Information Theoretic Perspectives on 5G Systems and Beyond

Experience a guided tour of the key information theoretic principles that underpin the design of next-generation cellular systems with this invaluable reference. Written by experts in the field, the text encompasses principled theoretical guidelines for the design and performance analysis of network architectures, coding and modulation schemes, and communication protocols. Presenting an extensive overview of the most important information-theoretical concepts underpinning the development of future wireless systems, this is the perfect tool for researchers and graduate students in the fields of information theory and wireless communications, as well as the practitoners in the telecommunications industry.

Ivana Marić is a research scientist with the Prematurity Research Center at the Department of Pediatrics, Stanford University School of Medicine.

Shlomo Shamai (Shitz) is a distinguished professor at the Department of Electrical Engineering, Technion, Israel Institute of Technology. He is an IEEE life fellow, an URSI fellow, a member of the Israeli Academy of Sciences and Humanities, and a foreign member of the US National Academy of Engineering.

Osvaldo Simeone is Professor of Information Engineering at the Centre for Telecommunications Research at the Department of Engineering, King's College London, and Fellow of the IEEE and of the IET.

Information Theoretic Perspectives on 5G Systems and Beyond

Edited by

IVANA MARIĆ
Stanford University, California

SHLOMO SHAMAI (SHITZ)
Technion, Israel Institute of Technology

OSVALDO SIMEONE
King's College London

CAMBRIDGE
UNIVERSITY PRESS

University Printing House, Cambridge CB2 8BS, United Kingdom

One Liberty Plaza, 20th Floor, New York, NY 10006, USA

477 Williamstown Road, Port Melbourne, VIC 3207, Australia

314–321, 3rd Floor, Plot 3, Splendor Forum, Jasola District Centre, New Delhi – 110025, India

103 Penang Road, #05–06/07, Visioncrest Commercial, Singapore 238467

Cambridge University Press is part of the University of Cambridge.

It furthers the University's mission by disseminating knowledge in the pursuit of education, learning, and research at the highest international levels of excellence.

www.cambridge.org
Information on this title: www.cambridge.org/9781108416474
DOI: 10.1017/9781108241267

© Cambridge University Press 2022
Cover illustration © Minya Mikic

First published 2022

Printed in the United Kingdom by TJ Books Limited, Padstow Cornwall

A catalogue record for this publication is available from the British Library.

Library of Congress Cataloging-in-Publication Data
Names: Shamai, Shlomo, 1958– author. | Simeone, Osvaldo, author. | Marić, Ivana, author.
Title: Information theoretic perspectives on 5G systems and beyond / Shlomo Shamai (Shitz), Osvaldo Simeone and Ivana Marić.
Other titles: Information theoretic perspectives on 5G systems and beyond
Description: New York : Cambridge University Press, 2020. | Includes bibliographical references and index.
Identifiers: LCCN 2019037656 | ISBN 9781108416474 (hardback) | ISBN 9781108241267 (epub)
Subjects: LCSH: 5G mobile communication systems.
Classification: LCC TK5103.25 .S53 2020 | DDC 621.3845/9–dc23
LC record available at https://lccn.loc.gov/2019037656

ISBN 978-1-108-41647-4 Hardback

Contents

Contributors

Navid Naderi Alizadeh
Intel Labs

Erdal Arıkan
Bilkent University

Salman Avestimehr
University of Southern California

Paolo Banelli
University of Perugia

Giuseppe Caire
Technische Universität Berlin

Giulio Colavolpe
University of Parma

Zhiguo Ding
Lancaster University

Giuseppe Durisi
Chalmers University of Technology

Michael Gastpar
Ecole polytechnique fédérale de Lausanne

Geordie George
Fraunhofer Institute

Thorkild B. Hansen
Seknion Inc.

Song-Nam Hong
Ajou University

Young-Han Kim
University of California, San Diego

Gerhard Kramer
Technical University of Munich

Lifeng Lai
University of California Davis

Erik G. Larsson
Linköping University

Yingbin Liang
The Ohio State University

Gianluigi Liva
German Aerospace Center

Sung Hoon Lim
Korean Institute of Ocean Science and Technology

Jaime Llorca
Nokia Bell Lab

Angel Lozano
Universitat Pompeu Fabra

Mohammad Ali Maddah-Ali
Nokia Bell Labs

Ivana Marić
Stanford University

Thomas L. Marzetta
New York University

Ratheesh K. Mungara
Ericsson

Bobak Nazer
Boston University

Ayfer Ozgur
Stanford University

Seok-Hwan Park
Chonbuk National University

Haim Permuter
Ben Gurion University

Yury Polyanskiy
Massachusetts Institute of Technology

H. Vincent Poor
Princeton University

Petar Popovski
Aalborg University

Luca Rugini
University of Perugia

Aydin Sezgin
Ruhr Universität Bochum

Shlomo Shamai (Shitz)
Technion, Israel Institute of Technology

Dor Shaviv
Lyft, Level

Osvaldo Simeone
King's College London

Anelia Somekh-Baruch
Bar-Ilan University

Čedomir Stefanović
Aalborg University

Ravi Tandon
University of Arizona

Kasper F. Trillingsgaard
Aalborg University

Antonia Tulino
DIETI, University of Naples Federico II

Alessandro Ugolini
University of Parma

Michèle Wigger
Telecom ParisTech

Peng Xu
University of Electronics Science & Technology of China

Benjamin M. Zaidel
Bar-Ilan University

Shaofeng Zou
University at Buffalo, the State University of New York

1 Introduction

Shlomo Shamai (Shitz), Osvaldo Simeone, and Ivana Marić

Evolving across the previous four generations, wireless cellular technology has transformed the way people communicate and acquire information. This transformation has taken place gradually since the 1990s to the first decade of the twenty-first century, with the first and second generation of cellular systems supporting mobile voice transmission, and the third and fourth generation enabling mobile internet access.

With its fourth generation (4G), wireless cellular technology has arguably completed an arc that was started by the work of Claude Shannon at the Bell Labs in the 1940s. In his seminal 1948 paper, Shannon derived a theoretical upper bound on the amount of information that can be conveyed on a communication link between two endpoints. His achievability proof was famously non-constructive, and hence it left open the problem of engineering efficient systems for the encoding and decoding of information at transmitter and receiver, respectively. The problem was essentially solved by the discovery of turbo and low-density parity check (LDPC) codes and corresponding decoders, all of which are included in the 4G standard.

As suggested by the discussion above, the path followed by the first four generations of cellular technology was one inspired by the goal of reaching the Shannon limit for communication over point-to-point links. In the process, communication engineers strived, and eventually succeeded, to design practical solutions that matched the theoretical results developed by Shannon. In contrast, the next generations of wireless systems, starting with the fifth (5G), face a new design landscape that lacks the strong theoretical guiding principles of Shannon's analysis of point-to-point communications.

In fact, 5G systems aim at providing not only an enhanced version of the 4G broadband communication service, known as enhanced mobile broadband (eMBB), but also new services that operate in the novel regimes of ultra-reliable, low-latency, and massive-access communications. Specifically, the 3rd Generation Partnership Project (3GPP), a collaboration between groups of telecommunications standards associations, has identified, beside eMBB, the services of ultra-reliable low-latency communications (URLLC) and massive machine-type communication (mMTC). The former is meant to support applications that require high reliability and low delays, such as for industrial control tasks or autonomous driving, while the latter enables the connectivity from a massive number of Internet of Things (IoT) devices.

The lack of a strong theoretical framework for the design of wireless systems in the novel regimes calls for a renewed effort by information theorists to extend the scope of Shannon's theory. With the evolution from 4G to 5G, the main design question has in fact shifted from one concerning "how to do" – namely, how to reach the optimal point-to-point performance promised by Shannon – to "what to do," as effective coding schemes and protocols for the new requirements are unknown. The latter is indeed the central question tackled by information theory through a mathematical lens. Despite its theoretical focus, information theory has proved to be able to offer valuable design insights, even when, as it is often the case in network scenarios, optimal solutions remain elusive the exact information theoretic characterizations, in terms of capacity regions of most relevant network problems for cellular communications, such as the Broadcast, Interference and Relay channels remain unsolved for decades. Yet the theoretical insights on network information theory, gained through the years, carry basic and important practical implications [1].

Broadly speaking, the development of an information theory for 5G requires a shift of focus from point-to-point channels with long transmission blocks to networks of devices with delay and reliability constraints. It is the goal of this book to offer an updated picture of the state of the art on information theoretic results that are relevant for the design of the next generations of wireless systems.

We organize the discussion around three main areas: network architecture, coding and modulation, and protocols. As will be seen in the text, the network architecture of 5G systems and beyond includes new elements that pose novel theoretical challenges, such as device-to-device links; high-speed backhaul links that support a decentralized and heterogeneous infrastructure; caches at the base stations that store popular content; and fronthaul links that allow for the centralization of baseband processing in cloud and fog-based architectures. Furthermore, the design of coding and modulation schemes needs to be revisited in order to account for novel aspects such as massive antenna arrays at the base stations; short-transmission blocks; non-orthogonal and uncoordinated multiple access; and asynchronous transmissions. Finally, communication protocols can benefit from an integration with breakthrough techniques inspired by information theory at the lower layers of the protocol stack, such as interference alignment; cooperative cellular transmission; physical-layer secrecy; and cognitive transmission.

The excitement and importance of 5G developments is attested by the extensive literature on the topic. Several books describe 5G architecture and envisioned use cases, and provide an overview of various techniques that will be deployed in 5G system [2–11]. Described techniques include multiple-access, device-to-device communications, spectrum sharing, mm-wave communications, Massive MIMO (multiple-input, multiple-output), interference management and relaying, among others.

Several books and monographs address 5G-related topics from the information-theoretic view, including Massive MIMO [12], interference alignment [13], security [14], cooperative cellular systems [15], and cloud radio access networks (C-RAN) [16–18]. These books highlight the impact information theory had and is yet to have on 5G and systems that will follow beyond.

Numerous research papers analyze in detail important 5G technologies and approaches, including multiple access [19, 20], modulation and waveforms [21], Massive MIMO [22], mm-wave [23], decentralized processing [24], heterogeneous networks

[25–27], C-RAN [28, 29], cooperation [30], cognitive cooperation [31, 32], polar coding [33], code design for short packets [34], nonorthogonal multiple access (NOMA) [35], caching [36], network coding [37], physical layer security [38], scheduling [39], mMTC [40], and implementation aspects [41]. Many surveys of 5G communications exist [42–60]. Overviews of 5G technologies can be found in the literature [61–67].

While several topics covered in this book appear also in many other references and books (e.g., [3, 4, 12]), the focus of this book is on the information theoretic framework and theoretical foundations that will make 5G and future wireless systems possible. The approach taken in this book is grounded in the discussed premise that information theory has played, and will play, a central role in developing wireless systems. The book starts with a detailed introductory chapter on network information theory that provides the background and the foundation for later chapters. The chapter covers first the point-to-point communications channel and its capacity. It then presents the main building blocks of communication networks, namely multiple access, broadcast, interference and relay channels, as well as coding schemes developed for each of these channels. Following the discussion above, the rest of the book is divided into three parts. The first part focuses on network architecture, and covers topics of device-to-device communications, caching, backhaul and fronthaul design, and energy harvesting. The second part covers coding and modulations schemes, including polar coding, Massive MIMO, short-packet transmission, NOMA, compute-forward strategies and waveform design. Finally, the third part presents protocols, and it encompasses chapters on 5G protocols, interference management, content delivery, and cooperative, cognitive, and confidential communications.

The first chapter in Part I covers device-to-device communications (D2D) that allow users in close proximity to establish direct communication with each other. In order to understand the fundamental limits of D2D critical for developing practical D2D strategies, this chapter analyzes the spectral efficiency of D2D communication, while relying on a novel modeling approach for the interference.

Chapter 4 describes backhaul architectures for small cells, in which backhaul is the main bottleneck. The chapter presents approaches in which access nodes not only deliver service to end-users but also act as relays carrying traffic to/from other access nodes, thereby establishing a multihop wireless connection for backhaul traffic.

Chapter 5 presents content caching, a promising direction to substantially improve the capacity of wireless networks by exploiting memories across the network.

Chapter 6 presents an information theoretic analysis of C-RAN, as well as of an extension that includes caches at the edge nodes of the network, also known as Fog-RAN (F-RAN), in the presence of capacity-limited fronthaul links.

The last chapter in the network architecture part of the book is concerned with the design of communication systems that contain wireless devices capable of harvesting energy they need for communication from the natural resources in the environment.

Part II of the book starts with a chapter on polar coding, presenting its general overview, capacity, and practical considerations.

The following chapter presents the most important forms of Massive MIMO and then discusses possible new directions for the physical layer that may go beyond Massive MIMO.

Chapter 10 reviews the most recent advances in finite-blocklength information theory. It provides theoretical principles governing the transmission of short packets, which are expected to be present in large volumes in machine-type communications.

A novel multiple-access technique, NOMA, that uses the power domain for achieving multiple access is presented in Chapter 11 in this part of the book.

Chapter 12 presents a novel way of treating interference in a communications strategy called compute-forward. In compute-forward, instead of treating interference as noise or decoding and canceling it, a receiver decodes a linear combination of simultaneously transmitted codewords, thereby improving the performance.

Chapter 13 provides a review of some of the most compelling waveforms for 5G networks. By employing a general signal-processing framework coupled with an information theoretic approach, the spectral efficiency of different techniques is compared.

Part III of the book starts with a chapter that discusses how to bridge the gap between information theory and protocol design.

Interference is one of the key obstacles in achieving higher spectral efficiency in wireless networks. Chapter 15 describes the most important interference management techniques.

Chapter 16 presents an overview of advanced cooperative communication techniques. and the benefits they attain over standard non-cooperative techniques.

Chapter 17 covers recent advances around the design of algorithms and protocols for the end-to-end optimization of real-time computation services over cloud-integrated 5G networks.

Chapter 18 presents physical layer security and, based on information theoretic ideas, characterizes the ability of the physical channel to achieve secure communication.

Chapter 19 analyzes settings in which transmitting and/or receiving nodes have acquired information about other users and/or interference in the network. As presented in this chapter, such knowledge allows for cooperation and interference mitigation, thereby improving spectral efficiency.

While work on 5G is still ongoing at the time of the writing, researchers have already started to consider technologies that may power the next generation of cellular systems (see, e.g., [75]–[79], [86]). Among the most promising directions for research, we mention here terahertz communication [80], reconfigurable intelligent surfaces [81], super-directive Massive MIMO [82, 83], and the application of artificial intelligence (AI) techniques for network management and operation [84, 85]. Information theory will play an essential role in understanding the fundamental benefits and limitations of all these technologies [87, 88]. Furthermore, as a complement to the analytical approach and fundamental insights obtained by means of information theory, recent activity has advocated the use of data-driven methods in order to overcome modeling or algorithmic deficits of existing solutions. Several important connections between Information Theory and Data Science are explored in detail in [89]. Deep learning is increasingly finding its application in wireless networks, including areas of mobile data analysis, signal processing, network security [67], data analytics in IoT [69], and big data [70]. We refer to [71–74] for further discussion on this expanding line of work.

Acknowledgments

This book is a joint project of many researchers in the field of information theory. We would like to thank all of the authors for contributing chapters to this book. We would also like to thank Cambridge University Press editors Sarah Strange and Julie Lancashire for their continuing and patient support throughout the preparation of this book.

We are thankful to the European Union's Horizon 2020 (ERC) Research and Innovation Programme, which supported the chapters composed by S. Shamai and O. Simeone, under grant agreement nos 694630 and 725731.

We acknowledge also the Heron and the Wireless Intelligent Networks (WIN) consortia of the Israel Ministry of Economy and Science, focused on development of essential building blocks for the 5G/6G next-generation mobile technology, supporting the relevant work of S. Shamai.

References

[1] A. El Gamal and Y.-H. Kim, *Network Information Theory,*. Cambridge: Cambridge University Press, 2011.

[2] A. Osseiran, J. Monserrat, and P. Marsch, Eds., *5G Mobile and Wireless Communications Technology*. Cambridge: Cambridge University Press, 2015.

[3] M. Vaezi, Z. Ding, and H. Poor, Eds., *Multiple Access Techniques for 5G Wireless Networks and Beyond*. New York: Springer, 2018.

[4] V. W. S. Wong, R. Schober, D. W. K. Ng, and L.-C. Wang, Eds., *Emerging Technologies for 5G Wireless Systems*. Cambridge: Cambridge University Press, 2017.

[5] F.-L. Luo and C. Zhang, Eds., *Signal Processing for 5G: Algorithms and Implementations*. Chichester: Wiley, 2016.

[6] J. Rodriguez, Ed., *Fundamentals of 5G Mobile Networks*. Chichester: Wiley, 2015.

[7] F. Hu, Ed., *Opportunities in 5G Networks*. Boca Raton, FL: CRC Press, 2016.

[8] M. Chiang, B. Balasubraman, and F. Bonomi, *FoG for 5G and IoT*. Chichester: Wiley, 2017.

[9] L. Yang and W. Zhang, *Interference Coordination for 5G Cellular Networks*. New York: Springer, 2015.

[10] W. Xiang, K. Zheng, and X. Shen, Eds., *5G Mobile Communications*. New York: Springer, 2017.

[11] R. Prasad, Ed., *5G Outlook: Innovations and Applications*. Aalborg: River Publishers, 2016.

[12] T. L. Marzetta, E. G. Larsson, H. Yang, and H. Q. Ngo, *Fundamentals of Massive MIMO*. Cambridge: Cambridge University Press, 2016.

[13] S. Jafar, "Interference alignment: A new look at signal dimensions in a communication network," *Found. Trends Commun. Inform. Theory*, vol. 7, no. 1, pp. 1–134, 2011.

[14] Y. Liang, V. Poor, and S. Shamai (Shitz), "Information theoretic security," *Found. Trends Commun. Inform. Theory*, vol. 5, pp. 355–580, 2009.

[15] O. Simeone, N. Levy, A. Sanderovich, et al., "Cooperative wireless cellular systems: An information theoretic view," *Found. Trends Commun. Inform. Theory*, vol. 8, no. 12, 2012.

[16] T. Quek, M. Peng, O. Simeone, and W. Yu, Eds., *Cloud Radio Access Networks: Principles, Technologies and Applications*. Cambridge: Cambridge University Press, 2017.

[17] H. Venkataraman and R. Trestian, Eds., *5G Radio Access Networks: Centralized RAN, Cloud-RAN and Virtualization of Small Cells*. Boca Raton, FL: CRC Press, 2017.

[18] M. Ghorbanzadeh, A. Abdelhadi, and C. Clancy, *Cellular Communications Systems in Congested Environments*. New York: Springer, 2017.

[19] N. Mokari, M. Javan, M. Moltafet, H. Saeedi, and H. Pishro-Nik, "A new multiple access technique for 5G: Power domain sparse code multiple access (PSMA)," *arxiv:1706.06439v1*, 2017.

[20] H. Kim, Y.-G. Lim, C.-B. Chae, and D. Hong, "Multiple access for 5G new radio: Categorization, evaluation, and challenges," *arXiv:1703.09042*, 2017.

[21] M. Nekovee, Y. Wang, M. Tesanovic, et al., "Overview of 5G modulation and waveforms candidates," *J. Commun. Inform. Netw.*, vol. 1, no. 1, pp. 44–60, Jun. 2016.

[22] G. Interdonato, E. Björnson, H. Ngo, P. Frenger, and E. Larsson, "Ubiquitous cell-free massive MIMO communications," *arXiv:1804.03421v1*, 2018.

[23] L. Lianming, N. Xiaokang, C. Yuan, et al., "The path to 5G: mmWave aspects," *J. Commun. Inform. Netw.*, vol. 1, no. 2, pp. 1–18, Aug. 2016.

[24] A. Sanderovich, S. Shamai (Shitz), Y. Steinberg, and G. Kramer, "Communication via decentralized processing," *IEEE Trans. Inf. Theory*, vol. 54, no. 7, pp. 3008–3023, Jul. 2008.

[25] H. Leifang, M. Wei, and Z. Shenghua, "A novel approach for radio resource management in multi-dimensional heterogeneous 5G networks," *J. Commun. Inform. Netw.*, vol. 1, no. 2, pp. 77–83, Aug. 2016.

[26] J. Riihijrvi, P. Mahonen, and M. Petrova, "What will interference be like in 5G hetnets?" *Physical Commun.*, vol. 18, pp. 85–94, Mar. 2016.

[27] K. Khawam, S. Lahoud, M. Ibrahim, et al., "Radio access technology selection in heterogeneous networks," *Physical Commun.*, vol. 18, pp. 125–139, Mar. 2016.

[28] S.-H. Park, O. Simeone, O. Sahin, and S. Shamai (Shitz), "Fronthaul compression for cloud radio access networks," *Special Issue on Signal Processing for the 5G Revolution*, vol. 31, no. 6, pp. 69–79, Nov. 2014.

[29] Z. Guizani and N. Hamdi, "CRAN, H-CRAN, and F-RAN for 5G systems: Key capabilities and recent advances," *Netw. Manag.*, vol. 27, Sep. 2017. DOI: /10.1002/nem.1973.

[30] M. Wigger, R. Timo, and S. Shamai (Shitz), "Conferencing in Wyner's asymmetric interference network: Effect of number of rounds," *IEEE Trans. Inf. Theory*, vol. 63, no. 2, Feb. 2017.

[31] R. Kolte, A. Ozgur, and H. Permuter, "Cooperative binning for semideterministic channels," *IEEE Trans. Inf. Theory*, vol. 62, no. 3, pp. 1193–1205, Mar. 2016.

[32] H. Permuter, S. Shamai (Shitz), and A. Somekh-Baruch, "Message and state cooperation in a multiple access channel," *IEEE Trans. Inf. Theory*, vol. 57, no. 10, pp. 6379–6396, Oct. 2011.

[33] V. Bioglio, C. Condo, and I. Land, "Design of polar codes in 5G new radio," *arXiv:1804.04389*, 2018.

[34] C. Liva, L. Gaudio, T. Ninacs, and T. Jerkovits, "Code design for short blocks: A survey," *arXiv:1610.00873*, 2016.

[35] M. Le, G. Ferrante, G. Caso, L. D. Nardis, and M.-G. D. Benedetto, "On information-theoretic limits of code-domain NOMA for 5G," *IET Commun.*, vol. 12, no. 15, pp. 1864–1871, Jul. 2018.

[36] Y. Fadlallah, D. B. A. Tulino, G. Vettigli, and J. Llorca, "Coding for caching in 5G networks," *IEEE Commun. Mag.*, vol. 55, no. 2, pp. 106–113, 2017.

[37] D. Szabo, F. Nemeth, B. Sonkoly, A. Gulys, and F. Fitzek, "Towards the 5G revolution: A software defined network architecture exploiting network coding as a service," *ACM SIGCOMM Comput. Commun. Rev. SIGCOMM'15*, vol. 45, no. 4, pp. 105–106, Oct. 2015.

[38] Y. Wu, A. Khisti, C. Xiao, et al., "A survey of physical layer security techniques for 5G wireless networks and challenges ahead," *IEEE J. Sel. Areas Commun.*, vol. 36, pp. 679–695, 2018.

[39] C. Gueguen, M. Ezzaouia, and M. Yassin, "Inter-cellular scheduler for 5G wireless networks," *Physical Commun.*, vol. 18, pp. 113–124, Mar. 2016.

[40] C. Bockelmann, N. Pratas, G. Wunder, et al., "Towards massive connectivity support for scalable mMTC communications in 5G networks," *IEEE Commun. Mag.*, vol. 6, pp. 28969–289922, 2018.

[41] L. Li, D. Wang, X. Niu, et al., "mmWave communications for 5G: Implementation challenges and advances," *Sci. China Inform. Sci.*, vol. 61, Feb. 2018. DOI: 10.1007/s11432-017-9262-8.

[42] J. G. Andrews, S. Buzzi, W. Choi, et al., "What will 5G be?" *IEEE J. Sel. Areas Commun.*, vol. 32, no. 6, pp. 1065–1082, Sep. 2014.

[43] M. Agiwal, A. Roy, and N. Saxena, "Next generation 5G wireless networks: A comprehensive survey," *IEEE Commun. Surv. Tutorials*, vol. 18, no. 3, pp. 1617–1655, 2016.

[44] F. Boccardi, R. W. Heath Jr., A. Lozano, T. L. Marzetta, and P. Popovski, "Five disruptive technology directions for 5G," *IEEE Commun. Mag.*, vol. 52, no. 2, pp. 74–80, Feb. 2014.

[45] Z. Ma, Z.-Q. Zhang, Z.-G. Ding, P.-Z. Fan, and H.-C. Li, "Key techniques for 5G wireless communications: Network architecture, physical layer, and MAC layer perspectives," *Sci. China Inform. Sci.*, vol. 8, no. 4, pp. 1–20, Apr. 2015.

[46] A. Gupta and R. K. Jha, "A survey of 5G network: Architecture and emerging technologies," *IEEE Access*, vol. 3, pp. 1206–1232, Aug. 2015.

[47] S.-Y. Lien, K.-C. Chen, Y.-C. Liang, and Y. Lin, "Cognitive radio resource management for future cellular networks," *IEEE Wireless Commun.*, vol. 21, no. 1, pp. 70–79, Feb. 2014.

[48] A. Gohil, H. Modi, and S. Patel, "5G technology of mobile communication: A survey," in *2013 Int. Conf. Intell. Syst. Signal Processing (ISSP2013)*, Gujarat, Mar. 2013, pp. 1–2.

[49] Q. Li, G. Li, D. Mazzarese, B. Clerckx, and Z. Li, "MIMO techniques in WiMAX and LTE: A feature overview," *IEEE Commun. Mag.*, vol. 62, no. 3, pp. 86–92, May 2010.

[50] M. Renzo, H. Haas, and P. M. Grant, "Spatial modulation for multiple-antenna wireless systems: A survey," *IEEE Commun. Mag.*, vol. 49, no. 12, pp. 182–191, Dec. 2011.

[51] G. Araniti, M. Condoluci, P. Scopelliti, A. Molinaro, and A. Iera, "Multicasting over emerging 5G networks: Challenges and perspectives," *IEEE Netw.*, vol. 31, no. 2, pp. 80–89, 2017.

[52] D. Sahinel, C. Akpolat, M. Khan, F. Sivrikaya, and S. Albayrak, "Beyond 5G vision for IOLITE community," *IEEE Commun. Mag.*, vol. 55, no. 1, pp. 41–47, 2017.

[53] Y. Liu, H.-H. Chen, and L. Wang, "Physical layer security for next generation wireless networks: Theories, technologies, and challenges," *IEEE Commun. Surv. Tutorials*, vol. 19, no. 1, pp. 347–376, 2017.

[54] P. Dat, A. Kanno, N. Yamamoto, and T. Kawanishi, "5G transport networks: The need for new technologies and standards," *IEEE Commun. Mag.*, vol. 54, no. 9, pp. 74–80, Sep. 2016.

[55] T. Olwal, K. Djouani, and A. Kurien, "A survey of resource management toward 5G radio access networks," *IEEE Commun. Surv. Tutorials*, vol. 3, no. 3, pp. 1656–1686, 2016.

[56] S. Sharma, T. Bogale, L. Le, et al., "Dynamic spectrum sharing in 5G wireless networks with full-duplex technology: Recent advances and research challenges," *IEEE Commun. Surv. Tutorials*, vol. 20, no. 1, pp. 674–707, 2018.

[57] Z. Wei, J. Yuan, D. Ng, M. Elkashlan, and Z. Ding, "A survey of downlink non-orthogonal multiple access for 5G wireless communication networks," *arXiv: 1609.01856*, 2016.

[58] H. Murata, E. Okamoto, M. Mikami, et al., "R&D activities for 5G in IEICE technical committee on radio communication systems," in *2015 21st Asia-Pacific Conf. Commun. (APCC)*, Kyoto, Oct. 2015, pp. 250–256.

[59] N. Luong, P. Wang, D. Niyato, et al., "Applications of economic and pricing models for resource management in 5G wireless networks: A survey," *arXiv:1710.04771*, Oct. 2017.

[60] Y. Niu, Y. Li, D. Jin, L. Su, and A. Vasilakos, "Survey of millimeter wave (mmwave) communications for 5G: Opportunities and challenges," *arXiv:1502.07228*, Feb 2015.

[61] M. Shafi, A. Molisch, P. Smith, et al., "5G: A tutorial overview of standards, trials, challenges, deployment and practice," *IEEE J. Sel. Areas Comm.*, vol. 35, no. 6, pp. 1201–1221, Jun. 2017.

[62] Y. Huo, X. Dong, and W. Xu, "5G cellular user equipment: From theory to practical hardware design," *IEEE Access*, vol. 5. DOI: 10.1109/ACCESS.2017.2727550.

[63] S.-Y. Lien, S.-L. Shieh, Y. Huang, et al., "5G new radio: Waveform, frame structure, multiple access, and initial access," *IEEE Commun. Mag.*, vol. 55, no. 6, pp. 64–71, Jun. 2017.

[64] A. Taufique, M. Jaber, A. Imran, Z. Dawy, and F. E. Yacoub, "Planning wireless cellular networks of future: Outlook, challenges and opportunities," *IEEE Access*, vol. 5, pp. 4821–4845, Mar. 2017.

[65] Q. Wu, G. Li, W. Chen, D. Ng, and R. Schober, "An overview of sustainable green 5G networks," *arXiv:1609.09773*, 2016.

[66] S. Shi, W. Yang, J. Zhang, and Z. Chang, "Review of key technologies of 5G wireless communication system," *MATEC Web Conf.*, vol. 20, 2015.

[67] Q. Nadeem, A. Kammoun, and M.-S. Alouini, "Elevation beamforming with full dimension MIMO architectures in 5G systems: A tutorial," *arXiv:1805.00225*, May 2018.

[68] C. Zhang, P. Patras, and H. Haddadi, "Deep learning in mobile and wireless networking: A survey," *IEEE Comm. Surv. Tutorials*, vol. 21, no. 3, pp. 2224–22387, 2018.

[69] M. Mohammadi and A. Al-Fuqaha, "Deep learning for IoT big data and streaming analytics: A survey," *IEEE Comm. Surv. Tutorials*, vol. 20, no. 4, pp. 2923–2960, 2018.

[70] J. Qiu, Q. Wu, G. Ding, Y. XuS, and H. Feng, "A survey of machine learning for big data processing," *EURASIP J. Adv. Signal Processing*, Dec. 2016. DOI: 10.1186/s13634-016-0355-x.

[71] S. Shea and J. Hoydis, "An introduction to deep learning for the physical layer," *IEEE Trans. Cogn. Commun. Netw.*, vol. 3, no. 4, pp. 563–575, Dec. 2017.

[72] X. You, C. Zhang, X. Tan, S. Jin, and H. Wu, "AI for 5G: Research directions and paradigms," *Sci. China Inform. Sci.*, Jul. 2018. DOI: 10.1007/s11432-018-9596-5.

[73] N. Kato, Z. Fadlullah, B. Mao, et al., "The deep learning vision for heterogeneous network traffic control: Proposal, challenges, and future perspective," *IEEE Wireless Commun.*, vol. 24, no. 3, pp. 146–153, Jun. 2017.

[74] O. Simeone, "A very brief introduction to machine learning with applications to communication systems," *arXiv:1808.02342*, Aug. 2018.

[75] Z. Zhang, Y. Xiao, Z. Ma, et al., "6G wireless networks: Vision, requirements, architecture, and key technologies," *IEEE Veh. Technol. Mag.*, vol. 14, no. 3, pp. 28–41, Sept. 2019.

[76] L. Zhang, Y.-C. Liang, and D. Niyato, "6G visions: Mobile ultra-broadband, super Internet-of-Things, and artificial intelligence," *China Commun.*, vol. 16, no. 8, Aug. 2019.

[77] Y. Yuan, Y. Zhao, B. Zong, and S. Parolari, "Potential key technologies for 6G mobile communications," *arXiv:1910.00730*, Oct. 2019.

[78] P. Popovski, O. Simeone, F. Boccardi, D. Gunduz, and O. Sahin, "Semantic-effectiveness filtering and control for post-5G wireless connectivity," *arXiv:1907.02441*, Jul. 2019.

[79] E.C. Strinati, S. Barbarossa, J.L. Gonzalez-Jimenez, et al. "6G: The next frontier," *arXiv:1901.03239*, Jan. 2019.

[80] H. Sarieddeen, N. Saeed, T.Y. Al-Naffouri, and M.S. Alouini. "Next generation terahertz communications: A rendezvous of sensing, imaging and localization," *arXiv:1909.10462*, Sept. 2019.

[81] J. Zhao, "A survey of intelligent reflecting surfaces (IRSs): Towards 6G wireless communication networks with massive MIMO," *arXiv:1907.04789*, Aug. 2019.

[82] E. Björnson, L. Sanguinetti, H. Wymeersch, J. Hoydis, and T.L. Marzetta. "Massive MIMO is a reality: What is next? Five promising research directions for antenna arrays." *Digital Signal Process.*, vol. 94, pp. 3–20, Jun. 2019.

[83] J. Zhang, E. Björnson, M. Matthaiou, et al., "Multiple antenna technologies for beyond 5G," *arXiv:1910.00092*, Sept. 2019.

[84] K.J. Ray Liu and B. Wang, *Wireless AI: Wireless Sensing, Positioning, IoT, and Communications*. Cambridge: Cambridge University Press, 2019.

[85] K.B. Letaief, W. Chen, Y. Shi, J. Zhang, and Y.-J.A. Zhang, "The roadmap to 6G: AI empowered wireless networks," *IEEE Commun. Mag.*, pp. 84–49, Aug. 2019.

[86] S.-H. Zhang, J.-H. Zhang, Y. Chen, and J.-K. Zhu, "Wireless big data enabled emerging technologies for beyond 5G system[J]." *J. Beijing Uni. Posts Telecomms.*, vol. 41, no. 5, pp. 52–61, 2018.

[87] M. Medard, "Is 5 just what comes after 4?," *Nature Electronics*, vol. 3, pp. 2-4, Jan. 2020.

[88] M. Vaezi, "*Cognitive Radio Networks: An Information Theoretic Perspective*," NOVA Science Publishers, 2020. <arXiv:2001.09261>

[89] Miguel R.D. Rodrigues and Yonina Eldar, *Information-Theoretic Methods in Data Science*. Cambridge: Cambridge University Press, 2021.

2 Information Theory for Cellular Wireless Networks

Young-Han Kim and Gerhard Kramer

Network information theory studies the fundamental limits on information flow in networks, and optimal coding schemes, protocols, and architectures that achieve these limits. Although a complete theory of information flow over networks with multiple sources and multiple destinations remains in terra incognita, many elegant results and techniques have been developed over the past 50 years.

This chapter provides a short introduction to network information theory through a few canonical network models that we believe are most relevant to the physical layer design of next-generation wireless systems. Slightly deviating from the usual emphasis in the literature on the determination of capacity, our main focus will be on conceptual building blocks for modulation and coding schemes, and for transceiver architectures for these network models, some of which have already proven useful to real-world applications, and many of which have great potential to do so. We hope that inspired engineers can someday translate these conceptual building blocks to practical communication techniques.

We start our discussion in Section 2.1 with the simplest model of point-to-point communication channels. After reviewing Shannon's channel coding theorem and the capacity of a few basic channels, we present coded modulation and multiple-input–multiple-output (MIMO) transmission schemes using elementary ideas from network information theory such as *superposition coding* and *successive cancellation decoding*. Sections 2.2 and 2.3 discuss multiple access channels and broadcast channels as canonical models for uplink and downlink communications in a single cell. Highlights are *simultaneous decoding* for uplink communications and *Marton coding* (writing on dirty paper) for downlink communications. In Section 2.4, we study interference channels as a simple platform to investigate different approaches to mitigate intercell interference. We present several coding schemes that combine basic coding techniques thus far developed for point-to-point, multiple access, and broadcast communications. Section 2.5 is devoted to two-hop network models in which dedicated relays facilitate communications between sources and destinations. We present a few canonical relaying schemes such as *decode–forward* and *compress–forward*, and discuss their use cases for wireless networks.

We remark that the topics covered in this chapter comprise only a small fraction of network information theory. For a more comprehensive treatment on the subject, we refer the interested reader to [1–3].

We remind the reader that the *entropy* of a discrete random variable X with probability mass function (pmf) $p(x)$ on an alphabet $\mathcal{X}$ is defined as

$$H(X) = \sum_{x \in \mathcal{X}, p(x) > 0} p(x) \log \frac{1}{p(x)}$$

$$= \mathsf{E}\left[\log \frac{1}{p(X)}\right],$$

and that the *mutual information* of a pair of discrete random variables (X, Y) with joint pmf $p(x, y)$ and marginal pmfs $p(x)$ and $p(y)$ is defined as

$$I(X; Y) = \mathsf{E}\left[\log \frac{p(X, Y)}{p(X)p(Y)}\right].$$

For key properties of these information measures and their extensions to continuous random variables, the reader is referred to standard textbooks on information theory [4–7].

2.1 Point-to-Point Communication

Consider the communication model with one sender and one receiver as depicted in Figure 2.1. A sender wishes to communicate a message $m \in [2^{nR}] = \{1, 2, \ldots, 2^{nR}\}$ over a channel by encoding the message into a codeword $x^n = (x_1, x_2, \ldots, x_n) \in \mathcal{X}^n$ of length n, and then sending the codeword as the channel input sequence. The receiver forms an estimate $\hat{m}$ of the message m by decoding the channel output sequence $y^n \in \mathcal{Y}^n$.

As the simplest model, suppose that the channel is *discrete*, i.e., the input and output alphabets $\mathcal{X}$ and $\mathcal{Y}$ are finite, and *memoryless* with the probability of receiving an output symbol y from the transmitted input symbol x specified by the conditional probability $p_{Y|X}(y|x)$. Then the conditional probability of receiving the channel output y^n when the message m is communicated factors as

$$p(y^n | m) = \prod_{i=1}^{n} p_{Y|X}(y_i | x_i(m)).$$

The indices $i \in [n] = \{1, 2, \ldots, n\}$ can refer to time, frequency, spatial location, and so on.

The encoder mapping $x^n(m)$ and the decoder mapping $\hat{m}(y^n)$ are collectively referred to as a $(2^{nR}, n)$ *code*, where R is called the *rate* of the code (measured in bits per transmission) and n is called its codeword length or *blocklength*. The set $\mathcal{C} = \{x^n(m) : m \in [2^{nR}]\}$ is referred to as the *codebook* associated with the $(2^{nR}, n)$ code.

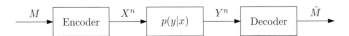

Figure 2.1 Point-to-point communication system.

The performance of a given code over a given channel is measured by the probability that the decoder fails to recover the message sent. Let the message $M \sim \text{Unif}[2^{nR}]$ be drawn uniformly at random from the message set and $\hat{M} = \hat{m}(Y^n)$ be the message estimate from the random output Y^n of the channel. Then the *average error probability* of the $(2^{nR}, n)$ code is defined as

$$P_e^{(n)} = \mathsf{P}(\hat{M} \neq M).$$

A natural engineering question is which code minimizes the error probability for given R and n, or equivalently, what is the optimal tradeoff between R, n, and $P_e^{(n)}$. This problem seems to be extremely difficult for most channels, however. Shannon [8] simplified the problem by considering instead the tradeoff between R and $P_e^{(n)}$ without any restriction on n, and asking when $P_e^{(n)}$ can be made arbitrarily small. More formally, we say that a rate R is *achievable* (for a given channel) if, for any target error probability $\epsilon > 0$ and sufficiently large n, there exists a $(2^{nR}, n)$ code with $P_e^{(n)} \leq \epsilon$. This require-ment is often stated in asymptotic terms by postulating a sequence of codes, i.e., R is achievable if there exists a sequence of $(2^{nR}, n)$ codes such that $\lim_{n \to \infty} P_e^{(n)} = 0$. In words, reliable communication is possible at an achievable rate R.

The *capacity* C of a given channel is the supremum of all achievable rates for the channel. Thus, if $R < C$, then reliable communication is possible and $P_e^{(n)}$ can be made arbitrarily small. Conversely, if $R > C$, then no code of rate R can provide reliable com-munication and $P_e^{(n)}$ is bounded away from zero. In fact, an even stronger result holds when one is interested in a single message and the channel is discrete and memoryless – if $R > C$, then $\lim_{n \to \infty} P_e^{(n)} = 1$ for every sequence of $(2^{nR}, n)$ codes [9]. Thus, the capacity C is the critical rate at which the error probability makes an asymptotically sharp transition from 0 to 1, a behavior similar to phase transitions in statistical physics.

2.1.1   Random Coding and Joint Typicality Decoding

As mentioned before, the optimal $(2^{nR}, n)$ code, i.e., the code that achieves the minimum error probability, is usually very difficult to find. A capacity-achieving code, however, can be constructed surprisingly easily.

Let $p(x)$ be given. We construct a codebook $\mathcal{C}$ of rate R and blocklength n randomly by generating its entries independently and identically distributed (i.i.d.) according to $p(x)$, i.e., the n symbols in each codeword are i.i.d. and the 2^{nR} codewords in the codebook are i.i.d. Each realization of this random ensemble defines a codebook. Intuitively speaking, although two codewords in a randomly generated codebook can be hard to distinguish from each other (they may even be identical!), most codewords are distinguishable from the rest as long as there are not too many of them. This intuition can be made precise by considering a specific decoder and analyzing its decoding error probability.

For a given codebook $\mathcal{C} = \{x^n(m) : m \in [2^{nR}]\}$, the error probability for a uniformly distributed message M is minimized by the *maximum likelihood (ML) decoder*,

$$\hat{m}^*(y^n) = \arg \max_m p(y^n | m).$$

Although optimal, the ML decoder is somewhat cumbersome to analyze. As a simpler alternative, we instead use the *joint typicality decoder*. To describe this decoder, we first give some definitions. For a random variable X on a finite alphabet $\mathcal{X}$, we say that an n-sequence x^n is ϵ-*typical* (or simply *typical*) with respect to X [10, 11] if

$$(1 - \epsilon)p(x) \leq \pi(x|x^n) \leq (1 + \epsilon)p(x), \quad x \in \mathcal{X},$$

where $\pi(x|x^n)$ denotes the empirical frequency of symbol x in the sequence x^n. Thus, x^n is typical if its empirical distribution is close to the distribution of X. Similarly, we say that x^n and y^n are *jointly typical* with respect to (X, Y) if

$$(1 - \epsilon)p(x, y) \leq \pi(x, y|x^n, y^n) \leq (1 + \epsilon)p(x, y), \quad (x, y) \in \mathcal{X} \times \mathcal{Y}.$$

It can be verified [11] that an i.i.d. sequence X^n is typical with high probability (w.h.p.) for every $\epsilon > 0$ and sufficiently large n, and that a pair of statistically independent i.i.d. sequences X^n and Y^n are jointly typical with probability at most $2^{-n(I(X;Y)-\delta)}$ for some small δ that vanishes with ϵ.

Now back to the decoding problem. The joint typicality decoder declares that $\hat{m}$ was sent if it is the unique message such that $x^n(\hat{m})$ and y^n are jointly typical. The average performance of this decoder for a randomly generated codebook is simple to analyze. Assume without loss of generality that $m = 1$ was sent. On the one hand, the corresponding (random) codeword $X^n(1)$ and its noisy observation Y^n are jointly typical w.h.p. On the other hand, each of the remaining $2^{nR} - 1$ codewords $X^n(m)$, $m \neq 1$, is independent of Y^n and thus jointly typical with Y^n with probability at most $2^{-n(I(X;Y)-\delta)}$. By the union-of-events bound, this implies that the probability that *some* wrong codeword $X^n(m)$, $m \neq 1$, is jointly typical is upper bounded as

$$(2^{nR} - 1)2^{-n(I(X;Y)-\delta)} \leq 2^{-n(I(X;Y)-\delta-R)},$$

which tends to zero as $n \to \infty$ if $R < I(X; Y)-\delta$. Having both types of error – the correct codeword being not typical and some wrong codeword being typical – accounted for, joint typicality decoding of a random codebook of rate R less than $I(X; Y) - \delta$ generated from $p(x)$ has vanishing error probability *on average*, and thus that there *exists* at least one $(2^{nR}, n)$ code with the same or smaller error probability. In paraphrase, the rate $R < I(X; Y)-\delta$ is achievable (for simplicity, assume that $I(X; Y)-\delta$ is positive). Finally, by taking $\epsilon > 0$ arbitrarily small in joint typicality decoding and optimizing $p(x)$ in random coding, we conclude that any rate

$$R < \max_{p(x)} I(X; Y)$$

is achievable for the discrete memoryless channel $p(y|x)$, or equivalently, the capacity of the channel $p(y|x)$ is *lower bounded as*

$$C \geq \max_{p(x)} I(X; Y). \tag{2.1}$$

2.1.2   Channel Coding Theorem

Standard information theoretic inequalities [4, sec. 7.9] imply that any achievable rate R must be no greater than $\max_{p(x)} I(X;Y)$, or equivalently, we have

$$C \leq \max_{p(x)} I(X;Y). \tag{2.2}$$

Combining (2.1) and (2.2), we have Shannon's channel coding theorem [8]:

THEOREM 2.1   *The capacity of the discrete memoryless channel $p(y|x)$ is*

$$C = \max_{p(x)} I(X;Y). \tag{2.3}$$

On the one hand, the left-hand side (LHS) of (2.3) is a physical, or operational, quantity – the maximum rate of reliable communication via an arbitrary number of transmissions over a given channel. On the other hand, the right-hand side (RHS) of (2.3) is a mathematical quantity – the maximum mutual information of a single input–output pair over a given channel. Shannon's channel coding theorem shows that these two quantities are the same, providing a simple mathematical expression for the critical rate of phase transition.

Broadly speaking, Shannon's channel coding theorem establishes two important principles of the optimal design of communication systems. First, and more directly, it establishes a fundamental limit of point-to-point communication as a computable expression. Equipped with the capacity expression (2.3) as a benchmark, a communication engineer can make an educated judgment on the performance of a given code and channel. In this context, we point out a few refinements of Shannon's channel coding theorem. First, by allowing errors in the message represented by a sequence of independent and uniformly distributed bits, one can show [5, thm. 9.5.1] that the asymptotically optimal rate under a given bit-error rate P_b is

$$C(P_b) = \frac{C}{1 - H(P_b)}, \tag{2.4}$$

where $H(P_b) = -P_b \log_2 P_b - (1 - P_b) \log_2 (1 - P_b)$ is the binary entropy function. Setting $P_b = 0$ in (2.4) recovers the channel coding theorem in (2.3). Second, by analyzing the error probability more carefully, one can establish *second-order asymptotics* [12–14] on the maximum number of bits $k^*(n, \epsilon)$ that can be communicated with blocklength n and block error probability ϵ as

$$k^*(n, \epsilon) = nC - \sqrt{nV}Q^{-1}(\epsilon) + O(\log n), \tag{2.5}$$

where $Q^{-1}(\epsilon)$ is the inverse of the standard Q-function (i.e., the tail probability of the standard Gaussian distribution) and

$$V = \text{Var}\left(\log \frac{p(X,Y)}{p(X)p(Y)}\right) \tag{2.6}$$

is called the *dispersion* of the channel, which is evaluated under $p(x)$ that minimizes the variance in (2.6) among all capacity-attaining input distributions, i.e., those that maximize the average

$$I(X;Y) = \mathsf{E}\left[\log \frac{p(X,Y)}{p(X)p(Y)}\right].$$

The second-order term $\sqrt{nV}Q^{-1}(\epsilon)$ in (2.5) provides a fine adjustment to the first, linear capacity term, and vanishes when normalized by n to compute the asymptotic rate $R = k/n$. Nonasymptotic bounds on $k^*(n,\epsilon)$ can be found in work by Polyanskiy et al. [14]. Third, by focusing on the probability of error $P_e^{(n)}$ at a rate R strictly below the capacity, we can study the *reliability function*

$$E(R) = \lim_{n\to\infty} -\frac{1}{n}\log \min_{(nR,n)\text{ codes}} P_e^{(n)},$$

which characterizes the largest *error exponent* of codes of rate R. In particular, $E(R) > 0$ if $R < C$. Upper and lower bounds exist on $E(R)$ that coincide for sufficiently large R [5, ch. 5]. For rates above the capacity, a similar exponential behavior of the minimum probability of error

$$E(R) = \lim_{n\to\infty} -\frac{1}{n}\log \max_{(nR,n)\text{ codes}} (1 - P_e^{(n)}).$$

can be characterized [15, 16].

The second engineering principle from the channel coding theorem is more subtle. Although the aforementioned random coding argument proves only the *existence* of a good code (as with other instances of the *probabilistic method* [17]), the performance of a randomly generated codebook concentrates around the average. Therefore, we may well say that *most* codes are good. This statement can be refined for *linear codes* over a finite field [5, sec. 6.2]. In addition, capacity-achieving codes share several important characteristics with a random code [18]. These observations lead to the design challenge of finding "random-code-like" codes with low encoding and decoding complexity, which has been successfully met for important classes of channels by convolutional, low-density parity check, turbo, raptor, spatially coupled, and polar codes [19–28].

In the forthcoming sections, we attempt to replicate these engineering principles of the channel coding theorem – performance bounds and optimal coding schemes – for network models beyond point-to-point communication.

2.1.3   Channel Capacity Examples

The *binary symmetric channel* (BSC) is one of the simplest examples of a discrete memoryless channel. As depicted in Figure 2.2, the input and output are binary, and the input

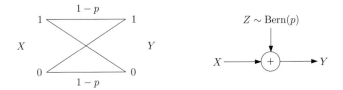

Figure 2.2 The BSC with crossover probability p and its equivalent representation.

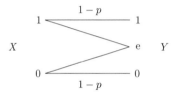

Figure 2.3 The BEC with erasure probability p.

symbol is flipped with probability p. Equivalently, the channel output Y corresponding to the input X can be expressed as

$$Y = X \oplus Z,$$

where the binary additive noise $Z \sim \text{Bern}(p)$ is independent of X. The capacity of the BSC with crossover probability p is

$$C = \max_{p(x)} I(X; Y) = 1 - H(p),$$

where the maximum mutual information is attained by $X \sim \text{Bern}(1/2)$.

The *binary erasure channel* (BEC) is another simple example of a discrete memoryless channel. As depicted in Figure 2.3, the input X is binary, and each input symbol is received intact with probability $1 - p$ and erased (mapped into a third, erasure symbol e) with probability p. The receiver keeps track of which transmission is lost, unlike the more complicated *deletion* channel [29] that loses symbols as well as their precise locations. If the encoder were to know before communication which transmissions will be erased, approximately $n(1 - p)$ bits would be transmitted reliably. The channel coding theorem shows that

$$C = \max_{p(x)} I(X; Y) = 1 - p$$

is again achieved by $X \sim \text{Bern}(1/2)$. Thus, *forward error correction* (FEC) can achieve the same asymptotic performance without any foresight on the erasure locations.

2.1.4 Gaussian Channel

The *additive white Gaussian noise* (AWGN) channel (or the *Gaussian* channel in short), as depicted in Figure 2.4, is the most popular channel model for baseband point-to-point communication in wireless or wired systems. Here and henceforth, we use a complex channel model to account for passband communications. The channel output Y corresponding to the input X is

$$Y = hX + Z, \tag{2.7}$$

where h is a complex channel coefficient and the additive noise $Z \sim \text{CN}(0, 1)$ is complex, Gaussian, circularly symmetric (and thus zero-mean), unit variance, and independent of X. Thus, at transmission time $i \in [n]$, the channel output is

$$Y_i = hX_i + Z_i, \tag{2.8}$$

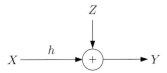

Figure 2.4 Additive white Gaussian noise channel.

where $\{Z_i\}$ is a complex, circularly symmetric, white Gaussian noise process with unit average power. Without any constraint on the code, the capacity of the Gaussian channel is unbounded, which is achieved by spending an unbounded amount of input power.

We now consider a more interesting problem in which there is a *block power constraint*, i.e., we have

$$\frac{1}{n} \sum_{i=1}^{n} |x_i(m)|^2 \leq P, \quad m \in [2^{nR}]. \tag{2.9}$$

The channel coding theorem for discrete channels can be extended to discrete or continuous channels with input cost constraint $(1/n) \sum_{i=1}^{n} b(x_i) \leq B$ to yield the capacity–cost function

$$C(B) = \max_{F(x): \, \mathrm{E}[b(X)] \leq B} I(X; Y), \tag{2.10}$$

where $F(x)$ is the cumulative distribution function of X. Specializing $C(B)$ for the Gaussian channel with block power constraint $(b(x) = |x|^2)$, we obtain the celebrated Gaussian capacity formula [8]

$$C(P) = \max_{F(x): \, \mathrm{E}[|X|^2] \leq P} I(X; Y) = \log(1 + S), \tag{2.11}$$

where the received signal-to-noise ratio (SNR) is $S = |h|^2 P$, and where the maximum mutual information is attained by $X \sim \mathrm{CN}(0, P)$.

In many practical systems, a specific modulation scheme restricts the input signals that can be transmitted. Such a restriction can often be incorporated into the channel model by setting the input alphabet $\mathcal{X}$ to be the corresponding constellation of the modulation scheme. For example, the input alphabet $\mathcal{X} = \{-\sqrt{P}, +\sqrt{P}\}$ can represent binary phase shift keying (BPSK), and the Gaussian channel with this input alphabet is sometimes referred to as the *binary-input additive white Gaussian noise* (BI-AWGN) channel. The capacity of this channel is again expressed as

$$C_{\mathrm{BPSK}}(P) = \max_{p(x)} I(X; Y), \tag{2.12}$$

where the maximum is attained by the uniform input distribution on $\mathcal{X} = \{-\sqrt{P}, +\sqrt{P}\}$. Unlike (2.11), there is no closed-form expression for (2.12), but it can be evaluated numerically. The maximum mutual information, or the mutual information under the uniform input distribution, can be similarly computed for other constellations, thus providing the fundamental limit of communication under specific modulation schemes. Since the codewords from a given constellation satisfy the average power constraint

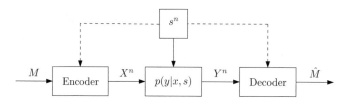

Figure 2.5 Point-to-point communication system with state.

symbol by symbol, the capacities of discrete-input Gaussian channels are upper bounded by the Gaussian capacity expression in (2.11).

2.1.5   Channels with State

In many communication systems, the channel statistics are not fully known or vary over transmissions. For example, for the Gaussian channel (2.8), only an estimate of the channel gain h may be available at the receiver, but not at the sender. To incorporate such uncertainty, we can extend the standard point-to-point communication channel model to a state-dependent channel model

$$p(y^n|x^n, s^n) = \prod_{i=1}^{n} p_{Y|X,S}(y_i|x_i, s_i),$$

as depicted in Figure 2.5. In this model, the channel output y^n depends on the channel input x^n as well as the channel state sequence s^n, which may be available at the sender and/or the receiver.

If the channel state is constant over the transmissions, but unknown to the sender and receiver, then the worst-case guarantee of reliable communication is studied as a *compound channel* [30–32], the capacity of which will be discussed in Section 2.3.1.

If the channel state is random and varies in a *memoryless* fashion over the transmissions, then the capacity depends on whether and when the state information is available at the sender and receiver; see, for example, [33, 34] and [2, ch. 7]. An interesting case is when the state information is available only at the sender, which was studied in [35–37]. A concrete example will be discussed in Section 2.3.3, where the sender can adapt its transmission with respect to known interference.

The channel gain h (or state h) in the Gaussian channel (2.8) may vary over time due to user mobility and multipath propagation. This model is often referred to as the *Gaussian fading channel*. When the channel gain h varies rapidly relative to the duration of communication, then the tradeoff between the rate and the error probability has a sharp phase transition and the capacity, which was defined earlier as the maximum achievable rate, is still meaningful as in the memoryless state case mentioned above. If the receiver has *channel state information* (CSI), namely, can incorporate the channel coefficient h into decoding, then the capacity can be expressed as the average of the (random) capacity under each h, and thus is commonly referred to as the *ergodic capacity*. If the channel variation is slow, then there is no phase transition and the performance can be measured by the *outage capacity*, the largest rate achievable at a given confidence level measured by the probability of failure or outage. For both fast and slow fading, these notions of ergodic and outage capacities depend on the availability of CSI at the sender, or receiver, or both, or neither; refer to [2, ch. 23] for details.

2.1.6   Coded Modulation

The random coding argument in Section 2.1.1 can generate a capacity-achieving code in any finite input alphabet. Most practical codes, however, are binary and the binary codewords should be mapped to sequences of channel input symbols. A systematic method for accomplishing this mapping is often referred to as *coded modulation*. For the BI-AWGN channel, the mapping can be simple and straightforward, say, $0 \mapsto \sqrt{P}$ and $1 \mapsto -\sqrt{P}$. For other constellations, there are many alternatives, among which we discuss three here. Our focus will be achievable rates, and the analysis will form a basis for such rates for network communication problems discussed in subsequent sections.

To be concrete, consider the uniformly spaced 4-PAM constellation

$$\mathcal{X} = \left\{ -\frac{3}{\sqrt{5}}, -\frac{1}{\sqrt{5}}, \frac{1}{\sqrt{5}}, \frac{3}{\sqrt{5}} \right\} \tag{2.13}$$

under the average power constraint $P = 1$ (see Figure 2.6). Note that each symbol $x \in \mathcal{X}$ can be expressed as a function of two BPSK symbols (or *layers*) u_1 and u_2 taking values in $\{-1, 1\}$ as

$$x = \frac{1}{\sqrt{5}} u_1 + \frac{2}{\sqrt{5}} u_2. \tag{2.14}$$

Note that this is not the unique mapping from two BPSK symbols to a 4-PAM symbol. For example, x can be generated from a Gray mapping

$$x = \frac{1}{\sqrt{5}} u_1 u_2 + \frac{2}{\sqrt{5}} u_2.$$

In both cases, if U_1 and U_2 are chosen independently and uniformly at random from $\{-1, 1\}$, then X is uniformly distributed over the 4-PAM constellation $\mathcal{X}$. We remark that, under the cost constraint (2.9), a uniform distribution $p(x)$ is suboptimal. We discuss a few approaches to shaping the distribution $p(x)$ shortly.

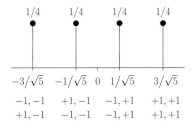

Figure 2.6 4-PAM constellation with uniform spacing, uniform probabilities, and two sets of labels for u_1 and u_2.

We now consider three alternatives for mapping binary codewords into a sequence x^n of transmitted symbols in $\mathcal{X}$. Figure 2.7 illustrates these three schemes. Common to all three schemes is *superposition coding* in which the input symbol is decomposed into multiple layers, and codewords are mapped into these layers and superimposed into x^n. The terminology *horizontal*, *vertical*, and *diagonal* is taken from the wireless communications literature [38, sec. 20.2.8].

In the first, *horizontal* superposition coding scheme, the message $m \in [2^{nR}]$ is split into two parts $m_1 \in [2^{nR_1}]$ and $m_2 \in [2^{nR_2}]$, where $R = R_1 + R_2$, that are encoded into respective codewords $u_1^n(m_1)$ and $u_2^n(m_2)$. These two codewords are then superimposed according to a specified symbol-by-symbol mapping, e.g., (2.14), to generate

$$x^n(m_1, m_2) = \frac{1}{\sqrt{5}} u_1^n(m_1) + \frac{2}{\sqrt{5}} u_2^n(m_2),$$

which is then transmitted over the channel. The receiver decodes y^n to recover m_1 and m_2 successively. It first finds the estimate $\hat{m}_1$ of m_1 from y^n (using, for example, ML or joint typicality decoding, or any low-complexity alternative) while treating $u_2^n(m_2)$ as noise (in other words, the decoder treats $u_2^n(m_2)$ as an i.i.d. sequence instead of a codeword, but it uses the correct distribution for the individual binary symbols). Assuming that the correct $\hat{m}_1$ was found, the receiver then incorporates $u_1^n(\hat{m}_1)$ into decoding and finds the estimate $\hat{m}_2$ of m_2 from $(y^n, u_1^n(\hat{m}_1))$.

When random coding is used, the performance of this *successive cancellation decoding* scheme can be characterized in a simple expression. Suppose that codewords $u_1^n(m_1)$ and $u_2^n(m_2)$ are generated independently and randomly according to $p(u_1)$ and $p(u_2)$, respectively. Following a similar analysis as presented in Section 2.1.1, decoding in the first step is successful w.h.p. if

$$R_1 < I(U_1; Y)$$

and decoding in the second step is successful w.h.p. if

$$R_2 < I(U_2; Y, U_1) = I(U_2; Y|U_1),$$

where the last identity follows since U_1 and U_2 are drawn from a product distribution $p(u_1)p(u_2)$. By allocating the rates R_1 and R_2 accordingly, the scheme can achieve the total rate

$$R < I(U_1; Y) + I(U_2; Y|U_1) = I(U_1, U_2; Y) = I(X; Y),$$

which is exactly what a random code on the original nonbinary input alphabet $\mathcal{X}$ would achieve. Note that each step in successive cancellation decoding involves a single

Figure 2.7 Illustration of three coded modulation schemes using horizontal, vertical, and diagonal superposition coding.

(binary) codeword. Multilevel coding (MLC) with multistage decoding (MSD) [39] implements this scheme in practical settings.

In the second, *vertical* superposition coding scheme, the message m is encoded into a long binary codeword $u^{2n}(m)$ of length $2n$. The first half $u^n(m)$ and the second half $u_{n+1}^{2n}(m)$ of the codeword are mapped to the u_1 and u_2 layers, respectively, and superimposed, for example, into

$$x^n(m) = \frac{1}{\sqrt{5}} u^n(m) + \frac{2}{\sqrt{5}} u_{n+1}^{2n}(m).$$

The receiver finds the unique message $\hat{m}$ such that *both* $u^n(m)$ and y^n are jointly typical (while treating $u_{n+1}^{2n}(m)$ as noise), *and* $u_{n+1}^{2n}(m)$ and y^n are jointly typical (while treating $u^n(m)$ as noise). The standard random coding analysis shows that decoding is successful w.h.p. if

$$R < I(U_1; Y) + I(U_2; Y), \tag{2.15}$$

which is in general less than $I(X; Y)$, with the gap depending on the symbol-by-symbol mapping. For the Gaussian channel, this gap can be made very small [39]. Bit-interleaved coded modulation (BICM) [40] implements this scheme, along with interleaving of the long codeword $u^{2n}(m)$, in practical settings.

The third, *diagonal* superposition scheme [41] communicates a stream of messages $m(1), m(2), \ldots$. As in BICM, each message $m(b)$ is encoded into a long codeword $u^{2n}(m(b))$. Blocks of n transmissions are used to communicate these codewords in a staggered manner. The first half $u^n(m(b))$ of the codeword is transmitted in block $b + 1$ using the u_1 layer, while the second half $u_{n+1}^{2n}(m(b))$ is transmitted in block b using the u_2 layer. Thus, in block b the transmitted sequence conveys information about both $m(b - 1)$ and $m(b)$ as

$$x^n(b) = \frac{1}{\sqrt{5}} u^n(m(b - 1)) + \frac{2}{\sqrt{5}} u_{n+1}^{2n}(m(b)).$$

The receiver finds $\hat{m}(1), \hat{m}(2), \ldots$ successively over decoding windows of two blocks. Assuming that $\hat{m}(1), \ldots, \hat{m}(b - 1)$ are correctly recovered, it finds the unique $\hat{m}(b)$ such that $u_{n+1}^{2n}(\hat{m}(b))$ and $(u^n(\hat{m}(b - 1)), y^n(b))$ are jointly typical, and $u^n(\hat{m}(b))$ and $y^n(b + 1)$ are jointly typical. This decoding method is called *sliding-window decoding* [42–44]. The standard random coding analysis shows that decoding for each message $m(b)$ is successful w.h.p. if

$$R < I(U_1; Y) + I(U_2; Y, U_1) = I(X; Y),$$

provided that the previous decoding steps were successful. Sliding-window coded modulation (SWCM) [45] implements this scheme in practical settings.

The three coded modulation approaches we have discussed so far used uniformly spaced constellations with uniform distributions. We now discuss techniques to *shape* the probability distribution of X beyond uniform spacing and the uniform distribution. We consider four approaches: superpose i.i.d. layers, many-to-one mapping, geometric shaping, and probabilistic shaping.

The first approach is conceptually simple [46]: superpose L layers $(u_1, \ldots, u_L)$ as in (2.14) but weight the u_i, $i = 1, 2, \ldots, L$, with the same factor $\sqrt{1/L}$. For example, for $L = 2$ the weights are $1/2$ rather than $1/\sqrt{5}$ and $2/\sqrt{5}$ as in (2.14). Observe that X becomes Gaussian for large L by the central limit theorem. One disadvantage of this approach is that it does not permit a fine control of $p(x)$. Another disadvantage is that the mapping from the L layers to X is a many-to-one mapping, e.g., for $L = 2$ the pairs $(u_1, u_2) = (-1, +1)$ and $(u_1, u_2) = (+1, -1)$ both map to $X = 0$.

The second approach to shaping uses many-to-one mappings more general than the ones resulting from superposition, as proposed by Gallager [5, p. 208]. For example, a Bern(1/4) input distribution can be generated by mapping $(u_1, u_2) = (+1, +1)$ to $x = 1$ and mapping the other three pairs to $x = 0$. In addition to the inherent disadvantage of many-to-one mappings, the size of the underlying alphabet (or equivalently, the number of layers) should grow very large in order to permit a fine control of $p(x)$.

The third approach to shaping, called geometric shaping, is to carefully place the constellation points so that a uniform distribution mimics a Gaussian distribution [47]. For example, instead of using the constellation (2.13), one could use the nonuniformly spaced constellation

$$\mathcal{X} = \left\{ -\frac{\sqrt{17}}{3}, -\frac{1}{3}, \frac{1}{3}, \frac{\sqrt{17}}{3} \right\},$$

as shown in Figure 2.8. One disadvantage of geometric shaping is that the effective number of bits (ENOB) of an analog-to-digital converter (ADC) with a linear uniform quantizer is reduced as compared to a uniformly spaced constellation. Moreover, geometric shaping is not as flexible as, and does not perform as well as, probabilistic shaping in general [48].

The fourth approach to shaping, called probabilistic shaping, generally uses a uniformly spaced constellation as in (2.13) but nonuniform probabilities, as shown in Figure 2.9. Observe that the nonuniform distribution lets one stretch the spacing so that the constellation is

$$\mathcal{X} = \left\{ -\frac{3}{\sqrt{3}}, -\frac{1}{\sqrt{3}}, \frac{1}{\sqrt{3}}, \frac{3}{\sqrt{3}} \right\}.$$

One can create nonuniform probabilities by using a *distribution matching* algorithm that takes as input independent and uniformly distributed bits and puts out symbols from $\mathcal{X}$

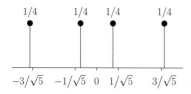

Figure 2.8 4-PAM constellation with geometric shaping.

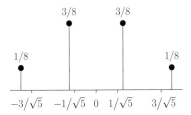

Figure 2.9 4-PAM constellation with probabilistic shaping.

with a desired distribution [49]. For example, a common approach in industrial applications is to use *shell mapping* [50] and more recently *constant composition distribution matching* [51]; the methods are compared by Schulte and Steiner [52]. Distribution matching algorithms add redundancy that reduces the rate from L bits per symbol (or more generally $\sum_{i=1}^{L} H(U_i)$ bits per symbol) to $H(U_1, \ldots, U_L)$ bits per symbol. If the receiver decodes as for vertical superposition by treating layers as noise, then the rate (2.15) for $L = 2$ layers changes to

$$R < H(U_1, U_2) - H(U_1|Y_1) - H(U_2|Y_2) \tag{2.16a}$$
$$= I(U_1; Y_1) + I(U_2; Y_2) - I(U_1; U_2), \tag{2.16b}$$

assuming that the expressions on the RHS of (2.16) are positive. In other words, distribution matching induces a rate loss of $I(U_1; U_2)$, a phenomenon that we will encounter again in *Marton coding* for broadcast channels (cf. (2.51) in Section 2.3.3). The key insight here, as for Marton coding, is that we gain more through the dependence of U_1 and U_2 than we lose by sacrificing the rate $I(U_1; U_2)$. For AWGN channels, the gain is in power efficiency (see the stretched spacing in Figure 2.9), while for broadcast channels the gain is because the transmitter can control interference more carefully (see Section 2.3.3 for details). Finally, we remark that the preferred approach is to apply shaping before channel coding. A layered, rate adaptive, and systematic architecture for this approach was recently proposed [49] and implemented in commercial systems [53, 54].

2.1.7 Multiple-Input Multiple-Output Systems

The Gaussian vector channel

$$\mathbf{Y} = \mathsf{H}\mathbf{X} + \mathbf{Z} \tag{2.17}$$

is a simple model for MIMO wireless communication systems (see Figure 2.10). Here, $\mathbf{Y}$ is an r-dimensional complex column vector, $\mathbf{X}$ is a t-dimensional complex column vector, $\mathbf{Z} \sim \mathrm{CN}(0, \mathsf{K_Z})$ is a r-dimensional complex and circularly symmetric noise column vector with covariance matrix $\mathsf{K_Z}$, and H is an $r \times t$ constant complex channel matrix with its element H_{jk} representing the channel coefficient from transmitter antenna k to receiver antenna j. For simplicity, we assume that the entries of the noise vector are statistically independent, and thus that the noise covariance matrix is identity ($\mathsf{K_Z} = \mathsf{I}_r$).

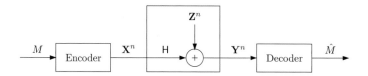

Figure 2.10 MIMO point-to-point communication system.

We consider a block power constraint P on every codeword $\mathbf{x}^n(m) = (\mathbf{x}_1(m), \ldots, \mathbf{x}_n(m))$, i.e., we have

$$\frac{1}{n} \sum_{i=1}^{n} \mathbf{x}_i^\dagger(m)\mathbf{x}_i(m) \le P, \quad m \in [2^{nR}], \tag{2.18}$$

where $\mathbf{x}^\dagger$ is the complex-conjugate transpose of $\mathbf{x}$.

The capacity of the Gaussian vector channel is

$$C = \max_{\mathsf{K_X} \succeq 0: \mathrm{tr}(\mathsf{K_X}) \le P} \log |\mathsf{I}_r + \mathsf{H}\mathsf{K_X}\mathsf{H}^\dagger|, \tag{2.19}$$

which is derived by evaluating the capacity under the input cost constraint in (2.10) as in the single-antenna case. Suppose that the channel matrix H has rank $d \le \min(t, r)$ and singular value decomposition $\mathsf{H} = \Phi\Gamma\Psi^\dagger$, where Φ is a complex $r \times d$ matrix with $\Phi^\dagger\Phi = \mathsf{I}_d$, Ψ is a complex $t \times d$ matrix with $\Psi^\dagger\Psi = \mathsf{I}_d$, and $\Gamma = \mathrm{diag}(\gamma_1, \ldots, \gamma_d)$ is a $d \times d$ diagonal matrix of d (real and positive) singular values. Then the optimization of the input covariance matrix $\mathsf{K_X}$ can be carried out in terms of these d singular values as

$$C = \max_{\Pi} \log |\mathsf{I}_d + \Gamma\Pi\Gamma^\dagger| \tag{2.20a}$$

$$= \max_{\pi_1, \ldots, \pi_d} \sum_{j=1}^{d} \log(1 + \gamma_j^2 \pi_j), \tag{2.20b}$$

where $\Pi = \Psi^\dagger\mathsf{K_X}\Psi$ in the first maximization is a diagonal covariance matrix satisfying $\mathrm{tr}(\Pi) \le P$, and where $\pi_1, \ldots, \pi_d$ in the second maximization are the diagonal entries of Π satisfying $\sum_{j=1}^{d} \pi_j \le P$. The optimal values $\pi_1^*, \ldots, \pi_d^*$ can be expressed [55, sec. IX] as

$$\pi_j^* = \max\left(0, \lambda - \frac{1}{\gamma_j^2}\right), \tag{2.21}$$

where λ satisfies $\sum_{j=1}^{d} \pi_j = P$. This solution has a water-filling interpretation as illustrated in Figure 2.11.

The capacity expressions in (2.19) and (2.20) suggest a simple coding scheme through *transmit beamforming* (also referred to as *precoding*) with right singular vectors $\boldsymbol{\psi}_1, \ldots, \boldsymbol{\psi}_d$ of the channel matrix H (namely, the columns of Ψ), and *receive beamforming* (also referred to as *linear equalization*) with left singular vectors $\boldsymbol{\phi}_1, \ldots, \boldsymbol{\phi}_d$ of H (namely, the columns of Φ). For a d-dimensional column vector $\mathbf{U}$, let $\mathbf{X} = \Psi\mathbf{U}$ and

$$\mathbf{V} = \Phi^\dagger\mathbf{Y} = \Phi^\dagger(\Phi\Gamma\Psi^\dagger(\Psi\mathbf{U}) + \mathbf{Z})$$
$$= \Gamma\mathbf{U} + \mathbf{W},$$

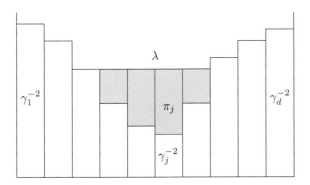

Figure 2.11 Water-filling interpretation of optimal $\pi_1^*, \ldots, \pi_d^*$.

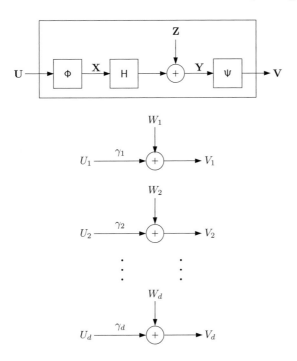

Figure 2.12 Transmit and receive beamforming and the orthogonal subchannels formed by them.

where $\mathbf{W} = \Phi^\dagger \mathbf{Z} \sim \mathrm{CN}(0, \mathsf{I}_d)$. Since Γ is diagonal, each complex symbol $U_j, j \in [d]$, transmitted in the direction of the transmit beamforming vector $\boldsymbol{\psi}_j$, can be received free of interference by projecting the channel output in the direction of the receive beamforming vector $\boldsymbol{\phi}_j$. Figure 2.12 illustrates the overall transmit and receive beamforming scheme that transforms the original Gaussian vector channel into d orthogonal subchannels from U_j to $V_j, j \in [d]$, thus bringing in new spatial degrees of freedom.

Similar to coded modulation discussed in Section 2.1.6, there are several alternatives to transmitting codewords (or more precisely, sequences of complex symbols generated from codewords via coded modulation) over the subchannels thus formed. Identifying each subchannel input U_j as a symbol layer as in coded modulation leads naturally to the following three transmission schemes. First, we can communicate d sequences of length n, say, $u_1^n, \ldots, u_d^n$ separately over the d layers. This is analogous to horizontal superposition coding or MLC. Second, we can communicate a single long sequence u^{nd} across the d layers vertically, essentially forming an n-sequence of d-dimensional complex vector symbols. This is analogous to vertical superposition coding or BICM. Third, we can communicate a stream of long sequences $u^{nd}(1), u^{nd}(2), \ldots$ across the d layers in a staggered manner. This is analogous to diagonal superposition coding or SWCM. Under optimal beamforming and power allocation, all three schemes can achieve the capacity of the Gaussian vector channel in (2.19) and (2.20), provided that scalar channel coding and coded modulation are perfect.

These transmission schemes provide canonical MIMO transmission architectures even when optimal beamforming is not feasible due to fading and inaccurate or limited CSI at the sender. In this case, symbols $X_1, \ldots, X_t$ at the transmitter antennas are typically used as signal layers, interfering with each other at the receiver. Bell Labs layered space-time (BLAST) architectures [56, 57] implement the three transmission schemes in practical settings.

The capacity gain from using multiple antennas without beamforming can still be significant when compared to the single-antenna counterpart. For fast fading, the capacity for a random channel matrix H with CSI at the receiver is

$$C = \sum_{j=1}^{\min(t,r)} \mathsf{E}\left[\log\left(1 + \frac{P}{t}\Lambda_j\right)\right], \tag{2.22}$$

where $\Lambda_1, \ldots, \Lambda_{\min(t,r)}$ are the random nonzero eigenvalues of $\mathsf{HH}^\dagger$. This capacity is achieved without beamforming, i.e., the best covariance matrix is $\mathsf{K_X} = (P/t)\mathsf{I}_t$. In a rich scattering environment, such as Rayleigh fading (say, $\mathsf{H}_{jk} \sim \mathrm{CN}(0,1)$ and the channel coefficients are independent), the capacity in (2.22) scales linearly in r at low SNR and linearly in $\min(t,r)$ at high SNR [56, 58].

2.2   Multiple Access Channels

The multiple access channel (MAC) provides a simple model for uplink communication in which multiple senders (user devices) communicate to a common receiver (base station). We start our discussion with the two-sender MAC depicted in Figure 2.13. Each sender $j = 1, 2$ wishes to communicate its message $m_j \in [2^{nR_j}]$ over a shared communication channel by encoding the message into a codeword x_j^n and transmitting the codeword over the channel. Transmissions from the senders take place simultaneously in general, with the output symbol y from each transmitted input symbol pair (x_1, x_2) drawn according to the conditional probability $p_{Y|X_1,X_2}(y|x_1, x_2)$. Thus, the conditional probability of receiving the output sequence y^n when m_1 and m_2 are communicated is

$$p(y^n|m_1, m_2) = \prod_{i=1}^{n} p_{Y|X_1,X_2}(y_i|x_{1i}(m_1), x_{2i}(m_2)).$$

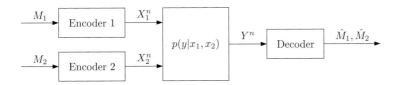

Figure 2.13 Multiple access channel with two senders.

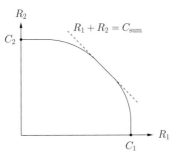

Figure 2.14 The capacity region $\mathscr{C}$ of a MAC.

The receiver forms a pair of estimates $(\hat{m}_1, \hat{m}_2)$ by decoding y^n.

The encoder mappings $x_1^n(m_1)$ and $x_2^n(m_2)$ and the decoder mappings $\hat{m}_1(y^n)$ and $\hat{m}_2(y^n)$ are referred to as a $(2^{nR_1}, 2^{nR_2}, n)$ code. The average error probability of the $(2^{nR_1}, 2^{nR_2}, n)$ code is

$$P_e^{(n)} = \mathsf{P}(M_1 \neq \hat{M}_1 \text{ or } M_2 \neq \hat{M}_2),$$

where we assume that M_1 and M_2 are independent and uniformly distributed. The optimal tradeoff between the rates R_1 and R_2 of reliable communication is our primary concern. More precisely, we say that a rate pair (R_1, R_2) is *achievable* if, for any target error probability $\epsilon > 0$ and sufficiently large n, there exists a $(2^{nR_1}, 2^{nR_2}, n)$ code with $P_e^{(n)} \leq \epsilon$. As for point-to-point channels, this can be stated in terms of a sequence of codes, i.e., (R_1, R_2) is achievable if there exists a sequence of $(2^{nR_1}, 2^{nR_2}, n)$ codes such that $\lim_{n \to \infty} P_e^{(n)} = 0$. The *capacity region* $\mathscr{C}$ of the MAC is the closure of the set of achievable (R_1, R_2), and it has the form sketched in Figure 2.14. Sometimes we are interested in the sum-capacity

$$C_{\text{sum}} = \max\{R_1 + R_2 : (R_1, R_2) \in \mathscr{C}\},$$

namely, the largest achievable sum-rate. Information theory for MACs was developed in work by Ahlswede and by Liao [59, 60]; see also [61, sec. 17]. A useful collection of classical results can also be found [62–64].

2.2.1   Basic Signaling Schemes

Let C_1 be the capacity from sender 1 to the receiver, when sender 2 is not communicating anything (which, in our model, is equivalent to transmitting a fixed symbol, say, zero). By Shannon's channel coding theorem in Section 2.1.2, we have

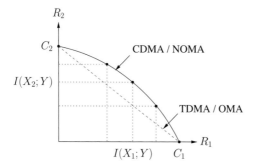

Figure 2.15 Rate regions achieved with TDMA/OMA, and CDMA/NOMA when treating interference as noise. For different channel coefficients, the TDMA/OMA curve may also lie above the CDMA/NOMA curve.

$$C_1 = \max_{p(x_1), x_2} I(X_1; Y | X_2 = x_2). \tag{2.23a}$$

Similarly, let

$$C_2 = \max_{p(x_2), x_1} I(X_2; Y | X_1 = x_1) \tag{2.23b}$$

be the individual capacity of sender 2. Then by having sender 1 transmit a fraction α of the time, and sender 2 transmit a fraction $1 - \alpha$ of the time, the senders can avoid interference and achieve rate pairs (R_1, R_2) satisfying

$$R_1 < \alpha C_1$$
$$R_2 < (1 - \alpha)C_2.$$

The rate pairs achievable by this *time sharing* or *time division multiple access* (TDMA) are sketched in Figure 2.15. More generally, any *orthogonal multiple access* (OMA) scheme that divides orthogonal resources such as frequency among the senders achieves the same rates.

Another simple signaling method is for the senders to transmit $x_1^n(m_1)$ and $x_2^n(m_2)$ simultaneously, i.e., they use *non-orthogonal multiple access* (NOMA) signaling. A conceptually simple receiver decodes y^n to recover the message m_1 by treating the signal $x_2^n(m_2)$ as noise, and similarly recovers m_2 by treating $x_1^n(m_1)$ as noise. This approach is usually used with spread-spectrum or code division multiple access (CDMA) signals. When codewords are generated randomly and independently according to $p(x_1)$ and $p(x_2)$, decoding is successful w.h.p. if

$$R_1 < I(X_1; Y)$$
$$R_2 < I(X_2; Y).$$

One may now choose different $p(x_1)$ and $p(x_2)$ to sweep out a set of rate points, as shown by the solid line in Figure 2.15.

2.2.2   Successive Cancellation Decoding

It turns out that the NOMA rates can be improved, in general, if the receiver uses a decoder that does not simply treat interference as noise. We illustrate this point by discussing the achievable rate region of a successive cancellation decoder, where the receiver decodes y^n to recover m_1 and m_2 successively as in MLC (see Section 2.1.6). The receiver first finds the estimate $\hat{m}_1$ of m_1 from y^n while treating $x_2^n(m_2)$ as part of the noise, i.e., we assume that $x_2^n(m_2)$ is an i.i.d. sequence. As above, when codewords are generated randomly and independently according to $p(x_1)$ and $p(x_2)$, decoding is successful w.h.p. if

$$R_1 < I(X_1; Y). \tag{2.24a}$$

Assuming the correct $\hat{m}_1$ was found, the receiver then incorporates $x_1^n(\hat{m}_1)$ into decoding and finds the estimate $\hat{m}_2$ of m_2 from $(y^n, x_1^n(\hat{m}_1))$. Decoding in the second step is successful w.h.p. if

$$R_2 < I(X_2; Y, X_1) = I(X_2; Y|X_1), \tag{2.24b}$$

where the last identity follows since X_1 and X_2 are drawn from a product distribution $p(x_1)p(x_2)$. Thus, any rate pair (R_1, R_2) satisfying (2.24) is achievable, namely, there exists a $(2^{nR_1}, 2^{nR_2}, n)$ code of rates R_1 and R_2 arbitrarily close to $I(X_1; Y)$ and $I(X_2; Y|X_1)$, respectively, with vanishingly small probability of error.

By changing the order of decoding, the receiver can recover m_2 and m_1 successively. Decoding in this case is successful w.h.p. if

$$\begin{aligned} R_1 &< I(X_1; Y|X_2) \\ R_2 &< I(X_2; Y). \end{aligned} \tag{2.25}$$

The rate pairs achievable by the two successive cancellation decoding schemes ($m_1 \rightarrow m_2$ and $m_2 \rightarrow m_1$) are sketched in Figure 2.16.

We can now apply time sharing as for TDMA/OMA in Section 2.2.1. Suppose that, for the first fraction α of a transmission block, we use a code that achieves the rates R_1' and R_2' reliably, while for the second fraction $1 - \alpha$ of this transmission block, we use a code that achieves the rates R_1'' and R_2'' reliably. Then the overall achievable rate pair

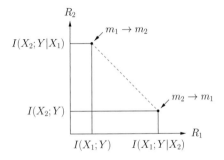

Figure 2.16 Rate pairs achieved with two successive cancellation decoding schemes.

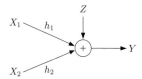

Figure 2.17 Gaussian MAC.

(R_1, R_2) of the combined code is between two rate pairs (R'_1, R'_2) and (R''_1, R''_2), i.e., we have

$$R_1 = \alpha R'_1 + (1 - \alpha) R''_1$$
$$R_2 = \alpha R'_2 + (1 - \alpha) R''_2.$$

That is, the final rate pair is a *convex combination* of the original rate pairs. In this way, we can achieve all rate pairs on the dashed line in Figure 2.16.

We remark that both successive cancellation decoding schemes achieve the sum-rate as high as

$$I(X_1, X_2; Y) = I(X_1; Y) + I(X_2; Y|X_1) = I(X_1; Y|X_2) + I(X_2; Y),$$

which improves upon time division in general. In fact, this sum-rate can be shown to be the sum-capacity

$$C_{\text{sum}} = \max_{p(x_1)p(x_2)} I(X_1, X_2; Y) \tag{2.26}$$

of the MAC when the choice of $p(x_1)$ and $p(x_2)$ is optimized.

We illustrate successive cancellation decoding through the following concrete example. The (complex alphabet) Gaussian multiple access channel

$$Y = h_1 X_1 + h_2 X_2 + Z$$

is a simple model for wireless uplink communication systems (see Figure 2.17). Here, Y is the output of the channel, X_1 and X_2 are the inputs, $Z \sim \text{CN}(0, 1)$ is the noise, and h_1 and h_2 are the channel coefficients. We consider the block transmit power constraints

$$\frac{1}{n} \sum_{i=1}^{n} |x_{ji}(m_j)|^2 \le P_j, \quad m_j \in [2^{nR_j}]$$

for each sender $j = 1, 2$. As in the point-to-point case, we denote the received SNRs as $S_j = |h_j|^2 P_j$, $j = 1, 2$. The individual capacities are $C_1 = \log(1 + S_1)$ and $C_2 = \log(1 + S_2)$.

Successive cancellation decoding for the Gaussian MAC is depicted in Figure 2.18. The receiver first finds $\hat{m}_1$ by decoding

$$y^n = h_1 x_1^n(m_1) + h_2 x_2^n(m_2) + z^n,$$

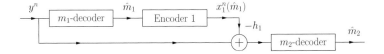

Figure 2.18 Successive cancellation decoding for the MAC.

while treating $h_2 x_2^n(m_2)$ as part of the noise. For random coding under the input distributions $X_1 \sim CN(0, P_1)$ and $X_2 \sim CN(0, P_2)$ that achieve the individual capacities, decoding in the first step is successful w.h.p. if

$$R_1 < I(X_1; Y) = \log\left(1 + \frac{S_1}{1 + S_2}\right), \tag{2.27}$$

where $S_1/(1 + S_2)$ can be recognized as the signal-to-interference-noise-ratio (SINR). The receiver then subtracts $h_1 x_1^n(\hat{m}_1)$ from y^n and finds $\hat{m}_2$ by decoding

$$y^n - h_1 x_1^n(\hat{m}_1) = h_2 x_2^n(m_2) + z^n + h_1(x_1^n(m_1) - x_1^n(\hat{m}_1)),$$

where the last term is zero if m_1 was correctly recovered as $\hat{m}_1$. Decoding in the second step is thus successful w.h.p. if

$$R_2 < I(X_2; Y|X_1) = I(X_2; Y - h_1 X_1) = \log(1 + S_2). \tag{2.28}$$

Combining (2.27) and (2.28), successive cancellation decoding achieves the sum-rate

$$\log\left(1 + \frac{S_1}{1 + S_2}\right) + \log(1 + S_2) = \log(1 + S_1 + S_2),$$

which is the sum-capacity of the Gaussian MAC under block transmit power constraints. Achievable rates for discrete inputs can be characterized similarly as the mutual information expressions in (2.27) and (2.28), which can be computed numerically as in the point-to-point case.

2.2.3　Capacity Region

Suppose that we use time sharing between the two successive cancellation decoding schemes ($m_1 \to m_2$ and $m_2 \to m_1$) that achieve the rate regions in (2.24) and (2.25). The combined code with $(\alpha, 1 - \alpha)$ time sharing can achieve any rate pair (R_1, R_2) satisfying

$$R_1 < \alpha I(X_1; Y) + (1 - \alpha)I(X_1; Y|X_2) \tag{2.29a}$$
$$R_2 < \alpha I(X_2; Y|X_1) + (1 - \alpha)I(X_2; Y). \tag{2.29b}$$

This region, when evaluated over all possible values of $\alpha \in [0, 1]$, is equivalent to the pentagon $\mathscr{R}(X_1, X_2)$ depicted in Figure 2.19 that consists of nonnegative rate pairs (R_1, R_2) satisfying

$$R_1 < I(X_1; Y|X_2) \tag{2.30a}$$
$$R_2 < I(X_2; Y|X_1) \tag{2.30b}$$
$$R_1 + R_2 < I(X_1, X_2; Y). \tag{2.30c}$$

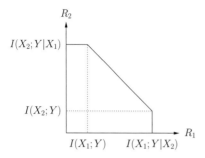

Figure 2.19 The region $\mathscr{R}(X_1, X_2)$ for a typical MAC.

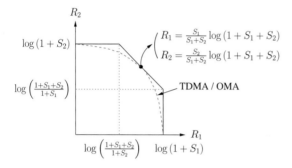

Figure 2.20 The capacity region of the Gaussian MAC is the solid pentagon. TDMA/OMA with block power constraints can achieve the rates shown by the dashed line.

For the Gaussian MAC, the region $\mathscr{R}(X_1, X_2)$ for $X_1 \sim CN(0, P_1)$ and $X_2 \sim CN(0, P_2)$ is the set of (R_1, R_2) such that

$$R_1 < \log(1 + S_1) \tag{2.31a}$$
$$R_2 < \log(1 + S_2) \tag{2.31b}$$
$$R_1 + R_2 < \log(1 + S_1 + S_2). \tag{2.31c}$$

These three bounds match the individual capacities and sum-capacity in (2.23) and (2.26), and thus (the closure of) this region is the capacity region of the Gaussian MAC; see Figure 2.20.

We remark that TDMA/OMA can achieve higher rates than those suggested by Figure 2.15, as long as the transmitters have block power constraints. The reason is that sender 1 can transmit with power P_1/α during its time slot of length αn, and sender 2 can transmit with power $P_2/(1 - \alpha)$ during its time slot of length $(1 - \alpha)n$. The rates are then

$$R_1 < \alpha \log\left(1 + \frac{S_1}{\alpha}\right)$$
$$R_2 < (1 - \alpha)\log\left(1 + \frac{S_2}{1 - \alpha}\right),$$

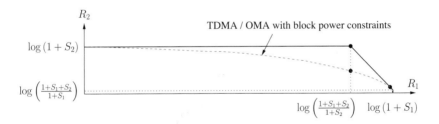

Figure 2.21 The capacity region of the Gaussian MAC, and rates achieved by TDMA/OMA with block power constraints.

and one can sweep out the dashed curve in Figure 2.20 by varying $\alpha \in [0, 1]$. In particular, the choice $\alpha = S_1/(S_1 + S_2)$ gives a point that achieves the sum-capacity. Figure 2.20 suggests that TDMA/OMA might be "good enough" for most applications, as the TDMA/OMA rates almost fill out the entire capacity region. However, the plot is misleading because the SNRs S_1 and S_2 were chosen to be similar, while for cellular wireless applications the SNRs are often very different. For example, suppose that $S_1 \gg S_2$, in which case the rate regions are as shown in Figure 2.21. The interesting point is the corner point of $\mathscr{C}$ where $R_2 = \log(1 + S_2)$, which can be achieved by NOMA with successive cancellation decoding. Here, sender 1 has given up a small fraction of its rate while boosting the rate of sender 2 by approximately a factor of two as compared to using TDMA/OMA.

Returning to a generic MAC, such as the Gaussian *vector* MAC discussed in Section 2.2.8, the capacity region is the convex closure of multiple $\mathscr{R}(X_1, X_2)$ regions over all $p(x_1)p(x_2)$ as depicted in Figure 2.22. At a conceptual level, achieving the capacity region is fairly straightforward. Any rate pair in a single $\mathscr{R}(X_1, X_2)$ region can be achieved by time sharing between the point-to-point codes that respectively achieve the two corner points of the pentagon as illustrated in Figure 2.16. Any rate pair that is not covered by some $\mathscr{R}(X_1, X_2)$ region can be achieved by time sharing between *two* $\mathscr{R}(X_1, X_2)$ regions. In summary, a proper combination of good point-to-point channel codes, successive cancellation decoding, and time sharing suffices to achieve the entire capacity region.

Each $\mathscr{R}(X_1, X_2)$ region in (2.30) for the MAC $p(y|x_1, x_2)$ plays the role of the interval $[0, I(X; Y))$ for the point-to-point channel $p(y|x)$. So far, we have seen that $\mathscr{R}(X_1, X_2)$ can be achieved by successive cancellation decoding and time sharing. In the next three sections, we present alternative coding schemes that achieve all points in the $\mathscr{R}(X_1, X_2)$ region without time sharing, either by employing a more sophisticated decoding scheme, or by introducing a new encoding scheme.

2.2.4 Simultaneous Decoding

Suppose that, rather than decoding for one message at a time, the receiver decodes for both messages at the same time. Intuitively, such a *simultaneous* decoding scheme, albeit

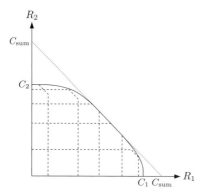

Figure 2.22 The capacity region of a generic MAC. The individual capacities C_1 and C_2, and the sum-capacity C_{sum} are characterized in (2.23) and (2.26).

more complex than successive cancellation decoding, can achieve higher rates since neither codeword is treated as noise. To characterize the achievable rates, it is instructive to consider joint typicality decoding, in which the receiver declares that the message pair $(\hat{m}_1, \hat{m}_2)$ was sent if it is a unique pair such that $x_1^n(\hat{m}_1), x_2^n(\hat{m}_2)$, and y^n are jointly typical. If $(m_1, m_2) = (1, 1)$, then the corresponding randomly generated codewords $X_1^n(1), X_2^n(1)$ are jointly typical with Y^n w.h.p. To compute the probability that wrong codewords are jointly typical with the channel output, we consider three separate cases:

1. a wrong codeword $X_1^n(m_1)$, $m_1 \neq 1$, is jointly typical with the correct codeword $X_2^n(1)$ and Y^n;
2. a wrong codeword $X_2^n(m_2)$, $m_2 \neq 1$, is jointly typical with the correct codeword $X_1^n(1)$ and Y^n;
3. wrong codewords $X_1^n(m_1)$ and $X_2^n(m_2)$, $m_1 \neq 1$, $m_2 \neq 1$, are jointly typical with Y^n.

Following a similar analysis as for the point-to-point case in Section 2.1.1, the probabilities of these error events vanish for large n if the rate pair (R_1, R_2) satisfy the bounds in (2.30). Thus, each rate pair in $\mathcal{R}(X_1, X_2)$, and in particular an arbitrary tradeoff between individual rates that give the sum-capacity C_{sum}, can be achieved by using a single point-to-point code at each sender without time sharing. In fact, by slightly modifying the encoding structure through a *coded time sharing* technique [2, sec. 4.5.3], each rate pair in the entire *capacity region* in Figure 2.22 can be achieved by a single code without time sharing.

Unlike successive cancellation decoding, in which the decoder handles a single codeword in each step, simultaneous decoding involves multiple codewords at a time. For most channel codes, there is no known low-complexity implementation for simultaneous decoding. Two closely related methods to approach the performance of simultaneous decoding are soft interference cancellation [65, 66] and message passing on factor graphs [67, 68, sec. 5.5]. Furthermore, polar codes [69] and spatially coupled codes [70] let one implement simultaneous decoding to achieve rates comparable to $\mathcal{R}(X_1, X_2)$.

2.2.5 Offset Transmission

Consider the transmission scheme in Figure 2.23, in which streams of messages $m_1(1), m_2(1), m_1(2), m_2(2), \ldots$ are communicated by staggering the transmission of the codewords $x_1^n(m_1(1)), x_2^n(m_2(1)), x_1^n(m_1(2)), x_2^n(m_2(2)), \ldots$ by alternating offsets of $(1 - \alpha)n$ and αn symbols.

The receiver finds $\hat{m}_1(1), \hat{m}_2(1), \hat{m}_1(2), \hat{m}_2(2), \ldots$ successively by sliding a decoding window of length n by $(1 - \alpha)n$ and αn symbols. For example, after observing the first n symbols y^n, the receiver decodes for $\hat{m}_1(1)$ while treating $x_2^{\alpha n}(m_2(1))$ as noise. Hence, decoding is successful w.h.p. if (2.29a) holds. The estimate $\hat{m}_1(b)$ thus recovered can then be incorporated into decoding for $\hat{m}_2(b)$, which is successful w.h.p. if (2.29b) holds. In summary, each rate pair (R_1, R_2) satisfying (2.29) can be achieved by using a single point-to-point code at each sender (instead of time sharing between two codes), and so can the sum-capacity C_{sum} with an arbitrary tradeoff between individual rates. Practical implementations of this offset transmission scheme have been reported [71]. Offset transmission is also useful for other channel models with multiple senders [72, 73].

2.2.6 Rate Splitting

Consider the encoding structure in Figure 2.24. The message $m_1 \in [2^{nR_1}]$ is split into two parts $m_{11} \in [2^{nR_{11}}]$ and $m_{12} \in [2^{nR_{12}}]$ with $R_1 = R_{11} + R_{12}$ that are encoded separately into two codewords $u_{11}^n(m_{11})$ and $u_{12}^n(m_{12})$. As in superposition coding for MLC (see Section 2.1.6), these codewords are mapped to the channel input x_1^n by a symbol-by-symbol mapping $x_1(u_{11}, u_{12})$, which is then transmitted.

The receiver uses successive cancellation decoding to find $\hat{m}_{11}, \hat{m}_2$, and $\hat{m}_{12}$ in order; see also Figure 2.25. First, it finds $\hat{m}_{11}$ by treating $x_2^n(m_2)$ and $u_{12}^n(m_{12})$ as noise. Second, it finds $\hat{m}_2$ by incorporating $u_{11}^n(\hat{m}_{11})$ into decoding while treating $u_{12}^n(m_{12})$ as noise. Finally, it finds $\hat{m}_{12}$ by incorporating $u_{11}^n(\hat{m}_{11})$ and $x_2^n(\hat{m}_2)$ into decoding. Following a

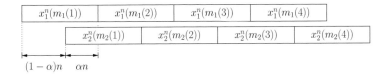

Figure 2.23 Offset transmission of codewords.

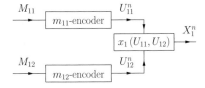

Figure 2.24 Rate-splitting at encoder 1.

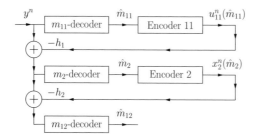

Figure 2.25 Successive cancellation decoder for rate-splitting and a Gaussian MAC.

similar analysis of successive cancellation decoding for two codewords in Section 2.2.2, when a random code generated from the product distribution $p(u_{11})p(u_{12})p(x_2)$ is used, decoding is successful w.h.p. if

$$R_{11} < I(U_{11};Y)$$
$$R_2 < I(X_2;Y|U_{11})$$
$$R_{12} < I(U_{12};Y|U_{11},X_2).$$

By combining the split rates R_{11} and R_{12} and choosing different superposition functions and code distributions, the scheme can achieve the rate pairs (R_1, R_2) satisfying

$$R_1 < I(U_{11};Y) + I(U_{12};Y|U_{11},X_2) \tag{2.32a}$$
$$R_2 < I(X_2;Y|U_{11}) \tag{2.32b}$$

for any $x_1(u_{11}, u_{12})$ and $p(u_{11})p(u_{12})p(x_2)$.

As a simple example, consider the Gaussian MAC in Figure 2.17; the successive cancellation decoder is shown in Figure 2.25. We choose $U_{11} \sim CN(0, \alpha P_1)$, $U_{12} \sim CN(0, (1 - \alpha)P_1)$, and $X_2 \sim CN(0, P_2)$, independent of each other, for some $\alpha \in [0, 1]$. We also choose the symbol-by-symbol mapping as $X_1 = U_{11} + U_{12} \sim CN(0, P_1)$. Thus the transmitted sequence $x_1^n = u_{11}^n(m_{11}) + u_{12}^n(m_{12})$ is a simple superposition of two codewords. The achievable rate region in (2.32) then becomes

$$R_1 < \log\left(1 + \frac{\alpha S_1}{1 + (1 - \alpha)S_1 + S_2}\right) + \log(1 + (1 - \alpha)S_1)$$
$$R_2 < \log\left(1 + \frac{S_2}{1 + (1 - \alpha)S_1}\right).$$

By varying $\alpha \in [0, 1]$, this region sweeps the entire capacity region for the Gaussian MAC in (2.31).

More generally, this *rate-splitting multiple access* scheme [74, 75] can achieve any $\mathcal{R}(X_1, X_2)$ rate region for a general MAC. This can be seen by the sum-rate

$$I(U_{11};Y) + I(U_{12};Y|U_{11},X_2) + I(X_2;Y|U_{11}) = I(X_1,X_2;Y)$$

in (2.32), which can be traded off between individual rates by choosing the input distribution $p(u_{11})p(u_{12})$ and the symbol-by-symbol mapping $x_1(u_{11}, u_{12})$ appropriately. In practice, however, the choice of feasible code distributions is somewhat limited and not every rate pair (R_1, R_2) in $\mathscr{R}(X_1, X_2)$ is achievable.

2.2.7   Multiple Access Channels with Many Senders

The capacity region for the MAC, as well as basic coding schemes we have discussed so far, such as time sharing, successive cancellation decoding, simultaneous decoding, offset transmission, and rate splitting, can be readily generalized to more than two senders [76, sec. 7.1]. For example, successive cancellation decoding over a k-sender MAC $p(y|x_1, x_2, \dots, x_k)$ for random codes generated from $p(x_1)p(x_2) \cdots p(x_k)$ can achieve any rate tuple $(R_1, R_2, \dots, R_k)$ such that

$$R_1 < I(X_1; Y)$$
$$R_2 < I(X_2; Y|X_1)$$
$$R_3 < I(X_3; Y|X_1, X_2)$$
$$\vdots$$
$$R_k < I(X_k; Y|X^{k-1}).$$

Similarly, the capacity region is the convex closure of all $\mathscr{R}(X_1, \dots, X_k)$ regions, each of which consists of all rate tuples $(R_1, \dots, R_k)$ such that

$$\sum_{j \in \mathcal{J}} R_j < I(X(\mathcal{J}); Y|X(\mathcal{J}^c)), \quad \mathcal{J} \subseteq [k], \tag{2.33}$$

where $X(\mathcal{J}) = (X_j : j \in \mathcal{J})$ and $[k] = \{1, \dots, k\}$. In general, each $\mathscr{R}(X_1, \dots, X_k)$ region is of a *polymatroidal* shape, namely, a multidimensional generalization of a pentagon. Figure 2.26 illustrates a typical $\mathscr{R}(X_1, X_2, X_3)$ for $k = 3$, which is characterized by $7 = 2^3 - 1$ inequalities:

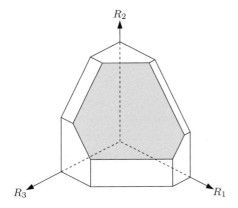

Figure 2.26 A typical $\mathscr{R}(X_1, X_2, X_3)$ region of a polymatroidal shape.

$$R_1 < I(X_1; Y|X_2, X_3)$$
$$R_2 < I(X_2; Y|X_1, X_3)$$
$$R_3 < I(X_3; Y|X_1, X_2)$$
$$R_1 + R_2 < I(X_1, X_2; Y|X_3)$$
$$R_1 + R_3 < I(X_1, X_3; Y|X_2)$$
$$R_2 + R_3 < I(X_2, X_3; Y|X_1)$$
$$R_1 + R_2 + R_3 < I(X_1, X_2, X_3; Y).$$

This region has $6 = 3!$ corner points, each corresponding to different orders for successive cancellation decoding.

The k-sender Gaussian MAC with received SNRs $S_1, \ldots, S_k$ has the capacity region consisting of rate tuples satisfying

$$\sum_{j \in \mathcal{J}} R_j \leq \log\left(1 + \sum_{j \in \mathcal{J}} S_j\right), \quad \mathcal{J} \subseteq [k],$$

and the sum-capacity is

$$C_{\text{sum}} = \log\left(1 + \sum_{j=1}^{k} S_j\right).$$

Unfortunately, this shows that the sum-capacity C_{sum} scales only logarithmically in the number of senders k if each sender has the same block power constraint. Furthermore, C_{sum} does not scale at all with k if there is a constraint on the total received power, and such a constraint might be necessary to permit electromagnetic compatibility with different receiving devices. Thus, when there are many senders, an average sender can communicate only at a vanishing rate. This poor capacity scaling can be remedied in two ways. As discussed in the next section, using sufficiently many antennas at the receiver can essentially provide separate channels to each sender and thus achieve a linear scaling of sum-capacity. Alternatively, as discussed in Section 2.5.6, having multiple access points or remote radio heads that serve as relays and communicate cooperatively to the receiver can again achieve a linear scaling of sum-capacity, even with relay-to-receiver links of limited capacity.

2.2.8   Uplink Multiuser MIMO Systems

The Gaussian vector MAC

$$\mathbf{Y} = \sum_{j=1}^{k} \mathsf{H}_j \mathbf{X}_j + \mathbf{Z}$$

provides a simple model for uplink multiuser MIMO systems; see Figure 2.27 for the two-sender case. As for point-to-point MIMO, $\mathbf{Y}$ is an r-dimensional output vector, $\mathbf{X}_j$ is a t_j-dimensional input vector from sender j, H_j is an $r \times t_j$ channel matrix from sender j, and $\mathbf{Z} \sim \text{CN}(0, \mathsf{I}_r)$ is an r-dimensional noise vector. We consider block power constraints of the form (2.18) on every codeword $\mathbf{x}_j^n(m_j) = (\mathbf{x}_{j1}(m_j), \ldots, \mathbf{x}_{jn}(m_j))$.

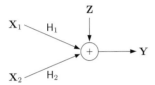

Figure 2.27 Uplink multiuser MIMO system with $k = 2$ senders.

The capacity region of the Gaussian vector MAC is the union of the polymatroidal regions consisting of all rate pairs $(R_1, \ldots, R_k)$ such that

$$\sum_{j \in \mathcal{J}} R_j \leq \log \left| I_r + \sum_{j \in \mathcal{J}} H_j K_j H_j^\dagger \right|, \quad \mathcal{J} \subseteq [k] \tag{2.34}$$

for some $K_1, \ldots, K_k \succeq 0$ with $\mathrm{tr}(K_j) \leq P_j, j \in [k]$. Each polymatroidal region in (2.34) corresponds to the $\mathcal{R}(X_1, \ldots, X_k)$ region in (2.33) for $X_j \sim CN(0, K_j), j \in [k]$. If $k = 2$, then the capacity region is the union of the pentagons characterized by the inequalities

$$R_1 \leq \log |I_r + H_1 K_1 H_1^\dagger|$$
$$R_2 \leq \log |I_r + H_2 K_2 H_2^\dagger|$$
$$R_1 + R_2 \leq \log |I_r + H_1 K_1 H_1^\dagger + H_2 K_2 H_2^\dagger|$$

for some $K_1, K_2 \succeq 0$ with $\mathrm{tr}(K_1) \leq P_1$ and $\mathrm{tr}(K_2) \leq P_2$. The sum-capacity of the k-sender Gaussian vector MAC is

$$C_{\mathrm{sum}} = \max_{K_1, \ldots, K_k \succeq 0 : \mathrm{tr}(K_j) \leq P_j} \log \left| I_r + \sum_{j=1}^{k} H_j K_j H_j^\dagger \right|.$$

The optimal covariance matrices $K_1^*, \ldots, K_k^*$ can be found efficiently by convex optimization [77], and even in a distributed manner by iteratively optimizing each covariance matrix while treating the other covariance matrices as part of the noise covariance matrix [78]. If the channel matrices $H_j = \Phi_j \Gamma_j \Psi_j^\dagger, j \in [k]$, have the same left singular vectors (say, $\Phi_j = \Phi, j \in [k]$), then the Gaussian vector MAC can be factorized as orthogonal Gaussian scalar MACs by optimal input beamforming with Ψ_j at each sender j and output beamforming with Φ, and the capacity region can be achieved by communicating multiple codewords over these orthogonal MACs in parallel. In general, however, the channel cannot be factorized into orthogonal channels. As in the point-to-point Gaussian vector channel without input beamforming (see Section 2.1.7), codewords from multiple antennas of multiple senders interfere with each other at the receiver.

If the transmitters do not have knowledge of the channel matrices, then a simple approach is to transmit independent codewords over different antennas at each sender, which means that $K_j = (P_j/t_j)I_{t_j}, j \in [k]$. The Gaussian vector MAC can then be viewed as a point-to-point Gaussian vector channel

$$\mathbf{Y} = \mathsf{H}\mathbf{X} + \mathbf{Z}$$

with $r \times (\sum_j t_j)$ channel matrix $\mathsf{H} = [\mathsf{H}_1 \cdots \mathsf{H}_k]$ and $(\sum_j t_j)$-dimensional input vector $\mathbf{X} = [\mathbf{X}_1 \cdots \mathbf{X}_k]^T$. As in point-to-point MIMO (see (2.22)), the achievable sum-rate scales linearly in $\min(r, \sum_j t_j)$ at high SNR in a rich scattering environment. Therefore, if the number of receiver antennas r is greater than the number of senders k, then the sum-capacity scales at least linearly in k, overcoming the capacity bottleneck of single-antenna single-receiver multiple access communication.

Finally, assuming that the receiver can estimate the channel matrices H_j, $j \in [k]$, a conceptually simple receiver design is to separate the input signals with *receive beamforming*, also known as *linear equalization* or *user separation* or *multiuser detection* [79]. To explain the idea, suppose that the senders have one antenna each, i.e., we have $t_j = 1$ for all $j \in [k]$. The matrices H_j are thus r-dimensional row vectors that we write as $\mathbf{h}_j^T$, $j \in [k]$, and H is an $r \times k$ matrix. We now project the received vector $\mathbf{Y}$ onto k *beams* $\mathbf{w}_j^T$, $j \in [k]$, each of which is an r-dimensional row vector. The receiver then uses a *mismatched decoder* and considers $\tilde{Y}_j = \mathbf{w}_j^T \mathbf{Y}$ to be the output of an AWGN channel (2.7) whose SNR is the SINR of $\tilde{Y}_j$.

To design the beams, we collect the new outputs into the vector $\tilde{\mathbf{Y}} = [\tilde{Y}_1 \cdots \tilde{Y}_k]^T$, where $\tilde{\mathbf{Y}} = \mathsf{W}\mathbf{Y}$ with the $k \times r$ *beamforming* matrix

$$\mathsf{W} = \begin{bmatrix} \mathbf{w}_1^T \\ \vdots \\ \mathbf{w}_k^T \end{bmatrix}.$$

The usual goal is to maximize the uplink SINR for each input X_k (which is equivalent to maximizing the rate achieved by treating interference as noise), and the solution to this problem is the minimum mean squared error (MMSE) beamforming matrix, or *Wiener filter*,

$$\mathsf{W}_{\mathrm{MMSE}} = \Pi\,\mathsf{H}^\dagger (\mathsf{I}_r + \mathsf{H}\,\Pi\,\mathsf{H}^\dagger)^{-1} \tag{2.35a}$$

$$= (\mathsf{I}_k + \Pi\,\mathsf{H}^\dagger\mathsf{H})^{-1}\Pi\,\mathsf{H}^\dagger, \tag{2.35b}$$

where Π is the diagonal covariance matrix of the input vector $\mathbf{X}$, which typically has the diagonal entries equal to the power constraint values P_j. The identity (2.35b) follows by the matrix identity

$$\mathsf{A}(\mathsf{I} + \mathsf{B}\mathsf{A})^{-1} = (\mathsf{I} + \mathsf{A}\mathsf{B})^{-1}\mathsf{A}. \tag{2.36}$$

For example, for small power constraint values, the input covariance matrix Π has small entries, and from (2.35) we recover the *maximum ratio combining (MRC)* matrix

$$\mathsf{W}_{\mathrm{MRC}} = \Pi\mathsf{H}^\dagger \tag{2.37}$$

that maximizes the powers of $\tilde{Y}_j$, $j \in [k]$. For large power constraint values, Π has large entries, and from (2.35b) we recover the *zero-forcing* matrix

$$\mathsf{W}_{\mathrm{ZF}} = (\mathsf{H}^\dagger\mathsf{H})^{-1}\mathsf{H}^\dagger, \tag{2.38}$$

where we have assumed that $r \geq k$ and H has full rank, i.e., H has rank $\min(r,k) = k$. The name zero-forcing emphasizes that W_{ZF} is the *Moore–Penrose pseudoinverse* of H (i.e., $\mathsf{W}_{ZF}\mathsf{H} = \mathsf{I}$) and cancels the interference completely as

$$\tilde{\mathbf{Y}} = \mathsf{W}_{ZF}\mathbf{Y} = \mathbf{X} + \mathsf{W}_{ZF}\mathbf{Z}$$

or equivalently, each input signal can be separated as

$$\tilde{Y}_j = X_j + \tilde{Z}_j, \quad j \in [k],$$

where $\tilde{Z}_j$ is the jth component of the filtered noise. For example, if $r = k$ and H has full rank, then we obtain the *inverse filter* $\mathsf{W}_{ZF} = \mathsf{H}^{-1}$.

2.3   Broadcast Channels

The broadcast channel (BC) is a simple model for downlink communication in which a sender communicates messages to multiple receivers. We start our discussion with the two-receiver BC depicted in Figure 2.28. Here, the sender wishes to reliably communicate private messages $m_1 \in [2^{nR_1}]$ and $m_2 \in [2^{nR_2}]$ to their respective receiver and a common message $m_0 \in [2^{nR_0}]$ to both receivers over a shared communication channel by encoding (m_0, m_1, m_2) into a codeword x^n and broadcasting it over the channel. The output symbols y_1 and y_2 from each transmitted input symbol x are drawn according to the conditional probability $p_{Y_1,Y_2|X}(y_1, y_2|x)$, and the conditional probability of receiving the output sequences y_1^n and y_2^n at receivers 1 and 2, respectively, when m_0, m_1, and m_2 are communicated is

$$p(y_1^n, y_2^n|m_0, m_1, m_2) = \prod_{i=1}^{n} p_{Y_1,Y_2|X}(y_{1i}, y_{2i}|x_i(m_0, m_1, m_2)).$$

By decoding y_j^n, receiver $j = 1, 2$ forms the estimates $\hat{m}_j$ of m_j and $\hat{m}_{0j}$ of m_0.

The encoder mapping $x^n(m_0, m_1, m_2)$ and the decoder mappings $(\hat{m}_1(y_1^n), \hat{m}_{01}(y_1^n))$ and $(\hat{m}_2(y_2^n), \hat{m}_{02}(y_2^n))$ are referred to as a $(2^{nR_0}, 2^{nR_1}, 2^{nR_2}, n)$ code. The average error probability of this code is

$$P_e^{(n)} = \mathsf{P}((M_1, M_0) \neq (\hat{M}_1, \hat{M}_{01}) \text{ or } (M_2, M_0) \neq (\hat{M}_0, \hat{M}_{02})),$$

where we assume that M_0, M_1, M_2 are independent and uniformly distributed. In order to understand the optimal tradeoff between the rates, we study the capacity region of the BC. We say that a rate triple (R_0, R_1, R_2) is achievable if, for any target error probability

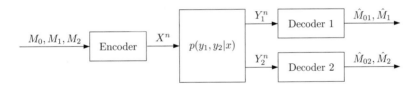

Figure 2.28 Two-receiver broadcast communication system.

$\epsilon > 0$ and sufficiently large n, there exists a $(2^{nR_0}, 2^{nR_1}, 2^{nR_2}, n)$ code with $P_e^{(n)} \leq \epsilon$. As usual, this can be stated in terms of a sequence of codes, i.e., (R_0, R_1, R_2) is achievable if there exists a sequence of $(2^{nR_0}, 2^{nR_1}, 2^{nR_2}, n)$ codes such that $\lim_{n \to \infty} P_e^{(n)} = 0$. The capacity region $\mathscr{C}$ of the BC is the closure of the set of achievable rate triples (R_0, R_1, R_2). The capacity region of a generic BC is sketched in Figure 2.29. This region is not known in general, even when the common message is not considered (i.e., $R_0 = 0$). Nonetheless, $\mathscr{C}$ can be characterized for several important classes of BCs, including many canonical models for downlink communication.

Our main focus will be on optimal coding schemes that achieve the capacity region for these BCs. For simplicity of presentation, we discuss primarily the special cases in which there is only the common message or there are only private messages. Information theory for BCs was developed in [80–83]. Surveys of key results are given in [62, 84, 85]. For example, one useful property is that the capacity region depends only on the marginal channel distributions

$$p(y_1|x) = \sum_{y_2} p(y_1, y_2|x) \quad \text{and} \quad p(y_2|x) = \sum_{y_1} p(y_1, y_2|x).$$

2.3.1 Common-Message Broadcasting

Suppose that the sender wishes to communicate a common message $M_0 \in [2^{nR_0}]$ to both receivers, as depicted in Figure 2.30. Let C_0 denote the *common-message capacity*, namely, the largest rate R_0 at which the communication can be reliable.

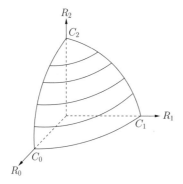

Figure 2.29 The capacity region $\mathscr{C}$ of a broadcast channel.

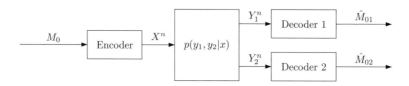

Figure 2.30 Broadcast channel with a common message.

When codewords are generated randomly and independently according to $p(x)$, then by the random coding analysis in Section 2.1.1, decoding at both receivers is successful w.h.p. if

$$R < \min(I(X; Y_1), I(X; Y_2)).$$

It can be shown [31, 32] that the common-message capacity can be achieved by optimizing over $p(x)$ and we have

$$C_0 = \max_{p(x)} \min(I(X; Y_1), I(X; Y_2)). \tag{2.39}$$

Let

$$C_1 = \max_{p(x)} I(X; Y_1) \tag{2.40}$$

be the capacity of the channel $p(y_1|x) = \sum_{y_2} p(y_1, y_2|x)$ from the sender to receiver 1. Similarly, let

$$C_2 = \max_{p(x)} I(X; Y_2) \tag{2.41}$$

be the individual capacity of receiver 2. Since the maximin is in general less than the minimax, we have

$$C_0 \le \min(C_1, C_2).$$

The upper bound is not achievable in general, since the maximum in (2.39)–(2.41) may not be attained by the same input distribution $p(x)$. We will discuss an important example in Section 2.3.5.

For a few interesting broadcast channel models, however, the optimal distribution $p(x)$ is the same for both receivers and $C_0 = \min(C_1, C_2)$. For instance, the *binary symmetric BC* consists of two BSCs (see Figure 2.2) with crossover probabilities $p_1, p_2 \le 1/2$ and individual capacities $C_j = 1 - H(p_j), j = 1, 2$. Since both individual capacities are attained by $X \sim \text{Bern}(1/2)$, we have

$$C_0 = \min(C_1, C_2) = 1 - H(\max(p_1, p_2)).$$

A second example is the *binary erasure BC* that consists of two BECs (see Figure 2.3) with erasure probabilities p_1 and p_2. Once again, the individual capacities $C_j = 1 - p_j$, $j = 1, 2$, are both attained by $X \sim \text{Bern}(1/2)$ and

$$C_0 = 1 - \max(p_1, p_2).$$

As a third example, consider the Gaussian BC depicted in Figure 2.31 that provides a simple model for wireless downlink communication systems. Here the channel outputs corresponding to the input X are

$$Y_1 = h_1 X + Z_1$$
$$Y_2 = h_2 X + Z_2,$$

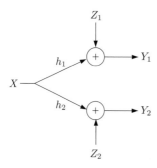

Figure 2.31 Gaussian broadcast channel.

where h_1 and h_2 are the channel coefficients, and $Z_1 \sim \text{CN}(0,1)$ and $Z_2 \sim \text{CN}(0,1)$ are noise components, whose correlation is irrelevant to the capacity. As before, we consider the block transmit power constraint

$$\frac{1}{n} \sum_{i=1}^{n} |x_i(m_0, m_1, m_2)|^2 \le P, \quad (m_0, m_1, m_2) \in [2^{nR_0}] \times [2^{nR_1}] \times [2^{nR_2}]$$

and denote the received SNRs as $S_j = |h_j|^2 P$, $j = 1, 2$. The individual capacities are $C_1 = \log(1 + S_1)$ and $C_2 = \log(1 + S_2)$, both of which are attained by $X \sim \text{CN}(0, P)$. Hence, the common-message capacity is

$$C_0 = \min(C_1, C_2) = \log(1 + \min(S_1, S_2)).$$

As discussed in Section 2.1.4, the BC with a common message is closely related the compound channel. For example, the problem of broadcasting over the Gaussian BC discussed above can be viewed as a problem in which the sender is unaware of whether the true received SNR is S_1 or S_2, but still would like to communicate reliably in either case. Unlike the receivers in the BC, the receiver in the compound channel does not have the channel state information a priori, but this subtle difference does not affect the capacity defined in asymptotics. Hence, similar to (2.39) the capacity of a compound channel is characterized as

$$C = \max_{p(x)} \min_{s} I(X; Y_s),$$

where Y_s is the channel output when the channel state is s.

As an alternative to guaranteeing the worst-case reliable communication in the same blocklength, we can communicate a very long (or infinite) stream of coded symbols and allow each receiver to stop listening at its own pace [85]. By slightly adapting the random coding analysis for point-to-point communication in Section 2.1.1, the effective rate of $I(X; Y_s)$ can be achieved at each receiver Y_s for any input distribution $p(x)$. Such *rateless codes* have many applications in content broadcasting, especially over erasure channels. Raptor codes [22] provide practical implementations of low-complexity rateless codes.

2.3.2   Superposition Coding

We now focus attention on the BC with private messages as depicted in Figure 2.32. This problem can be seen as the dual of the MAC.

Let C_1 and C_2 be individual capacities defined in (2.40) and (2.41). Then time division, or any scheme that divides orthogonal resources between the receivers, achieves any rate pair (R_1, R_2) satisfying

$$R_1 < \alpha C_1$$
$$R_2 < (1 - \alpha)C_2.$$

As in the MAC case, the sender can communicate reliably at higher rates by transmitting the messages simultaneously, or non-orthogonally, rather than separately. The most basic non-orthogonal coding scheme is (horizontal) superposition coding, as discussed in Sections 2.1.6 and 2.2.6, in which the signal space is decomposed into multiple layers and superimposed into the channel input.

Consider the encoding structure in Figure 2.33. The messages m_1 and m_2 are encoded separately into two codewords $u_1^n(m_1)$ and $u_2^n(m_2)$. These codewords generate the channel input $x^n(m_1, m_2) = x(u_1^n(m_1), u_2^n(m_2))$ by a symbol-by-symbol mapping $x(u_1, u_2)$, which is then transmitted. A geometric interpretation of the codewords is shown in Figure 2.34.

Suppose that each receiver recovers its own message while treating the other codeword as noise (see Section 2.2.1). When codewords are generated randomly and independently according to $p(u_1)$ and $p(u_2)$, decoding at the receivers is successful w.h.p. if

$$R_1 < I(U_1; Y_1) \tag{2.42a}$$
$$R_2 < I(U_2; Y_2). \tag{2.42b}$$

Suppose now that receiver 1 is "better" than receiver 2 and can recover m_2 as well as receiver 2 does. Then it may as well recover m_2 first and then recover m_1 by successive

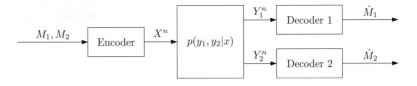

Figure 2.32 Broadcast channel with private messages.

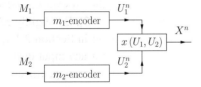

Figure 2.33 Superposition coding for the BC.

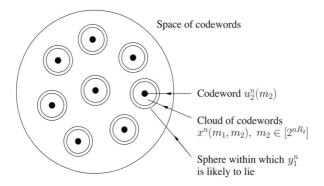

Figure 2.34 Space of codewords for superposition coding.

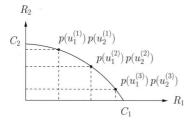

Figure 2.35 The capacity region of a typical two-receiver BC with one receiver less noisy than the other. Each rectangle corresponds to a different choice of $p(u_1)p(u_2)$.

cancellation decoding, whereas receiver 2 continues to treat $u_1^n(m_1)$ as noise. As in the earlier analysis of successive cancellation decoding in Section 2.2.2, decoding at both receivers is successful w.h.p. if

$$R_1 < I(U_1; Y_1|U_2) = I(X; Y_1|U_2) \tag{2.43a}$$

$$R_2 < I(U_2; Y_1) \tag{2.43b}$$

$$R_2 < I(U_2; Y_2). \tag{2.43c}$$

If receiver 1 is "better" than receiver 2 in the precise sense that $I(U_2; Y_1) \geq I(U_2; Y_2)$, then the decoding condition for m_2 at receiver 1 in (2.43b) is automatically satisfied if the decoding condition for m_2 at receiver 2 in (2.43c) is met. Hence, the rate region in (2.43) simplifies to

$$R_1 < I(U_1; Y_1|U_2) = I(X; Y_1|U_2) \tag{2.44a}$$

$$R_2 < I(U_2; Y_2). \tag{2.44b}$$

In fact, if receiver 1 is "better" than receiver 2 for all choices of code distributions $p(u_1)p(u_2)$ and superposition mappings $x(u_1, u_2)$, a condition for which receiver 1 is called *less noisy* [86], then the union of the rectangular regions in (2.44) over all $p(u_1)p(u_2)$ and $x(u_1, u_2)$ is the capacity region (see Figure 2.35). In general, this capacity region is strictly larger than the rate region achievable by time division.

A straightforward condition for receiver 1 to be less noisy than receiver 2 is that the channel $p(y_2|x)$ for the latter has the same conditional distribution as a further degradation of the channel $p(y_1|x)$ for the former, i.e., we have

$$p(y_2|x) = \sum_{y_1} p(y_1|x)\tilde{p}(y_2|y_1)$$

for some conditional distribution $\tilde{p}(y_2|y_1)$. A broadcast channel satisfying this condition is commonly referred to as a *degraded* BC. An even more restrictive condition is when Y_2 is a noisy version of Y_1 so that $p(y_2|y_1,x) = p(y_2|y_1)$ for all x, y_1, y_2. A BC satisfying this condition is called a *physically degraded* BC. A degraded BC has the same marginal distributions as a physically degraded BC, so when studying coding and capacity of a degraded BC, one may as well study coding and capacity of its physically degraded counterpart.

We illustrate how superposition coding improves upon time division through the Gaussian BC in Figure 2.31. This channel is an example of a degraded BC, as we show in the following. We assume without loss of generality that receiver 1 has a better channel than receiver 2, i.e., $S_1 \geq S_2$. Superposition coding for the Gaussian BC is depicted in Figure 2.36. Two codewords $u_1^n(m_1)$ and $u_2^n(m_2)$ are superimposed as the transmitted codeword

$$x^n(m_1, m_2) = u_1^n(m_1) + u_2^n(m_2).$$

Receiver 2 recovers m_2 by decoding $y_2^n = h_2 u_2^n(m_2) + h_2 u_1^n(m_1) + z_2^n$ while treating $h_2 u_1^n(m_1)$ as part of noise. When random coding with $U_1 \sim \mathrm{CN}(0, \alpha P)$ and $U_2 \sim \mathrm{CN}(0, (1-\alpha)P)$ is used, decoding at receiver 2 is successful w.h.p. if

$$R_2 < I(U_2; Y_2) = \log\left(1 + \frac{(1-\alpha)S_2}{\alpha S_2 + 1}\right). \tag{2.45}$$

Receiver 1 uses successive cancellation decoding. It first recovers m_2 from $y_1^n = h_1 u_2^n(m_2) + h_1 u_1^n(m_1) + z_1^n$ while treating $h_1 u_1^n(m_1)$ as noise. Since $S_1 \geq S_2$ and thus $(1-\alpha)S_1/(\alpha S_1 + 1) \geq (1-\alpha)S_2/(\alpha S_2 + 1)$, we have

$$I(U_2; Y_1) = \log\left(1 + \frac{(1-\alpha)S_1}{\alpha S_1 + 1}\right)$$

$$\geq \log\left(1 + \frac{(1-\alpha)S_2}{\alpha S_2 + 1}\right) = I(U_2; Y_2)$$

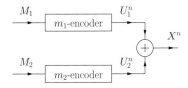

Figure 2.36 Superposition coding for the Gaussian BC.

and decoding at receiver 1 is successful w.h.p. if (2.45) is satisfied. Receiver 1 then recovers m_1 by decoding

$$y_1^n - h_1 u_2^n(\hat{m}_2) = h_1 u_1^n(m_1) + z_1^n + h_1(u_2^n(m_2) - u_2^n(\hat{m}_2)).$$

Assuming that m_2 was correctly recovered as $\hat{m}_2$, decoding in this step is successful w.h.p. if

$$R_1 < I(U_1; Y_1|U_2) = I(U_1; Y_1 - h_1 U_2) = \log(1 + \alpha S_1). \qquad (2.46)$$

The rectangular region characterized by (2.45) and (2.46) is sketched in Figure 2.37. The union of these regions, evaluated over all possible values of $\alpha \in [0, 1]$, is the capacity region of the Gaussian BC [81].

For the Gaussian BC, the channel output Y_2 at the worse receiver given the input X has the same conditional distribution as

$$Y_2' = \frac{h_2}{h_1} Y_1 + Z_2',$$

where the noise $Z_2' \sim \text{CN}(0, 1 - |h_2|^2/|h_1|^2)$ is independent of Y_1. Hence, the Gaussian BC (for any input alphabet) is degraded.

Consider next the binary symmetric BC with $0 \le p_1 \le p_2 \le 1/2$. Now Y_2 has the same conditional distribution as

$$Y_2' = Y_1 \oplus Z_2',$$

where the additive binary noise $Z_2' \sim \text{Bern}((p_2 - p_1)/(1 - 2p_1))$ is independent of Y_1. Hence, the BC consisting of two BSCs is degraded. The capacity region of the binary symmetric BC is shown in Figure 2.38 [82], and it is the set of rate pairs (R_1, R_2) such that

$$R_1 \le H(\alpha * p_1) - H(p_1)$$
$$R_2 \le 1 - H(\alpha * p_2)$$

for some $\alpha \in [0, 1/2]$, which is in general larger than the rate region achievable by time division. Here $\alpha * p = \alpha(1 - p) + (1 - \alpha)p$ and $H(p)$ is the binary entropy function of $p \in [0, 1]$.

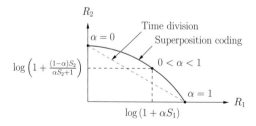

Figure 2.37 Superposition coding and capacity region for the Gaussian BC.

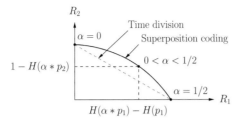

Figure 2.38 Superposition coding and capacity region for the binary symmetric BC.

Similarly, for the binary erasure BC with erasure probabilities $p_1 \leq p_2$, Y_2 has the same conditional distribution as Y_2' that is generated by further erasing Y_1 with probability $p_2' = (p_2 - p_1)/(1 - p_1)$. Hence, the BC consisting of two BECs is degraded. The capacity region of the binary erasure BC is the set of rate pairs (R_1, R_2) such that

$$R_1 \leq \alpha(1 - p_1)$$
$$R_2 \leq (1 - \alpha)(1 - p_2)$$

for some $\alpha \in [0, 1]$, which is equivalent to the rate region achievable by time division.

Superposition coding is also useful for broadcasting a common message and a private message. Consider the broadcast channel in Figure 2.28, where there is no private message M_2 (i.e., $R_2 = 0$). Thus, receiver 1 recovers both m_0 and m_1, whereas receiver 2 recovers only m_0. Superposition coding of m_0 and m_1, along with simultaneous decoding at receiver 1, can achieve any rate pair (R_0, R_1) satisfying

$$R_0 < I(U_2; Y_2)$$
$$R_1 < I(U_1; Y_1|U_2)$$
$$R_0 + R_1 < I(U_1, U_2; Y_1) = I(X; Y_1).$$

The closure of this region over all $p(u_1)p(u_2)$ and $x(u_1, u_2)$ is the capacity region [87].

2.3.3   Marton Coding

As discussed in the previous section, several interesting BCs are degraded. There are, however, many other interesting channel models that are *not* degraded, including the Gaussian vector BC for downlink MIMO discussed in Section 2.3.5. Since neither receiver is "better" than the other, successive cancellation decoding may be worse than merely treating the other codeword as noise that achieves the rate region in (2.42).

Marton coding [88] provides a fundamentally different approach to broadcast communication. Unlike superposition coding with two independent codewords, Marton coding generates two *dependent* codewords even if the messages themselves are independent. This dependence is achieved by *multicoding* that generates multiple codewords per message, as depicted in Figure 2.39.

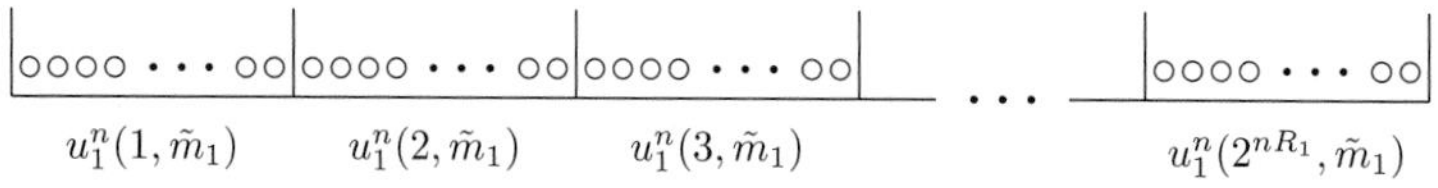

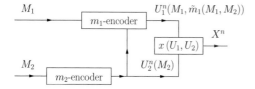

$$u_1^n(1,\tilde{m}_1) \qquad u_1^n(2,\tilde{m}_1) \qquad u_1^n(3,\tilde{m}_1) \qquad\qquad u_1^n(2^{nR_1},\tilde{m}_1)$$

Figure 2.39 Multicoding for the BC. Each $m_1 \in [2^{nR_1}]$ is represented by multiple codewords $u_1^n(m_1,\tilde{m}_1)$ indexed by $\tilde{m}_1 \in [2^{n\tilde{R}_1}]$.

Figure 2.40 Encoding structure for Marton coding.

More concretely, fix a joint distribution $p(u_1,u_2)$ with marginal distributions

$$p(u_1) = \sum_{u_2} p(u_1,u_2) \quad \text{and} \quad p(u_2) = \sum_{u_1} p(u_1,u_2)$$

and a superposition function $x(u_1,u_2)$. We first generate a codeword $u_2^n(m_2)$ for each $m_2 \in [2^{nR_2}]$ i.i.d. according to the marginal distribution $p(u_2)$ as in superposition coding. For each $m_1 \in [2^{nR_1}]$, however, we generate multiple codewords $u_1^n(m_1,\tilde{m}_1)$, indexed by $\tilde{m}_1 \in [2^{n\tilde{R}_1}]$, i.i.d. according to the marginal distribution $p(u_1)$. So far, $u_1^n(m_1,\tilde{m}_1)$ and $u_2^n(m_2)$ are independent for each $m_1,\tilde{m}_1,m_2$, generated from the product distribution $p(u_1)p(u_2)$, not the joint distribution $p(u_1,u_2)$. The desired dependence can be, however, induced by choosing $\tilde{m}_1$ dependent on $u_2^n(m_2)$. For each m_1 and m_2, we pick an index $\tilde{m}_1 = \tilde{m}_1(m_1,m_2)$ such that $u_1^n(m_1,\tilde{m}_1)$ and $u_2^n(m_2)$ are jointly typical. Then, by the definition of joint typicality in Section 2.1.1, it can be shown that $u_1^n(m_1,\tilde{m}_1)$ and $u_2^n(m_2)$ are jointly typical with probability at least $2^{-n(I(U_1;U_2)+\delta)}$ for n sufficiently large. Consequently, if there are sufficiently many $\tilde{m}_1$ indices, or more precisely, if

$$\tilde{R}_1 > I(U_1;U_2) + \delta, \tag{2.47}$$

then *joint typicality encoding* is successful, namely, there exists at least one index $\tilde{m}_1(m_1,m_2)$ such that $u_1^n(m_1,\tilde{m}_1(m_1,m_2))$ and $u_2^n(m_2)$ are jointly typical w.h.p. This pair of jointly typical codewords generates the channel input $x^n(m_1,m_2) = x(u_1^n(m_1,\tilde{m}_1(m_1,m_2)),u_2^n(m_2))$ by the symbol-by-symbol mapping $x(u_1,u_2)$, as shown in Figure 2.40, which is then transmitted to communicate (m_1,m_2).

Receiver 2 recovers m_2 by treating $u_1^n(m_1,\tilde{m}_1)$ as noise, which is successful w.h.p. if

$$R_2 < I(U_2;Y_2). \tag{2.48}$$

Receiver 1 recovers $(m_1,\tilde{m}_1)$ by treating $u_2^n(m_2)$ as noise, which is successful w.h.p. if

$$R_1 + \tilde{R}_1 < I(U_1;Y_1). \tag{2.49}$$

Observe that $\tilde{R}_1$ represents a rate loss for message 1; this is the price we pay for creating dependence between u_1^n and u_2^n. Combining the inequalities in (2.47)–(2.49) and setting $\tilde{R}_1$ as small as possible, Marton coding achieves any rate pair (R_1, R_2) satisfying

$$R_1 < I(U_1; Y_1) - I(U_1; U_2) \tag{2.50a}$$

$$R_2 < I(U_2; Y_2). \tag{2.50b}$$

Apparently, this rate region looks smaller than the rate region in (2.42) due to the penalty term of $\tilde{R}_1 = I(U_1; U_2)$. However, when evaluated under a product distribution $p(u_1)p(u_2)$, the rate region in (2.50) simplifies to the rate region in (2.42). Since there is a richer collection of joint distributions $p(u_1, u_2)$ that can be used for random codebook generation, Marton coding can achieve higher rates than superposition coding. Furthermore, by time sharing between the region in (2.50) and its counterpart achieved by changing the roles of U_1 and U_2, Marton coding can achieve any rate pair (R_1, R_2) satisfying

$$R_1 < I(U_1; Y_1) \tag{2.51a}$$

$$R_2 < I(U_2; Y_2) \tag{2.51b}$$

$$R_1 + R_2 < I(U_1; Y_1) + I(U_2; Y_2) - I(U_1; U_2). \tag{2.51c}$$

This pentagonal region is sketched in Figure 2.41. One can, of course, achieve the union of such regions, the union being taken over all choices of $p(u_1, u_2)$. Similar to simultaneous decoding for multiple access, the same region can also be achieved by generating multiple codewords $u_1^n(m_1, \tilde{m}_1)$ and $u_2^n(m_2, \tilde{m}_2)$ for both messages and adapting them simultaneously.

We revisit the Gaussian BC to provide a more concrete illustration of Marton coding. Let the joint distribution $p(u_1, u_2)$ be chosen as $U_2 \sim \text{CN}(0, (1 - \alpha)P)$ and

$$U_1 = \beta U_2 + X_1, \tag{2.52}$$

where

$$\beta = \alpha S_1 / (1 + \alpha S_1) \tag{2.53}$$

is the coefficient of the MMSE estimate of X_1 given $X_1 + Z_1$, and $X_1 \sim \text{CN}(0, \alpha P)$ is independent of U_2. As described above, codewords $u_1^n(m_1, \tilde{m}_1)$ and $u_2^n(m_2)$ are generated

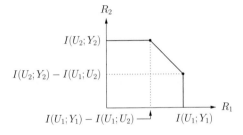

Figure 2.41 Marton coding region for the BC with a fixed choice of $p(u_1, u_2)$ and $x(u_1, u_2)$. The two corner points are achieved by multicoding for m_1 and for m_2, respectively.

independently according to $p(u_1)$ and $p(u_2)$, respectively. The auxiliary index $\tilde{m}_1 = \tilde{m}_1(m_1, m_2)$ is chosen such that $u_1^n(m_1, \tilde{m}_1)$ and $u_2^n(m_2)$ are jointly typical, or roughly speaking, by (2.52), $x_1^n(m_1, m_2) = u_1^n(m_1, \tilde{m}_1) - \beta u_2^n(m_2)$ "looks like" an i.i.d. $CN(0, \alpha P)$ sequence. This step can be viewed as an extension of Harashima–Tomlinson precoding [89, 90] to a high-dimensional signal space under the block power constraint. This joint typicality encoding step is successful w.h.p. if

$$\tilde{R}_1 > I(U_1; U_2) = \log\left(1 + \frac{\beta^2(1 - \alpha)}{\alpha}\right). \tag{2.54}$$

Let the superposition function $x(u_1, u_2)$ be chosen as $X = X_1 + X_2$ with the two independent signal layers $X_1 = U_1 - \beta U_2$ and $X_2 = U_2$ for receivers 1 and 2, respectively. These layers carry the sequences

$$x_1^n(m_1, m_2) = u_1^n(m_1, \tilde{m}_1(m_1, m_2)) - \beta u_2^n(m_2)$$
$$x_2^n(m_2) = u_2^n(m_2).$$

The superposition of these sequences $x^n(m_1, m_2) = x_1^n(m_1, m_2) + x_2^n(m_2)$ is transmitted, as depicted in Figure 2.42. Note that, by our choice of $p(u_1, u_2)$ in (2.52), signals x_1^n and x_2^n in these two layers are essentially independent of each other, provided that joint typicality encoding was successful. Hence, the block power constraint for $x^n = x_1^n + x_2^n$ is satisfied.

The channel to receiver 2 can now be viewed as

$$Y_2 = h_2 X + Z_2 = h_2 X_2 + h_2 X_1 + Z_2, \tag{2.55}$$

with input $X_2 = U_2$, interference $h_2 X_1$, and noise Z_2. Hence, by treating interference as noise, receiver 2 can recover m_2 (carried by $u_2^n(m_2)$ on the X_2 layer) successfully w.h.p. if

$$R_2 < I(U_2; Y_2) = \log\left(1 + \frac{(1 - \alpha)S_2}{1 + \alpha S_2}\right). \tag{2.56}$$

The channel to receiver 1 can be viewed as

$$Y_1 = h_1 X + Z_1 = h_1 X_1 + h_1 X_2 + Z_1, \tag{2.57}$$

with input $X_1 = U_1 - \beta U_2$, interference $h_1 X_2$, and noise Z_1. Unlike the other channel in (2.55), the message m_1 is carried by the codeword $u_1^n(m_1, \tilde{m}_1)$ adapted to the interfering codeword $u_2^n(m_2)$. Hence, receiver 1 can recover m_1 and $\tilde{m}_1$ successfully w.h.p. if

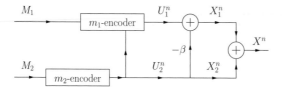

Figure 2.42 Multicoding for the Gaussian BC.

$$R_1 + \tilde{R}_1 < I(U_1; Y_1) = \log\left((1 + \alpha S_1)\left(1 + \frac{\beta^2(1-\alpha)}{\alpha}\right)\right). \tag{2.58}$$

When combined with the encoding condition in (2.54), the decoding condition in (2.58) yields

$$R_1 < I(U_1; Y_1) - I(U_1; U_2) = \log(1 + \alpha S_1), \tag{2.59}$$

which is equal to the rate that is achievable for the channel (2.57) if the interference $h_1 X_2$ were not present. The resulting region in (2.56) and (2.59) is exactly the capacity region of the Gaussian BC in (2.45) and (2.46).

The development above may feel like overkill for the simple Gaussian BC, the capacity of which can be achieved by point-to-point codes, superposition coding, and successive cancellation decoding. But in the Marton coding scheme presented above, we did not rely on the relative strengths of the channels, while the superposition coding scheme presented earlier required $S_1 \geq S_2$. This flexibility is particularly useful for the Gaussian vector channel for which each receiver is not necessarily better than the other.

Note that the channel in (2.57) and the achievable rate in (2.59) have the interesting feature that known interference at the encoder can be canceled by adapting codewords to the interference. This feature continues to hold for multiple antennas. Consider the following variant of the Gaussian vector channel in (2.17):

$$\mathbf{Y} = \mathbf{HX} + \mathbf{S} + \mathbf{Z}, \tag{2.60}$$

where $\mathbf{S}$ is the known interference. This is a special case of the channel with state discussed in Section 2.1.5, in which the state sequence $\mathbf{s}^n$ is available only at the sender [36]. By communicating codewords $\mathbf{u}^n(m, \tilde{m})$ with $\tilde{m}(m, \mathbf{s}^n)$ adapted to the interference $\mathbf{s}^n$, it can be shown that the capacity of this channel is

$$\begin{aligned} C &= \max_{F(\mathbf{u}|\mathbf{s}),\, \mathbf{x}(\mathbf{u},\mathbf{s}):\, E[\mathbf{x}^\dagger(\mathbf{U},\mathbf{S})\mathbf{x}(\mathbf{U},\mathbf{S})] \leq P} \big(I(\mathbf{U}; \mathbf{Y}) - I(\mathbf{U}; \mathbf{S})\big) \\ &= \max_{F(\mathbf{x}):\, E[\mathbf{X}^\dagger \mathbf{X}] \leq P} I(\mathbf{X}; \mathbf{HX} + \mathbf{Z}) \\ &= \max_{\mathbf{K_X} \succeq 0:\, \mathrm{tr}(\mathbf{K_X}) \leq P} \log|\mathbf{I}_r + \mathbf{HK_X H}^\dagger|, \end{aligned} \tag{2.61}$$

which is equal to the capacity of the Gaussian vector channel in (2.17) without interference $\mathbf{S}$. Thus, adapting the codeword under the optimal choice of the code distribution $F(\mathbf{u}|\mathbf{s})$ and the mapping $\mathbf{x}(\mathbf{u}, \mathbf{s})$, the effect of the additive interference $\mathbf{S}$ can be canceled completely. By taking the analogy of writing $\mathbf{X}$ on a paper with dirt $\mathbf{S}$, this channel model and optimal coding scheme are often referred to as "writing on dirty paper" [91] and "dirty paper coding," respectively. Implementations of dirty paper coding and its variations are investigated for more concrete codes, such as lattice codes [92], turbo codes [93, 94], and low-density parity check codes [95]. Dirty paper coding plays an important role for Gaussian vector BCs, as discussed in Section 2.3.5.

2.3.4   Broadcast Channels with Many Receivers

Superposition coding and Marton coding can be generalized to BCs with more than two receivers. If the k-receiver BC $p(y_1, y_2, \ldots, y_k|x)$ is *degraded*, namely, $p(y_j|x)$ is a further degradation of $p(y_{j-1}|x), j = 2, \ldots, k$, then superposition coding and successive cancellation decoding can achieve any rate tuple $(R_1, \ldots, R_k)$ such that

$$R_1 < I(U_1; Y_1|U_2, \ldots, U_k)$$
$$R_2 < I(U_2; Y_2|U_3, \ldots, U_k)$$
$$\vdots$$
$$R_k < I(U_k; Y_k).$$

The capacity region is achieved by considering all $p(u_1) \cdots p(u_k)$ and $x(u_1, \ldots, u_k)$.

The k-receiver Gaussian BC with received SNRs $S_1 \geq S_2 \geq \cdots \geq S_k$ has a capacity region that is the set of rate tuples satisfying

$$R_1 \leq \log(1 + \alpha_1 S_1)$$

$$R_j \leq \log\left(1 + \frac{\alpha_j S_j}{(\sum_{i=1}^{j-1} \alpha_i) S_j + 1}\right), \quad j = 2, \ldots, k,$$

where $\alpha_1, \ldots, \alpha_k \geq 0$ with $\sum_{i=1}^{k} \alpha_i = 1$, and the sum-capacity is

$$C_{\text{sum}} = \log(1 + S_1),$$

which is independent of k and achieved by serving only the best receiver. Thus, as in the MAC case, as the number of receivers grows, even an "average" receiver can be served only at a vanishing rate. We can improve this poor capacity scaling by either broadcasting from multiple antennas (see the next section) or coordinating transmission from multiple relays (see Section 2.5.6).

2.3.5   Downlink Multiuser MIMO Systems

The Gaussian vector BC

$$\mathbf{Y}_j = \mathsf{H}_j \mathbf{X} + \mathbf{Z}_j, \quad j \in [k]$$

provides a simple model for downlink multiuser MIMO systems; see Figure 2.43 for the two-receiver case. In this dual structure to the Gaussian vector MAC, $\mathbf{X}$ is a t-dimensional input vector and $\mathbf{Y}_j, j \in [k]$, is an r_j-dimensional output vector at receiver j, $\mathsf{H}_j, j \in [k]$, is an $r_j \times t$ channel matrix to receiver j, and $\mathbf{Z}_j \sim \text{CN}(0, \mathsf{I}_{r_j})$ is an r_j-dimensional noise vector. As before, we consider the block power constraint (2.18) on every codeword $\mathbf{x}^n(m_1, \ldots, m_k)$.

Unlike the Gaussian scalar BC, the Gaussian vector BC is not degraded in general, so that superposition coding is not necessarily optimal. However, before considering

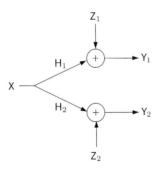

Figure 2.43 MIMO broadcast communication system with two receivers.

Marton coding, or more precisely dirty paper coding, we consider the superposition coding methods that are usually used in practice. For convenience, we collect the channel outputs into one vector

$$\mathbf{Y} = \mathsf{H}\mathbf{X} + \mathbf{Z},$$

where

$$\mathbf{Y} = \begin{bmatrix} \mathbf{Y}_1 \\ \vdots \\ \mathbf{Y}_k \end{bmatrix}, \quad \mathsf{H} = \begin{bmatrix} \mathsf{H}_1 \\ \vdots \\ \mathsf{H}_k \end{bmatrix}, \quad \text{and} \quad \mathbf{Z} = \begin{bmatrix} \mathbf{Z}_1 \\ \vdots \\ \mathbf{Z}_k \end{bmatrix}.$$

Defining $r = \sum_{j=1}^{k} r_j$, the outputs vector $\mathbf{Y}$ and the noise vector $\mathbf{Z}$ are r-dimensional vectors, and H is an $r \times t$ channel matrix.

Superposition coding for the Gaussian vector BC is usually implemented by *linear precoding*, which is closely related to the linear equalization at the receiver for the Gaussian vector MAC discussed in Section 2.2.8. For simplicity, consider receivers with one antenna, i.e., we have $r_j = 1$ for all $j \in [k]$ and therefore $r = k$. We encode k messages $m_1, m_2, \ldots, m_k$ into codewords represented by the random variables $U_1, U_2, \ldots, U_k$, respectively, and we transmit the vector $\mathbf{U} = [U_1, U_2, \ldots, U_k]^T$ via the linear transformation

$$\mathbf{X} = \mathsf{W}\mathbf{U} = \sum_{j=1}^{k} \mathbf{w}_j U_j,$$

where W is a $t \times k$ precoding matrix with columns $\mathbf{w}_j, j \in [k]$. The jth column $\mathbf{w}_j$ of W is called the *transmit beamforming vector* for the data stream represented by U_j.

One simple choice for W is known as *maximum ratio transmission* (MRT) with

$$\mathsf{W}_{\mathrm{MRT}} = \mathsf{H}^{\dagger}\Lambda, \tag{2.62}$$

where Λ is a $k \times k$ real, nonnegative, and diagonal matrix that distributes the transmit power to optimize rates and fulfill a transmit power constraint similar to (2.18). This choice of precoding matrix maximizes the received signal powers (cf. the uplink MRC matrix in (2.37)), but it does not necessarily maximize the SINRs. Suppose next that

H has full rank and $t \geq k$. A common choice for W is known as the *zero-forcing* (ZF) matrix with

$$W_{ZF} = H^\dagger (HH^\dagger)^{-1} \Lambda. \tag{2.63}$$

This choice of precoding matrix removes the interference across the data streams, but it again does not necessarily maximize the SINRs (cf. the uplink zero-forcing matrix in (2.38)).

The above two precoding matrices are special cases of a more general approach that maximizes the SINRs of the receivers according to a *utility function* that is strictly increasing in each of the k individual SINRs. The resulting precoding matrix has the general form [96]

$$W_{MMSE} = H^\dagger (I_k + \Pi HH^\dagger)^{-1} \Lambda \tag{2.64a}$$

$$= (I_t + H^\dagger \Pi H)^{-1} H^\dagger \Lambda, \tag{2.64b}$$

where Π is a $k \times k$ real, nonnegative, and diagonal matrix that should be optimized for the given utility function under the transmit power constraint. We remark that Π also appears in (2.35), but now the entries can be optimized under a *sum* power constraint. This relationship is a particular type of the MAC–BC *duality* result, which will be discussed again in Figure 2.45. Note that the identity (2.64b) follows by the matrix identity presented in (2.36). For small P the matrix Π has small values, and from (2.64a) or (2.64b) we recover the MRT matrix in (2.62). For large P, the matrix Π has large values, and from (2.64a) we recover the ZF matrix in (2.63).

We now return to the general problem with $r_j \geq 1$ for $j \in [k]$ and study dirty paper coding. For simplicity, we focus on the case of $k = 2$ receivers and consider the multicoding scheme shown in Figure 2.42, where the encoder for one message cancels the interference due to the other message. The variables x_1^n, x_2^n, and x^n are now sequences of t-dimensional vectors that we write as $\mathbf{x}_1^n$, $\mathbf{x}_2^n$, and $\mathbf{x}^n$, respectively. We choose the superposition function as $\mathbf{X} = \mathbf{X}_1 + \mathbf{X}_2$ where $\mathbf{X}_1 \sim CN(0, K_1)$ and $\mathbf{X}_2 \sim CN(0, K_2)$ are independent, and where the $t \times t$ covariance matrices K_1 and K_2 satisfy the transmit power constraint $\mathrm{tr}(K_1 + K_2) \leq P$. We further choose

$$U_1 = BU_2 + X_1$$
$$U_2 = X_2,$$

where $B = K_1 H_1^\dagger (I_{r_1} + H_1 K_1 H_1^\dagger)^{-1}$ is a $t \times t$ matrix that generalizes β in (2.53).

To send the message m_2 to receiver 2, consider the channel

$$\mathbf{Y}_2 = H_2 \mathbf{X}_2 + H_2 \mathbf{X}_1 + \mathbf{Z}_2$$

with input $\mathbf{X}_2 = \mathbf{U}_2$, interference $H_2 \mathbf{X}_1$, and noise $\mathbf{Z}_2$. By treating the interference $H_2 \mathbf{X}_1$ as noise, m_2 can be communicated reliably at receiver 2 if

$$R_2 < I(\mathbf{U}_2; \mathbf{Y}_2) = \log \frac{|I_{r_2} + H_2 K_2 H_2^\dagger + H_2 K_1 H_2^\dagger|}{|I_{r_2} + H_2 K_1 H_2^\dagger|}.$$

To send the message m_1 to receiver 1, consider the channel

$$\mathbf{Y}_1 = \mathsf{H}_1 \mathbf{X}_1 + \mathsf{H}_1 \mathbf{X}_2 + \mathbf{Z}_1$$

with input $\mathbf{X}_1 = \mathbf{U}_1 - \mathsf{B}\mathbf{U}_2$, *known* interference $\mathsf{H}_1 \mathbf{X}_2$, and noise $\mathbf{Z}_1$. By the vector writing on dirty paper result (2.61), m_1 can be sent reliably if

$$R_1 < I(\mathbf{U}_1; \mathbf{Y}_1) - I(\mathbf{U}_1; \mathbf{U}_2) = \log \|\mathbf{I}_{r_1} + \mathsf{H}_1 \mathsf{K}_1 \mathsf{H}_1^\dagger|.$$

We thus achieve the region $\mathscr{R}_1$ consisting of all rate pairs (R_1, R_2) satisfying

$$R_1 < \log \|\mathbf{I}_{r_1} + \mathsf{H}_1 \mathsf{K}_1 \mathsf{H}_1^\dagger|$$

$$R_2 < \log \frac{\|\mathbf{I}_{r_2} + \mathsf{H}_2 \mathsf{K}_2 \mathsf{H}_2^\dagger + \mathsf{H}_2 \mathsf{K}_1 \mathsf{H}_2^\dagger|}{\|\mathbf{I}_{r_2} + \mathsf{H}_2 \mathsf{K}_1 \mathsf{H}_2^\dagger|}$$

for some $K_1, K_2 \succeq 0$ with $\mathrm{tr}(K_1 + K_2) \leq P$. Similarly, by reversing the roles of receivers 1 and 2, we can achieve the region $\mathscr{R}_2$ consisting of all rate pairs (R_1, R_2) satisfying

$$R_1 < \log \frac{\|\mathbf{I}_{r_1} + \mathsf{H}_1 \mathsf{K}_1 \mathsf{H}_1^\dagger + \mathsf{H}_1 \mathsf{K}_2 \mathsf{H}_1^\dagger|}{\|\mathbf{I}_{r_1} + \mathsf{H}_1 \mathsf{K}_2 \mathsf{H}_1^\dagger|}$$

$$R_2 < \log \|\mathbf{I}_{r_2} + \mathsf{H}_2 \mathsf{K}_2 \mathsf{H}_2^\dagger|$$

for some $K_1, K_2 \succeq 0$ with $\mathrm{tr}(K_1 + K_2) \leq P$. The convex closure of $\mathscr{R}_1$ and $\mathscr{R}_2$, which can be achieved by time sharing, turns out to be the capacity region of the Gaussian vector BC [97] (see Figure 2.44).

This capacity region can be characterized efficiently without convex closure through a MAC–BC *duality* result. Given the Gaussian vector BC in Figure 2.43 with channel matrices H_1 and H_2, consider a Gaussian vector MAC with channel matrices $\mathsf{H}_1^\dagger$ and $\mathsf{H}_2^\dagger$ (referred to as the *dual MAC*) as depicted in Figure 2.45.

As discussed in Section 2.2.8, for $r_1 \times r_1$ and $r_2 \times r_2$ covariance matrices K_1 and K_2 for senders 1 and 2, respectively, the rate region $\mathscr{R}(\mathsf{K}_1, \mathsf{K}_2)$ consisting of all rate pairs satisfying

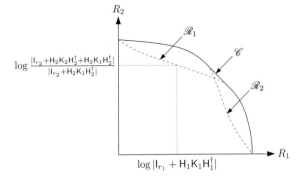

Figure 2.44 The capacity region of a Gaussian vector BC is the convex closure of the two regions $\mathscr{R}_1$ and $\mathscr{R}_2$ achieved by dirty paper coding with encoding orders $m_2 \to m_1$ and $m_1 \to m_2$, respectively.

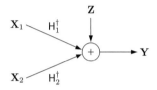

Figure 2.45 The dual MAC to the Gaussian vector BC in Figure 2.43.

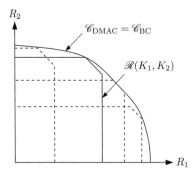

Figure 2.46 The capacity region $\mathscr{C}_{\mathrm{DMAC}}$ of the dual MAC, which is equivalent to the capacity region $\mathscr{C}_{\mathrm{BC}}$ of the Gaussian vector BC depicted in Figure 2.44.

$$R_1 < \log \|I_t + H_1^{\dagger} K_1 H_1\| \tag{2.65a}$$

$$R_2 < \log \|I_t + H_2^{\dagger} K_2 H_2\| \tag{2.65b}$$

$$R_1 + R_2 < \log \|I_t + H_1^{\dagger} K_1 H_1 + H_2^{\dagger} K_2 H_2\| \tag{2.65c}$$

is achievable. In fact, by (2.34), the capacity region $\mathscr{C}_{\mathrm{DMAC}}$ of the dual MAC under the *sum* power constraint is the closure of

$$\bigcup_{K_1, K_2 \succeq 0: \, \mathrm{tr}(K_1) + \mathrm{tr}(K_2) \leq P} \mathscr{R}(K_1, K_2).$$

The general form of the region $\mathscr{C}_{\mathrm{DMAC}}$ is depicted in Figure 2.46. Remarkably, this region turns out to be the same as the capacity region $\mathscr{C}_{\mathrm{BC}}$ of the Gaussian vector BC [98]. Since the rate bounds in (2.65) are convex in K_1 and K_2, the capacity region $\mathscr{C}_{\mathrm{DMAC}} = \mathscr{C}_{\mathrm{BC}}$ can be computed efficiently. This result extends to more than two receivers.

2.4   Interference Channels

The interference channel (IC) is a model that includes effects of interference and lack of cooperation among multiple cells. We begin with the two sender–receiver pair communication system depicted in Figure 2.47, where each sender (base station/user device) wishes to communicate a message to its respective receiver (user device/base station)

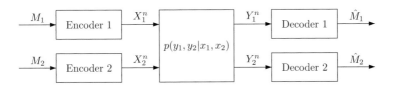

Figure 2.47 Two sender–receiver pair communication system.

over a shared communication channel. Senders 1 and 2 wish to reliably communicate their messages $m_1 \in [2^{nR_1}]$ and $m_2 \in [2^{nR_2}]$ to receivers 1 and 2, respectively. The output symbol pair (y_1, y_2) from each transmitted input symbol pair (x_1, x_2) is drawn according to the conditional probability $p_{Y_1,Y_2|X_1,X_2}(y_1, y_2|x_1, x_2)$, and the conditional probability of receiving the output sequences y_1^n and y_2^n at receivers 1 and 2, respectively, is

$$p(y_1^n, y_2^n|m_1, m_2) = \prod_{i=1}^{n} p_{Y_1,Y_2|X_1,X_2}(y_{1i}, y_{2i}|x_{1i}(m_1), x_{2i}(m_2)).$$

By decoding y_j^n, receiver $j = 1, 2$ forms the estimate $\hat{m}_j$ of m_j.

The encoder mappings $x_1^n(m_1)$ and $x_2^n(m_2)$ and the decoder mappings $\hat{m}_1(y_1^n)$ and $\hat{m}_2(y_2^n)$ are referred to as a $(2^{nR_1}, 2^{nR_2}, n)$ code. The average error probability of this code is

$$P_e^{(n)} = \mathsf{P}(M_1 \neq \hat{M}_1 \text{ or } M_2 \neq \hat{M}_2),$$

where we assume that M_1 and M_2 are independent and uniformly distributed. As usual, we say that a rate pair (R_1, R_2) is achievable if, for any target error probability $\epsilon > 0$ and sufficiently large n, there exists a $(2^{nR_1}, 2^{nR_2}, n)$ code with $P_e^{(n)} \leq \epsilon$. Alternatively, (R_1, R_2) is achievable if there exists a sequence of $(2^{nR_1}, 2^{nR_2}, n)$ codes such that $\lim_{n\to\infty} P_e^{(n)} = 0$. The capacity region $\mathscr{C}$ of the IC is the closure of the set of achievable rate pairs (R_1, R_2).

The capacity region $\mathscr{C}$ of a generic IC has the same convex form as that of a generic MAC (see Figure 2.14). However, as for the BC, $\mathscr{C}$ is not known in general. As one might expect, $\mathscr{C}$ can be characterized for several important classes of ICs, but it turns out that we still do not know $\mathscr{C}$ for the interesting class of Gaussian ICs that have moderate interference. A survey of key results is in [99–103] and [2, ch. 6]. For example, as for the BC, $\mathscr{C}$ depends on $p(y_1, y_2|x_1, x_2)$ only through the conditional marginal distributions $p(y_1|x_1, x_2)$ and $p(y_2|x_1, x_2)$.

As a concrete example, consider the two-user Gaussian IC depicted in Figure 2.48, which is a simple model for a wireless interference channel or a DSL cable bundle. The channel outputs corresponding to the inputs X_1 and X_2 are

$$Y_1 = h_{11}X_1 + h_{12}X_2 + Z_1$$
$$Y_2 = h_{21}X_1 + h_{22}X_2 + Z_2,$$

where $h_{jk}, j, k = 1, 2$, is the channel coefficient from sender k to receiver j, and $Z_1 \sim \mathrm{CN}(0, 1)$ and $Z_2 \sim \mathrm{CN}(0, 1)$ are noise variables, whose correlation is irrelevant. We consider the block transmit power constraints

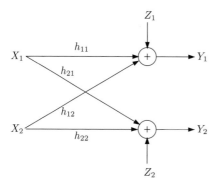

Figure 2.48 Gaussian interference channel.

$$\frac{1}{n}\sum_{i=1}^{n}|x_{ji}(m_j)|^2 \le P_j, \quad m_j \in [2^{nR_j}], \quad j = 1, 2.$$

We further define the received SNRs as $S_j = |h_{jj}|^2 P_j, j = 1, 2$, and we define the received *interference-to-noise ratios* (INRs) as $I_1 = |h_{12}|^2 P_2$ and $I_2 = |h_{21}|^2 P_1$.

2.4.1   Basic Signaling Schemes

We develop several coding schemes for the IC, which are basically the same as those for the MAC (see Section 2.2). The simplest scheme is to use TDMA/OMA where either sender 1 or sender 2 transmits information. By Shannon's channel coding theorem, the individual capacities are

$$C_1 = \max_{p(x_1), x_2} I(X_1; Y_1 | X_2 = x_2) \tag{2.66a}$$

$$C_2 = \max_{p(x_2), x_1} I(X_2; Y_2 | X_1 = x_1). \tag{2.66b}$$

Then by having sender 1 transmit a fraction α of the time, and sender 2 transmit a fraction $1 - \alpha$ of the time, the senders can avoid interference and achieve rate pairs (R_1, R_2) satisfying

$$R_1 < \alpha C_1 \tag{2.67a}$$

$$R_2 < (1 - \alpha)C_2. \tag{2.67b}$$

These rate pairs are sketched in Figure 2.49 as the dashed line, which is the same as in Figure 2.15 except that Y is replaced by Y_1 or Y_2.

We can alternatively let the senders transmit using CDMA/NOMA where each receiver treats the interfering signal as noise (TIN). When codewords are generated randomly according to $p(x_1)p(x_2)$, decoding is successful w.h.p. if

$$R_1 < I(X_1; Y_1) \tag{2.68a}$$

$$R_2 < I(X_2; Y_2). \tag{2.68b}$$

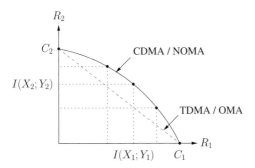

Figure 2.49 Rate regions achieved with TDMA/OMA, and CDMA/NOMA when treating interference as noise (TIN). For different channel coefficients, the TDMA/OMA curve may also lie above the CDMA/NOMA curve (see Figure 2.51).

One may now choose different $p(x_1)p(x_2)$ to sweep out a set of rate points, as shown by the solid line in Figure 2.49. This curve is the same as the NOMA curve in Figure 2.15 except that Y is replaced by Y_1 or Y_2. For the Gaussian IC with sufficiently small INRs, the exact condition of which is referred to as *weak* interference, NOMA with TIN achieves the sum-capacity [104–106]. Further results on NOMA with TIN can be found in [107].

The NOMA region plotted in Figure 2.49 is not convex in general and can be improved by time sharing across different choices of $p(x_1)p(x_2)$. This can be achieved by introducing a *time sharing* random variable Q that takes on different values to represent different choices of distributions via $p(x_1|q)p(x_2|q)$. Given a choice of $p(q)p(x_1|q)p(x_2|q)$, a codebook now consists of a time sharing sequence q^n generated according to $p(q)$ and codewords generated according to $p(x_1|q)p(x_2|q)$ conditioned on q^n [2, sec. 4.5.3]. This technique was referred to as *coded time sharing* in Section 2.2.4. We can write the resulting rates for NOMA with TIN as

$$R_1 < I(X_1; Y_1|Q) = \sum_q p(q)I(X_1; Y_1|Q = q) \tag{2.69a}$$

$$R_2 < I(X_2; Y_2|Q) = \sum_q p(q)I(X_2; Y_2|Q = q), \tag{2.69b}$$

which can be achieved under different choices of $p(q)p(x_1|q)p(x_2|q)$. Observe that, although we have not explicitly included conditions such as $X_2 = x_2$ and $X_1 = x_1$ as in (2.66), these conditions may be included implicitly via the $Q = q$ in (2.69). For example, $Q = 1$ might imply that $X_2 = x_2$ and $Q = 2$ might imply that $X_1 = x_1$.

2.4.2 NOMA with Advanced Decoding Schemes

We continue by mimicking the MAC development. The NOMA rates in (2.69) can be improved if the receivers do not necessarily use TIN. For example, if the interference at receiver 1 is strong, e.g., $|h_{12}| \geq |h_{22}|$ for the Gaussian IC, then intuition suggests that receiver 1 should first decode for m_2 while treating $x_1^n(m_1)$ as noise. Assuming the correct message $\hat{m}_2 = m_2$ was found, receiver 1 then decodes for m_1 from $(y_1^n, x_2^n(\hat{m}_2))$. Decoding is successful in both steps w.h.p. if

$$R_1 < I(X_1; Y_1 | X_2, Q) \tag{2.70a}$$
$$R_2 < I(X_2; Y_1 | Q). \tag{2.70b}$$

If receiver 2 also uses successive cancellation decoding, then we have the rate bounds

$$R_1 < I(X_1; Y_2 | Q) \tag{2.71a}$$
$$R_2 < I(X_2; Y_2 | X_1, Q). \tag{2.71b}$$

The resulting rate region is the intersection of (2.70) and (2.71) over all distributions $p(q)p(x_1|q)p(x_2|q)$.

If the interference is so strong that

$$I(X_1; Y_1 | X_2, Q) \le I(X_1; Y_2 | Q)$$
$$I(X_2; Y_2 | X_1, Q) \le I(X_2; Y_1 | Q)$$

for all interesting $p(q)p(x_1|q)p(x_2|q)$, then the IC is said to have *very strong* interference. The region above then coincides with the capacity region [108, 109], which consists of the rate bounds in (2.70a) and (2.71b) evaluated over all $p(q)p(x_1|q)p(x_2|q)$. For example, the Gaussian IC with very strong interference has

$$|h_{22}|^2 \le \frac{|h_{12}|^2}{1 + |h_{11}|^2} \quad \text{and} \quad |h_{11}|^2 \le \frac{|h_{21}|^2}{1 + |h_{22}|^2}$$

and the capacity region simplifies to the interference-free rates

$$R_1 < \log(1 + S_1)$$
$$R_2 < \log(1 + S_2),$$

which is achieved under the choice $X_1 \sim CN(0, P_1)$ and $X_2 \sim CN(0, P_2)$ without time sharing. Thus, having very strong interference is effectively the same as having no interference.

We next consider simultaneous decoding, which performs better than both TIN and successive cancellation decoding in general; see [110] for similar performance improvement for detection. Suppose that receiver 1 decodes for both messages by using joint typicality decoding as described in Section 2.2.4. However, since only m_1 must be recovered correctly, m_2 does not have to be unique or even correct.

The decoding error for message m_1 may occur if both of the following two error events occur [111]. First, a wrong codeword $x_1^n(\tilde{m}_1)$, $\tilde{m}_1 \ne m_1$, is jointly typical with a pair $(x_2^n(\tilde{m}_2), y_1^n)$, $\tilde{m}_2 \in [2^{nR_2}]$. This error has a vanishing probability if the rates are in the MAC pentagon as in (2.30), except that the bound (2.30b) is not required; see [102, 112, 113]. Second, a wrong codeword $x_1^n(\tilde{m}_1)$ is jointly typical with y_1^n. This error has a vanishing probability if the rate bounds for TIN in (2.69a) are satisfied. Hence, decoding at receiver 1 is successful w.h.p. if the rates satisfy either condition, or equivalently, if the rates are in $\mathscr{R}_{1,\text{SD}} = \mathscr{R}_{1,\text{TIN}} \cup \mathscr{R}_{1,\text{MAC}}$, where

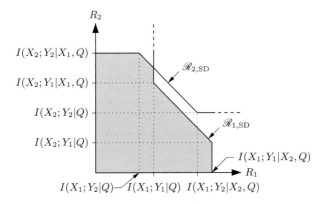

Figure 2.50 The achievable rate region of simultaneous decoding for a given $p(q)p(x_1|q)p(x_2|q)$ is the intersection of the rate region $\mathscr{R}_{1,\text{SD}}$ at receiver 1 and the rate region $\mathscr{R}_{2,\text{SD}}$ at receiver 2.

$$\mathscr{R}_{1,\text{TIN}} = \left\{ (R_1, R_2) : \begin{array}{l} 0 \leq R_1 < I(X_1; Y_1|Q) \\ 0 \leq R_2 \end{array} \right\} \tag{2.72a}$$

$$\mathscr{R}_{1,\text{MAC}} = \left\{ (R_1, R_2) : \begin{array}{l} R_1 < I(X_1; Y_1|X_2, Q) \\ R_1 + R_2 < I(X_1, X_2; Y_1|Q) \end{array} \right\}, \tag{2.72b}$$

as shown in Figure 2.50. When receiver 2 uses simultaneous decoding as well with the corresponding rate bounds $\mathscr{R}_{2,\text{SD}}$, any rate pair in the intersection of $\mathscr{R}_{1,\text{SD}}$ and $\mathscr{R}_{2,\text{SD}}$ is achievable. Note that this region includes the rate region (2.69) for TIN, and the rate region (2.70) and (2.71) by successive cancellation decoding.

We remark that achieving the rate pairs in $\mathscr{R}_{1,\text{TIN}}$ requires a simpler decoder that treats interference as noise, instead of the more complicated simultaneous decoder that achieves all rate pairs in $\mathscr{R}_{1,\text{SD}}$. The choice of decoder will thus usually depend on the system operating point (R_1, R_2). We further remark that $\mathscr{R}_{1,\text{SD}}$ can be expressed more compactly as nonnegative rate pairs satisfying

$$R_1 < I(X_1; Y_1|X_2, Q) \tag{2.73a}$$

$$R_1 + \min\left(R_2, I(X_2; Y_1|X_1, Q)\right) < I(X_1, X_2; Y_1|Q). \tag{2.73b}$$

As pointed out by Bandemer [111], the minimum term in (2.73b) can be interpreted as the effective rate of the interfering signal $x_2^n(m_2)$ at receiver 1, which is a monotone increasing function of R_2 until it saturates at the maximum possible rate $I(X_2; Y_1|X_1, Q)$. Thus, if R_2 is small, then $x_2^n(m_2)$ is distinguishable and the effective rate is R_2; if R_2 is large, then $x_2^n(m_2)$ is not distinguishable from other codewords, and one may treat $x_2^n(m_2)$ as noise. A similar phenomenon appears for Marton coding [114].

When the IC has *strong* interference [115, 116], the simultaneous decoding rate region simplifies to $\mathscr{R}_{1,\text{MAC}} \cap \mathscr{R}_{2,\text{MAC}}$; see (2.72b). Taking the union over all $p(q)p(x_1|q)p(x_2|q)$ results in the capacity region under strong interference. For the Gaussian IC, the strong interference condition is

$$|h_{12}| \geq |h_{22}| \quad \text{and} \quad |h_{21}| \geq |h_{11}|$$

and the capacity region is the intersection of the two Gaussian MAC pentagons (see (2.31)), which is characterized by the rate bounds

$$R_1 \leq \log(1 + S_1) \tag{2.74a}$$

$$R_2 \leq \log(1 + S_2) \tag{2.74b}$$

$$R_1 + R_2 \leq \log(1 + \min(S_1 + I_1, S_2 + I_2)) \tag{2.74c}$$

and achieved under the choice of $X_1 \sim \mathrm{CN}(0, P_1)$ and $X_2 \sim \mathrm{CN}(0, P_2)$ without time sharing.

2.4.3 Comparison of OMA and NOMA for the Gaussian IC

We illustrate the above rate regions for the Gaussian IC. First, for TDMA/OMA, using Gaussian symbols with maximum power is optimal, and (2.67) becomes

$$\mathscr{R}_{\mathrm{TDMA/OMA}} = \bigcup_{0 \leq \alpha \leq 1} \left\{ (R_1, R_2): \begin{array}{l} 0 \leq R_1 \leq \alpha \log\left(1 + \frac{S_1}{\alpha}\right) \\ 0 \leq R_2 \leq (1 - \alpha) \log\left(1 + \frac{S_2}{1-\alpha}\right) \end{array} \right\}.$$

Next, if we use NOMA with Gaussian symbols, which is not necessarily optimal, and TIN, then the region based on (2.68) is

$$\mathscr{R}_{\mathrm{NOMA,TIN}} = \bigcup_{0 \leq \beta_1, \beta_2 \leq 1} \left\{ (R_1, R_2): \begin{array}{l} 0 \leq R_1 \leq \log\left(1 + \frac{\beta_1 S_1}{\beta_2 I_1 + 1}\right) \\ 0 \leq R_2 \leq \log\left(1 + \frac{\beta_2 S_2}{\beta_1 I_2 + 1}\right) \end{array} \right\}.$$

One can check that the union can be restricted to $(\beta_1, \beta_2) \in \{(1, 0), (0, 1), (1, 1)\}$ if we subsequently permit time sharing. For simultaneous decoding, again with Gaussian symbols, we achieve the convex hull $\mathscr{R}_{\mathrm{NOMA,SD}}$ of the rates in the union of $\mathscr{R}_{\mathrm{NOMA,TIN}}$ and the pentagon in (2.74).

The performance of these schemes is shown in Figure 2.51 for an IC with symmetric powers $P_1 = P_2$, real channel gains $h_{11} = h_{22} = 1$ and $h_{12} = h_{21}$, and increasing values of $h_{12} = h_{21}$. Observe that TIN works well when the crosstalk coefficients are small, TDMA/OMA works well when the crosstalk coefficients are intermediate, and decoding interference is best when the crosstalk coefficients are large (or small). The TDMA/OMA rates are not a function of the crosstalk coefficients. Observe also that the outer bound is loose when the crosstalk coefficients are intermediate.

2.4.4 Rate Splitting and the Han–Kobayashi Coding Scheme

Non-orthogonal multiple access rates under simultaneous decoding discussed in Section 2.4.2 cannot be improved by any other decoding method [111]. Higher rates can be achieved, however, by a different encoder structure. Recall rate splitting and horizontal superposition coding discussed in Sections 2.1.6 and 2.2.6. Rate splitting at sender 1 involves splitting the message m_1 into two parts m_{11} and m_{12} that are encoded to $u_1^n(m_{11})$ and $v_1^n(m_{12})$, respectively. The message m_{11} has rate R_{11} and the message m_{12} has rate

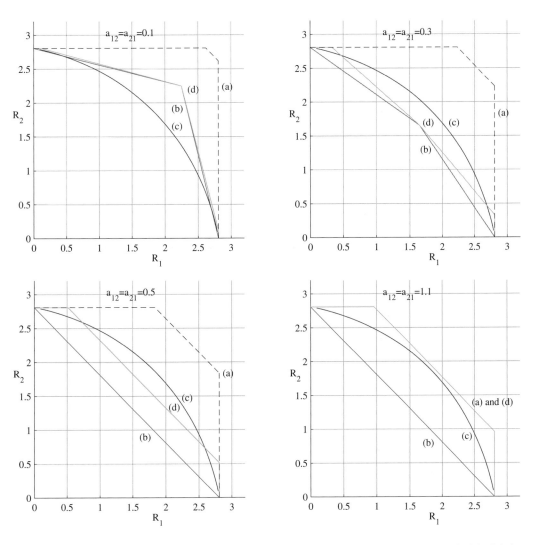

Figure 2.51 Rate regions for $S_1 = S_2 = 6$ and increasing $a_{12} = a_{21} = I_1/S_1 = I_2/S_2$. The labels are for (a) an outer bound on $\mathscr{C}$; (b) $\mathscr{R}_{\text{NOMA,TIN}}$ achieved by TIN with time sharing; (c) $\mathscr{R}_{\text{TDMA/OMA}}$ achieved using TDMA/OMA; (d) $\mathscr{R}_{\text{NOMA,SD}}$ achieved by decoding interference. The rate units are bits per channel symbol.

R_{12} so that $R_1 = R_{11} + R_{12}$. Similarly, the message m_2 is split into m_{21} and m_{22} of rates R_{21} and R_{22}, encoded to $u_2^n(m_{21})$ and $v_2^n(m_{22})$. This encoding structure was introduced by Carleial [100].

If the receivers use *successive cancellation decoding* [100] as in MLC (Section 2.1.6) or rate splitting multiple access (Section 2.2.6), then the resulting rate region is in general smaller than the simultaneous decoding region without rate splitting (2.72) [41]. Thus, under successive cancellation decoding, the flexibility from rate splitting does not improve the rates at all.

In the coding scheme by Han and Kobayashi [116], *simultaneous decoding* is used instead. In particular, decoder 1 finds the unique pair $(\hat{m}_{11}, \hat{m}_{12})$ such that $(u_1^n(\hat{m}_{11}), v_1^n(\hat{m}_{12}), u_2^n(m_{21}), y_1^n)$ is jointly typical for some m_{21}. Similarly, decoder 2 finds the unique $(\hat{m}_{21}, \hat{m}_{22})$ such that $(u_2^n(\hat{m}_{21}), v_2^n(\hat{m}_{22}), u_1^n(m_{11}), y_2^n)$ is jointly typical for some m_{11}. The achievable rates $R_1 = R_{11} + R_{12}$ and $R_2 = R_{21} + R_{22}$ of this coding scheme can be simplified as [102, 117]

$$R_1 < I(X_1; Y_1 | U_2, Q) \tag{2.75a}$$

$$R_2 < I(X_2; Y_2 | U_1, Q) \tag{2.75b}$$

$$R_1 + R_2 < I(X_1, U_2; Y_1 | Q) + I(X_2; Y_2 | U_1, U_2, Q) \tag{2.75c}$$

$$R_1 + R_2 < I(X_2, U_1; Y_2 | Q) + I(X_1; Y_1 | U_1, U_2, Q) \tag{2.75d}$$

$$R_1 + R_2 < I(X_1, U_2; Y_1 | U_1, Q) + I(X_2, U_1; Y_2 | U_2, Q) \tag{2.75e}$$

$$2R_1 + R_2 < I(X_1, U_2; Y_1 | Q) + I(X_1; Y_1 | U_1, U_2, Q) + I(X_2, U_1; Y_2 | U_2, Q) \tag{2.75f}$$

$$R_1 + 2R_2 < I(X_2, U_1; Y_2 | Q) + I(X_2; Y_2 | U_1, U_2, Q) + I(X_1, U_2; Y_1 | U_1, Q) \tag{2.75g}$$

for any distribution $p(q)p(u_1, x_1 | q)p(u_2, x_2 | q)$. The cardinalities $|\mathcal{U}_1|$, $|\mathcal{U}_2|$, $|\mathcal{Q}|$ of the respective random variables U_1, U_2, Q can be chosen to satisfy $|\mathcal{U}_1| \leq |\mathcal{X}_1| + 4$, $|\mathcal{U}_2| \leq |\mathcal{X}_2| + 4$, $|\mathcal{Q}| \leq 7$. This region includes all rate regions in Sections 2.4.1 and 2.4.2 as special cases. For example, it reduces to (2.69) by taking degenerate U_1 and U_2. At the other extreme, it recovers all points in Figure 2.50 by time sharing between $U_1 = X_1$ with degenerate U_2, and $U_2 = X_2$ with degenerate U_1.

As a concrete example, consider the Gaussian IC with $h_{11} = h_{22} = 1$, $h_{12} = h_{21} = 0.05$, and $P_1 = P_2 = 5000$. If $U_j \sim \text{CN}(0, (1 - \beta_j)P_j)$ and $V_j \sim \text{CN}(0, \beta_j P_j)$ are independent of each other, and $X_j = U_j + V_j$ for $j = 1, 2$, then the Han–Kobayashi region (2.75) without time sharing is a heptagonal region. Figure 2.52 shows the resulting heptagon for $\beta_1 = \beta_2 = 0.05$.

More generally, we can choose $X_j(q) = U_j(q) + V_j(q)$, where $U_j(q)$ and $V_j(q)$ are independent Gaussian random variables with respective variances $(1 - \beta_j(q))P_j(q)$ and $\beta_j(q)P_j(q)$. The power constraints are thus $\sum_q p(q)P_j(q) \leq P_j$, $j = 1, 2$. The region (2.75), restricted to such Gaussian distributions, is thus the union over $p(q)$, and $\{(\beta_j(q), P_j(q)) : j = 1, 2, q \in \mathcal{Q}\}$ satisfying the power constraints, of the (R_1, R_2) satisfying the bounds in (2.75), where

$$I(X_1; Y_1 | U_2, Q)$$

$$= \sum_q p(q) \log \left(1 + \frac{P_1(q)|h_{11}|^2}{\beta_2(q)P_2(q)|h_{12}|^2 + 1} \right)$$

$$I(X_1, U_2; Y_1 | Q)$$

$$= \sum_q p(q) \log \left(1 + \frac{P_1(q)|h_{11}|^2 + (1 - \beta_2(q))P_2(q)|h_{12}|^2}{\beta_2(q)P_2(q)|h_{12}|^2 + 1} \right)$$

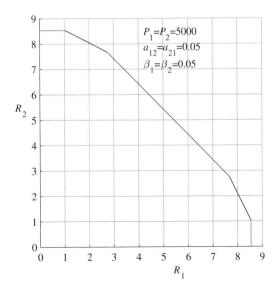

Figure 2.52 Rate region for $S_1 = S_2 = 5000$ and $a_{12} = a_{21} = I_1/S_1 = I_2/S_2 = 0.05$. Seven bounds are active. The rate units are bits per channel symbol.

$$I(X_1; Y_1 | U_1, U_2, Q)$$
$$= \sum_q p(q) \log \left(1 + \frac{\beta_1(q) P_1(q) |h_{11}|^2}{\beta_2(q) P_2(q) |h_{12}|^2 + 1} \right)$$

$$I(X_1, U_2; Y_1 | U_1, Q)$$
$$= \sum_q p(q) \log \left(1 + \frac{\beta_1(q) P_1(q) |h_{11}|^2 + (1 - \beta_2(q)) P_2(q) |h_{12}|^2}{\beta_2(q) P_2(q) |h_{12}|^2 + 1} \right)$$

and similarly for the other mutual information expressions. Since this region includes all the rate regions in Sections 2.4.1 and 2.4.2, it is sum-rate optimal for weak interference, and optimal for strong and very strong interference. In fact, for any channel coefficients and power constraints, this region is within 1 bit per dimension of the capacity region of the Gaussian IC [118, 119].

Simultaneous decoding plays a pivotal role in achieving the rates in (2.73) and (2.75) and no low-complexity implementation exists for most codes (see Section 2.2.4). By employing *diagonal*, instead of horizontal, superposition coding (Section 2.1.6), low-complexity successive cancellation decoding can replace simultaneous decoding without any rate loss [41, 45]. Several alternatives to practical implementation and their performances are discussed by Nam et al. [120].

2.4.5 MIMO Systems with Intercell Interference

The Gaussian vector IC

$$\mathbf{Y}_1 = \mathbf{H}_{11} \mathbf{X}_1 + \mathbf{H}_{12} \mathbf{X}_2 + \mathbf{Z}_1$$
$$\mathbf{Y}_2 = \mathbf{H}_{21} \mathbf{X}_1 + \mathbf{H}_{22} \mathbf{X}_2 + \mathbf{Z}_2$$

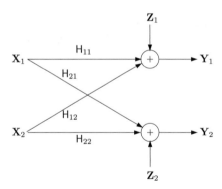

Figure 2.53 MIMO systems with intercell interference.

provides a simple model for uplink and downlink MIMO systems in two neighboring cells that communicate over the same time and frequency resources; see Figure 2.53. As in the MIMO MAC and BC, $\mathbf{Y}_j$ is an r_j-dimensional output vector at receiver j, $\mathbf{X}_j$ is a t_j-dimensional input vector from sender j, H_{jk} is an $r_j \times t_k$ channel matrix from sender k to receiver j, and $\mathbf{Z}_j \sim \mathrm{CN}(0, \mathsf{I}_{r_j})$ is an r_j-dimensional noise vector. As before, we consider block power constraints on every codeword $\mathbf{x}_j^n(m_j)$.

Similar to the single-antenna case, the Han–Kobayashi coding scheme operates within r_1 and r_2 bits, respectively, of any rate pair (R_1, R_2) in the capacity region [119, 121]. At high SINR, a much simpler transceiver structure can be employed. For simplicity, suppose that $t_1 = t_2 = t$, $r_1 = r_2 = r$, and $t = r = 2s$. Suppose that sender $j = 1, 2$ uses a $t \times s$ precoding matrix V_j and transmits $\mathbf{X}_j = \mathsf{V}_j\mathbf{U}_j$, where $\mathbf{U}_j = [U_{j1}, \dots, U_{js}]^T$ as in the Gaussian vector BC in Section 2.3.5. Receiver $j = 1, 2$ uses an $s \times r$ beamforming matrix W_j and the resulting channel outputs are

$$\tilde{\mathbf{Y}}_1 = \mathsf{W}_1\mathbf{Y}_1 = \mathsf{W}_1(\mathsf{H}_{11}\mathsf{V}_1\mathbf{U}_1 + \mathsf{H}_{12}\mathsf{V}_2\mathbf{U}_2 + \mathbf{Z}_1)$$
$$\tilde{\mathbf{Y}}_2 = \mathsf{W}_2\mathbf{Y}_2 = \mathsf{W}_2(\mathsf{H}_{21}\mathsf{V}_1\mathbf{U}_1 + \mathsf{H}_{22}\mathsf{V}_2\mathbf{U}_2 + \mathbf{Z}_2).$$

If the beamforming matrices W_1 and W_2 are chosen such that the zero-forcing condition

$$\mathsf{W}_1\mathsf{H}_{12}\mathsf{V}_2 = \mathsf{W}_2\mathsf{H}_{21}\mathsf{V}_1 = 0$$

is satisfied, then both sender–receiver pairs can enjoy interference-free communication. For example, W_1 can be taken as the first s rows of the pseudoinverse of $\mathsf{H}_{1*} = [\mathsf{H}_{11}\mathsf{V}_1 \quad \mathsf{H}_{12}\mathsf{V}_2]$ as in the zero-forcing matrix (2.38) for the MAC. Similarly, precoding matrices V_j can be taken as the first s columns of the pseudoinverse of

$$\mathsf{H}_{*1} = \begin{bmatrix} \mathsf{W}_1\mathsf{H}_{11} \\ \mathsf{W}_2\mathsf{H}_{21} \end{bmatrix}$$

as in the zero-forcing matrix (2.63) for the BC. As P_1 and P_2 tend to infinity, the achievable rates scale asymptotically as $R_j = s\log(P_j)$, that is, the achievable *degrees of freedom* equal half the number of antennas.

The simple zero-forcing transceiver structure has a far-reaching extension for ICs with more than two sender–receiver pairs. To illustrate the basic idea, consider the beamforming scheme above for the three-user Gaussian vector IC with the same number of antennas at each sender/receiver, which results in

$$\tilde{\mathbf{Y}}_1 = \mathbf{W}_1(\mathbf{H}_{11}\mathbf{V}_1\mathbf{U}_1 + \mathbf{H}_{12}\mathbf{V}_2\mathbf{U}_2 + \mathbf{H}_{13}\mathbf{V}_3\mathbf{U}_3 + \mathbf{Z}_1)$$
$$\tilde{\mathbf{Y}}_2 = \mathbf{W}_2(\mathbf{H}_{21}\mathbf{V}_1\mathbf{U}_1 + \mathbf{H}_{22}\mathbf{V}_2\mathbf{U}_2 + \mathbf{H}_{23}\mathbf{V}_3\mathbf{U}_3 + \mathbf{Z}_2)$$
$$\tilde{\mathbf{Y}}_3 = \mathbf{W}_3(\mathbf{H}_{31}\mathbf{V}_1\mathbf{U}_1 + \mathbf{H}_{32}\mathbf{V}_2\mathbf{U}_2 + \mathbf{H}_{33}\mathbf{V}_3\mathbf{U}_3 + \mathbf{Z}_3).$$

As before, the beamforming matrices $\mathbf{W}_j$, $j = 1, 2, 3$, are chosen such that the interference at each receiver is forced to zero, i.e., we have

$$\mathbf{W}_1(\mathbf{H}_{12}\mathbf{V}_2 + \mathbf{H}_{13}\mathbf{V}_3) = \mathbf{W}_2(\mathbf{H}_{21}\mathbf{V}_1 + \mathbf{H}_{23}\mathbf{V}_3) = \mathbf{W}_3(\mathbf{H}_{31}\mathbf{V}_1 + \mathbf{H}_{32}\mathbf{V}_2) = 0.$$

Note that in order to satisfy this zero-forcing condition, the row-dimension s of $\mathbf{W}_j$ cannot be larger than the difference between the full dimension r and the dimension of the combined interference (such as $\mathbf{H}_{12}\mathbf{V}_2 + \mathbf{H}_{13}\mathbf{V}_3$ at receiver 1). For arbitrary precoding matrices, s can be no more than $r/3$. By judiciously optimizing over the precoding matrices, however, the combined interference at each receiver can be *aligned* to occupy a lower-dimensional subspace simultaneously. In particular, as $t = r \to \infty$, a solution to the zero-forcing condition exists if $s/r \to 1/2$, which continues to hold for any number of sender–receiver pairs. Thus, as in the two-user case, each user in a k-sender/receiver IC can achieve degrees of freedom equal to half the number of antennas.

This *interference alignment* technique was introduced in [122] and [123]. The degrees of freedom of $s = r/2$ can be achieved even for a small number of antennas by extending the spatial dimension to multiple transmission times, the extreme of which is the *ergodic* interference alignment scheme in [124]. Other dimensions beyond space and time have been utilized, such as *real* interference alignment [125]. For more on interference alignment, we refer the interested reader to [126].

2.5   Two-Hop Relay Networks

Relay networks provide simple models for multihop communication systems in which some nodes, referred to as *relays*, do not have messages of their own to send and cooperate with other nodes to facilitate communication between senders and receivers. We present a few simple two-hop relay network models for wireless communication and discuss basic coding schemes such as symbol relaying, decode–forward, and compress–forward. Useful surveys of cooperative communication are presented in [127, 128].

2.5.1   Relay Channels

The relay channel (RC) is the simplest multihop communication system model with a single relay, as shown in Figure 2.54. The message $m \in [2^{nR}]$ is communicated by the

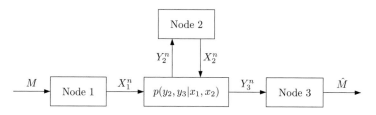

Figure 2.54 Point-to-point communication system with a relay.

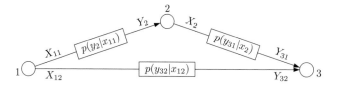

Figure 2.55 A network of point-to-point channels.

sender (node 1 that transmits x_1) with the help of a relay (node 2 that transmits x_2 and receives y_2) to the receiver (node 3 that receives y_3), where the output symbols y_2 and y_3 from transmitted input symbols x_1 and x_2 are drawn according to the conditional probability $p_{Y_2,Y_3|X_1,X_2}(y_2,y_3|x_1,x_2)$.

The encoder mapping $x_1^n(m)$, the relay mapping $x_{2i}(y_2^{i-1})$, $i \in [n]$, and the decoder mapping $\hat{m}(y_3^n)$ are referred to as a $(2^{nR}, n)$ code, which induces a joint probability distribution of the random variables that factors as

$$p(m, x_1^n, x_2^n, y_2^n, y_3^n, \hat{m})$$
$$= p(m)\, p(x_1^n|m) \left[\prod_{i=1}^{n} p(x_{2i}|y_2^{i-1}) p_{Y_2 Y_3|X_1 X_2}(y_{2i}, y_{3i}|x_{1i}, x_{2i}) \right] p(\hat{m}|y_3^n),$$

where $p(x_1^n|m)$, $p(x_{2i}|y_2^{i-1})$, and $p(\hat{m}|y_3^n)$ take on the values 0 and 1 only because they represent functions. The average error probability, achievability, and capacity are defined as in the point-to-point communication channel without a relay.

The RC model includes a wide variety of practical problems. For example, consider the network of point-to-point channels shown in Figure 2.55. The channel input of the sender is a vector $X_1 = (X_{11}, X_{12})$, where X_{11} is the input of the point-to-point channel from the sender (node 1) to the relay (node 2), and X_{12} is the input of the point-to-point channel from the sender (node 1) to the receiver (node 3). The input of the point-to-point channel from the relay (node 2) to the receiver (node 3) is X_2. The relay observes Y_2 and the receiver observes the vector $Y_3 = (Y_{31}, Y_{32})$. The channel probability distribution thus factors as

$$p(y_2, y_{31}, y_{32}|x_{11}, x_{12}, x_2) = p(y_2|x_{11}) p(y_{31}|x_2) p(y_{32}|x_{12}). \tag{2.76}$$

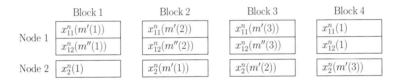

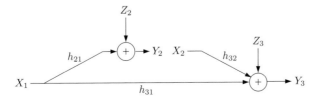

Figure 2.56 Block transmission scheme for the relay network of Figure 2.55.

Figure 2.57 Gaussian relay channel.

As a special instance of such a network, suppose that X_{11}, X_{12}, and X_2 are binary, and that

$$Y_2 = X_{11}, \quad Y_{31} = X_2, \quad \text{and} \quad Y_{32} = X_{12}. \tag{2.77}$$

The capacity is known to be 2 bits per transmission, as follows from the max-flow min-cut theorem [129, 130]. A block-oriented transmission scheme is shown in Figure 2.56, where the sender communicates a stream of messages $m(1), m(2), \dots$, each of $2n$ bits. The message $m(b)$ is split into two parts $m'(b)$, $m''(b)$ with n bits each. The first part $m'(b)$ is *routed* via the relay node 2, and the second part $m''(b)$ is sent via the direct link to the receiver node 3. It is clear that one cannot achieve more than 2 bits per transmission because node 1 can send (and node 3 can receive) at most 2 bits per transmission. Observe also that the first message part is delayed by one extra block due to the longer path through the relay; see [131] for an introduction to network flow problems.

Another example of an RC is the Gaussian RC depicted in Figure 2.57. The channel is defined by

$$Y_2 = h_{21}X_1 + Z_2 \tag{2.78a}$$
$$Y_3 = h_{31}X_1 + h_{32}X_2 + Z_3, \tag{2.78b}$$

where h_{21}, h_{31}, and h_{32} are channel coefficients, and $Z_2 \sim CN(0, 1)$ and $Z_3 \sim CN(0, 1)$ are independent noise variables. For example, a common situation is when $|h_{31}|$ is very small, in which case *two-hop* relaying can be used for range extension. We consider the block power constraints

$$\frac{1}{n}\sum_{i=1}^{n}|x_{1i}(m)|^2 \le P_1, \quad m \in [2^{nR}]$$

and

$$\frac{1}{n}\sum_{i=1}^{n}|x_{2i}(y_2^{i-1})|^2 \le P_2.$$

Observe that, as for the Gaussian MAC, the model (2.78) has the transmissions taking place synchronously, i.e., only x_{1i} and x_{2i} affect the channel outputs y_{1i}, y_{2i} at time i. Furthermore, the model permits the relay to react with a short delay of one time unit, whereas in practice the relay might require more time to develop its response. It turns out, however, that the capacity of the RC is unaffected by the relay delay as long as it is one or more time units. For example, if the delay is two time units, then one can split transmission into odd and even times and operate the RC as if there is a delay of one time unit; the delay increases but the capacity does not change. We remark that there are practical situations in which the relay delay can be modeled as zero, or even less than zero, in which case new coding possibilities arise [132, 133]. We do not consider such models here, however.

We remark that (2.78) is implicitly a *full-duplex* model in which the relay can transmit and receive at the same time in the same frequency band. Furthermore, the channel output at the relay is a noisy function of X_1 only, i.e., the channel slightly simplifies as

$$p(y_2, y_3 | x_1, x_2) = p(y_2 | x_1) p(y_3 | x_1, x_2, y_2). \qquad (2.79)$$

Such a model is reasonable for a relay with two antennas, one for reception and one for transmission, that are spaced sufficiently far apart. However, if there is insufficient space, or there is only one antenna, then the transmission and reception will interfere. A common approach to treat such problems is to use either time- or frequency-division to separate the physical resources for receiving and transmitting signals at the relay. Such relays are called *half-duplex*, and several models have been considered in the literature. For example, a simple model for half-duplex relays [134] adds the following constraint to (2.78):

$$Y_2 = 0 \text{ if } X_2 \neq 0.$$

Alternatively, suppose the relay uses frequency-division with two non-overlapping frequency bands labeled as $f = 1, 2$. If the relay receives $Y_2^{(1)}$ in frequency band 1 and transmits $X_2^{(2)}$ in frequency band 2, then a reasonable model might be

$$Y_2^{(1)} = h_{21}^{(1)} X_1^{(1)} + Z_2^{(1)}$$
$$Y_3^{(1)} = h_{31}^{(1)} X_1^{(1)} + Z_3^{(1)}$$
$$Y_3^{(2)} = h_{31}^{(2)} X_1^{(2)} + h_{32}^{(2)} X_2^{(2)} + Z_3^{(2)},$$

where $X_1^{(f)}, Y_3^{(f)}, Z_3^{(f)}, f = 1, 2$, are the various signals at the sender and receiver, the $h_{ij}^{(f)}, i = 2, 3, j = 1, 2, f = 1, 2$ are channel coefficients, and $Z_2^{(1)}$ is noise at the relay in frequency band 1. Of course, when computing the rate, one should normalize by the total time and frequency resources consumed.

A simpler time-division scheme for half-duplex relays is shown in Figure 2.58, where the sender and relay alternate in communicating a message. In odd-numbered blocks $b = 1, 3, \ldots$, the sender transmits $x_1^n(m((b + 1)/2))$ and the relay decodes its output y_2^n to $\tilde{m}((b + 1)/2)$. Assuming that the relay decoded correctly, we have $\tilde{m}((b + 1)/2) = m((b + 1)/2)$. In even-numbered blocks $b = 2, 4, \ldots$, the relay transmits $x_1^n(\tilde{m}(b/2))$ and the destination decodes its output y_3^n to $\hat{m}(b/2)$. This transmission

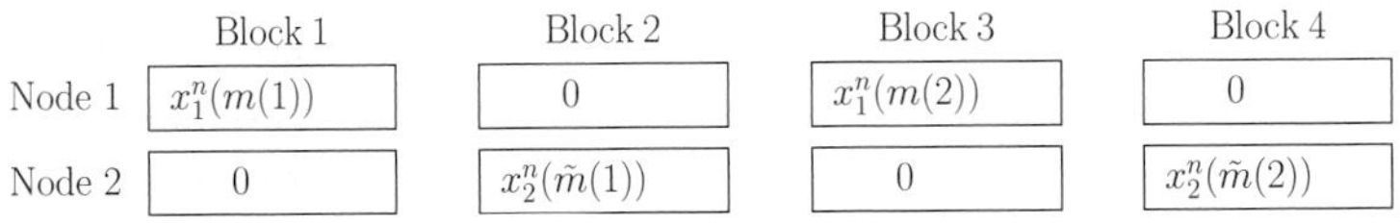

	Block 1	Block 2	Block 3	Block 4
Node 1	$x_1^n(m(1))$	0	$x_1^n(m(2))$	0
Node 2	0	$x_2^n(\tilde{m}(1))$	0	$x_2^n(\tilde{m}(2))$

Figure 2.58 Block transmission scheme for a half-duplex RC.

scheme is called *multihopping*. In fact, multihopping is suboptimal for half-duplex RCs for several reasons. First, the sender could transmit in both odd- and even-numbered blocks; second, the receiver should decode by using its channel outputs from both the odd- and even-numbered blocks; and finally, the relay could communicate information to the receiver node by modulating the times it chooses to receive and transmit [134].

The RC was studied early on by van der Meulen [135]. The capacity of the RC is still not known in general. We will develop four coding schemes and show that these can sometimes achieve capacity. The first scheme is *symbol relaying* that includes methods such as *amplify–forward* (AF) and *demodulate–forward*. The second scheme uses a technique called *block Markov superposition encoding* and is often called *decode–forward* (DF). The third scheme adds partial decoding, and the fourth scheme, called *quantize–forward* (QF) or *compress–forward* (CF), uses data compression. The second, third, and fourth schemes are due to Cover and El Gamal [136].

2.5.2 Symbol Relaying

As the simplest relaying scheme, the relay can transmit $x_{2i} = f(y_{2,i-1})$ for some symbol-relaying function $f(y)$. For the Gaussian RC, the receiver observes

$$Y_{3i} = h_{31}X_{1i} + h_{32}f(h_{21}X_{1,i-1} + Z_{2,i-1}) + Z_{3i}, \quad i \in [n].$$

We should now design $p(x_1^n)$ to maximize the end-to-end mutual information $I(X_1^n; Y_3^n)$. Note that the channel from x_1^n to y_3^n has unit memory.

For half-duplex relays, a common approach is to use a line network model [137] with two channels $Y_2 = h_{21}X_1 + Z_2$ and $Y_3 = h_{32}X_2 + Z_3$ that are used in turn. In this case, the received signals are

$$Y_{3i} = h_{32}f(h_{21}X_{1,i-1} + Z_{2,i-1}) + Z_{3i}, \quad i = 2, 4, 6, \ldots \tag{2.80}$$

and the end-to-end channel is effectively memoryless (shift the time i by 1 for X_{1i} and Z_{2i}).

Finding the best symbol-relaying function is a rather difficult optimization problem [137]. We discuss a few common choices. The simplest choice is the linear function

$$f(y) = c\,y \tag{2.81}$$

for some complex constant c chosen to satisfy the power constraint at the relay. This relaying method is called *amplify–forward*, which has been studied extensively [138–140].

For the half-duplex model (2.80), we have the effective channel

$$Y_3 = c\,h_{32}h_{21}X_1 + (c\,h_{32}Z_2 + Z_3)$$

that has the capacity

$$C = \log\left(1 + \frac{|c|^2|h_{32}h_{21}|^2 P_1}{|c|^2|h_{32}|^2 + 1}\right)$$

and we can choose c to satisfy

$$|c|^2(|h_{21}|^2 P_1 + 1) \le P_2. \tag{2.82}$$

The best coefficient c satisfies (2.82) with equality so that

$$C = \log\left(1 + \frac{|h_{32}h_{21}|^2 P_1 P_2}{|h_{21}|^2 P_1 + |h_{32}|^2 P_2 + 1}\right). \tag{2.83}$$

Note that if P_1 or P_2 are large, then (2.83) is close to the capacity

$$\log\left(1 + \min(|h_{21}|^2 P_1, |h_{32}|^2 P_2)\right)$$

of the line network model without a half-duplex constraint. However, if P_1 and P_2 are small, then the rate $\log(1 + |h_{32}|^2|h_{21}|^2 P_1 P_2)$ can be far from capacity.

The poor performance of AF for strong noise motivates two alternatives. First, use *bursty* signaling, i.e., under a block power constraint we conserve energy for long intervals, and transmit with high energy in short intervals to mimic high-SNR. Second, choose nonlinear functions that limit the adverse effects of large noise values [137, 141]. For example, if the sender uses BPSK then one might choose the binary decision

$$f(y) = \begin{cases} -\sqrt{P_2} & \text{if } y < 0 \\ \sqrt{P_2} & \text{if } y \ge 0. \end{cases} \tag{2.84}$$

More generally, the relaying function can be the scaled *hard decision* for the symbol from the sender alphabet. This relaying method is sometimes called *demodulate–forward* or *scalar–quantize–forward*. Yet another nonlinear relaying function is

$$f(y) = c\,y \quad \text{mod } [-A, A), \tag{2.85}$$

where a mod $[-A, A) = a - k \cdot 2A$ and k is the unique integer so that $-A \le a - k \cdot 2A < A$.

2.5.3   Decode–Forward

We next describe the *decode–forward* relaying scheme, originally introduced by Cover and El Gamal [136], that provides a digital interface at the relay. Figure 2.59 depicts the block structure. As in diagonal superposition coding in Section 2.1.6, and in the routing scheme of Figure 2.56, the sender communicates a stream of messages $m(1), m(2), \ldots,$ each of nR bits. In block b, the relay transmits the codeword $x_2^n(\tilde{m}(b-1))$ that carries the message estimate $\tilde{m}(b-1)$ from block $b-1$. At the same time, the sender transmits

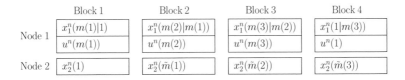

Figure 2.59 Decode–forward block transmission scheme.

the superposition of the codeword $u^n(m(b))$ carrying the new message and the codeword $x_2^n(m(b-1))$ via a symbol-by-symbol mapping $x_1(u, x_2)$ as

$$x_1^n(m(b)|m(b-1)) = x_1(u^n(m(b)), x_2^n(m(b-1))).$$

Thus, if the estimate $\tilde{m}(b-1)$ at the relay is correct, then the sender and relay transmit the same codeword $x_2^n(m(b-1))$ coherently in block b, while the sender superposes the additional codeword $u^n(m(b))$.

At the end of block b, the relay decodes its received signal $y_2^n(b)$ to the $\tilde{m}(b-1)$ of the next block. Assuming that $\tilde{m}(b-1)$ was recovered correctly, the relay can incorporate $x_2^n(\tilde{m}(b-1))$ into decoding as in successive cancellation decoding for the MAC (see Section 2.2.2). When codewords are generated randomly according to $p(u)$ and $p(x_2)$, decoding at the relay is successful w.h.p. if

$$R < I(U; Y_2|X_2) = I(X_1; Y_2|X_2). \tag{2.86}$$

Since the transmission of $m(b)$ is over two blocks b and $b+1$, carried by $u^n(m(b))$ and $x_2^n(m(b))$, respectively, the receiver uses *sliding-window decoding* [42–44]. Assuming that $\hat{m}(b-1)$ was recovered correctly, the receiver finds the unique $\hat{m}(b)$ such that $u^n(\hat{m}(b))$ and $(x_2^n(\hat{m}(b-1)), y_3^n(b))$ are jointly typical, and $x_2^n(\hat{m}(b))$ and $y_3^n(b+1)$ are jointly typical. As in diagonal superposition coding, decoding at the receiver is successful w.h.p. if

$$R < I(U; Y_3|X_2) + I(X_2; Y_3) = I(U, X_2; Y_3) = I(X_1, X_2; Y_3). \tag{2.87}$$

By combining the rate bounds in (2.86) and (2.87), and choosing the optimal input distribution $p(u)p(x_2)$ and symbol-by-symbol function $x_1(u, x_2)$, the DF scheme achieves the rate

$$R_{\text{DF}} = \max_{\substack{p(u)p(x_2) \\ x_1(u,x_2)}} \min\left(I(X_1; Y_2|X_2), I(X_1, X_2; Y_3)\right)$$

$$= \max_{p(x_1,x_2)} \min\left(I(X_1; Y_2|X_2), I(X_1, X_2; Y_3)\right). \tag{2.88}$$

We remark that this rate can be achieved with several different encoding and decoding methods, as described in [2, 44]. Decode–forward can be extended to general multihop relay networks [44, 142].

For the Gaussian RC in Figure 2.57, the maximum in (2.88) is attained by jointly Gaussian symbols $X_1 \sim \mathcal{CN}(0, P_1)$ and $X_2 \sim \mathcal{CN}(0, P_2)$ with correlation coefficient $\rho = E[X_1 X_2^*]/\sqrt{P_1 P_2}$. As a concrete example, suppose that the relay is placed along

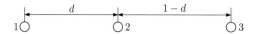

Figure 2.60 The relay (node 2) between the sender (node 1) and receiver (node 3).

the line between the sender and receiver, as depicted in Figure 2.60. If the path loss exponent is α, then the channel outputs are

$$Y_2 = \frac{e^{j\phi_{21}} X_1}{d^{\alpha/2}} + Z_2 \tag{2.89a}$$

$$Y_3 = e^{j\phi_{31}} X_1 + \frac{e^{j\phi_{32}} X_2}{(1-d)^{\alpha/2}} + Z_3, \tag{2.89b}$$

where $d \in (0,1)$, and where ϕ_{21}, ϕ_{31}, and ϕ_{32} are phases corresponding to the carrier frequency and propagation delays of the paths. The DF rate is here optimized by choosing $\rho = \tilde{\rho}\, e^{-j(\phi_{31}-\phi_{32})}$, where $\tilde{\rho}$ is real and nonnegative, so that (2.88) becomes

$$R_{\mathrm{DF}} = \max_{0 \le \tilde{\rho} \le 1} \min\left(\log\left(1 + \frac{P_1}{d^\alpha}(1 - \tilde{\rho}^2)\right), \right.$$
$$\left. \log\left(1 + \frac{P_1}{d^\alpha} + \frac{P_2}{(1-d)^\alpha} + 2\tilde{\rho}\sqrt{\frac{P_1 P_2}{d^\alpha(1-d)^\alpha}}\right)\right). \tag{2.90}$$

The rates R_{DF} are plotted against d in Figure 2.61, where we have chosen $\alpha = 3$ and $P_1 = P_2 = 10$. For instance, suppose that $d \to 1$, in which case the best achievable rate is $R_{\mathrm{DF}} = \log(1+10) \approx 3.5$ bits per transmission. This is the same rate achieved without a relay. However, as $d \to 0$, we have $R_{\mathrm{DF}} \to \log(1 + 40) \approx 5.4$ bits per transmission, which is a substantial boost from the rate without using the relay. As expected, the relay is most useful when it is roughly halfway between the sender and receiver. For small d, R_{DF} is very close to a capacity upper bound called the *cutset bound* [136]. In fact, R_{DF} is within 1 bit from the cutset bound (and thus from the capacity) for all channel coefficients and power constraints for the Gaussian RC.

To compare with multihopping, consider the scheme shown in Figure 2.58, and suppose that we optimize the transmit and receive times of the relay. The resulting rate is

$$R_{\mathrm{MH}} = \max_{0 \le \beta \le 1} \min\left(\beta \log\left(1 + \frac{P_1}{\beta d^\alpha}\right), \right.$$
$$\left. (1 - \beta) \log\left(1 + \frac{P_2}{(1-\beta)(1-d)^\alpha}\right)\right), \tag{2.91}$$

where β specifies the fraction of time the sender transmits and the relay listens. The rates R_{MH} are plotted in Figure 2.58, and we see that this multihopping scheme performs poorly. The reason is that time sharing incurs a rate loss (a pre-logarithmic factor) that is difficult to compensate with the power gain due to smaller distances (a logarithmic factor). One can improve R_{MH} substantially by having the receiver decode by using both the odd- and even-numbered output blocks in Figure 2.58. The resulting rates are shown as the dash-dotted curve labeled R_{MH2} in Figure 2.61.

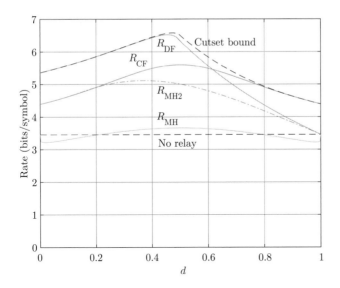

Figure 2.61 Comparison of the achievable rates for the Gaussian RC using DF, multihopping, and CF against the cutset upper bound (upper dashed curve) for varying distance d between the relay and the sender.

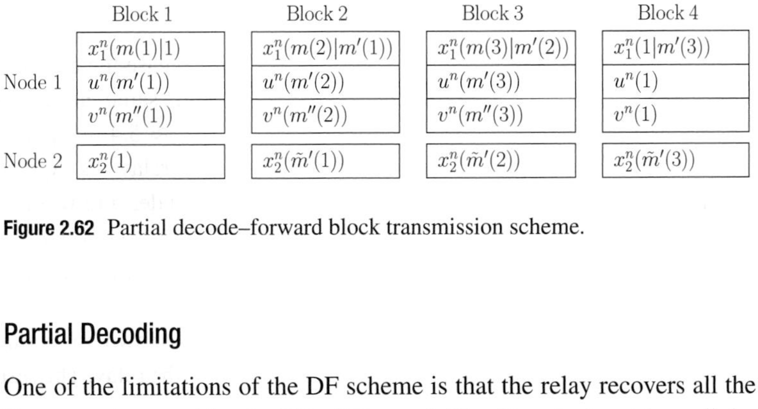

	Block 1	Block 2	Block 3	Block 4
Node 1	$x_1^n(m(1)\|1)$	$x_1^n(m(2)\|m'(1))$	$x_1^n(m(3)\|m'(2))$	$x_1^n(1\|m'(3))$
	$u^n(m'(1))$	$u^n(m'(2))$	$u^n(m'(3))$	$u^n(1)$
	$v^n(m''(1))$	$v^n(m''(2))$	$v^n(m''(3))$	$v^n(1)$
Node 2	$x_2^n(1)$	$x_2^n(\tilde{m}'(1))$	$x_2^n(\tilde{m}'(2))$	$x_2^n(\tilde{m}'(3))$

Figure 2.62 Partial decode–forward block transmission scheme.

2.5.4   Partial Decoding

One of the limitations of the DF scheme is that the relay recovers all the message bits. For example, consider the RC of Figure 2.55 with noiseless binary links (2.77). In this case, R_{DF} in (2.88) is only 1 bit per transmission because the relay can decode and forward at most 1 bit at a time.

To circumvent this problem, we proceed as for the routing scheme of Figure 2.56 and split each message $m(b)$ into two parts $m'(b)$ and $m''(b)$ with respective rates R' and R'', and require that the relay decode and forward only $m'(b)$. Such a *partial decode–forward* (PDF) or *multipath* scheme can be designed by allowing an additional superposition layer at the sender.

The message part $m'(b)$ is cooperatively transmitted by the sender and relay via the codeword $u^n(m'(b))$ in block b and the codeword $x_2^n(\tilde{m}'(b))$ in block $b + 1$ (see Figure 2.62). On top of that transmission, $m''(b)$ is transmitted at the sender via the codeword $v^n(m''(b))$. Thus, in block b the sender transmits the superposition of $u^n(m'(b))$,

$v^n(m''(b))$, and $x_2(m'(b-1))$ via a symbol-by-symbol mapping $x_1(u, v, x_2)$, and in the same block the relay transmits $x_2^n(\tilde{m}'(b-1))$.

As before, the relay decodes for $m'(b)$ at the end of block b, which is successful w.h.p. if

$$R' < I(U; Y_2|X_2).$$

The receiver finds $m'(b)$ using sliding-window decoding, which is successful w.h.p. if

$$R' < I(U, X_2; Y_3).$$

The receiver then finds $m''(b)$ using successive cancellation decoding (assuming $m'(b)$ was recovered correctly), which is successful w.h.p. if

$$R'' < I(V; Y_3|U, X_2).$$

By combining the rate bounds for $R = R' + R''$ and choosing the optimal input distribution and superposition function, the PDF scheme achieves

$$R_{\text{PDF}} = \max_{\substack{p(u)p(v)p(x_2) \\ x_1(u,v,x_2)}} \left[\min\left(I(U; Y_2|X_2),\ I(U, X_2; Y_3)\right) + I(V; Y_3|U, X_2) \right]. \qquad (2.92)$$

By rearranging the mutual information terms and expanding the set of distributions (which does not incur any rate change), the achievable rate can be rewritten as

$$R_{\text{PDF}} = \max_{p(u,x_1,x_2)} \min\left(I(X_1; U, Y_3|X_2) - I(X_1; U|X_2, Y_2),\ I(X_1, X_2; Y_3)\right). \qquad (2.93)$$

The second term in the minimum in (2.93) is reminiscent of the rate for a MIMO system with two transmit antennas and one receive antenna. The first term is reminiscent of the rate for a MIMO system with one transmit antenna and two receive antennas. After all, the sender conveys the message through Y_3 as well as U recovered at the relay. There is, however, a cost of setting up U at the relay, which is reflected in the rate loss $I(X_1; U|X_2, Y_2)$.

Consider the RC of (2.77) and set $U = X_{11}$ and $V = X_{12}$ in (2.92). Then, the rate expression becomes

$$R_{\text{PDF}} = \max_{p(x_{11})p(x_{12})p(x_2)} \left[\min\left(H(X_{11}),\ H(X_{12}) + H(X_2)\right) + H(X_{12}) \right] = 2,$$

which is the capacity. Partial decode–forward achieves capacity for several other classes of RCs [143, 144]. Moreover, PDF can be extended to *distributed decode–forward* (DDF) [145] for general relay networks.

2.5.5   Compress–Forward

We next develop a relaying scheme that avoids decoding at the relay, and instead requires the relay to forward a digital summary of its received signal [136]. This scheme is known by several names, including *quantize–forward* and *compress–forward*.

As in DF, the sender communicates a stream of messages $m(1), m(2), \ldots$, each of nR bits. The encoding at the sender is much simpler, however, and involves only transmitting the codeword $x_1^n(m(b))$ in block b.

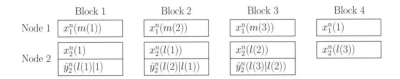

Figure 2.63 Compress–forward block transmission scheme.

At the end of block b, the relay finds a sequence $\hat{y}_2^n(l(b))$, $l(b) \in [2^{n\tilde{R}}]$, from a codebook of "quantization points" to represent its received signal $y_2^n(b)$. The index $l(b)$ of the quantized sequence with $n\tilde{R}$ bits is then transmitted as $x_2^n(l(b))$ in block $b+1$. For higher quantization efficiency, the codeword $x_2^n(l(b-1))$ and the quantization sequence $\hat{y}_2^n(l(b)|l(b-1))$ are designed jointly (see Figure 2.63). More precisely, the codewords $x_2^n(l(b-1))$, $l(b-1) \in [2^{n\tilde{R}}]$, are generated randomly according to $p(x_2)$ and for each $l(b-1)$, the sequences $\hat{y}_2^n(l(b)|l(b-1))$, $l(b) \in [2^{n\tilde{R}}]$, are generated according to $p(\hat{y}_2|x_2)$ conditioned on each $x_2^n(l(b-1))$. Using joint typicality encoding as in Section 2.3.3, the index $l(b)$ is found such that the quantization sequence $\hat{y}_2^n(l(b)|l(b-1))$ is jointly typical with the received signal $y_2^n(b)$ and the transmitted codeword $x_2^n(l(b-1))$ in block b, which is successful w.h.p. if

$$\tilde{R} > I(Y_2; \hat{Y}_2|X_2). \tag{2.94}$$

We remark that the rate (2.94) can be achieved by different encoding and decoding methods, as described in [2, 44, 146]. For example, if there is no self-interference and channel decomposes as (2.79), then there is no need to design the codeword x_2^n and the quantization sequence $\hat{y}_2^n$ jointly. Compress–forward can be extended to *network compress–forward* [44] and *noisy network coding* [73, 147] for general multihop relay networks. Different encoding and decoding methods for noisy network coding are described by Hou and Kramer [148].

Since $m(b)$ is communicated over two blocks – in block b as $x_1^n(m(b))$ and in block $b+1$ as $x_2^n(l(b))$ – the receiver uses sliding-window decoding over the two blocks. Assuming that $\hat{l}(b-1)$ is recovered correctly, the receiver finds the unique $(\hat{m}(b), \hat{l}(b))$ such that $(x_1^n(m(b)), x_2^n(\hat{l}(b-1)), \hat{y}_2^n(\hat{l}(b)|\hat{l}(b-1)))$, and $y_3^n(b)$ are jointly typical, and $x_2^n(\hat{l}(b))$, and $y_3^n(b+1)$ are jointly typical. The standard random coding analysis shows that decoding is successful w.h.p. if

$$R < I(X_1; \hat{Y}_2, Y_3|X_2) \tag{2.95a}$$

$$\tilde{R} < I(\hat{Y}_2; X_1, Y_3|X_2) + I(X_2; Y_3) \tag{2.95b}$$

$$R + \tilde{R} < I(X_1, X_2; Y_3) + I(X_1, Y_3; \hat{Y}_2|X_2). \tag{2.95c}$$

Combining the bounds in (2.94) and (2.95), we can achieve

$$R < \min\left(I(X_1; \hat{Y}_2, Y_3|X_2), I(X_1, X_2; Y_3) - I(Y_2; \hat{Y}_2|X_1, X_2, Y_3)\right)$$

for any distribution $p(x_1)p(x_2)p(\hat{y}_2|x_2, y_2)$ satisfying

$$I(\hat{Y}_2; Y_2|X_1, X_2, Y_3) < I(X_2; Y_3). \tag{2.96}$$

By adding a time sharing random variable Q [149], and noting that the constraint (2.96) is inactive at the optimum, we conclude that the CF scheme achieves

$$R_{CF} = \max \min \left\{ \begin{array}{l} I(X_1; \hat{Y}_2, Y_3 | X_2, Q), \\ I(X_1, X_2; Y_3 | Q) - I(Y_2; \hat{Y}_2 | X_1, X_2, Y_3, Q) \end{array} \right\}, \tag{2.97}$$

where the maximum is over all $p(q)p(x_1|q)p(x_2|q)p(\hat{y}_2|x_2, y_2, q)$.

The first mutual information term in (2.97) is reminiscent of the rate for a MIMO system with one transmit antenna and two receive antennas. After all, the receiver sees both Y_3 and an approximation $\hat{Y}_2$ of Y_2. The second term in the minimum is reminiscent of the rate for a MIMO system with two transmit antennas and one receive antenna, but there are several differences: X_1 and X_2 are independent, $\hat{Y}_2 \neq Y_2$ in general, and thus there is a compression loss $I(Y_2; \hat{Y}_2 | X_1, X_2, Y_3, Q)$. It is interesting to note the dual relationship between R_{CF} in (2.97) and R_{PDF} in (2.93), where $\hat{Y}_2$ and U, respectively, play the role of Y_2 in a MIMO system.

As a concrete example, consider the Gaussian RC in Figure 2.57 with parameters in (2.89). We use CF with Gaussian X_1 and X_2, and $\hat{Y}_2 = Y_2 + \hat{Z}_2$, where $\hat{Z}_2 \sim CN(0, N)$ is independent of all other random variables. Under this choice of the jointly Gaussian distribution, which is suboptimal in general, the CF rate (2.97) under the best value of N becomes

$$R_{CF} = \log \left(1 + P_1 + \frac{P_1 P_2}{(1-d)^{\alpha} P_1 + d^{\alpha} P_2 + d^{\alpha}(1-d)^{\alpha}(1+P_1)} \right).$$

The resulting rates against d are plotted in Figure 2.61. As $d \to 1$, CF achieves the capacity. In fact, it can be shown that R_{CF} is within 1 bit from the capacity for all channel coefficients and power constraints for the Gaussian RC.

2.5.6   Cloud Radio Access Networks

We conclude our discussion on relay networks by considering cloud radio access networks (C-RANs) as simple wireless two-hop networks with multiple relays (see Figures 2.64 and 2.67). We will extend the DF and CF schemes to these network models.

We first consider the uplink C-RAN shown in Figure 2.64, where the first hop is the wireless channel $p(y^l | x^k)$ from the k senders (user equipment) to the l relays (radio heads), and the second hop consists of orthogonal links (fronthaul) of capacities $C_1, \ldots, C_l$ bits per transmission from the relays to the receiver (central processor).

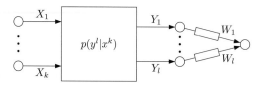

Figure 2.64 The uplink cloud radio access network.

The second hop is decoupled from the first hop, and the orthogonal links are decoupled from each other. For simplicity, the channel output at the receiver is assumed to be $(W_1, \ldots, W_l)$, where $W_i \in [2^{nC_i}]$ is communicated from relay i to the receiver reliably over n transmissions. When the link capacities are unbounded, the uplink C-RAN can be viewed as a MAC where the receiver observes a vector output $\mathbf{Y} = [Y_1, \ldots, Y_l]^T$. As a special instance of an uplink C-RAN, suppose that the first hop is a Gaussian channel with

$$\mathbf{Y} = \mathsf{H}\mathbf{X} + \mathbf{Z}, \tag{2.98}$$

where H is the $l \times k$ channel matrix with coefficients h_{ij} from sender j to relay i, $\mathbf{X} = [X_1, \ldots, X_k]^T$ and $\mathbf{Z} = [Z_1, \ldots, Z_l]^T$ are the vectors of channel inputs and noise variables, respectively, and $Z_i \sim \mathrm{CN}(0, 1)$, $i \in [l]$, are mutually independent of each other and $\mathbf{X}$. As in the MAC and IC, we consider the block transmit power constraints $P_1, \ldots, P_k$.

The highest rate of decoding-based schemes, whether using DF or PDF, is achieved by using only the best relay. Hence, we turn to compression-based schemes. A naive approach is to compress the channel output y_i^n at relay i into quantization points $\hat{y}_i^n(w_i)$ with nC_i bits, which requires

$$I(Y_i; \hat{Y}_i) < C_i, \quad i \in [l] \tag{2.99}$$

under random coding and joint typicality encoding. The overall channel can then be treated as a MAC with k inputs $X_1, \ldots, X_k$ and a single vector output $\hat{Y}^l = (\hat{Y}_1, \ldots, \hat{Y}_l)$ with achievable rate tuple $(R_1, \ldots, R_k)$ satisfying

$$\sum_{j \in \mathcal{J}} R_j < I(X(\mathcal{J}); \hat{Y}^l | X(\mathcal{J}^c)), \quad \mathcal{J} \in [k] \tag{2.100}$$

as in the MAC with multiple senders (see Section 2.2.7).

The CF scheme presented in Section 2.5.5 has more flexibility in that the compression index generated at the relay is decoupled from its own transmission. Hence, compression rates higher than (2.99) can be used at the relays, while only nC_i bits of the compression indices are sent to the receiver, which then decodes for the remaining bits of the compression indices, along with the messages. This scheme was developed by Sanderovich et al. [150, 151], and can achieve rate tuples satisfying

$$\sum_{j \in \mathcal{J}} R_j < I(X(\mathcal{J}); \hat{Y}(\mathcal{I}^c) | X(\mathcal{J}^c)) + \sum_{i \in \mathcal{I}} (C_i - I(Y_i; \hat{Y}_i | X^k)), \quad \mathcal{I} \in [l], \mathcal{J} \in [k] \tag{2.101}$$

for any distribution $p(x_1) \cdots p(x_k) p(\hat{y}_1 | y_1) \cdots p(\hat{y}_l | y_l)$. The same rate region can also be achieved by specializing network CF [44] or noisy network coding [147] for general multihop relay networks.

As a concrete example, suppose that $k = 1$ sender is placed on the line between $l = 2$ relays at distances d and $1 - d$, respectively (see Figure 2.65). For the Gaussian uplink C-RAN, the channel outputs are

Figure 2.65 C-RAN with $k = 1$ sender between $l = 2$ relays.

$$Y_1 = \frac{e^{j\phi_1}X}{d^{\alpha/2}} + Z_1$$

$$Y_2 = \frac{e^{j\phi_2}X}{(1-d)^{\alpha/2}} + Z_2,$$

where ϕ_1 and ϕ_2 are phases similar to those in (2.89). Partial decode–forward under the optimal Gaussian input distribution $X \sim \text{CN}(0, P)$ achieves

$$R_{\text{PDF}} = \max \left\{ \begin{array}{c} \min \left[\log\left(1 + \frac{P}{d^\alpha}\right), C_1 \right], \\ \min \left[\log\left(1 + \frac{P}{(1-d)^\alpha}\right), C_2 \right] \end{array} \right\}, \tag{2.102}$$

which is always less than the link capacity of a single relay. The achievable rate of the naive compression-based scheme in (2.100) is

$$R_{\text{NCF}} = \log \left(1 + \frac{(2^{C_1} - 1)P}{P + 2^{C_1}d^\alpha} + \frac{(2^{C_2} - 1)P}{P + 2^{C_2}(1-d)^\alpha} \right) \tag{2.103}$$

under the joint Gaussian distribution $X \sim \text{CN}(0, P)$, $\hat{Y}_i = Y_i + \hat{Z}_i$, where $\hat{Z}_i = \text{CN}(0, N_i)$, as in Section 2.5.5, and N_i is chosen to satisfy (2.99). In comparison, the achievable rate of CF in (2.101) under the joint Gaussian distribution is

$$R_{\text{CF}} = \max_{N_1, N_2} \min \left\{ \begin{array}{c} \log\left(1 + \frac{P}{d^\alpha(N_1+1)} + \frac{P}{(1-d)^\alpha(N_2+1)}\right), \\ \log\left(1 + \frac{P}{d^\alpha(N_1+1)}\right) + C_2 - \log\left(1 + \frac{1}{N_2}\right), \\ \log\left(1 + \frac{P}{(1-d)^\alpha(N_2+1)}\right) + C_1 - \log\left(1 + \frac{1}{N_1}\right), \\ C_1 - \log\left(1 + \frac{1}{N_1}\right) + C_2 - \log\left(1 + \frac{1}{N_2}\right) \end{array} \right\}. \tag{2.104}$$

Figure 2.66 plots these rates against d for $\alpha = 3$, $P = 1$, and $C_1 = C_2 = 2$. Compress–forward performs consistently better than naive compression since it allows for higher-resolution quantization of the received signals at the relays.

For the Gaussian uplink C-RAN with k senders and l receivers in (2.98), R_{CF} is within $\log(2el)$ bits per dimension from the cutset bound (and hence from the capacity region) for any channel coefficients and power constraints, regardless of k [152]. Therefore, if the fronthaul capacities are $\log(2el)$ bits above the wireless capacities, then CF can approximately achieve rate tuples as in the centralized Gaussian vector MAC in Section 2.2.8 with one antenna per sender and l antennas at the receiver. In particular, the sum-capacity scales as the number of users if l is larger than the number of users k.

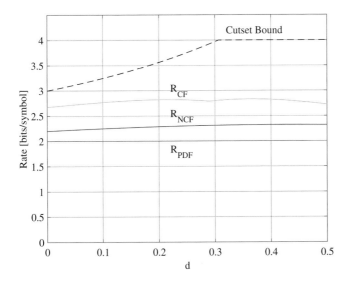

Figure 2.66 Comparison of the achievable rates for the Gaussian uplink C-RAN using PDF, (R_{PDF}), naive compression (R_{NCF}), and CF (R_{CF}) against the cutset bound (dashed curve) for varying position d of the sender between the two relays.

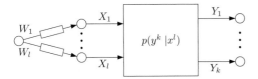

Figure 2.67 The downlink cloud radio access network.

We now consider the downlink C-RAN with l relays and k receivers, depicted in Figure 2.67. In this dual model to the uplink C-RAN, the first hop consists of orthogonal links (backhaul) of capacities $C_1, \ldots, C_l$ bits per transmission from the sender (central processor) to the relays (radio heads), decoupled from the second hop and from each other, and the second hop is the wireless channel $p(y^k | x^l)$ from the relays to the receivers (user equipment). As a special instance of a downlink C-RAN, suppose that the second hop is a Gaussian channel with

$$\mathbf{Y} = \mathbf{H}\mathbf{X} + \mathbf{Z}, \tag{2.105}$$

where $\mathbf{H}$ is the $k \times l$ channel matrix with coefficients h_{ji} from relay i to receiver j, $\mathbf{X} = [X_1, \ldots, X_l]^T$ and $\mathbf{Z} = [Z_1, \ldots, Z_k]^T$ are the vectors of channel inputs and noise variables, respectively, and $Z_j \sim \mathrm{CN}(0, 1)$, $j \in [k]$, are mutually independent of each other and $\mathbf{X}$. We consider the block transmit power constraints $P_1, \ldots, P_l$ at the relays.

When there is only a single ($k = 1$) receiver (and thus a single message), the sender can split the message into l independent parts and send them to the relays. The relays can then communicate the parts as in a MAC. This simple scheme achieves the *split–forward* (SF) rate

$$R_{SF} = \max_{p(x_1)\cdots p(x_l)} \min_{\mathcal{I}} \left(I(X(\mathcal{I}); Y | X(\mathcal{I}^c)) + \sum_{i \in \mathcal{I}^c} C_i \right),$$

which is close to the capacity in general. For example, for the Gaussian downlink C-RAN in (2.105), the SF scheme achieves the capacity within $\log l$ bits per transmission for all channel parameters and power constraints. Specialized capacity results for $k = 1$ appear in [153–155].

When there is more than one message, however, rate splitting becomes infeasible. We can use another version of PDF called *distributed decode–forward* (DDF) in this case. Instead of explicit rate splitting, the sender uses precoding as in Marton coding (see Section 2.3.3) to prepare auxiliary codewords $u_j^n(m_j, l_j)$, $m_j \in [2^{nR_j}]$, $l_j \in [2^{n\tilde{R}_j}]$, together with $x_i(w_i)$, $w_i \in [2^{nC_i}]$, and find appropriate $(l_1, \ldots, l_j)$ and $(w_1, \ldots, w_l)$. This encoding step is successful w.h.p. if

$$\sum_{j \in \mathcal{J}} \tilde{R}_j + \sum_{i \in \mathcal{I}} C_i > I^*(U(\mathcal{J}), X(\mathcal{I})), \quad \mathcal{I} \subseteq [l], \mathcal{J} \subseteq [k] \tag{2.106}$$

under random coding with respect to $p(u^k, x^l)$, where

$$I^*(U(\mathcal{J}), X(\mathcal{I})) = \sum_{j \in \mathcal{J}} H(U_j) + \sum_{i \in \mathcal{I}} H(X_i) - H(U(\mathcal{J}), X(\mathcal{I}))$$

is the multivariate mutual information of $U_j, j \in \mathcal{J}$, and $X_i, i \in \mathcal{I}$. The chosen indices $w_1, \ldots, w_l$ are communicated to the relays via the sender-to-relay links and relay i transmits $x_i^n(w_i)$. Finally, receiver j decodes y_j^n to recover m_j and l_j, which is successful w.h.p. if

$$R_j + \tilde{R}_j < I(U_j; Y_j), \quad j \in [k]. \tag{2.107}$$

Combining the bounds in (2.106) and (2.107), we achieve rate tuples satisfying

$$\sum_{j \in \mathcal{J}} R_j < I(X(\mathcal{I}); U(\mathcal{J}) | X(\mathcal{I}^c)) + \sum_{i \in \mathcal{I}^c} C_i - \sum_{j \in \mathcal{J}} I(U_j; X^l | Y_j) \tag{2.108}$$

for any $p(x_1) \cdots p(x_l) p(u_1 | x^l) \cdots p(u_k | x^l)$. This rate region is similar to that of the uplink C-RAN in (2.101), but in a dual form as we have seen in the MAC–BC duality, and in the relationship between PDF (2.93) and CF (2.97). This DDF scheme was analyzed by Ganguly and Kim [152] as a special case of a more general scheme of Lim et al. [145].

For the Gaussian downlink C-RAN with l relays and k receivers in (2.105), the rate region in (2.108) is within $\log(ekl)$ bits per dimension of the capacity region for all channel parameters and power constraints. Therefore, if the fronthaul capacities are $\log(ekl)$ bits above the wireless capacities, then DDF can approximately achieve rate tuples as in the centralized Gaussian vector BC in Section 2.3.5 with one antenna per receiver and l antennas at the sender. Once again, the sum-capacity scales as the number of users if l is larger than the number of users k.

References

[1] I. Csiszár and J. Körner, *Information Theory: Coding Theorems for Discrete Memoryless Systems*, 2nd ed. Cambridge: Cambridge University Press, 2011.

[2] A. El Gamal and Y.-H. Kim, *Network Information Theory*. Cambridge: Cambridge University Press, 2011.

[3] G. Kramer, "Topics in multi-user information theory," *Found. Trends Comm. Inf. Theory*, vol. 4, no. 4/5, pp. 265–444, 2007.

[4] T. M. Cover and J. A. Thomas, *Elements of Information Theory*, 2nd ed. New York: Wiley, 2006.

[5] R. G. Gallager, *Information Theory and Reliable Communication*. New York: Wiley, 1968.

[6] D. J. C. MacKay, *Information Theory, Inference and Learning Algorithms*. Cambridge: Cambridge University Press, 2003.

[7] R. W. Yeung, *Information Theory and Network Coding*. New York: Springer, 2008.

[8] C. E. Shannon, "A mathematical theory of communication," *Bell Syst. Tech. J.*, vol. 27, no. 3, pp. 379–423, 1948.

[9] J. Wolfowitz, "The coding of messages subject to chance errors," *Illinois J. Math.*, vol. 1, no. 4, pp. 591–606, 1957.

[10] J. L. Massey, *Lecture Notes for Applied Digital Information Theory (1980–1998)*. Zurich, Switerland: ETH Zurich, 1998.

[11] A. Orlitsky and J. R. Roche, "Coding for computing," *IEEE Trans. Inf. Theory*, vol. 47, no. 3, pp. 903–917, 2001.

[12] V. Strassen, "Asymptotische Abschätzungen in Shannon's informationstheorie," in *Trans. 3rd Prague Conf. Inf. Theory, Statist. Decision Functions, Random Processes (Liblice, Prague, 1962)*. Prague: Czechoslovak Academy of Sciences, 1962, pp. 689–723.

[13] M. Hayashi, "Information spectrum approach to second-order coding rate in channel coding," *IEEE Trans. Inf. Theory*, vol. 55, no. 11, pp. 4947–4966, Nov. 2009.

[14] Y. Polyanskiy, H. V. Poor, and S. Verdú, "Channel coding rate in the finite blocklength regime," *IEEE Trans. Inf. Theory*, vol. 56, no. 5, pp. 2307–2359, May 2010.

[15] S. Arimoto, "On the converse to the coding theorem for discrete memoryless channels (corresp.)," *IEEE Trans. Inf. Theory*, vol. 19, no. 3, pp. 357–359, 1973.

[16] G. Dueck and J. Körner, "Reliability function of a discrete memoryless channel at rates above capacity," *IEEE Trans. Inf. Theory*, vol. 25, no. 1, pp. 82–85, 1979.

[17] N. Alon and J. H. Spencer, *The Probabilistic Method*, 3rd ed. Hoboken, NJ: Wiley, 2008.

[18] S. Shamai and S. Verdú, "The empirical distribution of good codes," *IEEE Trans. Inf. Theory*, vol. 43, no. 3, pp. 836–846, May 1997.

[19] P. Elias, "Coding for noisy channels," in *IRE Int. Conv. Rec.*, vol. 3, part 4, 1955, pp. 37–46.

[20] R. G. Gallager, *Low-Density Parity-Check Codes*. Cambridge, MA: MIT Press, 1963.

[21] C. Berrou, A. Glavieux, and P. Thitimajshima, "Near Shannon limit error-correcting coding and decoding: Turbo codes," in *Proc. IEEE Int. Conf. Commun.*, Geneva, May 1993, pp. 1064–1070.

[22] A. Shokrollahi and M. Luby, "Raptor codes," *Found. Trends Comm. Inf. Theory*, vol. 6, no. 3–4, pp. 213–322, 2011.

[23] M. Lentmaier, A. Sridharan, D. J. Costello, and K. S. Zigangirov, "Iterative decoding threshold analysis for LDPC convolutional codes," *IEEE Trans. Inf. Theory*, vol. 56, no. 10, pp. 5274–5289, Oct. 2010.

[24] S. Kudekar, T. J. Richardson, and R. L. Urbanke, "Threshold saturation via spatial coupling: Why convolutional LDPC ensembles perform so well over the BEC," *IEEE Trans. Inf. Theory*, vol. 57, no. 2, pp. 803–834, Feb. 2011.

[25] N. Stolte, "Rekursive Codes mit der Plotkin-Konstruktion und ihre Decodierung," PhD dissertation, TU Darmstadt, Germany, 2002.

[26] E. Arıkan, "Channel polarization: A method for constructing capacity-achieving codes for symmetric binary-input memoryless channels," *IEEE Trans. Inf. Theory*, vol. 55, no. 7, pp. 3051–3073, Jul. 2009.

[27] E. Arikan, "On the origin of polar coding," *IEEE J. Sel. Areas Commun.*, vol. 34, no. 2, pp. 209–223, Feb. 2016.

[28] D. J. Costello and G. D. Forney, "Channel coding: The road to channel capacity," *Proc. IEEE*, vol. 95, no. 6, pp. 1150–1177, June 2007.

[29] V. I. Levenshtein, "Binary codes capable of correcting deletions, insertions and reversals," *Soviet Physics: Doklady*, vol. 10, pp. 707–710, Feb. 1966.

[30] R. L. Dobrushin, "Optimum information transmission through a channel with unknown parameters," *Radio Eng. Electron.*, vol. 4, no. 12, pp. 1–8, 1959.

[31] D. Blackwell, L. Breiman, and A. J. Thomasian, "The capacity of a class of channels," *Ann. Math. Statist.*, vol. 30, no. 4, pp. 1229–1241, 1959.

[32] J. Wolfowitz, *Coding Theorems of Information Theory*. Berlin: Springer, 1961.

[33] A. Lapidoth and P. Narayan, "Reliable communication under channel uncertainty," *IEEE Trans. Inf. Theory*, vol. 44, no. 6, pp. 2148–2177, 1998.

[34] G. Keshet, Y. Steinberg, and N. Merhav, "Channel coding in the presence of side information," *Found. Trends Comm. Inf. Theory*, vol. 4, no. 6, pp. 445–586, 2008.

[35] C. E. Shannon, "Channels with side information at the transmitter," *IBM J. Res. Develop.*, vol. 2, no. 4, pp. 289–293, 1958.

[36] S. I. Gelfand and M. S. Pinsker, "Coding for channel with random parameters," *Probl. Control Inf. Theory*, vol. 9, no. 1, pp. 19–31, 1980.

[37] C. Heegard and A. El Gamal, "On the capacity of computer memories with defects," *IEEE Trans. Inf. Theory*, vol. 29, no. 5, pp. 731–739, 1983.

[38] A. F. Molisch, *Wireless Communications*, 2nd ed. Chichester: Wiley, 2011.

[39] U. Wachsmann, R. F. H. Fischer, and J. B. Huber, "Multilevel codes: Theoretical concepts and practical design rules," *IEEE Trans. Inf. Theory*, vol. 45, no. 5, pp. 1361–1391, 1999.

[40] G. Caire, G. Taricco, and E. Biglieri, "Bit-interleaved coded modulation," *IEEE Trans. Inf. Theory*, vol. 44, no. 3, pp. 927–946, May 1998.

[41] L. Wang, Y.-H. Kim, C.-Y. Chen, H. Park, and E. Sasoglu, "Sliding-window superposition coding: Two-user interference channels," 2017, preprint available at http://arxiv.org/abs/1701.02345.

[42] A. B. Carleial, "Multiple-access channels with different generalized feedback signals," *IEEE Trans. Inf. Theory*, vol. 28, no. 6, pp. 841–850, Nov. 1982.

[43] L.-L. Xie and P. R. Kumar, "A network information theory for wireless communication: Scaling laws and optimal operation," *IEEE Trans. Inf. Theory*, vol. 50, no. 5, pp. 748–767, 2004.

[44] G. Kramer, M. Gastpar, and P. Gupta, "Cooperative strategies and capacity theorems for relay networks," *IEEE Trans. Inf. Theory*, vol. 51, no. 9, pp. 3037–3063, Sep. 2005.

[45] K. T. Kim, S.-K. Ahn, Y.-S. Kim, J. Park, C.-Y. Chen, and Y.-H. Kim, "Interference management via sliding-window coded modulation for 5G cellular networks," *IEEE Commun. Mag.*, vol. 52, no. 5, pp. 52–60, May 2014.

[46] L. Duan, B. Rimoldi, and R. Urbanke, "Approaching the AWGN channel capacity without active shaping," in *Proc. IEEE Int. Symp. Inf. Theory*, June 1997, p. 374.

[47] F.-W. Sun and H. C. A. van Tilborg, "Approaching capacity by equiprobable signaling on the Gaussian channel," *IEEE Trans. Inf. Theory*, vol. 39, no. 5, pp. 1714–1716, Sep. 1993.

[48] F. Steiner and G. Böcherer, "Comparison of geometric and probabilistic shaping with application to ATSC 3.0," in *Int. ITG Conf. Systems, Commun. Coding*, Hamburg, Feb. 2017, pp. 1–6.

[49] G. Böcherer, F. Steiner, and P. Schulte, "Bandwidth efficient and rate-matched low-density parity-check coded modulation," *IEEE Trans. Commun.*, vol. 63, no. 12, pp. 4651–4665, Dec. 2015.

[50] S. A. Tretter, *Constellation Shaping, Nonlinear Precoding, and Trellis Coding for Voice-band Telephone Channel Modems with Emphasis on ITU-T Recommendation V. 34*. Norwell, MA: Kluwer, 2002.

[51] P. Schulte and G. Böcherer, "Constant composition distribution matching," *IEEE Trans. Inf. Theory*, vol. 62, no. 1, pp. 430–434, Jan. 2016.

[52] P. Schulte and F. Steiner, "Shell mapping for distribution matching," 2018, preprint available at http://arxiv.org/abs/1803.03614.

[53] F. Buchali, F. Steiner, G. Böcherer, L. Schmalen, P. Schulte, and W. Idler, "Rate adaptation and reach increase by probabilistically shaped 64-QAM: An experimental demonstration," *J. Light. Technol.*, vol. 34, no. 7, pp. 1599–1609, Apr. 2016.

[54] J. Cho, X. Chen, S. Chandrasekhar, et al., "Trans-atlantic field trial using high spectral efficiency probabilistically shaped 64-QAM and single-carrier real-time 250-Gb/s 16-QAM," *J. Light. Technol.*, vol. 36, no. 1, pp. 103–113, Jan. 2018.

[55] C. E. Shannon, "Communication in the presence of noise," *Proc. IRE*, vol. 37, no. 1, pp. 10–21, Jan. 1949.

[56] G. J. Foschini, "Layered space-time architecture for wireless communication in a fading environment when using multi-element antennas," *Bell Labs Tech. J.*, vol. 1, no. 2, pp. 41–59, 1996.

[57] G. J. Foschini, D. Chizhik, M. J. Gans, C. Papadias, and R. A. Valenzuela, "Analysis and performance of some basic space-time architectures," *IEEE J. Select. Areas Commun.*, vol. 21, no. 3, pp. 303–320, Apr. 2003.

[58] İ. E. Telatar, "Capacity of multi-antenna Gaussian channels," *Euro. Trans. Telecomm.*, vol. 10, no. 8, pp. 585–595, 1999.

[59] R. Ahlswede, "Multiway communication channels," in *Proc. 2nd Int. Symp. Information Theory*, Tsahkadsor, Armenian S.S.R., 1971, pp. 23–52.

[60] H. H. J. Liao, "Multiple access channels," PhD Thesis, University of Hawaii, Honolulu, HI, Sep. 1972.

[61] C. E. Shannon, "Two-way communication channels," in *Proc. 4th Berkeley Symp. Math. Statist. Probab.* Berkeley, CA: University of California Press, 1961, vol. I, pp. 611–644.

[62] E. C. van der Meulen, "A survey of multi-way channels in information theory: 1961–1976," *IEEE Trans. Inf. Theory*, vol. 23, no. 1, pp. 1–37, 1977.

[63] E. C. van der Meulen, "Recent coding theorems and converses for multi-way channels – II: The multiple access channel (1976–1985)," Katholieke Universiteit Leuven, Leuven, Belgium, Department Wiskunde, 1985.

[64] N. Abramson, Ed., *Multiple Access Communications: Foundations for Emerging Technologies*. Piscataway, NJ: IEEE Press, 1993.

[65] M. C. Reed, C. B. Schlegel, P. D. Alexander, and J. A. Asenstorfer, "Iterative multiuser detection for CDMA with FEC: Near-single-user performance," *IEEE Trans. Commun.*, vol. 46, no. 12, pp. 1693–1699, Dec. 1998.

[66] X. Wang and H. V. Poor, "Iterative (turbo) soft interference cancellation and decoding for coded CDMA," *IEEE Trans. Commun.*, vol. 47, no. 7, pp. 1046–1061, Jul. 1999.

[67] F. Kschischang, B. Frey, and H.-A. Loeliger, "Factor graphs and the sum-product algorithm," *IEEE Trans. Inf. Theory*, vol. 47, no. 2, pp. 498–519, Feb. 2001.

[68] T. Richardson and R. Urbanke, *Modern Coding Theory*. Cambridge: Cambridge University Press, 2008.

[69] L. Wang and E. Şaşoğlu, "Polar coding for interference networks," 2014, preprint available at http://arxiv.org/abs/1401.7293.

[70] A. Yedla, P. Nguyen, H. Pfister, and K. Narayanan, "Universal codes for the Gaussian MAC via spatial coupling," in *Proc. 49th Ann. Allerton Conf. Comm. Control Comput.*, Monticello, IL, Sep. 2011, pp. 1801–1808.

[71] J. Hou, J. Smee, H. Pfister, and S. Tomasin, "Implementing interference cancellation to increase the EV-DO Rev A reverse link capacity," *IEEE Commun. Mag.*, vol. 44, no. 2, pp. 58–64, Feb. 2006.

[72] L. Sankar, G. Kramer, and N. B. Mandayam, "Offset encoding for multiple-access relay channels," *IEEE Trans. Inf. Theory*, vol. 53, no. 10, pp. 3814–3821, 2007.

[73] M. H. Yassaee and M. R. Aref, "Slepian–Wolf coding over cooperative relay networks," *IEEE Trans. Inf. Theory*, vol. 57, no. 6, pp. 3462–3482, 2011.

[74] R. Urbanke and B. Rimoldi, "Lattice codes can achieve capacity on the AWGN channel," *IEEE Trans. Inf. Theory*, vol. 44, no. 1, pp. 273–278, 1998.

[75] A. J. Grant, B. Rimoldi, R. L. Urbanke, and P. A. Whiting, "Rate-splitting multiple access for discrete memoryless channels," *IEEE Trans. Inf. Theory*, vol. 47, no. 3, pp. 873–890, 2001.

[76] D. Slepian and J. K. Wolf, "A coding theorem for multiple access channels with correlated sources," *Bell Syst. Tech. J.*, vol. 52, no. 7, pp. 1037–1076, Sep. 1973.

[77] L. Vandenberghe, S. Boyd, and S.-P. Wu, "Determinant maximization with linear matrix inequality constraints," *SIAM J. Matrix Anal. Appl.*, vol. 19, no. 2, pp. 499–533, 1998.

[78] W. Yu, W. Rhee, S. Boyd, and J. M. Cioffi, "Iterative water-filling for Gaussian vector multiple-access channels," *IEEE Trans. Inf. Theory*, vol. 50, no. 1, pp. 145–152, 2004.

[79] S. Verdú, *Multiuser Detection*. Cambridge: Cambridge University Press, 1998.

[80] T. M. Cover, "Broadcast channels," *IEEE Trans. Inf. Theory*, vol. 18, no. 1, pp. 2–14, Jan. 1972.

[81] P. P. Bergmans, "Random coding theorem for broadcast channels with degraded components," *IEEE Trans. Inf. Theory*, vol. 19, no. 2, pp. 197–207, 1973.

[82] A. D. Wyner, "A theorem on the entropy of certain binary sequences and applications: II," *IEEE Trans. Inf. Theory*, vol. 19, no. 6, pp. 772–777, 1973.

[83] R. G. Gallager, "Capacity and coding for degraded broadcast channels," *Probl. Inf. Transm.*, vol. 10, no. 3, pp. 3–14, 1974.

[84] E. C. van der Meulen, "Recent coding theorems and converses for multi-way channels – I: The broadcast channel (1976–1980)," in *New Concepts in Multi-User Communication*, J. K. Skwyrzinsky, Ed. Alphen aan den Rijn: Sijthoff & Noordhoff, 1981, pp. 15–51.

[85] T. M. Cover, "Comments on broadcast channels," *IEEE Trans. Inf. Theory*, vol. 44, no. 6, pp. 2524–2530, 1998.

[86] J. Körner and K. Marton, "General broadcast channels with degraded message sets," *IEEE Trans. Inf. Theory*, vol. 23, no. 1, pp. 60–64, 1977.

[87] J. Körner and K. Marton, "Images of a set via two channels and their role in multi-user communication," *IEEE Trans. Inf. Theory*, vol. 23, no. 6, pp. 751–761, 1977.

[88] K. Marton, "The capacity region of deterministic broadcast channels," in *Théorie de L'information: Développements récents et Applications (Cachan, 1977)*, C. F. Picard and P. Camion, Eds. Paris: CNRS Editions, 1979, pp. 243–248.

[89] M. Tomlinson, "New automatic equaliser employing modulo arithmetic," *Electron. Lett.*, vol. 7, no. 5, pp. 138–139, 1971.

[90] H. Harashima and H. Miyakawa, "Matched-transmission technique for channels with intersymbol interference," *IEEE Trans. Commun.*, vol. 20, no. 4, pp. 774–780, 1972.

[91] M. H. M. Costa, "Writing on dirty paper," *IEEE Trans. Inf. Theory*, vol. 29, no. 3, pp. 439–441, 1983.

[92] U. Erez, S. Shamai, and R. Zamir, "Capacity and lattice strategies for canceling known interference," *IEEE Trans. Inf. Theory*, vol. 51, no. 11, pp. 3820–3833, 2005.

[93] U. Erez and S. ten Brink, "A close-to-capacity dirty paper coding scheme," *IEEE Trans. Inf. Theory*, vol. 51, no. 10, pp. 3417–3432, 2005.

[94] Y. Sun, M. Uppal, A. D. Liveris, et al., "Nested turbo codes for the Costa problem," *IEEE Trans. Comm.*, vol. 56, no. 3, 2008.

[95] G. Shilpa, A. Thangaraj, and S. Bhashyam, "Dirty paper coding using sign-bit shaping and LDPC codes," in *Proc. IEEE Int. Symp. Inf. Theory*, Austin, TX, Jun. 2010, pp. 923–927.

[96] E. Björnson, M. Bengtsson, and B. Ottersten, "Optimal multiuser transmit beamforming: A difficult problem with a simple solution structure," *IEEE Signal Proc. Mag.*, vol. 31, no. 4, pp. 142–148, Jul. 2014.

[97] H. Weingarten, Y. Steinberg, and S. Shamai, "The capacity region of the Gaussian multiple-input multiple-output broadcast channel," *IEEE Trans. Inf. Theory*, vol. 52, no. 9, pp. 3936–3964, Sep. 2006.

[98] S. Vishwanath, N. Jindal, and A. J. Goldsmith, "Duality, achievable rates, and sum-rate capacity of Gaussian MIMO broadcast channels," *IEEE Trans. Inf. Theory*, vol. 49, no. 10, pp. 2658–2668, 2003.

[99] R. Ahlswede, "The capacity region of a channel with two senders and two receivers," *Ann. Probab.*, vol. 2, no. 5, pp. 805–814, 1974.

[100] A. B. Carleial, "Interference channels," *IEEE Trans. Inf. Theory*, vol. 24, no. 1, pp. 60–70, 1978.

[101] E. C. van der Meulen, "Some reflections on the interference channel," in *Communications and Cryptography: Two Sides of One Tapestry*, R. E. Blahut, D. J. Costello, U. Maurer, and T. Mittelholzer, Eds. Boston, MA: Kluwer, 1994, pp. 409–421.

[102] G. Kramer, "Review of rate regions for interference channels," in *Int. Zurich Seminar Commun.*, Zurich, Feb. 2006, pp. 162–165.

[103] X. Shang, "Two-user Gaussian interference channels: An information theoretic point of view," *Found. Trends Comm. Inf. Theory*, vol. 10, no. 3, pp. 247–378, 2013.

[104] V. S. Annapureddy and V. V. Veeravalli, "Gaussian interference networks: Sum capacity in the low interference regime and new outer bounds on the capacity region," *IEEE Trans. Inf. Theory*, vol. 55, no. 7, pp. 3032–3050, Jul. 2009.

[105] A. S. Motahari and A. K. Khandani, "Capacity bounds for the Gaussian interference channel," *IEEE Trans. Inf. Theory*, vol. 55, no. 2, pp. 620–643, Feb. 2009.

[106] X. Shang, G. Kramer, and B. Chen, "A new outer bound and the noisy-interference sum-rate capacity for Gaussian interference channels," *IEEE Trans. Inf. Theory*, vol. 55, no. 2, pp. 689–699, Feb. 2009.

[107] A. Dytso, D. Tuninetti, and N. Devroye, "Interference as noise: Friend or foe?" *IEEE Trans. Inf. Theory*, vol. 62, no. 6, pp. 3561–3596, Jun. 2016.

[108] A. B. Carleial, "A case where interference does not reduce capacity," *IEEE Trans. Inf. Theory*, vol. 21, no. 5, pp. 569–570, 1975.

[109] H. Sato, "The capacity of the Gaussian interference channel under strong interference," *IEEE Trans. Inf. Theory*, vol. 27, no. 6, pp. 786–788, Nov. 1981.

[110] J. Lee, D. Toumpakaris, and W. Yu, "Interference mitigation via joint detection," *IEEE J. Select. Areas Commun.*, vol. 29, no. 6, pp. 1172–1184, Jun. 2011.

[111] B. Bandemer, A. El Gamal, and Y.-H. Kim, "Optimal achievable rates for interference networks with random codes," *IEEE Trans. Inf. Theory*, vol. 61, no. 12, pp. 6536–6549, Dec. 2015.

[112] H.-F. Chong, M. Motani, and H. K. Garg, "A comparison of two achievable rate regions for the interference channel," in *Proc. UCSD Inf. Theory Appl. Workshop*, San Diego, CA, Feb. 2006.

[113] C. Nair and A. El Gamal, "The capacity region of a class of three-receiver broadcast channels with degraded message sets," *IEEE Trans. Inf. Theory*, vol. 55, no. 10, pp. 4479–4493, Oct. 2009.

[114] I. Marić, A. Goldsmith, G. Kramer, and S. Shamai, "On the capacity of interference channels with one cooperating transmitter," *Eur. Trans. Telecommun.*, vol. 19, pp. 405–420, Apr. 2008.

[115] H. Sato, "On the capacity region of a discrete two-user channel for strong interference," *IEEE Trans. Inf. Theory*, vol. 24, no. 3, pp. 377–379, May 1978.

[116] T. S. Han and K. Kobayashi, "A new achievable rate region for the interference channel," *IEEE Trans. Inf. Theory*, vol. 27, no. 1, pp. 49–60, 1981.

[117] H.-F. Chong, M. Motani, H. K. Garg, and H. El Gamal, "On the Han–Kobayashi region for the interference channel," *IEEE Trans. Inf. Theory*, vol. 54, no. 7, pp. 3188–3195, Jul. 2008.

[118] R. Etkin, D. N. C. Tse, and H. Wang, "Gaussian interference channel capacity to within one bit," *IEEE Trans. Inf. Theory*, vol. 54, no. 12, pp. 5534–5562, Dec. 2008.

[119] İ. E. Telatar and D. N. C. Tse, "Bounds on the capacity region of a class of interference channels," in *Proc. IEEE Int. Symp. Inf. Theory*, Nice, Jun. 2007, pp. 2871–2874.

[120] W. Nam, D. Bai, J. Lee, and I. Kang, "Advanced interference management for 5G cellular networks," *IEEE Commun. Mag.*, vol. 52, no. 5, pp. 52–60, May 2014.

[121] S. Karmakar and M. K. Varanasi, "The capacity region of the MIMO interference channel and its reciprocity to within a constant gap," *IEEE Trans. Inf. Theory*, vol. 59, no. 8, pp. 4781–4797, 2013.

[122] M. A. Maddah-Ali, A. S. Motahari, and A. K. Khandani, "Communication over MIMO X channels: Interference alignment, decomposition, and performance analysis," *IEEE Trans. Inf. Theory*, vol. 54, no. 8, pp. 3457–3470, 2008.

[123] V. Cadambe and S. A. Jafar, "Interference alignment and degrees of freedom of the K-user interference channel," *IEEE Trans. Inf. Theory*, vol. 54, no. 8, pp. 3425–3441, Aug. 2008.

[124] B. Nazer, M. Gastpar, S. A. Jafar, and S. Vishwanath, "Ergodic interference alignment," in *Proc. IEEE Int. Symp. Inf. Theory*, Seoul, Jun./Jul. 2009, pp. 1769–1773.

[125] A. S. Motahari, S. Oveis-Gharan, M.-A. Maddah-Ali, and A. K. Khandani, "Real interference alignment: Exploiting the potential of single antenna systems," *IEEE Trans. Inf. Theory*, vol. 60, no. 8, pp. 4799–4810, 2014.

[126] S. A. Jafar, "Interference alignment: A new look at signal dimensions in a communication network," *Found. Trends Comm. Inf. Theory*, vol. 7, no. 1, pp. 1–134, 2011.

[127] G. Kramer, I. Marić, R. D. Yates, et al., "Cooperative communications," *Found. Trends Comm. Inf. Theory*, vol. 1, no. 3–4, pp. 271–425, 2007.

[128] O. Simeone, N. Levy, A. Sanderovich, et al. "Cooperative wireless cellular systems: An information-theoretic view," *Found. Trends Comm. Inf. Theory*, vol. 8, no. 1–2, pp. 1–177, 2012.

[129] P. Elias, A. Feinstein, and C. E. Shannon, "A note on the maximum flow through a network," *IRE Trans. Inf. Theory*, vol. 2, no. 4, pp. 117–119, Dec. 1956.

[130] L. R. Ford, Jr. and D. R. Fulkerson, "Maximal flow through a network," *Canad. J. Math.*, vol. 8, no. 3, pp. 399–404, 1956.

[131] R. K. Ahuja, T. L. Magnanti, and J. B. Orlin, "Network flows," in *Optimization*, G. L. Nemhauser, A. H. G. Rinnooy Kan, and M. J. Todd, Eds. Amsterdam: Elsevier, 1989, vol. 1, pp. 211–369.

[132] A. El Gamal, N. Hassanpour, and J. Mammen, "Relay networks with delays," *IEEE Trans. Inf. Theory*, vol. 53, no. 10, pp. 3413–3431, Oct. 2007.

[133] G. Kramer, "Information networks with in-block memory," *IEEE Trans. Inf. Theory*, vol. 60, no. 4, pp. 2105–2120, Apr. 2014.

[134] G. Kramer, "Models and theory for relay channels with receive constraints," in *Proc. 42nd Ann. Allerton Conf. Comm. Control Comput.*, Monticello, IL, Sep./Oct. 2004, pp. 1312–1321.

[135] E. C. van der Meulen, "Three-terminal communication channels," *Adv. Appl. Prob.*, vol. 3, pp. 120–154, 1971.

[136] T. M. Cover and A. El Gamal, "Capacity theorems for the relay channel," *IEEE Trans. Inf. Theory*, vol. 25, no. 5, pp. 572–584, Sep. 1979.

[137] C. A. Desoer, "Communication through channels in cascade," PhD dissertation, Massachusetts Institute of Technology, Cambridge, MA, 1953.

[138] B. E. Schein, "Distributed coordination in network information theory," PhD thesis, Massachusetts Institute of Technology, Cambridge, MA, Sep. 2001.

[139] J. N. Laneman, "Cooperative diversity in wireless networks: Algorithms and architectures," PhD dissertation, Massachusetts Institute of Technology, Cambridge, MA, 2002.

[140] M. Gastpar and M. Vetterli, "On the capacity of large Gaussian relay networks," *IEEE Trans. Inf. Theory*, vol. 51, no. 3, pp. 765–779, 2005.

[141] M. N. Khormuji and M. Skoglund, "On instantaneous relaying," *IEEE Trans. Inf. Theory*, vol. 56, no. 7, pp. 3378–3394, Jul. 2010.

[142] L.-L. Xie and P. R. Kumar, "An achievable rate for the multiple-level relay channel," *IEEE Trans. Inf. Theory*, vol. 51, no. 4, pp. 1348–1358, 2005.

[143] A. El Gamal and M. R. Aref, "The capacity of the semideterministic relay channel," *IEEE Trans. Inf. Theory*, vol. 28, no. 3, p. 536, May 1982.

[144] A. El Gamal and S. Zahedi, "Capacity of a class of relay channels with orthogonal components," *IEEE Trans. Inf. Theory*, vol. 51, no. 5, pp. 1815–1817, 2005.

[145] S. H. Lim, K. T. Kim, and Y.-H. Kim, "Distributed decode–forward for relay networks," *IEEE Trans. Inf. Theory*, vol. 63, no. 7, pp. 4103–4118, 2017.

[146] H.-F. Chong, M. Motani, and H. K. Garg, "Generalized backward decoding strategies for the relay channel," *IEEE Trans. Inf. Theory*, vol. 53, no. 1, pp. 394–401, 2007.

[147] S. H. Lim, Y.-H. Kim, A. El Gamal, and S.-Y. Chung, "Noisy network coding," *IEEE Trans. Inf. Theory*, vol. 57, no. 5, pp. 3132–3152, May 2011.

[148] J. Hou and G. Kramer, "Short message noisy network coding with a decode–forward option," *IEEE Trans. Inf. Theory*, vol. 62, no. 1, pp. 89–107, Jan. 2016.

[149] A. El Gamal, M. Mohseni, and S. Zahedi, "Bounds on capacity and minimum energy-per-bit for AWGN relay channels," *IEEE Trans. Inf. Theory*, vol. 52, no. 4, pp. 1545–1561, 2006.

[150] A. Sanderovich, S. Shamai, Y. Steinberg, and G. Kramer, "Communication via decentralized processing," *IEEE Trans. Inf. Theory*, vol. 54, no. 7, pp. 3008–3023, Jul. 2008.

[151] A. Sanderovich, O. Somekh, H. V. Poor, and S. Shamai, "Uplink macro diversity of limited backhaul cellular network," *IEEE Trans. Inf. Theory*, vol. 55, no. 8, pp. 3457–3478, Aug. 2009.

[152] S. Ganguly and Y.-H. Kim, "On the capacity of cloud radio access networks," in *Proc. IEEE Int. Symp. Inf. Theory*, Aachen, 2017, pp. 2063–2067.

[153] W. Kang, N. Liu, and W. Chong, "The Gaussian multiple access diamond channel," *IEEE Trans. Inf. Theory*, vol. 61, no. 11, pp. 6049–6059, Nov. 2015.

[154] S. Saeedi Bidokhti and G. Kramer, "Capacity bounds for diamond networks with an orthogonal broadcast channel," *IEEE Trans. Inf. Theory*, vol. 62, no. 12, pp. 7103–7122, Dec. 2016.

[155] S. Saeedi Bidokhti, G. Kramer, and S. Shamai, "Capacity bounds on the downlink of symmetric, multi-relay, single receiver C-RAN networks," *Entropy*, vol. 19, no. 11, pp. 1–14, Nov. 2017.

Part I

Architecture

3 Device-to-Device Communication

Ratheesh K. Mungara, Geordie George, and Angel Lozano

3.1 Introduction

Device-to-device (D2D) communication allows users in close proximity to establish direct communication [1–3]. For wireless traffic exhibiting sufficient locality, such D2D connectivity may enable direct communication in emergency situations [4], proximity-aware internetworking [5, 6], vehicle-to-vehicle communication [7], and media content dissemination [8, 9], among other applications. However, the incorporation of D2D onto cellular networks also introduces new challenges, such as additional user-to-user interference and signaling overheads. Understanding the fundamental limits of D2D is critical to derive practical D2D strategies for managing interference and balancing overheads.

A seminal paper by Gupta and Kumar [10] on the capacity scaling laws of wireless ad-hoc networks inspired many subsequent works [11–13]. While such analyses yield clean expressions that capture the scaling of capacity with respect to the number of links or the network size, these expressions fall short of informing about the achievable spectral efficiencies. Addressing this limitation, Hong and Caire [14] derived explicit expressions for the achievable user spectral efficiencies in a dense D2D network with users placed on a regular grid.

Recognizing the irregular network topology and behavior of D2D users, more recent efforts [15–18] have applied stochastic geometry tools in order to study D2D communication by modeling the user locations via PPP (Poisson point process) distributions. Some of the existing analyses [15–17] postulate a block-fading model for the small-scale variations of the channel and characterize the outage probability (i.e., the probability that a desirable communication rate is not feasible). However, this metric has limited applicability in modern systems featuring link adaptation as well as broad bandwidths, whereby outages due to small-scale fading are either avoided or essentially averaged out. A more appropriate figure of merit for such systems is the ergodic spectral efficiency, extensively analyzed in the context of cellular networks [19, 20].

This chapter extends the ergodic spectral efficiency analysis to the realm of D2D communication, further relying on a novel modeling approach for the interference that is based on the following arguments [18].

- The observed signal at the receiver is subject to both large-scale (distance-dependent pathloss and shadowing) and small-scale channel effects. Over the span of each signal codeword, the large-scale features are regarded as constant while the small-scale fading may or may not remain constant, depending on the fading coherence and on how codewords are arranged in time and/or frequency.

- Under the foregoing separation of scales between small- and large-scale channel features, as mentioned, the ergodic spectral efficiency involving expectations over the small-scale fading is arguably the most representative information theoretic metric in modern wireless systems where codewords can be interspersed over wide bandwidths, across hybrid-ARQ (automatic-repeat-request) repetitions, and possibly on multiple antennas, while being subject to scheduling and link adaptation [21].
- At each receiver, the fading of its intended signal can be tracked, but not those of the interference components, and the decoder treats interference as additional Gaussian noise.

Both *underlay* or *overlay* options are possible, where respectively the D2D communication reuses the existing uplink or is segregated on a swath of dedicated spectrum. A key distinguishing feature of the framework in [18] is a variable degree of spatial averaging, which allows characterizing the performance both for specific user locations and on average over all possible locations.

We begin the chapter, in Section 3.2, with a description of integrated cellular and D2D networks including signal and propagation models. In Section 3.3, we formulate the spectral efficiency and present the interference modeling approach. We subsequently characterize the system-wide distributions of the local-average SIR (signal-to-interference ratio) and of the ergodic spectral efficiency, in Section 3.4, and of the spatially averaged spectral efficiency in Section 3.5. Next, Section 3.6 presents various examples to evaluate the potential of D2D in both overlay and underlay settings. Then, Section 3.7 dwells extensively on various interference management schemes, demonstrating their effectiveness in improving spatial spectrum reuse. In Section 3.8, we briefly review the idea of cache-aided D2D communication. Section 3.9 concludes the chapter.

3.2   Integrated Cellular and D2D Networks

3.2.1   Network Geometry

We consider interference-limited cellular networks where the base stations (BSs) are regularly placed on a hexagonal grid. (It would also be possible to model the BS locations stochastically, but in any event the emphasis in this chapter is on the location of the D2D users.) At each BS, cellular transmissions are orthogonalized while multiple D2D links share each time–frequency signaling resource. Transmitters and receivers have a single antenna and each receiver knows the fading of only its own link, be it cellular or D2D. Our focus is on a given time–frequency resource, where one cellular uplink and/or (underlay/overlay) multiple D2D links are active within each cell.

To facilitate the readability of the equations, we utilize distinct fonts for the cellular and D2D variables.

The locations of the transmitters, both cellular and D2D, are modeled relative to the location of a given receiver under consideration. For the cellular uplink, the receiver under consideration is a BS, whereas for the D2D link it is a user. In either case, and without loss of generality, we place such receiver at the origin and index the intended transmitter with zero. All other transmitters, acting as interferers, are indexed in order of increasing distance within each class (cellular and D2D). For a cellular uplink, the intended transmitter is always the closest cellular transmitter while, for a D2D link, the intended transmitter need not be the closest D2D transmitter.

Cellular Uplink

To study this link, we place a receiving BS at the origin and locate an intended cellular transmitter uniformly within the cell associated with that BS (see Figure 3.1), which is circular with radius R and denoted by $\mathcal{B}(0, R)$. There is one and only one cellular transmitter within $\mathcal{B}(0, R)$, and its distance to the BS at the origin is denoted by r_0.

The cellular interferers from other cells are outside $\mathcal{B}(0, R)$, modeled via a PPP Φ with density $\lambda = \frac{1}{\pi R^2}$. With this density made to coincide with the number of BSs per unit area, this has been shown to be a fine model for a network of BSs on a hexagonal grid with one cellular transmitter per cell [15, 22]. Then,

$$R = \sqrt{\frac{\sqrt{3}}{2\pi}} D, \tag{3.1}$$

with D the inter-BS distance.

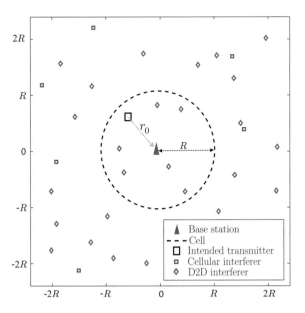

Figure 3.1 Cellular uplink with D2D. Located at the origin is a receiving BS and shown with a square marker in the surrounding circle is its intended cellular transmitter; shown with square markers outside the circle are the cellular interferers; shown with diamond markers are the D2D interferers.

The D2D interferer locations form another independent PPP Φ with density $\lambda = K\lambda$ such that there are, on average, K active D2D links per cell.

D2D Link

To study this link, we place a D2D receiver at the origin and locate its intended D2D transmitter at a distance r_0. In the absence of empirical data on whether and how r_0 depends on the user density, various canonical characterizations have been entertained for r_0: a Rayleigh distribution [15, 23, 24], a distribution corresponding to uniform placements on a circle [25], or an inverse function of the link density [18, 26]. In this chapter, we adopt the latter one,

$$r_0 = \frac{r_0}{K^\beta}, \tag{3.2}$$

with $r > 0$ and $\beta \geq 0$. Through the parameters r and β, this model can be tuned, which renders it more flexible than a fixed distribution. For strictly positive β, the link length shrinks as the user density intensifies – a behavior that is intuitively reasonable – whereas, for $\beta = 0$, we obtain $r_0 = r$ independently of the user density.

The cellular and D2D interferers conform to Φ and Φ, respectively. Note that there may be cellular interferers arbitrarily close to a D2D receiver, a point whose implications are discussed later.

3.2.2   Received Signal

We denote by P and P the (fixed) transmit powers of cellular and D2D users, respectively, both measured at 1 m from the transmitter and with their ratio being

$$\mu = \frac{P}{P}. \tag{3.3}$$

Unit-gain antennas are featured at the users. The BSs, in turn, feature unsectorized antennas whose precise gains are immaterial to the cellular uplink because, in interference-limited conditions, they affect signal and interference equally.

In order to present the results in the most general fashion, we define a binary parameter $\alpha \in \{0, 1\}$ that distinguishes between underlay and overlay as follows:

$$\begin{cases} \alpha & = 1 \qquad \text{underlay} \\ \alpha & = 0 \qquad \text{overlay.} \end{cases} \tag{3.4}$$

Cellular Uplink

The BS at the origin observes at symbol n

$$y[n] = \sqrt{P r_0^{-\eta}} \, H_0[n] \, s_0[n] + z[n], \tag{3.5}$$

where the first term is the signal from the intended cellular user while the second term represents the interference

$$z[n] = \sum_{k=1}^{\infty} \sqrt{P r_k^{-\eta}} \, H_k[n] \, s_k[n] + \alpha \sum_{j=1}^{\infty} \sqrt{P r_j^{-\eta}} \, H_j[n] \, s_j[n], \tag{3.6}$$

whose first summation spans the other-cell cellular users in $\Phi \setminus \mathcal{B}(0, R)$ and whose second summation spans all the D2D transmitters in Φ. In turn, $\eta > 2$ is the pathloss exponent for cellular links, r_k represents the distance between the kth cellular transmitter and the BS at the origin, H_k denotes the corresponding fading, and s_k is the symbol transmitted by the kth cellular transmitter. Similarly, r_j represents the distance between the jth D2D transmitter and the BS at the origin, H_j denotes the corresponding fading, and s_j is the symbol transmitted by the jth D2D transmitter. The fading coefficients H_k and H_j are independent and of unit power, but otherwise arbitrarily distributed. For their part, $s_k \sim \mathcal{N}_{\mathbb{C}}(0, 1)$ and $\mathsf{s}_j \sim \mathcal{N}_{\mathbb{C}}(0, 1)$, a choice that is justified later.

Conditioned on the link distances $\{r_k\}_{k=0}^{\infty}$ and $\{\mathsf{r}_j\}_{j=1}^{\infty}$, which are fixed as far as the small-scale modeling is concerned, the instantaneous SIR enjoyed by the BS at the origin is, at symbol n,

$$\mathrm{SIR}[n] = \frac{r_0^{-\eta} \, |H_0[n]|^2}{\sum_{k=1}^{\infty} r_k^{-\eta} \, |H_k[n]|^2 + \alpha \, \mu \sum_{j=1}^{\infty} \mathsf{r}_j^{-\eta} \, |\mathsf{H}_j[n]|^2}. \tag{3.7}$$

D2D Link

To analyze this link, we shift the origin to the D2D receiver of interest, which observes at symbol n

$$\mathsf{y}[n] = \sqrt{\mathsf{P} \, \mathsf{r}_0^{-\eta}} \, \mathsf{H}_0[n] \, \mathsf{s}_0[n] + \mathsf{z}[n], \tag{3.8}$$

where the first term is the signal from the intended D2D transmitter while the second term is the interference

$$\mathsf{z}[n] = \alpha \sum_{k=1}^{\infty} \sqrt{P \, r_k^{-\eta}} \, H_k[n] \, s_k[n] + \sum_{j=1}^{\infty} \sqrt{\mathsf{P} \, \mathsf{r}_j^{-\eta}} \, \mathsf{H}_j[n] \, \mathsf{s}_j[n] \tag{3.9}$$

received from the D2D transmitters in Φ and the cellular transmitters in Φ. For these user-to-user signals, we consider a different pathloss exponent $\eta > 2$.

As for the cellular uplink, conditioned on the link distances $\{\mathsf{r}_j\}_{j=0}^{\infty}$ and $\{r_k\}_{k=1}^{\infty}$, the instantaneous SIR enjoyed by the D2D receiver at the origin is, at symbol n,

$$\mathsf{SIR}[n] = \frac{\mathsf{r}_0^{-\eta} \, |\mathsf{H}_0[n]|^2}{\frac{\alpha}{\mu} \sum_{k=1}^{\infty} r_k^{-\eta} \, |H_k[n]|^2 + \sum_{j=1}^{\infty} \mathsf{r}_j^{-\eta} \, |\mathsf{H}_j[n]|^2}. \tag{3.10}$$

3.3   Interference Modeling

As indicated, a key differentiating feature of our analytical framework is the novel interference modeling approach that facilitates the spectral efficiency analysis. This approach is motivated and described in this section.

3.3.1   Spectral efficiency

Cellular Uplink

Assuming that the receiver is privy to $H_0[1], \ldots, H_0[N]$, the mutual information (in bits/symbol) over codewords spanning N symbols is

$$\frac{1}{N} I\left(s_0[1], \ldots, s_0[N]; y[1], \ldots, y[N] \,\big|\, H_0[1], \ldots, H_0[N], \{r_k\}_{k=0}^{\infty}, \{r_j\}_{j=1}^{\infty}\right), \qquad (3.11)$$

which, with i.i.d. codeword symbols, becomes

$$\frac{1}{N} \sum_{n=1}^{N} I\left(s_0[n]; y[n] \,\big|\, H_0[n], \{r_k\}_{k=0}^{\infty}, \{r_j\}_{j=1}^{\infty}\right). \qquad (3.12)$$

With spans of a few thousand symbols, codewords in contemporary broadband systems are long enough and arranged in such a way – interspersed in time, frequency, and increasingly across antennas – to experience sufficiently many fading swings for an effective averaging of the mutual information over the small-scale fading. As such fading is essentially stationary and ergodic over the codeword, the averaging in (3.12) becomes an expectation and bestows the significance of the ergodic spectral efficiency (in bits/s/Hz)

$$C_{\text{exact}} = \mathbb{E}_{H_0}\left[I\left(s_0; y \,\big|\, H_0, \{r_k\}_{k=0}^{\infty}, \{r_j\}_{j=1}^{\infty}\right)\right] \qquad (3.13)$$

$$= \mathbb{E}_{H_0}\left[I\left(s_0; \sqrt{P\, r_0^{-\eta}}\, H_0\, s_0 + z \,\Big|\, H_0, \{r_k\}_{k=0}^{\infty}, \{r_j\}_{j=1}^{\infty}\right)\right]. \qquad (3.14)$$

D2D Link

Similarly, the ergodic spectral efficiency of the D2D link is

$$\mathsf{C}_{\text{exact}} = \mathbb{E}_{\mathsf{H}_0}\left[I\left(\mathsf{s}_0; \mathsf{y} \,\big|\, \mathsf{H}_0, \{r_j\}_{j=0}^{\infty}, \{r_k\}_{k=1}^{\infty}\right)\right] \qquad (3.15)$$

$$= \mathbb{E}_{\mathsf{H}_0}\left[I\left(\mathsf{s}_0; \sqrt{P\, r_0^{-\eta}}\, \mathsf{H}_0\, \mathsf{s}_0 + \mathsf{z} \,\Big|\, \mathsf{H}_0, \{r_j\}_{j=0}^{\infty}, \{r_k\}_{k=1}^{\infty}\right)\right]. \qquad (3.16)$$

The quantities C_{exact} and $\mathsf{C}_{\text{exact}}$, which serve as baselines in the sequel, do not admit explicit expressions. Rather, their evaluation demands computationally very intensive simulations involving lengthy Monte-Carlo histograms [20, app. A].

3.3.2   Local Distribution of z and z

Conditioned on the link distances $\{r_k, r_j\}$, let us examine the local distribution (around the respective transceivers) of the aggregate interference terms z and z as defined, respectively, in (3.6) and (3.9). Without further conditioning on $\{H_k\}_{k=1}^{\infty}$ and $\{\mathsf{H}_j\}_{j=1}^{\infty}$, i.e., without the receiver knowing the fading coefficients from *all* interfering transmitters, these distributions are generally not Gaussian. In fact, given only the link distances, the distributions of z and z are highly involved. For instance, in Rayleigh fading, they comprise products of Gaussians. The non-Gaussianity of the interference is irrelevant to the SIR as only the variance of z or z matters in that respect. However, it is relevant to information theoretic derivations and especially that of the spectral efficiency, which does depend on the distribution of the interference.

3.3.3 Conventional Modeling Approach

Cellular Uplink

Customarily, C_{exact} in (3.14) is analyzed in the form it would have if $\{H_k\}_{k=1}^{\infty}$ and $\{H_j\}_{j=1}^{\infty}$ were actually known by the receiver at the origin and z were consequently Gaussian, namely the form

$$C_{\text{ub}} = \mathbb{E}_{\{H_k\}_{k=0}^{\infty},\{H_j\}_{j=1}^{\infty}} \left[\log_2 \left(1 + \frac{r_0^{-\eta} |H_0[n]|^2}{\sum_{k=1}^{\infty} r_k^{-\eta} |H_k[n]|^2 + \alpha \mu \sum_{j=1}^{\infty} r_j^{-\eta} |H_j[n]|^2} \right) \right],$$
(3.17)

where the redefinition of z as Gaussian, rarely made explicit, is unmistakable from $I(s_0; \sqrt{\gamma} s_0 + z) = \log_2(1 + \gamma)$, which is applicable only when s_0 and z are Gaussian. Besides being analytically convenient, a Gaussian modeling of z turns out to be sensible because, if a decoder is designed for Gaussian interference-plus-noise (either by choice or because the distribution thereof is unknown), it can achieve the spectral efficiency obtained by rendering the interference-plus-noise Gaussian [27]. Furthermore, with z taken to be Gaussian, the capacity-achieving signal distribution is also Gaussian, validating our choice for s_0. Meanwhile, the gifting of $\{H_k\}_{k=1}^{\infty}$ and $\{H_j\}_{j=1}^{\infty}$ as extra side information to the receiver renders (3.17) an upper bound to C_{exact}, hence the denomination C_{ub}.

D2D Link

Similarly, $\mathsf{C}_{\text{exact}}$ would typically be analyzed via the corresponding upper bound

$$\mathsf{C}_{\text{ub}} = \mathbb{E}_{\{H_j\}_{j=0}^{\infty},\{H_k\}_{k=1}^{\infty}} \left[\log_2 \left(1 + \frac{\mathsf{r}_0^{-\eta} |\mathsf{H}_0[n]|^2}{(\alpha/\mu) \sum_{k=1}^{\infty} r_k^{-\eta} |H_k[n]|^2 + \sum_{j=1}^{\infty} \mathsf{r}_j^{-\eta} |\mathsf{H}_j[n]|^2} \right) \right].$$
(3.18)

Although much more tractable than C_{exact} and $\mathsf{C}_{\text{exact}}$, the forms of C_{ub} and C_{ub} have the strain of depending not only on H_0 and H_0, respectively, but also on $\{H_k\}_{k=1}^{\infty}$ and $\{H_j\}_{j=1}^{\infty}$. As a consequence, their analysis remains considerably cluttered.

3.3.4 A Novel Interference Modeling Approach

A Gaussian Model for the Interference Distribution

We propose to go one step further in the Gaussian modeling of the interference, forgoing the small-scale variations in the power of z and z so as to model them respectively as

$$z \sim \mathcal{N}_{\mathbb{C}}\left(0, \sigma^2\right)$$
(3.19)

with

$$\sigma^2 = \sum_{k=1}^{\infty} P\, r_k^{-\eta} + \alpha \sum_{j=1}^{\infty} \mathsf{P}\, \mathsf{r}_j^{-\eta},$$
(3.20)

and

$$z \sim \mathcal{N}_{\mathbb{C}}\left(0, \sigma^2\right) \tag{3.21}$$

with

$$\sigma^2 = \alpha \sum_{k=1}^{\infty} P\, r_k^{-\eta} + \sum_{j=1}^{\infty} \mathsf{P}\, r_j^{-\eta}. \tag{3.22}$$

The closeness between these Gaussian distributions for the interference and their originals (see (3.6) and (3.9)) has been tightly bounded [28, 29].

The rendering of z and z as Gaussian without having to grant the receivers with $\{H_k\}_{k=1}^{\infty}$ and $\{\mathsf{H}_j\}_{j=1}^{\infty}$ yields gratifying analytical benefits.

Partition of the Interference Terms

Note that (3.20) and (3.22) contain an infinite number of terms, of which a handful largely dominate the total interference power because of the distance-dependent pathloss. Recognizing this, the interfering transmitters can be partitioned into two sets depending on whether they lie inside or outside a circle – we term it *averaging circle* – surrounding the receiver of interest. The terms within the circle are explicitly modeled, while the aggregate interference emanating from outside this circle is replaced by its expected (over the interference locations) value. The introduction of the averaging circle facilitates establishing the performance for specific locations of the users within the circle, which are the dominant interferers.

The radius of the averaging circle becomes a modeling parameter that should be chosen to balance simplicity and accuracy. A natural and very safe choice is to have the size of the circle coincide with that of a cell, $\mathcal{B}(0, R)$. Unless otherwise stated, such is the size of the averaging circle in this text, whereby the interference power in (3.20) can be rewritten as

$$\sigma^2 = \underbrace{\alpha \sum_{j=1}^{K'} \mathsf{P}\, r_j^{-\eta}}_{\sigma_{\mathrm{in}}^2} + \underbrace{\alpha \sum_{j=K'+1}^{\infty} \mathsf{P}\, r_j^{-\eta} + \sum_{k=1}^{\infty} P\, r_k^{-\eta}}_{\sigma_{\mathrm{out}}^2}, \tag{3.23}$$

where σ_{in}^2 corresponds to the K' D2D transmitters in $\Phi \cap \mathcal{B}(0, R)$ for the given network realization, whereas σ_{out}^2 corresponds to the transmitters in $\Phi \backslash \mathcal{B}(0, R)$ and $\Phi \backslash \mathcal{B}(0, R)$. Since $\mathbb{E}[K'] = K$, the expectation of σ_{out}^2 over the PPPs equals

$$\overline{\sigma_{\mathrm{out}}^2} = \alpha\, \mathbb{E}_{\Phi}\left[\sum_{j=K'+1}^{\infty} \mathsf{P}\, r_j^{-\eta}\right] + \mathbb{E}_{\Phi}\left[\sum_{k=1}^{\infty} P\, r_k^{-\eta}\right] \tag{3.24}$$

$$= \alpha \int_{R}^{\infty} 2\pi K\lambda\, \mathsf{P}\, r^{1-\eta} dr + \int_{R}^{\infty} 2\pi \lambda\, P\, r^{1-\eta} dr \tag{3.25}$$

$$= \frac{2\,(\alpha K \mathsf{P} + P)}{(\eta - 2) R^{\eta}}, \tag{3.26}$$

where (3.25) follows from the application of Campbell's theorem [30, thm. 4.1] to expect over the locations of all those interferers whose distances are greater than R

while (3.26) is obtained by evaluating the integrals and substituting $\lambda = \frac{1}{\pi R^2}$. Then, we may approximate the interference power in the cellular uplink as

$$\sigma^2 \approx \sigma_{\text{in}}^2 + \overline{\sigma_{\text{out}}^2}. \tag{3.27}$$

Similarly, for the D2D link, considering $\mathcal{B}(0, R)$ around the D2D receiver at the origin, the interference power in (3.22) can be rewritten as

$$\sigma^2 = \underbrace{\sum_{j=1}^{K'} \mathsf{P}\,\mathsf{r}_j^{-\eta} + \alpha \sum_{k=1}^{K''} \mathsf{P}\,\mathsf{r}_k^{-\eta}}_{\sigma_{\text{in}}^2} + \underbrace{\sum_{j=K'+1}^{\infty} \mathsf{P}\,\mathsf{r}_j^{-\eta} + \alpha \sum_{k=K''+1}^{\infty} \mathsf{P}\,\mathsf{r}_k^{-\eta}}_{\sigma_{\text{out}}^2}, \tag{3.28}$$

where σ_{in}^2 corresponds to the K' D2D transmitters in $\Phi \cap \mathcal{B}(0, R)$ and the K'' uplink cellular transmitters in $\Phi \cap \mathcal{B}(0, R)$, whereas σ_{out}^2 corresponds to the transmitters in $\Phi \setminus \mathcal{B}(0, R)$ and $\Phi \setminus \mathcal{B}(0, R)$. Noting that $\mathbb{E}[K'] = K$ and $\mathbb{E}[K''] = 1$, the expectation of σ_{out}^2 over the PPPs, computed by applying steps similar to (3.24)–(3.26), is

$$\overline{\sigma_{\text{out}}^2} = \frac{2(K\mathsf{P} + \alpha\,\mathsf{P})}{(\eta - 2)R^\eta}. \tag{3.29}$$

Then, we may approximate the interference power in the D2D link as

$$\sigma^2 \approx \sigma_{\text{in}}^2 + \overline{\sigma_{\text{out}}^2}. \tag{3.30}$$

3.3.5 Spectral Efficiency for Specific Network Geometries

Cellular Uplink

With the redefinition of the interference distribution in (3.19), the instantaneous SIR of the cellular uplink becomes

$$\text{SIR} = \frac{P\,r_0^{-\eta}\,|H_0|^2}{\sigma^2} \tag{3.31}$$

$$= \rho\,|H_0|^2, \tag{3.32}$$

where ρ is the local-average SIR,

$$\rho = \frac{r_0^{-\eta}}{\sum_{k=1}^{\infty} r_k^{-\eta} + \alpha\,\mu \sum_{j=1}^{\infty} r_j^{-\eta}} \tag{3.33}$$

$$\approx \frac{r_0^{-\eta}}{\alpha\,\mu \sum_{j=1}^{K'} r_j^{-\eta} + \frac{2(\alpha\mu K + 1)}{(\eta - 2)R^\eta}} \tag{3.34}$$

$$= \frac{a_0^{-\eta}}{\alpha\,\mu \sum_{j=1}^{K'} a_j^{-\eta} + \frac{2(\alpha\mu K + 1)}{(\eta - 2)}}, \tag{3.35}$$

with the approximation in (3.34) following from the application of (3.27) while (3.35) is obtained by normalizing all the terms by $R^{-\eta}$; with that, $a_0 = \frac{r_0}{R}$, $a_k = \frac{r_k}{R}$, and

$a_j = \frac{r_j}{R}$ are normalized distances. For a user uniformly located in the cell, we can write the cumulative distribution function (CDF) of a_0 as

$$F_{a_0}(a) = \begin{cases} a^2 & 0 \le a \le 1 \\ 1 & a > 1 \end{cases} \tag{3.36}$$

and its probability density function (PDF) as

$$f_{a_0}(a) = 2a \quad 0 \le a \le 1. \tag{3.37}$$

For specific locations of the transmitters, i.e., conditioned on all the link distances, ρ is determined and the conditional distribution of the instantaneous SIR is then given directly by that of $|H_0|^2$ while the ergodic spectral efficiency of the cellular uplink with local-average SIR ρ is characterized as

$$C = \mathbb{E}_{H_0}\left[\log_2\left(1 + \rho |H_0|^2\right)\right] \tag{3.38}$$

$$= \int_0^\infty \log_2(1 + \rho \zeta)\, dF_{|H_0|^2}(\zeta). \tag{3.39}$$

D2D Link
Similarly, for the D2D link, with the interference distributed as in (3.21), the instantaneous SIR becomes

$$\mathrm{SIR} = \frac{P\, r_0^{-\eta}\, |H_0|^2}{\sigma^2} \tag{3.40}$$

$$= \varrho\, |H_0|^2, \tag{3.41}$$

where ϱ is the local-average SIR,

$$\varrho = \frac{r_0^{-\eta}}{\frac{\alpha}{\mu}\sum_{k=1}^{\infty} r_k^{-\eta} + \sum_{j=1}^{\infty} r_j^{-\eta}} \tag{3.42}$$

$$\approx \frac{r_0^{-\eta}}{\sum_{j=1}^{K'} r_j^{-\eta} + \frac{\alpha}{\mu}\sum_{k=1}^{K''} r_k^{-\eta} + \frac{2(K+\alpha/\mu)}{(\eta-2)R^\eta}} \tag{3.43}$$

$$= \frac{\left(\frac{a}{K^\beta}\right)^{-\eta}}{\sum_{j=1}^{K'} a_j^{-\eta} + \frac{\alpha}{\mu}\sum_{k=1}^{K''} a_k^{-\eta} + \frac{2(K+\alpha/\mu)}{\eta-2}}, \tag{3.44}$$

with (3.43) following from the application of (3.30) while (3.44) is obtained by normalizing all the terms by $R^{-\eta}$; with that, $\frac{a}{K^\beta} = a_0 = \frac{r_0}{R}$, $a_k = \frac{r_k}{R}$, and $a_j = \frac{r_j}{R}$ are normalized distances.

The corresponding spectral efficiency is

$$C = \mathbb{E}_{H_0}\left[\log_2\left(1 + \varrho |H_0|^2\right)\right] \tag{3.45}$$

$$= \int_0^\infty \log_2(1 + \varrho \zeta)\, dF_{|H_0|^2}(\zeta). \tag{3.46}$$

EXAMPLE 3.1   In Rayleigh fading, $F_{|H_0|^2}(\zeta) = F_{|H_0|^2}(\zeta) = 1 - e^{-\zeta}$, from which

$$C(\rho) = e^{1/\rho}\, \mathcal{E}_1\left(\frac{1}{\rho}\right) \log_2 e \tag{3.47}$$

and

$$\mathsf{C}(\varrho) = e^{1/\varrho}\, \mathcal{E}_1\left(\frac{1}{\varrho}\right) \log_2 e, \tag{3.48}$$

where $\mathcal{E}_n(x) = \int_1^\infty t^{-n} e^{-xt}\, dt$ is an exponential integral.

For fading distributions other than Rayleigh (see Example 3.2), or with MIMO or other features, corresponding forms can be obtained for the ergodic spectral efficiency, always with the key property of these being a function of the local-average SIR and not of the instantaneous fading coefficients.

EXAMPLE 3.2   In Nakagami- m fading ($m \geq 1/2$),

$$F_{|H_0|^2}(\zeta) = F_{|H_0|^2}(\zeta) = 1 - \frac{\Gamma(m, m\,\zeta)}{\Gamma(m)}, \tag{3.49}$$

with $\Gamma(u) = \int_0^\infty t^{u-1} e^{-u}\, dt$ and $\Gamma(u, v) = \int_v^\infty t^{u-1} e^{-u}\, dt$ being, respectively, the gamma function and the upper incomplete gamma function. Then, the ergodic spectral efficiencies become

$$C(\rho) = \left(\frac{m}{\rho}\right)^m \mathcal{I}_m\left(\frac{m}{\rho}\right) \frac{\log_2 e}{\Gamma(m)} \tag{3.50}$$

and

$$\mathsf{C}(\varrho) = \left(\frac{m}{\varrho}\right)^m \mathcal{I}_m\left(\frac{m}{\varrho}\right) \frac{\log_2 e}{\Gamma(m)}, \tag{3.51}$$

where $\mathcal{I}_n(u) = \int_0^\infty x^{n-1} \log_e(1 + x) e^{-ux}\, dx$.

Note that setting $m = 1$ in (3.50) and (3.51) returns (3.47) and (3.48), respectively.

EXAMPLE 3.3   Consider a D2D link in an overlay system ($\alpha = 0$) with $K = 3$ links per cell on average, with $\beta = 0$ (i.e., with link lengths that are independent of the user density) and with a pathloss exponent $\eta = 4.5$. Draw around the receiver an averaging circle of size R such that only the locations of the inner interferers are conditioned upon. Three different situations are considered, with the interferers respectively placed at the interior, the middle, and the edge of the averaging circle (see Figure 3.2). Compared in Figure 3.3 are the spectral efficiency in (3.48) and the exact mutual information under the non-Gaussian interference z as per (3.9), with such mutual information numerically computed through lengthy Monte-Carlo histograms and averaged over many fading realizations for 10 different snapshots of the interferer locations outside the averaging circle (recall C_{exact} in (3.15)). It can be observed, under each of the three situations, that the simulation results for the different (out-of-circle) interference snapshots are tightly clustered around the analytical counterpart involving their spatial average. In fact, for the interior and middle situations, the results for multiple snapshots fully overlap.

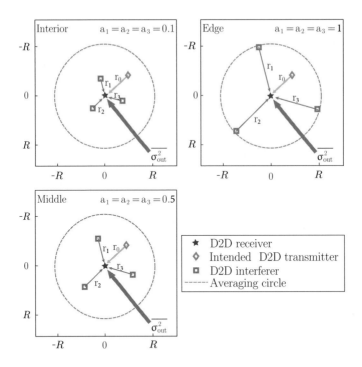

Figure 3.2 The three situations considered in Example 3.3.

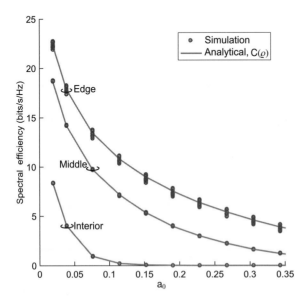

Figure 3.3 D2D link spectral efficiency in an overlay system as a function of the normalized link distance a_0 with $\beta = 0$ and $K = 3$: analysis with our interference model vs. simulation with the interference modeled as per (3.9). The pathloss exponent $\eta = 4.5$.

Similarly, good agreements are observed for other overlay settings and also for underlay settings.

3.4   System-Wide Distributions

3.4.1   Distribution of the Local-Average SIR

The spatial distribution of the transmitter locations induces distributions of their own for ρ and ϱ, i.e., large-scale distributions for the local-average SIRs, which are derived next. As the exact distribution of the local-average aggregate interference generally does not admit closed forms, certain approximate characterizations have been proposed [20, 31–33]. Here, we adapt the approach in [33]: Obtain the distributions of $1/\rho$ and $1/\varrho$ exactly in the Laplace domain, and then express the Laplace numerical inverse via an Euler series expansion.

Cellular Uplink
The CDF of ρ is given by

$$F_\rho(\theta) = \mathbb{P}[\rho \leq \theta] \tag{3.52}$$

$$= \mathbb{P}\left[\frac{1}{\rho} \geq 1/\theta\right] \tag{3.53}$$

$$= 1 - F_{1/\rho}(1/\theta). \tag{3.54}$$

At this point, we invoke the numerical inversion of the Laplace transform of $F_{1/\rho}(\cdot)$, which yields an accurate approximation in a series form for $F_\rho(\cdot)$ [33, 34]:

$$F_\rho(\theta) \approx 1 - \theta\, \frac{e^{A/2}}{2^B} \sum_{b=0}^{B} \binom{B}{b} \sum_{q=0}^{Q+b} \frac{(-1)^q}{D_q} \Re\{\mathcal{L}_{F_{1/\rho}}(\theta\, t)\}, \tag{3.55}$$

where $\mathcal{L}_{F_{1/\rho}}(\cdot)$ is the Laplace transform of $F_{1/\rho}(\cdot)$ and $t = \frac{A+j2\pi q}{2}$, with j the imaginary unit; in turn, $D_0 = 2$ and $D_q = 1$ for $q \geq 1$. The parameters A, B, and Q control the accuracy, with suggested values for multiple-digit precision being $A = 18.4$, $B = 11$, and $Q = 15$ [34]. For our purposes, $A = 9.21$, $B = 5$, and $Q = 8$ yield a more than sufficient accuracy [35].

Next, from the relationship [33, remark 5]

$$\mathcal{L}_{F_{1/\rho}}(\upsilon) = \frac{1}{\upsilon} \mathcal{L}_{1/\rho}(\upsilon), \tag{3.56}$$

where $\mathcal{L}_{1/\rho}(\cdot)$ is the Laplace transform of $1/\rho$, we can rewrite (3.55) as

$$F_\rho(\theta) \approx 1 - \theta\, \frac{e^{A/2}}{2^B} \sum_{b=0}^{B} \binom{B}{b} \sum_{q=0}^{Q+b} \frac{(-1)^q}{D_q} \Re\left\{\frac{\mathcal{L}_{1/\rho}(\theta\, t)}{\theta\, t}\right\}. \tag{3.57}$$

The Laplace transform of $1/\rho$ can be computed over the spatial locations of all interferers in the network, without invoking any a-priori averaging circle (or, equivalently, taking its size to infinity, which can only sharpen the model).

The derivation steps of $\mathcal{L}_{1/\rho}(\cdot)$ are

$$\mathcal{L}_{1/\rho}(\upsilon) = \mathbb{E}\left[e^{-\upsilon/\rho}\right] \tag{3.58}$$

$$= \mathbb{E}\left[\exp\left(-\upsilon\frac{\sum_{k=1}^{\infty} r_k^{-\eta} + \alpha\,\mu \sum_{j=1}^{\infty} r_j^{-\eta}}{r_0^{-\eta}}\right)\right] \tag{3.59}$$

$$= \mathbb{E}_{r_0}\left[\mathbb{E}_\Phi\left\{\prod_{k=1}^{\infty} e^{-\upsilon\,r_0^\eta\,r_k^{-\eta}}\right\}\cdot\mathbb{E}_\Phi\left\{\prod_{j=1}^{\infty} e^{-\alpha\,\mu\,\upsilon\,r_0^\eta\,r_j^{-\eta}}\right\}\right] \tag{3.60}$$

$$= \mathbb{E}_{r_0}\left[\exp\left\{-2\pi\lambda\int_R^\infty \left(1 - e^{-\upsilon\,r_0^\eta\,x^{-\eta}}\right)x\,dx\right\}\right.$$
$$\left.\cdot\exp\left\{-2\pi\lambda\int_0^\infty \left(1 - e^{-\alpha\,\mu\,\upsilon\,r_0^\eta\,x^{-\eta}}\right)x\,dx\right\}\right] \tag{3.61}$$

$$= \mathbb{E}_{r_0}\left[\exp\left\{2\pi\lambda\left[\frac{R^2}{2} + \frac{1}{\eta}r_0^2\,\upsilon^{2/\eta}\,\bar{\Gamma}\left(-\frac{2}{\eta}, R^{-\eta}\,r_0^\eta\,\upsilon\right)\right]\right\}\right.$$
$$\left.\cdot\exp\left\{2\pi\lambda\alpha\,\frac{1}{\eta}r_0^2\,(\mu\,\upsilon)^{2/\eta}\,\Gamma\left(-\frac{2}{\eta}\right)\right\}\right] \tag{3.62}$$

$$= \mathbb{E}_{a_0}\left[\exp\left\{1 + \frac{2}{\eta}a_0^2\,\upsilon^{2/\eta}\,\bar{\Gamma}\left(-\frac{2}{\eta}, a_0^\eta\,\upsilon\right)\right\}\right.$$
$$\left.\cdot\exp\left\{\frac{2}{\eta}K\alpha\,a_0^2\,(\mu\,\upsilon)^{2/\eta}\,\Gamma\left(-\frac{2}{\eta}\right)\right\}\right] \tag{3.63}$$

$$= 2\int_0^1 \exp\left\{1 + a^2\,\upsilon^{2/\eta}\,\frac{2}{\eta}\left[\bar{\Gamma}\left(-\frac{2}{\eta}, a^\eta\,\upsilon\right) + \alpha\,K\,\mu^{2/\eta}\,\Gamma\left(-\frac{2}{\eta}\right)\right]\right\}a\,da, \tag{3.64}$$

where (3.58) follows from the definition of $\mathcal{L}_{1/\rho}$; (3.59) follows from invoking the expression for ρ in (3.33); (3.60) follows from the fact that the locations of cellular uplink transmitters and D2D transmitters are two independent PPPs; (3.61) follows by separately invoking the definition of the probability generating functionals of the PPP [30] for the cellular and D2D transmitter processes Φ and Φ; (3.62) follows from the variable changes $x' = \upsilon\,r_0^\eta\,x^{-\eta}$ and $x' = \alpha\,\mu\,\upsilon\,r_0^\eta\,x^{-\eta}$; (3.63) follows from $\lambda = \frac{1}{\pi R^2}$, $\lambda = \frac{K}{\pi R^2}$, and $a_0 = \frac{r_0}{R}$; (3.64) follows by expecting over the PDF of a_0 in (3.37); and $\bar{\Gamma}(\cdot)$ is the lower incomplete gamma function. Plugging (3.64) into (3.57) yields a close approximation of $F_\rho(\cdot)$, which can be numerically evaluated using any standard software package.

EXAMPLE 3.4   In Figure 3.4, the CDF in (3.57) is contrasted against its simulation brethren, obtained numerically by means of the corresponding local-average SIR CDF. Both underlay and overlay are considered with $K = 10$, $\mu = -10$ dB, and $\eta = 4$. An excellent match is observed, supporting the validity of the Euler series expansion of the inverse Laplace transform.

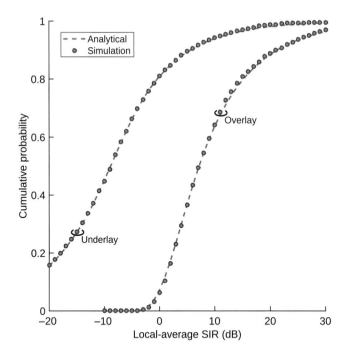

Figure 3.4 CDF of ρ for $K = 10$, $\mu = -10$ dB, $\eta = 4$.

While the expression in (3.57) is general and can be solved for either instantaneously or within seconds using software packages such as Mathematica or MATLAB, it is not simple enough to provide immediate insight. Alternatively, we invoke the averaging circle approximation and obtain relatively compact expressions separately for underlay and overlay. Though the analytical form of the distribution of ρ is unwieldy for underlay ($\alpha = 1$) and arbitrary η, we can derive a closed form for $\eta = 4$. Specifically, for this derivation it is more convenient to retain an averaging circle (equal to a cell size) only for the cellular interferers, while not applying it to D2D interferers. Then, the local-average SIR with underlay ($\alpha = 1$) becomes

$$\rho = \frac{r_0^{-\eta}}{\mu \sum_{j=1}^{\infty} r_j^{-\eta} + \frac{2}{(\eta-2)R^\eta}}. \tag{3.65}$$

Conditioned on the summation in the denominator of (3.65), denoted hereafter by $\mathcal{Y}_1 = \sum_{j=1}^{\infty} r_j^{-\eta}$, the CDF of ρ can be expressed as

$$F_{\rho|\mathcal{Y}_1}(\theta) = \mathbb{P}\left[\frac{r_0^{-\eta}}{\mu \mathcal{Y}_1 + \frac{2}{(\eta-2)R^\eta}} < \theta \,\middle|\, \mathcal{Y}_1 \right] \tag{3.66}$$

$$= \mathbb{P}\left[a_0 > \theta^{-\frac{1}{\eta}} \left(\mu \mathcal{Y}_1 R^\eta + \frac{2}{\eta - 2} \right)^{-\frac{1}{\eta}} \,\middle|\, \mathcal{Y}_1 \right] \tag{3.67}$$

$$= 1 - F_{a_0} \left[\theta^{-\frac{1}{\eta}} \left(\mu \, \mathcal{Y}_1 R^\eta + \frac{2}{\eta - 2} \right)^{-\frac{1}{\eta}} \right] \tag{3.68}$$

$$= 1 - \theta^{-\frac{2}{\eta}} \left(\mu \, \mathcal{Y}_1 R^\eta + \frac{2}{\eta - 2} \right)^{-\frac{2}{\eta}} \qquad \theta \geq \frac{\eta - 2}{\mu \, \mathcal{Y}_1 R^\eta (\eta - 2) + 2}. \tag{3.69}$$

Expecting $F_{\rho | \mathcal{Y}_1}(\theta)$ over the distribution of $\mathcal{Y}_1$, the unconditional CDF of ρ can be obtained. For $\eta = 4$ [36, sec. V],

$$f_{\mathcal{Y}_1}(v) = \frac{\pi}{2} \lambda \, v^{-\frac{3}{2}} e^{\frac{-\pi^3 \lambda^2}{4v}} \qquad v > 0. \tag{3.70}$$

Averaging $F_{\rho | \mathcal{Y}_1}(\cdot)$ over the distribution of $\mathcal{Y}_1$, we obtain

$$F_\rho(\theta) = \begin{cases} \displaystyle\int_{\frac{1-\theta}{R^4 \mu \theta}}^{\infty} \left(f_{\mathcal{Y}_1}(v) - \frac{f_{\mathcal{Y}_1}(v)}{\sqrt{\theta (\mu \, v R^4 + 1)}} \right) dv & 0 < \theta < 1 \\[4mm] 1 - \displaystyle\int_0^\infty \frac{f_{\mathcal{Y}_1}(v)}{\sqrt{\theta (\mu \, v R^4 + 1)}} \, dv & \theta \geq 1, \end{cases} \tag{3.71}$$

where the limits of the integrals follow from $v \geq \frac{1-\theta}{R^4 \mu \theta}$ and $v > 0$.

Inserting (3.70) into (3.71), substituting $\lambda = \frac{K}{\pi R^2}$ and solving the two integrals, we obtain

$$F_\rho(\theta) = \begin{cases} \dfrac{e^{\kappa^2} \left[\mathrm{erf}(\kappa) - \mathrm{erf}\left(\frac{\kappa}{\sqrt{1-\theta}} \right) \right]}{\sqrt{\theta}} + \mathrm{erf}\left(\frac{\kappa \sqrt{\theta}}{\sqrt{1-\theta}} \right) & 0 < \theta < 1 \\[5mm] 1 - \dfrac{e^{\kappa^2} \mathrm{erfc}(\kappa)}{\sqrt{\theta}} & \theta \geq 1, \end{cases} \tag{3.72}$$

where $\kappa = \frac{\sqrt{\pi \mu} K}{2}$ while $\mathrm{erf}(v) = \frac{2}{\sqrt{\pi}} \int_0^v e^{-x^2} dx$ (the error function) and $\mathrm{erfc}(v) = 1 - \mathrm{erf}(v)$.

With overlay ($\alpha = 0$), the cellular link is not subject to D2D interference and the local-average SIR, $\rho = \frac{\eta - 2}{2 a_0^\eta}$, depends only on a_0. Thus, the CDF of ρ can be expressed as

$$F_\rho(\theta) = 1 - F_{a_0} \left[\left(\frac{\eta - 2}{2 \theta} \right)^{1/\eta} \right]. \tag{3.73}$$

Applying (3.36) to (3.73) yields a CDF that is simpler and general in η, namely

$$F_\rho(\theta) = 1 - \left(\frac{\eta - 2}{2 \theta} \right)^{2/\eta} \qquad \theta \geq \frac{\eta - 2}{2}. \tag{3.74}$$

D2D Link

By leveraging the derivation of its cellular uplink counterpart (see (3.52)–(3.57)), the CDF of the D2D link local-average SIR can be expressed as

$$F_\varrho(\theta) \approx 1 - \theta \frac{e^{A/2}}{2^B} \sum_{b=0}^{B} \binom{B}{b} \sum_{q=0}^{Q+b} \frac{(-1)^q}{D_q} \Re \left\{ \frac{\mathcal{L}_{1/\varrho}(\theta \, t)}{\theta \, t} \right\}, \tag{3.75}$$

where the Laplace transform $\mathcal{L}_{1/\varrho}(\cdot)$ is derived as (see (3.58)–(3.64))

$$\mathcal{L}_{1/\varrho}(\upsilon) = \exp\left\{-\frac{a^2}{K^{2\beta}}\upsilon^{2/\eta}\,\Gamma\left(1-\frac{2}{\eta}\right)\left[K+\frac{\alpha}{\mu^{2/\eta}}\right]\right\}. \tag{3.76}$$

In this case, we may also obtain a direct expression in an infinite series form as follows. Again, the distribution of ϱ can be computed over the spatial locations of all interferers in the network, which is tantamount to taking the size of the averaging circle to infinity. Thereby replacing the average interference power $\overline{\sigma^2_{\text{out}}}$ with σ^2_{out} in the definition of ϱ in (3.44),

$$\varrho = \frac{r_0^{-\eta}}{\sum_{j=1}^{\infty} r_j^{-\eta} + \frac{\alpha}{\mu}\sum_{k=1}^{\infty} r_k^{-\eta}}. \tag{3.77}$$

The denominator of (3.77), denoted hereafter by $\mathcal{Y}_2 = \sum_{j=1}^{\infty} r_j^{-\eta} + \frac{\alpha}{\mu}\sum_{k=1}^{\infty} r_k^{-\eta}$, has the characteristic function

$$\phi_{\mathcal{Y}_2}(\omega) = \mathbb{E}\left[e^{j\omega\mathcal{Y}_2}\right] \tag{3.78}$$

$$= \exp\left\{-\pi\left(\lambda+\frac{\alpha}{\mu^{2/\eta}}\lambda\right)\Gamma\left(1-\frac{2}{\eta}\right)e^{-\frac{j\pi}{\eta}}\omega^{\frac{2}{\eta}}\right\} \qquad \omega \geq 0. \tag{3.79}$$

The expression for $\phi_{\mathcal{Y}_2}(\omega)$ in (3.79) is obtained as the product of the characteristic functions of the first and second summations in $\mathcal{Y}_2$, which are computed as illustrated in [36, sec. V]. Then, the density of $\mathcal{Y}_2$ can be obtained by taking the inverse Fourier transform of $\phi_{\mathcal{Y}_2}(\omega)$ and the corresponding CDF is

$$F_{\mathcal{Y}_2}(v) = 1 - \frac{1}{\pi}\sum_{k=1}^{\infty}\left[\pi\left(\lambda+\frac{\alpha}{\mu^{2/\eta}}\lambda\right)\Gamma\left(1-\frac{2}{\eta}\right)v^{\frac{-2}{\eta}}\right]^k \frac{\Gamma\left(\frac{2k}{\eta}\right)}{k!}\sin\left[k\pi\left(1-\frac{2}{\eta}\right)\right] \tag{3.80}$$

which, for $\eta = 4$, equals

$$F_{\mathcal{Y}_2}(v) = 1 - \operatorname{erf}\left[\frac{\pi^{3/2}\left(\lambda+\frac{\alpha}{\sqrt{\mu}}\lambda\right)}{2\sqrt{v}}\right]. \tag{3.81}$$

The CDF of ϱ is given by

$$F_{\varrho}(\theta) = 1 - F_{\mathcal{Y}_2}\left(\frac{r_0^{-\eta}}{\theta}\right). \tag{3.82}$$

Using (3.80) and (3.81) in (3.82), and further substituting $r_0 = \frac{r}{K^{\beta}}$ and $\lambda = \frac{K}{\pi R^2}$, we obtain the CDF of ϱ, for arbitrary η, as

$$F_{\varrho}(\theta) = \frac{1}{\pi}\sum_{k=1}^{\infty}\left[\frac{\theta^{2/\eta}a^2}{K^{2\beta}}\left(K+\frac{\alpha}{\mu^{2/\eta}}\right)\Gamma\left(1-\frac{2}{\eta}\right)\right]^k \frac{\Gamma\left(\frac{2k}{\eta}\right)}{k!}\sin\left[k\pi\left(1-\frac{2}{\eta}\right)\right] \tag{3.83}$$

and, for $\eta = 4$, as

$$F_\varrho(\theta) = \mathrm{erf}\left[\frac{\sqrt{\pi}\,\theta\,\mathsf{a}^2}{2\,K^{2\beta}}\left(K + \frac{\alpha}{\sqrt{\mu}}\right)\right].\tag{3.84}$$

3.4.2   Distribution of the Spectral Efficiency

The randomness that ρ and ϱ acquire once the large-scale features are randomized is then inherited by $C(\rho)$ or $\mathsf{C}(\varrho)$, and the respective CDFs provide complete descriptions of the ergodic spectral efficiencies achievable over all possible network locations. Provided that an invertible form for C or C is available, the CDF of C is

$$F_C(\gamma) = F_\rho\left[C^{-1}(\gamma)\right]\tag{3.85}$$

whereas the CDF of C is

$$F_\mathsf{C}(\gamma) = F_\varrho\left[\mathsf{C}^{-1}(\gamma)\right].\tag{3.86}$$

If no invertible forms are available, by mapping the applicable functions $C(\rho)$ and $\mathsf{C}(\varrho)$ onto the expressions for $F_\rho(\cdot)$ and $F_\varrho(\cdot)$ put forth in the previous section, $F_C(\cdot)$ and $F_\mathsf{C}(\cdot)$ are readily characterized.

EXAMPLE 3.5   Under Rayleigh fading, the ergodic spectral efficiency of a cellular uplink is given by (3.47) and its CDF equals

$$F_C(\gamma) = \mathbb{P}\left[e^{1/\rho}\,\mathcal{E}_1\left(\frac{1}{\rho}\right)\log_2 e < \gamma\right].\tag{3.87}$$

Invoking [37],

$$e^v \mathcal{E}_1(v)\log_2 e \approx 1.4\log\left(1 + \frac{0.82}{v}\right),\tag{3.88}$$

we can approximate (3.87) as

$$F_C(\gamma) \approx F_\rho\left(\frac{e^{\frac{\gamma}{1.4}} - 1}{0.82}\right),\tag{3.89}$$

which is validated Example 3.6. Similarly, from the D2D link local-average SIR ϱ, we can compute its CDF as

$$F_\mathsf{C}(\gamma) \approx F_\varrho\left(\frac{e^{\frac{\gamma}{1.4}} - 1}{0.82}\right).\tag{3.90}$$

EXAMPLE 3.6   In Figure 3.5, the approximated CDFs in (3.89) and (3.90) are contrasted against the ones obtained numerically by means of the corresponding local-average SIR CDFs (see (3.72), (3.74), and (3.83)) and the link spectral efficiency equations, and further against the ones obtained completely through Monte-Carlo. Setting $\mathsf{a}_0 = 0.12/K^\beta$, $K = 10$, $\mu = -10$ dB, $\eta = 4$, and $\eta = 4$, underlay with $\beta = 0.2$ and overlay with $\beta = 0.4$ are considered. Very good agreements are observed, again validating our interference modeling approach.

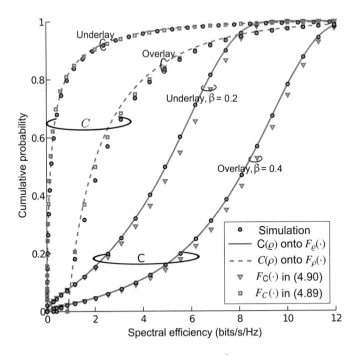

Figure 3.5 CDFs of C and C for $a_0 = 0.12/K^\beta$, $K = 10$, $\mu = -10$ dB, $\eta = 4$, and $\eta = 4$.

3.5   Spatially Averaged Spectral Efficiency

As an alternative to the characterization for specific network geometries, we can characterize the average spectral efficiency over all possible network geometries. Although the quantities thus obtained are less informative, they do allow gauging system-level benefits.

3.5.1   Cellular Uplink

With underlay, the cellular uplink spectral efficiency averaged over all network geometries is

$$\bar{C} = \mathbb{E}_\rho \left[C(\rho) \right] \tag{3.91}$$

$$= \mathbb{E}_\rho \left[\mathbb{E}_{H_0} \left[\log_2 \left(1 + \rho |H_0|^2 \right) \right] \right] \tag{3.92}$$

$$= \mathbb{E}_\rho \left[\int_0^\infty \mathbb{P} \left[\log_2 \left(1 + \rho |H_0|^2 \right) > \gamma \, \Big| \, \rho \right] d\gamma \right] \tag{3.93}$$

$$= \mathbb{E}_\rho \left[\int_0^\infty \frac{\log_2 e}{\zeta + 1} \left(1 - F_{|H_0|^2}(\zeta/\rho) \right) d\zeta \right] \tag{3.94}$$

$$= \int_0^\infty \frac{\log_2 e}{\zeta + 1} \left(1 - \mathbb{E}_\rho \left[F_{|H_0|^2}(\zeta/\rho) \right] \right) d\zeta, \tag{3.95}$$

where (3.94) follows from the variable change $\gamma = \log_2(1 + \zeta)$.

EXAMPLE 3.7   In Rayleigh fading, $\bar{C}$ in (3.95) can be rewritten as

$$\bar{C} = \int_0^\infty \frac{\log_2 e}{\zeta + 1} \mathbb{E}_\rho \left[e^{-\frac{\zeta}{\rho}} \right] d\zeta. \tag{3.96}$$

Conditioned on r_0 and with ρ as given in (3.65),

$$\mathbb{E}\left[e^{-\zeta/\rho} \,\middle|\, r_0 \right] = \exp\left(-\zeta \frac{r_0^\eta}{R^\eta} \frac{2}{\eta - 2} \right) \mathbb{E}_\Phi \left[\prod_{j=1}^\infty \exp\left(-\zeta\,\alpha\mu\,r_0^\eta\,r_j^{-\eta} \right) \,\middle|\, r_0 \right] \tag{3.97}$$

$$= 1 - \exp\left(-\zeta \frac{r_0^\eta}{R^\eta} \frac{2}{\eta - 2} \right) \exp\left\{ -2\pi\lambda \int_0^\infty \left(1 - e^{-\zeta\,\alpha\mu\,r_0^\eta\,x^{-\eta}} \right) x\,dx \right\} \tag{3.98}$$

$$= 1 - \exp\left\{ -\zeta \frac{r_0^\eta}{R^\eta} \frac{2}{\eta - 2} - \zeta^{\frac{2}{\eta}} \frac{r_0^2}{R^2} K \frac{2}{\eta} \int_0^\infty \left(1 - e^{-\alpha\mu u} \right) \frac{1}{u^{1+2/\eta}}\,du \right\}, \tag{3.99}$$

where (3.98) follows from the definition of the probability generating functional [30] and (3.99) follows from the variable change $\zeta\,r_0^\eta\,x^{-\eta} = u$ and the relation $\pi\lambda = K/R^2$. Applying integration by parts in (3.99) and invoking $a_0 = \frac{r_0}{R}$, we obtain

$$\mathbb{E}\left[e^{-\zeta/\rho} \,\middle|\, a_0 \right] = 1 - \exp\left\{ -\zeta\,a_0^\eta \frac{2}{\eta - 2} - \alpha\,(\zeta\mu)^{\frac{2}{\eta}}\,a_0^2\,K\,\Gamma\left(1 - \frac{2}{\eta} \right) \right\}. \tag{3.100}$$

Expecting (3.100) over the PDF of a_0 in (3.37) and substituting the resulting expression in (3.96) yields

$$\bar{C} = 2 \int_0^\infty \frac{\log_2 e}{\zeta + 1} \int_0^1 a\,\exp\left\{ -\zeta \frac{2\,a^\eta}{\eta - 2} - (\zeta\mu)^{\frac{2}{\eta}} a^2\,K\,\Gamma\left(1 - \frac{2}{\eta} \right) \right\} da\,d\zeta \tag{3.101}$$

which, for $\eta = 4$, simplifies to

$$\bar{C} = \frac{\sqrt{\pi}\,e^{-\pi\mu K^2/4}}{2\log_e 2} \int_0^\infty \frac{\mathrm{erf}\left(\sqrt{\zeta} + \frac{\sqrt{\pi\mu K}}{2} \right) - \mathrm{erf}\left(\frac{\sqrt{\pi\mu K}}{2} \right)}{\sqrt{\zeta}\,(1 + \zeta)}\,d\zeta. \tag{3.102}$$

With overlay ($\alpha = 0$), a compact result involving only the Meijer-G function

$$G_{p,q}^{m,n}\left(z \,\middle|\, \begin{array}{c} a_1, \ldots, a_n, a_{n+1}, \ldots, a_p \\ b_1, \ldots, b_m, b_{m+1}, \ldots, b_q \end{array} \right) \tag{3.103}$$

and the cellular pathloss exponent η is obtained, namely

$$\bar{C} = \frac{2\log_2 e}{\eta}\,G_{2,3}^{2,2}\left(\frac{2}{\eta - 2} \,\middle|\, \begin{array}{c} 0, \frac{\eta-2}{\eta} \\ 0, 0, \frac{-2}{\eta} \end{array} \right). \tag{3.104}$$

The versatility of our interference modeling approach is on display here, facilitating this expression for a quantity that had previously been obtained only in integral form.

3.5.2   D2D link

By leveraging the derivation of its cellular uplink counterpart (see (3.91)–(3.95)), the D2D link spectral efficiency averaged over all network geometries can be seen to equal

$$\bar{\mathsf{C}} = \int_0^\infty \frac{\log_2 e}{\zeta + 1} \left(1 - \mathbb{E}_\varrho\left[F_{|\mathsf{H}_0|^2}(\zeta/\varrho)\right]\right) d\zeta. \tag{3.105}$$

EXAMPLE 3.8   In Rayleigh fading, recalling the Laplace transform

$$\mathcal{L}_{1/\varrho}(\zeta) = \mathbb{E}_\varrho\left[e^{-\frac{\zeta}{\varrho}}\right]$$

in (3.76), (3.105) can be expressed as

$$\bar{\mathsf{C}} = \int_0^\infty \frac{\log_2 e}{\zeta + 1} \mathbb{E}_\varrho\left[e^{-\frac{\zeta}{\varrho}}\right] d\zeta \tag{3.106}$$

$$= \int_0^\infty \frac{\log_2 e}{\zeta + 1} \exp\left\{-\zeta^{2/\eta} \frac{\mathsf{a}^2}{K^{2\beta}}\left(K + \frac{\alpha}{\mu^{2/\eta}}\right)\Gamma\left(1 - \frac{2}{\eta}\right)\right\} d\zeta, \tag{3.107}$$

which, for $\eta = 4$, reduces to

$$\bar{\mathsf{C}} = 2\left[\sin\left(\mathcal{K}\,\mathsf{a}^2\right)\mathrm{si}\left(\mathcal{K}\,\mathsf{a}^2\right) - \cos\left(\mathcal{K}\,\mathsf{a}^2\right)\mathrm{ci}\left(\mathcal{K}\,\mathsf{a}^2\right)\right]\log_2 e, \tag{3.108}$$

where $\mathcal{K} = \frac{\sqrt{\pi}}{K^{2\beta}}(K + \frac{\alpha}{\sqrt{\mu}})$ while $\mathrm{si}(x) = \int_x^\infty \frac{\sin t}{t}\,dt$ and $\mathrm{ci}(x) = -\int_x^\infty \frac{\cos t}{t}\,dt$.

EXAMPLE 3.9   For an overlay system ($\alpha = 0$) with $\mathsf{a} = 0.1$ and $K = 10$, Figure 3.6 shows the average spectral efficiencies in (3.104) and (3.107) alongside the respective

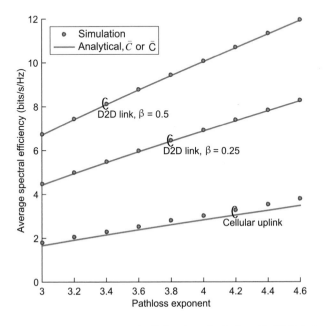

Figure 3.6 Average spectral efficiencies of uplink and D2D link for varying η and η with overlay.

numerically computed values with z as in (3.6) and z as in (3.9). The match is once again excellent.

3.6   Benefits of D2D

In order to assess the benefits of D2D and to demonstrate the usefulness of the framework presented in earlier sections, we next provide some additional examples. Unless otherwise specified, the pathloss exponents are set to $\eta = 3.5$ and $\eta = 4.5$.

3.6.1   Overlay D2D

EXAMPLE 3.10   Consider an overlay system with fixed D2D link distance $\mathsf{a}_0 = \mathsf{a}$, i.e., with $\beta = 0$, and with an average of $K = 10$ D2D links per cell. To render the system as typical as possible, K' is set to its expected value and, for $j = 1, \dots, K'$, a_j is set to the expected value of the normalized distance to the jth nearest neighboring point in a PPP with density λ [38]. In such a setup, equating the link spectral efficiencies $C(\rho)$ and $\mathsf{C}(\varrho)$, respectively from (3.47) and (3.48), we obtain

$$\frac{(3.5-2)\,a_0^{-3.5}}{2} = \frac{\mathsf{a}_0^{-4.5}}{\sum_{j=1}^{10}\left(\frac{\Gamma(0.5+j)}{\sqrt{10}\,\Gamma(j)}\right)^{-4.5} + \frac{2\times10}{4.5-2}}, \qquad (3.109)$$

which leads to

$$a_0 = 5.12\,\mathsf{a}_0^{\frac{4.5}{3.5}}, \qquad (3.110)$$

for which a contour plot is shown in Figure 3.7. Within the unshaded region, the D2D link has a better spectral efficiency than a corresponding uplink transmission from the same user would have, and thus D2D is advantageous. The share of geometries for which $\mathsf{C}(\varrho) > C(\rho)$ for a given a_0 is $\mathbb{P}[\mathsf{a}_0 > a] = 1 - a^2$ where a is the corresponding x-axis value of the contour. Some such shares are displayed, e.g., to have D2D be preferable in over 90% of situations, the D2D links must satisfy $\mathsf{a}_0 < 0.11$.

Example 3.10 shows how, from a link vantage, D2D is very often preferable to communicating via the BS even if only the uplink is considered, let alone if both uplink and downlink are considered.

EXAMPLE 3.11   Considering overlay and $\mathsf{a}_0 = 0.1/K^\beta$, the CDFs of $C(\rho)$ and $\mathsf{C}(\varrho)$ are plotted in Figure 3.8 for various β and K. Even for high values of K, with plenty of D2D interference, thanks to their short range many D2D links enjoy higher spectral efficiencies than the respective cellular uplinks. From the CDFs of $\mathsf{C}(\varrho)$ we observe the following.

- For $\beta < 1/2$, $\mathsf{C}(\varrho)$ worsens with increasing K because a_0 shrinks more slowly than the interferer distances.
- For the boundary value of $\beta = 1/2$, $\mathsf{C}(\varrho)$ is independent of K.

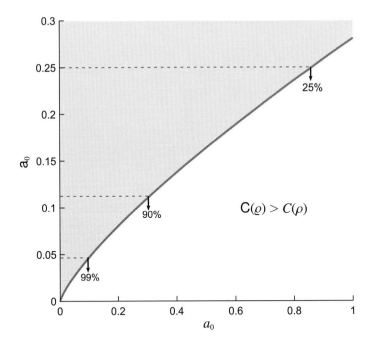

Figure 3.7 Contour plot for the relationship in (3.110). Within the unshaded region, $C(\varrho) > C(\rho)$.

- For $\beta > 1/2$, $C(\varrho)$ improves with increasing K because a_0 shrinks faster than the interferer distances.

The number of D2D links that can coexist on a given signaling resource is large, and to better appreciate the benefits of such dense spectral reuse we next turn our attention to the system spectral efficiency (bits/s/Hz per cell).

EXAMPLE 3.12   Since there are K active D2D links per cell on average, the average system spectral efficiency of the D2D traffic is $K\bar{C}$, whereas for the cellular uplink the average system spectral efficiency is $\bar{C}$ as there is only one active cellular user per cell. Shown in Figure 3.9 is the comparison of these quantities as a function of K, for various β. As K grows beyond the range of values shown in the plot, the D2D link curves for $\beta = 0$ and $\beta = 0.1$ eventually fall below the cellular uplink curve. The following is observed for a variety of such settings.

- For each $\beta < 1/2$, there is an optimum "load" K.
- When $\beta \geq 1/2$, the D2D system spectral efficiency increases monotonically with K.
- Even when not monotonic in K, the D2D system spectral efficiency is generally much higher than its cellular counterpart.

3.6.2   Underlay D2D

EXAMPLE 3.13   In Figure 3.10, the average system spectral efficiency achieved by underlaid D2D with $a_0 = 0.12$ is plotted until its peak value by varying K, for different

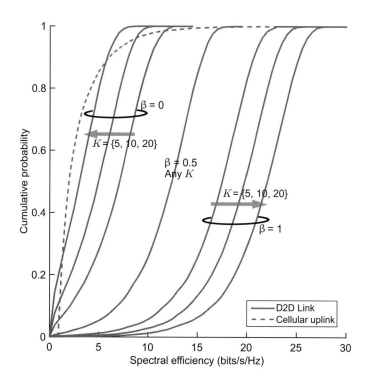

Figure 3.8 CDFs of cellular and D2D link spectral efficiencies in an overlay system with $a_0 = 0.1/K^\beta$ and different β.

values of μ. Suppose that we want the average uplink system spectral efficiency to satisfy $\bar{C} \geq \nu \bar{C}|_{K=0}$ where $\bar{C}|_{K=0}$ denotes its value without D2D and $\nu > 0$ parametrizes its degradation; for instance, $\nu = 0.8$ means a degradation of less than 20%. The maximum average system spectral efficiencies of D2D for different values of ν are as indicated in the figure.

The strong dependence on ν in Example 3.13 indicates that, with underlay, the cellular uplink spectral efficiency is affected substantially by D2D interference; this encourages research looking into ways of protecting the cellular uplink from the D2D interference, and that is precisely the focus of the next section.

3.7   Interference Management

In the context of underlay D2D, [15, 16] proposed various resource allocation strategies to efficiently partition the time and frequency dimensions between cellular and D2D users, while [17, 39] tackled the excessive interference from D2D users to cellular users by means of power control. Another way to reduce the interference seen in the uplink with underlay D2D is to have exclusion regions around the BSs wherein no co-channel D2D transmitters are allowed. This idea has been explored in [18, 40, 41],

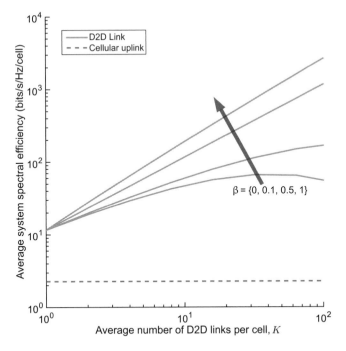

Figure 3.9 Average system spectral efficiency in an overlay system with $a_0 = 0.1/K^\beta$.

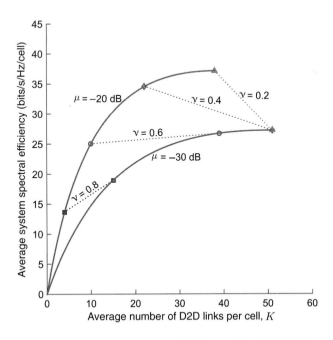

Figure 3.10 Average system spectral efficiency of underlaid D2D with $a_0 = 0.12$ for different values of K, μ, and v.

where it is shown that exclusion regions alleviate the surge in D2D interference seen at the BS. This, however, results in a sparser reuse of spectrum among D2D links, thereby compromising the system spectral efficiency. The emergence of Massive MIMO BSs is expected to also help relieve the problem of D2D interference onto co-channel cellular links [42]. In any case, substantial changes in the standards are required to accommodate underlay D2D and, as a result, there is growing interest in infrastructure-assisted overlay alternatives [1, 43, 44].

In overlay mode, there is no cellular interference, but a certain share of the links do experience strong interference from other unintended D2D transmissions. And, as the density increases, this interference could progressively clog the network. One way to address this problem is to parse D2D transmissions into noninterfering sets to be allocated to separate channels, a process classically termed dynamic channel allocation [45] and that for the D2D context we shorten to *channelization*.

By leveraging stochastic geometry tools, channelization schemes have been studied in the context of wireless ad-hoc networks, where fixed infrastructure is absent [24, 46–50], and for cognitive networks [51, 52]. With the growing interest in overlaid D2D, new schemes are being discussed that can befit networks with an infrastructure-supported control plane and the ensuing ability to synchronize transmissions, discover neighbors, and disseminate side information.

A first such scheme, termed FlashLinQ, was formulated in [53, 54] and experimentally demonstrated. A subsequent scheme, referred to as ITLinQ, was proposed in [26] and evaluated by means of Monte-Carlo simulations. This latter scheme is enticing because it is underpinned by information theoretic optimality notions, with the subsequent advantages:

- It is more suited to analysis [55], with the ensuing broader generality and with the possibility of optimizing controllable parameters.
- It is more apt to provide insight and understanding on the mechanisms exercised to manage interference.

Recognizing that both FlashLinQ and ITLinQ schemes operate by enforcing, through various parameters and thresholds, exclusion regions around transmitters and receivers, the remainder of this section probes the idea of signal-strength-based exclusion regions and demonstrates its benefits through examples adapted from [56].

3.7.1   Channelization via Exclusion Regions

User Spatial Distribution

Let us consider a D2D-only network with the following spatial characteristics:

- Φ is, as usual, the uniform PPP with density λ representing the locations of all existing transmitters.
- Each transmitter has its intended receiver at a fixed distance r_0, independent of the link density and at a random angle uniformly distributed in $[0, 2\pi)$.

- Each link is endowed with a random mark m, uniformly distributed in $[0, 1]$, which may represent a time stamp or the priority of that link.

Description of the Scheme

The exclusion regions around the transmitters or receivers have a radius $\delta \cdot r_0$, where $\delta \geq 0$ is the ratio of the exclusion radius to the intended link distance r_0. A link is allowed on the channel under consideration only when the following two conditions are met:

1. No transmitter with a lower mark is present inside a circle of radius $\delta \cdot r_0$ around the link receiver.
2. No receiver with a lower mark is present inside a circle of radius $\delta \cdot r_0$ around the link transmitter.

If a link is not allowed on a given channel, then the link is served on another channel and thus we can regard the analysis herein as corresponding to the channel where the link under consideration is being served.

Co-channel Link Density

Denoting by $\tilde{K}$ the average number of cochannel links per cell, we use the following approach to compute $\tilde{K}$. Consider a link with mark m. Since the marks are uniformly distributed in $[0, 1]$, the density of links with a mark lower than m is $m\lambda$ [30]. Condition 1 is based on the interfering transmitter locations relative to the given link, and Condition 2 is based on the interfered receiver locations, which also depend on the corresponding transmitter locations through the link distance r_0. Combining all that, we can identify the transmitter locations that would violate Condition 2 and delineate an area, denoted by A_c, where no other transmitters must be present for both conditions to be satisfied. Such area is [56]

$$A_c = \pi \, \xi(\delta) \, r_0^2, \tag{3.111}$$

where

$$\xi(\delta) = \delta(\delta + 1) - \frac{2}{\pi^2} \int_{|\delta-1|}^{\delta+1} r \arcsin^2 \left(\frac{r^2 + 1 - \delta^2}{2r} \right) dr. \tag{3.112}$$

The probability of allowing a link with random mark m is

$$p = e^{-m\lambda A_c} \tag{3.113}$$

$$= e^{-mK \, \xi(\delta) \, a_0^2}, \tag{3.114}$$

which is the void probability of a PPP with density $m\lambda$ in an area A_c [30], and recall that $a_0 = r_0/R$ and $\lambda = \frac{K}{\pi R^2}$. Then, the average number of co-channel links per cell equals

$$\tilde{K} = K \int_0^1 e^{-mK \, \xi(\delta) \, a_0^2} \, dm \tag{3.115}$$

$$= \frac{1 - e^{-K\xi(\delta)a_0^2}}{\xi(\delta) \, a_0^2}, \tag{3.116}$$

where (3.116) follows from integration via (3.111). As evidenced by (3.116), $\tilde{K}$ is monotonically increasing in K and, as $K \to \infty$,

$$\tilde{K} \to \frac{1}{\xi(\delta)\, a_0^2}. \tag{3.117}$$

Interference Modeling

Recalling the first step of the interference modeling approach in Section 3.3.4 and invoking $\alpha = 0$ in (3.22), the distribution of z, conditioned on $\{r_j\}$, is modeled as $z \sim \mathcal{N}_{\mathbb{C}}\left(0, \sigma^2\right)$ where

$$\sigma^2 = \sum_{j=1}^{\infty} P\, r_j^{-\eta}. \tag{3.118}$$

The analysis is not straightforward because exclusion regions introduce dependencies (specifically, repulsions) across the locations of co-channel transmitters. The statistical properties of similar repulsive processes such as Matérn processes have been investigated [49, 57] and it has been shown that, when the separation between any two transmitters is at least $\delta \cdot r_0$, the locations of co-channel transmitters within $2\delta \cdot r_0$ are correlated while those of co-channel transmitters beyond $2\delta \cdot r_0$ behave essentially like a PPP. Applying this insight, we can borrow a modeling assumption introduced in [24]: the locations of cochannel transmitters outside the receiver's exclusion region belong to a homogeneous PPP with scaled-down density $\tilde{\lambda} = \frac{\tilde{K}}{\pi R^2}$. This assumption allows deriving compact expression for the link spectral efficiency, which does exhibit a satisfactory agreement with respect to the exact value in various settings [46, 55, 56].

System-Wide Distributions

Invoking the overlay setting ($\alpha = 0$) that befits a D2D-only network, the local-average SIR in (3.42) specializes to

$$\varrho = \frac{r_0^{-\eta}}{\sum_{j=1}^{\infty} r_j^{-\eta}}. \tag{3.119}$$

By leveraging the derivation of its unchannelized D2D network counterpart (see Section 3.4.1), the CDF of the local-average SIR and the spectral efficiency of the channelized D2D network can be computed by means of (3.75), with the Laplace transform of $1/\varrho$ being

$$\mathcal{L}_{1/\varrho}(\upsilon) \approx \exp\left\{\tilde{K} a_0^2 \left[\delta^2 + \frac{2}{\eta}\upsilon^{2/\eta}\bar{\Gamma}\left(-2/\eta, \upsilon/\delta^{\eta}\right)\right]\right\}, \tag{3.120}$$

where $\tilde{K}$ depends on K and δ as per (3.116).

Consequently, the CDF $F_{\mathsf{C}}(\cdot)$ can be evaluated via the mapping of $\mathsf{C}(\varrho)$ onto $F_\varrho(\cdot)$.

Example 3.14 validates the goodness of the Poisson approximation of the co-channel transmitter locations and the Gaussian model for the interference.

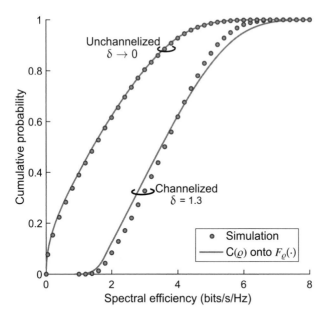

Figure 3.11 CDF of spectral efficiency for $K = 40$, $a_0 = 0.1$, and $\eta = 4.5$.

EXAMPLE 3.14   Let $K = 40$ with a normalized intended link distance of $a_0 = 0.1$ and with $\delta = 1.3$ and $\eta = 4.5$. In Figure 3.11, the spectral efficiency CDFs obtained by mapping $\mathsf{C}(\varrho)$ onto the CDF $F_\varrho(\cdot)$ are contrasted against their counterparts obtained via Monte Carlo. In the simulations, the locations of interferers outside the exclusion circle no longer conform to a PPP.

Similarly good agreements are observed for a variety of other settings.

Spatially Averaged Spectral Efficiency

For given K and δ, under Rayleigh fading, the link spectral efficiency averaged over all network geometries is

$$\bar{\mathsf{C}}(K, \delta) \approx \log_2(e) \int_0^\infty \frac{1}{\zeta + 1} \exp\left\{ \tilde{K} a_0^2 \left[\delta^2 + \frac{2}{\eta} \zeta^{2/\eta} \bar{\Gamma}\left(-2/\eta, \zeta/\delta^\eta \right) \right] \right\} d\zeta, \quad (3.121)$$

where $\tilde{K}$ depends on K and δ as per (3.116) while (3.121) follows from the application of (3.120) to (3.106).

By taking $\lim_{\delta \to 0} \bar{\mathsf{C}}(K, \delta)$, we can recover the spatially averaged link spectral efficiency for an unchannelized network, i.e., a network where all links are co-channel. This baseline,

$$\bar{\mathsf{C}}(K, 0) \approx \log_2(e) \int_0^\infty \frac{1}{\zeta + 1} \exp\left\{ -K a_0^2 \zeta^{2/\eta} \Gamma\left(1 - 2/\eta \right) \right\} d\zeta, \quad (3.122)$$

is useful to establish the benefits of channelization.

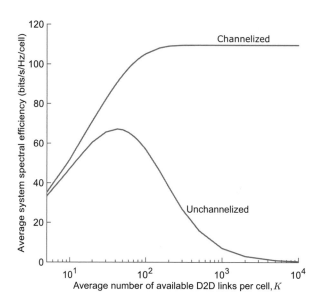

Figure 3.12 Average system spectral efficiency (bits/s/Hz per cell) as a function of K, for $a_0 = 0.1$ and $\eta = 4.5$, with optimized δ.

Furthermore, from (3.121) we can also obtain the spatially averaged system spectral efficiency by scaling the spatially averaged link spectral efficiency $\bar{\mathsf{C}}(K, \delta)$ by the average number of co-channel links per cell $\tilde{K}$. This gives

$$\bar{C}(K, \delta) = \tilde{K}\,\bar{\mathsf{C}}(K, \delta), \tag{3.123}$$

from which the average benefits of channelization over all possible network geometries can be assessed. Moreover, the parameter δ can be tuned as a function of the user density and link distance in order to maximize the average system spectral efficiency, i.e., to obtain

$$\bar{C}^{\star}(K) = \max_{\delta}\ \bar{C}(K, \delta). \tag{3.124}$$

EXAMPLE 3.15   Figure 3.12 plots $\bar{C}^{\star}(K)$, obtained numerically for $a_0 = 0.1$ and $\eta = 4.5$. The figure also shows the average system spectral efficiency without channelization, $\bar{C}(K, 0)$.

By repeating Example 3.15 for other values of a_0 and η, the following observations are found to hold qualitatively:

- Channelization yields a significant improvement in average system spectral efficiency relative to the unchannelized baseline, with more pronounced performance advantages at higher link densities.
- For any fixed intended link distance, the co-channel link density $\tilde{\lambda} = \frac{\tilde{K}}{\pi R^2}$ increases with growing λ and eventually saturates. This, in turn, leads to saturation in the average system spectral efficiency.

3.8   Cache-Aided D2D Communication

Cache-aided communication is one of the most promising ideas to reduce the wireless backhaul load in exchange for cache memory – an inexpensive network resource compared to bandwidth – at the BSs and/or devices [58, 59]. The idea in cache-aided communication is to store the most popular files in memory at the BSs and/or devices, delivering content on demand to users in close proximity. Various forms of caching strategies are proposed, which can be classified into two categories:

1. Infrastructure-based caching, where small BSs with low-rate backhaul links but large storage capacities are deployed at the network edge to serve the users' requests, e.g., femtocaching [60] and coded multicasting [61, 62].
2. Infrastructureless caching, where content is disseminated through direct D2D links by utilizing cached files at the user devices [63].

There is a growing interest in the latter option due to the possibility of higher density of caching and higher spectral efficiency over short-range D2D links [9, 59, 63–66].

In [65], the throughput-outage scaling laws were established for a D2D one-hop caching network where N users are placed on a square grid with a minimum distance $1/\sqrt{N}$ and users precache files of size f from a fixed library of size F. A request from a user for a random file is served by its neighboring user if the file is not available in its own cache. When $F = \mathcal{O}(N)$ and the total cache size satisfies $fN \gg F$, the per-user throughput is proportional to f/F and independent of the number of users N for any fixed outage probability. This implies that D2D caching achieves a system-level performance that is scalable with the cache capacity f of the user devices, which makes it beneficial in dense environments with a large number of users requesting a relatively small library of popular files. These observations have been verified to hold qualitatively for limited link ranges, and even with realistic propagation conditions [9, 67].

3.9   Summary and Outlook

This chapter has shown how to analytically characterize the D2D spectral efficiencies of specific network geometries as well as their average, for both underlay and overlay, with easy-to-evaluate expressions that in some cases are even in closed form. From these expressions, and from the supporting examples, we can distill the following insights:

- For local traffic, direct D2D is better than uplink–downlink communication in a vast majority of situations, upwards of 80% for relatively long D2D links (15% of cell radius) and upwards of 98% for shorter D2D links (5% of cell radius).
- Tens of D2D links can be packed in each cell with acceptable mutual interference and, given a properly sized exclusion region, with only a minor effect (order of 10–20%) on the cellular uplinks in that cell.
- For local traffic, the average system spectral efficiency with D2D can be between one and three orders of magnitude larger than without D2D.

Altogether then, D2D communication offers a prime opportunity for network densification in the face of local traffic [68–70]. Both overlay and underlay options can enhance the system spectral efficiency as long as there exists a reasonable protection against the co-channel interference by means of exclusion regions or more sophisticated channel allocation schemes.

Research directions of further interest include:

- Throughout the chapter, the power has been considered fixed at each D2D transmitter. Due to the scarcity of power for battery-operated devices, it is desirable to have power control policies [71, 72] whereby transmit power levels are adjusted to achieve a specified target SINR at the receivers or to maximize system performance. Power control, an intensely researched topic in the context of wireless networks, can improve spatial frequency reuse by helping to control the interference. An improved version of ITLinQ has been formulated, which allows for a tighter packing of co-channel D2D links [73]. This scheme, termed ITLinQ+, implements both channel allocation and power control in a distributed fashion. ITLinQ+ is shown (through Monte-Carlo simulations) to improve the system spectral efficiency by 5–20% over ITLinQ in a typical D2D network. The analytical characterization of ITLinQ+ could constitute an interesting follow-up within the context of channel allocation schemes.
- Given that a cluster-centric placement is favorable in the context of cache-aided D2D communication, where the content is placed in each cluster to improve the collective performance of all the devices therein, it would be desirable to extend the analytical framework to encompass a Poisson cluster process for the user locations [74].

Acknowledgments

The work reported in this chapter was supported by the Intel/Verizon University Research Program "5G: Transforming the Wireless User Experience" and by Project TEC 2015-66228-P (MINECO/FEDER, UE), as well as by the European Research Council under the H2020 Framework Programme/ERC grant agreement 694974. The authors greatly appreciate the feedback from the anonymous reviewer.

References

[1] S. Andreev, A. Pyattaev, K. Johnsson, O. Galinina, and Y. Koucheryavy, "Cellular traffic offloading onto network-assisted device-to-device connections," *IEEE Commun. Mag.*, vol. 52, no. 4, pp. 20–31, Apr. 2014.

[2] D. Wu, J. Wang, R. Q. Hu, Y. Cai, and L. Zhou, "Energy-efficient resource sharing for mobile device-to-device multimedia communications," *IEEE Trans. Veh. Technol.*, vol. 63, no. 5, pp. 2093–2103, Jun. 2014.

[3] X. Bao, U. Lee, I. Rimac, and R. R. Choudhury, "Data spotting: Offloading cellular traffic via managed device-to-device data transfer at data spots," *ACM SIGMOBILE Mobile Comput. Commun. Rev.*, vol. 14, no. 3, pp. 37–39, Jul. 2010.

[4] 3GPP TR 22.803 V1.0.0, "Feasibility study for proximity services (ProSe) (Release 12)," 3rd Generation Partnership Project 3GPP, www.3gpp.org, Tech. Rep., Aug. 2012.

[5] M. S. Corson, R. Laroia, J. Li, T. Richardson, and G. Tsirtsis, "Toward proximity-aware internetworking," *IEEE Wireless Commun. Mag.*, vol. 17, no. 6, pp. 26–33, Dec. 2010.

[6] M. S. Corson, R. Laroia, J. Li, et al. "Flashlinq: Enabling a mobile proximal internet," *IEEE Wireless Commun. Mag.*, vol. 20, no. 5, pp. 110–117, Oct. 2013.

[7] 3GPP TR 22.885 V14.0.0, "Study on LTE support for vehicle to everything (V2X) services (Release 14)," 3rd Generation Partnership Project 3GPP, www.3gpp.org, Tech. Rep., Dec. 2015.

[8] N. Golrezaei, A. Molisch, A. G. Dimakis, and G. Caire, "Femtocaching and device-to-device collaboration: A new architecture for wireless video distribution," *IEEE Commun. Mag.*, vol. 51, no. 4, pp. 142–149, Apr. 2013.

[9] N. Golrezaei, P. Mansourifard, A. F. Molisch, and A. G. Dimakis, "Base-station assisted device-to-device communications for high-throughput wireless video networks," *IEEE Trans. Wireless Commun.*, vol. 13, no. 7, pp. 3665–3676, Jul. 2014.

[10] P. Gupta and P. R. Kumar, "The capacity of wireless networks," *IEEE Trans. Inform. Theory*, vol. 46, no. 12, pp. 388–404, Mar. 2000.

[11] A. Ozgur, O. Leveque, and D. N. C. Tse, "Hierarchical cooperation achieves optimal capacity scaling in ad hoc networks," *IEEE Trans. Inform. Theory*, vol. 53, no. 10, pp. 3549–3572, Oct. 2007.

[12] M. Franceschetti, D. Migliore, and P. Minero, "The capacity of wireless networks: Information-theoretic and physical limits," *IEEE Trans. Inform. Theory*, vol. 55, no. 8, pp. 3413–3424, Aug. 2009.

[13] A. Ozgur, O. Leveque, and D. Tse, "Spatial degrees of freedom of large distributed MIMO systems and wireless ad hoc networks," *IEEE J. Select. Areas Commun.*, vol. 31, no. 2, pp. 202–214, Feb. 2013.

[14] S. N. Hong and G. Caire, "Beyond scaling laws: On the rate performance of dense device-to-device wireless networks," *IEEE Trans. Inform. Theory*, vol. 61, no. 9, pp. 4735–4750, Sep. 2015.

[15] X. Lin, J. G. Andrews, and A. Ghosh, "Spectrum sharing for device-to-device communication in cellular networks," *IEEE Trans. Wireless Commun.*, vol. 13, no. 12, pp. 6727–6740, Dec. 2014.

[16] Q. Ye, M. Al-Shalash, C. Caramanis, and J. G. Andrews, "Resource optimization in device-to-device cellular systems using time-frequency hopping," *IEEE Trans. Wireless Commun.*, vol. 13, no. 10, pp. 5467–5480, Oct. 2014.

[17] N. Lee, X. Lin, J. G. Andrews, and R. W. Heath Jr., "Power control for D2D underlaid cellular networks: Modeling, algorithms and analysis," *IEEE J. Select. Areas Commun.*, vol. 33, no. 1, pp. 1–13, Jan. 2015.

[18] G. George, R. K. Mungara, and A. Lozano, "An analytical framework for device-to-device communication in cellular networks," *IEEE Trans. Wireless Commun.*, vol. 14, no. 11, pp. 6297–6310, Nov. 2015.

[19] J. G. Andrews, F. Baccelli, and R. K. Ganti, "A tractable approach to coverage and rate in cellular networks," *IEEE Trans. Commun.*, vol. 59, no. 11, pp. 3122–3134, Nov. 2011.

[20] G. George, R. K. Mungara, A. Lozano, and M. Haenggi, "Ergodic spectral efficiency in MIMO cellular networks," *IEEE Trans. Wireless Commun.*, vol. 16, no. 5, pp. 2835–2849, May 2017.

[21] A. Lozano and N. Jindal, "Are yesterday's information-theoretic fading models and performance metrics adequate for the analysis of today's wireless systems?" *IEEE Commun. Mag.*, vol. 50, no. 11, pp. 210–217, Nov. 2012.

[22] T. D. Novlan, H. S. Dhillon, and J. G. Andrews, "Analytical modeling of uplink cellular networks," *IEEE Trans. Commun.*, vol. 12, no. 6, pp. 2669–2679, Jun. 2013.

[23] M. Haenggi, "Local delay in Poisson networks with and without interference," in *Proc. Annual Allerton Conf. Commun., Cont., Computing*, Sep. 2010, pp. 1482–1487.

[24] F. Baccelli, J. Li, T. Richardson, et al. "On optimizing CSMA for wide area ad-hoc networks," in *Proc. Int. Symp. Modell. Opt. Mobile, Ad-hoc Wireless Netw.*, May 2011, pp. 354–359.

[25] N. Lee, F. Baccelli, and R. W. H. Jr., "Spectral efficiency scaling laws in dense random wireless networks with multiple receive antennas," *IEEE Trans. Inform. Theory*, vol. 62, no. 3, pp. 1344–1359, Mar. 2016.

[26] N. Naderializadeh and A. S. Avestimehr, "ITLinQ: A new approach for spectrum sharing in device-to-device communication systems," *IEEE J. Select. Areas Commun.*, vol. 32, no. 6, pp. 1139–1151, Jun. 2014.

[27] A. Lapidoth and S. Shamai, "Fading channels: How perfect need "perfect side information" be?" *IEEE Trans. Inform. Theory*, vol. 48, no. 5, pp. 1118–1134, May 2002.

[28] A. Giorgetti and M. Chiani, "Influence of fading on the Gaussian approximation for BPSK and QPSK with asynchronous cochannel interference," *IEEE Trans. Wireless Commun.*, vol. 4, no. 3, pp. 384–389, Mar. 2005.

[29] S. Ak, H. Inaltekin, and H. V. Poor, "Gaussian approximation for the downlink interference in heterogeneous cellular networks," in *Proc. IEEE Int. Symp. Inform. Theory*, Barcelona, Jul. 2016, pp. 1611–1615.

[30] M. Haenggi, *Stochastic Geometry for Wireless Networks*. Cambridge: Cambridge University Press, 2012.

[31] C. C. Chan and S. V. Hanly, "Calculating the outage probability in a CDMA network with spatial Poisson traffic," *IEEE Trans. Veh. Technol.*, vol. 50, no. 1, pp. 183–204, Jan. 2001.

[32] A. Ghasemi and E. S. Sousa, "Interference aggregation in spectrum-sensing cognitive wireless networks," *IEEE J. Select. Topics Signal Processing*, vol. 2, no. 1, pp. 41–56, Feb. 2008.

[33] B. Blaszczyszyn and M. Karray, "Spatial distribution of the SINR in Poisson cellular networks with sector antennas," *IEEE Trans. Wireless Commun.*, vol. 15, no. 1, pp. 581–593, Jan. 2016.

[34] J. Abate and W. Whitt, "Numerical inversion of Laplace transforms of probability distributions," *ORSA J. Compt.*, vol. 7, no. 1, pp. 36–43, 1995.

[35] C. A. O'Cinneide, "Euler summation for Fourier series and Laplace transform inversion," *Communications in Statistics. Stochastic Models*, vol. 13, no. 2, pp. 315–337, 1997.

[36] E. S. Sousa and J. A. Silvester, "Optimum transmission ranges in a direct-sequence spread-spectrum multihop packet radio network," *IEEE J. Select. Areas Commun.*, vol. 8, no. 5, pp. 762–771, Jun. 1990.

[37] S. Catreux, P. F. Driessen, and L. J. Greenstein, "Data throughputs using multiple-input multiple-output (MIMO) techniques in a noise-limited cellular environment," *IEEE Trans. Wireless Commun.*, vol. 1, no. 2, pp. 226–235, Apr. 2002.

[38] M. Haenggi, "On distances in uniformly random networks," *IEEE Trans. Inform. Theory*, vol. 51, no. 10, pp. 3584–3586, Oct. 2005.

[39] A. Memmi, Z. Rezki, and M.-S. Alouini, "Power control for D2D underlay cellular networks with channel uncertainty," *IEEE Trans. Wireless Commun.*, vol. 16, no. 2, pp. 1330–1343, Feb. 2017.

[40] M. Ni, L. Zheng, F. Tong, J. Pan, and L. Cai, "A geometrical-based throughput bound analysis for device-to-device communications in cellular networks," *IEEE J. Select. Areas Commun.*, vol. 33, no. 1, pp. 100–110, Jan. 2015.

[41] Z. Syu and C. Lee, "Spatial constraints of device-to-device communications," in *IEEE First Int. Black Sea Conf. on Commun. and Networking (BlackSeaCom)*, Jul. 2013, pp. 94–98.

[42] X. Lin, R. W. H. Jr., and J. G. Andrews, "The interplay between massive MIMO and underlaid D2D networking," *IEEE Trans. Wireless Commun.*, vol. 14, no. 6, pp. 3337–3351, Jun. 2015.

[43] B. Kaufman and B. Aazhang, "Cellular networks with an overlaid device to device network," in *Proc. Annual Asilomar Conf. Signals, Syst., Comp.*, Oct. 2008, pp. 1537–1541.

[44] G. Fodor, E. Dahlman, G. Mildh, et al. "Design aspects of network assisted device-to-device communications," *IEEE Commun. Mag.*, vol. 50, no. 3, pp. 170–177, Mar. 2012.

[45] A. Lozano and D. C. Cox, "Integrated dynamic channel assignment and power control in TDMA mobile wireless communication systems," *IEEE J. Select. Areas Commun.*, vol. 17, pp. 2031–2040, Nov. 1999.

[46] A. Hasan and J. G. Andrews, "The guard zone in wireless ad hoc networks," *IEEE Trans. Wireless Commun.*, vol. 6, no. 3, pp. 897–906, Mar. 2007.

[47] D. Torrieri and M. C. Valenti, "Exclusion and guard zones in DS-CDMA ad hoc networks," *IEEE Trans. Commun.*, vol. 61, no. 6, pp. 2468–2476, Jun. 2013.

[48] G. Alfano, M. Garetto, and E. Leonardi, "New directions into the stochastic geometry analysis of dense CSMA networks," *IEEE Trans. Mobile Comput.*, vol. 13, no. 2, pp. 324–336, Feb. 2014.

[49] M. Haenggi, "Mean interference in hard-core wireless networks," *IEEE Commun. Lett.*, vol. 15, no. 8, pp. 792–794, Aug. 2011.

[50] Y. Zhong, W. Zhang, and M. Haenggi, "Stochastic analysis of the mean interference for the RTS/CTS mechanism," in *Proc. IEEE Int. Conf. Commun.*, Jun. 2014, pp. 1996–2001.

[51] C. H. Lee and M. Haenggi, "Interference and outage in Poisson cognitive networks," *IEEE Trans. Wireless Commun.*, vol. 11, no. 4, pp. 1392–1401, Apr. 2012.

[52] S. Cho and W. Choi, "Relay cooperation with guard zone to combat interference from an underlaid network," in *Proc. IEEE Global Commun. Conf.*, Dec. 2011, pp. 1–5.

[53] X. Wu, S. Tavildar, and S. Shakkottai, et al. "FlashLinQ: A synchronous distributed scheduler for peer-to-peer ad hoc networks," in *Proc. Annual Allerton Conf. Commun., Cont., Computing*, Sep. 2010, pp. 514–521.

[54] X. Wu, S. Tavildar, and S. Shakkottai, et al. "FlashLinQ: A synchronous distributed scheduler for peer-to-peer ad hoc networks," *IEEE/ACM Trans. Networking*, vol. 21, no. 4, pp. 1215–1228, Aug. 2013.

[55] R. K. Mungara, X. Zhang, A. Lozano, and R. W. Heath Jr., "Analytical characterization of ITLinQ: Channel allocation for device-to-device communication networks," *IEEE Trans. Wireless Commun.*, vol. 15, no. 5, pp. 3603–3615, May 2016.

[56] G. George, R. K. Mungara, and A. Lozano, "Optimum exclusion regions for interference protection in device-to-device wireless networks," in *Proc. Int. Workshop Device-to-Device Commun. WiOpt*, May 2015, pp. 102–109.

[57] B. Cho, K. Koufos, and R. Jäntti, "Bounding the mean interference in Matérn type II hard-core wireless networks," *IEEE Wireless Commun. Lett.*, vol. 2, no. 5, pp. 563–566, Oct. 2013.

[58] N. Golrezaei, A. G. Dimakis, A. F. Molisch, and G. Caire, "Wireless video content delivery through distributed caching and peer-to-peer gossiping," in *Proc. Annual Asilomar Conf. Signals, Syst., Comp.*, Nov. 2011, pp. 1177–1180.

[59] A. F. Molisch, G. Caire, D. Ott, et al., "Caching eliminates the wireless bottleneck in video aware wireless networks," *Adv. Elect. Eng.*, vol. 2014, Nov. 2014, article ID. 261930.

[60] K. Shanmugam, N. Golrezaei, A. G. Dimakis, A. F. Molisch, and G. Caire, "FemtoCaching: Wireless content delivery through distributed caching helpers," *IEEE Trans. Inform. Theory*, vol. 59, no. 12, pp. 8402–8413, Dec. 2013.

[61] M. A. Maddah-Ali and U. Niesen, "Fundamental limits of caching," *IEEE Trans. Inform. Theory*, vol. 60, no. 5, pp. 2856–2867, May 2014.

[62] M. Bayat, R. K. Mungara, and G. Caire, "Achieving spatial scalability for coded caching via coded multipoint multicasting," *IEEE Trans. Wireless Commun.*, vol. 18, no. 1, pp. 227–240, Jan. 2019.

[63] M. Ji, G. Caire, and A. F. Molisch, "Fundamental limits of caching in wireless D2D networks," *IEEE Trans. Inform. Theory*, vol. 62, no. 2, pp. 849–869, Feb. 2016.

[64] A. Altieri, P. Piantanida, L. R. Vega, and C. G. Galarza, "On fundamental trade-offs of device-to-device communications in large wireless networks," *IEEE Trans. Wireless Commun.*, vol. 14, no. 9, pp. 4958–4971, Sep. 2015.

[65] M. Ji, G. Caire, and A. F. Molisch, "The throughput–outage tradeoff of wireless one-hop caching networks," *IEEE Trans. Inform. Theory*, vol. 61, no. 12, pp. 6833–6859, Dec. 2015.

[66] M. Ji, G. Caire, and A. F. Molisch, "Wireless device-to-device caching networks: Basic principles and system performance," *IEEE J. Select. Areas Commun.*, vol. 34, pp. 176–189, Jan. 2016.

[67] P. Kyösti, J. Meinilá, L. Hentila and X. Zhao, "Winner II channel models, Deliverable D1.1.2," Eur. Union, Brussels, Belgium, Tech. Rep., Sep. 2007. [Online]. Available: http://www.ist-winner.org/WINNER2-Deliverables/D1.1.2v1.1.pdf

[68] F. Boccardi, R. W. Heath Jr., A. Lozano, T. Marzetta, and P. Popovski, "Five disruptive technology directions for 5G," *IEEE Commun. Mag.*, vol. 52, no. 2, pp. 74–80, Feb. 2014.

[69] J. G. Andrews, S. Buzzi, W. Choi, et al., "What will 5G be?" *IEEE J. Select. Areas Commun.*, vol. 32, no. 6, pp. 1065–1082, Jun. 2014.

[70] B. Bangerter, S. Talwar, R. Arefi, and K. Stewart, "Networks and devices for the 5G era," *IEEE Commun. Mag.*, vol. 52, no. 2, pp. 90–96, Feb. 2014.

[71] G. J. Foschini and Z. Miljanic, "A simple distributed autonomous power control algorithm and it's convergence," *IEEE Trans. Veh. Technol.*, vol. 42, no. 4, pp. 641–646, Apr. 1993.

[72] P. Hande, S. Rangan, M. Chiang, and X. Wu, "Distributed uplink power control for optimal SIR assignment in cellular data networks," *IEEE/ACM Trans. Networking*, vol. 16, no. 6, pp. 1420–1433, Dec. 2008.

[73] X. Yi and G. Caire, "Optimality of treating interference as noise: A combinatorial perspective," *IEEE Trans. Inform. Theory*, vol. 62, no. 8, pp. 4654–4673, Aug. 2016.

[74] M. Afshang, H. S. Dhillon, and P. H. J. Chong, "Fundamentals of cluster-centric content placement in cache-enabled device-to-device networks," *IEEE Trans. Commun.*, vol. 64, no. 6, pp. 2511–2526, Jun. 2016.

4 Multihop Wireless Backhaul for 5G

Song-Nam Hong and Ivana Marić

4.1 Introduction

To be able to support the exponential increase in mobile data traffic, cellular networks are undergoing fundamental change from a well-planned deployment of tower-mounted base stations (BSs) to a capacity-driven, possibly ad hoc, deployment of smaller and low-power BSs forming *small cells*. As the amount of traffic to be carried by the network increases, the major bottleneck for such network architecture that can limit the throughput and deployment cost [1] is the backhaul. Conventional BSs are typically connected to a high-capacity point-to-point backhaul network. However, the same will not be possible for small cell BSs due to their dense deployment at often more adverse locations. This problem is further aggravated in systems planned to operate at high frequencies (i.e., millimeter-waves), such as 5G. In such scenarios, due to the limited range of radio signals at high frequencies [2], the use of relays is a promising technique that can extend network coverage and improve network throughput [3]. By allowing access nodes (ANs) to not only deliver service to end-users but also to act as relays (see Figure 4.1), a multihop wireless connection can be established between each AN and an aggregation node (AgN), i.e., a node at which an optical wired or high-speed radio link connection is available. Multihop relaying can be instrumental to implement a wireless backhaul able to overcome non-line-of-sight propagation, providing a cost-effective and rapidly deployable alternative to the conventional backbone wired network [4].

The main obstacle to achieving high network throughput in wireless multihop backhaul network is interference, owing to the broadcast nature of the wireless medium. An interference-aware routing scheme for wireless backhaul networks was proposed by Hui and Axnas [5] to address this problem. The proposed routing algorithm attempts to minimize the interpath interference by choosing paths that cause minimal interference to each other. However, this approach incurs significant limitations on network throughput at high load [5]. This behavior is expected since avoiding all interpath interference at high load where there is a large number of interfering paths is nearly impossible. Instead, in such scenarios, more advanced coding schemes that can efficiently manage (or harness) strong interference instead of simply treating it as noise should be considered. This chapter presents such schemes for 5G wireless backhaul.[1]

---

[1] The presented approach also applies to the uplink of cloud radio access networks (C-RANs) with a multihop backhaul network [6] where sources, relays, and destination (see Figure 4.9) correspond to mobile stations, radio units, and control units, respectively.

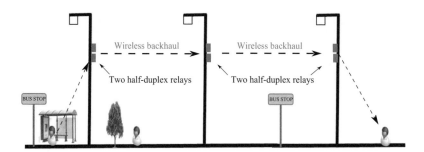

Figure 4.1 Multihop wireless backhaul network.

Several solutions for 5G backhaul, as well as specifically for multihop wireless back-haul, have been proposed and analyzed both in academia and industry. Following an information theoretic approach, capacity and scalability analysis of a backhaul networks in which the number of antennas on a BS increases proportionally with the network size can be found in [7]. Information theoretic upper bound to the capacity, under the same assumption on scaling the number of antennas, has been presented in [8]. A multihop wireless backhaul solution that focuses on energy efficiency has been proposed in [9]. Different multihop solutions focusing on several important metrics including deploying multiple disjoint routes [10]; adapting to traffic dynamics [11]; and performing joint path selection and rate allocation [12] have been proposed. User-level performance is inves-tigated in [13]. Various 5G backhaul requirements and technologies are overviewed and discussed in a survey [14].

The approach taken in this chapter is motivated by the view of the wireless back-haul network as a multiple-multicast relay network. There has been a lot of progress in information theory on developing efficient relaying strategies, starting with the intro-duction of the relay channel [15] and development of key relaying schemes in [16], reviewed in Chapter 2. Finding an optimal transmission strategy for the relay channel, and consequently also for wireless multihop backhaul networks, is still an open prob-lem. An efficient cooperative scheme called compute-and-forward (aka, physical layer network coding) has been developed more recently for these networks [17, 18]. The compute-and-forward transmission scheme requires channel coefficients to have inte-ger values, which is not the case in practice and for that reason the performance of compute-and-forward is limited by the non-integer penalty. It was shown in [19, 20] that due to the impact of the non-integer penalty, this scheme suffers from performance degradation that increases as the number of nodes (or hops) grows. In such networks, the communication scheme that achieves the best known performance is quantize-map-and-forward (QMF) [21], and its generalizations of noisy network coding (NNC) [22] and short message NNC (SNNC) [23]. In this scheme, all network nodes quantize and forward received source signals, achieving the cut-set upper bound within a constant gap that grows *linearly* with the number of relay nodes. This gap, however, may not be negligible for the system with multihop transmissions such as wireless backhaul net-work operating at high frequencies. Furthermore, there are several assumptions implicit in this strategy that need to be addressed and resolved for practical implementation of this strategy:

1. It was assumed that relays operate in full-duplex mode [21–23]. However, isolating the radio front-end when transmitting and receiving in the same frequency band and during the same time slot [24] is difficult to realize in practice, and therefore implementation of full-duplex radios is challenging.
2. It was assumed in [21, 22] that all nodes in the network quantize and forward even when: (1) signals that nodes receive from sources are very noisy, or (2) signals that are transmitted by these nodes are received poorly at the destination. In practice, an algorithm that selects relay nodes and their sequence of transmissions is needed.
3. A decoder that performs joint decoding of signals received from a large number of network nodes has prohibitively high complexity.

We next present a practical cooperative scheme for wireless backhaul that address the above limitations. We start by addressing the first challenge and assume that relays operate in half-duplex instead of full-duplex mode. Relaxing the full-duplex constraint increases the complexity of the considered problem because the transmit/receive mode of each relay for a given time slot now needs to be determined. We address this by separating the analysis into two parts as follows. In Section 4.2, we present an efficient communication scheme for half-duplex relays. We consider a multihop *virtual* full-duplex relay channel shown in Figure 4.2. In this configuration, each relay stage is formed of at least two half-duplex relays, used alternatively in transmit and receive nodes, such that, while one relay transmits its signal to the next stage, the other relay receives a new signal from the previous stage. The role of the relays is exchanged at the end of each time interval. This relaying operation is known as *successive relaying* [25]. This approach allows the source to send a new message to the destination at every time slot as if full-duplex relays were used. Every two consecutive source messages will travel across two alternate disjoint paths of relays. For a two-hop model of such network, it was shown that dirty paper coding (DPC) achieves the performance of ideal full-duplex relay [26]. In this case, the source can completely eliminate the "known" interference at the intended receiver. However, DPC cannot be used in the same manner in a multihop network (shown in Figure 4.2) because a transmit relay has no knowledge on interference signals at other stages [27]. In fact, determining an optimal strategy for the multihop case remains an open problem. In Section 4.2, we present a coding scheme

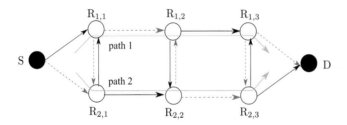

Figure 4.2 Virtual full-duplex relay network consisting of half-duplex relays.

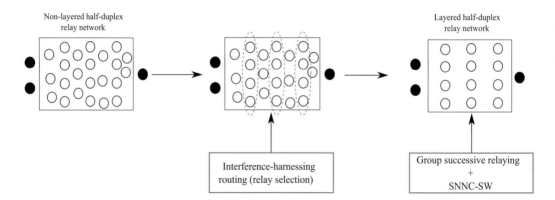

Figure 4.3 Steps in the transmission scheme for wireless backhaul: relay selection, group relaying, and SNNC-SW. Black nodes represent the sources and the destination. The white nodes are the relays.

for this network that does not require sending a "long" message and joint decoding at the receiver required by the scheme proposed in [21, 22], and later on relaxed in SNNC. It is shown that the performance of this scheme degrades *logarithmically* with the number of relay stages, instead of linearly as when the "background noise-level" quantization is used [21, 22].

After developing the results of Section 4.2, we consider a more general network with L sources communicating with a destination using half-duplex relays. We present a relaying scheme that consists of the following (see Figure 4.3):

1. A routing (relay selection) algorithm that chooses which nodes should serve as relays. The relay selection algorithm establishes a $(K + 1)$-hop network with $2L$ relays at each hop. We refer to such a network as a *layered* network.
2. A scheduling algorithm that assigns the transmit/receive mode of each half-duplex relay for a given time slot. We refer to this scheme as *group successive relaying*.
3. A cooperative relaying scheme optimized for the resulting layered half-duplex relay network. We refer to this scheme as *SNNC with sliding-window decoding* (SNNC-SW).

Cooperative strategies for layered networks under the assumption that the layered network has been established a priori have been analyzed in several related works (see [28] and references therein). Specifically, NNC in a full-duplex layered network has been analyzed [28], improving the performance gap to the capacity from linear to logarithmic growth as the number of layers increases. The presented SNNC-SW scheme outperforms existing techniques for half-duplex layered networks [29].

Some of the existing routing algorithms (e.g., interference-aware routing in [5]) could be used to form a (not necessary efficient) layered network. Instead, we present a relay selection algorithm named interference-harnessing routing that is *specifically* optimized for SNNC-SW. In interference-harnessing routing, the routing criterion contrasts the routing criterion of interference-aware routing of [30–32]: the routing metric is chosen to *harness* interference, instead to avoid it, as will be described in Section 4.3. Specifically, there are two very fundamental differences between interference-harnessing and more conventional routing: (1) Each relay chosen by the interference harnessing algorithm will relay information for multiple sources. Or, viewed in another way, there are several relays per each hop. Thus, we are not establishing routes as in common routing algorithms, rather a layered network. (2) The selection of relays is very different, in that interference harnessing chooses relays that are *as close to each other as possible*. This is a key ingredient that establishes a layered network that is optimized for SNNC-SW, enabling it to exploit interference.

After presenting interference-harnessing routing, we consider group successive relaying that generalizes the approach of Section 4.2. In the approach, $2L$ relays at each network stage are divided into two groups; while L relays in one group transmit their messages to the next stage, the relays in the other group receive signals from the previous stage. The role of each group is swapped at each time interval (see Figure 4.2). In this way, the sources send L new messages to the destination at every time slot. The group successive relaying is a simple extension of successive relaying [27, 33–35] to multilevel networks, and is a necessary part of the overall transmission scheme. As will be shown, by using group successive relaying, the developed SNNC-SW ensures high data rates.

Finally, after establishing the layered network with group successive relaying, we present the SNNC-SW transmission scheme. For *any* given layered network, an achievable rate region of the proposed SNNC-SW scheme will be derived. It will be shown that SNNC-SW achieves better rate-scaling than QMF, NNC, and SNNC; that is, that the performance degrades *logarithmically* with K, thereby achieving a larger performance gain as the number of hops grows. Furthermore, SNNC-SW has a lower decoding complexity and delay compared to QMF and NNC, since a joint decoding is performed only per each hop (i.e., very small network) instead of for the whole network.

The idea of sliding window decoding was originally proposed for the multiple access channel with generalized feedback (MAC-GF) in [36]. Although the two schemes have completely different encoding (and consequently decoding), resulting in two very different overall transmission schemes, the SNNC-SW scheme uses the sliding window idea, and we chose the name accordingly.

At the end of Section 4.3, the performance of the interference-harnessing routing that establishes a specific layered network is evaluated. As we will see, the interference-harnessing routing with SNNC-SW provides a significantly higher rate than interference-aware routing in which relays use store-and-forward, which is a solution currently proposed for wireless multihop networks (we refer to this scheme as *multihop routing (MR)*). The performance gap again increases as the network becomes denser (or load becomes higher). Finally, we present a practical, low-complexity stage-by-stage successive MIMO decoding for the proposed SNNC-SW scheme. Such a decoder can be implemented by using conventional MIMO decoding (e.g., zero-forcing, linear MMSE, and integer-forcing receivers) instead of using a complicated joint decoder. It is shown that this practical scheme provides a significant gain over multihop routing. Furthermore, the performance gap increases as a network becomes denser.

We start by considering multihop virtual full-duplex relay network.

4.2   Multihop Virtual Full-Duplex Relay Networks

We consider a virtual full-duplex relay channel with K relay stages shown in Figure 4.2. Encoding/decoding operations are performed over time slots consisting of n channel uses of a discrete memoryless channel. As explained in Section 4.1, successive relaying [19] is assumed in a multihop fashion such that, at each time slot t, the source transmits a new message $\underline{\mathbf{w}}_t \in \{1, 2, \ldots, 2^{nr_i}\}$ where $i = 1$ for odd time slot t and $i = 2$ for even time slot t, and the destination decodes a new message $\underline{\mathbf{w}}_{t-K}$. We define two message rates r_1 and r_2 since the odd-indexed and even-indexed messages are conveyed to the destination via two disjoint paths, i.e.,

$$\text{path 1: } (S, R_{1,1}, \ldots, R_{1,K}, D) \tag{4.1}$$

$$\text{path 2: } (S, R_{2,1}, \ldots, R_{2,K}, D). \tag{4.2}$$

The role of relays is alternatively reversed in successive time slots (Figure 4.2). During $N + K$ time slots, the destination decodes the $N/2$ messages from each path. Thus, the achievable rate of the messages via path i is given by $\frac{r_i N}{2(N+K)}$. By letting $N \to \infty$, the rate $r_i/2$ is achievable, provided that the error probability vanishes with n. As in standard relay channels (see for example [21, 22]), we take first the limit for $n \to \infty$ and then for $N \to \infty$, and focus on the achievable rate r_i. For ease of exposition, we use the notation $\bar{i}$ to indicate the complement of i, i.e., $\bar{i} = 2$ if $i = 1$ and $\bar{i} = 1$ if $i = 2$. As illustrated in Figure 4.2, it is assumed that each receiver can only receive signals from relays in the previous and the current relay stages. Given the multistage structure of the model topology, the transition probability distribution then factors as the product as given by (4.3):

$$\prod_{k=1}^{\lceil K/2 \rceil} p(y_{\bar{i},2k-1}|x_{i,2k-1}, x_{\bar{i},2k-2}) \prod_{k=1}^{\lfloor K/2 \rfloor} p(y_{i,2k}|x_{i,2k-1}, x_{\bar{i},2k}) p(y_D|x_{i,K}), \tag{4.3}$$

where $i = 1$ for odd t and $i = 2$ for even t, and where $x_{i,k}$ and $y_{i,k}$ denote the respective output and input at relay $R_{i,k}$, and $x_{1,0}$ and $x_{2,0}$ denote the source outputs. Note that there are two types of channels, where one is used for odd time slots and the other is used for even time slots.

4.2.1   Mixed Encoding Scheme for Multihop Virtual Full-Duplex Relay Channels

We next present a *mixed* scheme for a multihop virtual full-duplex relay channel. In the scheme, each relay can perform either QMF or DF (decode-forward) to better adapt to channel coefficients. This idea and its gains have been demonstrated for general networks in [23]. Each DF relay decodes its incoming message that is either coming from the source or is a quantization index sent by a QMF relay. As explained in [19], the destination explicitly decodes messages from the relays as well as the source message and exploits the decoded relays' messages as side-information in the next time slot. Furthermore, we allow each relay to incorporate rate-splitting into its encoding (QMF or DF) to enable partial decoding of its own signal, as it is interference at other relays.

Specifically, relay $R_{i,k}$ employs rate-splitting if $R_{\bar{i},k}$ performs DF. On the other hand, when $R_{\bar{i},k}$ performs QMF, the rate-splitting is not used because a QMF relay does not need to decode any message (unlike a DF relay) and because the destination decodes a message with full knowledge of the interference.

For decoding, we assume that joint decoding (JD) is performed only at DF relays associated with the DF-only stage (namely, a stage in which both relays perform DF). For all other relays, we assume a simpler successive decoding (SD). This assumption reduces the decoding complexity while achieving almost optimal performance [37]. Furthermore, this assumption yields an achievable rate that is easily computable and also allows finding the optimal relay configuration. The destination and the other DF relays perform SD and, accordingly, it is assumed that quantize-forward (QF) relays employ Wyner–Ziv quantization so that the destination can perform SD as in [19]. The details and the analysis of the scheme are presented next.

We next introduce the following notation. We let

$$\mathcal{V}_i = \{k_{i,1}, \ldots, k_{i,|\mathcal{V}_i|}\} \subseteq \{1, \ldots, K\}$$

denote the index subset containing the indices of QMF relays in the path i, where $k_{i,1} < k_{i,2} < \cdots < k_{i,|\mathcal{V}_i|}$. For a given $\mathcal{V}_i$, let $\mathcal{I}_{i,\ell} = \{k_{i,\ell}, \ldots, k_{i,\ell+1} - 1\}$ for $\ell = 0, \ldots, |\mathcal{V}_i|$ with $k_{i,0} = 0$ and $k_{i,|\mathcal{V}_i|+1} = K + 1$. Note that $\mathcal{I}_{i,\ell}$ contains all DF relays that transmit the message sent by QMF relay $R_{i,k_{i,\ell}}$ and $\{\mathcal{I}_{i,\ell}\}_{\ell=0}^{|\mathcal{V}_i|}$ forms a partition of $\{1, \ldots, K\}$. We define a mapping: $g_i(k) = k_{i,\ell}$ if $k \in \mathcal{I}_{i,\ell}, \ell = 0, \ldots, |\mathcal{V}_i|$.

DEFINITION 4.1 *According to the mode of a receiving relay, we define:*

$$I_{i,k} = \begin{cases} I(X_{i,k}; Y_{i,k+1}|U_{\bar{i},k+1}), & k+1 \in \mathcal{V}_i^c \\ I(X_{i,k}; \hat{Y}_{i,k+1}|X_{\bar{i},k+1}), & k+1 \in \mathcal{V}_i \end{cases} \tag{4.4}$$

$$I_{i,k1} = \begin{cases} I(X_{i,k}; Y_{i,k+1}|U_{\bar{i},k+1}, U_{i,k}), & k+1 \in \mathcal{V}_i^c \\ I(X_{i,k}; \hat{Y}_{i,k+1}|X_{\bar{i},k+1}, U_{i,k}), & k+1 \in \mathcal{V}_i \end{cases}, \tag{4.5}$$

where $Y_{i,K+1} = Y_D$, $X_{i,K+1} = \phi$, and $U_{i,k}$ denotes an auxiliary random variable to be used for superposition coding.

By letting $r_{i,k}$ denote the rate of the relay $R_{i,k}$, we have:

THEOREM 4.2 *For a $(K+1)$-hop virtual full-duplex relay channel, the achievable rate region of the mixed scheme with SD is the set of all rate pairs $(r_1/2, r_2/2)$ that satisfy:*

$$r_i \leq \min\left\{\min_{k \in \mathcal{I}_{i,0}} I_{i,k}, \min_{k \in \mathcal{I}_{i,0} \cap \mathcal{V}_i^c} I(U_{i,k}; Y_{\bar{i},k}) + I_{i,k1}\right\} \tag{4.6}$$

and for $k \in \mathcal{V}_i$ with $g_i(k) = k_{i,\ell}$,

$$I(\hat{Y}_{i,k}; Y_{i,k}|X_{\bar{i},k}) = \min\left\{\min_{k' \in \mathcal{I}_{i,\ell}} I_{i,k'}, \min_{k' \in \mathcal{I}_{i,\ell} \cap \mathcal{V}_i^c} I(U_{i,k'}; Y_{\bar{i},k'}) + I_{i,k'1}\right\}, \tag{4.7}$$

for any index subset $\mathcal{V}_i \subseteq \{1, \ldots, K\}, i = 1, 2,$ *and any joint distributions that factors as*

$$\prod_{i=1}^{2} p(x_{i,0}) \prod_{k \in \mathcal{V}_i} p(x_{i,k}) \prod_{k \in \mathcal{V}_i^c} p(u_{i,k}) p(x_{i,k}|u_{i,k}) \prod_{k \in \mathcal{V}_i} p(\hat{y}_{i,k}|y_{i,k}). \tag{4.8}$$

We can further improve the performance by using JD at DF relays that belong to DF-only stages, yielding:

THEOREM 4.3   *For a* $(K + 1)$*-hop virtual full-duplex relay channel, the achievable rate region of the mixed scheme with JD (at DF-only stages) is the set of all rate pairs* $(r_1/2, r_2/2)$ *that satisfy:*

$$r_i \leq \min\{I_{i,k} : k \in \mathcal{I}_{i,0}\}, \tag{4.9}$$

and for $k \in \mathcal{V}_1^c \cap \mathcal{V}_2^c,$

$$r_{1,g_1(k)} + r_{2,g_2(k)} \leq \min\{I(U_{1,k}, X_{2,k-1}; Y_{2,k}) + I_{1,k1}, I(U_{2,k}, X_{1,k-1}; Y_{1,k}) + I_{2,k1}\}, \tag{4.10}$$

and for $k \in \mathcal{V}_i$ *with* $g_i(k) = k_{i,\ell},$

$$I(\hat{Y}_{i,k}; Y_{i,k}|X_{\bar{i},k}) = r_{i,k} \tag{4.11}$$

$$r_{i,k} \leq \min\{I_{i,k'} : k' \in \mathcal{I}_{i,\ell}\} \tag{4.12}$$

$$r_{i,k} \leq I(U_{i,k}; Y_{\bar{i},k}) + I_{i,k1}, k \in \mathcal{V}_{\bar{i}}^c, \tag{4.13}$$

for any subset $\mathcal{V}_i \subseteq \{1, \ldots, K\}$ *and any joint distribution defined in* (4.8), *where* $r_{i,0} = r_i.$

Proofs are provided in [37].

4.2.2   Gaussian Channels

We next consider a $(K + 1)$-hop Gaussian virtual full-duplex relay channels for three different scenarios referred to as a *symmetric channel*, *per-path symmetric channel*, and *per-hop symmetric channel*.

Symmetric Channels

In this scenario, each path has the same channel gain, denoted by SNR (signal-to-noise ratio), and each inter-relay interference level is equal to INR (interference-to-noise ratio) $= \gamma^2$SNR where γ denotes the relative interference level. We first consider an upper bound on capacity. This bound is valid for any relay scheduling, including the successive relaying considered in this chapter.

LEMMA 4.4   *If* SNR ≥ 1, *the capacity of the* $(K + 1)$*-hop virtual full duplex relay network is upper bounded by*

$$R_{\text{upper}}^{(K)} = \log(1 + \text{SNR}). \tag{4.14}$$

Notice that the upper bound is independent of the interference level γ.

We next analyze the performance of communication schemes of DF, amplify-and-forward (AF), compute-and-forward (CF), and CF with power allocation (CF-P), which are reviewed in Chapter 1 and in [19]. Also, we consider the proposed scheme with only QMF relays in which, due to the symmetry of the channel, it is not necessary to employ a mixed encoding scheme. The corresponding achievable rates are obtained as follows:

THEOREM 4.5   *For a symmetric $(K + 1)$-hop Gaussian virtual full-duplex relay network, the following rates are achievable:*

$$R_{\text{DF}}^{(K)} = \min \left\{ \max \left\{ \log \left(1 + \frac{\text{SNR}}{1 + \gamma^2 \text{SNR}} \right), \right. \right.$$
$$\left. \left. \frac{1}{2} \log(1 + (1 + \gamma^2)\text{SNR}) \right\}, \log(1 + \text{SNR}) \right\} \tag{4.15}$$

$$R_{\text{AF}}^{(K)} = \log \left(1 + \left(\frac{1 + \text{SNR}}{1 + (1 + \gamma^2)\text{SNR}} \right)^K \frac{\text{SNR}^{K+1}}{(1 + \text{SNR})^{K+1} - \text{SNR}^{(K+1)}} \right) \tag{4.16}$$

$$R_{\text{CoF}}^{(K)} = \min \left\{ \log^+ \left(\frac{\text{SNR}}{\mathbf{b}^H (\text{SNR}^{-1}\mathbf{I} + \mathbf{h}\,\mathbf{h}^H)^{-1}\mathbf{b}} \right), \log(1 + \beta_2^2 \text{SNR}) \right\} \tag{4.17}$$

$$R_{\text{CoF-P}}^{(K)} = \log(\text{SNR}) - K \log(1/\gamma_{max}^2) \tag{4.18}$$

$$R_{\text{QMF}}^{(K)} = \log \left(1 + \frac{\text{SNR}^{K+1}}{(1 + \text{SNR})^{K+1} - \text{SNR}^{K+1}} \right) \tag{4.19}$$

for some $\mathbf{b} \neq 0 \in \mathbf{Z}^2[j]$, $\mathbf{beta} = (\beta_1, \beta_2)$ with $|\beta_i| \leq 1$, where $\mathbf{h} = [\beta_1, \beta_2\gamma]^T$, and where $\gamma_{max} = \max\{\gamma/\lceil\gamma\rceil, \lfloor\gamma\rfloor/\gamma\}$.

COROLLARY 4.6   *With the high-SNR condition (i.e., SNR $\gg$ 1), the performance degradations according to the number of relay stages K are given by*

$$R_{\text{DF}}^{(1)} - R_{\text{DF}}^{(K)} = 0 \tag{4.20}$$

$$R_{\text{AF}}^{(1)} - R_{\text{AF}}^{(K)} = (K - 1) \log(1 + \gamma^2) \tag{4.21}$$

$$R_{\text{CoF-P}}^{(1)} - R_{\text{CoF-P}}^{(K)} = (K - 1) \log(1/\gamma_{max}^2) \tag{4.22}$$

$$R_{\text{QMF}}^{(1)} - R_{\text{QMF}}^{(K)} = \log \left(\frac{K + 1}{2} \right). \tag{4.23}$$

We observe a significant improvement compared to the previous schemes, i.e., that the gap of the proposed QMF from the cut-set upper bound grows logarithmically with the number of stages, and not linearly as in the case of "noise-level" quantization as performed in [21, 22]. The major difference between the proposed QMF and the conventional QMF (or NNC) is that in the former, the quantization levels are optimized, while in the latter they are simply set by noise level. This demonstrates that the optimization of quantization levels can significantly improve the performance. The performance gap becomes larger as the number of stages increases. We have the following corollary.

COROLLARY 4.7   *When the inter-relay interference level is equal to direct channel gain (i.e., $\gamma = 1$), CF achieves the upper bound within 0.5 bit.*

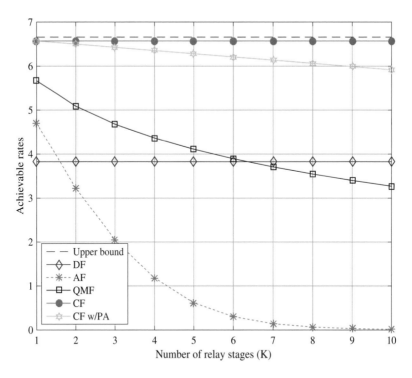

Figure 4.4 SNR $= 20$ dB. Achievable *ergodic* rates of different coding schemes averaging over $\gamma^2 \sim \text{Unif}(0.9, 1.1)$.

Corollary 4.7 demonstrates that CF is almost optimal for the Gaussian multihop virtual full-duplex relay channel provided that the inter-relay interference and the direct channel gains are balanced. However, this result does not capture the impact of the non-integer penalty, which may potentially greatly degrade the performance of CF. In order to demonstrate the actual performance of CF in this setting, we next consider Monte-Carlo averaging over the inter-relay interference level γ. Results are shown in Figures 4.4 and 4.5. When γ^2 is close to 1 (i.e., $\gamma^2 \sim \text{Unif}(0.9, 1.1)$), CF performance is close to the upper bound and it generally outperforms the other coding schemes, especially as K increases. Figure 4.5 shows that even if γ^2 is not close to 1, CF achieves the best performance for a sufficiently large number of relay stages (in this case, $K > 3$). Also, for $K \leq 3$, CF–P outperforms the other schemes. Therefore, CF (with or without PA [power allocation]) appears to be a strong candidate for the practical implementation of multihop virtual full-duplex relay networks, especially when the relative power of the interfering and direct links can be tuned by node placement and line-of-sight propagation, making the channel coefficients essentially deterministic. Proofs are provided in [19].

Per-Path Symmetric Channel

The assumption in this channel model is that each path has the same channel gain. In particular, SNR_i denotes the direct channel gain from $R_{i,k-1}$ to $R_{i,k}$ for all k, and

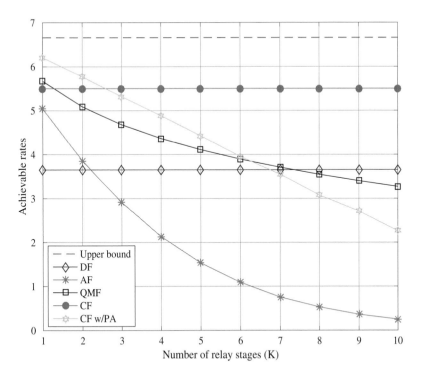

Figure 4.5 SNR = 20 dB. Achievable *ergodic* rates of different coding schemes averaging over $\gamma^2 \sim \text{Unif}(0.5, 1)$.

INR$_i$ denotes the interference channel gain R$_{\bar{i},k}$ to R$_{i,k}$ for all k. For this channel, we assume that relays on path 1 perform QMF and relays on path 2 perform DF, which yields:

COROLLARY 4.8   *For a $(K + 1)$-hop virtual full-duplex relay channel, the achievable rate region of the mixed scheme is the set of all rate pairs $(r_1/2, r_2/2)$ such that*

$$r_1 \leq \log(1 + \text{SNR}_1/(1 + \hat{\sigma}_1^2)) \tag{4.24}$$

$$r_2 \leq \log(1 + \text{SNR}_2/(1 + \theta\text{INR}_2)), \tag{4.25}$$

for some power-splitting parameter $\theta \in [0, 1]$, where $\hat{\sigma}_1^2$ is computed by the following recursive equation:

$$\hat{\sigma}_k^2 = \max\left\{(1 + \hat{\sigma}_{k+1}^2)\frac{1 + \text{SNR}_1}{\text{SNR}_1}, \frac{1 + \text{SNR}_1}{\phi_k - 1}\right\}, \tag{4.26}$$

with initial value $\hat{\sigma}_{K+1}^2 = 0$, where

$$\phi_k = \left(1 + \frac{\theta\text{SNR}_1}{1 + \hat{\sigma}_{k+1}^2}\right)\left(1 + \frac{(1 - \theta)\text{INR}_2}{1 + \text{SNR}_2 + \theta\text{INR}_2}\right). \tag{4.27}$$

Achievable average rates $(r_1 + r_2)/2$ of the proposed schemes are numerically evaluated for different values of K (Figure 4.6). For QMF, three quantization methods are

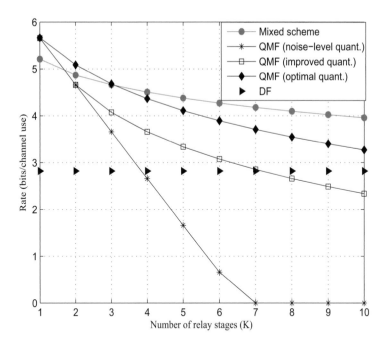

Figure 4.6 The 3-hop full-duplex relay channels where $SNR_1 = 10$, $SNR_2 = 20$ dB, and $INR_1 = INR_2 = 5$ dB.

considered: noise-level quantization [21], improved quantization [28], and the proposed optimal quantization. From Figure 4.6, we observe that the mixed scheme outperforms the QMF-only schemes especially for $K > 3$, since the mixed scheme reduces the impact of propagation of quantization noise thereby improving capacity scaling. This demonstrates that the mixed scheme may be the preferred choice when considering multihop transmission and operating at finite SNR.

Per-Hop Symmetric Channel

In this channel, each hop has the same channel gain. Let SNR_k denote the direct channel gain from $R_{i,k-1}$ to $R_{i,k}$ for all i, and INR_k denote the interference channel gain from $R_{\bar{i},k}$ to $R_{i,k}$ for all i. This scenario can be related to the wireless backhaul network shown in Figure 4.1, where two half-duplex relays operate as a distributed version of a full-duplex relay. Due to the symmetric channel of each stage, we assume that each stage employs the same relaying scheme, i.e., $\mathcal{V}_1 = \mathcal{V}_2 = \mathcal{V} = \{k_1, \ldots, k_{|\mathcal{V}|}\}$ and also assume that $r_{1,k} = r_{2,k}$ for all k. From Definition 4.1, we obtain:

$$I_k = \log\left(1 + \frac{SNR_{k+1}}{1 + (1 - \theta_{k+1})INR_{k+1} + \hat{\sigma}_{k+1}^2}\right) \tag{4.28}$$

$$I_{k1} = \log\left(1 + \frac{(1 - \theta_k)SNR_{k+1}}{1 + (1 - \theta_{k+1})INR_{k+1} + \hat{\sigma}_{k+1}^2}\right), \tag{4.29}$$

where $\theta_{k+1} = 1$ if $k + 1 \in \mathcal{V}$ and $\hat{\sigma}_{k+1}^2 = 0$ if $k + 1 \in \mathcal{V}^c$. In this section, we focus on the symmetric achievable rate of r with $r = r_1 = r_2$. Then, we obtain:

COROLLARY 4.9   *For a $(K + 1)$-hop Gaussian virtual full-duplex relay channel, the achievable symmetric rate of the mixed scheme with SD (or JD) is given by*

$$r = \min\{\min\{I_k : k \in \mathcal{I}_0\}, \min\{I_k' : k \in \mathcal{I}_0 \setminus \{0\}\}\} \tag{4.30}$$

$$r_{k_\ell} = \min\{\min\{I_k : k \in \mathcal{I}_\ell\}, \min\{I_k' : k \in \mathcal{I}_\ell \setminus \{k_\ell\}\}\} \tag{4.31}$$

$$\hat{\sigma}_{k_\ell}^2 = (1 + SNR_{k_\ell})/(2^{r_{k_\ell}} - 1), \ell = 1, \ldots, |\mathcal{V}|, \tag{4.32}$$

for any subset $\mathcal{V} \subseteq \{1, \ldots, K\}$ and any $\theta_k \in [0, 1]$ with $\theta_k = 1$ for $k \in \mathcal{V}$, where

$$I_k' = \begin{cases} \log\left(1 + \frac{\theta_{k+1} INR_{k+1}}{1 + SNR_{k+1}}\right) + I_{k1}, & SD \\ \frac{1}{2}\log\left(1 + \frac{SNR_{k+1} + \theta_{k+1} INR_{k+1}}{1 + (1 - \theta_{k+1}) INR_{k+1}}\right) + \frac{1}{2} I_{k1}, & JD. \end{cases} \tag{4.33}$$

In Corollary 4.9, we need to find an optimal relay configuration among 2^K possible configurations, with respect to maximizing a symmetric rate. To achieve this, we present a simple greedy algorithm (Algorithm 4.1). The optimality of this algorithm is proved in [37]. In the algorithm, given the relay modes at stages $K, K - 1, \ldots, k + 1$, we determine the relay mode at stage k. According to the relay mode at stage k, the message of the relay at stage $k - 1$ should be decoded at either the relay at stage k or the destination. Since the relay modes at stages $K, K - 1, \ldots, k + 1$ have already been determined, we can compute the rate-constraint such that the message is decoded at the next-hop relay, denoted by $r_{DF,k}$, where the optimal power allocation (i.e., rate-splitting) is applied. We can also compute the rate constraint such that the message is decoded at the destination, denoted by $r_{QMF,k}$. With these definitions, the greedy algorithm proceeds as shown in Algorithm 4.1.

Algorithm 4.1 Relay Mode Selection

Determine the mode of each stage k for $k = K, K - 1, \ldots, 1$, in descending order.

- Given the relay modes at stages $K, K - 1, \ldots, k + 1$, determine the relay mode at stage k such as

$$\begin{cases} QMF, & \text{if } r_{QMF,k} > r_{DF,k} \\ DF, & \text{otherwise.} \end{cases} \tag{4.34}$$

- Repeatedly perform the above procedure until all relay modes are chosen.

The achievable symmetric rate of the presented scheme is numerically evaluated for different values of K and shown in Figures 4.7 and 4.8. The optimal relay configuration is obtained from Algorithm 4.1. Results show that the mixed scheme outperforms QMF schemes with various quantization methods presented in [19, 21, 28]. Furthermore, Figure 4.7 shows that JD can improve the performance compared to SD. This gain becomes larger in strong inter-relay interference since, in weak interference, interference can be treated as noise.

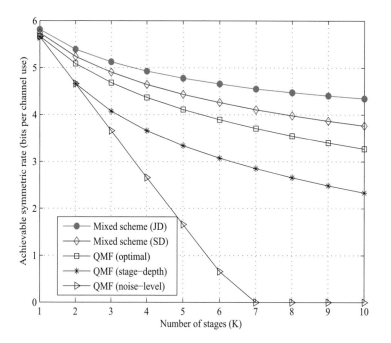

Figure 4.7 SNR $= 20\,$dB and INR$_k = $ SNR$^{\alpha_k}$ where $\alpha_k \sim$ Unif$[1, 2]$.

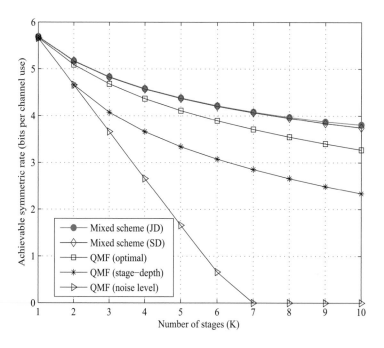

Figure 4.8 SNR $= 20$dB and INR$_k = $ SNR$^{\alpha_k}$ where $\alpha_k \sim$ Unif$[0, 1]$.

4.3   Wireless Backhaul Networks

We next consider a general half-duplex multihop relay network in which L sources wish to send their messages to a destination with the help of half-duplex relays. As explained in Section 4.1, the presented communication scheme consists of three parts (Figure 4.3): (1) a routing algorithm that establishes a $(K + 1)$-hop $2L$-layer network; (2) a relay scheduling scheme referred to as *group successive relaying* that assigns the transmit/receive mode to each relay for each time slot – such scheduling maximizes the number of messages simultaneously transmitted/received, because it enables destination to receive L messages in every time slot; and (3) a cooperative relaying scheme referred to as SNNC-SW that achieves a high data rate.

4.3.1   Routing (Relay Selection)

A routing algorithm establishes routes for L sources, as shown in Figure 4.3. An important difference from classical routing is that each relay at any route will simultaneously forward information of multiple sources by using SNNC-SW. As shown in Figure 4.9, the number of hops (or relay stages K) of each route is chosen according to the communication range and source–destination distance. For ease of explanation, it is assumed

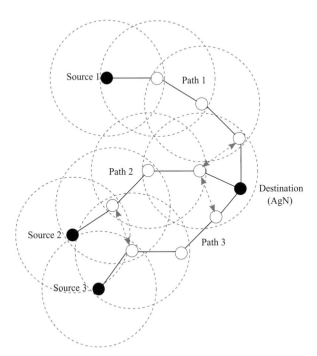

Figure 4.9 Routes obtained via a routing algorithm in a network with multiple sources and a single destination. A dashed circle represents a communication range whose radius can be determined as a function of a transmit power. Dashed lines indicate strong interference. Any signal outside the communication range is relatively small and captured as additive Gaussian noise.

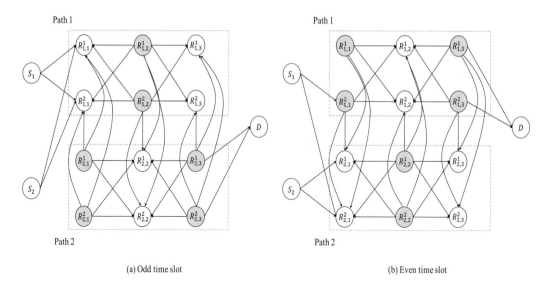

Figure 4.10 Group successive relaying for layered half-duplex relay networks when $L = 2$ and $K = 3$. In each time slot, relays that transmit are shown in gray circles and relays that receive are shown in white circles. Sources transmit and the destination receives in every time slot. Each path contains L routes for L sources.

that each route is composed of K relay stages (i.e., $(K+1)$ hops). However, the presented scheme can be extended into an *asymmetric* layered network where the routes from different sources have different numbers of hops due to the various source–destination distances. This case will be discussed in Section 4.4. As in the previous section, each stage of a route contains two relays per source in order to overcome the half-duplex constraint. These relays are again used in alternated transmit and receive modes, such that while one relay transmits its signal to the next stage, the other relay receives a signal from the previous stage. The routing algorithm establishes two routes per source. Therefore, a $(K+1)$-hop $2L$-layer network is produced as shown in Figure 4.10 for $L = 2$ and $K = 3$.

As a result of a routing algorithm, we obtain a layered network with associated capacity (equivalently, a channel transition probability). We then develop a transmission scheme in Sections 4.3.2 and 4.3.3 to achieve this capacity. Therefore, we need to optimize a routing algorithm so that the capacity of the layered network is maximized. This problem will be addressed in Section 4.3.5. In Section 4.3.3, an achievable rate region of the SNNC-SW scheme is derived as a function of channel transition probability, that is, for any layered network.

4.3.2   Group Successive Relaying

In group successive relaying, relays are divided into two groups depending on their transmit/receive mode such that

$$\mathcal{G}_1 = \left\{ \{R^j_{1,2k}\}^{K_1}_{k=1}, \{R^j_{2,2k-1}\}^{K_2}_{k=1} : j = 1, \dots, L \right\} \tag{4.35}$$

$$\mathcal{G}_2 = \left\{ \{R^j_{1,2k-1}\}^{K_2}_{k=1}, \{R^j_{2,2k}\}^{K_1}_{k=1} : j = 1, \ldots, L \right\}, \tag{4.36}$$

where $R^j_{i,k}$ denotes the relay at stage k on the ith route of source j, and $K_1 = \lfloor (K-1)/2 \rfloor$ and $K_2 = \lceil (K-1)/2 \rceil$. At odd (resp. even) time slot (consisting of n channel uses), the relays in $\mathcal{G}_1$ (resp. $\mathcal{G}_2$) operate in transmit mode and the relays in $\mathcal{G}_2$ (resp. $\mathcal{G}_1$) operate in receive mode. In this way, at every time slot t, each source j transmits $\underline{w}^j_t \in \{1, \ldots, 2^{nr_{j,i}}\}$ at rate $r_{j,i}$ and the destination decodes L new messages $(\underline{w}^1_{t-K}, \ldots, \underline{w}^L_{t-K})$, where $i = 1$ for odd time slot t and $i = 2$ for even time slot t. We used two different rates $r_{j,1}$ and $r_{j,2}$ per source since the odd-indexed and even-indexed messages are conveyed to the destination via two disjoint paths:

$$\text{path 1:} \left((S_1, \ldots, S_L), (R^1_{1,1}, \ldots, R^1_{1,K}), \ldots, (R^L_{1,1}, \ldots, R^L_{1,K}), D \right)$$

$$\text{path 2:} \left((S_1, \ldots, S_L), (R^1_{2,1}, \ldots, R^1_{2,K}), \ldots, (R^L_{2,1}, \ldots, R^L_{2,K}), D \right),$$

where we note that each path contains L routes for L sources. As shown in Figure 4.10, it is assumed that each transmit signal is received only at the neighboring relays in the received mode, namely, relays at the same stage on the other path and relays at the previous and next stages on the same path. This assumption may be reasonable since the strengths of signals from the outside of neighboring relays are relatively low, thereby being able to be captured by additive Gaussian noise (see Figure 4.9).

Incorporating the group successive relaying into a channel, the discrete memoryless channel (DMC) is described by transition probabilities as

$$\prod_{j=1}^{L} \prod_{k=1}^{\lceil K/2 \rceil} p(y^j_{i,2k-1} | x^j_{i,2k-2}, x^1_{i,2k-1}, \ldots, x^L_{i,2k-1}, x^1_{i,2k}, \ldots, x^L_{i,2k})$$

$$\times \prod_{k=1}^{\lfloor K/2 \rfloor} p(y^j_{i,2k} | x^j_{i,2k-1}, x^1_{i,2k}, \ldots, x^L_{i,2k}, x^1_{i,2k+1}, \ldots, x^L_{i,2k+1}) \times p(y_D | x^1_{i,K}, \ldots, x^L_{i,K}), \tag{4.37}$$

where $i = 1$ for an odd time slot and $i = 2$ for an even time slot, and where $x^j_{i,k}$ and $y^j_{i,k}$ denote respective input and output at relay $R^j_{i,k}$, and $x^j_{i,0}$ denotes source j's inputs.

4.3.3 SNNC-SW

We next assume that group successive relaying is used for relay scheduling and present a coding scheme for a layered network. Accordingly, the channel model in (4.37) is considered. The *time-expanded* network model will be used to consider decoding of QF-based schemes in order to deal with cycles in the original network [21]. Figure 4.11 shows a *time-expanded* network for the case of $L = 2$ and $K = 2$. Before giving a detailed explanation of proposed SNNC-SW, we compare it with other QF-based schemes, i.e., QMF, NNC, and SNNC.

1. QMF and NNC in [21, 22] consist of message-repetition encoding (i.e., one long message sent via repetitive encoding), signal quantization at relay, and simultaneous

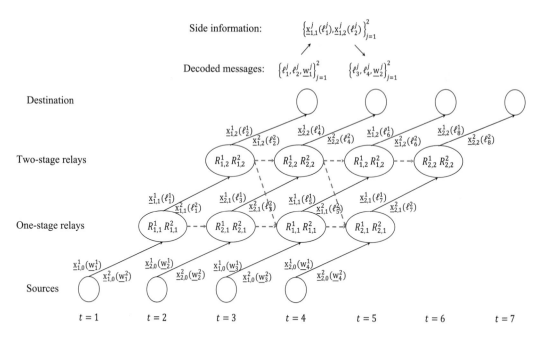

Figure 4.11 Time-expanded multihop network with $L = 2$ and $K = 2$ where relays in transmit mode are shown at each time slot. Dashed lines stand for the "known" interferences at the destination and each arrow actually represents the wireless MIMO channel with two inputs and two outputs.

joint decoding of the received signals from all time slots (say, $t = 1, 2, \ldots, T$) without explicitly decoding relays' messages (i.e., quantization indices).

2. SNNC in [23] overcomes the long delay of NNC by transmitting many short messages in time slots rather than using one long message with repetitive encoding. Instead of simultaneous joint decoding, the destination performs *backward* decoding where, after T time slots, the destination starts decoding the source message $\{\underline{\mathbf{w}}_t^\ell\}_{\ell=1}^L$ and relays' messages in the slot t, for $t = T, T-1, \ldots, 1$, in descending order. Here, joint decoding is separately performed for each time slot, instead of jointly over all time slots as in QMF and NNC.

3. SNNC-SW follows the encoding and relaying operation of SNNC but the destination performs *sliding window decoding* instead of backward decoding, where, after t time slots, the destination can decode the source message $\{\underline{\mathbf{w}}_t^\ell\}_{\ell=1}^L$ without waiting for the whole T time slots. Thus, SNNC-SW has a lower decoding delay than SNNC in [23]. The SW decoding proceeds as follows. The time-expanded network is partitioned into T subnetworks, each of which is denoted by $\mathcal{S}_t$ for $t = 1, \ldots, T$, as shown in Figure 4.11. After time slot $K + 1$, the destination decodes the messages (i.e., source message $\{\underline{\mathbf{w}}_1^\ell\}_{\ell=1}^L$ and relays' messages) associated with subnetwork $\mathcal{S}_1$, and then, after time slot $K + 2$, decodes the messages associated with subnetwork $\mathcal{S}_2$ by exploiting the previously decoded messages as side information. Due to the use of group successive relaying, the destination completely knows the interference seen at relays (depicted

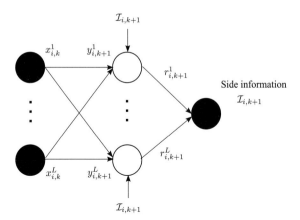

Figure 4.12 Multiple access relay channel (MARC) consisting of L sources, L intermediate relays, and one destination. Here, each relay is connected to the destination via noiseless digital link.

by arrows in Figure 4.11) since they are fully determined by the previously decoded relays' messages. In this way, the destination can decode all source messages $\underline{\mathbf{w}}_t$ for $t = 1, \ldots, T$, in that order.

For ease of exposition, we introduce notation $\mathcal{I}_{i,K-1} = (\underline{\mathbf{x}}^1_{i,K-1}, \ldots, \underline{\mathbf{x}}^L_{i,K-1}, \underline{\mathbf{x}}'^1_{i,K}, \ldots, \underline{\mathbf{x}}'^L_{i,K})$ that denotes known interference (i.e., side information) at the destination. Namely, the destination completely knows $(\mathcal{I}_{i,1}, \ldots, \mathcal{I}_{i,K})$ when decoding source messages conveyed via path i.

Before presenting the achievable rate of SNNC-SW, we provide the following useful lemma and definition.

LEMMA 4.10 *Consider the multiple access relay channel (MARC) in Figure 4.12, defined as*

$$\left(\prod_{j=1}^{L} p(y^j_{i,k+1}|x^1_{i,k}, \ldots, x^L_{i,k}, \mathcal{I}_{i,k+1}), (r^1_{i,k+1}, \ldots, r^L_{i,k+1}) \right),$$

where the first part represents a transition probability of the wireless channel induced by sources and relays, and the second part represents the noiseless wired links with given rates $(r^1_{i,k+1}, \ldots, r^L_{i,k+1})$ between relays and destination. When $\mathcal{I}_{i,k+1}$ is known to the destination, a rate tuple $(r^1_{i,k}, \ldots, r^L_{i,k})$ is achievable if for all $\mathcal{L} \subseteq \{0, \ldots, L-1\}$,

$$\sum_{\ell \in \mathcal{L}} r^\ell_{i,k} \leq \min_{\mathcal{S} \subseteq [1:L]} \sum_{j \in \mathcal{S}} \left[r^j_{i,k+1} - I\left(Y^j_{i,k+1}; \hat{Y}^j_{i,k+1} | X^1_{i,k}, \ldots, X^L_{i,k}, \mathcal{I}_{i,k+1} \right) \right]$$
$$+ I\left((X^\ell_{i,k} : \ell \in \mathcal{L}); (\hat{Y}^j_{i,k+1} : j \in \mathcal{S}^c) | (X^j_{i,k} : j \in \mathcal{L}^c), \mathcal{I}_{i,k+1} \right),$$

for any input distribution that factors into

$$\prod_{j=1}^{L} p(x^j_{i,k}) \prod_{j=1}^{L} p(\hat{y}^j_{i,k}|y^j_{i,k}).$$

DEFINITION 4.11   *We denote the rate region of Lemma 4.10 by*

$$\mathcal{R}\Big(\prod_{j=1}^{L} p(y_{i,k+1}^{j}|x_{i,k}^{1},\ldots,x_{i,k}^{L},\mathcal{I}_{k+1}),(r_{i,k+1}^{1},\ldots,r_{i,k+1}^{L})\Big).$$

We can now obtain the achievable rate of SNNC-SW:

THEOREM 4.12   *For the layered half-duplex relay network in Figure 4.10, a rate tuple* $(r_{1}^{1},\ldots,r_{1}^{L},r_{2}^{1},\ldots,r_{2}^{L})$ *is achievable if*

$$(r_{i,k}^{1},\ldots,r_{i,k}^{L})\in$$

$$\mathcal{R}\Big(\prod_{j=1}^{L} p(y_{i,k+1}^{j}|x_{i,k}^{1},\ldots,x_{i,k}^{L},\mathcal{I}_{i,k+1}),(r_{i,k+1}^{1},\ldots,r_{i,k+1}^{L})\Big),$$

for $k = K, K-1,\ldots,0$, *with initial value* $r_{i,K+1}^{j} = \infty$, $j = 1,\ldots,L$, *for any input distribution defined as*

$$\prod_{i=1}^{2}\prod_{k=0}^{K}\prod_{j=1}^{L} p(x_{i,k}^{j})p(\hat{y}_{i,k}^{j}|y_{i,k}^{j}). \tag{4.38}$$

The proof can be found in [29].

In Theorem 4.12, an achievable rate region of the presented scheme was derived for any given layered half-duplex relay network. Since the layered half-duplex network is constructed as a consequence of the routing algorithm, the rate region can be further increased by optimizing a routing algorithm. That is, to maximize a sum-rate or symmetric user rate, we need to jointly optimize a routing algorithm and relays' transmission rates $\{r_{i,k}^{\ell}\}$ subject to their achievable rate region in Theorem 4.12. This is a nontrivial combinatorial optimization problem. This problem is partially solved in Sections 4.3.4 and 4.3.5. In Section 4.3.4, relays' transmission rates are optimized for some interesting networks, namely dense and sparse networks. The obtained solution provides a useful intuition to optimize a routing algorithm. Based on this, in Section 4.3.5 a routing algorithm suitable for the proposed SNNC-SW scheme is presented.

4.3.4   Gaussian Networks

We consider a Gaussian layered half-duplex relay network where destination is equipped with L receiver antennas and, accordingly, the L-dimensional observations are denoted by $(\underline{y}_{i,K+1}^{1},\ldots,\underline{y}_{i,K+1}^{L})$. Let $\mathbf{H}_{i,k}$ denote the $L \times L$ channel matrix with inputs $(\underline{x}_{i,k}^{1},\ldots,\underline{x}_{i,k}^{L})$ and outputs $(\underline{y}_{i,k+1}^{1},\ldots,\underline{y}_{i,k+1}^{L})$ for $k = 0, 1,\ldots,K$. Let $\mathbf{G}_{i,k}$ denote the $L \times L$ *intrapath interference* channel matrix with inputs $(\underline{x}_{i,k+1}^{1},\ldots,\underline{x}_{i,k+1}^{L})$ and outputs $(\underline{y}_{i,k}^{1},\ldots,\underline{y}_{i,k}^{L})$ for $k = 1,\ldots,K-1$. Also, let $\mathbf{S}_{i,k}$ denote the $L \times L$ *interpath interference* channel matrix with inputs $(\underline{x}_{i,k}^{1},\ldots,\underline{x}_{i,k}^{L})$ and outputs $(\underline{y}_{i,k}^{1},\ldots,\underline{y}_{i,k}^{L})$ for $k = 1,\ldots,K$. From Theorem 4.12, we can see that the performance of the proposed scheme is independent from the intrapath and interpath interference matrices $\mathbf{G}_{i,k}$ and $\mathbf{S}_{i,k}$. For the Gaussian channel, Lemma 4.10 is simplified as

LEMMA 4.13   *For the Gaussian MARC defined by* $(\mathbf{H}_{i,k}, (r_{i,k+1}^1, \ldots, r_{i,k+1}^L))$, *a rate tuple* $(r_{i,k}^1, \ldots, r_{i,k}^L)$ *is achievable if for all* $\mathcal{L} \subseteq \{0, \ldots, L-1\}$,

$$\sum_{\ell \in \mathcal{L}} r_{i,k}^\ell \leq \min_{\mathcal{S} \subseteq [1:L]} \sum_{j \in \mathcal{S}} \left[r_{i,k+1}^j - \log\left(1 + 1/Q_{i,k}^j\right) \right]$$

$$+ \log \det\left(\mathbf{I} + \mathbf{D}_{i,k}\mathbf{H}_{i,k}(\mathcal{S}^c)\mathbf{H}_{i,k}(\mathcal{S}^c)^H\right),$$

for some quantization level $Q_{i,k}^j \geq 0$, *where* $\mathbf{H}_{i,k}(\mathcal{S})$ *represents the channel submatrix to contain the rows of* $\mathbf{H}_{i,k}$ *with their indices belong to* $\mathcal{S} \subseteq [1:L]$, $\mathbf{D}_{i,k}$ *represents a diagonal matrix whose jth diagonal element is* $P_{tx}/(1+Q_{i,k}^j)$, *and* P_{tx} *denotes a transmit power at each node.*

We evaluate the performance of the scheme for two extreme scenarios of network structure called *sparse* and *dense* networks. They are classified according to the structure of $\mathbf{H}_{i,k}$. In a sparse network, the number of interfering nodes is constant as the number of nodes L grows (i.e., $\mathbf{H}_{i,k}$ forms a band matrix) while in a dense network it increases linearly with L (i.e., $\mathbf{H}_{i,k}$ forms a "full" matrix). Although we limit our attention to Gaussian networks, the computation of the achievable rate in Lemma 4.13 is generally difficult since it involves a complicated combinatorial optimization. In order to make a problem manageable, we consider the particular structures of $\mathbf{H}_{i,k}$ that capture the features of sparse and dense networks. They are described in the following. Further, we let $L \rightarrow \infty$ (i.e., asymptotic analysis) and focus on the achievable symmetric rate of r.

Sparse Networks

In this section, it is assumed that each transmit signal is received at its desired relay and two neighboring relays, i.e., $\mathbf{H}_{i,k}$ has the form of a tri-diagonal structure with $(\alpha, 1, \alpha)$, where α denotes the relative interference strength. Namely, other interferences are relatively weak and are captured by additive Gaussian noise. This model is well known as the Wyner model [38]. Obviously, this model is the form of sparse network in the sense that the number of interfering nodes is fixed to 2, independently from the number of users L. The transmit power at each node is assumed to be $P_{tx} = \text{SNR}$. Hereafter, we restrict ourselves to choose relays' rates as $r_k = r_{i,k}^j$ for all i and j, which is reasonable due to the symmetric structure of $\mathbf{H}_{i,k}$. Then, by symmetry and concavity, this limits the sum-rate inequality to be dominant in Lemma 4.13, and by choosing the same quantization levels at each relay (i.e., $Q_k = Q_{i,k}^j$ for all i and j), we have:

$$r_{k-1} = \frac{1}{L} \min_{\mathcal{S} \subset [1:L]} |\mathcal{S}| \left(r_k - \log\left(1 + \frac{1}{Q_k}\right)\right)$$

$$+ \log \det\left(\mathbf{I} + \left(\frac{P_{tx}}{1+Q_k}\right) \mathbf{H}_{i,k}(\mathcal{S}^c)\mathbf{H}_{i,k}^H(\mathcal{S}^c)\right). \tag{4.39}$$

A remarkable result of [39] is that in the limit of $L \rightarrow \infty$, (4.39) can be simplified to (4.40). Consequently, we get the following:

COROLLARY 4.14   *The proposed scheme achieves the symmetric rate (per user)*
r = r_0 to satisfy:

$$r_{k-1} = F(x^*), \tag{4.40}$$

for k = K + 1, ..., 1, where

$$F(x) = \int_0^1 \log\left(1 + \mathrm{SNR}(1 - 2^{-x})(1 + 2\gamma \cos(2\pi\theta))^2)\right) d\theta, \tag{4.41}$$

and x^* is the solution of the equation $F(x) = r_k - x$ with the initial value $r_{K+1} = \infty$.

For the comparison, we consider the multihop relaying (MR) where each relay
decodes its desired message (forwarded by its own route) by treating all other signals as
noise. In order to derive its achievable rate, we assume that $\mathbf{G}_{i,k} = 0$ and $\mathbf{S}_{i,k} = 0$. Thus,
the obtained achievable rate of the MR is optimistic since, in practice, these may not be
zero matrices. Under this assumption, the achievable symmetric rate of the MR is given
by

$$r_{\mathrm{MR}} = \log\left(1 + \frac{\mathrm{SNR}}{1 + 2\alpha^2 \mathrm{SNR}}\right). \tag{4.42}$$

Figure 4.13 shows the achievable symmetric rate of our proposed scheme for various
quantization levels at relays. We first observe that using optimal quantization outperforms the other cases as background noise-level (i.e., $Q_k = 1$) [21], stage-depth (i.e.,

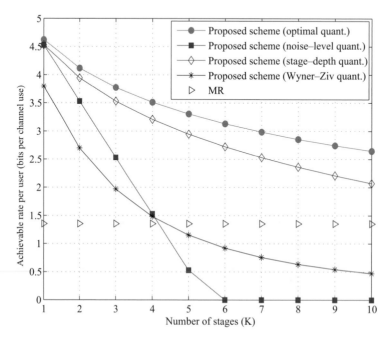

Figure 4.13 SNR = 20 dB and INR = 15 dB (e.g., $\alpha = \sqrt{\mathrm{INR}/\mathrm{SNR}} = 0.56$). Performance of the
proposed scheme for various quantization levels as a function of the number of stages K.

$Q_k = K$) [28], and Wyner–Ziv quantization (i.e., $Q_k = \frac{(1+2\alpha^2)\text{SNR}}{2^{r_k}-1}$) [40]. As anticipated, the optimal and stage-depth quantization provide a larger performance gain due to the improved rate-scaling. When $K \geq 4$, the noise-level and Wyner–Ziv quantization show worse performance than the MR and hence optimizing a quantization level plays an important role in improving performance of multihop networks.

Dense Networks

In this section it is assumed that $\mathbf{H}_{i,k}$ is an i.i.d. matrix whose entries have a complex Gaussian distribution with zero mean and unit variance. Note that the transmit power at each node is assumed to be $P_{\text{tx}} = \text{SNR}/L$. If we choose $P_{\text{tx}} = \text{SNR}$ as in the sparse network, the received power at each relay goes to infinity, thereby yielding an infinite rate. As done in the previous section, we limit our attention to the symmetric choice of $r_k = r_{i,k}^j$ for all j and i. With the same argument as there, the sum-rate inequality in Lemma 4.13 is dominant and is represented as in (4.39) with $P_{\text{tx}} = \text{SNR}/L$. A closed-form expression of (4.39) has been derived in [20] by using the fact that as $L \to \infty$, the problem becomes symmetric although the channel is "full" and non-tri-diagonal as in the Wyner model. The closed-form expression was obtained from the asymptotic random matrix theory and the submodular structure of the rate expression, which is equal to (4.43). Consequently, we get the following:

COROLLARY 4.15   *The proposed scheme achieves the symmetric rate (per user)* $r = r_0$ *to satisfy:*

$$r_{k-1} = \min\left\{ r_k - \log\left(1 + \frac{1}{Q_k}\right), \mathcal{C}\left(\frac{\text{SNR}}{1+Q_k}\right)\right\}, \tag{4.43}$$

for $k = K+1, \ldots, 1$, *with initial value* $Q_{K+1} = 0$ *and* $r_{K+1} = \infty$, *where*

$$\mathcal{C}(x) = 2\log\left(\frac{1+\sqrt{1+4x}}{2}\right) - \frac{\log e}{4x}(\sqrt{1+4x}-1)^2. \tag{4.44}$$

Since the achievable rate in (4.43) is the minimum of two terms, where the first is an increasing function of Q_k and the second is a decreasing function of Q_k, the optimal value of Q_k is attained by solving

$$r_k - \log(1 + 1/Q_k) = \mathcal{C}(\text{SNR}/(1+Q_k)). \tag{4.45}$$

By letting

$$f(Q_k) = r_k - \log(1 + 1/Q_k) - \mathcal{C}(\text{SNR}/(1+Q_k)), \tag{4.46}$$

we can find $Q_{k,\min} = 1/(2^{r_k}-1)$ and $Q_{k,\max} = (1+\text{SNR})/(2^{r_k}-1)$ such that $f(Q_{k,\min}) \leq 0$ and $f(Q_{k,\max}) \geq 0$. This is because $Q_{k,\min}$ makes the first term in the minimum in (4.43) zero and $Q_{k,\max}$ is the Wyner–Ziv quantization, which makes the second term attain the minimum in (4.43). Using the bisection method, we can quickly find an optimal quantization level $Q_{k,\text{opt}}$. This value will be used to plot the performance of the proposed scheme.

Figure 4.14 shows the achievable symmetric rate of the presented scheme for various quantization levels. The performance trend is similar as in the case of a sparse network.

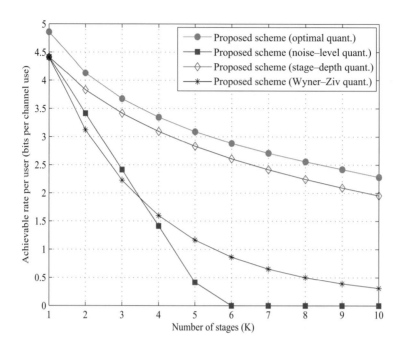

Figure 4.14 LSNR $= 20$ dB. Performance of the proposed scheme for various quantization levels as a function of the number of stages K. Notice that the performance of the MR goes to zero as $L \to \infty$ due to the impact of severe interference.

The most remarkable observation is that the proposed scheme can achieve almost the same performance as in Figure 4.13 with a lower transmit power (per node) as $L \to \infty$.

4.3.5   Interference-Harnessing Routing

The presented transmission scheme can be applied together with any multihop routing algorithm (e.g., interference-aware routing [5]). In this section, we provide a routing criterion that further improves the performance of the scheme. From Figures 4.13 and 4.14, we observe that the presented scheme can achieve almost the same performance in both dense and sparse networks. In dense networks, however, the transmit power consumption is reduced proportionally to the number of users L (i.e., $P_{tx} = \text{SNR}/L$). In other words, the presented scheme has a *higher power efficiency* achieving a desired rate performance with a lower transmission power as the network becomes denser. The improved power efficiency comes from enabling each relay (by using SNNC-SW) to collect all signals sent by transmitters. Any interfering signal that is received at the relay will be forwarded through SNNC-SW and treated as a useful signal at the destination. For this reason, SNNC-SW performs better when the network is dense and interference is stronger. This motivates *interference-harnessing* routing in which the routing criterion contrasts the routing criterion used in the interference-aware routing. Interference-harnessing routing refers to a family of routing algorithms in which the metric is chosen to exploit interference. This can be done by, for example, choosing the metric that maximizes an achievable rate between every two consecutive relay stages. For any choice of relays at the two stages, this rate corresponds to an achievable rate for a MIMO channel (see Section 4.3.6) and can thus be calculated, as we will specify in this section. The performance gain of the interference-harnessing algorithm compared to the MR is demonstrated via a simulation result in Section 4.3.6.

For interference-harnessing routing, one efficient algorithm can be developed using the *iterative algorithm* in [5] by properly modifying the routing criterion. In [5], the algorithm establishes one route at a time while keeping other previously established routes fixed, and repeats the process until negligible improvements in the sum through-put can be made. The routing criterion maximizes each link-capacity on the route, which is computed by taking into account interference from all other routes. For the interference-harnessing routing, on the other hand, we select a relay (at stage k) to max-imize the MIMO capacity defined by two consecutive stages $k - 1$ and k. We provide a description of this algorithm for the network with two sources using Figure 4.15. Generalization to the case with more sources is straightforward. We use the notation $\mathcal{C}_{\mathrm{MIMO}}(\{R_1, R_2\}, \{R_3, R_4\})$ to denote the MIMO capacity induced by two transmitters $\{R_1, R_2\}$ and two receivers $\{R_3, R_4\}$. Without loss of generality, we focus on path 1 (i.e., $i = 1$ in Figure 4.15).

The interference-harnessing routing algorithm proceeds as follows:

1. **(Initial route)** Establish the route from S_1 to D so that the number of hops is min-imized and each link-capacity along the route is maximized (namely, the received power is maximized). This process closely follows the step that establishes the initial route in [5].

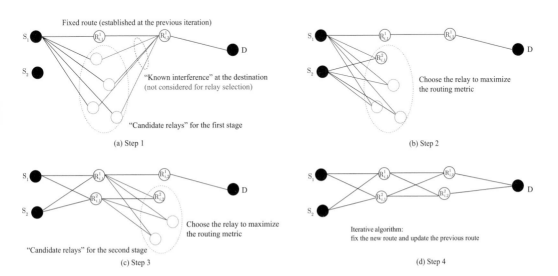

Figure 4.15 Illustration of an interference-harnessing routing to establish the two routes in path i where "candidate relays" denotes the relays within the communication range from a transmitter.

2. For the fixed route from S_1 to D, establish a new route from S_2 to D, as illustrated by the steps in Figure 4.15 and described below:

    a. Find the relays within a communication range from S_2 that have not already been chosen for other routes. The communication range is a design parameter to be determined as a function of a transmit power where the transmit power should be larger than a threshold to satisfy network connectivity [41]. Let $\mathcal{T}_1 = \{R_{1,1}, \ldots, R_{1,|\mathcal{T}_1|}\}$ denote the set of such relays (i.e., "candidate relays" of the first stage). For the jth relay in $\mathcal{T}_1$, we can compute the MIMO capacity (i.e., routing metric) $\mathcal{C}_{\mathrm{MIMO}}(\{S_1, S_2\}, \{R^1_{1,1}, R_{1,j}\})$ where $R^1_{1,1}$ denotes the first relay on the route from S_1 to D. With S_1, S_2, and $R^1_{1,1}$ fixed, we choose a relay $R_{1,j}$ to maximize the MIMO capacity:

$$j_1^\star = \mathrm{argmax}_{j\in[1:|\mathcal{T}_1|]}\mathcal{C}_{\mathrm{MIMO}}(\{S_1, S_2\}, \{R^1_{1,1}, R_{1,j}\}).$$

    b. Let $R^2_{1,1} = R_{1,j_1^\star}$. Find the candidate relays of the second stage, denoted by $\mathcal{T}_2 = \{R_{2,1}, \ldots, R_{2,|\mathcal{T}_2|}\}$. Then choose a relay $R_{2,j}$ to maximize the g metric:

$$j_2^\star = \mathrm{argmax}_{j\in[1:|\mathcal{T}_2|]}\mathcal{C}_{\mathrm{MIMO}}(\{R^1_{1,1}, R^2_{1,1}\}, \{R^1_{1,2}, R_{2,j}\}),$$

    where $R^1_{1,2}$ denotes the second relay on the route from S_1 to D.

    c. By setting $R^2_{1,2} = R_{2,j_2^\star}$, we can establish the route from S_2 to D.

3. For the fixed route from S_2 to D, update the route from S_1 to D following the above procedures in a–c.

4. Repeat process 2 and 3 until negligible improvements in the sum throughput can be made.

Note that since the performance of the proposed scheme is independent of the interpath interference, as explained in Section 4.3.3, we first apply the above procedure to find path 1 and then apply the procedure to find path 2 using only the relays not included in path 1.

4.3.6   Beyond Information-Theoretical Results

We focus on the practical implementation of the presented scheme and provide its numerical results for a finite network size.

Practical Construction

It was shown in Section 4.3.4 that the presented scheme outperforms MR. Recall that, in the presented scheme, destination must perform joint decoding to decode relays' messages as well as sources' messages. Joint decoding in this case is performed separately for each stage (i.e., window), instead of over entire messages as in [21, 22]. Therefore, the presented scheme has lower complexity. Nevertheless, the joint decoding typically requires high complexity, presenting an obstacle in designing a practical coding method. Although some progress towards practical joint decoding has been made in [42, 43] for small one-relay networks, extending theses previous works to the considered multihop network is a difficult problem.

In this section, we develop a low-complexity decoder referred to as *stage-by-stage successive MIMO decoder* where conventional MIMO decoding is applied to each stage (i.e., window) instead of joint decoding. First of all, we avoid the joint decoder at the destination by using the Wyner–Ziv quantization in which each relay $\mathsf{R}^j_{i,k}$ chooses the quantization level such as

$$Q^j_{i,k} = \frac{1 + \sum_{\ell=1}^L |h^{j,\ell}_{i,k}|^2 \mathrm{SNR}}{2^{r^j_{i,k}} - 1}, \tag{4.47}$$

where $h^{j,\ell}_{i,k}$ denotes the (j, ℓ)-th element of $\mathbf{H}_{i,k}$ and $r^j_{i,k}$ denotes the transmission rate of relay $\mathsf{R}^j_{i,k}$. As shown in [40], this particular choice of quantization level ensures that the destination can find a *unique* quantization sequence from the decoded relay's message $\ell^j_{i,k}$ and side information, and therefore we can avoid joint decoding over all quantized sequences associated with $\ell^j_{i,k}$.

Using the above fact, the destination performs the stage-by-stage successive MIMO decoding, i.e., the following procedures are repeated for $k = K + 1, K, \ldots, 1$, in that order:

MIMO decoding at the stage k: The destination has the side information $\mathcal{I}_{i,k}$ and $\{\ell^j_{i,k}\}_{j=1}^L$ (i.e., decoded relays' messages at stage $k + 1$).

- Using the side information $\{\ell^j_{i,k}\}_{j=1}^L$ and $\mathcal{I}_{i,k}$, the destination can find *unique* quantized sequences $\{\hat{\underline{y}}^j_{i,k}\}_{j=1}^L$:

$$
\hat{\underline{y}}^j_{i,k} = \sum_{\ell=1}^L h^{j,\ell}_{i,k-1} \underline{x}^\ell_{i,k-1}
$$

$$
+ \underbrace{\sum_{\ell=1}^L g^{j,\ell}_{i,k} \underline{x}^\ell_{i,k+1} + \sum_{\ell=1}^L s^{j,\ell}_{i,k} \underline{x}^\ell_{i,k}}_{\mathcal{I}_{i,k}} + \hat{\underline{z}}^j_{i,k} + \underline{z}^j_{i,k},
$$

where $\hat{\underline{z}}^j_{i,k}$ denotes an i.i.d. Gaussian random variable $\sim \mathcal{CN}(0, Q^j_{i,k})$ and $\underline{z}^j_{i,k}$ denotes an i.i.d. Gaussian random variable $\sim \mathcal{CN}(0, 1)$.

- After removing the known interference $\mathcal{I}_{i,k}$, the destination can decode relays' messages $(\ell^1_{i,k-1}, \ldots, \ell^L_{i,k-1})$ from the resulting "quantized MIMO MAC" channel, given by

$$
\begin{bmatrix} \tilde{\underline{y}}^1_{i,k} \\ \vdots \\ \tilde{\underline{y}}^L_{i,k} \end{bmatrix} = \mathbf{H}_{i,k} \begin{bmatrix} \underline{x}^1_{i,k-1}(\ell^1_{i,k-1}) \\ \vdots \\ \underline{x}^L_{i,k-1}(\ell^L_{i,k-1}) \end{bmatrix} + \begin{bmatrix} \hat{\underline{z}}^1_{i,k} + \underline{z}^1_{i,k} \\ \vdots \\ \hat{\underline{z}}^L_{i,k} + \underline{z}^L_{i,k} \end{bmatrix},
$$

where $\tilde{\underline{y}}^j_{i,k} = \hat{\underline{y}}^j_{i,k} - \mathcal{I}_{i,k}$. Equivalently, we have the matrix form as

$$\tilde{\mathbf{Y}}_{i,k} = \mathbf{H}_{i,k} \underline{\mathbf{X}}_{i,k-1} + \hat{\mathbf{Z}}_{i,k} + \underline{\mathbf{Z}}_{i,k}, \tag{4.48}$$

where each row of a matrix consists of n-dimensional vector.

Since each stage k consists of the *quantized* MIMO MAC defined in (4.48), many approaches to MIMO decoding can be applied. Clearly, the best performance can be attained by maximum likelihood (ML) decoding, which actually achieves the information-theoretical rates of using Wyner–Ziv quantization in Figures 4.13 and 4.14. The complexity of this approach, however, is exponential in the product of the coding blocklength n and the number of receiver antennas (i.e., L in our case). The complexity of ML decoding can be significantly reduced through the use of sphere decoding algorithms [44, 45]. Rather than processing all the observed signals from the antennas jointly, one simple and widely used approach is to separate out the transmitted data streams using linear equalization and then decode each data stream individually as in zero-forcing (ZF), linear minimum mean-squared error (MMSE), and integer-forcing (IF) receivers [46]. The ZF receiver (aka, decorrelator) inverts the channel matrix so that each data stream can be recovered via a single-user decoder. The MMSE receiver performs the same operation except with a *regularized* channel inverse that accounts for possible noise amplification. Both of these architectures permit the use of powerful point-to-point channel codes (e.g., turbo code, low-density parity check [LDPC] code, and polar code) that can achieve high data rates at practically relevant SNRs. Recently, an IF receiver was presented in [46] where the decoder first eliminates the noise by decoding a linear combination of interfering data streams and then eliminates interference between data streams in the digital domain. This scheme is based on the fact that each data stream is drawn from the same lattice codebook which ensures that any integer combination of codewords is itself a codeword, and thus decodable at high rates. Further, low-complexity coding frameworks based on quadrature amplitude modulation (QAM) modulation and non-binary linear codes haven been proposed in [19, 47].

Next, we attain the performance of the low-complexity scheme with linear receivers. In this case, the receiver (i.e., destination) applies the equalization matrix $\mathbf{B}_{i,k} \in \mathcal{C}^{L \times L}$ to obtain

$$\tilde{\mathbf{Y}}'_{i,k} = \mathbf{B}_{i,k}\underline{\mathbf{H}}_{i,k}\underline{\mathbf{X}}_{i,k} + \mathbf{B}(\hat{\underline{\mathbf{Z}}}_{i,k} + \underline{\mathbf{Z}}_{i,k}), \tag{4.49}$$

where $\mathbf{B}_{i,k}$ will be specifically defined according to linear schemes. From (4.49), we can derive an achievable rate $r^j_{i,k}$ as a function of $\mathbf{B}_{i,k}$:

$$r^j_{i,k} = \log\left(1 + \frac{\text{SNR}\left|\underline{\mathbf{b}}^j_{i,k}\,\mathbf{h}^j_{i,k}\right|^2}{\sum_{\ell=1}^{L}|b^{j,\ell}_{i,k}|^2(1 + Q^\ell_{i,k}) + \text{SNR}\sum_{\ell \neq j}\left|\underline{\mathbf{b}}^j_{i,k}\mathbf{h}^\ell_{i,k}\right|^2}\right),$$

where $\mathbf{h}^j_{i,k}$ denotes the jth column of $\mathbf{H}_{i,k}$ and $\underline{\mathbf{b}}^j_{i,k} = (b^{j,1}_{i,k}, \dots, b^{j,L}_{i,k})$ denotes the jth row of $\mathbf{B}_{i,k}$. With initial value $Q^\ell_{i,K+1} = 0$ for all i and ℓ, we can derive all relays' rates at stage k for $k = K, \dots, 1$ and sources' rates recursively. From [46], we classify the equalization matrix $\mathbf{B}_{i,k}$ according to linear schemes: (1) $\mathbf{B}_{i,k} = \mathbf{H}^{-1}_{i,k}$ for ZR receiver; (2) $\mathbf{B}_{i,k} = \mathbf{H}^H_{i,k}(\mathbf{Q}^{-1}_{i,k} + \mathbf{H}_{i,k}\,\mathbf{H}^H_{i,k})^{-1}$ for linear MMSE receiver, where $\mathbf{Q}_{i,k}$ denotes a diagonal matrix with $\text{SNR}/(1 + Q^j_{i,k})$ as its jth entry; (3) $\mathbf{B}_{i,k} = \mathbf{A}_{i,k}\mathbf{H}^H_{i,k}(\mathbf{Q}^{-1}_{i,k} + \mathbf{H}_{i,k}\,\mathbf{H}^H_{i,k})^{-1}$ for IF receiver, where $\mathbf{A}_{i,k} \in \mathbb{Z}[j]^{L \times L}$ is an integer full-rank matrix and is optimized as a function of channel matrix, and $\mathbf{Q}_{i,k}$ denotes a diagonal matrix with its jth diagonal element $\text{SNR}/(1 + Q_{i,k}^j)$.

If we choose $\mathbf{A} = \mathbf{I}$, this scheme reduces to the linear MMSE receiver. The key step underlying this approach is the selection of an integer matrix $\mathbf{A}$ to approximate the channel matrix $\mathbf{H}_{i,k}$. Although finding the optimal $\mathbf{A}_{i,k}$ has a worst-case complexity that is exponential in L, this search only needs to be performed once per coherence interval. In practice, efficient approximation algorithms (see [48] and [19] for details) can be used to find near-optimal $\mathbf{A}_{i,k}$ in polynomial time.

REMARK 4.16 *In this section, we developed a low-complexity decoder which is implementable in practice. In our analysis, it is assumed that the destination exactly knows the entire channel coefficients over the network, which can be possible by letting each relay send its local channel coefficients to the destination via a control channel. In practice, the relay cannot send exact channel coefficients via the control channel due to its limited capacity and hence may need to send quantized channel coefficients. The impact of this on the performance of any transmission scheme including the presented scheme is an interesting future work.*

Numerical Results

In this subsection, we evaluate the performance of the low-complexity scheme presented in wider section for nonasymptotic cases (i.e., small numbers of users). From this, we can identify the practical feasibility of the presented scheme. We first investigate the performance of the presented scheme for various receiver architectures, namely ML, ZF, MMSE, and IF receivers. In our simulation, we consider the dense network with i.i.d. channel matrix $\mathbf{H}_{i,k}$ where each entry has a complex Gaussian distribution with zero mean and unit variance. Notice that in Figure 4.16, the ML receiver achieves the information-theoretical achievable rate. Among the linear receivers, the IF receiver shows the best performance, achieving about 1-bit gain over the MMSE receiver. For $K < 3$, the IF receiver can provide a satisfactory performance within 1-bit from the ML performance, significantly reducing the decoding complexity over the ML receiver. Thus, the IF receiver can be a good candidate for a practical system where the number of stages is limited due to a delay constraint. For $K \geq 4$, however, a low-complexity ML receiver (e.g., sphere decoding) should be considered.

We next evaluate the performance of the interference-harnessing algorithm and compare it to the performance of interference-aware routing. In simulations, we consider the multihop wireless network depicted in Figure 4.17 with $L = 4$ sources and $K = 3$ stages of relays. We define the *cluster* (i.e., a square in Figure 4.17) whose area depends on the transmit power. It is assumed that each relay can only communicate with the relays within eight neighboring clusters and its own cluster (i.e., communication range). Any interference outside the communication range is captured by additive Gaussian noise. We further assume the symmetric channel model for which a channel coefficient (within the communication range) is defined as $\text{SNR} \exp(j\theta)$, where SNR captures the distance-dependent path-loss and transmit power, and $\theta \sim \text{Unif}[0, 2\pi)$ denotes a random i.i.d. phase. In our simulation, we assume that the whole network is divided into $4 \times K$ disjoint

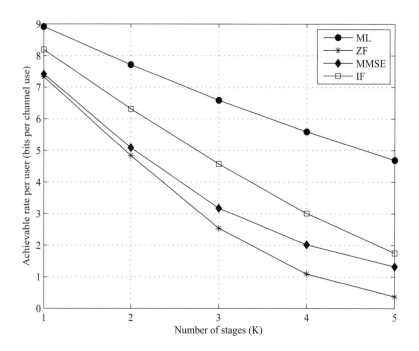

Figure 4.16 SNR $= 30\,$dB and $L = 4$. Performance comparison of the proposed low-complexity scheme for various receiver architectures as a function of the number of stages K.

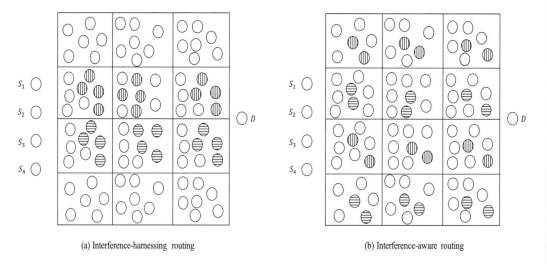

(a) Interference-harnessing routing   (b) Interference-aware routing

Figure 4.17 The opposite trend of the routes selected by interference-harnessing and interference-aware routing. Relays with vertical dashed lines establish path 1 and relays with horizontal dashed lines establish path 2.

clusters as shown in Figure 4.17 with $K = 3$, where K captures the source–destination distance (i.e., determines the network area). It is also assumed that each cluster contains n_c half-duplex relays with $n_c \geq L$. Notice that for a given K (i.e., network area), the network becomes denser as L grows, since the network includes more nodes. Hence, the parameters K and L control the network area and the network density, respectively.

For such a network, the interference-harnessing and interference-aware routing are used to establishes the L routes (per each path) for the proposed scheme and the MR, respectively. In the example of Figure 4.17, the relays chosen by the algorithm are shown in the circles. This can be easily extended to any L due to the assumption of $n_c \geq L$. To be specific, the interference-harnessing routing chooses the six clusters as in Figure 4.17(a) where each active cluster contains the L selected relays, while the interference-aware routing chooses the 12 clusters as in Figure 4.17(b) where each active cluster contains the $L/2$ selected relays. For the MR, there are two types of relays according to the number of received interference signals where the relays in the top or bottom route receive $(\frac{3}{2}L - 2)$ interference signals and the relays in the middle routes receive $(2L - 2)$ interference signals. Therefore, the achievable symmetric rate of the MR is determined by the minimum of all message rates, given by

$$r_{\mathrm{MR}} = \log\left(1 + \mathrm{SNR}/(1 + (2L - 2)\mathrm{SNR})\right). \tag{4.50}$$

Figure 4.18 shows that the presented scheme outperforms the MR, having a larger performance gain as SNR increases. This is because the strong interference limits the performance of the MR while it further improves the performance of the presented scheme that treats interference as a useful signal. From Figure 4.19, we observe that in contrast to the MR, the performance of the presented scheme is improved as

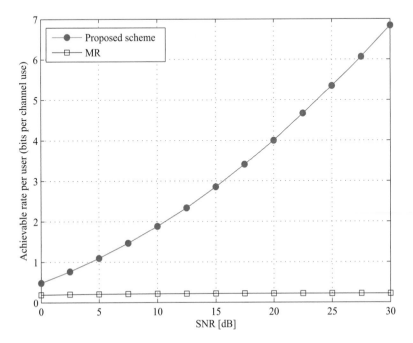

Figure 4.18 $L = 4$ and $K = 3$. Performance comparisons of the proposed scheme and multihop routing where the low-complexity decoding (with ML) is used for the proposed scheme.

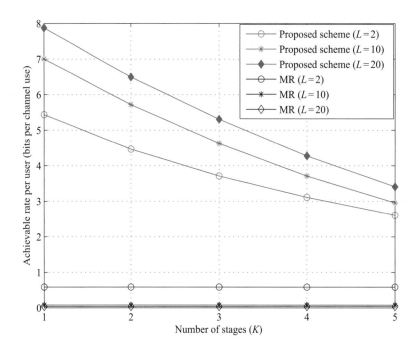

Figure 4.19 SNR $= 20$ dB. Performance comparisons of the proposed scheme and multihop routing where the low-complexity decoding (with ML) is used for the proposed scheme.

the number of users L increases, which is well-matched to the asymptotic result in Section 4.3.4.

4.4   Discussion and Concluding Remarks

This chapter presented a cooperative relaying scheme for half-duplex multihop relay networks. In contrast to routing, in the proposed scheme, a message from every source is simultaneously forwarded by all relays at every hop. This is realized by using the SNNC-SW relaying scheme in which a relay forwards a function of all source messages simultaneously. By optimizing both the relay selection (i.e., routing) as well as the cooperative relaying scheme, the performance can be significantly improved to outperform other state-of-the-art schemes, in particular QMF, NNC, and SNNC. Furthermore, the performance gain increases with the number of hops due to better rate-scaling.

Beyond an information-theoretical contribution, the proposed scheme can be made practical by using low-complexity stage-by-stage successive MIMO decoding to avoid prohibitively complex joint decoding. The proposed decoder can be implemented using a conventional MIMO receiver such as ML, ZF, linear MMSE, and IF receivers. By comparing various receiver architectures, it appears that, when the number of hops is fewer than four, the IF receiver provides a satisfactory performance that is within 1-bit from the ML performance. It also shows significantly better performance than multihop routing. In practical systems, the maximum number of hops is usually limited due to a delay constraint. In this case, the proposed scheme can provide a substantial gain over the multihop routing with similar encoding/decoding complexity, by deploying the IF linear receiver and a point-to-point channel code (e.g., turbo code, LDPC code, or polar code).

We note that the presented scheme can be extended to *asymmetric layered networks* and networks with multiple destinations, and bring further performance gains in comparison to MR for systems with delay constraints, as discussed in [29].

The mixed scheme was only developed for multihop virtual full-duplex relay channels with a single source and its extension to general L sources would be an interesting future work.

References

[1] Small Cell Forum, "Backhaul technologies for small cells," white paper, Dec. 2013.

[2] R. Daniels, R. Heath, J. Murdock, and T. Rappaport, *60 GHz Wireless Communication Systems*. Upper Saddle River, NJ: Prentice Hall Press, 2012.

[3] K. Doppler, C. Wijting, and K. Valkealahti "On the benefits of relays in a metropolitan area network," *IEEE Trans. Veh. Tech.* DOI: 10.1109/VETECS.2008.511.

[4] R. Baldemair, T. Irnich, K. Balachandran, et al., "Ultra-dense networks in millimeter-wave frequencies," *IEEE Commun. Mag,* vol. 53, no. 1, pp. 202–208, Jan. 2015.

[5] D. Hui and J. Axnas, "Joint routing and resource allocation for wireless self-backhaul in an indoor ultra-dense network," in *Proc. IEEE Int. Symp. Personal, Indoor and Mobile Radio Commun.*, London, Sep. 2013, pp. 3083–3088.

[6] H. S. Dhillon and G. Caire, "Wireless backhaul networks: Capacity bound, scalability analysis and design guidelines," *IEEE Trans. Inf. Theory,* vol. 14, no. 11, pp. 6043–6056, Nov. 2015.

[7] H.S. Dhillon and G. Caire, "Information theoretic upper bound on the capacity of wireless backhaul networks," in *Proc. IEEE Int. Symp. Inf. Theory (ISIT)*, Honolulu, HI, Jun./Jul. 2014, pp. 251–255.

[8] Y. Liang, T. Song and T. Li, "Energy efficient multihop wireless backhaul in heterogeneous cellular networks," in *Proc. IEEE Global Conf. Sig. Info. Process. (GlobalSIP)*, Washington, DC, Dec. 2016, pp. 625–629.

[9] S. Saadat, D. Chen, and T. Jiang, "Multipath multihop mmWave backhaul in ultra-dense small-cell network," *Digital Commun. Netw.*, vol. 4, no. 2, pp. 111–117, Apr. 2018.

[10] B. P. S. Sahoo, C.-H. Yao, and H-Y. Wei, "Millimeter-wave multi-hop wireless backhauling for 5G cellular networks," in *Proc. IEEE Veh. Tech. Conf.,* Sydney, Jun. 2017, pp. 1–5.

[11] T. K. Vu, M. Bennis, M Debbah, and M. Latva-aho, "Joint path selection and rate allocation framework for 5G self-backhauled mmWave networks," *arXiv:1805.07743*.

[12] A. Drozdy, J. Kapanen, and J. Manner, "User level performance analysis of multi-hop in-band backhaul for 5G," *Wireless Netw.*, pp. 1-15.

[13] M. Jaber, M.A. Imran, and R. Tafazolli, "5G backhaul challenges and emerging research directions: A survey," *IEEE Access,* vol. 4, pp. 1743–1766, Apr. 2016.

[14] S.-H. Park, O. Simeone, O. Sahin, and S. Shamai, "Multihop backhaul compression for the uplink of cloud radio access networks," *IEEE Trans. Veh. Tech.,* vol. 65, no. 5, pp. 3185–3199, May 2016.

[15] E. C. van der Meulen, "Three-terminal communication channels," *Adv. Appl. Prob.,* vol. 3, pp. 120–154, Jun. 1971.

[16] T. Cover and A. El Gamal, "Capacity theorems for the relay channel," *IEEE Trans. Inf. Theory,* vol. 25, no. 5, pp. 572–584, Sep. 1979.

[17] B. Nazer and M. Gastpar, "Compute-and-forward: Harnessing interference through structured codes," *IEEE Trans. Inf. Theory,* vol. 57, no. 10, pp. 6463–6486, Oct. 2011.

[18] B. Nazer and M. Gastpar, "Reliable physical layer network coding," *IEEE Special Issue on Network Coding,* vol. 99, no. 3, pp. 438–460, Mar. 2011.

[19] S.-N. Hong and G. Caire, "Compute-and-forward strategies for cooperative distributed antenna systems," *IEEE Trans. Inf. Theory,* vol. 59, no. 9, pp. 5227–5243, Aug. 2013.

[20] S.-N. Hong and G. Caire, "Beyond scaling law: On the rate performance of dense device-to-device wireless networks," *IEEE Trans. Inf. Theory,* vol. 61, no. 9, pp. 4735–4750, Sep. 2015.

[21] S. Avestimehr, S. Diggavi, and D. Tse, "Wireless network information flow: A deterministic approach," *IEEE Trans. Inf. Theory,* vol. 57, no. 4, pp. 1872–1905, Apr. 2011.

[22] S. Lim, Y. H. Kim, A. E. Gamal, and S. Chung, "Noisy network coding," *IEEE Trans. Inf. Theory,* vol. 57, no. 5, pp. 3132–3152, May 2011.

[23] J. Hou and G. Kramer, "Short message noisy network coding with a decode–forward option," *IEEE Trans. Inf. Theory,* vol. 62, no. 1, pp. 89–107, Aug. 2013.

[24] S. W. Peters and R. W. Heath, Jr., "The future of WiMAX: Multi-hop relaying with IEEE 802.16j," *IEEE Commun. Mag.* vol. 47, no. 1, pp. 104–111, Jan. 2009.

[25] S. S. C. Rezaei, S. O. Gharan, and A. K. Khandani, "Cooperative strategies for the half-duplex Gaussian parallel relay channel: Simultaneous relaying versus successive relaying," in *Proc. Allerton Conf. Comm., Control Comput.,* Monticello, Sep. 2008, pp. 1309–1316.

[26] H. Bagheri, A. S. Motahari, and A. K. Khandani, "On the capacity of the half-duplex diamond channel," in *Proc. IEEE Int. Symp. Inf. Theory (ISIT),* Austin, TX, Jun. 2010, pp. 649–653.

[27] K. Jain, J. Padhye, V. Padmanabhan, and L. Qiu, "The impact of interference on multi-hop wireless network performance," in *Proc. Annual Int. Conf. Mobile Comp. Net. (MobiCom),* Miami, Sep. 2013, pp. 66–80.

[28] R. Draves, J. Padhye, and B. Zill, "Routing in multi-radio, multi-hop wireless mesh networks," in *Proc. Annual Int. Conf. Mobile Comp. Net. (MobiCom),* Maui, HF, Sep. 2014, pp. 114–128.

[29] G. Parissidis, M. Karaliopoulos, T. Spyropoulos, and B. Plattner, "Interference-aware routing in wireless multihop networks," *IEEE Trans. Mobile Computing,* vol. 10, no. 5, pp. 716–733, May 2011.

[30] B. Rankov and B. A. Wittneben, "Spectral efficient protocols for half-duplex fading relay channels," *IEEE J. Sel. Areas Commun.,* vol. 25, no. 2, pp. 379–389, Feb. 2007.

[31] W. Chang, S.-Y. Chung, and Y.-H. Lee, "Capacity bounds for alternating two-path relay channels," in *Proc. Allerton Conf. Commun. Control Comput.,* Monticello, IL, Oct. 2007, pp. 1149–1155.

[32] B. Muthuramalingam, S. Bhashyam, and A. Thangaraj, "A decode and forward protocol for two-stage Gaussian relay networks," *IEEE Trans. Commun.,* vol. 60, no. 1, pp. 68–73, Jan. 2012.

[33] S.-N. Hong and G. Caire, "Virtual full-duplex relaying with half-duplex relays," *IEEE Trans. Inf. Theory,* vol. 61, no. 9, pp. 4700–4720, Sept. 2015.

[34] S.-N. Hong, I. Marić, and D. Hui, "Short message noisy network coding with sliding-window decoding for half-duplex multihop relay networks," *IEEE Trans. Wireless Commun.*, vol. 14, no. 8, pp. 6679–6689, Aug. 2015.

[35] N. Laneman and G. Kramer, "Window decoding for the multi-access channel with generalized feedback," in *Proc. IEEE Int. Symp. Inf. Theory (ISIT),* Chicago, IL, Jun./Jul. 2004, p. 281.

[36] A. Sanderovich, O. Somekh, H. V. Poor, and S. Shamai (Shitz), "Uplink macro diversity of limited backhaul cellular network," *IEEE Trans. Inf. Theory,* vol. 55, no. 8, pp. 3457–3478, Aug. 2009.

[37] A. D. Wyner, "Shannon-theoretic approach to a Gaussian cellular multiple-access channel," *IEEE Trans. Inf. Theory,* vol. 40, no. 6, pp. 1713–1727, Nov. 1994.

[38] R. Kolte, A. Ozgur, and A. E. Gamal, "Capacity approximations for Gaussian relay networks," *IEEE Trans. Inf. Theory,* vol. 61, no. 9, pp. 4721–4734, Sep. 2015.

[39] T. Cover and J. Thomas, *Elements of Information Theory*. Chichester: Wiley, 1991.

[40] P. Gupta and P. R. Kumar, "Critical power for asymptotic connectivity," in *Proc. IEEE Conf. Decision Control,* Tampa, FL, Dec. 1998, pp. 1106–1110.

[41] V. Nagpal, I.-H. Wang, M. Jorgovanović, D. Tse, and B. Nikolić, "Coding and system design for quantize-map-and-forward relaying," *IEEE J. Sel. Areas Commun.,* vol. 31, no. 8, pp. 1423–1435, Aug. 2013.

[42] M. Duarte, A. Sengupta, S. Brahma, C. Fragouli, and S. Diggavi, "Quantize-map-forward (QMF): An experimental study," in *Proc. ACM Int. Symp. Mobile Ad Hoc Net. Comput (MobiHoc),* New York, NY, 2013, pp. 227–236.

[43] E. Viterbo and J. Boutros, "A universal lattice decoder for fading channels," *IEEE Trans. Inf. Theory,* vol. 45, no. 5, pp. 1639–1642, Jul. 1999.

[44] M. O. Damen, H. E. Gamal, and G. Caire, "On maximum-likelihood detection and the search for the closest lattice point," *IEEE Trans. Inf. Theory,* vol. 49, no. 10, pp. 2389–2402, Oct. 2003.

[45] J. Zhan, B. Nazer, U. Erez, and M. Gastpar, "Integer-forcing linear receivers," *IEEE Trans. Inf. Theory*, vol. 60, no. 12, pp. 7661–7685, Dec. 2014.

[46] C. Feng, D. Silva, and F. Kschischang, "An algebraic approach to physical-layer network coding," *IEEE Trans. Inf. Theory,* vol. 59, no. 11, pp. 7576–7596, Nov. 2013.

[47] A. K. Lenstra, H. W. Lenstra, and L. Lovász, "Factoring polynomials with rational coefficients," *Mathematische Annalen*, vol. 261, no. 4, pp. 515–534, 1982.

[48] A. Ozgur, O. Lévêque, and D. Tse, "Operating regimes of large wireless networks," *Found. Trends Netw*, 2011.

[49] S.-N. Hong, I. Marić, D. Hui, and G. Caire, "On the achievable rates of virtual full-duplex relay channel," *IEEE Trans. Inf. Theory,* vol. 65, no. 1, pp. 354–367 Sept. 2018.

5 Edge Caching

Navid Naderi Alizadeh, Mohammad Ali Maddah-Ali, and Salman Avestimehr

5.1 Introduction

With the explosion of wireless data traffic over the past few years, many opportunities are being studied in development of 5G wireless networks to avoid the wireless spectrum crunch. One of the promising directions is to use the memory across the network and at the users to *cache* popular content during off-peak hours in order to increase the capacity and improve users' quality of service. This trend is particularly fueled by two trends: (1) storage memory is the fastest growing and cheapest network resource that can be made widely available in communication devices; (2) "cacheable" content (e.g., wireless video on demand) accounts for the majority (~70%) of the predicted 10,000% wireless traffic demand increase [1].

In this chapter, we will provide a cohesive view of state-of-the-art research for utilizing content caching for increasing the capacity of future wireless networks. In particular, we focus on the physical layer and illustrate the opportunities that caching can provide for enabling new communication and interference management strategies that can substantially improve the capacity of wireless networks. We will also highlight several exciting and timely research problems in this area for future studies.

5.2 Cache-Enabled Opportunities in the Physical Layer

In this chapter our goal is to review some of the opportunities that caching can provide and discuss and evaluate their gains in overall rate. In particular, we investigate the gain of the following opportunities in different scenarios:

- *cooperative interference cancellation*
- *interference alignment*
- *coded multicasting or cancelling known interference.*

Here, we briefly explain each of these opportunities over a wireless interfering channel with two transmitters (access points or base stations) and two users. In this system, we have a database of two equal-size files (e.g., videos), called file A and file B, somewhere in the core of the network. Each transmitter has a cache of size M_T files, and each user has a cache of size M_R files. The system has two phases, (1) a *placement phase*, where each of caches is populated, without knowing the future requests of the users, (2) a *delivery phase*, where each user will request a file from the database, and the transmitter has to deliver the requests.

5.2.1   Cooperative Interference Cancellation

Consider an extreme case, where each transmitter can store the entire database (i.e., $M_T = 2$), while the cache at the receiver is zero (i.e., $M_R = 0$). In this case, in the delivery phase, no matter which files are requested, the two transmitters can fully collaborate in delivering the files. For example, if user 1 asks for A and user 2 asks for B, then since both transmitters already have both A and B in their caches, they can collaboratively cancel interference, using techniques such as zero-forcing, in delivering the two data streams. Therefore, we can send two interference-free data streams at the same time and same bandwidth over the network. Roughly speaking, the number of interference-free data streams that the system can support at the same time and bandwidth is called degrees of freedom (DoF). Indeed, DoF is the first-order approximation of the rate, i.e., $R = \mathsf{DoF} \log(\mathsf{SNR}) + o(\mathsf{SNR})$, where SNR is the signal-to-noise ratio and R is the rate in terms of bits per second per hertz.

In this case, cache at the transmitters enables the gain of collaboration for interference cancellation to achieve two DoF.

Clearly, if $M_T < N$, then full collaboration is not feasible; still, judicious content placement would allow the system to exploit the gain of collaboration to some extent.

5.2.2   Interference Alignment

Consider the case where $M_T = 1$ and $M_R = 0$. In this case, each transmitter can store only one file. To clarify the gain of interference alignment, enabled by caching, let us compare two different placement approaches:

- Scenario 1: In this case, transmitter 1 caches file A and transmitter 2 caches file B. Let us assume that in the delivery phase, user 1 asks for file A and user 2 asks for file B, then the system turns into an interference channel. It is very well known that a two-user interference channel can support only one interference-free data stream with one DoF. Indeed, in order to have interference-free communication, we need to schedule the two data streams to user 1 and user 2 in two non-overlapping time slots or some other orthogonal subchannels.
- Scenario 2: In this case, in the placement phase, we split each file into two equal-size non-overlapping subfiles. In particular, we split file A into two equal-size subfiles A_1 and A_2. Similarly, we split file B into two equal-size subfiles B_1 and B_2. Then, transmitter 1 caches subfiles A_1 and B_1 and transmitter 2 caches subfiles A_2 and B_2. Now consider an arbitrary demand, say user 1 asks for file A and user 2 asks for file B. In this case, the channel turns into an X channel, where each transmitter has a message for each receiver. In particular, transmitter 1 has A_1 for user 1 and B_1 for user 2. Similarly, transmitter 2 has A_2 for user 1 and B_2 for user 2. In this case, we can use a technique called interference alignment to minimize the contribution of interference

at each receiver. In this approach, we can arrange the signaling such that interference of B_1 and B_2 at receiver 1 are aligned and interference of A_1 and A_2 are aligned at receiver 2. In other words, interference of B_1 and B_2 are such that from user 1's perspective there is only one source of interference, instead of 2. Similarly, interference of A_1 and A_2 are such that from user 2's perspective, there is only one source of interference, instead of two. In this case, one-third of communication dimensions at receiver 1 is wasted for interference of B_1 and B_2 and two-thirds are used for desired data A_1 and A_2. The same statement is true for receiver 2. Thus, overall, the system can support 4/3 interference-free data streams (i.e., $\mathsf{DoF} = \frac{4}{3}$), instead of one in the first scenario. Therefore, caching with careful placement can enable and increase the gain of interference alignment. Interference alignment is explored further in [2–4].

5.2.3   Coded Multicasting or Cancelling Known Interference

Now consider a system with only one transmitter and two users, where the cache size at the transmitter is $M_T = 2$ and the cache size at each receiver is $M_R = 1$. Assume that there are only two files A and B and therefore the transmitter can cache both. For the placement phase at the receivers, again, we split each file into two equal-size subfiles, i.e., $A = (A_1, A_2)$ and $B = (B_1, B_2)$. Assume that user 1 caches (A_1, B_1) and user 2 caches (A_2, B_2). Let us assume that user 1 asks for file A and user 2 asks for file B. We note that user 1 misses A_2 and user 2 misses B_1. To deliver the missing parts, the transmitter can multicast $A_2 \oplus B_1$ over the wireless link to both users. Thus user 1 can decode A_2 from $A_2 \oplus B_1$, because B_1 is already in its cache. Similarly user 2 can decode B_1 from $A_2 \oplus B_1$, because A_2 is already in its cache. The gain of sending one coded packet that is simultaneously useful for more than one user is called the gain of coded multicasting. In this example, the gain of coded multicasting increases the throughput by a factor of two. In other words, we can support two independent interference-free data streams, and thus $\mathsf{DoF} = 2$.

The main objective now is to evaluate the gains of exploiting these opportunities in overall throughput and to investigate whether these gains scale with the size of the networks. In what follows, we talk about different scenarios and provide formulas for each gain.

5.3   Canonical Cache Network

As mentioned in the previous section, one of the gains of cache in networks is the gain of coded multicasting. In this section, we evaluate this gain over a single-cell wireless network. In this network we have a single transmitter (base station), and K_R users. We have a database of N files $\{\mathcal{W}_n\}_{n=1}^{N}$ each of size F packets. The cache at the base station has space to store all N files (i.e., $M_T = N$), and each user has an isolated cache memory of size M_R files, for some real number $M_R \in [0, N]$. The coding rate at the transmitter has been chosen such that all the receivers can decode the transmitted message with very

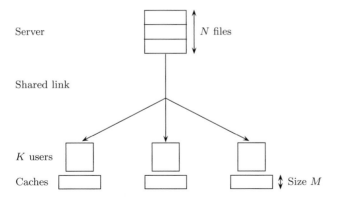

Server — N files

Shared link

K users

Caches — Size M

Figure 5.1 Canonical cache network from [5]: A server containing N files of size F packets each is connected through a shared link to K users each with an isolated cache of size MF bits. In the figure, $N = K = 3$ and $M = 1$.

small probability of error. For simplicity, we assume $N \geq K$ in this chapter. Figure 5.1 shows a schematic diagram of the network.

This system operates in two phases:

1. *Placement phase:* In this phase, each of the caches is populated as a function of the content, up to its size. For example, each user can store some of the files or some fraction of some files, or any other function of the content. This phase is carried out when or where the traffic is low. In this phase, the requests of the users have not yet been revealed.
2. *Delivery phase:* In this phase, which may happen at the peak of the traffic, each user reveals its requests for one of the files. The base station needs to transmit some signals such that each user can recover its requested file, using the received signal and also its cache content.

The objective is to minimize the maximum load of the base station, i.e., the number of bits that it needs to send to all users.

We note that each user has space to store an M_R/N fraction of each file. Therefore, in the worst case, the base station needs to deliver the missing $1 - M_R/N$ fraction of the requested file of each user. Therefore, in total the base station needs to deliver $K_R(1 - M_R/N)$. The factor of $1 - M_R/N$ appears because an M_R/N fraction of each file is locally cached at each user. We call the factor $1 - M_R/N$ the *local caching gain*. In what follows, we show that by exploiting the gain of coded multicasting, one can reduce this load by another factor of $1 + \frac{K_R M_R}{N}$.

We explain the main idea with two examples from [5].

EXAMPLE 5.1 (*Coded caching $K_R = N = 3$*)   Consider a system with $K_R = 3$ users, each with a cache large enough to store M_R files. Assume that the base station has $N = 3$ files, $\mathcal{W}_1 = A, \mathcal{W} = B$, and $\mathcal{W}_3 = C$. For this setup, we focus on two cases of $M = 1$ and $M = 2$.

- Case $M_R = 1$:

    In this case, for the placement phase, we split each file into three subfiles of equal size, i.e., $A = (A_1, A_2, A_3)$, $B = (B_1, B_2, B_3)$, and $C = (C_1, C_2, C_3)$. Then, user k caches (A_k, B_k, C_k) as shown in Figure 5.2. Note that each subfile has one-third of the size of a whole file, the size of (A_k, B_k, C_k) is equal to one file, satisfying the memory constraint of $M_R = 1$.

    In the delivery phase, we consider a generic case in which user 1 requests file A, user 2 requests file B, and user 3 requests file C. Recall that user 1 misses A_2 and A_3, user 2 misses B_1 and B_3, and user 3 misses C_1 and C_2. Therefore, overall, the base station needs to deliver six subfiles each of size one-third of a file, or equivalently two files. However, using the gain of coded multicasting, we can reduce this to one file as follows.

    We note that user 2 has stored A_2, which is needed by user 1, and user 1 has stored B_1, which is needed by user 2. These two users would like to swap these two subfiles, but cannot since their caches are isolated. Instead the base station can exploit this situation by transmitting $A_2 \oplus B_1$ to both users. Then, user 1 can recover A_2 by XORing the received message $A_2 \oplus B_1$ and stored subfile B_1. Similarly, user 2 can recover B_1 by XORing the received message $A_2 \oplus B_1$ and stored subfile A_2. Similarly, the base station transmits $A_3 \oplus C_1$ over the wireless link to deliver A_3 to user 1 and C_1 to user 3, and transmits $B_3 \oplus C_2$ to deliver B_3 to user 2 and C_2 to user 3 (see Figure 5.2). Since each server transmission is simultaneously useful for two users, the load of the wireless link is reduced by a factor of 2, which represents the gain of coded multicasting. We note that this gain can be exploited in any possible demand.

- Case $M = 2$:

    In this case, for the placement phase, we again split each file into three subfiles of equal size; however, for simplicity of explanation, we use different labeling, namely $A = (A_{12}, A_{13}, A_{23})$, $B = (B_{12}, B_{13}, B_{23})$, and $C = (C_{12}, C_{13}, C_{23})$. Then, user k stores all the subfiles, for which k is their index set, as shown in Figure 5.3. For the delivery phase, let us assume user 1 asks for file A, user 2 requests file B, and user 3 asks for file C. We note that each user has already two-thirds of the requested file, and misses one-third. In particular, users 1, 2, and 3 need A_{23}, B_{13}, and C_{12}. However,

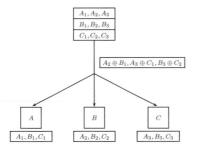

Figure 5.2 Coded multicasting opportunity for $K_R = 3$ users, $N = 3$ files, and cache size $M_R = 1$. Each file is split into three subfiles of size $1/3$, e.g., $A = (A_1, A_2, A_3)$. In the delivery phase, each coded packet is useful for two users.

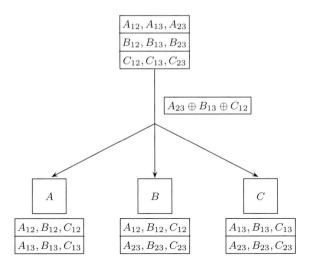

Figure 5.3 Coded multicasting for $K_R = 3$ users, $N = 3$ files, and cache size $M_R = 2$. Each file is split into three subfiles of size $1/3$, e.g., $A = (A_{12}, A_{13}, A_{23})$. In the delivery phase, the coded combination transmitted by the base station is useful for all three users.

Algorithm 5.1 Coded caching [5]

---

procedure PLACEMENT$(\mathcal{W}_1, \ldots, \mathcal{W}_N)$
    $t \leftarrow M_R K_R / N$
    $\mathfrak{T} \leftarrow \{\mathcal{T} \subset [K_R] : |\mathcal{T}| = t\}$
    for $n \in [N]$ **do**
        split $\mathcal{W}_n$ into $(\mathcal{W}_{n,\mathcal{T}} : \mathcal{T} \in \mathfrak{T})$ of equal size
    end for
    for $k \in [K_R]$ **do**
        $Z_k \leftarrow (\mathcal{W}_{n,\mathcal{T}} : n \in [N], \mathcal{T} \in \mathfrak{T}, k \in \mathcal{T})$
    end for
end procedure

procedure DELIVERY$(\mathcal{W}_1, \ldots, \mathcal{W}_N, d_1, \ldots, d_{K_R})$
    $\mathfrak{S} \leftarrow \{\mathcal{S} \subset [K_R] : |\mathcal{S}| = t + 1\}$
    $X_{(d_1, \ldots, d_{K_R})} \leftarrow (\oplus_{k \in \mathcal{S}} W_{d_k, \mathcal{S} \setminus \{k\}} : \mathcal{S} \in \mathfrak{S})$
end procedure

---

each missing part has been stored in the cache of both other users. In this case, the base station can transmit $A_{23} \oplus B_{13} \oplus C_{12}$ to all three users. Since each user misses one term in this addition, and has the other two, it can recover the missing part. Therefore, this combination is simultaneously useful for all three users, and therefore the load of communication has been reduced by a factor of 3.

The general placement of the delivery phase is been detailed in Algorithm 5.1.

Example 5.1 shows that as we increase the size of the cache, the gain of coded multi-casting increases. In general, in [5], it is shown that coded multicasting can achieve the gain of $1 + K_R M_R / N$.

This result can be interpreted from two related perspectives:

- *Degrees of freedom:* Since at each time we support $1 + K_R M_R / N$ users, this can translate to achieve $\mathsf{DoF}_{\mathrm{sum}} = 1 + K_R M_R / N$, which is very surprising. This is because the base station has only one antenna, and therefore the DoF of the system without cache is at most 1. Here, with caching, we can surprisingly increase the DoF linearly with K_R.
- *Load of the transmitter:* Another perspective is to evaluate the load of the transmitter. This is stated in the following theorem.

THEOREM 5.2 ([5]) *For $N \in \mathbb{N}$ files and $K_R \in \mathbb{N}$ users each with cache of size M_R, the maximum load of the transmitter, denoted by $R(M_R)$, is*

$$R(M_R) \triangleq K_R \frac{1 - \frac{M_R}{N}}{1 + \frac{K_R M_R}{N}}$$

in terms of number of files, for $M_R \in \{0, N/K_R, 2N/K_R, \ldots, N\}$. For general $0 \leq M_R \leq N$, the lower convex envelope of these points is achievable.

Figure 5.4 illustrates the gain of coded multicasting for a system with $K = 30$ users. For example, if each user has space to store a fraction of 0.2 of the total database, in the

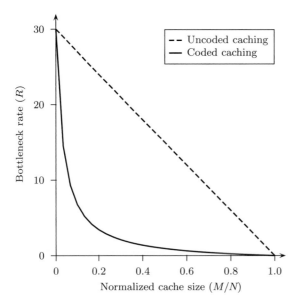

Figure 5.4 Load of the transmitter R in the delivery phase exploiting the gain of coded multicasting as a function of normalized memory size $M_R/N = M/N$ for $K_R = 30$ users from [5] as compared with the scheme that does not exploit the gain of coded multicasting.

uncoded scheme the load of the transmitter is 24 files, while in the coded scheme the load of the transmitter is less than 2.5 files, a factor of 7 improvement.

It has been shown that $R(M_R)$ is within a factor 12 of the optimum [5]. Later this bound is sharpened to within a factor 2 for general placement [6]. If we restrict the placement to the uncoded schemes, it is shown that $R(M_R)$ is optimum [7].

5.4 Cache-Aided Interference Channels

In this section we consider a more general wireless network setting, as illustrated in Figure 5.5, with K_T transmitters and K_R receivers, and we show how we can improve the physical layer performance of the network by using caches at both transmitters and receivers for interference management. We assume the system contains a library of N files, where each file consists of F packets (each packet is defined as a vector of B individual bits). Each transmitter is assumed to have a cache of size M_T files, while each receiver is assumed to have a cache of size M_R files. The system operation is decomposed into two phases, namely the prefetching phase and the delivery phase. In the prefetching phase, each transmitter and each receiver caches a subset of packets from the library up to their cache sizes, oblivious to any future requests of the receivers. The receivers' requests are revealed in the subsequent delivery phase, in which the transmitters need to cooperate in order to deliver the uncached parts of each requested file to the requesting receiver. We intend to maximize the delivery throughput for any possible set of receiver demands through an end-to-end design of the prefetching and delivery phases.

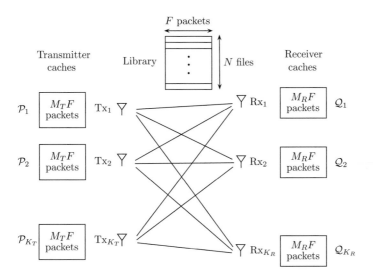

Figure 5.5 Wireless network with K_T transmitters and K_R receivers, where each transmitter and each receiver caches up to $M_T F$ packets and $M_R F$ packets, respectively, from a library of N files, each composed of F packets [8].

In such a setting, we are able to characterize the one-shot linear sum degrees-of-freedom (sum-DoF) of the network, i.e., the number of receivers that can be served at the same time interference-free via a one-shot linear delivery strategy, within a constant multiplicative factor of 2. In fact, we demonstrate that the following one-shot linear sum-DoF is achievable [8]:

$$\text{DoF} = \left\{ \frac{K_T M_T + K_R M_R}{N}, K_R \right\}, \tag{5.1}$$

and it is also within a factor of 2 of the optimum one-shot linear sum-DoF.

Our achievable scheme consists of a careful pattern of filling up the caches at the transmitters and receivers so as to create a content redundancy of a factor of $\frac{K_T M_T}{N}$ at the transmitters' caches and $\frac{K_R M_R}{N}$ at the receivers' caches. Such a redundancy creates two of the interference management opportunities explained in Section 5.2, namely collaborative interference cancellation by the transmitters and canceling known interference by the receivers. Through a particular one-shot linear delivery strategy that will be detailed shortly, we are able to serve min $\left\{ \frac{K_T M_T + K_R M_R}{N}, K_R \right\}$ receivers at each time without any inter-user interference, for any set of receiver demands.

5.4.1   Problem Formulation

We consider a discrete-time additive white Gaussian noise (AWGN) channel, as shown in Figure 5.5, with K_T transmitter nodes $\{\text{Tx}_i\}_{i=1}^{K_T}$ and K_R receiver nodes $\{\text{Rx}_i\}_{i=1}^{K_R}$. At time t, the received signal by each receiver Rx_j can be written as

$$Y_j(t) = \sum_{i=1}^{K_T} h_{ji} X_i(t) + Z_j(t), \tag{5.2}$$

where $X_i(t) \in \mathbb{C}$ denotes the signal transmitted by $\text{Tx}_i, i \in [K_T] \triangleq \{1, \ldots, K_T\}$ (subject to power constraint P), $h_{ji} \in \mathbb{C}$ denotes the channel gain from Tx_i to Rx_j, and $Z_j(t) \sim \mathcal{CN}(0, 1)$ denotes the additive white Gaussian noise at Rx_j at time slot t.

We assume a library of N files $\{\mathcal{W}_n\}_{n=1}^{N}$, where each file $\mathcal{W}_n$ includes F packets $\{\mathbf{w}_{n,f}\}_{f=1}^{F}$, with each packet containing B bits, i.e., $\mathbf{w}_{n,f} \in \mathbb{F}_2^B$. We assume that each transmitter is equipped with a cache memory of size $M_T F$ packets and each receiver is equipped with a cache memory of size $M_R F$ packets, which can be utilized to prefetch parts of the library before communication begins.

The network is assumed to operate in two phases, *the prefetching phase* and *the delivery phase*, detailed below:

Prefetching phase: In this phase, each of the transmitter and receiver nodes stores arbitrary pieces of content from the library in their caches, unaware of what the requests of the receivers will be in the future. Specifically, each transmitter Tx_i (resp., each receiver Rx_j) stores a subset $\mathcal{P}_i$ (resp., $\mathcal{Q}_j$) of size at most $M_T F$ (resp., $M_R F$) packets from the library. Throughout this section, we assume *uncoded prefetching* at all transmitter/receiver caches, which implies that storing coded combinations of multiple packets in the library is not permitted.

Delivery phase: In this phase, the receivers will reveal their demands for arbitrary files from the library. In particular, for any $j \in [K_R]$, we let $\mathcal{W}_{d_j}$ denote the file requested by receiver Rx_j. Upon revealing the requests, the transmitters will need to deliver parts of the requested files that are not available at the requesting receivers. We assume that $K_T M_T \geq N$, which guarantees that any packet in the library can be available in the cache of at least one of the transmitters in the network.

To deliver the requested packets to the receivers, each packet is first coded using a random Gaussian coding scheme of rate $\log P + o(\log P)$ into a *coded packet* consisting of PHY complex symbols. In particular, for each packet $\mathbf{w}_{n,f}$ in the library, we use $\tilde{\mathbf{w}}_{n,f}$ to denote its coded version. This suggests that each coded packet carries 1 (DoF).

We assume the delivery of the requested packets to the receivers occurs over H communication *blocks*. In each block, the transmitters cooperate to deliver a subset of requested packets to a subset of receivers, where each packet is intended for exactly one of the selected receivers. Specifically, in the *mth* block, a subset $\mathcal{D}_m$ of requested packets is delivered to a subset $\mathcal{R}_m$ of receivers. We assume that the delivery scheme at each block m is a one-shot linear scheme, in which each transmitter sends a linear combination of the coded packets $\mathcal{D}_m$ which it has already cached, and each receiver $Rx_j \in \mathcal{R}_m$ forms a linear combination $\mathcal{L}_{j,m}(\mathbf{y}_j(m), \tilde{\mathcal{Q}}_j)$ of its received signal $\mathbf{y}_j(m)$ and its (coded) cache contents $\tilde{\mathcal{Q}}_j$. The communication on block m is said to be *successful* if for all $Rx_j \in \mathcal{R}_m$, there exists a linear combination such that [8]

$$\mathcal{L}_{j,m}(\mathbf{y}_j(m), \tilde{\mathcal{Q}}_j) = \tilde{\mathbf{w}}_{d_j,f} + \mathbf{z}_j(m), \tag{5.3}$$

where $\mathbf{w}_{d_j,f} \in \mathcal{D}_m$ is the packet intended to receiver Rx_j in block m. Equation (5.3) hence implies that the desired packets at all receivers in $\mathcal{R}_m$ can be decoded with arbitrarily low error probabilities for infinitely high blocklengths. As mentioned before, the delivery phase will continue for H successful blocks until all receivers can recover their desired files.

Using the aforementioned scheme consisting of multiple successful blocks of communication, it is clear that at each block m, a one-shot linear sum-DoF of $|\mathcal{D}_m|$ can be achieved. This suggests that the entire one-shot linear sum-DoF of $\frac{\left| \bigcup_{m=1}^{H} \mathcal{D}_m \right|}{H}$ can be achieved in the entire delivery phase consisting of H communication blocks. Hence, for any arbitrary prefetching scheme which results in the transmitter/receiver cache contents $\left(\{\mathcal{P}_i\}_{i=1}^{K_T}, \{\mathcal{Q}_j\}_{j=1}^{K_R} \right)$, the one-shot linear sum-DoF is defined as [8]

$$\mathrm{DoF}_{\mathrm{L,sum}}^{\left(\{\mathcal{P}_i\}_{i=1}^{K_T}, \{\mathcal{Q}_j\}_{j=1}^{K_R} \right)} = \inf_{\mathbf{d}} \sup_{H, \{\mathcal{D}_m\}_{m=1}^{H}} \frac{\left| \bigcup_{m=1}^{H} \mathcal{D}_m \right|}{H}, \tag{5.4}$$

where $\mathbf{d}$ denotes the vector of receiver demands. This definition is based on the best achievable one-shot linear sum-DoF for the worst case receiver demands. Optimizing over the cache contents, we therefore define the one-shot linear sum-DoF of the network as the highest achievable one-shot linear sum-DoF achieved by the best caching scheme, i.e., [8]

$$\mathsf{DoF}^*_{\mathrm{L,sum}}(N, M_T, M_R) = \sup_{\{\mathcal{P}_i\}_{i=1}^{K_T}, \{\mathcal{Q}_j\}_{j=1}^{K_R}} \mathsf{DoF}_{\mathrm{L,sum}}^{\left(\{\mathcal{P}_i\}_{i=1}^{K_T}, \{\mathcal{Q}_j\}_{j=1}^{K_R}\right)} \tag{5.5}$$

$$\text{s.t.} \quad |\mathcal{P}_i| \le M_T F, \ \forall i \in [K_T] \tag{5.6}$$

$$|\mathcal{Q}_j| \le M_R F, \ \forall j \in [K_R]. \tag{5.7}$$

5.4.2   Main Result and Its Implications

In this section, we present our main result and a few remarks on its significance.

THEOREM 5.3   *For a network with a library of N files, each containing F packets, where F is sufficiently large, and cache size of M_T and M_R files at each transmitter and each receiver, respectively, the one-shot linear sum-DoF of the network satisfies [8]*

$$\min \left\{ \frac{K_T M_T + K_R M_R}{N}, K_R \right\} \le \mathsf{DoF}^*_{\mathrm{L,sum}}(N, M_T, M_R)$$

$$\le \min \left\{ 2 \frac{K_T M_T + K_R M_R}{N}, K_R \right\}. \tag{5.8}$$

Observations
- Comparing the upper and lower bounds in (5.8) reveals that the one-shot linear sum-DoF of a cache-aided wireless network can be characterized to within a constant multiplicative factor of 2, regardless of the system parameters.
- Theorem 5.3 demonstrates that the performance of the cache-aided wireless network scales with the entire cache size that is available in the network, both at the transmitter nodes and at the receiver nodes. As will be shown in Section 5.4.3, the transmitter caches can be used for collaborative zero-forcing (leading to the term $\frac{K_T M_T}{N}$ in (5.8)), whereas the receiver caches can be utilized for subtracting the interference from already-cached packets (leading to the term $\frac{K_R M_R}{N}$ in (5.8)).
- For the special case of a cache-aided fully connected interference channel, where $K_T = K_R = K$, Theorem 5.3 indicates that the sum-DoF growth is proportional to the number of users (i.e., transmitter–receiver pairs). This is in stark contrast to a regular interference channel without caches, where the one-shot linear sum-DoF is bounded by 2 [9].
- In Section 5.3 it was shown that a global caching gain, hence a sum-DoF, of $\min \left\{ 1 + \frac{K_R M_R}{N}, K_R \right\}$ is achievable for the special case of a single server having access to the entire library. Hence, our result in Theorem 5.3 subsumes the result of Section 5.3 by extending it to multiple cooperating transmitters.

5.4.3   Achievable Scheme

In this section, we present an algorithm to achieve the sum-DoF lower bound in (5.8). Our algorithm uses a careful pattern of prefetching the contents from the library in the prefetching phase, which maximizes the gains of interference mitigation at both the transmitter and receiver sides in the delivery phase. We first introduce our scheme through a simple example.

EXAMPLE 5.4   Consider a system with a library of $N = 3$ files, $W_1 = A$, $W_2 = B$, and $W_3 = C$, each comprising F packets. Moreover, assume that there exist $K_T = 3$ transmitters in the network, each with a cache of size $M_T = 2$ files, and $K_R = 3$ receivers each with a cache of size $M_R = 1$ file. For this network, we will detail our proposed prefetching and delivery schemes in the following:

Prefetching phase: In this phase, each file in the library is partitioned into $\binom{3}{2}\binom{3}{1} = 9$ disjoint subfiles, each of size $F/9$ packets. We denote the subfiles of each file W_n as $W_{n,\mathcal{T},\mathcal{R}}$ for any $\mathcal{T} \subseteq [K_T] = [3]$ and $\mathcal{R} \subseteq [K_R] = [3]$ such that $|\mathcal{T}| = 2$ and $|\mathcal{R}| = 1$. This notation indicates that for any combination of the aforementioned transmitter and receiver subsets, the subfiles $\{W_{n,\mathcal{T},\mathcal{R}}\}_{n \in [3]}$ are cached by the transmitters in $\mathcal{T}$ and the receivers in $\mathcal{R}$. For example, file B is partitioned into the following nine subfiles:

$$B_{12,1}, B_{12,2}, B_{12,3}, B_{13,1}, B_{13,2}, B_{13,3}, B_{23,1}, B_{23,2}, B_{23,3},$$

where $B_{12,1}$ is cached by transmitters Tx_1 and Tx_2 and receiver Rx_1, while $B_{12,2}$ is stored at transmitters Tx_1 and Tx_2 and receiver Rx_2.

Using the aforementioned prefetching strategy, the number of distinct subfiles cached by each transmitter is equal to six, whereas the number of distinct subfiles cached by each receiver equals three. It is straightforward to check that these numbers satisfy the memory constraints at all nodes.

Delivery phase: In this phase, initially the receivers disclose their requests for their desired files from the library. For the sake of this example and without loss of generality, assume that files A, B, and C are requested by receivers Rx_1, Rx_2, and Rx_3, respectively. Using the caching scheme outlined in the prefetching phase, each receiver already has three out of the nine subfiles from its requested file in its own cache. Hence, after the requests are revealed, the transmitters are responsible for delivering the following subfiles to their requesting receivers:

$$A_{12,2}, A_{12,3}, A_{13,2}, A_{13,3}, A_{23,2}, A_{23,3} \text{ to receiver } Rx_1,$$
$$B_{23,3}, B_{13,1}, B_{12,3}, B_{23,1}, B_{13,3}, B_{12,1} \text{ to receiver } Rx_2, \qquad (5.9)$$
$$C_{13,1}, C_{23,2}, C_{23,1}, C_{12,2}, C_{12,1}, C_{13,2} \text{ to receiver } Rx_3.$$

Interestingly, as shown in Figure 5.6, the above subfiles can be partitioned into six subsets of size three, such that the subfiles in each subset can be delivered at the same time to the receivers, without any inter-user interference. As an example, in the step illustrated in Figure 5.6(a), subfiles $A_{12,2}, B_{23,3}, C_{13,1}$ are delivered to receivers Rx_1, Rx_2, and Rx_3 at the same time, respectively. Figure 5.7 demonstrates how the interference between all the transmitted subfiles can be effectively eliminated. Using a careful delivery scheme, the signals transmitted by Tx_1, Tx_2, and Tx_3 can be respectively written as

$$X_1 = -h_{32}\tilde{A}_{12,2} + h_{23}\tilde{C}_{13,1}$$
$$X_2 = \quad h_{31}\tilde{A}_{12,2} - h_{13}\tilde{B}_{23,3}$$
$$X_3 = -h_{21}\tilde{C}_{13,1} + h_{12}\tilde{B}_{23,3},$$

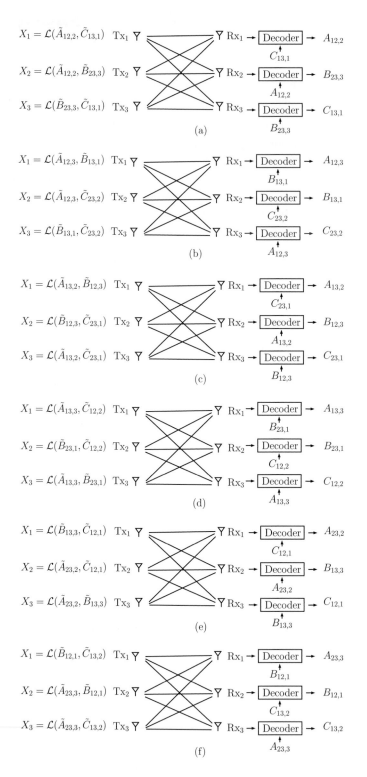

Figure 5.6 Delivery phase for Example 5.4 for files A, B, and C requested by receivers Rx_1, Rx_2, and Rx_3, respectively, where $\mathcal{L}(\mu, \lambda)$ represents a carefully chosen linear combination of μ and λ. In every step, each subset of two transmitters collaborate to zero-force the interference caused by a specific subfile at an unintended receiver. Furthermore, each receiver also utilizes its cached contents to subtract the known incoming interference contribution from interfering subfiles [8].

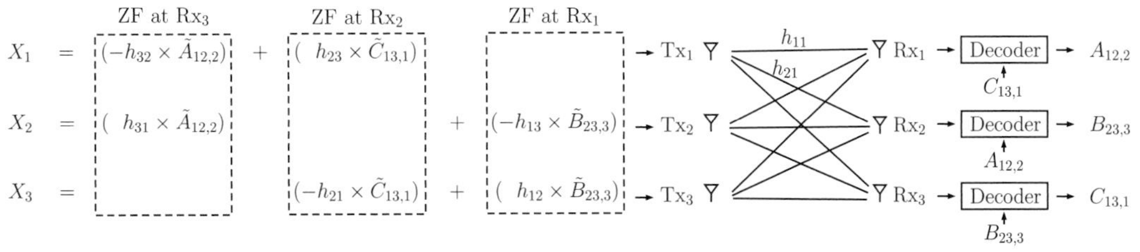

Figure 5.7 Detailed information regarding the encoding and decoding mechanisms used in the delivery phase step in Figure 5.6(a). In this step, Tx$_1$ and Tx$_2$ zero-force $A_{12,2}$ at Rx$_3$, Tx$_1$ and Tx$_3$ zero-force $C_{13,1}$ at Rx$_2$, and Tx$_2$ and Tx$_3$ zero-force $B_{23,3}$ at Rx$_1$. Moreover, Rx$_1$, Rx$_2$, and Rx$_3$ can, respectively, subtract the interference contribution from $C_{13,1}$, $A_{12,2}$, and $B_{23,3}$, since they have already cached the corresponding subfiles in the prefetching phase [8].

where for any subfile $W_{n,\mathcal{T},\mathcal{R}}$, $\tilde{W}_{n,\mathcal{T},\mathcal{R}}$ denotes its coded version. It is straightforward to derive the received signals by the receivers as

$$Y_1 = (h_{12}h_{31} - h_{11}h_{32})\tilde{A}_{12,2} + (h_{11}h_{23} - h_{13}h_{21})\tilde{C}_{13,1} + Z_1,$$
$$Y_2 = (h_{23}h_{12} - h_{22}h_{13})\tilde{B}_{23,3} + (h_{22}h_{31} - h_{21}h_{32})\tilde{A}_{12,2} + Z_2,$$
$$Y_3 = (h_{31}h_{23} - h_{33}h_{21})\tilde{C}_{13,1} + (h_{33}h_{12} - h_{32}h_{13})\tilde{B}_{23,3} + Z_3.$$

Now, it is clear that the interference due to subfiles $C_{13,1}$, $A_{12,2}$, and $B_{23,3}$ can be respectively cancelled at Rx$_1$, Rx$_2$, and Rx$_3$, since each subfile is already cached at the corresponding receiver. This shows how the inter-user interference can be completely mitigated in all the steps in Figure 5.6 so that all the receivers can decode their desired subfiles.

Regardless of the receivers' demands in the delivery phase, our prefetching phase is designed so as to maximize the gains of interference cancellation at both the transmitter and receiver sides. Hence, the sum-DoF of $\frac{18*F/9}{6*F/9} = 3 = \min\left\{\frac{K_T M_T + K_R M_R}{N}, K_R\right\}$ is achievable in this example.

The general achievable scheme is presented in Algorithm 5.2, using the following notation:

- For any set $\mathcal{S}$, $\Pi_\mathcal{S}$ denotes the set of all permutations of $\mathcal{S}$, and for any $t \in \{1, \ldots, |\mathcal{S}|\}$, $\Pi_{\mathcal{S},t}$ denotes the set of all permutations of all subsets of $\mathcal{S}$ of size t, i.e.,

$$\Pi_{\mathcal{S},t} = \bigcup_{\mathcal{A} \subseteq \mathcal{S}, |\mathcal{A}| = t} \Pi_\mathcal{A}.$$

- For a set $\mathcal{R}$, $\Pi_\mathcal{R}^{\mathrm{circ}}$ denotes the set of $(|\mathcal{R}| - 1)!$ circular permutations of $\mathcal{R}$.[1] Moreover, for a set $\mathcal{S}$, a permutation $\pi \in \Pi_\mathcal{S}$ and two integers i, j satisfying $j \geq i$, we define $\pi[i : j]$ as

$$\pi[i : j] = [\pi(i \oplus_{|\mathcal{S}|} 0) \ \pi(i \oplus_{|\mathcal{S}|} 1) \ \pi(i \oplus_{|\mathcal{S}|} 2) \ \ldots \ \pi(i \oplus_{|\mathcal{S}|} (j - i))],$$

[1] A circular permutation of a set $\mathcal{R}$ is a way of arranging the elements of $\mathcal{R}$ around a fixed circle. The number of distinct circular permutations of a set $\mathcal{R}$ equals $(|\mathcal{R}| - 1)!$.

Algorithm 5.2 Achievable scheme for Theorem 5.3 [8]

1: **procedure** PLACEMENT($\mathcal{W}_1, \ldots, \mathcal{W}_N$)
2:    $t_R \leftarrow M_R K_R / N$
3:    $t_T \leftarrow M_T K_T / N$
4:    **for** $n \in [N]$ **do**
5:        Partition $\mathcal{W}_n$ into $\binom{K_T}{t_T}\binom{K_R}{t_R}$ disjoint subfiles $\{\mathcal{W}_{n,\mathcal{T},\mathcal{R}}\}_{\mathcal{T} \subseteq [K_T], |\mathcal{T}| = t_T, \mathcal{R} \subseteq [K_R], |\mathcal{R}| = t_R}$ of equal sizes.
6:    **end for**
7:    **for** $i = 1, \ldots, K_T$ **do**
8:        **for** $i = 1, \ldots, K_T$
9:            Tx$_i$ caches all $\mathcal{W}_{n,\mathcal{T},\mathcal{R}}$ for which $i \in \mathcal{T}$.
10:   **end for**
11:   **for** $j = 1, \ldots, K_R$ **do**
12:        Rx$_j$ caches all $\mathcal{W}_{n,\mathcal{T},\mathcal{R}}$ for which $j \in \mathcal{R}$.
13:   **end for**
14: **end procedure**

15: **procedure** DELIVERY($\mathcal{W}_1, \ldots, \mathcal{W}_N, d_1, \ldots, d_{K_R}$)
16:    **for** $j \in [K_R]$ **do**
17:        **for** $\mathcal{T} \subseteq [K_T]$ s.t. $|\mathcal{T}| = t_T$ **do**
18:            **for** $\mathcal{R} \subseteq [K_R] \setminus \{j\}$ s.t. $|\mathcal{R}| = t_R$ **do**
19:                partition $\mathcal{W}_{d_j, \mathcal{T}, \mathcal{R}}$ to $\frac{t_R! [K_R - (t_R + 1)]!}{[K_R - (t_R + t_T)]!}$ disjoint subfiles $\{\mathcal{W}_{d_j, \mathcal{T}, \pi, \pi'}\}_{\substack{\pi \in \Pi_{\mathcal{R}} \\ \pi' \in \Pi_{[K_R] \setminus (\mathcal{R} \cup \{j\}), t_T - 1}}}$ of equal sizes.
20:            **end for**
21:        **end for**
22:    **end for**
23:    **for** $\mathcal{T} \subseteq [K_T]$ s.t. $|\mathcal{T}| = t_T$ **do**
24:        **for** $\mathcal{R} \subseteq [K_R]$ s.t. $|\mathcal{R}| = t_T + t_R$ **do**
25:            **for** $\pi \in \Pi_{\mathcal{R}}^{\text{circ}}$ **do**
26:                Each transmitter Tx$_i$ transmits a linear combination of the coded subfiles as in
$$X_i = \mathcal{L}_{i,\mathcal{T},\pi}\Bigg(\Big\{\tilde{W}_{d_{\pi(l)}, \mathcal{T} \oplus_{K_T} (l-1), \pi[l+1:l+t_R], \pi[l+t_R+1:l+t_R+t_T-1]} :$$
$$l \in [t_T + t_R], i \in \mathcal{T} \oplus_{K_T} (l-1)\Big\}\Bigg)$$
using a linear combination such that the subfiles
$$\Big\{\mathcal{W}_{d_{\pi(l)}, \mathcal{T} \oplus_{K_T} (l-1), \pi[l+1:l+t_R], \pi[l+t_R+1:l+t_R+t_T-1]} : l \in [t_T + t_R]\Big\}$$
are simultaneously delivered to the receivers in $\mathcal{R}$ interference-free.
27:            **end for**
28:        **end for**
29:    **end for**
30: **end procedure**

---

where for an integer m, $i \oplus_m j$ is defined as

$$i \oplus_m j = 1 + (i + j - 1 \mod m). \tag{5.10}$$

- For a set $\mathcal{T}$ and an integer j, $\mathcal{T} \oplus_m j$ represents entry-wise addition of elements of $\mathcal{T}$ with j modulo m, as defined in (5.10).

5.4.4   Other Results

In this section, the focus has been on exploiting the gain of collaborative interference cancellation and canceling known interference. As mentioned before, another gain to exploit is the gain of enabling interference alignment. Ideally, we would like to exploit all three gains together. However, combining all three gains is a very difficult task. This is done in [10] for three-user interference channels with caches only at the transmitter side. However, beyond the three-user case, it is very hard to handle the dependency among the resulting channel coefficients after applying zero-forcing. Due to this dependency, the available techniques of interference alignment (see [2–4]) is not applicable.

In [11], the focus is to exploit the gain of interference alignment and canceling known interference. In other words, the gain of collaborative interference cancellation has been completely ignored. Quite interestingly, it is shown that using this technique, we can achieve within a constant factor of optimum sum-DoF.

THEOREM 5.5   *For a network with a library of N files, each containing F packets, and cache size of M_T and M_R files at each transmitter and each receiver, respectively, the sum-DoF of the network, satisfies [11]*

$$d(K_T, K_R, M_T, M_R, N) \leq \mathsf{DoF}^*_{\mathrm{sum}}(N, M_T, M_R) \leq 13.5 \cdot d(K_T, K_R, M_T, M_R, N), \tag{5.11}$$

for sufficiently large F, where $d(K_T, K_R, M_T, M_R, N)$ is defined as

$$d(K_T, K_R, M_T, M_R, N) = \frac{K_T}{\left(1 - \frac{t_R}{K_R}\right)\left(K_T - 1 + \min\{\frac{K_R}{t_R+1}, N\}\right)}. \tag{5.12}$$

So far, we have assumed that the transmitters are isolated. Another scenario is to assume other than caches at the transmitters and at the receivers, there is a remote database that is connected to each of the transmitters with limited capacity. This scenario has been considered in [12].

5.5   Cache-Aided Interference Management in Wireless Cellular Networks

In this section, we consider how caching can be used to improve interference mitigation in wireless cellular networks, in which the receivers in the downlink direction do not receive signals or interference from base stations which are far from them. In particular, in a hexagonal cellular setting, we assume that each receiver only receives signals from its three adjacent base stations, while it receives signals from the rest of the base stations in the network below noise level; hence we ignore them for simplicity.

The aforementioned setting is distinct from the setting considered in the previous section, as the network is no longer fully connected. This implies that only a subset of transmitter–receiver links can be used to deliver the desired contents to the receivers. Furthermore, we make an additional assumption that in the considered cellular network, the locations of the receivers are not known during the prefetching phase, which is a reasonable assumption considering the users' mobility [13].

As will be shown later, we will extend the result of Section 5.4 and characterize the one-shot linear DoF that caching can achieve per cell in a cellular network to within an additive gap of 1 and a multiplicative gap of 2. Similar to the previous section, our result in this section highlights the importance of the aggregate cache size within a cell, as the per-cell one-shot linear DoF scales with the total cache size that is available at both the cell base station and the receivers inside the cell.

In order to make the prefetching phase independent of the user locations, we propose a randomized prefetching scheme for the receivers, inspired by [14], which is oblivious to their number, locations, and identities. In particular, each receiver stores a fraction μ_R of each file in the library uniformly at random, where μ_R is equal to the size of the receiver cache in terms of the fraction of the entire library that it can store. On the other hand, the base stations prefetch the library contents using a reuse pattern of three-cell clusters. The caching is done such that each packet in the library is cached by $3\mu_T$ base stations in each cluster of three adjacent base stations, where μ_T is equal to the maximum fraction of the entire library that each base station can store in its cache. After the receivers' requests are revealed in the delivery phase, the communication scheme depends on the cache sizes of the base stations. For the case of small cache sizes where there is no duplication in the base station cache contents, only one-third of the base stations are active at each time. However, for larger cache sizes, the base stations use the redundancy in their cache contents in order to collaboratively zero-force the interference due to their transmitted packets at some of the undesired receivers. The rest of the receivers will utilize their own cache contents to cancel the interference due to the undesired packets which they have already stored in their local caches.

5.5.1   System Model

Consider a wireless cellular network, comprising C hexagonal cells, each of which contains a base station located at the center of the cell and K_R receivers randomly located around the six corners of the cell, as illustrated in Figure 5.8. We denote the base station (BS) at the center of cell i by Tx_i. Due to path-loss effects, we assume that each receiver that is located at the intersection of three adjacent cells only receives signals from the three nearby base stations and does not receive signal or interference from the rest of the base stations. This implies that the received signal of $\text{Rx}_j, j \in [CK_R]$ at time t can be written as

$$Y_j(t) = \sum_{i \in \mathcal{N}_j} h_{ji} X_i(t) + Z_j(t), \tag{5.13}$$

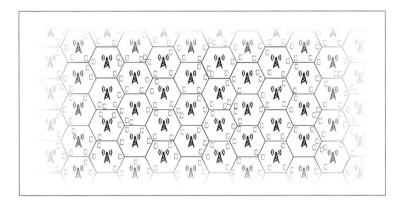

Figure 5.8 A cellular network model with $K_R = 4$ [16].

where $\mathcal{N}_j$ represents the set of three base stations whose signals can be "heard" by receiver Rx_j. In the above equation, $X_i(t) \in \mathbb{C}$ denotes the transmit signal of Tx_i at time slot t, subject to power constraint P, $h_{ji} \in \mathbb{C}$ denotes the channel gain from Tx_i to Rx_j, and $Z_j(t) \sim \mathcal{CN}(0, 1)$ denotes the additive white Gaussian noise at Rx_j at time slot t.

We assume that the system contains a library of N files, each containing F packets, as in Section 5.4.1. Furthermore, each BS is assumed to have a cache memory of size $\mu_T NF$ packets, while each receiver is equipped with a cache memory of size $\mu_R NF$ packets. The rest of the problem formulation, in terms of prefetching and one-shot linear delivery phases, is the same as Section 5.4.1, except for the fact that due to the partial connectivity of the cellular network topology, each piece of content in the library should be available in the immediate neighborhood of every receiver in the network. This implies that the base station cache sizes should satisfy $3\mu_T \geq 1$.

Let $\mathrm{DoF}^*_{\mathrm{L,sum}}(N, \mu_T N, \mu_R N, C)$ be the optimum sum-DoF of the network according to (5.5)–(5.7) for a system with N files, transmitter cache size of $\mu_T N$, receive cache-size of $\mu_R N$, and C cells. Then, we define the one-shot linear DoF per cell as

$$\mathrm{LDoF}^* = \lim_{C \to \infty} \lim_{N \to \infty} \frac{\mathrm{DoF}^*_{\mathrm{L,sum}}(N, \mu_T N, \mu_R N, C)}{C}. \qquad (5.14)$$

5.5.2 Main Result

In this section, we present our main result on the one-shot linear DoF per cell of the cache-aided cellular network and highlight two remarks on its significance.

THEOREM 5.6 *For a cellular network with fractional cache sizes of μ_T and μ_R at each base station and each receiver, respectively, the one-shot linear DoF per cell satisfies [15]*

$$\min\{\mu_T + K_R\mu_R - 1, K_R\} \leq \mathrm{LDoF}^* \leq \min\{2(\mu_T + K_R\mu_R), K_R\}. \qquad (5.15)$$

Remarks

- The upper and lower bounds in (5.15) are within an additive gap of 1 and a multiplicative gap of 2 for all system parameters. This indicates that the one-shot linear DoF per cell of a cache-aided wireless cellular network can be characterized within an additive gap of 1 and a multiplicative gap of 2.
- Note that the gain of caching, as stated in Theorem 5.6, scales linearly with the entire cache size inside each cell, hence with the number of receivers. This shows how caching can provide a scalable interference mitigation technique in cellular networks. A similar phenomenon was also observed in the previous section for the case of fully connected wireless networks. This implies that the partial connectivity of the network topology in the cellular case does not have a significant impact on the gains that caching can provide. This can be viewed as the tension between two conflicting effects: First, in the cellular case, content delivery from farther away base stations to a receiver is simply not possible as opposed to the fully connected case. This limits the capability of transmit zero-forcing and receiver interference cancellation. Second, it is clear that the local connectivity of the network topology also significantly reduces the amount of downlink interference from base stations to the receivers. Our result in Theorem 5.6 demonstrates that these two effects almost completely balance each other.

We will next prove the achievability of the lower bound in Theorem 5.6 by providing an achievable scheme in Section 5.5.3. The proof of the outer bound is relegated to [16].

5.5.3   Achievable Scheme

Prefetching phase at the receiver side: As explained in the system model, the locations of the receivers are assumed to be unknown during the prefetching phase. Hence, we take a randomized approach inspired by [14], where each receiver simply selects and caches $\mu_R F$ packets of each file in the library *uniformly at random*.

Prefetching phase at the base station side and delivery phase: For prefetching the contents at the base stations, we utilize the reuse pattern with clusters of size three, as shown in Figure 5.9. In this figure, the base stations with similar labels cache the exact same subset of packets from the library as each other.

The exact prefetching algorithm at the base stations and also the delivery phase depend on the size of the base station cache memories, leading to the following three cases that vary in terms of the level of collaboration among the BSs.[2]

$3\mu_T = 1$ (No BS Collaboration)

In this case, each file $W_n, n \in [N]$ in the library is partitioned into three disjoint subfiles $\{W_{n,1}, W_{n,2}, W_{n,3}\}$, each of size $\frac{F}{3}$ packets. Each base station with label $t \in [3]$ will then store all the subfiles in $\{W_{n,t} : n \in [N]\}$. This implies that the cache contents of the base

---

[2] If $3\mu_T$ is not an integer, the memory-sharing technique introduced in [5] can be utilized to achieve a convex combination of the DoFs per cell that can be achieved with base station fractional cache sizes $\mu_T' = \frac{1}{3}\lfloor 3\mu_T \rfloor$ and $\mu_T'' = \frac{1}{3}\lceil 3\mu_T \rceil$.

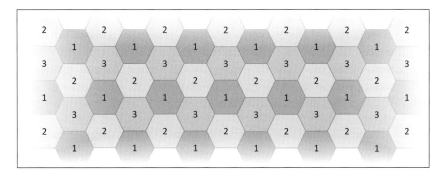

Figure 5.9 Prefetching pattern of the base stations [16].

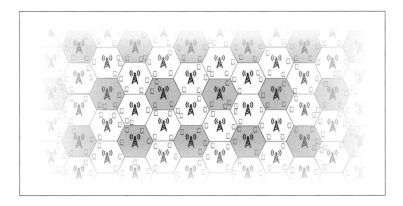

Figure 5.10 The delivery activation pattern for the case of $3\mu_T = 1$ [16].

stations with different labels are completely disjoint, hence no collaboration can exist among them.

Consequently, after the receivers' requests are revealed in the delivery phase, only one-third of the base stations which are labelled identically are activated at each time, while the rest of the base stations remain silent. One step of such a delivery phase is illustrated in Figure 5.10. We then use time-sharing to activate the rest of the base stations with different labels in turn. At each step, each active base station will transmit contents to all the receivers around its six corners. It is evident that such an activation pattern of the base stations incurs no interference among active cells.

For ease of notation, we use $\{1, 2, \ldots, \frac{C}{3}\}$ to denote the indices of active cells (base stations). For any cell $i \in [\frac{C}{3}]$, we use K_i to denote the number of receivers around its corners. It is easy to verify that $\sum_{i=1}^{\frac{C}{3}} K_i = CK_R$, which is the total number of receivers in the whole network. By the law of large numbers, the number of packets of each file in the library that are exactly stored in the caches of each subset of s receivers approximately equals

$$\mu_R^s (1 - \mu_R)^{K_i - s} F. \tag{5.16}$$

The *ith* active BS can then utilize a delivery scheme similar to the one proposed in [14] to identify packets that are stored at an equal number of receivers out of its surrounding K_i receivers and deliver them simultaneously. In particular, similar to the single-server scheme in Section 5.3, all the packets that are cached at s receivers can be delivered in groups of $s + 1$ to the receivers. We review the scheme in the following simple example.

EXAMPLE 5.7   Consider the case where the active base stations in Figure 5.10 are all labeled $t = 1$, as was illustrated in Figure 5.9. Consider one such active cell with three receivers around its cell corners, denoted by $\{Rx_1, Rx_2, Rx_3\}$. Without loss of generality, assume that these receivers will request files A, B, and C, respectively, during the delivery phase. Now, let A_1, B_1, and C_1 denote the subfiles of the aforementioned files that have been cached by this base station. Each of these subfiles can, in turn, be partitioned to smaller subfiles based on the cache contents of $\{Rx_1, Rx_2, Rx_3\}$. As an example, subfile B_1 can be partitioned to

$$\{B_{1,\emptyset}, B_{1,1}, B_{1,2}, B_{1,3}, B_{1,12}, B_{1,13}, B_{1,23}, B_{1,123}\}, \tag{5.17}$$

where for any $\mathcal{R} \subseteq \{1, 2, 3\}$, $B_{1,\mathcal{R}}$ represents the subfile of B_1 cached exclusively at the receivers in $\mathcal{R}$. Note that due to (5.16), the size of each subfile $B_{1,\mathcal{R}}$ approximately equals $\mu_R^{|\mathcal{R}|}(1 - \mu_R)^{3-|\mathcal{R}|}\frac{F}{3}$ packets. A_1 and C_1 can also be partitioned similarly.

Now, note that receiver Rx_2 has already stored $\{B_{1,1}, B_{1,12}, B_{1,13}, B_{1,123}\}$ in its cache, hence the base station should deliver the remaining four subfiles of B_1 to this receiver. Likewise, receivers Rx_1 and Rx_3 also need four subfiles of A_1 and C_1, respectively. The base station can now utilize the receivers' cache contents for creating coded multicasting opportunities to deliver the required subfiles. Two example steps of delivery using coded multicasting are illustrated in Figure 5.11, where for any subfile $W_{n,t,\mathcal{R}}$, $\tilde{W}_{n,t,\mathcal{R}}$ represents the corresponding coded subfile containing PHY coded symbols. In both cases, the receivers can use their cache contents to subtract the interference contributions due to the incoming undesired subfiles and recover their desired subfiles interference-free.

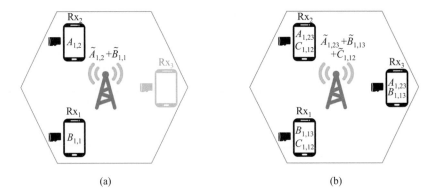

(a)   (b)

Figure 5.11 Two steps of delivery in Example 5.7, where the base station respectively delivers (a) subfiles $A_{1,2}$ and $B_{1,1}$ to receivers Rx_1 and Rx_2, and (b) subfiles $A_{1,23}$, $B_{1,13}$, and $C_{1,12}$ to receivers Rx_1, Rx_2, and Rx_3 at the same time [16].

Algorithm 5.3 Delivery phase for the *ith* base station with label $t \in [3]$, for the case $3\mu_T = 1$ [15]

1: **for** $s = K_i, K_i - 1, \ldots, 1$ **do**
2:     **for** $\mathcal{R} \subseteq [K_i]$ s.t. $|\mathcal{R}| = s$ **do**
3:         The base station transmits $\sum_{k \in \mathcal{R}} \tilde{W}_{d_k, t, \mathcal{R} \backslash \{k\}}$
4:     **end for**
5: **end for**

Algorithm 5.3 outlines the delivery scheme utilized by the *ith* active base station with label $t \in [3]$ for the receiver demands $[d_1 \ \ldots \ d_{K_i}]$. In this algorithm, for any $n \in [N]$ and $\mathcal{R} \subseteq [K_i]$, $W_{n,t,\mathcal{R}}$ represents the subfile of $W_{n,t}$ exclusively cached at receivers in $\mathcal{R}$ and $\tilde{W}_{n,t,\mathcal{R}}$ represents the corresponding PHY coded subfile.

This implies that the following network sum-DoF is achievable:

$$\text{DoF}^*_{\text{L,sum}}(N, \mu_T N, \mu_R N, C) \geq \sum_{i=1}^{\frac{C}{3}} \frac{1}{1 - \mu_R} \sum_{s=0}^{K_i - 1} \binom{K_i - 1}{s} \mu_R^s (1 - \mu_R)^{K_i - s} (s+1), \quad (5.18)$$

from which it can be shown that the achievable DoF per cell is lower bounded by

$$\text{LDoF}^* = \lim_{C \to \infty} \lim_{N \to \infty} \frac{\text{DoF}^*_{\text{L,sum}}(N, \mu_T N, \mu_R N, C)}{C} \geq \mu_T + K_R \mu_R - \frac{1}{3}. \quad (5.19)$$

$3\mu_T = 2$ (Partial BS Collaboration)

In this case, each file $W_n, n \in [N]$ in the library is partitioned into three disjoint subfiles $\{W_{n,12}, W_{n,13}, W_{n,23}\}$, each of size $\frac{F}{3}$ packets. Then, for any $\mathcal{T} \in \{\{1,2\}, \{1,3\}, \{2,3\}\}$, all subfiles $\{W_{n,\mathcal{T}}\}_{n \in [N]}$ are stored at all the base stations with labels $t \in \mathcal{T}$. It is straightforward to verify that such a prefetching scheme satisfies the cache size constraint at the base stations.

Using the aforementioned prefetching pattern, each packet in the library will be stored at two-thirds of the base stations in the network, leading to a partial zero-forcing opportunity at the base station side. Hence, at each time, two-thirds of the base stations will get activated, while the last one-third will be turned off. Such an activation pattern is illustrated in Figure 5.12.

The above activation pattern suggests that the packets that are cached exclusively at s receivers in the network can be delivered in groups of size $\min\{s + \frac{2C}{3}, CK_R\}$. The term $\frac{2C}{3}$ comes from the gain of collaborative zero-forcing of the interference by the base stations, which is possible because of the careful aforementioned prefetching phase. Once at least one receiver in each active cell is served, the number of transmit beamforming variables will suffice for providing such a transmit zero-forcing gain as in the fully connected setting. This gain is complementary to the s term in $s + \frac{2C}{3}$, which results from canceling known interference contributions at the receiver side. The details of the delivery scheme are demonstrated in Algorithm 5.4 for the demand vector

Figure 5.12 The delivery activation pattern for the case of $3\mu_T = 2$ [16].

$[d_1 \ldots d_{CK_R}]$. In this algorithm, for any $n \in [N]$ and $\mathcal{R} \subseteq [CK_R]$, $W_{n,\mathcal{T},\mathcal{R}}$ represents the subfile of $W_{n,\mathcal{T}}$ that is exclusively cached at the receivers in $\mathcal{R}$ and $\tilde{W}_{n,\mathcal{T},\mathcal{R}}$ represents the corresponding PHY-coded subfile. The rest of the notation is based on the notation defined in Section 5.4.3.

Therefore, our delivery scheme can achieve the following DoF per cell:

$$\text{LDoF}^* \geq \lim_{C \to \infty} \frac{1}{C} \cdot \frac{1}{1 - \mu_R} \tag{5.20}$$

$$\sum_{s=0}^{CK_R-1} \binom{CK_R - 1}{s} \mu_R^s (1 - \mu_R)^{CK_R-s} \min\left\{ s + \frac{2C}{3}, CK_R \right\}, \tag{5.21}$$

from which, we can show that

$$\text{LDoF}^* \geq \mu_T + K_R \mu_R - \frac{2}{3}. \tag{5.22}$$

$3\mu_T = 3$ (Full BS Collaboration)

In this case, each base station has the storage capacity to cache the entire library. Hence, all the base stations can be activated at the same time in order to deliver requested contents to the receivers. This implies that all the packets which are exclusively cached at a set of s receivers can be delivered in groups of size $\min\{s + C, CK_R\}$. Therefore, following similar arguments as the previous case with partial collaboration, it can be shown that the following DoF per cell can be achieved:

$$\text{LDoF}^* \geq \mu_T + K_R \mu_R - 1. \tag{5.23}$$

5.6   Future Research Directions

The results reviewed in the previous sections promise significant gains in using caches in the wireless systems, if one carefully exploits the gains that the cache memories

---

Algorithm 5.4 Delivery phase for base stations with labels $\mathcal{T} \subset [3], |\mathcal{T}| = 2$, for the case $3\mu_T = 2$ [15]

1: **for** $s = CK_R - \frac{2C}{3}, CK_R - \frac{2C}{3} - 1, \ldots, 0$ **do**

2:   **for** $j \in [CK_R]$ **do**

3:     **for** $\mathcal{R} \subseteq [CK_R] \setminus \{j\}$ s.t. $|\mathcal{R}| = s$ **do**

4:       Let $\mathscr{R}' = \{\mathcal{R}' \subseteq [CK_R] \setminus (\mathcal{R} \cup \{j\}) : |\mathcal{R}'| = \frac{2C}{3} - 1$

5:       and $\mathcal{R}' \cup \mathcal{R} \cup \{j\}$ contains at least one receiver in each active cell}.

6:       Let $\Pi' = \bigcup_{\mathcal{R}' \in \mathscr{R}'} \Pi_{\mathcal{R}'}$.

7:       partition $W_{d_j, \mathcal{T}, \mathcal{R}}$ to $s! |\Pi'|$ disjoint subfiles $\left\{ W_{d_j, \mathcal{T}, \pi, \pi'} \right\}_{\substack{\pi \in \Pi_{\mathcal{R}} \\ \pi' \in \Pi'}}$ of equal

  sizes.

8:     **end for**

9:   **end for**

10:   **for** $\mathcal{R} \subseteq [CK_R]$ s.t. $|\mathcal{R}| = s + \frac{2C}{3}$ and $\mathcal{R}$ contains at least one receiver in each active cell **do**

11:     **for** $\pi \in \Pi_{\mathcal{R}}^{\mathrm{circ}}$ **do**

12:       Each base station Tx_i transmits a linear combination of the coded subfiles as in

$$X_i = \mathcal{L}_{i, \mathcal{T}, \pi} \left(\left\{ \tilde{W}_{d_{\pi(l)}, \mathcal{T}, \pi[l+1:l+s], \pi[l+s+1:l+s+\frac{2C}{3}-1]} : l \in [s + \frac{2C}{3}] \right\} \right)$$

  such that the subfiles $\left\{ W_{d_{\pi(l)}, \mathcal{T}, \pi[l+1:l+s], \pi[l+s+1:l+s+\frac{2C}{3}-1]} : l \in [s + \frac{2C}{3}] \right\}$
  are simultaneously delivered to the receivers in $\mathcal{R}$ interference-free.

13:     **end for**

14:   **end for**

15: **end for**

---

offer. However, to materialize these gains in practice, there are many problems to be investigated. These problems often need major fundamental research to be addressed. Here we briefly mention some of these research directions.

5.6.1   System-Level Evaluation

To be able to evaluate the gain of caching, complete system-level simulations are needed, where more practical models for the macro and pico cells with different values for SNRs have been considered.

5.6.2   Effect of Uncertainty, Quantization, Delay, and Locality in Channel State Information

In previous sections, we assume the channel state information (CSI) is perfectly available at the receivers and/or the transmitters. However, in practice, this knowledge is available locally in some of the nodes, with some error due to noise in measurement and quantization, and with some delay due to the time needed to measure the channel and transfer the information. A major research initiative is needed to evaluate the effect of the uncertainty in CSI, and also to design the new signaling schemes suitable for these scenarios. Some results in this area have been already published [17–19].

5.6.3   Content with Different Popularity, Time Variation of Popularity, and Cache Updating

In this chapter, we mainly focus on the worst case analysis, where the popularity of the files is not relevant. In addition, we assumed that the content of the database is fixed. However, in reality, some files are more popular, and the popularity of the files changes constantly. As a result, the content of the caches has to be designed and updated accordingly. Some results have been published in this direction (see [20–25]); however our knowledge in these scenarios is very limited.

5.6.4   Other Network Topologies

In this chapter, we focus on some canonical topologies. Some other scenarios such as hierarchical cache networks [26] and device-to-device networks [27] have been investigated in the literature. However, there are many topologies relevant in practice that require theoretical investigation.

5.6.5   Delay-Limited Application and Video Streaming

In some applications, such as video streaming, there is a bound on the delay between the time of request and the time of delivery. If that bound is violated, the system feels that the network is congested and automatically switches to some lower-quality version of the video. In this scenario, the challenge is to exploit the gains of caching as much as possible, while respecting the delay constraint. In this direction, some basic results have been reported in [24, 28]; however, more work is needed.

References

[1]   Cisco, "The Zettabyte Era: trends and analysis," 2013.

[2]   M. A. Maddah-Ali, A. S. Motahari, and A. K. Khandani, "Communication over MIMO X channels: Interference alignment, decomposition, and performance analysis," *IEEE Trans. Inform. Theory*, vol. 54, no. 8, pp. 3457–3470, 2008.

[3]   V. R. Cadambe and S. A. Jafar, "Interference alignment and the degrees of freedom of wireless X networks," *IEEE Trans. Inform. Theory*, vol. 55, no. 9, pp. 3893–3908, 2009.

[4]   A. S. Motahari, S. Oveis-Gharan, M.-A. Maddah-Ali, and A. K. Khandani, "Real interference alignment: Exploiting the potential of single antenna systems," *IEEE Trans. Inform. Theory*, vol. 60, no. 8, pp. 4799–4810, 2014.

[5]   M. A. Maddah-Ali and U. Niesen, "Fundamental limits of caching," *IEEE Trans. Inform. Theory*, vol. 60, no. 5, pp. 2856–2867, 2014.

[6]   Q. Yu, M. A. Maddah-Ali, and A. S. Avestimehr, "Characterizing the rate–memory tradeoff in cache networks within a factor of 2," *arXiv:1702.04563*, 2017.

[7] Q. Yu, M. A. Maddah-Ali, and A. S. Avestimehr, "The exact rate–memory tradeoff for caching with uncoded prefetching," *IEEE Trans. Inform. Theory*, vol. 64, no. 2, pp. 1281–1296, Feb. 2018.

[8] N. Naderializadeh, M. A. Maddah-Ali, and A. S. Avestimehr, "Fundamental limits of cache-aided interference management," *IEEE Trans. Inform. Theory*, vol. 63, no. 5, pp. 3092–3107, 2017.

[9] M. Razaviyayn, G. Lyubeznik, and Z.-Q. Luo, "On the degrees of freedom achievable through interference alignment in a MIMO interference channel," *IEEE Trans. Signal Process.*, vol. 60, no. 2, pp. 812–821, 2012.

[10] M. A. Maddah-Ali and U. Niesen, "Cache-aided interference channels," in *2015 IEEE International Symposium on Information Theory (ISIT)*, Hong Kong, 2015, pp. 809–813.

[11] J. Hachem, U. Niesen, and S. Diggavi, "Degrees of freedom of cache-aided wireless interference networks," *IEEE Trans. Inform. Theory*, p. 1, 2018.

[12] A. Sengupta, R. Tandon, and O. Simeone, "Fog-aided wireless networks for content delivery: Fundamental latency tradeoffs," *IEEE Trans. Inform. Theory*, vol. 63, no. 10, pp. 6650–6678, 2017.

[13] R. Wang, X. Peng, J. Zhang, and K. B. Letaief, "Mobility-aware caching for content-centric wireless networks: Modeling and methodology," *IEEE Commun. Mag.*, vol. 54, no. 8, pp. 77–83, August 2016.

[14] M. A. Maddah-Ali and U. Niesen, "Decentralized coded caching attains order-optimal memory-rate tradeoff," *IEEE/ACM Trans. Networking (TON)*, vol. 23, no. 4, pp. 1029–1040, 2015.

[15] N. Naderializadeh, M. A. Maddah-Ali, and A. S. Avestimehr, "Cache-aided interference management in wireless cellular networks," in *IEEE Int. Conf. Commun. (ICC)*, Jul. 2017.

[16] N. Naderializadeh, M. A. Maddah-Ali, and A. S. Avestimehr, "Cache-aided interference management in wireless cellular networks," *IEEE Trans. Commun.*, vol. 67, no. 5, pp. 3376–3387, May 2019.

[17] J. Zhang and P. Elia, "Fundamental limits of cache-aided wireless BC: Interplay of coded-caching and CSIT feedback," *IEEE Trans. Inform. Theory*, vol. 63, no. 5, pp. 3142–3160, 2017.

[18] A. Ghorbel, M. Kobayashi, and S. Yang, "Cache-enabled broadcast packet erasure channels with state feedback," *arXiv:1509.02074*, 2015.

[19] R. Timo and M. Wigger, "Joint cache-channel coding over erasure broadcast channels," in *Proc. IEEE ISWCS*, Aug. 2015.

[20] U. Niesen and M. A. Maddah-Ali, "Coded caching with nonuniform demands," in *Proc. IEEE INFOCOM Workshops*, Apr. 2014, pp. 221–226.

[21] M. Ji, A. M. Tulino, J. Llorca, and G. Caire, "On the average performance of caching and coded multicasting with random demands," in *Proc. IEEE ISWCS*, Aug. 2014, pp. 922–926.

[22] J. Zhang, X. Lin, and X. Wang, "Coded caching under arbitrary popularity distributions," *IEEE Trans. Inform. Theory*, vol. 64, no. 1, pp. 349–366, 2018.

[23] R. Pedarsani, M. A. Maddah-Ali, and U. Niesen, "Online coded caching," *IEEE/ACM Trans. Networking (TON)*, vol. 24, no. 2, pp. 836–845, 2016.

[24] F. Rezaei and B. H. Khalaj, "Stability, rate, and delay analysis of single bottleneck caching networks," *IEEE Trans. Commun.*, vol. 64, no. 1, pp. 300–313, 2016.

[25] J. Hachem, N. Karamchandani, and S. Diggavi, "Effect of number of users in multi-level coded caching," in *2015 IEEE Int. Symp. Inform. Theory (ISIT)*.   IEEE, 2015, pp. 1701–1705.

[26] N. Karamchandani, U. Niesen, M. A. Maddah-Ali, and S. N. Diggavi, "Hierarchical coded caching," *IEEE Trans. Inform. Theory*, vol. 62, no. 6, pp. 3212–3229, 2016.

[27] M. Ji, G. Caire, and A. F. Molisch, "Fundamental limits of caching in wireless D2D networks," *IEEE Trans. Inform. Theory*, vol. 62, no. 2, pp. 849–869, 2016.

[28] U. Niesen and M. A. Maddah-Ali, "Coded caching for delay-sensitive content," in *2015 IEEE Int. Conf. Commun. (ICC)*, Hong Kong, 2015, pp. 5559–5564.

6 Cloud and Fog Radio Access Networks

Osvaldo Simeone, Ravi Tandon, Seok-Hwan Park, and Shlomo Shamai (Shitz)

6.1 Introduction

In the evolution towards 5G networks, wireless mobile cellular systems are undergoing a profound paradigm shift at both the architectural and functional levels [1, 2]. At the architectural level, the trend is toward a *fog architecture*, which, as illustrated in Figure 6.1, encompasses a wireless edge segment and a hierarchical cloud segment. The cloud segment hosts computing servers that are connected to the wireless edge segment either directly via an access network or through metro and core networks [1]. In contrast, the wireless edge segment is composed of wireless access devices, such as base stations and remote radio heads, which will be referred to as edge nodes (ENs). A key novel aspect of the fog architecture is the availability of computing and storage resources at both cloud servers and ENs [3, 4].

At a functional level, 5G networks are envisaged to be operated by leveraging *network function virtualization* (NFV). This allows the cloud servers and ENs to be programmed and reconfigured via software. With NFV, the network functions that make up a network service, or *network slice*, such as computing tasks and caching, can hence be flexibly and adaptively allocated on the software-controlled network elements in the cloud and edge segments [1, 2, 5, 6].

The reconfigurability of the fog architecture enabled by NFV contrasts with the rigid allocation of network functions in current cellular systems. In particular, in existing conventional deployments, the wireless communication protocol stack is fully implemented at the base stations. This flexibility is expected to allow operators to offer a variety of services, such as mobile broadband communication, content delivery, Internet of Things IoT, and machine-type communications, while reducing capital and operating costs.

However, the adaptability afforded by fog networking comes with the following novel *challenges and open problems*: (1) How should network functions be engineered and instantiated in the fog architecture so as to optimize the performance of a network slice, given the available physical resources for computing, storage, and communication? (2) What is the resulting optimal tradeoff between key performance indicators, such as throughput and latency, and the available physical resources? This chapter takes a network information theoretic viewpoint on the problem with the aim of obtaining fundamental theoretical insights and performance benchmarks.

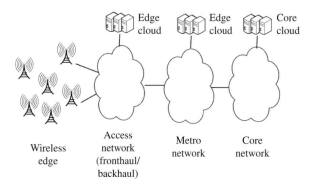

Figure 6.1 Next-generation cellular systems will be characterized by a fog network architecture in which software-controlled network elements can be adaptively assigned computing and caching network functions for different network services, or slices (adapted from [1]). Users (not shown) connect to the wireless edge.

6.1.1   Overview

At one extreme of the fog architecture resides the cloud radio access network (C-RAN), in which a cloud processor (CP) in the edge cloud (see Figure 6.1) implements the baseband signal processing functionalities of a set of ENs [7, 8]. The ENs of a C-RAN are low-complexity and low-cost devices that are connected to the CP by means of fronthaul links and implement only radio-frequency functionalities. At the other extreme lie conventional cellular systems, in which the wireless communication protocol functions are carried out at the ENs, which are full-fledged base stations. This type of architecture will be referred to as distributed RAN (D-RAN) [1]. In this chapter, we will first compare D-RAN and C-RAN for generic mobile broadband transmission, and then study D-RAN, C-RAN, and more flexible fog-based solutions for content delivery.

For mobile broadband communications, the typical key performance indicators of interest are the transmission rates that can be assigned to the users while satisfying reliability constraints. A useful and tractable approximation of the transmission rates is given by the high-signal to noise ratio (SNR) metric of the number of degrees of freedom (DoF). The DoF metric measures the number of interference-free channels that can be supported towards each user (see, e.g., [9]). It is noted that the high-SNR regime, with its focus on the impact of interference over the role played by background noise, is particularly well suited to describe dense and ultra-dense deployments that are expected to be typical in the next generation of wireless systems.

Content delivery, most notably video delivery, is predicted to account for 75% of the overall mobile data traffic by 2019 [10], hence playing a prominent role in the design of next-generation cellular systems [11–13]. A key performance indicator for content delivery is the coding latency required to deliver a requested fixed-size content to the users within a given transmission slot. A high-SNR measure of the delivery latency was recently introduced and studied [14–16] and is referred to as normalized delivery time (NDT). The NDT was adopted in a number of follow-up works [17–27].

In this chapter, we will first review the DoF analysis of mobile broadband communication considering D-RAN and C-RAN deployments in Section 6.2. We will then cover the NDT analysis of content delivery for D-RAN with edge caching, C-RAN and fog, or F-RAN, solutions, in Section 6.3. Finally, we will back off from the asymptotic analyses in Section 6.4 by considering the performance at finite SNR and finite blocklength of optimized delivery strategies, as well as by introducing various techniques that prove useful in nonasymptotic settings.

6.2   Degrees of Freedom Analysis of Mobile Broadband Communication in D-/C-RAN

We start by discussing the performance of D-RAN and C-RAN for the conventional network slice of mobile broadband communication, in which the key performance metric is the transmission rate. We use this section also to introduce the key modeling assumptions adopted throughout this chapter.

6.2.1   Main Assumptions

We consider the model in Figure 6.2, in which a CP, along with M cache-aided ENs, provide downlink broadband communication service to a set of K users over a shared wireless channel. In a C-RAN, all the ENs are connected to the CP in the edge cloud (see Figure 6.1) through an access network made of so-called *fronthaul* links. In a C-RAN, the CP carries out part or all of the encoding functions on behalf of the connected ENs, while in a D-RAN processing is performed locally at each EN. Strictly speaking, solutions in which part of the processing is carried out at the ENs and part at

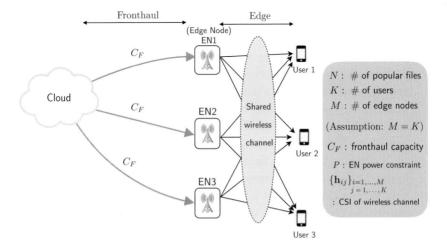

Figure 6.2 System model adopted for the study of mobile broadband communications in D-/C-RAN. This figure represents an instantiation of wireless edge and edge cloud of Figure 6.1.

the cloud could be classified as being fog-based approaches, but here we will reserve this label for the F-RAN system studied in the next section for content delivery. In order to simplify the discussion, in this chapter we focus on the case $K = M$, so that there is an equal number of ENs and users.

Traffic model: For mobile broadband applications, it is reasonable to assume that active users have backlogged queues of data yet to be received in the network. Accordingly, we assume that each user k can receive at any rate R_k in bits per channel use, or per second/Hz, of the wireless channel, that can be supported with high reliability. With an eye on fairness, we focus on equal achievable rates $R_k = R$. D-RAN and C-RAN models differ with respect to the availability of the data streams at the network nodes.

In a D-RAN, each data stream intended for a user k is available at a subset of ENs $\mathcal{M}_k \subseteq \{1, \ldots, M\}$, which act as encoders for this information. When explicitly stated, we will also consider the case in which each EN in the subset $\mathcal{M}_k$ has available a distinct equal fraction of the message intended for user k. In this case, rather than being shared by the ENs in $\mathcal{M}_k$, the message for user k is collectively stored across the ENs in $\mathcal{M}_k$. We emphasize that the definition of D-RAN given here is a generalization of the more conventional one in which ENs serve a disjoint set of users.

In contrast to D-RAN, in a C-RAN all data streams are available at the CP. The CP may encode all data streams or forward the data streams to suitable subsets of ENs for encoding. We note that this definition of C-RAN complies with current discussions on flexible functional splits across CP and ENs [8]. The definition generalizes more conventional C-RAN systems, in which all encoding is centralized at the CP and no local encoding is allowed at the ENs. We also emphasize that, in a C-RAN, delivering the message intended for user k to a subset of ENs entails an overhead due to transmission on the fronthaul links, which is not accounted for in a D-RAN.

Fronthaul model: The fronthaul connectivity between cloud and ENs is realized by means of dedicated links of capacity C_F (bits per symbol of the edge wireless channel). This model corresponds to most current implementations of cloud-assisted systems that leverage dedicated fronthaul links, such as fiber optic cables or microwave links (see, e.g., [7, 28]).

Edge model: The ENs are connected to the users over a shared wireless channel. In 5G systems, the radio interface will be based on multicarrier modulation, and hence a coding frame will occupy a given portion of the time–frequency plane as in the current generation of cellular systems. In particular, given a signaling bandwidth W, the number of symbols[1] in a frame of duration D equals approximately $T = D/W$. Considering the bandwidth W fixed, one can hence take the number of symbols T as a measure of the coding latency. In order to keep the discussion simple, we focus on a setup with single-antenna ENs and single-antenna users. Some discussion on the practically important case with multiantenna ENs is provided in Sections 6.4 and 6.5.

At each symbol, denoting the input of the mth edge node as X_m, the channel output at the kth user is given as

---

[1] We use the term "symbol" to indicate a channel use and not an OFDM symbol.

$$Y_k = \sum_{m=1}^{M} h_{mk} X_m + Z_k, \tag{6.1}$$

where h_{mk} is the channel coefficient between the mth edge node and the kth user, and Z_k additive Gaussian noise with power equal to 1. Each EN has a power constraint P. Unless stated otherwise, we assume that the channel coefficients have continuous distributions and that each coding frame contains a large number of channel coherence blocks in the time–frequency plane. This assumption can be considered a reasonable approximation for systems operating over sufficiently large bandwidths [29], as is the case, e.g., with carrier aggregation. In this chapter, we assume that full channel state information (CSI) about all channel coefficients $H = \{h_{km}\}$ across the coding frame is available at ENs and CP. It is observed that, while useful to obtain key insights into the optimal operation of the system, this assumption can affect the validity of the results in the more practical regime of imperfect CSI, especially at high SNR [30]. We will provide some additional discussion on this point below.

We finally note that model (6.1) assumes full connectivity, that is, it implies that all users are covered by all ENs. This is clearly only accurate in specific deployments such as locally ultra-dense networks. For a brief discussion on the effect of partial connectivity, refer to Section 6.5.

6.2.2 Degrees of Freedom

We now introduce the performance metric of per-user DoF and provide some discussion about its merits and limitations. The DoF measures the high-SNR behavior of the system under study, hence concentrating on the effect of interference. Other DoF analyses that focus on 5G systems include that by Lu and Wei [31].

In order to study the impact of the fronthaul links in the high-SNR regime, we parametrize the fronthaul capacity as $C_F = r \log(P)$. The rationale for this parametrization is that $\log(P)$ is approximately the high-SNR capacity of each EN-to-user wireless channel in the absence of interference. Therefore, the parameter r measures the relative capacity of each fronthaul link with respect to the mentioned reference wireless links. We observe that, when $r > 1$, each fronthaul link has a larger capacity than each reference interference-free wireless link and the opposite is true when $r < 1$.

For any achievable equal per-user rate $R(P, C_F)$ for an EN's power constraint of P and fronthaul capacity C_F, we define the per-user DoF as

$$\text{DoF}(r) = \lim_{P \to \infty} \frac{R(P, r \log P)}{\log P}, \tag{6.2}$$

and the maximum per-user DoF over all transmission strategies as DoF*. Note that a rate $R(P, C_F)$ is achievable if a communication strategy exists that guarantees an arbitrarily small probability of error as the communication blocklength grows large. The DoF (6.2) measures the number of interference-free data streams that can be sent reliably, i.e., with vanishing error probability, to the users at the maximum data rate of $\log P$. An important initial observation is that the maximum per user is one, since each single-antenna user can receive at most one data stream at capacity.

While useful to obtain analytical insights into the performance of different interference management schemes and hence widely studied, the DoF metric has a number of limitations, shared also by the NDT measures to be formalized below, that are worth discussing here. (1) By neglecting the impact of noise, the DoF metric does not capture the performance gains that can be obtained via techniques that improve the SNR, such as coherent beamforming. This will be further discussed in Section 6.4. (2) The DoF metric is quite sensitive to the assumed channel model. In particular, the number of channel coherence blocks per coding frame generally limits the achievable DoF. This will be further elaborated on below when describing the performance of specific schemes. As a special case, when the channel gains are constant through the coding frame, that is, with quasi-static fading, the maximum achievable DoF is a discontinuous function of the channel gains [32, 33]. We note that this dependence on the channel coefficients is less pronounced for multiantenna systems [34]. (3) The DoF metric is not able to capture differences between power levels on the direct link between a transmitter and the intended receiver and on the cross-link to an interfered user. This limitation can be alleviated by considering the metric of the generalized degrees of freedom (GDoF) (see, e.g., [35]). We will not discuss this aspect further in this chapter. (4) Finally, the DoF metric is sensitive to the assumption of perfect CSI. The impact of imperfect CSI will be further expanded on below.

In the following subsections, we describe various transmission strategy schemes and corresponding per-user DoFs for D-RAN and C-RAN.

6.2.3   DoF Analysis of D-RAN

In a D-RAN, the cloud connections are assumed to be disabled and, as discussed, the message to be delivered to user k is available at a subset $\mathcal{M}_k$ of ENs. In the following, we derive achievable per-user DoFs (6.2) for a number of baseline communication strategies for a D-RAN. The DoF metric is here denoted here as DoF by dropping the dependence on the fronthaul rate r, which is not relevant for D-RANs.

Interference avoidance (orthogonal multiplexing): Assume that each set $\mathcal{M}_k$ includes at least EN k, that is, $\mathcal{M}_k \subseteq \{k\}$. This means that the message intended for user k is available in full at least at EN k. Interference avoidance can then be realized by scheduling transmission from each EN k to user k in an orthogonal spectral resource. Since each user is only served for a fraction of time or bandwidth equal to $1/M$, the corresponding per-user DoF is given as

$$\text{DoF}_{\text{OR}} = \frac{1}{M}. \tag{6.3}$$

This DoF serves as a reference for more advanced interference management schemes.

Cooperative transmission: Assume now that the EN subsets $\mathcal{M}_k$ are all equal to the set of all ENs $\{1, \ldots, M\}$. In other words, we consider here that all ENs share all messages intended for the M users. In this case, cooperative beamforming can be performed across all ENs to null interference at each user via zero-forcing (ZF) beamforming. Zero-forcing beamforming converts the channel matrix into a multiple of the identity by pre-multiplying the transmitted signals by the inverse of the channel across all ENs. Note that the channel matrix is full-rank with high probability. Furthermore, we emphasize that this approach, unlike interference avoidance via orthogonal multiplexing, hinges on the assumption of full CSI at all ENs. Since each user can receive a data stream with no interference on the full time–frequency block, the resulting per-user DoF is the maximum possible, namely

$$\text{DoF}_{\text{COOP}} = 1. \tag{6.4}$$

To understand the dependence of (6.4) on the assumption of full CSI, it is useful to briefly review a result from [36]. There, each channel gain is assumed to be known at the ENs with an additive error of power $P^{-\beta}$ for some parameter $0 \le \beta \le 1$. It was shown that, under this assumption, the per-user DoF equals

$$\text{DoF}_{\text{COOP},\beta} = 1 - (1 - \beta)\left(1 - \frac{1}{M}\right). \tag{6.5}$$

This demonstrates that, in order to obtain the DoF (6.4), the CSI error must decrease as P^{-1}, while a finite-precision CSI, i.e., $\beta = 0$, yields a per-user DoF equal to interference avoidance (cf. (6.3)).

Interference alignment on an interference channel: Let us now consider again the setup in which we have $\mathcal{M}_k \subseteq \{k\}$ as in the case of interference avoidance. However, we consider here a more advanced transmission strategy whereby the ENs coordinate their transmission to manage mutual interference. Specifically, under the message assignment $\mathcal{M}_k = \{k\}$, the optimal interference management strategy is interference alignment (IA) [9].

The key idea of IA is to transmit linearly precoded symbols from each EN in such a way that the net interference caused by $M - 1$ ENs at each user aligns in a "small" subspace. Note that linear precoding operates across multiple transmitted symbols. It would be desirable that the fraction of the signal space that is occupied by interference be smaller than the fraction $1 - 1/M$ that would be occupied by each user with orthogonal multiplexing. Remarkably, the fraction of signal space occupied by the interference can be made to be equal to $1/2$ by means of IA. This achieves the following per-user DoF:

$$\text{DoF}_{\text{IA}} = \frac{1}{2}. \tag{6.6}$$

Again, it is important to point out the limitations of the result (6.6). A first observation is that, in a manner similar to the discussion above for the cooperative ZF scheme, finite precision CSI yields a DoF performance that is no better than interference avoidance [30]. In addition to needing full CSI, achieving the per-user $\text{DoF}_{\text{IA}} = 1/2$ with single antenna nodes also requires communication over a number of coherence blocks that grows exponentially with the number of users. More recent works have investigated the design of IA schemes with limited diversity [37, 38]. The key finding of these papers is that one can trade off a reduction in the number of coherence blocks over which precoding is done, and hence a reduction in complexity and delay, for a lower per-user DoF. In the extreme case in which channel diversity is not available, i.e., for static channels, nonlinear real interference alignment techniques can be used to also achieve the same per-user DoF $= 1/2$ under given conditions on the channel gains [32].

Interference alignment on an X channel: We assume here that the EN subsets $\mathcal{M}_k$ are all equal to the set of all ENs $\{1, \ldots, M\}$ as in the case of cooperative transmission, with the following caveat: Each EN has a distinct $1/M$-fraction of each message. As a result, each user must receive a distinct fraction of the message from all ENs. This configuration is conventionally referred to as an X-channel. As shown in [39] (also see [35]), through IA over a total of $2M - 1$ signal dimensions, the interference can be aligned, using linear precoding across multiple symbols, in a $(M - 1)$ dimensional space, and the remaining M dimensions can be used for decoding the desired signal at each user. This leads to the achievable per-user DoF

$$\text{DoF}_{\text{XIA}} = \frac{1}{2 - \frac{1}{M}}. \tag{6.7}$$

This IA scheme over the X channel also requires an exponential number of coherence blocks over which to perform precoding as well as full CSI. The study of static channels was carried in [32].

6.2.4   DoF Analysis of C-RAN

In this subsection, we consider the DoF performance of C-RAN systems. The operation of C-RAN can be first classified with respect to the way in which the fronthaul links connecting cloud and ENs are used. The two main operating modes for the fronthaul, which typically correspond also to distinct technological implementations, will be referred to as hard and soft transfer fronthauling.

Hard transfer calls for the transmission of messages, or fractions thereof, on the fronthaul links. Therefore, with hard transfer, encoding of the message takes place at the ENs. This mode of operation may be generally deployed over wired or wireless digital links with relatively loose synchronization and jitter requirements. We note that, when fronthaul links are operated by means of hard transfer, they are often also referred to as backhaul links. In contrast, soft transfer prescribes coding and precoding at the CP and the transmission of encoded and precoded messages over the fronthaul links. More specifically, the information sent on the fronthaul links amounts to a quantized and compressed version of the encoded and precoded messages. Quantization and compression are needed to ensure that the fronthaul capacity constraints are satisfied. Given that the ENs in this mode act as repeaters of the decompressed signals received on the fronthaul links, soft transfer generally needs digital fronthaul links with stringent synchronization and jitter requirements. We briefly observe that analog fronthauling solutions, whereby radio- or intermediate-frequency analog signals are transferred on the fronthaul links, are also deployed, e.g., over fiber optics or DSL cables (see, e.g., [40]).

We now provide a high-SNR analysis of soft and hard transfer modes in terms of DoF. We start by discussing the assumed orthogonal serial operation across fronthaul and edge channels.

Serial operation: To elaborate, we assume that fronthaul transmission is followed by wireless transmission in a serial manner, as seen in Figure 6.3. Note that we do not consider here the classical approach often invoked in information theory of enabling coding

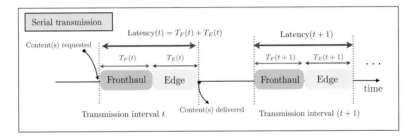

Figure 6.3 Serial transmission over cloud and edge segments. The overall latency is the sum of the latencies over both segments.

across many blocks, due to its scarce applicability to practical coding schemes that cannot afford extremely large latencies. Nevertheless, operating over multiple blocks would enable pipelining fronthaul and wireless transmissions. We refer to [16] for additional discussion on this point. With the approach adopted here, which is illustrated in Figure 6.3, transmission requires one block for the fronthaul and one for the wireless channel. The two blocks are generally of different durations.

We now review transmission strategies for C-RAN that build on the edge communication schemes presented above for D-RAN. We also discuss a simple upper bound on the achievable DoF that demonstrates the optimality of a soft transfer-based transmission strategy.

Hard transfer: With hard transfer, the cloud can provide each EN with a subset of messages, or fractions thereof, thereby ensuring that the message for each user k is available at a subset $\mathcal{M}_k$ of ENs. The transmission rate achievable on the wireless channel for the resulting message assignment was studied above in the context of D-RAN for a number of baseline strategies. The computation of the achievable rate in C-RAN, however, requires accounting also for the fronthaul transmission, as discussed next.

With hard transfer, if the fronthaul transmission has a per-user rate of R_F bits per symbol and a per-user rate of R_E bits per symbol is achievable on the wireless channel, the resulting per-user transmission rate is obtained as

$$R = \frac{1}{1/R_F + 1/R_E}. \tag{6.8}$$

This is because the transmission of L bits per user requires $L/R_F + L/R_E$ symbols, and hence the transmission rate is $L/(L/R_F + L/R_E)$. Furthermore, from (6.8) and (6.2), the per-user DoF is given as

$$\text{DoF}(r) = \frac{1}{1/\text{DoF}_F + 1/\text{DoF}_E}, \tag{6.9}$$

where we have defined the fronthaul DoF as $\text{DoF}_F = \lim_{P\to\infty} R_F/\log P$ and the edge DoF as $\text{DoF}_E = \lim_{P\to\infty} R_E/\log P$. As per (6.9), the DoF of C-RAN schemes depends on both fronthaul and edge transmission rates and hence on the corresponding DoFs. In order to apply orthogonal transmission over the edge channel, which achieves $\text{DoF}_E = 1/M$ (see (6.3)), each EN k requires only the message of its associated user k, which can be transmitted via hard transfer. The resulting fronthaul DoF is $\mathrm{DoF}_F = r$, since information can be delivered for each user at a rate $R_F = r \log P$ on the fronthaul links. Thus, the resulting per-user DoF is given as

$$\mathrm{DoF}_{\mathrm{HT+OR}}(r) = \frac{1}{1/r + M}. \tag{6.10}$$

As an alternative approach, the maximum edge DoF $\mathrm{DoF}_E = 1$ is achieved via cooperative ZF beamforming (see (6.4)), which requires the hard transfer of all messages to all ENs. This entails a fronthaul DoF equal to $\mathrm{DoF}_F = r/M$, since the message of each user can be transmitted on each fronthaul link at a rate $R_F = (\log P)/M$. This leads to the overall per-user DoF

$$\mathrm{DoF}_{\mathrm{HT+COOP}}(r) = \frac{1}{M/r + 1}. \tag{6.11}$$

It is also possible to leverage edge transmission strategies based on IA. As a first approach, one can employ conventional IA, which requires each EN to be informed about only one message. Following the same argument used above for the scheme based on orthogonal transmission and using (6.6), the per-user DoF

$$\mathrm{DoF}_{\mathrm{HT+IA}}(r) = \frac{1}{1/r + 2} \tag{6.12}$$

is achievable. As a second approach, the edge segment can implement an X-channel IA scheme, achieving the edge DoF (6.7). This requires each EN to be informed about a fraction $1/M$ of all messages. Therefore, the per-user transmission rate on the fronthaul is $R_F = (1/M)/(r \log P/M)$, achieving the per-user DoF

$$\mathrm{DoF}_{\mathrm{HT+XIA}}(r) = \frac{1}{1/r + 2 - 1/M}. \tag{6.13}$$

Soft transfer: With soft transfer fronthauling, as first proposed in [41], the CP implements ZF beamforming and quantizes the resulting encoded signals. Using a resolution of $\log P$ bits per downlink baseband sample, it can be shown that the effective SNR in the downlink scales proportionally to the power P [16, 41]. This implies that each user receives an interference-free channel with capacity scaling as $\log P$. Therefore, the number of channel uses needed to communicate to each user on the edge channel is approximately equal to $L/\log P$ in high SNR. Furthermore, each of the $L/\log P$ samples are quantized by the CP with $\log P$ bits, entailing a fronthaul transmission time equal to $(L/\log P \times \log P)/(r \log P) = L/(r \log P)$, given the fronthaul capacity $C_F = r \log P$. From the discussion above, the transmission rate per user equals the ratio between the L bits conveyed to each user and the overall time required to communicate on the fronthaul and edge channels, namely $L(1 + 1/r) \log P$. It follows that the following overall per-user DoF is achievable by means of soft transfer

$$\mathrm{DoF}_{\mathrm{ST}}(r) = \frac{1}{1/r + 1}. \tag{6.14}$$

As a practical note, we observe that a key difference between soft and hard transfer schemes is that soft transfer requires CSI at the cloud to enable centralized precoding, while hard transfer only requires CSI at the ENs.

We now argue that the DoF (6.14) achieved by soft transfer is in fact information-theoretically optimal. To see this, we can use the following cut-set argument. Consider an enhanced system in which the ENs can cooperate and are the final recipients of the messages. In this system, the time required to communicate from the cloud to the ENs would approximately equal $2L/(2r \log P) = L/(r \log P)$, since the cloud-to-EN connection has capacity $2C_F$. Note that this time matches that needed by soft transfer. Furthermore, the edge transmission time cannot be smaller than that of an interference-free system, namely $L/\log P$. Given that this also matches the time needed by soft transfer, we can conclude that there exists no scheme that can improve over the DoF of the discussed soft transfer-based scheme.

As a final remark, we emphasize that the optimality of soft transfer in the high-SNR regime does not necessarily translate into its dominance at lower values of the SNR. In fact, the optimality of soft transfer at high SNR can be intuitively attributed to the fact that the edge transmission packet tends to be small in this regime, which entails that quantizing this packet for fronthaul transmission yields an efficient use of the fronthaul resources. This point will be further discussed in Section 6.4.

6.2.5 Numerical Examples

In Figure 6.4, we compare the DoF performance of the schemes presented above for communication over a C-RAN model. We specifically plot the DoF as a function of the fronthaul rate r for $M = K = 10$, i.e., $K = 10$ users connected to $M = 10$ ENs. As argued above, soft transfer outperforms all other schemes, particularly for intermediate values of r. Note that for large values of r, both soft and hard transfer with cooperative transmission asymptotically achieve the upper bound of DoF equal to 1. However, we recall that the advantage of soft transfer comes at the cost of acquiring timely CSI at the cloud. We also observe that the relative behavior of hard transfer-based schemes is markedly different depending on the value of r. In particular, for smaller values of r (i.e., $r \in (0, 1]$), hard transfer-based IA for the X channel (HT + XIA) is to be preferred among all hard transfer-based schemes. Instead, for larger values of r, the hard transfer-based cooperative transmission strategy (HT+COOP) begins to dominate all other hard transfer-based schemes due to the reduction in latency for transmitting messages to all the ENs via hard transfer.

6.3 NDT Analysis of Content Delivery via D/C/F-RAN

In this section, we turn to the analysis of the performance of the network slice of content delivery in cloud and fog-assisted networks. We will first consider an offline caching scenario, which is more commonly studied in the information theoretic literature but is generally deemed to have narrow applicability in practical implementations. We will

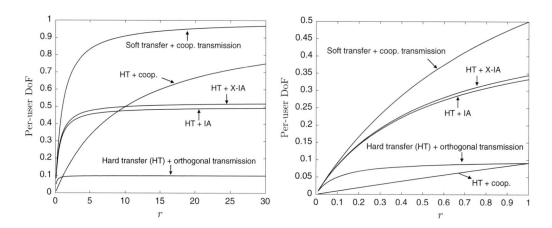

Figure 6.4 Per-user DoF achieved by soft and hard transfer-based schemes versus the fronthaul rate r. The right figure zooms in on the performance for the range $r \in (0, 1]$.

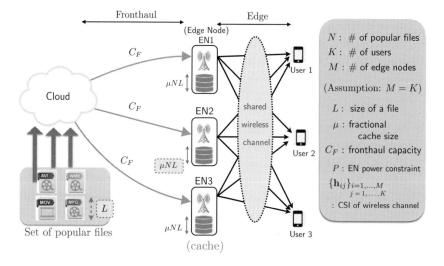

Figure 6.5 System model adopted for the study of content delivery in D-/C-/F-RAN.

then investigate the optimization of the system operation under an online caching setting, demonstrating the suitability of the insights offered by the offline caching analysis to guide the analysis of online methods.

6.3.1   Main Assumptions

Consider the fog-assisted model in Figure 6.5, in which a CP, which is directly connected to a content server, along with M cache-aided ENs, provide content-delivery service to a set of K users over a shared wireless channel. We refer to this system as being fog-assisted, since content delivery can take place by leveraging cloud computing and fronthaul transmission resources, as studied in the previous section, and also by leveraging the content cached at the ENs at the edge. We again focus here on the case $M = K$ to simplify the discussion, and we refer to [16] for technical details and extensions regarding the offline caching scenario and to [42] for the online case. We detail here the offline case and discuss separately the setup with online caching.

In an offline caching setting, the operation of the system is divided into two separate phases: a *prefetching*, or *caching*, phase in which the ENs' caches are populated on the basis of a static content library; and a *delivery* phase, in which users make requests from the content library. The caching phase should in practice take place when the network load is low, e.g., at night, so as not to be affected by fronthaul limitations. Instead, the delivery phase is carried out during the time of regular operation of the system, e.g., throughout the day, and consists of many transmission intervals.

Traffic model: Under an offline caching model, we assume the presence of a library $\mathcal{F} = \{F_1, F_2, \ldots, F_N\}$ of popular files, with $N \geq K$, where each file is of size L bits. This library represents the contents that may be requested by the users and is assumed to be invariant throughout the caching and delivery phases. In each transmission interval $t = 1, 2, \ldots$ of the delivery phase, users issue a vector of requests $D(t) = (D_1(t), \ldots, D_K(t))$, where $D_k(t) \in \{1, \ldots, N\}$ indicates that file $F_{D_k(t)}$ is requested by user k at time t. We will focus here on a *worst-case traffic* model, according to which the users' requests are arbitrary. Section 6.4 will also consider the performance under the *Independent Reference Model (IRM)*, in which each file F_k is requested independently according to a known popularity distribution, e.g., Zipf.

Offline caching model: Each EN can cache a fraction μ, with $0 \leq \mu \leq 1$, of the content library, that is, μNL bits. Caching is done offline in a prefetching phase based only on the knowledge of a set of popular files $\mathcal{F}$. Each content can be arbitrarily split prior to caching, with possible coding within each file, and stored across the caches of different ENs.

Fronthaul and edge models: The fronthaul capacity and the downlink received signal (6.1) are defined as discussed in the previous section.

For the content-delivery service of interest here, we will adopt the delivery coding latency as the performance indicator of interest, instead of the transmission rate. The focus on the latency is justified by the fact that, in content delivery, the amount of information to be delivered is fixed, rather than being allowed to depend on network state, as for mobile broadband communication. As a result, one typically wishes to minimize the time required for its transmission. The delivery coding latency measures the amount of time needed in a transmission interval in order to deliver the requested files reliably. Note that the delivery latency generally includes other contributions related to networking aspects, such as queueing and transmission of protocol control information. These terms depend on a number of specific technological details and, while important, will not be part of this study.

To evaluate the latency – the term used henceforth to refer to the coding latency only – we need to specify the timeline of transmission across the fronthaul and edge segments. To this end, we will assume here serial fronthaul–edge transmission, as illustrated in Figure 6.3, where the cloud first transmits to the ENs and then the ENs transmit to the users. The overall latency is hence the sum of the durations T_F and T_E of the fronthaul

and edge transmissions, respectively. For an analysis of pipelined transmission, we refer
to [16].

6.3.2   Normalized Delivery Latency

We next define the proposed performance metric of NDT by first introducing the
notion of delivery time per bit. A *delivery time per bit* $\Delta(\mu, C_F, P)$ is achievable if
there exists a sequence of feasible policies, each with fronthaul and edge transmission
times T_F and T_E (generally functions of the file size L), such that the following limit
holds

$$\Delta(\mu, C_F, P) = \limsup_{L \to \infty} \frac{T_F + T_E}{L}. \tag{6.15}$$

Note that the limit is over the file size, which is accordingly allowed to be arbitrarily large in order to obtain any desired level of accuracy as per Shannon's standard
framework. As discussed, the delivery time per bit accounts for the latency within each
transmission interval, as illustrated in Figure 6.3. In order to obtain a vanishing probability of error, the fronthaul and edge latencies T_F and T_E need to scale with L, and it
is this scaling that is measured by (6.15). We also observe that the definition of delivery time per bit in (6.15) is akin to the completion time studied in [43, 44] for standard
channel models, such as broadcast and multiple access channels.

We next define a more tractable metric that reflects the latency performance in the
high-SNR regime. As in the previous section, we let the fronthaul capacity scale with
the SNR parameter P as $C_F = r \log(P)$, where r is the fronthaul gain. For any achievable
$\Delta(\mu, C_F, P)$, with $C_F = r \log(P)$, the NDT, is defined as

$$\delta(\mu, r) = \lim_{P \to \infty} \frac{\Delta(\mu, r \log(P), P)}{1/\log P}. \tag{6.16}$$

In (6.16), the delivery time per bit (6.15) is normalized by the term $1/\log P$. The latter
is the delivery time per bit in the high-SNR regime for a reference system with no
interference and unlimited caching, in which each user can be served by a dedicated EN
which has locally stored all the files. An NDT of δ hence indicates that the worst-case
time required to serve any possible request vector is δ times larger than the time needed
by this reference system. Finally, for any given pair (μ, r), the minimum NDT is defined
as

$$\delta^*(\mu, r) = \inf \{\delta(\mu, r) : \delta(\mu, r) \text{ is achievable}\}. \tag{6.17}$$

The NDT in (6.17) can be seen to be equivalent to the inverse of the per-user DoF
metric introduced in the previous section. Specifically, we have the relationship $\delta =
1/\text{DoF}$ for any given sequence of achievable schemes. One of the key technical reasons
why it is convenient to work with the NDT metric is that, unlike the DoF metric, the
minimum NDT, $\delta^*(\mu, r)$, is a convex function of μ for every value of $r \geq 0$ [16]. In fact,
for any two feasible policies that require fractional cache capacity and fronthaul gain
pairs (μ_1, r) and (μ_2, r) and achieve NDTs $\delta(\mu_1, r)$ and $\delta(\mu_1, r)$, respectively, given an
F-RAN system with cache storage capacity $\mu = \alpha\mu_1 + (1 - \alpha)\mu_2$ for some $\alpha \in [0, 1]$

and fronthaul gain r, we can achieve the NDT $\delta(\mu, r) = \alpha\delta(\mu_1, r) + (1 - \alpha)\delta(\mu_2, r)$ by using the first policy for the first α-fraction of each file and the second for the remaining $1 - \alpha$ fraction. We refer to this approach as *file splitting*.

6.3.3 NDT of D-RAN via Edge Caching

We consider first the case in which delivery is carried out by leveraging only the cached contents via edge transmission. In this scenario, no cloud connectivity is used and hence one can set the fronthaul rate as $r = 0$ without loss of generality. We first describe caching and delivery strategies for two extreme regimes for the value of the fractional cache capacity μ, namely $\mu = 1$, in which full caching of the entire library is possible, and $\mu = 1/M$, which corresponds to the minimum value of the cache size that enables the storage of the library across the aggregate capacity of the ENs' caches. Caching and delivery strategies for intermediate values of μ can be obtained from the two extreme cases by means of file splitting. More precisely, defining as $\delta(1, 0)$ an NDT achievable with $\mu = 1$ and as $\delta(1/M, 0)$ an NDT achievable with $\mu = 1/M$, the NDT $\delta(\mu, 0) = \mu\delta(1, 0) + (1 - \mu)\delta(1/M, 0)$ is obtained by file splitting.

Full caching ($\mu = 1$): If $\mu = 1$, every EN can store the entire file library. Under these assumptions, given the worst-case request vector in which all M users request different files, the resulting system can be treated as a D-RAN with cooperative ENs, i.e., with $\mathcal{M}_k = \{1, \ldots, M\}$ for any user k. As explained in the previous section, by using ZF beamforming, a per-user DoF equal to (6.4) can be achieved and hence an NDT equal to $\delta_{\text{D-RAN}}(1, 0) = 1$ is obtained. In other words, the NDT performance is the same as for the reference system introduced above.

Minimum caching ($\mu = 1/M$): If $\mu = 1/M$, each EN can cache non-overlapping $1/M$-fractions of each file. Under this placement scheme, for the worst-case demand vector in which each user requests a different file, the edge transmission policy can follow the D-RAN interference alignment scheme for the X-channel discussed in the previous section, yielding an NDT of $\delta_{\text{D-RAN}}(1/M, 0) = 2 - 1/M$ (see (6.7)).

6.3.4 NDT of C-RAN

Cloud-aided policies that neglect the caches at the ENs follow directly the C-RAN approach described in the previous section. As a result, using soft transfer, which maximizes the DoF and hence minimizes the NDT, from (6.14), one can achieve the NDT

$$\delta_{\text{C-RAN}}(0, r) = \frac{1}{r} + 1. \tag{6.18}$$

6.3.5 NDT of F-RAN

By leveraging the cloud- and edge-assisted properties of the fog architecture, one can benefit from the advantages of both edge processing, namely reduced latency for the cached contents, and of cloud processing, notably centralized beamforming gains. A

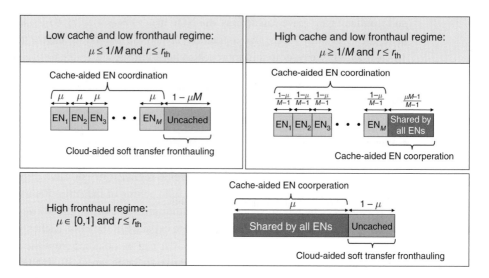

Figure 6.6 Caching and delivery schemes achieving the offline NDT $\delta_{\text{off,ach}}(\mu, r)$.

simple, and effective, way to do this is to perform file splitting between C-RAN and D-RAN strategies. Upon optimization of the file splitting ratios, detailed in [16], the resulting policy is illustrated in Figure 6.6 and can be summarized as follows.

In the regime of high fronthaul rate, i.e., when the fronthaul rate satisfies $r \geq 1$, the same fraction μ of each file is stored at all ENs. For any request vector, the cached fraction μ of the requested files is then transmitted using D-RAN operation via cooperation, while the uncached fraction $1 - \mu$ is delivered using C-RAN processing via soft-transfer fronthauling. In contrast, in the regime of low fronthaul rate, i.e., $r < 1$, placement is aimed at caching fractions of each file that are as distinct as possible at different ENs. The rationale of this choice is that one would like to minimize fronthaul usage in this regime. As a result, when $r < 1$, we need to distinguish the two cases represented in Figure 6.5 in which the μ-fractions of each file stored at different ENs are either fully disjoint, possibly leaving a portion of each file uncached, or partially overlapping. Delivery of the shared and uncached fractions takes place using D-RAN-based cooperation and C-RAN processing via soft transfer, respectively. Instead, the fractions cached at distinct ENs are delivered using D-RAN cooperation.

6.3.6 Minimum Offline NDT

In [16], it is shown that the F-RAN approach summarized above achieves the minimum NDT within a constant multiplicative gap equal to 2, independent of the problem parameters, for all regimes of fractional cache size μ and fronthaul gain r. Mathematically, we have the inequality

$$\frac{\delta_{\text{off,ach}}(\mu, r)}{\delta^*(\mu, r)} \leq 2, \qquad (6.19)$$

for $\mu \in [1/M, 1]$ when $r = 0$ and for $\mu \in [0, 1]$ when $r > 0$. The scheme is also shown to be exactly optimal in terms of NDT for a number of special cases including the scenario with $M = 2$. For this case, the minimum NDT is characterized as follows.

- Cache-only F-RAN ($r = 0$): For $M = 2$, the minimum NDT when $r = 0$ is given as

$$\delta^*(\mu, r) = 2 - \mu. \tag{6.20}$$

- Low fronthaul ($r \in (0, 1]$): Instead, when ($r \in (0, 1]$), we have

$$\delta^*(\mu, r) = \max \left(1 + \mu + \frac{1 - 2\mu}{r}, 2 - \mu \right). \tag{6.21}$$

- High fronthaul ($r > 1$): When $r > 1$, we have

$$\delta^*(\mu, r) = 1 + \frac{1 - \mu}{r}. \tag{6.22}$$

The optimal NDT for the case $M = 2$ is illustrated in Figure 6.7(b). We observe that, as per the discussion above, in the high fronthaul regime, the use of both cloud and caching resources are necessary to achieve the minimum NDT, while in the low fronthaul regime, if the cache size is sufficiently large, namely if $\mu \geq 1/2$, it is enough to leverage cache-only resources to achieve the minimum NDT.

6.3.7   Online Caching

Here, we describe the system model and some key results related to the analysis of fog-assisted cellular systems under online caching.

Main assumptions: The system model is defined as for the offline scenario studied above with the key caveat that the set of popular files, rather than being static, changes from one time slot to the next, yielding a time-varying set $\mathcal{F}_t$ of N popular files over the time slot index $t = 1, 2, \ldots$. As a result, there is no distinction between caching and delivery phases. Instead, in each time slot t, the signals sent by the cloud on the fronthaul links to the ENs can be used both to facilitate delivery, as in the offline case, and to update the caches. There is hence a tension between keeping the caches updated and reducing the delivery latency at each time slot t.

To state the *traffic model* more precisely, each user k requests a random file $F_{D_k(t)} \in \mathcal{F}_t$, chosen uniformly and without replacement as in [45]. The set of popular files $\mathcal{F}_t$ evolves as in the Markov model considered in [45]. Accordingly, given the popular set $\mathcal{F}_{t-1}$ at time $t - 1$, with probability $1 - p$ no new popular content is generated and we have $\mathcal{F}_t = \mathcal{F}_{t-1}$; while, with probability p, a new popular file is added to the set $\mathcal{F}_t$ by replacing a file selected uniformly at random from $\mathcal{F}_{t-1}$. One can generally distinguish two cases, namely: (1) *known popular set*: the cloud is informed about the set $\mathcal{F}_t$ at time t, e.g., by leveraging data external analytics tools; (2) *unknown popular set*: the set $\mathcal{F}_t$ may only be inferred via the observation of the users' requests. We note that the latter assumption is typically made in the networking literature.

As the performance criterion of interest, we account for the fact that caching decisions made in a given time slot t generally affect all future time slots by evaluating a long-term

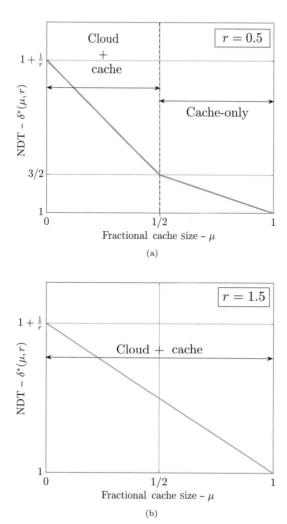

Figure 6.7 Minimum NDT for an F-RAN with $M = K = 2$: (a) low fronthaul regime, here $r = 0.25$; and (b) high fronthaul regime, here $r = 1.5$. The labels "Cache" and "Cloud" refer to the operation of the achievable schemes.

average of the per-slot NDT considered above. In particular, we define first the average achievable delivery time per bit in slot t for a given sequence of feasible policies as

$$\Delta_t(\mu, C_F, P) = \lim_{L \to \infty} \frac{1}{L} E(T_{F,t} + T_{E,t}), \qquad (6.23)$$

where the average is taken with respect to the distributions of the fronthaul and edge NDTs in each slot t. Note that, unlike (6.15), the average here is on the randomness of the popular file set. The NDT $\delta_t(\mu, r)$ in each time slot t is then defined as in (6.16). Finally, for any achievable sequence $\Delta_t(\mu, C_F, P)$ for $t = 1, 2, \ldots$, the NDT of time slot t is defined as

$$\bar{\delta}_{\text{on}}(\mu, r) = \limsup_{T \to \infty} \frac{1}{T} \sum_{t=1}^{T} \delta_t(\mu, r). \qquad (6.24)$$

We also denote the minimum long-term NDT over all feasible policies under the known popular set assumption as $\bar{\delta}^*_{\text{on,k}}(\mu, r)$, while $\bar{\delta}^*_{\text{on,u}}(\mu, r)$ denotes the minimum long-term NDT under the unknown popular set assumption. In the following, we describe an online caching strategy based on reactive caching.

Long-term NDT with reactive caching: A reactive online caching scheme updates the ENs' caches every time a new file is requested by any user. This scheme does not require knowledge of the popular set $\mathcal{F}_t$ and hence operates also for the case of unknown popular set. The reactive strategy considered here delivers a portion of the requested and uncached files to all ENs, which then cache these fractions by evicting from the caches a randomly selected file.

To elaborate, in a manner similar to [45], each EN stores a μ/α fraction of the same $N' = \alpha N$ files for some $\alpha > 1$. Note that the set of $N' > N$ cached files in the cached contents of all ENs m generally contains files that are no longer in the set $\mathcal{F}_t$ of N popular files. Caching $N' > N$ files is instrumental in keeping the intersection between the set of cached files and $\mathcal{F}_t$ from vanishing [45]. If Y_t requested files, with $0 \le Y_t \le K$, are not cached at the ENs, a μ/α fraction of each requested and uncached file is sent on the fronthaul link to each EN following the offline caching policy in Figure 6.6 with μ/α in lieu of μ. Delivery then takes place by following the achievable offline delivery strategy, with the only caveat that μ/α should replace μ. The overall NDT is hence the sum of the NDT $\delta_{\text{off,ach}}(\mu/\alpha, r)$ achievable by the offline delivery policy when the fractional cache size is μ/α and of the NDT due to the fronthaul transfer of the μ/α-fraction of each requested and uncached file on the fronthaul link. Details can be found in [42].

Reference [23] also shows that the minimum long-term NDT with online caching is proportional to the minimum NDT for offline caching, with an additive gap that is inversely proportional to the fronthaul rate r. This behavior is observed also for the reactive strategy presented above. To understand this result, note that, when $\mu \ge 1/M$ and hence the set of popular files can be fully stored across all the M ENs' caches, offline caching enables the delivery of all possible users' requests with a finite delay even when $r = 0$. In contrast, with online caching, the time variability of the set $\mathcal{F}_t$ of popular files implies that, with non-zero probability, some of the requested files cannot be cached at ENs and hence should be delivered by leveraging fronthaul transmission.

Numerical example: We close this section by describing a numerical experiment. We consider the long-term NDT achievable by a greedy scheme that minimizes only the current NDT at each time slot, along with the reactive scheme described above. The impact of the fronthaul rate r is considered in Figure 6.8. Here, we also plot for reference the achievable NDT for offline caching, and we assume random eviction for reactive caching. Parameters are set as $p = 0.5$, $\mu = 0.3$, $M = K = 5$, and $N = 10$. It is seen that reactive caching can significantly improve over greedy delivery by storing content for future slots. Furthermore, as the fronthaul rate r decreases to zero, the additive gap between online and offline caching is seen to grow without bound as per the

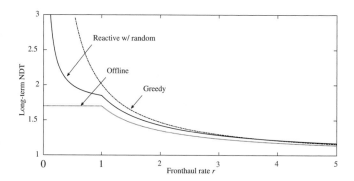

Figure 6.8 Achievable long-term NDT for greedy and reactive online caching with random eviction versus fronthaul rate r. For reference, the offline minimum NDT is also shown ($p = 0.5$, $\mu = 0.3$, $M = K = 5$, $N = 10$).

discussion above. Performance curves of reactive caching with other eviction strategies can be found in [23].

6.4   Backing Off from Asymptotics

The asymptotic regime assumed in the previous sections was seen to yield precise guidelines and performance benchmarks for the design of cloud and fog-aided systems. In this section, we first revisit the conclusions obtained from the asymptotic analysis of content delivery in Section 6.3 by considering the regime of limited SNR, as well as that of finite and fixed file sizes. In the corresponding subsections, we focus on signal processing aspects and concentrate on the techniques introduced in the previous section. We then provide a brief, but quantitative, discussion on advanced techniques for communication on fronthaul and wireless edge channels, whose benefits can be observed at finite SNR.

6.4.1   Finite SNR Performance

Backing off from the assumption of infinite SNR can be done by analyzing the performance of caching and delivery strategies via the integrated use of information theoretic tools for the evaluation of finite-SNR achievable rates and of nonconvex optimization strategies for the optimization of transmission parameters such as linear precoding matrices. Here, we focus on the joint design of fronthaul and edge transmission strategies for the delivery phase under the assumptions that the prefetching strategy, which determines the contents stored in the caches of ENs, is fixed. We refer to [46, 47] for extensions that include the design of the prefetching strategy.

To model arbitrary uncoded prefetching strategies, we split each file $F_k \in \mathcal{F}$ of size L bits into J equal-sized fragments, each of size $\tilde{L} = L/J$ bits, denoted as $(F_k, 1), \ldots, (F_k, J)$, and define binary caching variables $c_{k,j}^i \in \{0, 1\}$ that indicate whether the subfile (F_k,j) is cached at EN i ($c^i_{k,l} = 1$) or not ($c^i_{k,l} = 0$). We assume the standard IRM request model with a Zipf distribution (see Section 6.3). We define the set $\mathcal{D}_{req}(t) = \cup_{k\in\mathcal{K}}\{D_k(t)\}$ of the requested files at time slot t and remove the time index t throughout the rest of the subsection.

We first discuss the optimization of the delivery phase for latency minimization under the hard-transfer fronthauling scheme and then consider the design of the soft transfer scheme. Throughout, we focus on one-shot linear precoding strategies at the ENs. The term "one-shot" refers to the fact that linear precoding is applied independently on each symbol of the edge channel. We emphasize that "one-shot" linear precoding, or beamforming, is routinely implemented in practical systems.

Hard transfer: To evaluate the latency on the fronthaul link with the hard transfer fronthauling strategy, we define binary variables $d^i_{k,j}$ that take value 1 if a subfile (F_k,j) is transferred to EN i on the fronthaul link and value 0 otherwise. Then, the delivery latency T_F on the fronthaul link is given as

$$T_F = \max_{i\in\{1,\dots,M\}} T_{F,i}, \tag{6.25}$$

where $T_{F,i}$ represents the latency on the fronthaul link connected to the ith EN, which is given as

$$T_{F,i} = \frac{\sum_{f\in\mathcal{D}_{req}}\sum_{j\in\mathcal{J}} d^i_{f,j}\tilde{L}}{C_F}. \tag{6.26}$$

In (6.26), we defined the set $\mathcal{J} = \{1,\dots,J\}$ of fragments.

For given binary transfer variables $d^i_{f,j}$, the signal X_i transmitted by EN i can be written as

$$X_i = \sum_{f\in\mathcal{D}_{req}}\sum_{j\in\mathcal{J}} \left(1 - \bar{c}^i_{f,j}\bar{d}^i_{f,j}\right) V^i_{f,j}S_{f,j}, \tag{6.27}$$

where $S_{f,j}$ denotes the baseband signal that encodes the subfile (F_f,j), which is assumed to be taken from a complex Gaussian random codebook, i.e., $S_{f,j} \sim \mathcal{CN}(0,1)$; and $V^i_{f,j}$ represents the complex precoding coefficient for the signal $s_{f,j}$ at EN i. Note that the signal $S_{f,j}$ can be precoded at EN i only if it is cached, i.e., $c^i_{f,j} = 1$, or is transferred on the fronthaul link, i.e., $d^i_{f,j} = 1$.

Under the precoding model in (6.27), the signal Y_k received by the kth user is given as

$$Y_k = \mathbf{h}^H_k \sum_{f\in\mathcal{D}_{req}}\sum_{j\in\mathcal{J}} \mathbf{v}_{f,j}S_{f,j} + Z_k, \tag{6.28}$$

where we defined the channel vector $\mathbf{h}_k = [h^H_{1k}\dots h^H_{Mk}]^H$ from the ENs toward user k and the effective precoding vector $\mathbf{v}_{f,j} = [(1 - \bar{c}^1_{f,j}\bar{d}^1_{f,j})V^{1H}_{f,j}\dots(1 - \bar{c}^M_{f,j}\bar{d}^M_{f,j})V^{MH}_{f,j}]^H$ for subfile (f,j) across all the ENs.

We assume as in [48] that user k decodes the subfiles (F_{D_k},j), $j \in \mathcal{J}$, with a successive interference cancellation (SIC) approach, while treating the interference signals encoding the other files as additive noise. For any decoding order $S_{D_k,1} \rightarrow \dots \rightarrow S_{D_k,J}$, the achievable rate $R_{D_k,j}$, at which the subfile (F_{D_k},j) can be reliably transmitted to user k, is given as

$$R_{D_k,j} = q_{k,j}(\mathbf{v}) = I\left(S_{D_k,j}; Y_k | S_{D_k,1}, \ldots, S_{D_k,j-1}\right) \tag{6.29}$$

$$= \Phi\left(|\mathbf{h}_k^H \mathbf{v}_{D_k,j}|^2, \sum_{m=j+1}^{J} |\mathbf{h}_k^H \mathbf{v}_{D_k,m}|^2 + \sum_{f \in \mathcal{D}_{\mathrm{req}} \setminus \{D_k\}} \sum_{m \in \mathcal{J}} |\mathbf{h}_k^H \mathbf{v}_{f,m}|^2 + 1\right), \tag{6.30}$$

where we defined the function $\Phi(A, B) = \log_2(A + B) - \log_2(B)$.

For given delivery rates $R_{D_k,j}$, the delivery latency $T_{E,k}$ from the ENs to the kth user is given as

$$T_{E,k} = \max_{j \in \mathcal{J}} \frac{\tilde{L}}{R_{D_k,j}}, \tag{6.31}$$

and, as a result, the overall edge latency T_E across all the users amounts to

$$T_E = \max_{k \in \{1,\ldots,K\}} T_{E,k}. \tag{6.32}$$

Finally, the total latency T of the system can hence be written as

$$T = T_F + T_E. \tag{6.33}$$

The problem of minimizing the total delivery latency T can hence be stated as

$$\underset{T,\mathbf{v},\mathbf{R}}{\text{minimize}} \quad T_F + T_E \tag{6.34}$$

$$\text{s.t.} \quad T_F \geq T_{F,i}, \ i \in \{1,\ldots,M\}, \tag{6.35}$$

$$T_E \geq \frac{\tilde{L}}{R_{D_k,j}}, \ k \in \{1,\ldots,K\}, j \in \mathcal{J}, \tag{6.36}$$

$$R_{D_k,j} \leq q_{k,j}(\mathbf{v}), \ k \in \{1,\ldots,K\}, j \in \mathcal{J}, \tag{6.37}$$

$$\sum_{f \in \mathcal{D}_{\mathrm{req}}} \sum_{j \in \mathcal{J}} |V_{f,l}^i|^2 \leq P, \ i \in \{1,\ldots,M\}. \tag{6.38}$$

The above problem is nonconvex, but an efficient solution can be found by adopting the concave convex procedure (CCCP) approach to the problem obtained by change of variables $\mathbf{V}_{f,j} = \mathbf{v}_{f,j} \mathbf{v}_{f,j}^H$ and relaxing the rank constraints $\mathrm{rank}(\mathbf{V}_{f,j}) = 1$ (see, e.g., [48, 49] for the details of the CCCP approach).

Soft transfer: Unlike the hard-transfer scheme, as discussed, soft-transfer fronthauling prescribes that the uncached files (F_k, j) with $c_{k,j}^i = 0$ at EN i be encoded and precoded by the CP. The precoded signals are then quantized and transferred to the EN. Considering an additive model for the quantization noise, the signal X_i transmitted by EN i can be written as

$$X_i = \sum_{f \in \mathcal{D}_{\mathrm{req}}} \sum_{j \in \mathcal{J}} \left(\bar{c}_{f,j}^i U_{f,j}^i + c_{f,j}^i V_{f,j}^i\right) S_{f,j} + Q_i, \tag{6.39}$$

where $V^i_{f,j}$ and $U^i_{f,j}$ represent the precoding coefficients for subfile (F_f, j) applied at the EN i and the CP, respectively, and $Q_i \sim \mathcal{CN}(0, \omega_i)$ represents the additive quantization noise. We note that the additive model (6.39) can be justified by using either rate-distortion arguments as in, e.g., [50], or the Bussgang theorem (see, e.g., [51]). Note, however, that in the latter case the quantization noise would neither be Gaussian nor independent of the signal. It was shown in [50] that using standard rate-distortion theoretic arguments the number of bits that are needed to represent the quantized signal X_i can be written as

$$g_i(\mathbf{v}, \mathbf{u}, \boldsymbol{\omega}) = T_E \Phi \left(\sum_{f \in \mathcal{D}_{\text{req}}} \sum_{j \in \mathcal{J}} \bar{c}^i_{f,j} |U^i_{f,j}|^2, \omega_i \right). \tag{6.40}$$

Therefore, the fronthaul latency T_F with the soft-transfer scheme is given as

$$T_F = \max_{i \in \{1, \ldots, M\}} \frac{g_i(\mathbf{v}, \mathbf{u}, \boldsymbol{\omega})}{C_F}. \tag{6.41}$$

As for the hard-transfer scheme, the edge latency T_E with the soft-transfer fronthauling scheme is given in (6.32). If we assume that user k performs SIC decoding with the decoding order $S_{D_k,1} \to \ldots \to S_{D_k,J}$ while treating the interference signals as noise, the achievable rate $R_{D_k,j}$ is given as

$$R_{D_k,j} = q_{k,j}(\mathbf{v}, \mathbf{u}, \boldsymbol{\omega}) = I\left(S_{D_k,j}; Y_k | S_{D_k,1}, \ldots, S_{D_k,j-1}\right) \tag{6.42}$$

$$= \Phi \left(|\mathbf{h}_k^H \mathbf{v}_{D_k,j}|^2, \sum_{m=j+1}^{J} |\mathbf{h}_k^H \mathbf{v}_{D_k,m}|^2 \right.$$

$$\left. + \sum_{f \in \mathcal{D}_{\text{req}} \setminus \{D_k\}} \sum_{m \in \mathcal{J}} |\mathbf{h}_k^H \mathbf{v}_{f,m}|^2 + \mathbf{h}_k^H \Omega \mathbf{h}_k + 1 \right), \tag{6.43}$$

with the notations $\mathbf{v}_{f,j} = [(\bar{c}^1_{f,j} U^1_{f,j} + c^1_{f,j} V^1_{f,j})^H \ldots (\bar{c}^M_{f,j} U^M_{f,j} + c^M_{f,j} V^M_{f,j})^H]^H$ and $\Omega = \text{diag}(\{\omega_i\}_{i=1}^M)$.

The problem of minimizing the total latency T in (6.33) under the soft transfer fronthauling scheme is thus stated as

$$\underset{T, \mathbf{v}, \mathbf{u}, \boldsymbol{\omega}, \mathbf{R}}{\text{minimize}} \ T_F + T_E \tag{6.44}$$

$$\text{s.t.} \ T_F \geq \frac{g_i(\mathbf{v}, \mathbf{u}, \boldsymbol{\omega})}{C_F}, \ i \in \{1, \ldots, M\}, \tag{6.45}$$

$$T_E \geq \frac{\tilde{L}}{R_{D_k,j}}, \ k \in \{1, \ldots, K\}, j \in \mathcal{J}, \tag{6.46}$$

$$R_{D_k,j} \leq q_{k,j}(\mathbf{v}, \mathbf{u}, \boldsymbol{\omega}), \ k \in \{1, \ldots, K\}, j \in \mathcal{J}, \tag{6.47}$$

$$\sum_{f \in \mathcal{D}_{\text{req}}} \sum_{j \in \mathcal{J}} |V^i_{f,l}|^2 + \omega_i \leq P, \ i \in \{1, \ldots, M\}. \tag{6.48}$$

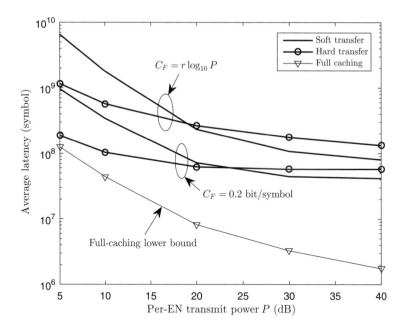

Figure 6.9 Average latency T_{total} versus the SNR for an F-RAN downlink with $(M, K) = (3, 3)$, $N = 6, J = 15, \mu = 1/3, L = 1$ Mbyte, and $C_F = 0.2$ or $C_F = r \log_{10} P$ bit/symbol with $r = 0.02$.

In a manner similar to problem (6.34) for the hard-transfer scheme, we can obtain an efficient solution to problem (6.44) by deriving a CCCP-based iterative algorithm with change of variables $\mathbf{V}_{fj} = \mathbf{v}_{fj} \mathbf{v}_{fj}^H$.

Numerical example: In Figure 6.9, we plot the average latency T versus the SNR for an F-RAN downlink with $(M, K) = (3, 3), N = 6, J = 15, \mu = 1/3, L = 1$ Mbyte. We consider both a fixed fronthaul capacity $C_F = 0.2$ bit/symbol and a fronthaul capacity scaling with the SNR as in the asymptotic analysis, namely $C_F = r \log_{10} P$ bit/symbol with fronthaul rate $r = 0.02$. We assume that the positions of the ENs and the users are uniformly distributed within a $500 \times 500 \, \text{m}^2$ square area. The channel coefficient h_{mk} is modeled as $h_{mk} = \sqrt{\rho_{mk}} \tilde{h}_{mk}$, where $\rho_{mk} = 1/(1 + (d_{mk}/50)^3)$ represents the path-loss, with d_{mk} denoting the distance between the EN m and user k, and $\tilde{h}_{mk} \sim \mathcal{CN}(0, 1)$ accounts for the small-scale Rayleigh fading. We also assume that each user k independently requests a file with the Zipf's distribution, i.e., $\Pr[F_{D_k} = F_n] = cn^{-\gamma}$, where c is a constant that satisfies $\sum_{n=1}^{N} cn^{-\gamma} = 1$, and the parameter γ is set to $\gamma = 0.5$. For the prefetching strategy, we consider the randomized fractional caching policy described in [52].

From Figure 6.9, it is observed that, in the high-SNR regime, the soft-transfer fronthauling scheme shows better latency performance than the hard-transfer fronthauling scheme, as predicted from the NDT analysis in the previous section. However, there exists a crossover point between the performance of soft-transfer and hard-transfer schemes as functions of the SNR. In fact, the hard-transfer fronthauling scheme can outperform the soft-transfer scheme when the SNR is relatively small as compared to the fronthaul link capacity. This is the case with both fixed and scaling fronthaul capacity. This results can be explained by noting that, at high SNR, the edge transmission time becomes small and hence the soft-transfer scheme uses the fronthaul capacity resources efficiently by quantizing the packet to be transmitted on the edge channel. Instead, at lower SNRs, it becomes more efficient to transmit the data messages rather than the compressed packet to be delivered on the edge channel.

6.4.2   Finite Blocklength Performance

Here, we alleviate the assumption of large file sizes made in the asymptotic analysis. To this end, we take a pragmatic approach that recognizes the complexity of extending the finite-blocklength capacity results of [53] to multiuser setups, as demonstrated by the limited literature on the subject [54–56]. We recall that reference [53] establishes that the number T of channel uses necessary to achieve a certain probability of block error ϵ when transmitting L bits at the capacity C on a point-to-point memoryless channel is closely approximated by the relation [53, eq. (54)]

$$L \approx TC - \sqrt{TV}Q^{-1}(\epsilon), \tag{6.49}$$

where V is the *dispersion* of the channel and Q^{-1} is the inverse Q-function. The approximation is accurate up to a term of order $O(\log T)$.

To evaluate the latency with finite file sizes, and hence blocklengths, we adopt the approximation (6.49) as extended in [57, eq. (15)] to account for channels in the presence of interference under the assumption of nearest-neighbor decoding. Accordingly, the edge latency $T_{E,k}$ for each user k is approximated as

$$T_{E,k} \approx \left(\frac{\sqrt{V_{k,j}}}{2R_{D_k,j}} Q^{-1}(\epsilon) + \sqrt{\frac{\tilde{L}}{R_{D_k,j}} + \frac{V_{k,j}}{4R_{D_k,j}^2}(Q^{-1}(\epsilon))^2} \right)^2, \tag{6.50}$$

where the dispersion $V_{k,j}$ for the jth fragment of the file F_{D_k} at user k is given according to [57, eq. (16)] as

$$V_{k,j} = \frac{P_{k,j}^2(\xi + 1) + 4P_{k,j}}{4(P_{k,j} + 1)^2}, \tag{6.51}$$

where $\xi = 3$ is the fourth moment of the interference-plus-noise signal that is normalized to have unit variance, and the effective normalized signal power $P_{k,j}$ is given as

$$P_{k,j} = \frac{|\mathbf{h}_k^H \mathbf{v}_{D_k,j}|^2}{1 + \theta \mathbf{h}_k^H \Omega \mathbf{h}_k + \sum_{m=j+1}^{J} |\mathbf{h}_k^H \mathbf{v}_{D_k,m}|^2 + \sum_{f \in \mathcal{D}_{\text{req}} \setminus \{D_k\}} \sum_{m \in \mathcal{J}} |\mathbf{h}_k^H \mathbf{v}_{f,m}|^2}, \tag{6.52}$$

with the parameter θ given as $\theta = 0$ and $\theta = 1$ for the hard-transfer and soft-transfer fronthauling strategies, respectively.

Numerical example: Figure 6.10 plots the average latency T, normalized by the file size L, versus L for an F-RAN downlink with the setup considered in Figure 6.9 and fixed

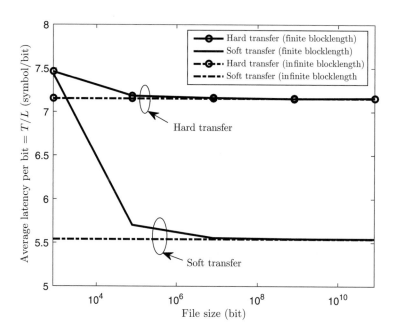

Figure 6.10 Average latency T, normalized by the file size L, versus L for an F-RAN downlink with $(M, K) = (3, 3)$, $N = 6$, $J = 15$, $\mu = 1/3$, $C_F = 0.2$ bit/symbol, SNR of 30 dB when accounting for the finite-blocklength capacity penalty ($\epsilon = 10^{-2}$).

fronthaul capacity $C = 0.2$ bit/symbol and SNR of 30 dB. The transmission parameters are obtained from the optimization described in the previous subsection that does not account for finite-blocklength effects. We set the parameter ϵ in (6.50) to $\epsilon = 10^{-2}$. Comparing the result with Figure 6.9, it is observed that the impact of finite blocklength is more deleterious for the soft-transfer fronthauling scheme than for the hard-transfer scheme. This is because the edge latency T_E per information bit tends to increase as L decreases for both the hard-transfer and soft-transfer schemes owing to the capacity penalty caused by finite-blocklength transmission. However, the increased edge latency affects the fronthaul latency T_F only for the soft transfer scheme as shown in (6.40) and (6.41). Nevertheless, we observe that the asymptotic results are closely approximated for files as small as a few megabits.

6.4.3   Advanced Fronthaul–Edge Transmission Techniques

This subsection discusses advanced fronthaul–edge transmission techniques that are inspired by network information theoretic concepts and have been demonstrated to be potentially beneficial in the nonasymptotic regime (see [58–62]). To focus solely on fronthaul and edge transmission, we consider here a conventional C-RAN downlink setup in which the ENs are not equipped with caching, i.e., $\mu = 0$. Furthermore, we concentrate on the conventional scenario in which the ENs operate as "oblivious" relays, in the sense that channel encoding and precoding are carried out only by the CP.

As discussed in the previous sections, in the presence of "oblivious" relays, the fronthaul links can be utilized in soft-transfer mode. The application of point-to-point (P2P) compression techniques for fronthaul transmission was proposed in [41]. This natural compression approach, which was implicitly assumed in the treatment of the previous sections, was found to be suboptimal in [63]. In particular, [63] demonstrated that multivariate, or joint, compression of the precoding output signals can outperform P2P compression under both linear and nonlinear precoding strategies. In fact, via joint compression, one can design the correlation among the quantization noise signals added to the precoded signals for different ENs in such a way that the signal spaces are least affected by the degradation caused by quantization. The idea was translated into practical quantization strategies in [64].

In the reviewed P2P and multivariate compression strategies, the quantization codebooks for fronthaul transmission are designed separately from the channel codebooks. In contrast, in the reverse compute-and-forward (RCoF) scheme [65], the CP employs the same lattice code for both channel coding and fronthaul quantization. As a result, no quantization noise is introduced due to fronthaul capacity constraints. On the flip side, the performance of RCoF is limited by the self-interference caused by the fact that the channel coefficients are not integers as required by the precoding scheme used by RCoF. We refer to [60, 62] for a brief summary of the compression-based and CoF schemes for the downlink and uplink of C-RAN systems.

Figure 6.11 plots per-user achievable rates on the edge channel (that is, R_E) for a C-RAN downlink with $M = K = 3$ and SNR $P = 20$ dB versus the fronthaul capacity C_F. The channel model is a simple deterministic Wyner-type downlink in which the channel coefficient h_{mk} is given as $h_{mk} = 1 \cdot 1(m = k) + \alpha \cdot 1(m \neq k)$, where $1(\cdot)$ is the indicator function and $\alpha \in [0, 1]$ represents the intercell channel gain. For this experiment, we set $\alpha = 0.6$. For reference, we plot two per-user rates that are achievable with D-RAN, namely: (1) orthogonal transmission, which achieves per-user rate $R_{OR} = 1/2 \log_2(1 + 2P)$ and in which users in odd and even cells are scheduled in orthogonal spectral resources; and (2) simultaneous transmission, which attains per-user rate $R_{SIM} = \log_2(1 + P/(1 + 2\alpha P))$ and in which ENs transmit simultaneously to their corresponding users while noise is treated as interference. We also plot the cut-set upper bound [59], which is given as $C_{cut} = \min\{R_{full}, MC_F\}/M$, where R_{full} is the sum-rate that can be achieved when the ENs can fully cooperate.

Before commenting on the figure, we remark that all the C-RAN based schemes surveyed above are able to achieve the same optimal DoF (6.14), while D-RAN with orthogonal transmission achieves (6.3) and simultaneous transmission yields a DoF of zero. As observed in Figure 6.11, the relative performance of the scheme demonstrates a more complex behavior for finite SNR.

In fact, from Figure 6.11, it is seen that multivariate compression shows significant performance gains as compared to P2P compression for both nonlinear precoding (dirty paper coding, or DPC) and linear precoding schemes. Also, in the regime of intermediate fronthaul capacity, RCoF outperforms the compression-based schemes, since the impact of the interference signals caused by non-integer channel coefficients is counterbalanced by the advantage of eliminating the quantization noise. We also

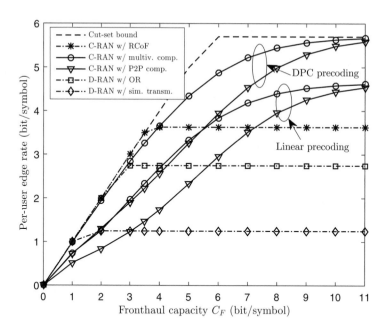

Figure 6.11 Average per-user edge rate versus the fronthaul capacity C_F for a Wyner-type C-RAN downlink with $M = K = 3$, $\mu = 0$, SNR of 20 dB.

note that the performance of the compression-based schemes combined with DPC approaches the cut-set upper bound when the fronthaul capacity is sufficiently large.

6.5   Outlook and Concluding Remarks

This chapter has presented a review of information theoretic results concerning the optimization of mobile broadband communications and content-delivery services on fog-assisted wireless networks. With the aim of highlighting key ideas and insights, the treatment has focused on the performance in the high-SNR regime. Among the main conclusions of this analysis, we underline the effectiveness of strategies based on cloud processing and fronthaul compression in achieving optimal or near-optimal performance in the high-SNR regime. The chapter has also provided a discussion on the limitations of the asymptotic analysis and on advanced fronthaul–edge transmission strategies.

We conclude the chapter with a brief overview of important aspects of the design and analysis of fog-aided networks that are not covered in the treatment presented in the previous sections. These aspects pertain to modeling and algorithmic considerations, and include a number of open problems.

First, while the presentation here has focused on fully connected networks with perfect CSI, assessing the impact of partial connectivity and imperfect CSI is of critical importance in order to evaluate the practical potential of the available schemes. Some discussion on partial CSI was provided in the text, and we refer to, e.g., [66] and references therein for a study that accounts for partial connectivity in D-RAN.

Second, the study of mobile broadband communication for the uplink of C-RAN is an active field of research. In this regard, we refer to [67–70] for studies of P2P and distributed fronthaul compression. In addition, [71, 72] studied the advantages of joint decompression and decoding (JDD) at the CP for the uplink of C-RAN, and it was shown in [73] that JDD is a capacity-achieving scheme under the assumption of oblivious processing at ENs.

Another aspect that was not covered in this chapter is the design of communications strategies in the presence of multiantenna ENs. Multiantenna ENs, possibly with massive antenna arrays, are expected to be commonplace in 5G systems. The spatial processing enabled by multiantenna ENs adds an important new dimension to the design problem and poses novel challenges for communications on both fronthaul and edge channels, as well as for optimized caching in F-RAN systems. As an example, massive antenna arrays call for the deployment of hybrid analog–digital processors at the ENs, which were studied in [74] in the context of C-RAN.

There are also a number of open theoretical questions, such as uplink and downlink duality in C-RAN and F-RAN [75] and the development of tight converse bounds [76].

Acknowledgments

This work was supported by the European Research Council (ERC) under the European Union's Horizon 2020 research and innovation programme (grant agreement no. 694630 and 725731). The work of R. Tandon was supported by the NSF grant CAREER 1651492.

References

[1] GPPP, "5G PPP view on 5G architecture," https://5g-ppp.eu/white-papers/, 2016.

[2] NGMN, "5G white paper," [online]. Available www.ngmn.org/5g-white-paper.html, 2015.

[3] H. Freeman and T. Zhang, "The emerging era of fog computing and networking [the president's page]," *IEEE Commun. Mag.*, vol. 54, no. 6, pp. 4–5, June 2016.

[4] M. Chiang, S. Ha, I. Chih-Lin, F. Risso, and T. Zhang, "Clarifying fog computing and networking: 10 questions and answers," *IEEE Commun. Mag.*, vol. 55, no. 4, pp. 18–20, 2017.

[5] B. Han, V. Gopalakrishnan, L. Ji, and S. Lee, "Network function virtualization: Challenges and opportunities for innovations," *IEEE Commun. Mag.*, vol. 53, no. 2, pp. 90–97, Feb. 2015.

[6] P. Rost, A. Banchs, I. Berberana, et al., "Mobile network architecture evolution toward 5G," *IEEE Commun. Mag.*, vol. 54, no. 5, pp. 84–91, May 2016.

[7] A. Checko, H. L. Christiansen, Y. Yan, et al., "Cloud RAN for mobile networks: A technology overview," *IEEE Commun. Surv. Tutorials*, vol. 17, no. 1, pp. 405–426, Sep. 2014.

[8]   O. Simeone, A. Maeder, M. Peng, O. Sahin, and W. Yu, "Cloud radio access network: Virtualizing wireless access for dense heterogeneous systems," *J. Commun. Netw.*, vol. 18, no. 2, pp. 135–149, Apr. 2016.

[9]   S. A. Jafar, *Interference Alignment: A New Look at Signal Dimensions in a Communication Network*. Now Publishers Inc., 2011.

[10] CVN. Index, "Forecast and methodology, 2014–2019 white paper," 2015.

[11] Sandvine, "Global internet phenomena report," [Online]. Available: https://www.sandvine.com/trends/global-internet-phenomena, 2016

[12] iGR Market Study, "A profile of the Netflix user: More devices, OTT services, and data," [Online]. Available: https://igr-inc.com/advisory-subscription-services/wireless-mobile-landscape, 2016

[13] C. Liang, F. R. Yu, and X. Zhang, "Information-centric network function virtualization over 5G mobile wireless networks," *IEEE New.*, vol. 29, no. 3, pp. 68–74, May 2015.

[14] R. Tandon and O. Simeone, "Cloud-aided wireless networks with edge caching: Fundamental latency trade-offs in fog radio access networks," in *IEEE Int. Symp. Inform. Theory (ISIT)*, Barcelona, 2016.

[15] R. Tandon and O. Simeone, "Harnessing cloud and edge synergies: Toward an information theory of fog radio access networks," *IEEE Commun. Mag.*, vol. 54, no. 8, pp. 44–50, Aug. 2016.

[16] A. Sengupta, R. Tandon, and O. Simeone, "Fog-aided wireless networks for content delivery: Fundamental latency tradeoffs," *IEEE Trans. Inform. Theory*, vol. 63, no. 10, pp. 6650–6678, Oct 2017.

[17] Y. Cao, M. Tao, F. Xu, and K. Liu, "Fundamental storage–latency tradeoff in cache-aided MIMO interference networks," *arXiv preprint arXiv:1609.01826*, 2016. [Online]. Available: http://arxiv.org/abs/1609.01826.

[18] X. Yi and G. Caire, "Topological coded caching," in *IEEE Int. Symp. Inform. Theory*, Barcelona 2016, pp. 2039–2043.

[19] F. Xu, K. Liu, and M. Tao, "Cooperative tx/rx caching in interference channels: A storage–latency tradeoff study," *CoRR*, vol. abs/1605.07729, 2016. [Online]. Available: http://arxiv.org/abs/1605.07729.

[20] A. M. Girgis, O. Ercetin, M. Nafie, and T. ElBatt, "Decentralized coded caching in wireless networks: Trade-off between storage and latency," *arXiv: 1701.06673*, Jan. 2017.

[21] J. S. Pujol Roig, F. Tosato, and D. Gündüz, "Interference networks with caches at both ends," *arXiv: 1703.04349*, Mar. 2017.

[22] S. M. Azimi, O. Simeone, and R. Tandon, "Content delivery in fog-aided small-cell systems with offline and online caching: An information-theoretic analysis," *Entropy*, vol. 19, no. 7, p. 366, 2017.

[23] S. M. Azimi, O. Simeone, A. Sengupta, and R. Tandon, "Online edge caching and wireless delivery in fog-aided networks with dynamic content popularity," *IEEE J. Sel. Areas Commun.*, vol. 36, no. 6, pp. 1189–1202, 2018.

[24] J. Goseling, O. Simeone, and P. Popovski, "Delivery latency trade-offs of heterogeneous contents in fog radio access networks," in *GLOBECOM 2017 IEEE Global Commun. Conf.*, 2017, pp. 1–6.

[25] R. Karasik, O. Simeone, and S. Shamai, "Fundamental latency limits for D2D-aided content delivery in fog wireless networks," *arXiv:1801.00754*, 2018.

[26] J. Zhang and O. Simeone, "Fundamental limits of cloud and cache-aided interference management with multi-antenna base stations," *arXiv:1712.04266*, 2017.

[27] J. Zhang and O. Simeone, "Cloud-edge non-orthogonal transmission for fog networks with delayed CSI at the cloud," *arXiv:1805.10024*, 2018.

[28] A. Pizzinat, P. Chanclou, F. Saliou, and T. Diallo, "Things you should know about fronthaul," *J. L. Technol.*, vol. 33, no. 5, pp. 1077–1083, 2015.

[29] A. Lozano and N. Jindal, "Are yesterday-s information-theoretic fading models and performance metrics adequate for the analysis of today's wireless systems?" *IEEE Commun. Mag.*, vol. 50, no. 11, 2012.

[30] A. G. Davoodi and S. A. Jafar, "Generalized DOF of the symmetric k-user interference channel under finite precision CSIT," in *2016 IEEE Int. Symp. Inform. Theory*, Barcelona 2016, pp. 1307–1311.

[31] Y. Lu and Z. Wei, *Interference Coordination for 5G Cellular Networks*. New York: Springer, 2015.

[32] A. S. Motahari, S. O-Gharan, M. A. Maddah-Ali, and A. K. Khandani, "Real interference alignment: Exploiting the potential of single antenna systems," *IEEE Trans. Inform. Theory*, vol. 60, no. 8, pp. 4799–4810, Aug. 2014.

[33] D. Stotz, S. A. Jafar, H. Bölcskei, and S. Shamai, "Canonical conditions for k/2 degrees of freedom," in *2016 IEEE Int. Symp. Inform. Theory (ISIT)*, Barcelona July 2016, pp. 1292–1296.

[34] D. Stotz and H. Bölcskei, "Degrees of freedom in vector interference channels," *IEEE Trans. Inform. Theory*, vol. 62, no. 7, pp. 4172–4197, July 2016.

[35] C. Huang, V. R. Cadambe, and S. A. Jafar, "Interference alignment and the generalized degrees of freedom of the X channel," *IEEE Trans. Inform. Theory*, vol. 58, no. 8, pp. 5130–5150, Aug. 2012.

[36] A. G. Davoodi, B. Yuan, and S. A. Jafar, "GDOF of the MISO BC: Bridging the gap between finite precision and perfect CSIT," *arXiv:1602.02203*, 2016.

[37] C. T. Li and A. Ozgur, "Channel diversity needed for vector space interference alignment," *IEEE Trans. Inform. Theory*, vol. 62, no. 4, pp. 1942–1956, Apr. 2016.

[38] G. Bresler, D. Cartwright, and D. Tse, "Feasibility of interference alignment for the MIMO interference channel," *IEEE Trans. Inform. Theory*, vol. 60, no. 9, pp. 5573–5586, Sep. 2014.

[39] V. R. Cadambe and S. A. Jafar, "Interference alignment and the degrees of freedom of wireless X networks," *IEEE Trans. Infor. Theory*, vol. 55, no. 9, pp. 3893–3908, Sep. 2009.

[40] S. H. R. Naqvi, A. Matera, L. Combi, and U. Spagnolini, "On the transport capability of LAN cables," *CoRR*, vol. abs/1702.03911, 2017. [Online]. Available: http://arxiv.org/abs/1702.03911.

[41] O. Simeone, O. Somekh, H. V. Poor, and S. Shamai, "Downlink multicell processing with limited-backhaul capacity," *EURASIP J. Adv. Sig. Proc.*, 2009. DOI: 10.1155/2009/840814.

[42] S. M. Azimi, O. Simeone, A. Sengupta, and R. Tandon, "Online edge caching and wireless delivery in fog-aided networks with dynamic content popularity," *IEEE J. Sel. Areas Communications*, 2018. DOI: 10.1109/jsac.2018.2844961.

[43] Y. Liu and E. Erkip, "Completion time in multi-access channel: An information theoretic perspective," in *IEEE Inform. Theory Workshop*, 2011.

[44] Y. Liu and E. Erkip, "Completion time in broadcast channel and interference channel," in *49th Ann. Allerton Conf. Commun. Control Comput.*, 2011.

[45] R. Pedarsani, M. A. Maddah-Ali, and U. Niesen, "Online coded caching," *IEEE/ACM Trans. Networking*, vol. 24, no. 2, pp. 836–845, 2016.

[46] J. Liu, B. Bai, J. Zhang, and K. B. Letaief, "Cache placement in fog-rans: From centralized to distributed algorithms," *Proc. IEEE Int. Conf. Commun.*, Kuala Lumpur May 2016.

[47] B. Dai and W. Yu, "Joint user association and content-placement for cache-enabled wireless access networks," *Proc. IEEE Int. Conf. Acoust., Speech and Sig. Proc.*, Shanghai Mar. 2016.

[48] S.-H. Park, O. Simeone, and S. Shamai, "Joint cloud and edge processing for latency minimization in fog radio access networks," in *2016 IEEE 17th Int. Workshop Signal Processing Adv. Wireless Commun. (SPAWC)*, Edinburgh, 2016, pp. 1–5.

[49] M. Tao, E. Chen, H. Zhou, and W. Yu, "Content-centric sparse multicast beamforming for cache-enabled cloud ran," *IEEE Trans. Wireless Comm.*, vol. 15, pp. 6118–6131, Sep. 2016.

[50] A. E. Gamal and Y.-H. Kim, *Network Information Theory*. Cambridge: Cambridge University Press, 2011.

[51] S. Jacobsson, G. Durisi, T. Goldstein, and C. Studer, "Quantized precoding for massive MU-MIMO," *arXiv:1610.07564*, Oct. 2016.

[52] S.-H. Park, O. Simeone, and S. Shamai, "Joint optimization of cloud and edge processing for fog radio access networks," *IEEE Trans. Wireless Commun.*, vol. 15, no. 11. pp, 7621–7632, 2016.

[53] Y. Polyanskiy, H. V. Poor, and S. Verdú, "Channel coding rate in the finite blocklength regime," *IEEE Trans. Inform. Theory*, vol. 56, no. 5, pp. 2307–2359, May 2010.

[54] E. MolavianJazi and J. N. Laneman, "A finite-blocklength perspective on Gaussian multi-access channels," *arXiv:1309.2343*, Sep. 2013.

[55] J. Scarlett and V. Y. F. Tan, "Second-order asymptotics for the Gaussian MAC with degraded message sets," in *2014 IEEE Int. Symp. Inform. Theory*, Honolulu, HI, Jun. 2014, pp. 461–465.

[56] S. Q. Le, V. Y. F. Tan, and M. Motani, "A case where interference does not affect the channel dispersion," *IEEE Trans. Inform. Theory*, vol. 61, no. 5, pp. 2439–2453, May 2015.

[57] J. Scarlett, V. Y. Tan, and G. Durisi, "The dispersion of nearest-neighbor decoding for additive non-Gaussian channels," *IEEE Trans. Inf. Theory*, vol. 63, no. 1, pp. 1380–1408, Jan. 2017.

[58] D. Gesbert, S. Hanly, H. Huang, et al., "Multi-cell MIMO cooperative networks: A new look at interference," *IEEE J. Sel. Areas Commun.*, vol. 28, no. 9, pp. 1380–1408, Dec. 2010.

[59] O. Simeone, N. Levy, A. Sanderovich, et al., Cooperative wireless cellular systems: An information-theoretic view. *Found. Trends Commun. Inf. Theory*, vol. 8, no. 1–2, pp. 1–177, 2011.

[60] S.-H. Park, O. Simeone, O. Sahin, and S. Shamai, "Fronthaul compression for cloud radio access networks," *IEEE Signal Process. Mag., Special Issue on Signal Processing for the 5G Revolution*, vol. 31, no. 6, pp. 69–79, Nov. 2014.

[61] M. Peng, C. Wang, V. Lau, and H. V. Poor, "Fronthaul-constrained cloud radio access networks: Insights and challenges," *IEEE Wireless Commun.*, vol. 22, no. 2, pp. 152–160, Apr. 2015.

[62] O. Simeone, S.-H. Park, O. Sahin, and S. Shamai, "Fronthaul compression for C-RAN," in *Cloud Radio Access Networks: Principles, Technologies, and Applications*. Cambridge: Cambridge University Press, 2017.

[63] S.-H. Park, O. Simeone, O. Sahin, and S. Shamai, "Joint precoding and multivariate backhaul compression for the downlink of cloud radio access networks," *IEEE Trans. Sig. Proc.*, vol. 61, no. 22, pp. 5646–5658, Nov. 2013.

[64] W. Lee, O. Simeone, J. Kang, and S. Shamai, "Multivariate fronthaul quantization for downlink C-RAN," *IEEE Trans. Signal Process.*, vol. 64, no. 19, pp. 5025–5037, Oct. 2016.

[65] S.-N. Hong and G. Caire, "Compute-and-forward strategies for cooperative distributed antenna systems," *IEEE Trans. Inform. Theory*, vol. 59, no. 9, pp. 5227–5243, 2013.

[66] A. El Gamal, "Topological interference management: Linear cooperation is not useful for Wyner's networks," *arXiv:1611.01278*, 2016.

[67] J. Hoydis, M. Kobayashi, and M. Debbah, "Optimal channel training in uplink network MIMO systems," *IEEE Trans. Sig. Proc.*, vol. 59, no. 6, pp. 2824–2833, Jun. 2011.

[68] A. del Coso and S. Simoens, "Distributed compression for MIMO coordinated networks with a backhaul constraint," *IEEE Trans. Wireless Comm.*, vol. 8, no. 9, pp. 4698–4709, Sep. 2009.

[69] S.-H. Park, O. Simeone, O. Sahin, and S. Shamai, "Robust and efficient distributed compression for cloud radio access networks," *IEEE Trans. Veh. Technol.*, vol. 62, no. 2, pp. 692–703, Feb. 2013.

[70] Y. Zhou, Y. Xu, W. Yu, and J. Chen, "On the optimal fronthaul compression and decoding strategies for uplink cloud radio access networks," *IEEE Trans. Inform. Theory*, vol. 62, no. 2, pp. 7402–7418, Dec. 2016.

[71] A. Sanderovich, O. Somekh, H. V. Poor, and S. Shamai, "Uplink macro diversity of limited backhaul cellular network," *IEEE Trans. Info. Theory*, vol. 55, no. 8, pp. 3457–3478, Aug. 2009.

[72] S.-H. Park, O. Simeone, O. Sahin, and S. Shamai, "Joint decompression and decoding for cloud radio access networks," *IEEE Sig. Proc. Lett.*, vol. 20, no. 5, pp. 503–506, May 2013.

[73] I. E. Aguerri, A. Zaidi, G. Caire, and S. S. (Shitz), "On the capacity of cloud radio access networks with oblivious relaying," *arXiv:1701.07237*, 2017.

[74] J. Kim, S.-H. Park, O. Simeone, I. Lee, and S. Shamai, "Joint design of digital and analog processing for downlink C-RAN with large-scale antenna arrays," *arXiv: 1704.00455*, Apr. 2017.

[75] W. He, B. Nazer, and S. Shamai, "Uplink–downlink duality for integer-forcing," in *IEEE Int. Symp. Inform. Theory (ISIT)*, Honolulu, HI, 2014, pp. 2544–2548.

[76] S. S. Bidokhti, G. Kramer, and S. Shamai, "Capacity bounds on the downlink of symmetric, multi-relay, single receiver c-ran networks," *arXiv:1702.01828*, 2017.

7 Communication with Energy Harvesting and Remotely Powered Radios

Dor Shaviv and Ayfer Ozgur

There has been significant recent interest in building wireless devices that do not have conventional batteries. Instead, such wireless devices harvest the energy they need for communication from the natural resources in their environment (e.g., ambient radio waves, light, mechanical vibrations, temperature variations, etc.) or are powered remotely via wireless energy transfer. These new means of powering wireless devices offer several important advantages. For example, eliminating the battery allows us to significantly reduce the size and cost of the wireless device, as batteries are relatively large and expensive components. Size is critical for many emerging medical applications, such as in-body health sensors, while reducing the cost of the wireless device is important for applications that envision massive deployments. The ability to harvest its own energy also enhances the mobility and the life span of the wireless device, and can enable new deployment models by eliminating the maintenance needed for replacing or recharging batteries. In principle, such self-powered wireless devices can operate perpetually and maintenance-free, their life spans limited by their hardware and not the size of their batteries.

From a communication and information theoretic perspective, this new form of powering wireless devices introduces a new paradigm – *energy dynamics* – into communication system design. While energy is central to the design of any engineering system, its associated dynamics so far has had minimal impact on the design of communication schemes for wireless systems. This is because conventionally, the battery and the encoder operate at two drastically different time scales and communication can be accurately modeled as constrained only in terms of average power. In contrast, in a harvesting system,[1] energy is continuously generated and consumed, and it is desirable for the wireless device to operate in an energy-neutral fashion where the incoming and outgoing energy processes are matched with minimal buffering in between. Moreover, due to inherent randomness in the harvesting process, the amount of energy available to the device at any given time becomes a random quantity. The communication system should now be designed by taking these *random energy dynamics* into account.

In this chapter, we present two different formulations for studying a communication system operating under such random energy dynamics. The first formulation leads to a

---

[1] Harvesting here refers to both passive and active harvesting, i.e., from natural resources and from a targeted energy transfer process respectively.

power control problem. It is relevant when the energy expenditure rate of the communication system needs to be adjusted over a time scale of the order of the codeword length. The optimal power control policy aims to maximize throughput under random energy availability when the transmitter is equipped with a finite energy buffer (we will often refer to this buffer as the battery even though it is not a conventional battery). In particular, available energy should not be consumed too fast, or transmission can be interrupted in the future due to an energy outage; on the other hand, if the energy consumption is too slow, it can result in the wasting of the harvested energy and missed recharging opportunities in the future due to an overflow in the battery capacity. This leads to an interesting online decision problem.

The second formulation we present is an information theoretic model for an energy harvesting transmitter operating under random energy availability. It replaces the classical average power constraint on the encoder for an additive white Gaussian noise (AWGN) channel with a dynamic and random energy constraint. This leads to a peculiar state-dependent channel with memory, where the state is causally known only at the transmitter. We discuss the capacity and optimal schemes for this channel, and how the information theoretic model relates to the power control problem mentioned earlier.

We finally present and analyze an information theoretic model for communication systems that are powered remotely via a targeted energy transfer process. These systems differ from passive harvesting systems that collect their energy from the natural resources in their environment in that, in this case, the energy arrival process can be (at least partially) controlled by the remote energy charger. Besides the theoretical analysis of the proposed models, throughout the chapter, we also emphasize the high-level principles that emerge from these theoretical studies for the optimal design of harvesting communication systems.

7.1   Power Control for Energy Harvesting Wireless Transmitters

We start with the *power control problem* for an energy harvesting transmitter. We consider a point-to-point single user discrete-time channel with AWGN (see Figure 7.1). The transmitter is equipped with a rechargeable battery of finite capacity $\bar{B}$, which can be used to buffer the harvested energy. Let E_t be the energy harvested at time t. We will refer to $\{E_t, t = 1, 2 \ldots\}$ as the energy harvesting or arrival process. A power control *policy* for an energy harvesting system is a sequence of mappings from energy arrivals

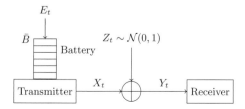

Figure 7.1 An energy harvesting transmitter communicating over an AWGN channel.

to a nonnegative number, which will denote the energy allocated for transmission at the corresponding time slot (we will also call it the power level for this time slot, since the energy normalized by the length of each time-step determines the power level for transmission). We will focus on two different types of policies: offline and online.

An online policy $\pi = \{\pi_t, t = 1, 2 \dots\}$ is a sequence of mappings

$$\pi_t : \mathcal{E}^t \to \mathbb{R}_+ \qquad , t = 1, 2, \dots, \tag{7.1}$$

such that the allocated power at time t is $g_t = \pi_t(E^t)$. An *offline policy* π^N over N time slots is a sequence of mappings

$$\pi_t : \mathcal{E}^N \to \mathbb{R}_+ \qquad , t = 1, \dots, N, \tag{7.2}$$

such that $g_t = \pi_t(E^N)$ gives the allocated power at time t. In words, in the online case the decision of how much energy to allocate to the channel at the current time t can depend only on the realizations of the energy harvesting process up to t. In the offline case, the energy arrivals for the whole horizon of the power control problem are assumed to be known ahead of time; therefore in the second case, the decision of how much energy to allocate to the channel at the current time t can also depend on the knowledge of future energy harvests. This scenario is applicable when the energy harvesting process is very predictable. In most applications, however, it is not possible to exactly predict future energy arrivals and the online scenario is a better fit. In the next subsection, we first discuss the offline case, as a prelude to the more sophisticated case of online power control. We then discuss the online case in detail.

Let b_t be the amount of energy available in the battery at time slot t. Any admissible online or offline policy must satisfy the following constraints for every possible harvesting sequence $\{E_t, t = 1, 2, \dots\}$:

$$0 \le g_t \le b_t, \qquad t = 1, 2, \dots, \tag{7.3}$$

$$b_t = \min(b_{t-1} - g_{t-1} + E_t, \bar{B}), \qquad t = 2, 3, \dots, \tag{7.4}$$

where b_1 is the initial battery level. In words, the energy allocated at time t cannot exceed the amount of energy b_t available in the battery, which in turn depends on the energy arrivals as well as the energy consumed in the previous time slots. Note that at any given time the amount of available energy in the battery cannot exceed the battery capacity $\bar{B}$; therefore any excessive energy arrivals are clipped at $\bar{B}$. By allocating power g_t at time t, we assume that we can achieve an instantaneous rate $r_t = \frac{1}{2} \log(1 + g_t)$, equal to the capacity of an AWGN channel with average power g_t (the noise variance is chosen to be 1 without loss of generality). This formulation implicitly assumes that there is an underlying transmission scheme operating at a finer time scale, such that each time slot constitutes a sufficiently long block for achieving the corresponding AWGN capacity.

In the online case, we will assume that the energy arrival process $\{E_t, t = 1, 2 \dots\}$ is a stochastic process with known distribution. In this case, we define the N-horizon total throughput for a policy π as

$$\mathcal{T}_N(\pi) = \sum_{t=1}^N \mathbb{E}\left[\frac{1}{2}\log(1 + g_t)\right],$$

where $g_t = \pi_t(E^t)$ and the expectation is over the random energy arrivals $E_1, \ldots, E_N$. The long-term average throughput is defined as

$$\mathscr{T}(\pi) = \liminf_{N \to \infty} \frac{1}{N} \mathscr{T}_N(\pi).$$

The goal of the online power control optimization problem is to maximize the long-term average throughput:

$$\Theta = \sup_\pi \mathscr{T}(\pi).$$

In the offline case, since $E_1, \ldots, E_N$ are known and fixed ahead of time, the power control problem can be solved optimally for any given sequence of future energy arrivals $E_1, \ldots, E_N$.

7.1.1 Offline Power Control

In this section we discuss the optimal solution for the N-horizon offline power control problem for any given and known sequence of future energy arrivals $E_1, \ldots, E_N$. We will assume that each $E_t \leq \bar{B}$. This is without loss of generality since whenever an energy arrival E_t is larger than $\bar{B}$, the battery will be fully charged to $\bar{B}$, and the remaining energy will be discarded as per (7.4).

The optimal energy allocation should ensure that energy is never wasted due to the finite storage capacity of the battery. This can be ensured in the offline scenario, since future energy arrivals are known ahead of time. Mathematically, this is simply the requirement

$$b_{t-1} - g_{t-1} + E_t \leq \bar{B}, \qquad t = 2, \ldots, N.$$

This leads to the following constraint, called *no battery overflow*, on the optimal energy allocations:

$$b_1 + \sum_{i=2}^{t} E_i - \sum_{i=1}^{t-1} g_i \leq \bar{B}, \qquad t = 2, \ldots, N. \tag{7.5}$$

When there is no battery overflow, the energy constraints in (7.3) and (7.4) reduce to the following simpler constraint, called the *energy causality* constraint. This constraint simply requires that total energy allocated up to a given time t does not exceed the total amount of energy harvested up to that time. Mathematically, this is written as

$$\sum_{i=1}^{t} g_i \leq b_1 + \sum_{i=2}^{t} E_i, \qquad t = 1, \ldots, N. \tag{7.6}$$

These two constraints are illustrated in Figure 7.2. To simplify the figure, we will assume that there are nonzero energy harvests only in some of the time slots, such that the cumulative harvested energy plot has a "staircase" form. The upper staircase plot is the cumulative harvested energy, which is the energy causality upper bound. The lower staircase plot is the "no battery overflow" constraint, which is a lower bound on the energy consumption. The *total energy consumption curve*, which is the sum of

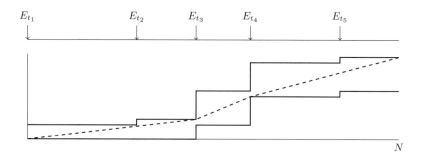

Figure 7.2 The feasibility tunnel (solid) along with the optimal energy consumption curve (dashed).

the energies allocated up to a given time, must lie between these two staircase curves, which is called the *feasibility tunnel*. The optimal energy consumption curve can be shown to be the tightest string between the bottom-left and top-right corners of the tunnel. This solution aims to keep energy consumption as uniform as possible over time, by allocating the longest stretches of constant energy allocations, while maintaining the energy causality and no battery overflow constraints and ensuring that all energy is consumed by the Nth time slot. The optimality of the uniform allocation follows from the concavity of the rate function.

7.1.2   Online Power Control: i.i.d. Energy Arrivals

While the offline case can be explicitly solved, the more interesting case is the *online* scenario, where future energy arrivals are random and unknown. In this section, we will focus on the online power control problem with *infinite horizon*, i.e., $t = 1, 2, \ldots$. We assume a random amount of energy E_t is harvested at time t, which is a nonnegative random variable drawn from a set $\mathcal{E} \subseteq \mathbb{R}_+$, such that $\mathbb{E}[E_t] > 0$. In this section we will focus on i.i.d. energy arrival processes; in the next section we will generalize to a *block* i.i.d. model.

Markov Decision Process Formulation

It can be observed that the model under consideration is a Markov decision process (MDP), with the state being the battery level b_t, the action g_t allowed to take values in the interval $[0, b_t]$, and the disturbance E_{t+1}. The state evolution equation is given by (7.4), and the stage reward is $r_t = \frac{1}{2} \log(1 + g_t)$. The online power control problem is exactly the maximization of the long-term average reward of this MDP.

It then follows by a well-known result in MDPs that the optimal policy is *Markovian*, i.e., it depends only on the current state: $\pi_t(E^t) = \pi_t(b_t)$. If the policy depends only on the current state and it is time-invariant, i.e., there is some function f such that $\pi_t(E^t) = f(b_t)$, we say it is *stationary*. The optimal throughput, along with the optimal policy, can be found by means of dynamic programming, which involves solving the Bellman equation:

THEOREM 7.1 (Bellman equation)   *If there exists a scalar $\lambda \in \mathbb{R}_+$ and a bounded function $h : [0, \bar{B}] \to \mathbb{R}_+$ that satisfy*

$$\lambda + h(b) = \sup_{0 \leq g \leq b} \left\{ \frac{1}{2} \log(1 + g) + \mathbb{E}[h(\min\{b - g + E_t, \bar{B}\})] \right\}, \quad 0 \leq b \leq \bar{B}, \quad (7.7)$$

then the optimal throughput is $\Theta = \lambda$. Furthermore, if $g^\star(b)$ attains the supremum in (7.7), then the optimal policy is given by $\pi_t(E^t) = g^\star(b_t(E^t))$.

The functional equation (7.7) is hard to solve explicitly, and requires an exact model for the statistical distribution of the energy arrivals E_t, which may be hard to obtain in practical scenarios. The equation can be solved numerically using the value iteration algorithm, but this can be computationally demanding, and the numerical solution does not provide much insight on the structure of the optimal online power control policy and the qualitative behavior of the resultant throughput, namely how it varies with the parameters of the problem. This kind of insight can be critical for design considerations, such as choosing the size of the battery to employ at the transmitter.

In what follows, we will see an explicit online power control policy and show that it is within a constant gap to optimality for all i.i.d. harvesting processes. This policy depends on the harvesting process only through its mean, and it also leads to a simple and insight-ful approximation of the achievable throughput. We first discuss a special case in which the optimal online solution can be explicitly found. This inspires the approximately optimal power control policy for general i.i.d. energy harvesting processes.

Bernoulli Energy Arrivals
Assume the energy arrivals are i.i.d. Bernoulli random variables, namely

$$E_t = \begin{cases} \bar{B} & , \text{w.p. } p, \\ 0 & , \text{w.p. } 1 - p. \end{cases} \quad (7.8)$$

At each time slot t, the battery is either fully charged to $\bar{B}$ with probability p (regardless of the amount of energy previously available in the battery), or no energy is harvested at all with probability $1 - p$.

Even in this simple case, the Bellman equation (7.7) is hard to solve analytically. Nevertheless, we can transform the problem into a deterministic form, which allows for an explicit solution. It can be shown that the optimal policy is stationary, i.e., it is given by some function $g_t = \pi^\star(b_t)$. Then the battery state behaves as follows:

$$b_t = \begin{cases} \bar{B} & , \text{w.p. } p, \\ b_{t-1} - \pi^\star(b_{t-1}) & , \text{w.p. } 1 - p. \end{cases}$$

Denote $b^{(1)} = \bar{B}$ and $b^{(i)} = b^{(i-1)} - \pi^\star(b^{(i-1)})$ for $i \geq 2$. Note that the sequence $\{b^{(i)}\}_{i=1}^{\infty}$ is deterministic, and can be computed given the function $\pi^\star$. The battery state process b_t will take the form of the sequence $b^{(i)}$, restarting from $b^{(1)}$ every time there is a battery recharge. A typical battery state process under Bernoulli energy arrivals is illustrated in Figure 7.3. The epoch durations L_k, $k = 1, 2, \ldots$ (the times between consecutive positive energy arrivals) are i.i.d. random variables, distributed according

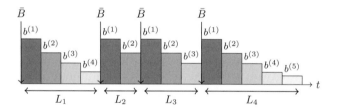

Figure 7.3 Battery state process for Bernoulli energy arrivals under the optimal stationary policy.

to a geometric distribution with parameter p. This unique behavior of the process b_t is called *regenerative*, formally defined below.

DEFINITION 7.2   *A stochastic process X_t is called* regenerative *if there exists a random time L such that the process $X_{L+1}, X_{L+2}, \ldots$ has the same distribution $X_1, X_2, \ldots$, and is independent of the past (L, X^L).*

Regenerative processes are composed of i.i.d. epochs with i.i.d. random duration L. At the beginning of each epoch, the process regenerates and all memory of the past is erased. Clearly the battery state process b_t is regenerative, which will be very useful in the analysis of the long-term average throughput, using the law of large numbers for regenerative processes.

THEOREM 7.3 (LLN for regenerative processes)   *Let X_t be a regenerative process with epoch duration L, and let f be some function. If $\mathbb{E}L < \infty$ and $\mathbb{E}[\sum_{t=1}^{L} |f(X_t)|] < \infty$, then:*

$$\frac{1}{N}\sum_{t=1}^{N} f(X_t) \xrightarrow{N\to\infty} \frac{\mathbb{E}\left[\sum_{i=1}^{L} f(X_i)\right]}{\mathbb{E}L} \quad a.s.$$

Applying the theorem to the long-term average throughput under the policy $\pi^\star$ (up to interchanging limit and expectation, which can be shown to hold):

$$\mathcal{T}(\pi^\star) = \frac{1}{\mathbb{E}L}\mathbb{E}\left[\sum_{i=1}^{L}\frac{1}{2}\log(1+\pi^\star(b^{(i)}))\right]. \tag{7.9}$$

Writing out the expectation over $L \sim$ Geometric(p):

$$\mathcal{T}(\pi^\star) = p\sum_{k=1}^{\infty} p(1-p)^{k-1}\sum_{i=1}^{k}\frac{1}{2}\log(1+\pi^\star(b^{(i)}))$$

$$= \sum_{i=1}^{\infty} p(1-p)^{i-1}\frac{1}{2}\log(1+\pi^\star(b^{(i)})),$$

where the second equality is by changing the order of summations followed by carrying out the summation over k.

Denote $\varepsilon_i = \pi^\star(b^{(i)})$. It follows from the energy constraints (7.3) and (7.4) that $\varepsilon_i \geq 0$ for all i and $\sum_{i=1}^{\infty} \varepsilon_i \leq \bar{B}$. The optimal throughput is therefore given by the solution to the following *deterministic* convex optimization problem:

$$\text{maximize} \quad \sum_{i=1}^{\infty} p(1-p)^{i-1} \frac{1}{2} \log(1 + \varepsilon_i)$$

$$\text{subject to} \quad \varepsilon_i \geq 0, \quad i = 1, 2, \ldots, \tag{7.10}$$

$$\sum_{i=1}^{\infty} \varepsilon_i \leq \bar{B}.$$

This can be readily solved using Karush–Kuhn–Tucker conditions, giving the following optimal policy.

THEOREM 7.4   *For the online power control problem with energy harvesting process given by (7.8), let $j_t(E^t)$ be the time of the last energy arrival, i.e.,*

$$j_t(E^t) = \{\sup \, \tau \leq t : E_\tau = \bar{B}\}.$$

Then the optimal policy is given by

$$\pi_t^\star(E^t) = \begin{cases} \frac{\tilde{N} + \bar{B}}{1 - (1-p)^{\tilde{N}}} p(1-p)^{t-j_t} - 1, & t - j_t < \tilde{N} \\ 0, & t - j_t \geq \tilde{N} \end{cases} \tag{7.11}$$

where $\tilde{N}$ is the smallest positive integer satisfying

$$1 > (1-p)^{\tilde{N}}[1 + p(\bar{B} + \tilde{N})].$$

It can be seen that this is a stationary policy, i.e., $\pi_t^\star(E^t)$ can be written as a time-invariant function of $b_t(E^t)$. Roughly speaking, the energy is allocated only to the first $\tilde{N}$ time slots after each battery recharge, and decays in an approximately exponential manner.

Approximately Optimal Policy for Bernoulli Energy Arrivals

Inspired by the exponentially decreasing structure of the optimal policy derived in the previous section, we suggest the following suboptimal but simpler policy for Bernoulli energy arrivals:

$$\varepsilon_i = \bar{B}p(1-p)^{i-1}, \quad i = 1, 2, \ldots. \tag{7.12}$$

Note that the policy decreases exactly exponentially with time since the last battery recharge, and observe that it satisfies the energy constraints since $\sum_{i=1}^{\infty} \varepsilon_i = \bar{B}$. An alternative way of writing this policy is via the stationary function

$$\pi(b) = p \cdot b. \tag{7.13}$$

That is, the policy simply uses a p fraction of the available energy in the battery.

This simplified policy can be intuitively motivated as follows: In a Bernoulli process, the time between two consecutive battery recharges is a geometric random variable with parameter p. Because the geometric distribution is memoryless and has mean $1/p$, at each time step the expected time to the next energy arrival is $1/p$. As we have seen for the offline case, due to the concavity of the rate function, uniform allocation of energy maximizes throughput. That is, if the current energy level in the battery is b_t, and we knew that the next battery recharge would be in exactly m channel uses, allocating b_t/m energy to each of the next m channel uses would maximize throughput. For the online case of interest here, we do not know m but we may instead use the expected time to the next energy arrival: Since at each time step the expected time to the next energy arrival is $1/p$, we always allocate a fraction p of the currently available energy in the battery.

This simple policy can be shown to be close to optimality up to a constant gap, which also provides an approximate formula for the optimal throughput Θ.

THEOREM 7.5   *For the online power control problem with energy harvesting process given by (7.8), the throughput of any policy is upper bounded by*

$$\Theta \leq \frac{1}{2}\log(1 + p\bar{B}), \tag{7.14}$$

while the throughput of the policy π defined in (7.13) is lower bounded by

$$\mathcal{T}(\pi) \geq \frac{1}{2}\log(1 + p\bar{B}) - \frac{1}{2}\log e, \tag{7.15}$$

$$\mathcal{T}(\pi) \geq \frac{1}{2} \cdot \frac{1}{2}\log(1 + p\bar{B}). \tag{7.16}$$

This suggests that the simplified policy (7.13) is always within $\frac{1}{2}\log e \approx 0.72$ bits per channel-use of optimality, and simultaneously within 50% of optimality; the former being especially useful when the mean energy arrival rate (or signal to noise ratio, SNR) $p\bar{B}$ is large, and the latter when $p\bar{B}$ is small. Numerical evaluations show, however, that the real gap to optimality is much smaller than the one given in Theorem 7.5. In fact, Figure 7.4 shows that the throughput obtained by the policy (7.13) is almost indistinguishable from the optimal throughput.

Proof of the Upper Bound

Fix an arbitrary stationary policy π, which is equivalently given by a sequence of power allocations $\{\varepsilon_i\}_{i=1}^{\infty}$. As in (7.10), the throughput can be expressed as follows:

$$\mathcal{T}(\pi) = \sum_{i=1}^{\infty} p(1-p)^{i-1}\frac{1}{2}\log(1 + \varepsilon_i)$$

$$\overset{(a)}{\leq} \frac{1}{2}\log\left(1 + \sum_{i=1}^{\infty} p(1-p)^{i-1}\varepsilon_i\right)$$

$$\overset{(b)}{\leq} \frac{1}{2}\log\left(1 + p\sum_{i=1}^{\infty}\varepsilon_i\right)$$

$$\overset{(c)}{\leq} \frac{1}{2}\log(1 + p\bar{B}),$$

where (a) is due to Jensen's inequality, (b) is because $(1-p)^{(i-1)} \leq 1$ for $i \geq 1$, and (c) follows from the constraint $\sum_{i=1}^{\infty}\varepsilon_i \leq \bar{B}$. Taking supremum over all policies completes the proof.

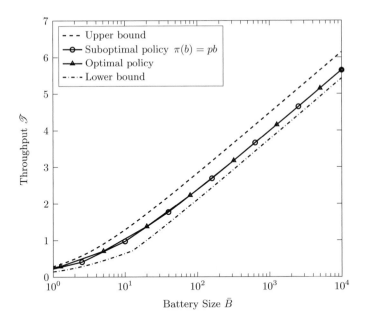

Figure 7.4 Optimal throughput and the throughput obtained by the policy (7.13), for $p = 0.5$, along with the upper and lower bounds suggested in Theorem 7.5.

Proof of the Lower Bounds

Let π be the policy in (7.13), which can be equivalently given in terms of the power allocation sequence (7.12). We can again express the throughput as in (7.10):

$$\mathcal{T}(\pi) = \sum_{i=1}^{\infty} p(1-p)^{i-1} \frac{1}{2} \log(1 + \varepsilon_i)$$

$$\stackrel{(a)}{=} \sum_{i=1}^{\infty} p(1-p)^{i-1} \frac{1}{2} \log(1 + \bar{B}p(1-p)^{i-1})$$

$$= \sum_{i=1}^{\infty} p(1-p)^{i-1} \left[\frac{1}{2} \log \left((1-p)^{-(i-1)} + p\bar{B} \right) + \frac{1}{2} \log(1-p)^{i-1} \right]$$

$$\geq \sum_{i=1}^{\infty} p(1-p)^{i-1} \left[\frac{1}{2} \log(1 + p\bar{B}) + (i-1)\frac{1}{2} \log(1-p) \right]$$

$$= \frac{1}{2} \log(1 + p\bar{B}) + \frac{1-p}{2p} \log(1-p)$$

$$\stackrel{(b)}{\geq} \frac{1}{2} \log(1 + p\bar{B}) - \frac{1}{2} \log e,$$

where (a) is by (7.12) and (b) is because $\frac{1-p}{2p} \log(1-p) \geq -\frac{1}{2} \log e$ for $0 \leq p \leq 1$.

To show the other lower bound, we start from the same throughput expression:

$$\mathcal{T}(\pi) = \sum_{i=1}^{\infty} p(1-p)^{i-1} \frac{1}{2} \log(1 + \bar{B}p(1-p)^{i-1})$$

$$\overset{(a)}{\geq} \sum_{i=1}^{\infty} p(1-p)^{i-1} \cdot (1-p)^{i-1} \frac{1}{2} \log(1 + p\bar{B})$$

$$= \frac{1}{2-p} \frac{1}{2} \log(1 + p\bar{B})$$

$$\geq \frac{1}{2} \cdot \frac{1}{2} \log(1 + p\bar{B}),$$

where (a) follows from the inequality $\log(1 + \alpha x) \geq \alpha \log(1 + x)$ for $0 \leq \alpha \leq 1$ and $x \geq 0$.

Approximately Optimal Policy for General Energy Arrival Distribution

Now assume E_t is an i.i.d. process with some known arbitrary distribution. We will present a natural generalization of the simplified policy (7.13), and show that it is approximately optimal with the same bounds as in the Bernoulli case.

Before presenting the policy, we make the following observation. Whenever an energy arrival E_t is larger than $\bar{B}$, the battery will be fully charged to $\bar{B}$ and the remaining energy will be discarded. Therefore, without loss of generality, we will replace the energy arrival process E_t with

$$\tilde{E}_t = \min(E_t, \bar{B})$$

in the sequel. We proceed to define our policy:

DEFINITION 7.6 (fixed fraction policy)   *Denote* $\mu = \mathbb{E}[\tilde{E}_t]$, *and let* $q = \mu/\bar{B}$. *The Fixed Fraction Policy is defined as*

$$\pi(b) = q \cdot b. \tag{7.17}$$

We use q here instead of p in the Bernoulli case; in that case, we also had $\mathbb{E}[\tilde{E}_t] = p\bar{B}$, *so this is a natural definition. Note that* $0 \leq q \leq 1$, *hence this policy satisfies the energy constraint* (7.3).

In the following theorem, we show that the fixed fraction policy is close to optimality within a constant additive gap and a constant multiplicative gap for any distribution of the energy arrivals. We prove this result by showing that under this policy, the Bernoulli harvesting process yields the *worst* performance compared to all other i.i.d. processes with the same mean μ. This implies that the lower bounds obtained for the throughput achieved under Bernoulli energy arrivals apply also to any i.i.d. harvesting process with the same mean.

THEOREM 7.7   *Let* E_t *be an i.i.d. energy harvesting process with* $\mu = \mathbb{E}[\min(E_t, \bar{B})]$. *The throughput obtained for any policy is upper bounded by*

$$\Theta \leq \frac{1}{2} \log(1 + \mu), \tag{7.18}$$

while the throughput obtained by the fixed fraction policy (7.17) is lower bounded by

$$\mathcal{T}(\pi) \geq \frac{1}{2}\log(1 + \mu) - \frac{1}{2}\log e, \tag{7.19}$$

$$\mathcal{T}(\pi) \geq \frac{1}{2} \cdot \frac{1}{2}\log(1 + \mu). \tag{7.20}$$

This theorem gives a simple approximation of the optimal throughput and its dependence on the energy harvesting process E_t and the battery size $\bar{B}$. It identifies two fundamentally different operating regimes for this channel, where the dependence of the average throughput on E_t and $\bar{B}$ is qualitatively different. Assume that E_t takes values in the interval $[0, \bar{E}]$. When $\bar{B} \geq \bar{E}$, we have $\Theta \approx \frac{1}{2}\log(1 + \mathbb{E}[E_t])$, and the throughput is approximately equal to the capacity of an AWGN channel with an average power constraint equal to the average energy harvesting rate. This is surprising given that the transmitter is limited by the additional constraints (7.3) and (7.4), and at finite $\bar{B}$ this can lead to part of the harvested energy being wasted due to overflows in the battery capacity. Note that in this large battery regime, the throughput depends only on the mean of the energy harvesting process – two energy harvesting profiles are equivalent as long as they provide the same energy on average – and is also independent of the battery size $\bar{B}$. In particular, choosing $\bar{B} \approx \bar{E}$ is approximately sufficient to achieve the throughput at infinite battery size.

When $\bar{B} < \bar{E}$, we can equivalently consider the distribution of E_t to be that in the right plot of Figure 7.5. Since every energy arrival with value $E_t \geq \bar{B}$ fully recharges the battery, this creates a point mass at $\bar{B}$ with value $\Pr(E_t \geq \bar{B})$. In this case, Theorem 7.7 reveals that the throughput is approximately determined by the mean of this modified distribution. This can be interpreted as the small battery regime of the channel. In particular, in this regime, the achievable throughput depends both on the shape of the distribution of E_t and the value of $\bar{B}$.

In Figure 7.6, we compare the actual performance of the fixed fraction policy (7.17) to the throughput achieved by two other common heuristic policies: the greedy policy, which always allocates all the energy available in the battery, i.e., $\pi(b) = b$; and the "constant" policy, which attempts to transmit at a constant power μ, and if there is

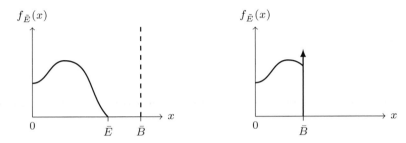

Figure 7.5 The distribution of $\tilde{E}_t = \min\{E_t, \bar{B}\}$ in the two battery regimes. The left plot corresponds to $\bar{B} \geq \bar{E}$, and the right plot depicts the case $\bar{B} < \bar{E}$.

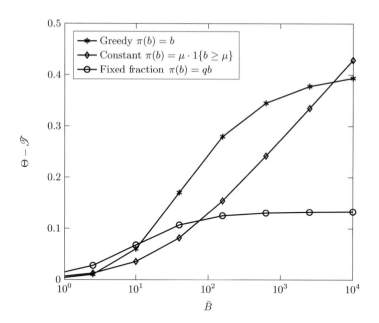

Figure 7.6 Plots of the additive gap between the optimal throughput Θ and achieved throughput $\mathscr{T}$ for different power allocation policies, when the energy arrivals are exponentially distributed, $E_t \sim \text{Exp}\left(\frac{1}{0.1\bar{B}}\right)$.

not enough energy in the battery, it simply waits until the battery is recharged again to a level at least μ. The figure depicts the additive gap to optimality obtained by each policy, i.e., the difference between the optimal throughput and the throughput obtained by the policy, when the energy arrivals follow an exponential probability distribution (this corresponds to the energy process arising from a Gaussian signal). Note that the gap for the fixed fraction policy remains small for all values of $\bar{B}$, while the gaps for the greedy and constant policies grow unboundedly as $\bar{B}$ increases.

Proof of the Upper Bound

Let π be an arbitrary policy, and let $\{g_t\}$ be the sequence of power allocations induced by this policy. Since $\{g_t\}$ satisfies the energy constraints (7.3) and (7.4):

$$g_N \overset{(a)}{\leq} \min(b_{N-1} - g_{N-1} + \tilde{E}_N, \bar{B})$$
$$\leq b_{N-1} - g_{N-1} + \tilde{E}_N,$$

where in (a) we replace E_t with $\tilde{E}_t$ according to the previous discussion. Repeating this step for b_{N-1}, followed by b_{N-2}, and so forth, we arrive at the inequality

$$\sum_{t=1}^{N} g_t \leq b_1 + \sum_{t=2}^{N} \tilde{E}_t. \tag{7.21}$$

The N-horizon average expected throughput obtained by this policy is

$$\frac{1}{N} \mathscr{T}_N(\pi) = \frac{1}{N} \sum_{t=1}^{N} \mathbb{E} \left[\frac{1}{2} \log(1 + g_t) \right]$$

$$\overset{(a)}{\leq} \frac{1}{N} \sum_{t=1}^{N} \frac{1}{2} \log(1 + \mathbb{E}[g_t])$$

$$\overset{(b)}{\leq} \frac{1}{2} \log \left(1 + \frac{1}{N} \sum_{t=1}^{N} \mathbb{E}[g_t] \right)$$

$$\overset{(c)}{\leq} \frac{1}{2} \log \left(1 + \frac{1}{N} \left(b_1 + \sum_{t=2}^{N} \mathbb{E}[\tilde{E}_t] \right) \right)$$

$$= \frac{1}{2} \log \left(1 + \frac{b_1}{N} + \frac{N-1}{N} \mu \right),$$

where (a) and (b) follow from Jensen's inequality, and (c) follows from (7.21). Taking the limit as $N \to \infty$:

$$\mathscr{T}(\pi) \leq \frac{1}{2} \log(1 + \mu).$$

Since the policy π was arbitrary, this concludes the proof of the upper bound.

Proof of the Lower Bounds

Observe that under the fixed fraction policy, the battery state is a Markov process:

$$b_{t+1} = \min((1 - q)b_t + \tilde{E}_{t+1}, \bar{B}).$$

For $x \in [0, \bar{B}]$, denote

$$J_N(x) = \sum_{t=1}^{N} \mathbb{E} \left[\frac{1}{2} \log(1 + qb_t) \Big| b_1 = x \right],$$

so that the N-horizon total throughput is given by $\mathscr{T}_N(\pi) = J_N(\bar{B})$.

Now define a new i.i.d. energy arrival process:

$$\hat{E}_t = \begin{cases} \bar{B} & , \text{w.p. } q, \\ 0 & , \text{w.p. } 1 - q. \end{cases}$$

Note that $\mathbb{E}[\hat{E}_t] = q\bar{B} = \mu$. The fixed fraction policy operating on this energy process induces the following battery state process:

$$\hat{b}_{t+1} = \min \left((1 - q)\hat{b}_t + \hat{E}_{t+1}, \bar{B} \right).$$

Similarly, we define the N-horizon expected throughput under this new energy arrival process when the initial battery state is $\hat{b}_1 = x$:

$$\hat{J}_N(x) = \sum_{t=1}^{N} \mathbb{E}\left[\frac{1}{2}\log(1 + q\hat{b}_t)\Big|\hat{b}_1 = x\right].$$

This is exactly the throughput obtained for the policy $\pi(b) = qb$ defined in (7.13) for Bernoulli energy arrivals. Therefore, it follows from Theorem 7.5:

$$\liminf_{N\to\infty} \frac{1}{N}\hat{J}_N(\bar{B}) \geq \frac{1}{2}\log(1 + q\bar{B}) - \frac{1}{2}\log e,$$

$$\liminf_{N\to\infty} \frac{1}{N}\hat{J}_N(\bar{B}) \geq \frac{1}{2}\cdot\frac{1}{2}\log(1 + q\bar{B}).$$

In Lemma 7.8, we claim that the N-horizon expected throughput for any distribution of i.i.d. energy arrivals is always better than the throughput obtained for i.i.d. Bernoulli energy arrivals with the same mean, for any N and any initial battery level x.

LEMMA 7.8   *For any $x \in [0, \bar{B}]$ and any $N \geq 1$:*

$$J_N(x) \geq \hat{J}_N(x).$$

An immediate result of this lemma is

$$\mathscr{T}(\pi) = \liminf_{N\to\infty} \frac{1}{N}J_N(\bar{B})$$

$$\geq \liminf_{N\to\infty} \frac{1}{N}\hat{J}_N(\bar{B})$$

$$\geq \frac{1}{2}\log(1 + q\bar{B}) - \frac{1}{2}\log e$$

$$= \frac{1}{2}\log(1 + \mu) - \frac{1}{2}\log e,$$

and similarly

$$\mathscr{T}(\pi) \geq \frac{1}{2}\cdot\frac{1}{2}\log(1 + \mu),$$

which concludes the proof of the lower bounds.

Proof of Lemma 7.8. We will give a proof by induction.
   Base case: Clearly for $N = 1$ we have

$$J_1(x) = \hat{J}_1(x) = \frac{1}{2}\log(1 + qx).$$

Observe that this is a non-decreasing and concave function of x. This will in fact be true for every $\hat{J}_N(x)$, $N \geq 1$, and we will use this fact in the induction step.
   Induction hypothesis: Assume that $J_N(x) \geq \hat{J}_N(x)$ for all $x \in [0, \bar{B}]$, and also that $\hat{J}_N(x)$ is monotonic non-decreasing and concave.
   Inductive step: We start by expressing $J_{N+1}(x)$ in terms of $J_N(x)$, using the fact that $\{b_t\}$ is a time homogeneous Markov chain:

$$J_{N+1}(x) = \sum_{t=1}^{N+1} \mathbb{E}\left[\frac{1}{2}\log(1+qb_t)\big|b_1 = x\right]$$

$$= \frac{1}{2}\log(1+qx) + \mathbb{E}[J_N(b_2)|b_1 = x]$$

$$= \frac{1}{2}\log(1+qx) + \mathbb{E}\left[J_N\left(\min((1-q)x+\tilde{E}_2,\bar{B})\right)\right]$$

$$\overset{(a)}{\geq} \frac{1}{2}\log(1+qx) + \mathbb{E}\left[\hat{J}_N\left(\min((1-q)x+\tilde{E}_2,\bar{B})\right)\right], \qquad (7.22)$$

where (a) is due to the induction hypothesis. Observe that if in the expression above we would have $\hat{E}_2$ instead of $\tilde{E}_2$, this would have been the desired result. Indeed, the following lemma will be used to further lower bound this expression using $\hat{E}_2$.

LEMMA 7.9   Let $f(z)$ be a concave function on the interval $[0,\bar{B}]$, and let Z be a random variable confined to the same interval, i.e., $0 \leq Z \leq \bar{B}$. Let $\hat{Z} \in \{0,\bar{B}\}$ be a Bernoulli random variable with $\Pr(Z = \bar{B}) = \frac{\mathbb{E}Z}{\bar{B}}$. Then

$$\mathbb{E}[f(Z)] \geq \mathbb{E}[f(\hat{Z})].$$

Proof.   By definition of concavity, for any $z \in [0,\bar{B}]$:

$$f(z) \geq \frac{z}{\bar{B}}f(\bar{B}) + \frac{\bar{B}-z}{\bar{B}}f(0).$$

Taking expectation:

$$\mathbb{E}[f(Z)] \geq \frac{\mathbb{E}Z}{\bar{B}}f(\bar{B}) + \frac{\bar{B}-\mathbb{E}Z}{\bar{B}}f(0) = \mathbb{E}[f(\hat{Z})].   \blacksquare$$

Consider the function

$$f(z) = \hat{J}_N\left(\min((1-q)x+z,\bar{B})\right).$$

By the induction hypothesis, $\hat{J}_N(x)$ is monotonic non-decreasing, hence

$$f(z) = \min\left(\hat{J}_N((1-q)x+z),\ \hat{J}_N(\bar{B})\right).$$

The first function in the min is concave (by the induction hypothesis). A minimum of concave functions is a concave function (the second function being simply a constant). Therefore, $f(z)$ is concave, and by application of Lemma 7.9 with $\tilde{E}_2$ and $\hat{E}_2$ instead of Z and $\hat{Z}$, respectively, we get

$$\mathbb{E}\left[\hat{J}_N\left(\min((1-q)x+\tilde{E}_2,\bar{B})\right)\right] \geq \mathbb{E}\left[\hat{J}_N\left(\min((1-q)x+\hat{E}_2,\bar{B})\right)\right].$$

Substituting in (7.22):

$$J_{N+1}(x) \geq \frac{1}{2}\log(1+qx) + \mathbb{E}\left[\hat{J}_N\left(\min((1-q)x+\hat{E}_2,\bar{B})\right)\right]$$

$$= \frac{1}{2}\log(1+qx) + \mathbb{E}[\hat{J}_N(\hat{b}_2)|\hat{b}_1 = x]$$

$$= \hat{J}_{N+1}(x).$$

This proves the first part of the induction hypothesis. It remains to show that $\hat{J}_{N+1}(x)$ is concave and non-decreasing. Expressing it in terms of $\hat{J}_N(x)$, and writing out the expectation over $\hat{E}_2$:

$$\hat{J}_{N+1}(x) = \frac{1}{2}\log(1 + qx) + q\hat{J}_N(\bar{B}) + (1 - q)\hat{J}_N((1 - q)x).$$

This is a weighted sum of concave and non-decreasing functions of x, hence $\hat{J}_{N+1}(x)$ is concave and non-decreasing.   ∎

7.1.3   Online Power Control: Block i.i.d. Energy Arrivals

So far we have assumed the energy arrivals are either perfectly known ahead of time, or are randomly generated from an i.i.d. process, such that the next energy arrival at each time instant is impossible to predict. However, most natural energy harvesting processes are neither of the two, but rather somewhere in between.

In what follows, we consider energy arrivals which follow a *block i.i.d.* model: The energy arrivals remain constant for a fixed duration of T time slots, and then change to an independent realization for the next T time slots. This can model, for example, a device which harvests RF energy from other transmitting devices in its environment. Such transmitting devices typically transmit continuously for certain periods of time and are silent for the remaining periods (as in time division multiple access [TDMA] for example), which warrants a block i.i.d. model. More importantly, the block i.i.d. model provides a simple way to study the impact of correlation in the harvesting process. Such models are popularly used in wireless communications to model correlations in the fading process. In this case, T is called the coherence time of the channel, which corresponds to the time duration over which the channel remains approximately constant. Analogously, we refer to T as the coherence time of the energy arrival process.

For a process $\{X_t\}$ with block structure, we will use the notation $X_j^{(i)}$ to denote the jth time slot of the ith block, that is

$$X_j^{(i)} = X_{(i-1)T+j}, \qquad j = 1, \ldots, T, \qquad i = 1, 2, \ldots.$$

The problem formulation is the same as before, except the harvesting process satisfies

$$E_t = E^{(i)}, \qquad t = (i - 1)T + 1, \ldots, iT, \qquad i = 1, 2, \ldots, \qquad (7.23)$$

where $\{E^{(i)}\}$ is an i.i.d. process generated from a given distribution P_E.

Due to the block structure of the energy arrivals, at the beginning of each block, upon observing $b_1^{(i)}$ and $E^{(i)}$, the transmitter can decide on the power allocations for the entire block $g_1^{(i)}, \ldots, g_T^{(i)}$. Hence the problem can be formulated as an MDP, where each MDP stage constitutes an entire block of T time slots. In fact, due to the concavity of the rate function and since it is suboptimal to have battery overflows inside the block, it can be shown that it is optimal to set $g_1^{(i)} = \ldots = g_{T-1}^{(i)}$, which simplifies the MDP (this is somewhat reminiscent of the results of the offline case). Note that the power allocated at the last time slot of the block, $g_T^{(i)}$, is different, since at this point the battery state at the next time slot is still unknown. (The easiest way to see that $g_T^{(i)}$ needs to be different is to consider a block Bernoulli energy arrivals process, where either $E^{(i)} = \bar{B}$ or $E^{(i)} = 0$. When $E^{(i)} = \bar{B}$, clearly $g_1^{(i)} = \ldots = g_{T-1}^{(i)} = \bar{B}$ since any energy allocation less than that will result in wasted energy. However, the optimal energy allocation for the last time slot needs to be smaller than $\bar{B}$; if we also allocate energy $\bar{B}$ and there is not any incoming energy over the next block, transmission will be interrupted which is likely to lead to low throughput.) This MDP is described formally in the following lemma.

LEMMA 7.10 *The online power control problem with a block i.i.d. harvesting process as defined in (7.23) can be formulated as an MDP with state pair $(b_1^{(i)}, E^{(i)})$, action pair $(g_1^{(i)}, g_T^{(i)})$, and disturbance $E^{(i+1)}$. The actions must satisfy the constraints*

$$0 \leq g_1^{(i)} \leq \min\left(E^{(i)} - \tfrac{E^{(i)} - b_1^{(i)}}{T-1}, \bar{B}\right),$$
$$0 \leq g_T^{(i)} \leq b_T^{(i)},$$

where $b_T^{(i)} = \min\{b_1^{(i)} + (T-1)(E^{(i)} - g_1^{(i)}), \bar{B}\}$. Furthermore, the policy for $g_T^{(i)}$ may depend only on $b_T^{(i)}$ instead of $(b_1^{(i)}, E^{(i)})$.
The state evolves according to the function

$$\left(b_1^{(i+1)}, E^{(i+1)}\right) = \left(\min\{b_T^{(i)} - g_T^{(i)} + E^{(i+1)}, \bar{B}\}, E^{(i+1)}\right),$$

and the stage reward is given by

$$r(b_1^{(i)}, E^{(i)}, g_1^{(i)}, g_T^{(i)}) = \tfrac{T-1}{T} C(g_1^{(i)}) + \tfrac{1}{T} C(g_T^{(i)}).$$

Extending the Fixed Fraction Policy to Block i.i.d. Arrivals

We next construct a heuristic policy for block i.i.d. arrivals which builds on the insights from the i.i.d. case. Recall that in the i.i.d. case, the fixed fraction policy was allocating a q fraction of the available amount of energy in the battery b_t. Here, at the beginning of block i, the amount of energy in the battery is $b_1^{(i)}$. In addition, we know we will harvest an amount of energy equal to $(T-1)E^{(i)}$ by the end of the block. Analogous to the i.i.d. case, let us take q fraction of the total energy, for some appropriately chosen $0 \leq q \leq 1$, and due to the concavity of the rate function, divide it equally across all T time slots. Consequently, we suggest the following policy:

$$g_j^{(i)} = \frac{q}{T}(b_1^{(i)} + (T-1)E^{(i)}), \qquad j = 1, \ldots, T.$$

If there is no battery overflow during the block, the battery state at the last time slot of the block is

$$b_T^{(i)} = b_1^{(i)} + (T-1)(E^{(i)} - g_1^{(i)})$$
$$= \frac{q + (1-q)T}{T}(b_1^{(i)} + (T-1)E^{(i)}).$$

We can therefore write the policy in a way that agrees with the optimal policy structure in Lemma 7.10:

$$g_1^{(i)} = \frac{q}{T}(b_1^{(i)} + (T-1)E^{(i)}),$$
$$g_T^{(i)} = \frac{q}{q + (1-q)T}b_T^{(i)}. \tag{7.24}$$

However, this extension of the fixed fraction policy to the block i.i.d. case can indeed lead to battery overflows inside the block when $E^{(i)}$ is large. In particular, battery overflow will occur if the battery level at the last time slot exceeds the battery size, i.e., $b_T^{(i)} > \bar{B}$, or equivalently

$$\frac{q + T(1 - q)}{T}(b_1^{(i)} + (T - 1)E^{(i)}) > \bar{B}.$$

If we assume the battery was empty at the end of the previous block, i.e., $b_1^{(i)} = E^{(i)}$, the battery overflow condition would reduce to $E^{(i)} > E_c$, where E_c is a *critical energy level* given by

$$E_c = \frac{\bar{B}}{q + T(1 - q)}. \tag{7.25}$$

Note that when $E^{(i)} > E_c$, battery overflow will occur regardless of the state of the battery at the end of the previous block. Since this is known at the beginning of the block, we would want to modify the policy to increase our energy allocations for such blocks. Specifically, when $E^{(i)} > E_c$, we increase the energy allocations in the first $T - 1$ time slots such that the battery is left fully charged at the end of the block, $b_T^{(i)} = \bar{B}$, and energy is not wasted:

$$
\begin{aligned}
g_1^{(i)} &= \min\left(E^{(i)} - \frac{\bar{B} - b_1^{(i)}}{T - 1}, \bar{B}\right), \\
g_T^{(i)} &= \frac{q}{q + (1 - q)T}\bar{B}.
\end{aligned} \tag{7.26}
$$

Note that the energy allocated at the last time slot follows the same policy as (7.24), since $b_T^{(i)} = \bar{B}$.

Choosing the Parameter q

Recall that in the i.i.d. case q was chosen to be $\frac{\mathbb{E}[E_t, \bar{B}]}{\bar{B}}$. Consider the following interpretation for the numerator: Whenever the energy arrival E_t is greater than $\bar{B}$, the battery will be fully charged, which constitutes a *regeneration* of the battery state process b_t. Let $p = \Pr(E_t > \bar{B})$. The epoch duration, i.e., the time between consecutive regeneration events, is $\tau = L$, where L is a geometric random variable with parameter p. The average energy in a single epoch is

$$\varepsilon = \bar{B} + (L - 1)\mathbb{E}[E_t | E_t \leq \bar{B}].$$

Note that this is a random amount, due to the randomness of the epoch length L. We can compute the average expected energy per time slot:

$$
\begin{aligned}
\frac{\mathbb{E}\varepsilon}{\mathbb{E}\tau} &= \frac{\bar{B} + (\frac{1}{p} - 1)\mathbb{E}[E_t | E_t \leq \bar{B}]}{1/p} \\
&= \mathbb{E}[\bar{B} \cdot 1\{E_t > \bar{B}\}] + \mathbb{E}[E_t \cdot 1\{E_t \leq \bar{B}\}] \\
&= \mathbb{E}[\min(E_t, \bar{B})].
\end{aligned}
$$

Therefore, q can be thought of as the average energy available in an epoch, divided by the size of the battery $\bar{B}$.

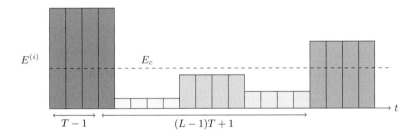

Figure 7.7 Illustration of an epoch of block i.i.d. energy arrivals. The first block constitutes a large arrival ($E^{(i)} > E_c$), and the subsequent $L - 1$ blocks are all "low-energy." The last block is again a large arrival, after which a new epoch begins.

Let us generalize this intuition to the block i.i.d. case. Note that when $E^{(i)} > E_c$, our policy will leave the battery fully charged by the end of the block, so this can be considered a regeneration event. Hence, let $p = \Pr(E^{(i)} > E_c)$, and imagine we put aside the first $T - 1$ time slots of the "large arrival" block (in which we abandon the fixed fraction policy), and instead concentrate only on the subsequent slots (where we do apply it). The average energy available for this period can be computed as

$$\varepsilon = \bar{B} + (L - 1)T\mathbb{E}[E^{(i)}|E^{(i)} \leq E_c],$$

where $L \sim \text{Geometric}(p)$ is the number of blocks between consecutive regeneration events $\{E^{(i)} > E_c\}$. This is because the battery is fully charged at the last time slot of the large arrival block, and at each one of the subsequent "low-energy" blocks the transmitter harvests an average amount of energy equal to $T\mathbb{E}[E^{(i)}|E^{(i)} \leq E_c]$. The duration for which this energy is available is $\tau = 1 + T(L - 1)$ slots. This is illustrated in Figure 7.7.

The average expected energy available per time slot during an epoch is given by

$$\frac{\mathbb{E}\varepsilon}{\mathbb{E}\tau} = \frac{p\bar{B} + (1 - p)T\mathbb{E}[E^{(i)}|E^{(i)} \leq E_c]}{p + (1 - p)T}.$$

Inspired by the i.i.d. case, given E_c we may want to choose $q = \frac{\mathbb{E}\varepsilon/\mathbb{E}\tau}{E_c}$. Recall, however, that given q, we want to choose E_c as in (7.25). These two desired relations for E_c and q, along with the identity

$$\mathbb{E}[\min(E^{(i)}, E_c)] = pE_c + (1 - p)\mathbb{E}[E^{(i)}|E^{(i)} \leq E_c],$$

can be solved to obtain the following equation:

$$TE_c - (T - 1)\mathbb{E}[\min(E^{(i)}, E_c)] = \bar{B}, \tag{7.27}$$

which can be solved for E_c for given T, $\bar{B}$, and P_E. Additionally, combining (7.25) and (7.27) yields the following simple formula for q, given E_c:

$$q = \frac{\mathbb{E}[\min(E^{(i)}, E_c)]}{E_c}. \tag{7.28}$$

Note that this is essentially the same expression for q as in the i.i.d. case, with $\bar{B}$ replaced by E_c. Indeed, when $T = 1$, (7.27) reduces to $E_c = \bar{B}$.

Approximate Throughput

To summarize, the online policy π for the block i.i.d. is given by

$$
g_1^{(i)} = \begin{cases} \frac{q}{T}(b_1^{(i)} + (T-1)E^{(i)}), & \text{if } E^{(i)} \le E_c, \\ E^{(i)} - \frac{\bar{B} - b_1^{(i)}}{T-1}, & \text{if } E^{(i)} > E_c, \end{cases}
$$
$$
g_T^{(i)} = \frac{q}{q + (1-q)T} b_T^{(i)}, \tag{7.29}
$$

where E_c and q are given by (7.27) and (7.28), respectively.

The throughput obtained by this policy can be bounded as stated in the following theorem, which is provided without proof.

THEOREM 7.11   *Let E_c be the unique solution of (7.27), let $p = \Pr(E^{(i)} > E_c)$, and let π be the policy defined in (7.29). The throughput obtained by this policy is bounded by*

$$
\bar{\Theta} - \frac{1}{2}\log e \le \mathscr{T}(\pi) \le \bar{\Theta} \le \bar{\Theta},
$$

where

$$
\bar{\Theta} = \frac{p(T-1)}{T} \mathbb{E}\left[\frac{1}{2}\log\left(1 + \min\left\{ E^{(i)} - \frac{\bar{B} - E^{(i)}}{T-1}, \bar{B} \right\}\right) \middle| E^{(i)} > E_c \right]
$$
$$
+ \frac{p + T(1-p)}{T} \frac{1}{2}\log\left(1 + \mathbb{E}[\min(E^{(i)}, E_c)]\right).
$$

The structure of the approximately optimal throughput $\bar{\Theta}$ has a natural interpretation in terms of the policy (7.29). The expression has two terms, corresponding to the two different operation modes of the policy. The first term corresponds to the throughput achieved in the first $T - 1$ time slots of a large energy arrival block. Note that these time slots correspond to a fraction $\frac{p(T-1)}{T}$ of the total time on average. In the remaining fraction of the time, we apply the fixed fraction policy, which, analogously to the i.i.d. case, achieves a throughput $\frac{1}{2}\log(1 + \mathbb{E}[\min(E^{(i)}, E_c)])$, where $\mathbb{E}[\min(E^{(i)}, E_c)] = \mathbb{E}\varepsilon/\mathbb{E}\tau = qE_c$ is the average available energy rate for a low-energy period.

Denote $\mu = \mathbb{E}[\min(E^{(i)}, \bar{B})]$. For all ergodic energy arrival processes (including block i.i.d.), the AWGN capacity $\frac{1}{2}\log(1 + \mu)$ is always an upper bound on the throughput, for any finite battery size. The following corollary provides a threshold for the battery size above which the AWGN capacity is nearly achievable in the block i.i.d. model.

COROLLARY 7.12   *If $\bar{B} \ge \mu + T(E^{(i)} - \mu)$ with probability 1, the approximate throughput reduces to*

$$
\bar{\Theta} = \frac{1}{2}\log(1 + \mu). \tag{7.30}
$$

Proof.   Choose $E_c = \mu + \frac{\bar{B} - \mu}{T}$, and observe that $\mathbb{E}[\min(E^{(i)}, E_c)] = \mu$. Therefore E_c is the solution to (7.27). It follows that $p = 0$, and $\bar{\Theta}$ in Theorem 7.11 reduces to $\frac{1}{2}\log(1 + \mu)$.   ■

We identify the case $\bar{B} \geq \mu + T(E^{(i)} - \mu)$ as the *large battery regime*. The threshold $\mu + T(E^{(i)} - \mu)$ can be intuitively interpreted as follows: When the battery size is infinite, it is straightforward to observe that the optimal policy is to allocate a constant amount of power equal to the mean energy arrival $g_t = \mu$. Assume we apply this policy for all time slots in block i. If no battery overflow occurs, the battery level at the jth slot of the block is given by $b_j^{(i)} = b_1^{(i)} + (j-1)(E^{(i)} - \mu)$. Assume further that the battery was empty prior to the beginning of the block, i.e., $b_1^{(i)} = E^{(i)}$. Then the battery level at the last time slot of the block is $b_T^{(i)} = \mu + T(E^{(i)} - \mu)$. This implies that we would need a battery size of at least $\mu + T(E^{(i)} - \mu)$ in order to not waste energy due to an overflow. However, the fact that we can nearly achieve the AWGN capacity as soon as the battery size is larger than this threshold is indeed surprising, since overflows can still happen with this finite battery size. Additionally, note that the statistical spread (or dispersion) of $E^{(i)}$ also comes into play in the threshold. When the probability distribution of $E^{(i)}$ is mostly concentrated around the mean, the quantity $E_{\max} - \mu$ would be small, and the dependence on T would be weaker. As an extreme example, if $E^{(i)} = \mu$ is constant (i.e., zero spread), then $E_{\max} = \mu$ and the threshold is simply μ, which does not depend on T, as expected.

The results suggest that for a fixed energy arrival distribution P_E and fixed battery size $\bar{B}$, the approximate throughput $\bar{\Theta}$ decreases with the block duration T. This can be seen by means of the following limiting cases: When the block duration is small, i.e., $T = 1$, the approximate throughput reduces to the expression given in Theorem 7.11, namely $\bar{\Theta} = C(\mathbb{E}[\min(E^{(i)}, \bar{B})])$. When the block duration is very large, $T \to \infty$, it can be seen from (7.27) that $E_c = 0$, and therefore the approximate throughput becomes $\bar{\Theta} = \mathbb{E}[C(\min(E^{(i)}, \bar{B}))]$, which is always less than $C(\mathbb{E}[\min(E^{(i)}, \bar{B})])$ due to Jensen's inequality.

7.2 Information Theoretic Capacity

So far, we have been concerned with the power control problem, which operates under the implicit assumption that each time slot is essentially long enough such that there exists an optimal code which enables achieving a rate of $\frac{1}{2}\log(1 + g_t)$ when allocating power g_t for communicating over the AWGN channel.

In the information theoretic view of the energy harvesting communication problem, the energy constraints are enforced at a per-symbol level. More specifically, the energy harvesting channel is composed of a single-user discrete-time AWGN channel, such that the output at time t is $Y_t = X_t + Z_t$, where N_t is i.i.d. additive Gaussian noise with unit variance, and $X_t \in \mathbb{R}$ is the input symbol. The transmitter is equipped with a battery of size $\bar{B}$, and the input symbol energy at each time slot is constrained by the amount of energy in the battery. The high-level system model remains the same as the one depicted in Figure 7.1; however, now the energy constraints are imposed on a per-symbol scale. At any time t, X_t needs to satisfy the following constraints:

$$X_t^2 \leq B_t, \tag{7.31}$$

$$B_{t+1} = \min(B_t - X_t^2 + E_{t+1}, \bar{B}). \qquad (7.32)$$

We assume E_t is a nonnegative i.i.d. stochastic process with $\mathbb{E}[E_t] > 0$. The transmitter observes the energy arrivals in a causal fashion. Note that this also implies causal knowledge of B_t at the transmitter.

Note that B_t can be regarded as the state of the system at time t. This state is random, has memory and is known causally at the transmitter but not at the receiver. For such channels, encoding cannot be done in the classical sense, by assigning a fixed codeword to each message ahead of time, since this codeword may not be transmittable during the course of the communications if energy arrivals are not sufficient. Instead, the transmitter fixes a sequence of coding functions corresponding to each message ahead of time, and the actual transmitted codeword is determined dynamically depending on the energy arrivals. A rate R code with blocklength n for this channel is defined as a set of encoding functions f_t^{enc} and a decoding function f^{dec}:

$$f_t^{\text{enc}} : \mathcal{M} \times \mathcal{E}^t \to \mathcal{X}, \qquad t = 1, \dots, n,$$
$$f^{\text{dec}} : \mathcal{Y}^n \to \mathcal{M},$$

where $\mathcal{X} = \mathcal{Y} = \mathbb{R}$ are the input and output alphabets, $\mathcal{E} \subset \mathbb{R}_+$ is the energy arrivals alphabet, and $\mathcal{M} = \{1, \dots, 2^{nR}\}$ is the message set. The capacity of the energy harvesting channel C is defined in the usual way as the supremum of all rates that are achievable with vanishing probability of error.

In the sequel, we will also be interested in the case when the energy arrivals are observed at the receiver in addition to the transmitter. In practice, this may be the case if the receiver itself is also harvesting energy from the same physical process as the transmitter. However, our main motivation for studying this case is that characterizing the capacity of the channel under receiver side information turns out to be a simpler, though still challenging, problem, and as we will see in the following, this capacity can be related to the case where the receiver does not have any side information. With energy arrival information at the receiver, we change the decoding function to $f^{\text{dec}} : \mathcal{Y}^n \times \mathcal{E}^n \to \mathcal{M}$. The capacity will be denoted by C_{RX}.

7.2.1   Infinite Battery

The only instance where the exact capacity is known is the case of infinite battery. In this case, the energy constraint (7.32) becomes

$$B_{t+1} = B_t - X_t^2 + E_{t+1}.$$

Assuming without loss of generality that $B_0 = 0$, this constraint along with (7.31) are equivalent to

$$\sum_{t=1}^{k} X_k^2 \le \sum_{t=1}^{k} E_k, \qquad k = 1, \dots, n.$$

In particular, the last constraint $\frac{1}{n}\sum_{t=1}^{n} X_t^2 \leq \frac{1}{n}\sum_{t=1}^{n} E_t$ implies, by the strong law of large numbers:

$$\limsup_{n \to \infty} \frac{1}{n}\sum_{t=1}^{n} X_t^2 \leq \mathbb{E}[E_t].$$

Therefore, the capacity of the energy harvesting channel is upper bounded by the classical AWGN capacity with power constraint equal to the expected value of the energy arrivals:

$$C \leq \frac{1}{2}\log(1 + \mathbb{E}[E_t]).$$

This upper bound can in fact be achieved, as stated in the Theorem 7.13.

THEOREM 7.13   *The capacity of the energy harvesting AWGN channel with an infinite battery and an i.i.d. energy arrivals process $\{E_t\}$ is given by*

$$C = \frac{1}{2}\log(1 + \mathbb{E}[E_t]).$$

We give a sketch of the proof, by describing a scheme that achieves the upper bound, called the *save-and-transmit* scheme. In this scheme, transmission is split into two phases. The first phase, which is called the *save* phase, takes place over the first $h(n)$ channel uses. In this phase the transmitter saves energy in the battery, i.e., it sets $X_t = 0$, in order to charge the battery to a sufficient level. Next, in the subsequent *transmit* phase, the transmitter sends a codeword from a codebook generated from i.i.d. samples of a Gaussian random variable with variance $\mathbb{E}[E_t]$.

By choosing $h(n)$ such that both $h(n) \to \infty$ and $\frac{h(n)}{n} \to 0$, such as $h(n) = \lceil \log n \rceil$, we can achieve the AWGN capacity. This is due to the following reasons: (1) if $h(n) \to \infty$, the energy saved during the save phase will be enough to prevent energy outages during the transmit phase of length $n - h(n)$ with high probability; and (2) if $\frac{h(n)}{n} \to 0$, the rate penalty incurred due to transmitting zeros during the save phase will vanish as $n \to \infty$.

7.2.2   Finite Battery

When the battery size is finite, no closed-form expression for capacity is known. In the spirit of the previous results for the online power control problem, in the following we will provide an *approximate* expression for capacity. As an intermediate result, we will see that the information theoretic capacity and the online power control problem, while seemingly two distinct formulations of the energy harvesting communication problem, are in fact highly related, and solving one can provide an approximate solution to the other.

Bernoulli Energy Arrivals

As in the power control problem, we start by solving the simpler case of Bernoulli energy arrivals, and later generalize to arbitrary energy arrival distributions.

Suppose the energy arrivals are given by

$$E_t = \begin{cases} \bar{B}, & \text{w.p. } p, \\ 0, & \text{w.p. } 1 - p. \end{cases}$$

We will also assume energy arrival observations are available at the receiver; namely we will be interested in the capacity C_{RX}.

Recall the regenerative structure imposed by the energy arrivals: Any time there is a positive energy arrival (i.e., $E_t = \bar{B}$), the battery will be fully recharged to $B_t = \bar{B}$, essentially erasing the memory of the system. We can therefore consider each epoch (the time between consecutive positive energy arrivals) to be an independent "channel use." Informally, capacity is given by the maximum number of bits we can transmit per epoch, divided by the expected epoch length. The latter is simply $\mathbb{E}L = 1/p$, where $L \sim$ Geometric(p) is the epoch length. To find the former, we define the *clipping channel*.

DEFINITION 7.14 (clipping channel) *The clipping channel, illustrated in Figure 7.8, is a memoryless channel which, at each channel use, admits as input an infinite sequence of real numbers $X^\infty = \{X_i\}_{i=1}^\infty$. The input sequence must satisfy the energy constraint*

$$\|X^\infty\|^2 = \sum_{i=1}^\infty X_i^2 \le \bar{B}.$$

Each channel use is associated with a state variable L, which is i.i.d., independent of the input, and follows a geometric distribution with parameter p. The state is known at the receiver but not at the transmitter. The channel output Y^L is a vector of length L given by

$$Y_i = X_i + Z_i, \qquad i = 1, \dots, L,$$

where $Z_i \sim \mathcal{N}(0, 1)$ is i.i.d. Gaussian noise.

In other words, at each channel use, the clipping channel generates a random length L, and outputs only the first L components of the infinite input sequence under additive Gaussian noise. Denote the capacity of this memoryless channel by C_{clipping}.

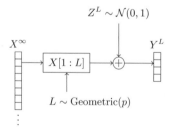

Figure 7.8 Schematic of the clipping channel. The input at each channel use is an infinite sequence X^∞. This is clipped to only the first L symbols, where $L \sim$ Geometric(p) is i.i.d. state known only to the receiver. This is then corrupted by additive white Gaussian noise and output to the receiver.

Each channel use of the clipping channel constitutes an entire epoch of the energy harvesting channel, where the i.i.d. state L is the random epoch length. Note that the epoch length is known at the receiver at the time of decoding, by our assumption that the energy arrivals are observed at the receiver. The transmitter, however, has no a-priori knowledge of the epoch length, and must produce an infinite input sequence to accommodate any possible epoch length. Since, by definition, there are no energy arrivals during an epoch, the total energy available for transmission is $\bar{B}$. It can be seen that any code for the energy harvesting channel can be converted into a code for the clipping channel, and vice versa. Combined with arguments along the lines of Theorem 7.3, it follows that

$$C_{RX} = \frac{C_{\text{clipping}}}{\mathbb{E}L} = p \cdot C_{\text{clipping}}. \tag{7.33}$$

The clipping channel is a memoryless channel with i.i.d. side information available at the receiver, hence capacity can be readily expressed in terms of maximal mutual information:

$$
\begin{aligned}
C_{\text{clipping}} &= \max_{\substack{p(x^\infty): \\ \|X^\infty\|^2 \leq \bar{B}}} I(X^\infty; Y^L, L) \\
&\overset{(a)}{=} \max_{\substack{p(x^\infty): \\ \|X^\infty\|^2 \leq \bar{B}}} I(X^\infty; Y^L | L) \\
&\overset{(b)}{=} \max_{\substack{p(x^\infty): \\ \|X^\infty\|^2 \leq \bar{B}}} \sum_{k=1}^{\infty} p(1-p)^{k-1} I(X^\infty; Y^k) \\
&= \max_{\substack{p(x^\infty): \\ \|X^\infty\|^2 \leq \bar{B}}} \sum_{k=1}^{\infty} p(1-p)^{k-1} I(X^k; X^k + Z^k),
\end{aligned}
$$

where (a) follows from the fact that the input is independent of the state L, and (b) follows by writing out the expectation over L. This yields the following infinite-letter capacity expression for the energy harvesting channel:

$$C_{RX} = \max_{\substack{p(x^\infty): \\ \|X^\infty\|^2 \leq \bar{B}}} \sum_{k=1}^{\infty} p^2 (1-p)^{k-1} I(X^k; X^k + Z^k). \tag{7.34}$$

While this expression is still hard to explicitly compute, we will use it to show a connection between the information theoretic capacity and the power control problem. Specifically, as stated in the following theorem, capacity can be approximated by the optimal solution to the online power control problem, up to a constant gap.

THEOREM 7.15 *The capacity of the energy harvesting channel with Bernoulli energy arrivals observed at the transmitter and the receiver is bounded by*

$$\Theta - \frac{1}{2} \log\left(\frac{\pi e}{2}\right) \leq C_{RX} \leq \Theta,$$

where Θ is the solution to the online power control problem (7.10).

Combining this result with the approximation of Θ in Theorem 7.5, we get the following approximation for capacity by a simple and insightful formula:

$$\frac{1}{2}\log(1+p\bar{B}) - \frac{1}{2}\log\left(\frac{\pi e^2}{2}\right) \le C_{RX} \le \frac{1}{2}\log(1+p\bar{B}). \tag{7.35}$$

Note that the gap $\frac{1}{2}\log\left(\frac{\pi e^2}{2}\right) \approx 1.77$ is independent of any of the parameters of the problem, namely p and $\bar{B}$.

Proof of the Upper Bound

We start with the capacity expression (7.34):

$$
\begin{aligned}
C_{RX} &= \max_{\substack{p(x^\infty):\\ \|X^\infty\|^2 \le \bar{B}}} \sum_{k=1}^{\infty} p^2(1-p)^{k-1} I(X^k; X^k + Z^k)\\[2mm]
&\overset{(a)}{\le} \max_{\substack{p(x^\infty):\\ \|X^\infty\|^2 \le \bar{B}}} \sum_{k=1}^{\infty} p^2(1-p)^{k-1} \sum_{i=1}^{k} I(X_i; X_i + Y_i)\\[2mm]
&\overset{(b)}{=} \max_{\substack{p(x^\infty):\\ \|X^\infty\|^2 \le \bar{B}}} \sum_{i=1}^{\infty} p(1-p)^{i-1} I(X_i; X_i + Y_i)\\[2mm]
&\overset{(c)}{\le} \max_{\substack{p(x^\infty):\\ \mathbb{E}\|X^\infty\|^2 \le \bar{B}}} \sum_{i=1}^{\infty} p(1-p)^{i-1} I(X_i; X_i + Y_i)\\[2mm]
&\overset{(d)}{\le} \max_{\substack{\{\varepsilon\}_{i=1}^{\infty}:\\ \varepsilon_i \ge 0\ \forall i\\ \sum_{i=1}^{\infty} \varepsilon_i \le \bar{B}}} \sum_{i=1}^{\infty} p(1-p)^{i-1} \frac{1}{2}\log(1+\varepsilon_i)\\[2mm]
&\overset{(e)}{=} \Theta,
\end{aligned}
$$

where (a) is because the underlying AWGN channel is memoryless, (b) is by changing the order of summation, (c) follows by relaxing the amplitude constraint $\sum_{i=1}^{\infty} X_i^2 \le \bar{B}$ to hold in expectation instead of almost surely, (d) is by denoting $\mathbb{E}X_i^2 = \varepsilon_i$ and because $I(X_i; X_i + Z_i) \le \frac{1}{2}\log(1+\mathbb{E}X_i^2)$ for any input distribution, and (e) is by (7.10).

Proof of the Lower Bound

To obtain a lower bound, we compute the mutual information expression in (7.34) under a specific input distribution, given as follows: Fix an arbitrary sequence of real numbers $\{\varepsilon_i\}_{i=1}^{\infty}$, satisfying $\varepsilon_i \ge 0\ \forall i$ and $\sum_{i=1}^{\infty} \varepsilon_i \le \bar{B}$. Let the X_i be independent random variables, where X_i is distributed uniformly over the interval $[-\sqrt{\varepsilon_i}, \sqrt{\varepsilon_i}]$. Note that this implies $X_i^2 \le \varepsilon_i$, and subsequently $\sum_{i=1}^{\infty} X_i^2 \le \sum_{i=1}^{\infty} \varepsilon_i \le \bar{B}$.

A simple lower bound on the capacity of the amplitude-constrained AWGN channel is given in the following lemma.

LEMMA 7.16   *Let* $X \sim [-\sqrt{\varepsilon}, \sqrt{\varepsilon}]$ *and let* $Z \sim \mathcal{N}(0, 1)$ *independent of X. The following holds:*

$$I(X; X + Z) \geq \frac{1}{2} \log(1 + \varepsilon) - \frac{1}{2} \log\left(\frac{\pi e}{2}\right).$$

Proof.   We make use of the entropy power inequality:

$$I(X; X + Z) = h(X + Z) - h(Z)$$

$$\geq \frac{1}{2} \log\left(2^{2h(X)} + 2^{2h(Z)}\right) - h(Z)$$

$$= \frac{1}{2} \log(4\varepsilon + 2\pi e) - \frac{1}{2} \log(2\pi e)$$

$$= \frac{1}{2} \log\left(\frac{\pi e}{2} + \varepsilon\right) - \frac{1}{2} \log\left(\frac{\pi e}{2}\right)$$

$$\geq \frac{1}{2} \log(1 + \varepsilon) - \frac{1}{2} \log\left(\frac{\pi e}{2}\right).$$

∎

We plug in the proposed input distribution to the expression in (7.34):

$$C_{\text{RX}} \overset{(a)}{\geq} \sum_{k=1}^{\infty} p^2 (1 - p)^{k-1} \sum_{i=1}^{k} I(X_i; X_i + Z_i)$$

$$\overset{(b)}{=} \sum_{i=1}^{\infty} p(1 - p)^{i-1} I(X_i; X_i + Z_i)$$

$$\overset{(c)}{\geq} \sum_{i=1}^{\infty} p(1 - p)^{i-1} \left[\frac{1}{2} \log(1 + \varepsilon_i) - \frac{1}{2} \log\left(\frac{\pi e}{2}\right)\right]$$

$$= \sum_{i=1}^{\infty} p(1 - p)^{i-1} \frac{1}{2} \log(1 + \varepsilon_i) - \frac{1}{2} \log\left(\frac{\pi e}{2}\right),$$

where (a) is because the X_i are independent, (b) is by changing the order of summation, (c) follows from Lemma 7.16. Since the LHS does not depend on $\{\varepsilon_i\}$ and the sequence was arbitrary, we can maximize over all such sequences. By (7.10), we obtain the desired lower bound.

General Energy Arrival Distribution

As in the power control problem, here as well we can extend the results for the Bernoulli case to general i.i.d. energy arrival processes. We do not pursue this direction here, but only state the main result.

THEOREM 7.17   *The capacity of the energy harvesting channel with i.i.d. energy arrivals observed at the transmitter and the receiver is bounded by*

$$\Theta - \frac{1}{2} \log\left(\frac{\pi e}{2}\right) \leq C_{\text{RX}} \leq \Theta,$$

where Θ is the solution to the online power control problem:

$$\Theta = \sup_{\pi} \liminf_{N \to \infty} \frac{1}{N} \sum_{t=1}^{N} \mathbb{E}\left[\frac{1}{2} \log(1 + g_t)\right].$$

Together with the results of the fixed fraction policy, namely Theorem 7.7, we have:

$$\frac{1}{2} \log(1 + \mu) - \frac{1}{2} \log\left(\frac{\pi e^2}{2}\right) \le C_{\mathrm{RX}} \le \frac{1}{2} \log(1 + \mu), \tag{7.36}$$

where $\mu = \mathbb{E}[\min(E_t, \bar{B})]$.

Capacity without Receiver Energy Arrival Observations

The capacity approximations discussed so far are only valid for the case when the receiver observes the energy harvesting process in addition to the transmitter. However, in most practical scenarios, the receiver does not necessarily have access to such energy arrival observations. While the capacity with side information at the receiver is always an upper bound to the capacity without it (since the receiver may simply choose not to use the side information), the lower bound must be corrected as follows:

$$C \ge \Theta - \frac{1}{2} \log\left(\frac{\pi e}{2}\right) - H(E_t).$$

This can be understood as follows: The extra gap $H(E_t)$ is roughly the amount of additional information needed to be conveyed to the receiver in order for the transmitter and the receiver to agree on the process E_t.

Unfortunately, this bound may be too loose in general, especially if E_t has a very large alphabet (or even a continuous alphabet). However, it can be shown that the bound can be tightened as follows: For any online policy π, let $\{g_t\}_{t=1}^{\infty}$ be the resulting sequence of power allocations. Note that this is a random process, induced by the randomness of the energy harvesting process $\{E_t\}$. Then

$$C \ge \liminf_{n \to \infty} \frac{1}{n}\left(\mathcal{T}_n(\pi) - H(g^n)\right) - \frac{1}{2} \log\left(\frac{\pi e}{2}\right), \tag{7.37}$$

where $\mathcal{T}_n(\pi) = \sum_{t=1}^{n} \mathbb{E}\left[\frac{1}{2} \log(1 + g_t)\right]$ is the n-horizon throughput of the policy π. Clearly, $\frac{1}{n} H(g^n) \le \frac{1}{n} H(E^n) = H(E_t)$ since g^n is a deterministic function of E^n, hence this is a tighter lower bound. This bound can be understood as follows: We use an input distribution composed of independent symbols, where each X_t is distributed uniformly over the interval $[-\sqrt{g_t}, \sqrt{g_t}]$ as before. In order for the transmitter and receiver to agree on the codebook of symbol X_t, it is enough for the receiver to know only $g_t(E^t)$ rather than the entire sequence E^t.

Using a different policy than the fixed fraction policy, one can obtain $\frac{1}{n} H(g^n) \le 1$ for any distribution of E_t. However, the lower bound on the throughput it achieves is slightly worse, giving a gap of 1.8, i.e. $\mathcal{T}(\pi) \ge \frac{1}{2} \log(1 + \mu) - 1.8$ (as compared to the gap in Theorem 7.7 which was $\frac{1}{2} \log e \approx 0.72$). Nevertheless, this yields a constant gap lower bound on capacity, stated in the following theorem.

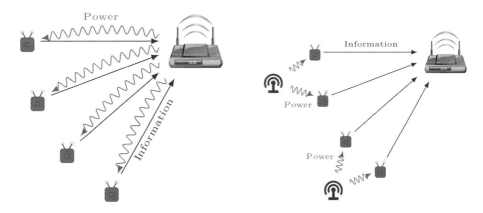

Figure 7.9 Remotely powered communication topologies. The left figure illustrates Topology 1, in which a sink node powers sensor nodes over the downlink, while the sensor nodes transmit over the uplink. The right figure illustrates Topology 2, in which the sensor nodes are powered by dedicated power beacons.

THEOREM 7.18   *The capacity of the energy harvesting channel with i.i.d. energy arrivals observed only at the transmitter is bounded by*

$$\frac{1}{2}\log(1+\mu) - 3.85 \leq C \leq \frac{1}{2}\log(1+\mu),$$

where $\mu = \mathbb{E}[\min(E_t, \bar{B})]$.

7.3   Remotely Powered Communication

In addition to systems harvesting energy from natural resources in their environment, wireless devices can also be actively powered by a targeted *wireless energy transfer* process. Indeed, RF power transfer is expected to be one of the dominant modes for powering wireless devices in the Internet of Things (IoT). Figure 7.9 illustrates two common topologies for deploying remotely powered wireless sensor nodes, which are described below.

Topology 1 Wireless sensors communicate to a central sink node, which gathers all the information and serves as a gateway to the cloud. The sink node has access to conventional power but the wireless sensors do not have any conventional batteries. They harvest the RF energy over the downlink channel and store it in a capacitor, which allows them to transmit over the uplink.

Topology 2 Dedicated power beacons that have access to conventional power are deployed to wirelessly charge nearby sensor nodes, while those communicate to the central sink node. Such power beacons do not require any backhaul links, and therefore their low cost can allow dense deployments.

The energy transfer in such settings can be organized in several simple ways: continuously transferring power to the sensor nodes; periodically releasing bursts of energy;

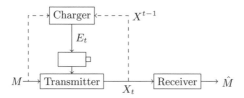

Figure 7.10 Model of remotely powered communication over a noiseless binary channel with binary energy alphabet.

or beamforming energy to different spatial clusters of sensor nodes at different times. However, the energy efficiency of the overall system can be maximized by taking advantage of the side information that can be naturally available to the powering device (sink node or power beacon) in certain cases.

To illustrate this point, consider the following simple example: Consider a noiseless binary channel, i.e., the input and output alphabets are $\mathcal{X} = \mathcal{Y} = \{0, 1\}$ and the output is given by $Y_t = X_t$. The transmitter is equipped with a battery of size 1. Assume that the energy cost of transmitting a "1" is 1, and transmitting a "0" does not expend energy. The transmitter's battery is being charged by a third device, called the *charger*. The charger can decide the amount of energy E_t to transfer to the transmitter's battery at time t. At each time instant, the charger can either charge one unit of energy or not charge at all, i.e., $E_t \in \{0, 1\}$. The model is illustrated in Figure 7.10.

Obviously, the simplest way to power the transmitter is to charge $E_t = 1$ every time slot. This will ensure the transmitter's battery is always full, and it can transmit at the maximal rate of $R = 1$. The energy cost is $\Gamma = 1$ units of energy per time slot. Note, however, that the transmitter uses only $\frac{1}{2}$ units per time slot on average, since only half of the time a 1 is transmitted.

Now suppose the charger has access to side information. Specifically, suppose the charger knows the message to be transmitted ahead of time. It could then charge the transmitter only when it intends to send a "1," thereby achieving rate $R = 1$ with only $\Gamma = 1/2$ units of energy on average.

Now consider the case where the charger does not observe the entire message, but only observes the transmitted signal in a strictly causal fashion, i.e., at time t it observes X^{t-1}. Here, the charger can charge only when the battery is empty, which occurs exactly after the transmitter sends a "1." This scheme achieves the same performance of rate $R = 1$ with average energy $\Gamma = 1/2$, even though the side information is significantly weaker.

System Model and Capacity

We next present an information theoretic model to study such remotely powered communication systems (see Figure 7.11). The physical channel is a discrete memoryless channel, with input $X_t \in \mathcal{X}$, output $Y_t \in \mathcal{Y}$, and channel transition probability distribution $p(y|x)$. The channel has an associated energy cost function $\phi : \mathcal{X} \to \mathbb{R}_+$, denoting the amount of energy used for transmission of each symbol. The transmitter has a battery

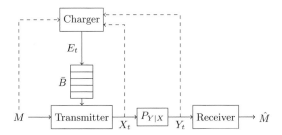

Figure 7.11 Energy harvesting communication system with a charger.

with finite capacity $\bar{B}$, and this battery is charged by an energy arrival $E_t \in \mathcal{E} \subset \mathbb{R}_+$ at each time slot t.

As in the energy harvesting channel, the input symbols must satisfy the energy constraints. Let B_t be the battery state at time t. The energy constraints are described by

$$\phi(X_t) \leq \min(B_t + E_t, \bar{B}), \tag{7.38}$$

$$B_{t+1} = \min(B_t + E_t, \bar{B}) - \phi(X_t). \tag{7.39}$$

Note that B_t is defined to be the battery state *before* charging, such that the amount of energy in the battery *after* charging which is available for transmission is $\min(B_t + E_t, \bar{B})$. This is different than the definition of B_t encountered so far; however, it does not change the underlying model.

To account for the energy efficiency of the communication system, we impose an average cost constraint on the energy emitted by the charger:

$$\frac{1}{n} \sum_{t=1}^{n} E_t \leq \Gamma. \tag{7.40}$$

The transmitter encoding functions and receiver encoding function are defined as in the energy harvesting channel:

$$f_t^{\text{TX}} : \mathcal{M} \times \mathcal{E}^t \to \mathcal{X}, \qquad t = 1, \ldots, n,$$
$$f^{\text{RX}} : \mathcal{Y}^n \to \mathcal{M}.$$

Additionally, a code for this channel must include a set of charger encoding functions f_t^{C}. These functions may vary depending on the side information available to the charger. We consider four cases of interest, motivated by different settings of practical interest.

Generic Charger

The charger does not observe any side information. The charger encoding function is

$$f^{\text{C}} : \emptyset \to \mathcal{E}^n.$$

In other words, the charger employs a predetermined fixed sequence e^n. The capacity in this case is denoted by $C_\emptyset$.

Receiver Charges Transmitter

This is the scenario described in Topology 1. The charger encoding functions are

$$f_t^C : \mathcal{Y}^{t-1} \to \mathcal{E}, \qquad t = 1, \dots, n.$$

Capacity is denoted by C_Y.

Charger Adjacent to Transmitter

This models the case when the charger is physically close to the transmitter, as in Topology 2. We can assume that the charger observes the transmitter's input noiselessly. The charger encoding functions are

$$f_t^C : \mathcal{X}^{t-1} \to \mathcal{E}, \qquad t = 1, \dots, n.$$

Capacity is denoted by C_X.

Fully Cognitive Charger

In this case we assume the charger knows the message to be transmitted ahead of time. Note that this does not correspond to the conventional average power constraint, since the transmitter still has a finite battery and must satisfy the battery constraints. We set

$$f^C : \mathcal{M} \to \mathcal{E}^n,$$

and denote capacity by C_M.

To state capacity expressions, we define the set of allowed transmitter–charger codeword pairs (i.e., such that satisfy the energy constraints):

$$\mathcal{A}_n(\Gamma) = \Big\{ x^n \in \mathcal{X}^n, \ e^n \in \mathcal{E}^n :$$

$$b_1 = \bar{B},$$

$$b_{t+1} = \min\{b_t + e_t, \bar{B}\} - \phi(x_t), \qquad t = 1, \dots, n-1,$$

$$\phi(x_t) \le \min\{b_t + e_t, \bar{B}\}, \qquad t = 1, \dots, n,$$

$$\sum_{t=1}^{n} e_t \le n\Gamma \qquad \qquad \Big\}.$$

By definition of the channel, we must have

$$(X^n, E^n) \in \mathcal{A}_n(\Gamma) \ \text{a.s.}$$

Using this condition, we can write n-letter capacity expressions for each of the four cases described above.

Generic Charger

$$C_\emptyset(\Gamma) = \lim_{n \to \infty} \frac{1}{n} \max_{\substack{p(X^n), \ e^n : \\ (X^n, e^n) \in \mathcal{A}_n(\Gamma) \ \text{a.s.}}} I(X^n; Y^n)$$

The maximization is over all distributions for X^n and deterministic charging sequences e^n s.t. the battery constraints are satisfied.

Receiver Charges Transmitter

$$C_Y(\Gamma) = \lim_{n\to\infty} \frac{1}{n} \max_{\substack{p(x^n\|e^n),\ \{e_t(y^{t-1})\}_{t=1}^n: \\ (X^n,e^n(Y^{n-1}))\in\mathcal{A}_n(\Gamma)\ \text{a.s.}}} I(X^n \to Y^n),$$

where

$$I(X^n \to Y^n) = \sum_{t=1}^n I(X^t; Y_t|Y^{t-1})$$

is directed information, and the maximization is over sets of function $\{e_t(y^{t-1})\}_{t=1}^n$ and causally conditioned input distributions

$$p(x^n\|e^n) = \prod_{t=1}^n p(x_t|x^{t-1}, e^t).$$

The reason for the emergence of directed information in this capacity expression is because the receiver can convey information to the transmitter through the energy sequence, thereby introducing feedback into the communication channel.

Charger Adjacent to Transmitter

$$C_X(\Gamma) = \lim_{n\to\infty} \frac{1}{n} \max_{\substack{p(x^n),\ \{e_t(x^{t-1})\}_{t=1}^n: \\ (X^n,e^n(X^{n-1}))\in\mathcal{A}_n(\Gamma)\ \text{a.s.}}} I(X^n; Y^n)$$

The maximization is over input distributions $p(x^n)$ and functions $\{e_t(x^{t-1})\}_{t=1}^n$. Note that the transmitter codeword does not need to explicitly depend on past energy arrivals, since these are given deterministically from past input symbols.

Fully Cognitive Charger

$$C_M(\Gamma) = \lim_{n\to\infty} \frac{1}{n} \max_{\substack{p(x^n,e^n): \\ (X^n,E^n)\in\mathcal{A}_n(\Gamma)\ \text{a.s.}}} I(X^n; Y^n)$$

Since the transmitter and charger both know the message, they can cooperate and choose the transmitter–charger codeword pair together. Therefore the maximization is over joint transmitter–charger input distributions.

While these n-letter expressions are in general hard to compute explicitly, in what follows we will see two interesting special cases where capacity can be computed exactly.

7.3.1 Special Case 1: Precision Charger

A channel with a *precision charger* is a channel in which the charger is finely tunable to different energy levels. This case is motivated by the observation that in practice the amount of transferred energy is controlled by the amplitude of the beamformed signal, which can be changed in a continuous fashion. Note that this may not be the case in general, since due to regulations or device limitations it may be necessary to restrict the energy alphabet $\mathcal{E}$.

DEFINITION 7.19 (Precision charger)   *A channel with a precision charger is such that* $\phi(\mathcal{X}) \subseteq \mathcal{E}$, *i.e., for every input symbol* $x \in \mathcal{X}$ *there exists an energy symbol* $e \in \mathcal{E}$ *s.t.* $e = \phi(x)$.

EXAMPLE 7.20   Consider the binary channel with a binary charger $\mathcal{X} = \mathcal{E} = \{0, 1\}$, $\phi(x) = x$, as in the example considered in the beginning of the section. Even with arbitrary battery size and noisy channel, this is a precision charger.

EXAMPLE 7.21   Consider the AWGN channel with energy cost function $\phi(x) = x^2$. Since $\phi(x) \leq \bar{B}$, we can restrict the input alphabet to the interval $\mathcal{X} = [-\sqrt{\bar{B}}, \sqrt{\bar{B}}]$. Let the energy alphabet be $\mathcal{E} = [0, \bar{B}]$. The condition in Definition 7.19 holds in this case.

In the example of the binary channel with binary charger, we saw that the performance when the charger observes the message (fully cognitive charger) can be achieved even if the charger has only strictly causal observations of the input (Charger Adjacent to Transmitter). This observation can be seen to hold in general for all channels with a precision charger, as seen in the following theorem.

THEOREM 7.22   *For a channel with a precision charger:*

$$C_X(\Gamma) = C_M(\Gamma) = \max_{\substack{p(x): \\ \mathbb{E}[\phi(X)] \leq \Gamma}} I(X; Y).$$

Note that the expression $\max_{p(x): \mathbb{E}[\phi(X)] \leq \Gamma} I(X; Y)$ is the capacity of the standard memoryless channel with average input constraint, which is always an upper bound regardless of the nature of side information available at the charger. This is because the average amount of energy the charger can provide to the transmitter is limited to Γ.

Proof.   Since C_M has more side information than C_X, we have

$$C_X(\Gamma) \leq C_M(\Gamma) \leq \max_{\substack{p(x): \\ \mathbb{E}[\phi(X)] \leq \Gamma}} I(X; Y).$$

It is left to prove $C_X(\Gamma) \geq \max_{p(x): \mathbb{E}[\phi(X)] \leq \Gamma} I(X; Y)$.

To show this, we find a code for the channel C_X which achieves the desired rate. Consider the optimal code for the standard input-constrained channel with capacity $\max_{p(x): \mathbb{E}[\phi(X)] \leq \Gamma} I(X; Y)$. This determines the transmitter encoding and receiver decoding functions; it remains to specify the charger function. We choose

$$f_t^C(X^{t-1}) = \phi(X_{t-1}).$$

In words, the charger just measures the energy of the last symbol and refills the battery with the same amount. By our definition of a precision charger, this is possible since the symbol $\phi(X_{t-1})$ always exists in $\mathcal{E}$. Assuming the battery starts full, this strategy will keep the battery full for the entire transmission. This means that the transmitter will always be able to transmit any symbol, and therefore any codeword will satisfy the energy constraints. The average energy constraint holds by construction of the code:

$$\frac{1}{n}\sum_{t=1}^{n}E_t = \frac{1}{n}\sum_{t=1}^{n-1}\phi(X_{t-1}) \leq \Gamma.$$

Clearly this scheme achieves the desired rate.   ∎

7.3.2   Special Case 2: Noiseless Channel

Another special case of interest is the *noiseless* channel, i.e., $Y_t = X_t$. This can typically approximate channels with high SNR. Note that in this case $C_X(\Gamma) = C_Y(\Gamma)$. It can be shown that the computation of the n-letter capacity in this case can be cast as maximization of the long-term average reward of a Markov decision process, and thereby can be readily computed by solving the Bellman equation or applying algorithms such as value iteration.

An interesting application of these results is the noiseless channel with ternary alphabet, i.e., $\mathcal{X} = \mathcal{Y} = \{0, 1, 2\}$. The cost function is $\phi(x) = x$, the battery capacity is $\bar{B} = 2$, and the energy alphabet is $\mathcal{E} = \{0, 2\}$. That is, we do not allow the charger to choose $e = 1$, otherwise this would fall under the category of the precision charger. In fact, this models a system in which the charger is inaccurate, and can only release large bursts of energy that completely charge the battery. Using the MDP formulations, closed-form expressions can be found for each one of the capacities $C_\emptyset(\Gamma)$, $C_X(\Gamma)$, and $C_M(\Gamma)$. It can be shown that

$$C_\emptyset(\Gamma) < C_X(\Gamma) < C_M(\Gamma) < C_{\mathrm{ub}}(\Gamma),$$

indicating that indeed the results of Theorem 7.22 do not hold in general.

7.4   Literature Sources

The offline power control solution is due to [1–3]. For the online case, there has been several solution approaches, such as the MDP formulation [4–6] or heuristic policies [7, 8]. A survey of many results in the area of power control for energy harvesting can be found in the work of Ulukus et al. [9].

The approximately optimal fixed fraction policy for i.i.d. energy arrivals, described in Section 7.1.2, was proposed by Shaviv and Özgür [10]. The extension to block i.i.d. energy arrivals in Section 7.1.3 was first developed for Bernoulli energy arrivals by the same authors [11] and later generalized to arbitrary block i.i.d. distributions [12]. The fixed fraction policy has been also generalized and applied to the multiple-access channel [13–17], the broadcast channel [18], and other multiuser networks [19], as well as to distortion minimization [20] and energy harvesting with a general utility function [21, 22].

The information theoretic formulation of the energy harvesting channel was first proposed by Ozel and Ulukus [23], where the infinite battery case was solved. Initial attempts at characterizing capacity in the finite battery case include [24], where upper and lower bounds are derived in the form of n-letter expressions; [25], where an n-letter capacity expression is given for the special case when the energy arrivals are constant; and [26], where the battery is of unit size and the channel is a noiseless bit pipe. Additionally, Shaviv et al. [27] show that feedback can increase the capacity of the energy harvesting channel. The approximate capacity for finite battery as described in Section 7.2.2 was developed in [28, 29] for Bernoulli energy harvesting processes and in [30] for general i.i.d. processes.

The information theoretic formulation and analysis of remotely powered communication systems in Section 7.3 is due to Shaviv et al. [31].

References

[1] J. Yang and S. Ulukus, "Optimal packet scheduling in an energy harvesting communication system," *IEEE Trans. Commun.*, vol. 60, no. 1, pp. 220–230, Jan. 2012.

[2] K. Tutuncuoglu and A. Yener, "Optimum transmission policies for battery limited energy harvesting nodes," *IEEE Trans. Wireless Commun.*, vol. 11, no. 3, pp. 1180–1189, Mar. 2012.

[3] O. Ozel, K. Tutuncuoglu, J. Yang, S. Ulukus, and A. Yener, "Transmission with energy harvesting nodes in fading wireless channels: Optimal policies," *IEEE J. Sel. Areas Commun.*, vol. 29, no. 8, pp. 1732–1743, Sep. 2011.

[4] M. Zafer and E. Modiano, "Optimal rate control for delay-constrained data transmission over a wireless channel," *IEEE Trans. Inform. Theory*, vol. 54, no. 9, pp. 4020–4039, Sep. 2008.

[5] C. K. Ho and R. Zhang, "Optimal energy allocation for wireless communications with energy harvesting constraints," *IEEE Trans. Signal Proces.*, vol. 60, no. 9, pp. 4808–4818, Sep. 2012.

[6] P. Blasco, D. Gündüz, and M. Dohler, "A learning theoretic approach to energy harvesting communication system optimization," *IEEE Trans. Wireless Commun.*, vol. 12, no. 4, pp. 1872–1882, Apr. 2013.

[7] R. Srivastava and C. E. Koksal, "Basic performance limits and tradeoffs in energy-harvesting sensor nodes with finite data and energy storage," *IEEE/ACM Trans. Networking*, vol. 21, no. 4, pp. 1049–1062, Aug. 2013.

[8] V. Sharma, U. Mukherji, V. Joseph, and S. Gupta, "Optimal energy management policies for energy harvesting sensor nodes," *IEEE Trans. Wireless Commun.*, vol. 9, no. 4, pp. 1326–1336, Apr. 2010.

[9] S. Ulukus, A. Yener, E. Erkip, et al., "Energy harvesting wireless communications: A review of recent advances," *IEEE J. Sel. Areas Commun.*, vol. 33, no. 3, pp. 360–381, Mar. 2015.

[10] D. Shaviv and A. Özgür, "Universally near optimal online power control for energy harvesting nodes," *IEEE J. Sel. Areas Commun.*, vol. 34, no. 12, pp. 3620–3631, Dec. 2016.

[11] D. Shaviv and A. Özgür, "Online power control for block i.i.d. Bernoulli energy harvesting channels," in *2017 IEEE Wireless Commun. Networking Conf. Workshops (WCNCW)*, San Francisco, CA, Mar. 2017, pp. 1–6.

[12] D. Shaviv and A. Özgür, "Online power control for block i.i.d. energy harvesting channels," *IEEE Trans. Inform. Theory*, vol. 64, no. 8, pp. 5920–5937, Aug. 2018.

[13] H. A. Inan and A. Özgür, "Online power control for the energy harvesting multiple access channel," in *2016 14th Int. Symp. Model. Optimization Mobile Ad Hoc and Wireless Netw. (WiOpt)*, Tempe, AZ, May 2016, pp. 1–6.

[14] H. A. Inan, D. Shaviv, and A. Özgür, "Capacity of the energy harvesting Gaussian MAC," in *2016 IEEE Int. Symp. Information Theory (ISIT)*, Barcelona Jul. 2016, pp. 2744–2748.

[15] H. A. Inan, D. Shaviv, and A. Özgür, "Capacity of the energy harvesting Gaussian MAC," *IEEE Trans. Inform. Theory*, vol. 64, no. 4, pp. 2347–2360, Apr. 2018.

[16] A. Baknina and S. Ulukus, "Online policies for multiple access channel with common energy harvesting source," in *2016 IEEE Int. Symp. Inform. Theory (ISIT)*, Barcelona, Jul. 2016, pp. 2739–2743.

[17] A. Baknina and S. Ulukus, "Energy harvesting multiple access channels: Optimal and near-optimal online policies," *IEEE Trans. Commun.*, vol. 66, no. 7, pp. 2904–2917, Jul. 2018.

[18] A. Baknina and S. Ulukus, "Optimal and near-optimal online strategies for energy harvesting broadcast channels," *IEEE J. Sel. Areas Commun.*, vol. 34, no. 12, pp. 3696–3708, Dec. 2016.

[19] B. Varan and A. Yener, "Online transmission policies for cognitive radio networks with energy harvesting secondary users," in *2017 IEEE Wireless Commun. Networking Conf. Workshops (WCNCW)*, San Francisco, Mar. 2017, pp. 1–6.

[20] A. Arafa and S. Ulukus, "Near optimal online distortion minimization for energy harvesting nodes," in *2017 IEEE Int. Symp. Inform. Theory (ISIT)*, Aachen Jun. 2017, pp. 1117–1121.

[21] D. Shaviv and A. Özgür, "Approximately optimal policies for a class of Markov decision problems with applications to energy harvesting," in *2017 15th Int. Symp. Model. Optimization Mobile Ad Hoc Wireless Netw. (WiOpt)*, Paris May 2017, pp. 1–8.

[22] A. Arafa, A. Baknina, and S. Ulukus, "Energy harvesting networks with general utility functions: Near optimal online policies," in *2017 IEEE Int. Symp. Inform. Theory (ISIT)*, Aachen Jun. 2017, pp. 809–813.

[23] O. Ozel and S. Ulukus, "Achieving AWGN capacity under stochastic energy harvesting," *IEEE Trans. Inform. Theory*, vol. 58, no. 10, pp. 6471–6483, Oct. 2012.

[24] W. Mao and B. Hassibi, "Capacity analysis of discrete energy harvesting channels," *IEEE Trans. Inform. Theory*, vol. 63, no. 9, pp. 5850–5885, Sep. 2017.

[25] V. Jog and V. Anantharam, "A geometric analysis of the AWGN channel with a (σ, ρ)-power constraint," *IEEE Trans. Inform. Theory*, vol. 62, no. 8, pp. 4413–4438, Aug. 2016.

[26] K. Tutuncuoglu, O. Ozel, A. Yener, and S. Ulukus, "The binary energy harvesting channel with a unit-sized battery," *IEEE Trans. Inform. Theory*, vol. 63, no. 7, pp. 4240–4256, Jul. 2017.

[27] D. Shaviv, A. Özgür, and H. Permuter, "Can feedback increase the capacity of the energy harvesting channel?" in *2015 IEEE Inform. Theory Workshop (ITW)*, Jerusalem Apr. 2015, pp. 1–5.

[28] D. Shaviv and A. Özgür, "Capacity of the AWGN channel with random battery recharges," in *2015 IEEE Int. Symp. Inform. Theory (ISIT)*, Aachen, Jun. 2015, pp. 136–140.

[29] D. Shaviv, A. Özgür, and H. H. Permuter, "A communication channel with random battery recharges," *IEEE Trans. Inform. Theory*, vol. 64, no. 1, pp. 38–56, Jan. 2018.

[30] D. Shaviv, P.-M. Nguyen, and A. Özgür, "Capacity of the energy-harvesting channel with a finite battery," *IEEE Trans. Inform. Theory*, vol. 62, no. 11, pp. 6436–6458, Nov. 2016.

[31] D. Shaviv, A. Özgür, and H. H. Permuter, "Capacity of remotely powered communication," *IEEE Trans. Inform. Theory*, vol. 63, no. 3, pp. 1364–1391, Mar. 2017.

Part II

Coding and Modulation

8 Polarization and Polar Coding

Erdal Arıkan

8.1 Introduction

Polarization is a technique that was introduced [1] to construct capacity-achieving polar codes for a certain class of channels (memoryless, symmetric, and having a binary-input alphabet). In their basic form, polar codes are a class of linear codes that follow the Plotkin construction [2] in a recursive manner, and in this respect are closely related to Reed–Muller codes [3, 4]. The recursive structure of polar codes makes it possible to encode and decode them at complexity $\mathcal{O}(N \log(N))$ in the code blocklength N. The recursiveness also renders the analysis of polar codes tractable and makes it possible to prove rigorously that they achieve Shannon limits with a probability of error decreasing to zero roughly as $2^{-\sqrt{N}}$.

Thanks to these properties, polar codes have attracted attention both for theoretical and practical reasons. From a theoretical viewpoint, polar codes were met with interest for being the first class of codes that achieve Shannon limits with an explicit construction and low-complexity construction, encoding, and decoding methods [5, 6]. Polar codes have also had an impact on engineering practice by being accepted into the emerging 5G standard [7, 8]. The goal of this chapter is to provide a survey of both the theoretical and practical aspects of this topic.

As the following survey will show, polarization is a very general phenomenon that can be exploited for code construction in virtually any information theoretic setting, from single-user to multiuser settings, and from lossy or lossless source coding to channel coding. The versatility of polar codes, combined with their low-complexity nature, makes them an enabling technology for turning sophisticated multiuser information theoretic ideas into practical coding schemes for future wireless systems. Hence, the survey covers polar coding for a broad range of multiuser settings which we believe will be important in the near future.

We had to leave out many topics from this survey to keep the length of the chapter within limits. In particular, we have left out papers on implementation of polar codes since it is a specialty subject that deserves a survey of its own. We have also omitted many papers on detailed implementation aspects such as puncturing/shortening of polar codes, hybrid-ARQ (automatic-repeat-request) with polar codes, etc., although there are many results of enduring value in this domain.

8.1.1   Outline of Chapter

The organization of topics is as follows. In Section 8.2, we state the basic results on polarization. This is followed in Section 8.3 by a discussion of polar coding for source and channel coding problems. Section 8.4 discusses the finite-length performance of polar codes and methods to improve it. In Section 8.5, a survey of generalizations of polarization and polar coding are given. These generalizations include extension of polar coding to non-binary code alphabets and/or use of alternative recursions (kernels). Section 8.6 consists of complementary material for extending polar codes to several important problems in single-user settings, such as rate-distortion coding. In Section 8.7 a survey of polar coding for multiuser settings is given. Section 8.8 reviews results on more applied aspects of polar coding. The chapter concludes in Section 8.9 with some complementary remarks.

8.2   Polarization

There are two types of channels for which the channel coding problem is trivial: noiseless channels and pure noise channels. If the channel is noiseless, information can be transmitted reliably without any channel coding. If the channel is a pure noise channel, no method of channel coding can transmit information reliably at any positive rate. Channel polarization is a method that exploits this observation and tries to create near-extreme channels out of multiple independent copies of non-extreme channels. This chapter aims to explain polarization and give a survey of its use in various source and channel coding problems. To keep the presentation simple, our setting will be that of binary-input memoryless channels and binary memoryless sources. The non-binary case will be mentioned only briefly in citing generalizations.

8.2.1   Problem Setting and Notation

The main ideas of polarization can be introduced abstractly, without explicitly mentioning source or channel coding. Source and channel coding are specific problems that can be solved by exploiting polarization. We will follow this approach and consider a pair of random variables (X, Y), where X takes values over $\mathcal{X} \triangleq \{0, 1\}$ (not necessarily from the uniform distribution) and Y takes values over some alphabet $\mathcal{Y}$ which is discrete but otherwise arbitrary. As usual $p_{X,Y}$ will denote the joint probability mass function (pmf) of (X, Y). We will refer to such a pair (X, Y) as a binary *source/observation* (s/o) pair, where X and Y are the source and observation variables, respectively.

In discussing polarization, we will consider a sequence of independent copies of a given binary s/o pair (X, Y) and denote the first N of them as (X^N, Y^N) or $\{(X_i, Y_i)\}_{i=1}^{N}$. Thus, for any integer $N \geq 1$, we will have

$$p_{X^N, Y^N}(x^N, y^N) = \prod_{i=1}^{N} p_{X,Y}(x_i, y_i),$$

where $X^N = (X_1, \ldots, X_N)$ and $Y^N = (Y_1, \ldots, Y_N)$.

Polarization results are expressed in terms of convergence of certain information measures. The primary information measures of interest are the *conditional entropy*

$$H(X|Y) \stackrel{\Delta}{=} - \sum_{x,y} p_{X|Y}(x|y) \log_2 p_{X|Y}(x|y),$$

the mutual information

$$I(X;Y) \stackrel{\Delta}{=} \sum_{x,y} p_{X,Y}(x,y) \log_2 \frac{p_{X,Y}(x,y)}{p_X(x)p_Y(y)},$$

and the Bhattacharyya parameter

$$Z(X|Y) \stackrel{\Delta}{=} 2 \sum_y P_Y(y)\sqrt{P_{X|Y}(0|y)P_{X|Y}(1|y)}.$$

The conditional entropy $H(X|Y)$ and mutual information $I(X;Y)$ represent Shannon limits on achievable rates in source and channel coding problems, respectively. The Bhattacharyya parameter $Z(X|Y)$ serves as a convenient upper bound on the probability of error for estimating X given Y. More precisely, if $\hat{X}(Y)$ denotes the maximum a-posteriori probability (MAP) estimate of X given Y, then

$$P\left(\hat{X}(Y) \neq X\right) \leq \frac{1}{2}Z(X|Y).$$

The goal of polarization is to begin with multiple i.i.d. copies of a binary s/o pair (X, Y) and turn them into near-extreme binary s/o pairs for which the source or channel coding problems are much simpler to solve. The required notion of extremeness in the setting of polarization is the following.

DEFINITION 8.1 *A binary source/observation pair (X, Y) is called extreme if $H(X|Y)$ equals 0 or 1.*

Note that since $H(X|Y) \leq H(X)$, $H(X|Y) = 1$ implies $H(X) = 1$. For an extreme binary s/o pair, the source and channel coding problems have trivial solutions. For example, in a source coding context, if $H(X|Y) = 0$, then X can be estimated perfectly at a decoder that has access to side information Y. So, the encoder can compress X to an empty string (meaning that nothing is sent to the decoder) and the decoder can still recover X with certainty.

We have defined the extremeness of a binary s/o pair in terms of the conditional entropy $H(X|Y)$. It turns out an s/o pair that is extreme with respect to $H(X|Y)$ is extreme also with respect to $Z(X|Y)$. This is because $H(X|Y)$ is near 0 or 1 if and only if $Z(X|Y)$ is near 0 or 1, respectively, as can be deduced from the following pair of inequalities from [9].

PROPOSITION 8.2 *For (X, Y) a binary source/observation pair as above, the following inequalities hold*

$$[Z(X|Y)]^2 \leq H(X|Y) \leq \log_2(1 + Z(X|Y)). \tag{8.1}$$

Either both inequalities are strict or both hold with equality. For equality to hold, it is necessary and sufficient that $H(X|Y)$ equals 0 or 1.

8.2.2 A Basic Polarization Result

Polarization consists of two steps: combining and splitting. The combining step receives as input a collection (X^N, Y^N) of N i.i.d. copies of a binary s/o pair (X, Y) and produces a new collection of random variable (U^N, Y^N) as output. The $U^N = (U_1, \ldots, U_N)$ here is a vector of length N over $\mathcal{X}$ and is obtained from X^N by calculating $U^N = G_N(X^N)$ where $G_N : \mathcal{X}^N \to \mathcal{X}^N$ is a one-to-one transform. The splitting step consists of splitting (U^N, Y^N) into a collection of N binary s/o pairs of the form $\{(U_i, Y^N U^{i-1})\}_{i=1}^N$, with $U^{i-1} = (U_1, \ldots, U_{i-1})$. The new collection of binary s/o pairs are no longer i.i.d. and tend to converge to extreme binary s/o pairs as N becomes large, provided the transform G_N has sufficient mixing properties. Below we will make these notions precise. Throughout, we will consider only linear transforms G_N and the transform operation $U^N = G_N(X^N)$ will be denoted as a matrix operation $U^N = X^N G_N$ over the vector space $\mathbb{F}_2^N$.

The basic polarization result of [1] and [9] is as follows.

THEOREM 8.3   *For each integer $n \geq 1$ and $N = 2^n$, let G_N be defined as*

$$G_N \triangleq F^{\otimes n}, \qquad F \triangleq \begin{bmatrix} 1 & 0 \\ 1 & 1 \end{bmatrix}, \tag{8.2}$$

where $F^{\otimes n}$ denotes the nth Kronecker power of F. Let $0 < \delta < 1$ be fixed. Let (X^N, Y^N) be N i.i.d. copies of a given binary s/o pair (X, Y). Let $U^N = X^N G_N$ and consider the binary s/o pairs $\{(U_i; Y^N, U^{i-1})\}_{i=1}^N$. This second set of binary s/o pairs become polarized in the sense that their conditional entropies $H_N^{(i)} \triangleq H(U_i|Y^N, U^{i-1})$ satisfy

$$\lim_{N \to \infty} \frac{\left| \{i \in \{1, \ldots, N\} : H_N^{(i)} \in (1 - \delta, 1]\} \right|}{N} \to H(X|Y) \tag{8.3}$$

and

$$\lim_{N \to \infty} \frac{\left| \{i \in \{1, \ldots, N\} : H_N^{(i)} \in [0, \delta)\} \right|}{N} \to 1 - H(X|Y). \tag{8.4}$$

Theorem 8.3 was first proved in [9] as an extension of a similar theorem in [1]. In essence, Theorem 8.3 states that the N binary s/o pairs $\{(U_i, Y^N U^{i-1})\}_{i=1}^N$ at the polar transform output approach extreme binary s/o pairs, with all but a vanishing fraction of the conditional entropies $\{H_N^{(i)}\}$ clustering around 0 or 1. For clarity, the theorem has been stated in a redundant form; clearly, the two statements (8.3) and (8.4) imply each other.

In the rest of this subsection, we wish to discuss various aspects of polarization and give pointers to topics that will be discussed in the rest of the chapter.

REMARK 8.4   *(The kernel F) The polar transform G_N in Theorem 8.3 is generated as Kronecker powers of a basic transform module F, which we will refer to as a kernel. Polar codes can be constructed using various other kernels as discussed in Section 8.5. The specific kernel F used in Theorem 8.3 will be referred to as the standard kernel. A circuit diagram for implementing the standard kernel is shown in Fig. 8.1.*

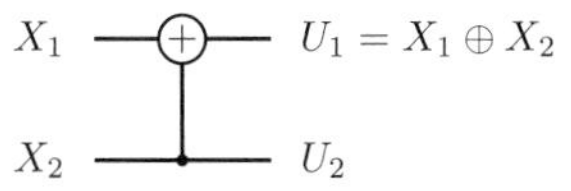

Figure 8.1 Circuit diagram for the standard kernel.

The $\oplus$ in the diagram implements addition mod-2 (XOR in binary logic). The operation $(X_1, X_2) \rightarrow (U_1, U_2)$ performed by the circuit preserves entropy, i.e.,

$$H(U_1, U_2 | Y_1, Y_2) = H(X_1, X_2 | Y_1, Y_2) = 2H(X|Y),$$

but is polarizing in the sense that

$$H(U_1 | Y_1, Y_2) \geq H(X|Y) \geq H(U_2 | Y_1, Y_2, U_1).$$

It is easy to show that equalities hold here if and only if $H(X|Y)$ equals 0 or 1. Thus, unless the binary s/o pair (X, Y) at the transform input is perfectly extreme, the s/o pairs (U_1, Y^2) and $(U_2, Y^2 U_1)$ will move away from each other in terms of conditional entropy and become more extreme relative to (X, Y). For a detailed discussion of how the polar transform G_N recursively extends the basic polarization step initiated by the kernel transform F, we refer to [9].

REMARK 8.5 *(Conservation of entropy) The polar transform preserves the conditional entropy. By the chain rule, we have $\sum_{i=1}^{N} H_N^{(i)} = H(U^N | Y^N)$. Since G_N is one-to-one, $H(U^N | Y^N) = H(X^N | Y^N)$, and since $\{(X_i, Y_i)\}_{i=1}^{N}$ are i.i.d. $\sim (X, Y)$, we have $H(X^N | Y^N) = NH(X|Y)$. Thus, the sum of the conditional entropies of the binary s/o pairs at the output of the polar transform equals the sum of the conditional entropies at its input.*

The entropy conservation is important in the context of polarization primarily to ensure that codes constructed on the basis of polarization achieve information theoretic limits. Entropy conservation also helps the analysis of polarization schemes by making it possible to set up a martingale and exploit results from martingale theory.

REMARK 8.6 *(Strong polarization) The theorem statement assumes that δ is fixed and independent of N. It turns out that this basic form of the polarization result is not strong enough to prove that source and channel codes constructed by using polarization are capacity-achieving. To prove coding theorems, a stronger form of polarization is needed in which δ is allowed to go to zero sufficiently fast as N grows. This question is investigated in Section 8.2.3.*

REMARK 8.7 *(Gap to capacity) For studying the finite-length performance of polar codes for source and channel coding, it is of interest to consider the rate at which the limits in (8.3) and (8.4) are reached. This question is addressed in Section 8.4.1 in the context of channel coding where this problem has been studied most extensively.*

REMARK 8.8 *(Polarization and random-coding) Polarization is a commonplace phenomenon. If a transformation G_N is constructed by choosing G_N at random from the class of all N-by-N transformations, then (8.3) and (8.4) will hold with high probability* *for all N large enough. The proof of this statement is essentially equivalent to proofs of the achievability of the Shannon limits in source and channel coding by random-coding methods. Conversely, any sequence of source or channel codes that achieves the Shannon limits has to exhibit polarization. In other words, polarization is an essential ingredient of all capacity-achieving schemes. The main novelty of Theorem 8.3 is that it exhibits an explicit and low-complexity method of polarization.*

8.2.3   Strong Polarization

As mentioned above, Theorem 8.3 is not strong enough to prove coding theorems. The Bhattacharyya parameter plays an essential role in the derivation of stronger polarization results because its evolution is easier to track through the steps of polarization compared to conditional entropy and mutual information.

The following result, sometimes referred to as strong polarization, states this fact in a form that is tailored for proving polar coding results.

THEOREM 8.9   *Consider the polarization scheme in Theorem 8.3. For shorthand, put $Z_N^{(i)} \triangleq Z(U_i|Y^N, U^{i-1})$. Let $\beta > 0$ be a fixed number. Then,*

$$\lim_{N\to\infty} \frac{\left|\{i \in \{1,\dots,N\}: Z_N^{(i)} < 2^{-N^\beta}\}\right|}{N} = \begin{cases} 1 - H(X|Y), & \text{if } 0 < \beta < \frac{1}{2}; \\ 0, & \text{if } \beta > \frac{1}{2}. \end{cases} \quad (8.5)$$

Theorem 8.9 was first proved in [10] using martingales and for the binary-input channels. Theorem 8.9 was refined and generalized later in [11] for the case of channel polarization. This generalization will be discussed in Section 8.3.2.

For self-contained alternative proofs of the strong polarization result, we refer to [12–14].

8.3   Polar Coding

8.3.1   Source Coding

In this section we describe a source coding method based on polarization. Let $\{(X_i, Y_i)\}_{i=1}^N$ be i.i.d. copies of a binary s/o pair (X, Y) for some integer $N \geq 1$. Recall that X takes values over $\mathcal{X} = \{0, 1\}$ and Y takes values over an arbitrary discrete alphabet $\mathcal{Y}$. A source code comprises an encoder $f : \mathcal{X}^N \to \{0, 1\}^K$ and a decoder $g : \{0, 1\}^K \times \mathcal{Y}^N \to \mathcal{X}^N$. The ratio K/N is called the compression ratio. Given a realization (x^N, y^N) of (X^N, Y^N), the encoder generates a codeword $v^K = f(x^N)$ and sends it to the decoder. The decoder receives v^K and y^N and computes $\hat{x}^N = g(v^K, y^N)$. An error is said to occur if $\hat{x}^N \neq x^N$. In terms of random variables, the encoder computes $V^K = f(X^N)$ and the decoder computes $\hat{X}^N = g(V^K, Y^N)$. The probability of error is defined as $P_e \triangleq \Pr(\hat{X}^N \neq X^N)$. Shannon's source coding theorem states that given any $\delta > 0$ and $\epsilon > 0$ and sufficiently large N and K, there exist encoder–decoder pairs (f, g) such that $K/N \leq H(X|Y) + \delta$ and $P_e < \epsilon$.

In [15, 16], the above source coding problem was solved for binary s/o pairs (X, Y), where X is $\mathrm{Ber}(\frac{1}{2})$ and $Y = X \oplus Z$, where Z is $\mathrm{Ber}(p)$ and independent of X. Their approach was to reduce the source coding problem to a channel coding problem by exploiting a duality relationship between the two problems. In [9], the solution was generalized to the case of arbitrary pairs (X, Y). The approach in [9] was direct and did not use duality of source and channel coding.

Details of the polar source coding solution are as follows. We will use the notation of Section 8.2. Suppose $\delta > 0$ and $\epsilon > 0$ are given and fixed. For the time being, we will leave the blocklength N of the target source code as a free parameter. We wish to construct a polar source code of length $N = 2^n$ for some $n \geq 1$ and a compression ratio $K/N \leq H(X|Y) + \delta$ such that $P_{\mathrm{e}} < \epsilon$. The blocklength N here is a free parameter that can be taken as large as necessary to meet the design requirements. For each candidate blocklength $N = 2^n$, let $K_N \triangleq \lfloor N[H(X|Y) + \delta] \rfloor$ and $\mathcal{A}_N \subset \{1, 2, \ldots, N\}$ be a subset of indices such that $|\mathcal{A}_N| = K_N$ (it has size K_N) and the sum

$$Z(\mathcal{A}_N) \triangleq \sum_{i \in \mathcal{A}_N} Z(U_i | Y^N, U^{i-1})$$

is as small as possible. Clearly, any set $\mathcal{A}_N$ that minimizes $Z(\mathcal{A}_N)$ must contain only the indices of the K_N smallest terms in the set $\{Z(U_i|Y^N, U^{i-1})\}_{i=1}^N$. (The solution may not be unique if the Bhattacharyya parameters are not all distinct. In case there are ties, they can be broken arbitrarily for purposes of this construction.)

Encoding: Given a realization (x^N, y^N) of (X^N, Y^N), the encoder computes $u^N = x^N G_N$ and sends out the subvector $u_{\mathcal{A}_N} \triangleq (u_i : i \in \mathcal{A}_N)$ as the compressed word. Note that the encoding operation does not require knowledge of y^N.

Decoding: The decoder receives $u_\mathcal{A}$ and y^N, and sequentially constructs an estimate $\hat{u}^N$ of u^N by the rule

$$\hat{u}_i = \begin{cases} u_i & \text{if } i \in \mathcal{A}_N \\ 0 & \text{if } i \in \mathcal{A}_N^c \text{ and } L_N^{(i)}(y^N, \hat{u}^{i-1}) \geq 1 \\ 1 & \text{else} \end{cases}$$

where

$$L_N^{(i)}(y^N, \hat{u}^{i-1}) = \frac{\Pr(U_i = 0 | Y^N = y^N, U^{i-1} = \hat{u}^{i-1})}{\Pr(U_i = 1 | Y^N = y^N, U^{i-1} = \hat{u}^{i-1})}$$

is a likelihood ratio. Following the construction of $\hat{u}^N$ in this manner, the decoder computes $\hat{x}^N = \hat{u}^N (G_N)^{-1}$ and outputs it as the estimate $\hat{x}^N$ of x^N. (The inverse transform here happens to be the same as the transform itself: $(G_N)^{-1} = G_N$.) In the following, we will regard $\hat{u}^N$ and $\hat{x}^N$ as realizations of the random vectors $\hat{U}^N$ and $\hat{X}^N$, respectively.

Performance: The probability of error has been defined above as $P_{\mathrm{e}} = \Pr(\hat{X}^N \neq X^N)$. Expressed in terms of the transforms, we have $P_{\mathrm{e}} = \Pr(\hat{U}^N \neq U^N)$. The decoder is guaranteed to provide correct reconstruction of the transform elements in the complement of $\mathcal{A}$; in other words, we will always have $\hat{U}_{\mathcal{A}_N^c} = U_{\mathcal{A}_N^c}$. So, $P_e = \Pr(\hat{U}_{\mathcal{A}_N} \neq U_{\mathcal{A}_N})$. Using the union bound, we obtain

$$P_e \leq \sum_{i \in \mathcal{A}_N} \Pr(\hat{U}_i \neq U_i) \leq \sum_{i \in \mathcal{A}_N} Z(U_i | Y^N U^{i-1}),$$

where $Z(U_i | Y^N U^{i-1})$ are the Bhattacharyya parameters (for details we refer to [1]). The following is now a simple corollary to Theorem 8.9.

THEOREM 8.10   *For any fixed $R > H(X|Y)$ and $\beta < \frac{1}{2}$, the probability of error for the above polar source coding method is bounded as $P_e = \mathcal{O}(2^{-N^\beta})$.*

Complexity: The complexity of encoding and decoding for the above source code are both $\mathcal{O}(N \log N)$. For details of complexity analysis, we refer to [9].

REMARK 8.11   *The above polar source coding method does not guarantee zero error probability. In order to achieve zero error probability (lossless) compression at rates approaching the entropy limit, it is necessary to consider a variable-length polar coding method. Such a method was developed in [17] and it was shown that the proposed scheme achieves the optimal compression rate asymptotically while retaining the low-complexity properties of polar coding.*

8.3.2   Channel Coding

In this part we consider polar coding schemes for binary-input discrete memoryless channels. Such a channel will be defined by a set of transition probabilities $\{W(y|x) : x \in \mathcal{X} = \{0, 1\}, y \in \mathcal{Y}\}$ where $\mathcal{X}$ is the input alphabet, $\mathcal{Y}$ is the output alphabet, and $W(y|x)$ is the conditional probability of receiving y at the channel output given that x is transmitted. A polar code for this type of channel can be constructed readily from the polarization results for a binary s/o pair (X, Y) with p_X uniform and $p_{Y|X}(y|x) = W(y|x)$. Such a polar code achieves the *symmetric* capacity of the channel W, defined as $I(W) \overset{\Delta}{=} I(X; Y)$ with X uniform on $\mathcal{X}$.

If the channel W has input–output symmetry, then the symmetric capacity $I(W)$ coincides with the true capacity. A channel W is called (input–output) symmetric if there exists a permutation $\pi : \mathcal{Y} \to \mathcal{Y}$ of the output alphabet such that $W(y|0) = W(\pi(y)|1)$ for every $y \in \mathcal{Y}$. Examples of symmetric channels include the binary symmetric channel (BSC) and the binary erasure channel (BEC). For the BSC, $\mathcal{Y} = \{0, 1\}$ and $W(0|0) = W(1|1) = 1 - p$ for some $p \in [0, 1/2]$. For the BEC, $\mathcal{Y} = \{0, 1, ?\}$ and $W(0|0) = W(1|1) = 1 - p$ and $W(?|0) = W(?|1) = p$. For the rest of this subsection, we will assume that the channel W under consideration is symmetric.

Recall that a channel code comprises an encoder $f : \{0, 1\}^K \to \mathcal{X}^N$ and a decoder $g : \mathcal{Y}^N \to \{0, 1\}^K$. The ratio K/N is called the rate (of information transmission). Given a realization v^K of a data word V^K (assumed uniformly distributed on $\{0, 1\}^K$), the encoder generates a codeword $x^N = f(v^K)$ and sends it to a receiver by N uses of a channel W. The receiver receives a channel output word y^N and decodes it to generate an estimate $\hat{v}^K = g(y^N)$ of the data word v^K. An error is said to occur if $\hat{v}^K \neq v^K$.

The probability of error, in terms of the corresponding random variables, is defined as $P_e \triangleq \Pr(\hat{V}^K \neq V^K)$. Shannon's channel coding theorem states that given any $\delta > 0$ and $\epsilon > 0$ and sufficiently large N and K, there exist encoder–decoder pairs (f, g) such that $K/N > I(X; Y) - \delta$ and $P_e < \epsilon$.

The polar channel coding solution to the channel coding problem is as follows. Let $\delta > 0$ and $\epsilon > 0$ be fixed. We will consider polar codes of blocklength $N = 2^n$ and dimension $K_N \triangleq \lceil N[I(X; Y) - \delta] \rceil$. Thus, all codes in the sequence will satisfy the rate constraint $R > I(X; Y) - \delta$. We will show that for N sufficiently large, the reliability constraint $P_e \leq \epsilon$ will be satisfied as well. For the length-N code, we will define an *information index set* $\mathcal{A}_N \subset \{1, 2, \ldots, N\}$ as a subset of indices such that $|\mathcal{A}_N| = K_N$ and the sum

$$Z(\mathcal{A}_N) \triangleq \sum_{i \in \mathcal{A}_N} Z(U_i | Y^N, U^{i-1})$$

is as small as possible.

Encoding: Given a data word v^K, the encoder prepares a transform input u^N by setting $u_{\mathcal{A}_N} = v^K$ and $u_{\mathcal{A}_N^c} = 0^{N-K}$ and computes $x^N = u^N(G_N)^{-1}$ which for the polar transform here is the same as computing $x^N = u^N G_N$.

Transmission: The codeword x^N is sent over the channel W and a channel output y^N is received.

Decoding: The decoder receives the channel output y^N and also knows that $u_{\mathcal{A}_N^c}$ was set equal to the all-zero vector 0^{N-K}. Based on this, the decoder constructs an estimate $\hat{u}^N$ of u^N using the same rule as in source coding. Following the construction of $\hat{u}^N$ in this manner, the decoder produces an estimate $\hat{v}^K$ of the data word v^K by setting $\hat{v}^K = \hat{u}_{\mathcal{A}_N}$.

Performance: The probability of error can be upper bounded as follows

$$P_e \triangleq \Pr(\hat{V}^K \neq V^K) \leq \sum_{i \in \mathcal{A}_N} \Pr(\hat{U}_i \neq U_i) \leq \sum_{i \in \mathcal{A}_N} Z(U_i | Y^N U^{i-1}). \tag{8.6}$$

The following is now a simple corollary to Theorem 8.9.

THEOREM 8.12   *For any fixed $R < I(X; Y) - \delta$ and $\beta < \frac{1}{2}$, the probability of error for the above polar channel coding method is bounded as $P_e = \mathcal{O}(2^{-N^\beta})$.*

Theorem 8.12 follows from Theorem 8.10 by noting that $R < I(X; Y) - \delta$ is equivalent to $1 - R > H(X|Y) + \delta$ (since $H(X) = 1$) and $1 - R$ is the compression rate in source coding.

The exponential error bound in Theorem 8.12 was proved in [10]. This result was later refined by [11], where a more precise rate-dependent form of the exponential bound on P_e was obtained. Further refinements of this result for nonasymptotic performance in N are discussed below in Section 8.4.1.

Complexity: The complexity of encoding and decoding for the above channel code is $\mathcal{O}(N \log N)$. For details of complexity analysis, we refer to [1].

8.3.3   Construction of Polar Codes

The construction problem for polar codes may be defined as follows. Given a channel W and a block-length N, rank the channels $\{W_N^{(i)}\}_{i=1}^N$ in terms of their quality as
measured, for example, by the Bhattaracharyya parameter $Z(W_N^{(i)})$. In [1], no efficient
algorithm was given for the construction problem. This problem was addressed by Mori
and Tanaka [18], who gave a construction method based on *density-evolution* and had
complexity $\mathcal{O}(N)$ floating-point operations.

A finite-precision implementation of density-evolution for polar code construction
was given by Tal and Vardy [19], who provided upper and lower bounds on the density
evolution parameters by using "upgrading" and "downgrading" procedures. Pedarsani
et al. [20, 21] showed that the complexity of finite-precision implementations of the
density-evolution method for polar code construction is of order $\mathcal{O}(N\text{poly}(\log N))$. In a
more recent study Guruswami and Xia [22] showed that the complexity of constructing
a polar code with rate $R > I(W) - \epsilon$ is bounded as $\text{poly}(\frac{1}{\epsilon}) + \mathcal{O}(N \log N)$.

The construction problem for polar codes can be alleviated somewhat by identifying
certain universal partial-order relations among the synthetic subchannels. This subject
has been studied extensively by Geiger and Geiger [23], Bardet et al. [24], Schurch
[25], and Mondelli et al. [26]; the last paper showed that, by exploiting the partial-
order relations, it is possible to construct polar codes in sublinear complexity in code
blocklength.

By showing that there exist efficient methods of polar code construction, in effect the
papers cited above establish that polar codes are *explicit* coding schemes.

In addition to the above precise methods of code construction, there exist some heuristic approximation techniques of lower computational complexity, such as the BEC
approximation [27], the Gaussian approximation [28] by Trifonov, and the 3GPP-NR
approximation [7]. The 3GPP-NR approximation aims to minimize storage space by
providing a single table that specifies a "mother" polar code from which the design of
multiple polar codes of various blocklengths are derived.

The above methods target polar code constructions for successive cancellation (SC)
decoding and might not be optimal for other decoding algorithms such as belief propagation (BP) or successive-cancellation list (SCL) decoding. This question was addressed
by Qin et al. [29], who proposed a design method based on the evolution of log-
likelihood ratios in the course of BP decoding and reported that improved performance
is observed not only under BP decoding, but also under SCL decoding.

8.4   Finite-Length Performance of Polar Codes

8.4.1   Gap to Capacity

In Theorem 8.12, the rate R is fixed and one is interested in the behavior of P_e as N
becomes large. It is also possible and meaningful to study the channel coding problem
by fixing P_e and studying the tradeoff between R and N. For polar codes, this problem

has been studied in detail by Hassani et al. [30], Guruswami and Xia [22], Błasiok et al. [5, 6] under the terms "scaling exponent" or "gap-to-capacity."

A scaling exponent μ for polar coding on a channel W under successive cancellation decoding is a hypothetical number such that

$$N = \Omega\left(\frac{\alpha}{(I(W) - R)^{\mu}}\right),$$

where α is a positive constant that depends on P_e and $I(W)$. The scaling exponent μ depends on the channel W. In general, it was shown in [30] that $3.55 < \mu < 7$ independent of the channel W. Goldin and Burshtein improved the upper bound to $\mu < 5.77$ [31].

In [22], the emphasis was on restricting the gap-to-capacity to a small value by requiring $0 < I(W) - R < \epsilon$ and studying the complexity of decoding and construction of polar codes as a function of $1/\epsilon$. It was shown that, for all binary-input symmetric memoryless channels, polar codes enable reliable communication at rates within an additive gap $\epsilon > 0$ to the Shannon capacity with a blocklength, construction complexity, and decoding complexity, all bounded by a polynomial in $1/\epsilon$. It was also mentioned that polar coding gives the first known explicit construction with rigorous proofs of all these properties. In [5, 6], these results were generalized to arbitrary kernels (such as those in Section 8.5) and to channels with a prime number of inputs.

As pointed out in [30], a classical result in information theory states that the best possible value for the scaling exponent (over all possible codes and under maximum likelihood [ML] decoding) is $\mu = 2$. Thus, polar codes require significantly longer blocklengths to satisfy a given P_e compared to optimal codes. In [32], Pfister and Urbanke studied the scaling problem for polar codes over large alphabets and showed that near-optimal finite-length scaling could be achieved.

8.4.2   Improvement of finite-length performance

Polar codes have been of interest for practical applications from the very start, given their close connection to Reed–Muller codes [1]. However, initial simulation studies revealed that polar codes are not strong enough to compete with the state-of-the-art low-density parity check (LDPC) and turbo codes. Part of the disappointing performance of polar codes is due to the suboptimality of the successive cancellation decoder (SCD), and part is due to the relatively poor minimum distance of polar codes.

Tal and Vardy provided a major impetus for the practicality of polar codes [33]. They fixed the deficiency of the SCD by combining it with list decoding and the minimum distance problem by using a concatenation scheme in which the polar code became the inner code and a cyclic redundancy check (CRC) code served as an outer code. Simulation studies presented in [33] showed that polar codes with these modifications can perform competitively against the state-of-the-art channel codes. Furthermore, it was shown in [33] that the list-of-L SCD can be implemented with complexity $\mathcal{O}(LN \log N)$ in time and $\mathcal{O}(LN)$ in space. List-decoding for polar codes was also studied by Chen et al. [34] and Niu and Chen [35].

While the above list-decoder carries out a breadth-first search in the polar code decoding tree, Niu and Chen studied a depth-first search (with backtracking) procedure [36], called successive cancellation stack (SCS) decoding. They provided simulation results showing that the SCS decoder achieves similar performance to SCL (in fact near ML performance) with time complexity much lower than that of SCL.

The above decoding methods for polar codes (SCD, SCL, SCS) all happen to be rediscoveries or adaptations of known methods studied in great detail for Reed–Muller codes. For a review of decoders from this perspective, we refer to the papers [37] and [38] by Dumer (see also [39]).

For a theoretical study of why serial concatenation of a polar code with a high-rate outer code (such as a CRC) improves the finite-length performance so remarkably, we refer to [40].

8.5   General Kernels

Until now, we considered polarization of a binary s/o pair (X, Y) using the standard kernel $F = \begin{bmatrix} 1 & 0 \\ 1 & 1 \end{bmatrix}$. The standard scheme achieves a polarization exponent $E = 1/2$ as stated in Theorem 8.9. It is of interest to seek alternative polarization schemes that improve the polarization exponent while retaining the low-complexity properties of polar codes. In this part, we summarize some results about this topic. The treatment is divided into three parts. Section 8.5.1 considers binary s/o pairs; Section 8.5.2 considers non-binary s/o pairs; and Section 8.5.3 reviews a number of different approaches to constructing polarization schemes.

8.5.1   Binary Kernels

In this section, we consider the polarization of a binary s/o pair (X, Y) as in Theorem 8.3, except here the kernel F is an ℓ-by-ℓ non-singular matrix over $\mathbb{F}_2$, where $\ell \geq 2$ is also arbitrary. The most basic question that needs to be answered first is whether a given kernel F of this type is polarizing in the sense that Theorem 8.3 holds with the transformation G_N modified accordingly so that $G_N = F^{\otimes n}$ and $N = \ell^n$.

In answer to this question, Korada et al. [41] showed that F is polarizing if and only if none of the column permutations of F is an upper triangular matrix. From now on we will restrict attention to *polarizing kernels*. Given that F is a polarizing kernel, a question of central importance is to determine the rate of polarization. It was shown in [41] that the rate of polarization by a kernel is governed by a parameter $E(F)$ defined as follows.

DEFINITION 8.13 (Polarization exponent)   *Let F be an ℓ-by-ℓ matrix over $\mathbb{F}_2$, for some $\ell \geq 2$. Let f_i denote the ith row of F, $i = 1, \ldots, \ell$. Let D_i be the ith partial distance defined as the minimum Hamming distance between f_i and any vector in the span of the vectors $f_{i+1}, \ldots, f_\ell$ for $i < \ell$; and, let D_ℓ be the Hamming weight of f_ℓ. The polarization exponent of F is defined as*

$$E(F) \overset{\Delta}{=} \frac{1}{\ell} \sum_{i=1}^{\ell} \log_\ell(D_i).$$

EXAMPLE 8.14   For the standard kernel, the partial distances are $(1, 2)$ and we have $E(F) = 1/2$. For $F = \begin{bmatrix} 1 & 0 & 0 \\ 1 & 1 & 0 \\ 1 & 0 & 1 \end{bmatrix}$, the partial distances are $(1, 2, 2)$ yielding $E(F) = \frac{1}{3} \log_3 2^2 = 0.4206$.

The rate of polarization result in [41] can now be stated as follows.

THEOREM 8.15   *For any kernel F, the limit in (8.5) in Theorem 8.9 converges to* $1 - H(X|Y)$ *if* $0 < \beta < E(F)$ *and to 0 if* $\beta > E(F)$.

Korada et al. [41] further showed that all kernels of size $\ell < 15$ have exponents $\leq \frac{1}{2}$, and presented a kernel of size $\ell = 16$ with exponent 0.51828. They showed that $E(F)$ could be made to approach 1 (the best possible value over any code) by increasing ℓ.

Presman et al. [42] and Lin et al. [43] provide lists of currently best available kernels. The latter reference gives explicit kernels with optimal exponents for each ℓ from 1 to 16. It is shown in [43] that one has to use a kernel of size at least $\ell = 15$ to get a polarization exponent above $\frac{1}{2}$; even then, the best possible exponent at size $\ell = 15$ has value 0.500651. These results show that, for binary alphabets, the standard two-by-two kernel is the only practically viable choice.

Polarization of binary s/o pairs was also studied using *nonlinear* kernels. Mori and Tanaka [44] gave a generalization of Theorem 8.15 for nonlinear kernels. In [42] and [43], nonlinear kernels with optimal polarization exponents were constructed for various dimensions. In particular, it was shown that, for ℓ equal to 14, 15, and 16, the exponents for nonlinear kernels are strictly better than those for linear kernels of the same dimension. However, it is disappointing that one has to use a kernel of size at least as large as $\ell = 14$ to improve the exponent 1/2 of the standard two-by-two (linear) kernel, and even then, the best exponent achievable at dimension $\ell = 14$ is 0.501940, which is negligibly better than 1/2.

8.5.2   Non-binary Kernels

To begin the discussion of non-binary s/o pairs, first we revisit some of the definitions given in Section 8.2.1 and restate them for the general q-ary s/o pair (X, Y). Since polarization is defined using matrix transformations, we will assume that the source alphabet is a finite field; accordingly, q will be a prime power, unless otherwise stated. In discussing the q-ary case, $H(X|Y)$ and $I(X; Y)$ will be defined using logarithms to base-q to ensure that these quantities take values in $[0, 1]$. The definition of the Bhattacharyya parameter will be modified as follows.

$$Z(X|Y) \overset{\Delta}{=} \frac{1}{q-1} \sum_y P_Y(y) \sum_{x, x'} \sqrt{P_{X|Y}(x|y) P_{X|Y}(x'|y)}.$$

We will say that a kernel F polarizes a q-ary source (X, Y) if Theorem 8.3 holds using transforms of the form $G_N = F^{\otimes n}$ with all matrix operations in $\mathbb{F}_q$.

The first question that needs to be answered is whether there is any transform that polarizes q-ary s/o pairs. Şaşoğlu et al. [45] showed that the standard kernel $F = \begin{bmatrix} 1 & 0 \\ 1 & 1 \end{bmatrix}$, interpreted as a matrix over $\mathbb{F}_q$, polarizes any q-ary s/o pair when q is a prime number. An example was also given in [45] showing that there exists a quaternary ($q = 4$) s/o pair that does not polarize under the standard kernel.

A general characterization of all kernels that polarize a q-ary s/o pair was given by Mori and Tanaka [46], which contained the result of [41] about binary s/o pairs mentioned above as a special case. The characterization in [46] implied that the kernel $F = \begin{bmatrix} 1 & 0 \\ \gamma & 1 \end{bmatrix}$, with γ not equal to 0 or 1, polarizes all q-ary s/o pairs for $q > 2$.

As for the rate of polarization, [44, 46] showed that Theorem 8.15 holds also for q-ary kernels and s/o pairs. Based on this result, [47, 48] showed that kernels over $\mathbb{F}_q$ can achieve significantly better polarization exponents than $\frac{1}{2}$. In particular, [47, 48] gave kernels of size $l < q$ that have polarization exponents equal to $\frac{1}{\ell} \log_\ell \ell!$, which is the best possible value over any kernel of size ℓ. For $\ell = q = 4$, such a kernel achieves a polarization exponent $E(F) = 0.57312$ which is significantly better than $\frac{1}{2}$. This demonstrated the usefulness of considering non-binary kernels for potential applications.

8.5.3   Other Constructions

For non-prime q, a counter example in [45] showed that the standard kernel may not achieve polarization. A remedy proposed in [45] is to reduce the non-prime case to the prime case by using multilevel polarization. For example, if $q = pr$ with p and r prime, a q-ary s/o pair (X, Y) can be written as $(X'X'', Y)$ where X' and X'' are (not-necessarily independent) random variables that take values over p-ary and r-ary alphabets, respectively. Standard polarization is applied first to the pair (X', Y), and then to the pair (X'', YX') where X' becomes part of the side information.

An interesting new phenomenon about polarization was reported and investigated by Park and Barg [49–52]. Specifically, it was observed that for a channel with an input alphabet of size $q = 2^r$, the standard kernel produces q-ary channels whose capacities cluster around the integer values $0, 1, 2, \ldots, r$ bits. This naturally led to a transmission scheme, which achieved the symmetric capacity of the original q-ary channel with low-complexity encoding and decoding, and had a probability of error bounded by the same exponential function as that in (8.6). This result shows that in the case of channels with q-ary inputs, extreme channels should be defined in a broad manner so that they include channels whose capacities may take integer values strictly between 0 and $\log_2 q$ bits. In other words, we should call a channel an extreme channel if the capacity of that channel can be achieved without coding.

In independent work, Sahebi and Pradhan [53] obtained similar results to the ones cited above for multilevel polarization. Channels produced by polarization were classified into three classes: useless, perfect, and partially perfect. The algebraic structure of the channel input alphabet and its relation to the partially perfect channels were investigated.

In [54], Presman et al. studied polarization using transforms G_N that are Kronecker products of several kernels of various sizes and alphabets. This type of construction provides greater flexibility in code construction in terms of choosing the code length N. A restricted version of mixing kernels was studied in [55] by Gabry et al. from a more practical viewpoint. The constructions in [55] target binary s/o pairs, so the kernels are restricted to $\mathbb{F}_2$ but they are allowed to have different sizes. Advantages of such constructions are cited as better performance compared to standard schemes that use puncturing or shortening to construct codes whose lengths are not a power of two.

8.6   Polar Codes in Other Single-User Settings

8.6.1   Rate-Distortion Problem

Korada et al. [41] and Korada and Urbanke [56] showed that polar codes can achieve the Shannon limit in lossy compression of a binary symmetric s/o pair using a low-complexity successive encoding algorithm. Based on this, they further showed the optimality of polar codes for various problems including the binary Wyner–Ziv and the binary Gelfand–Pinsker problem (see also Hussami et al. [15] and Korada [16, chapters 3 and 4]).

The above results were related to the case where the representation alphabet is binary. In [57], Karzand generalized these results to the case where the representation alphabet is q-ary, for q any prime number.

In [58], Ye and Barg showed that polar codes can be used to achieve the rate-distortion functions in the problem of hierarchical source coding also known as the successive refinement problem. Also, they analyzed the distributed version of this problem and constructed a polar coding scheme that achieved the rate distortion bound for successive refinement with side information.

8.6.2   Asymmetric Channels

Polar coding, as any other linear block code, uses the channel input symbols with equal frequency, and cannot achieve the true Shannon capacity if the capacity-achieving input distribution is nonuniform.

A pragmatic approach to fix this problem is to follow the approach in Gallager [59, p. 208], where the code is constructed in an extended alphabet with uniform distribution over the extended alphabet and the symbols of the extended alphabet are mapped to the channel input alphabet in a many-to-one manner so as to induce the desired channel input distribution either exactly or approximately. This approach was followed in [16, p. 74] and [45]. While this method provides a simple way of approximating the optimal input distribution, the decoding has to be carried out over the extended alphabet, which may increase the decoding complexity excessively.

Honda and Yamamoto [60] proposed an alternative approach in which the optimal nonuniform input distribution was generated by randomizing symbols in the frozen bits with an appropriate probability distribution. This approach combined source polarization and channel polarization and provided a solution fully within the framework of polar coding. Furthermore, as reported in [60], this method provided a complexity advantage over the alphabet-extension method.

8.6.3   Universal Polar Codes

Polar codes are not universal in the sense that the code construction depends on the channel transition probabilities. The construction is tailored for the optimal performance under successive cancellation decoding and a polar code designed for one channel may not work well under successive cancellation decoding on another channel of the same capacity. It is noted by Şaşoğlu [61] that this non-universality is an artifact of the suboptimality of the successive cancellation decoding method. Specifically, it was shown in [61, pp. 86–89] that a code that works well on a BSC under any type of decoder will work at least nearly as well under ML decoding on any other binary-input channel of the same capacity. It follows from this general result that there exist universal polar code constructions under ML decoding. However, due to practical reasons, it is still desirable to find polar code constructions that are universally good even under successive cancellation decoding.

In [62], Hassani and Urbanke presented two alternative polar coding schemes that are capable of achieving the compound capacity of the whole class of binary input memoryless symmetric channels with low complexity. In the first scheme, N polar code blocks of length N are stacked up on top of each other in such a way that each block is shifted with respect to the previous block so as to form a staircase. The columns of this staircase are encoded with a Reed–Solomon code. This scheme achieves the compound capacity using a successive cancellation decoder to decode the rows followed by a Reed–Solomon decoder to decode the columns. The second universal polar code construction scheme presented in [62] uses a method of chaining successive blocks of polar codes that achieves the compound capacity of any two channels. This second method is more in the spirit of polarization in that it uses only polar codes.

In [63], Şaşoğlu and Wang presented an alternative universal polar code construction which is based on combining channels of unequal capacities in each polarization step, as opposed to the standard method of combining identical channels. This yields a simple method to design universal polar codes for discrete memoryless channels.

While all three methods above achieve universality in channel coding with polar codes under successive cancellation decoding, the length of the overall construction is significantly increased relative to the case where there is only one channel.

In [64], Ye and Barg proposed a universal source polarization scheme that universally achieves the smallest possible compression rate for a class of s/o pairs with side information. They then applied this result to a joint source-channel coding problem over a broadcast channel.

8.6.4 Mismatched Decoder

In practice, there is usually some degree of mismatch between the true channel and the channel model used by a decoder. This mismatch may be due to quantization of channel outputs or other computational shortcuts in order to reduce complexity. The performance of polar codes under mismatched decoding has been studied by Alsan and Telatar [65] and Alsan [66]. An interesting finding in these references is that the mismatched decoding capacity is created by polarization in the sense that the sum of the mismatched decoder capacities of the synthetic channels $\{W_N^{(i)}\}_{i=1}^N$ created by polarization from N independent copies of a channel W is lower bounded by N times the mismatched decoding capacity of W.

8.7 Polar Coding in Multiuser Settings

There is already a large body of research on the possible use of polar coding for a wide range of scenarios involving multiple users. In this section, we will review some of those results. All polar coding schemes mentioned in this section have the low-complexity features of polar coding and provide reliable communication with a probability of error decreasing to zero roughly as $2^{-\sqrt{N}}$ in the blocklength N of the coding schemes. In this sense, polar coding is a candidate enabling technology for bridging the gap between theory and practice for a wide range of multiuser communication scenarios.

Due to space limitations, we will not describe details of the information theoretic problems discussed below; instead we will refer to the relevant chapter of El Gamal and Kim [67].

As general references on polar coding and their use in multiuser information theoretic settings, we refer to the theses by Korada [16], Şaşoğlu [61], Goela [68], Blasco-Serrano [69], Sahebi [70], Wang [71], Gulcu [72], and Mondelli [73].

8.7.1 Source Coding with Side Information

Lossless source coding with side information, also known as the Slepian–Wolf problem [74], [67, chapter 10], was first studied for polar codes by Hussami et al. in [15] (see also [16, pp. 69–72], [56]). These works showed that polar codes achieve optimum performance for the Slepian–Wolf problem in a restricted setting where the source X and side information Y are related to each other by $Y = X + Z$ (addition mod-2) where X and Z are independent 0–1 random variables with $X \sim \text{Ber}(\frac{1}{2})$ and $Z \sim \text{Ber}(p)$. In [9], this result was generalized to the case of arbitrary binary s/o pairs (X, Y) subject only to a symmetry constraint.

The above works achieved the Slepian–Wolf region by using time-sharing between the corner points of the achievable rate region. In [75], a method based on "monotone chain rules" was introduced that achieved the full admissible rate region in the Slepian–Wolf problem without time-sharing. A practical implementation of the monotone-chain-rule method was given by Onay [76].

In [77], Bhatt et al. applied the monotone-chain-rule technique to the multiple descriptions problem to show that polar coding can achieve the El Gamal–Cover inner bound. In an earlier work [78], Salamatian et al. used the monotone-chain-rule method to construct low-complexity Slepian–Wolf codes in a network setting.

8.7.2 Multi-Access Channel

The multi-access channel model [67, chapter 4] is a common scenario for wireless communications and is becoming more important as wireless systems move in the direction of nonorthogonal multi-access. In [79], Şaşoğlu et al. extended the polar coding method to the two-user multiple-access channel (MAC) and showed that if the polarization transform is applied to the two users separately, the resulting channels polarize to one of five possible extremals, on each of which uncoded transmission is optimal. In [80], Abbe and Telatar generalized the preceding result to the m-user MAC with binary inputs and showed that the polarization technique applied individually to each user polarizes any m-user binary input MAC into a finite collection of extremal MACs. In general, the *direct* polarization methods used in [79] and [80] were able to achieve the entire MAC capacity region only by time-sharing. In [75], a new method, which may be called "polar rate-splitting," was introduced that was capable of achieving the entire MAC capacity region without using time-sharing. In more recent work [81], Nasser and Telatar derived necessary and sufficient conditions that characterize all MACs that do not lose any part of their capacity region under the direct polarization method of [79] and [80].

8.7.3 Broadcast Channels

Broadcast channels are models for one-to-many communications, a natural scenario in a wireless setting. For details of various broadcast channel models and corresponding coding strategies, we refer to [67, chapters 5 and 8].

In [82], Goela et al. considered polar coding for discrete memoryless broadcast channels. For m-user deterministic broadcast channels, they showed that polar coding can achieve rates on the boundary of the private-message capacity region. For two-user noisy broadcast channels, they gave polar implementations of Cover's superposition codes and Marton's codes. They imposed degradedness constraints on the auxiliary and channel-input distributions in order to ensure proper alignment of polarization indices in the multiuser setting, and showed that polar codes could achieve rates on the capacity boundary of binary-input stochastically degraded broadcast channels.

In [83], Mondelli et al. were able to construct coding schemes that achieved any rate pair inside the region defined by Cover's superposition and Marton's binning strategies. Unlike [82], they did not make any degradedness assumptions, but instead considered the transmission of k blocks and employed chaining constructions that guaranteed the proper alignment of polarized indices. In [84], Mondelli et al. extended this method to achieve Marton's region with both common and private messages.

8.7.4   Relay Channels

In wireless networks, relaying may be employed to help increase the range of transmissions or improve the spectral efficiency per area by improving connectivity. As wireless networks become denser and transmit powers have to be curbed to control interference, it is likely that relays will play a greater role in the wireless network architecture. Relaying has been studied extensively in multiuser information theory and several coding strategies, such as decode-and-forward (DF) and compress-and-forward (CF), have been developed [67, chapter 16]. In this part, we mention several studies that show that polar coding can be useful for implementing relay coding strategies.

In [85], Andersson et al. considered polar coding schemes for symmetric discrete memoryless relay channels with orthogonal receiver components under the assumption that the source-to-destination channel is a physically degraded version of the source-to-relay channel. The channel inputs were assumed to be binary. They showed that DF relaying could be implemented by polar coding at low complexity and at rates approaching the information theoretic limits. The codes they constructed were nested polar codes in the sense that the source-to-destination polar code was a subcode of the source-to-relay node. The assumption of degradedness was used to help construct nested polar codes.

In [86], Blasco-Serrano et al. considered a relay model with orthogonal receiver components and with binary channel inputs. They described two polar coding schemes that implemented CF relaying. The first scheme was based on Slepian–Wolf coding, the second on Wyner–Ziv coding. They showed that the proposed schemes achieved the cut-set bound and the binary symmetric CF rate under some special conditions.

In a follow-up work [87], Blasco et al. generalized the results of [85] and [86] to relay channels with non-binary inputs.

In [88], Karzand considered polar coding for binary-input symmetric degraded relay channels. A novelty of [88] was the use of polar codes to implement block Markov coding at low complexity.

In more recent work, Wang [89] developed polar coding schemes for DF and CF relaying for the general relay channel (without any assumption about orthogonal receiver components and allowing q-ary channel inputs for arbitrary q). For DF, she applied a universal polarization technique from [90] to create the desired nested polar code structures. For CF, she generalized existing methods to allow arbitrary input distributions and channel statistics for the relay channel. Both schemes achieved full theoretical rates in general relay channels.

Recently, Mondelli et al. [91] presented a new coding scheme for relays with orthogonal receiver components. The new scheme combined CF and DF methods with a chaining construction over pairs of blocks. They showed that the proposed method could perform strictly better than both DF or CF alone. Furthermore, they showed that the method could be implemented with low-complexity polar coding schemes.

8.7.5   Interference Channel

As networks become denser, it is likely that multi-user interference will emerge as the main limiting factor of throughput per unit area. We refer to [67, chapter 6] for an information theoretic formulation of the interference channels and various coding schemes for such channels. Polar coding has been applied successfully to construct explicit and low-complexity versions of various coding ideas for interference channels. Notably, in [90], Wang and Şaşoğlu constructed a polar coding scheme that was able to achieve the Han–Kobayashi inner bound for two-user interference channels. In [92], Wei and Ulukus developed a polar coding scheme that was able to achieve the best-known inner bounds for interference channels with confidential messages.

8.7.6   Wiretap Channel

Privacy is a key concern, especially in a wireless setting. An information theoretic model for communication under secrecy constraints is the Wyner wiretap channel [67, section 22.1]. Koyluoglu and El Gamal [93], Andersson et al. [85], Hof and Shamai [94], and Mahdavifar and Vardy [95] exploited the nested property of polar codes to construct polar coding schemes that achieved the entire rate-equivocation region for degraded wiretap channels. In [92], Wei and Ulukus extended this result to general wiretap channels by using methods for universal polar coding and polar coding for asymmetric channels. Furthermore, they were able to construct polar codes that achieved the best-known inner bounds for the multiple-access wiretap channel, and the broadcast and interference channels with confidential messages. In [96], Chou and Yener studied the multiple-access wiretap channel without any degradation or symmetry assumptions and showed that any rate pair, known to be achievable with a random-coding like proof, is also achievable with a low-complexity polar coding scheme.

8.7.7   Interactive Computation

Often a communication network serves as an infrastructure to provide connectivity in a distributed system where the main goal of the system is to make decisions and take actions in response to data collected from various parts of the network. Given the growing demand for building intelligent systems using large amounts of data collected from various parts of a network, distributed and interactive computation subject to communications constraints is a subject that is likely to attract more attention. We refer to [67, chapter 21] for a formal definition of such problems from an information theoretic perspective. In [97], Gulcu and Barg studied polar coding for problems of interactive computation of functions by two terminals and interactive computation in a collocated network and showed that the rate regions for both of these problems can be achieved using several rounds of polar-coded transmissions.

8.8   Polar Coding for Various Applications

In this section we provide a brief survey of papers on polar coding for various specific applications. Polar codes do not have an error floor, which makes them attractive for applications that require extreme reliability, such as control channels, optical transport of bulk data, and storage. On the other hand, the sequential nature of decoding algorithms for polar codes make them less attractive where extreme throughput is required.

Here, we collect results about polar coding for applications. The emphasis is on finite-length performance for various application scenarios and the success criterion is performance compared to alternative codes, notably LDPC and turbo codes.

The section begins with a discussion of methods to improve the finite-length performance of polar codes, most notably, list-decoding for polar codes. We then consider polar coding in combination with modulation, in particular how to combine polar coding with multilevel modulation (MLM) and bit-interleaved coded modulation (BICM).

8.8.1   Modulation

Polar coding has been studied in connection with various modulation schemes, using both the Imai–Hirakawa [98] multilevel coding (MLC) and the BICM methods. In [99] and [100], Seidl et al. considered joint optimization of binary polar coding and 2^m-ary digital pulse-amplitude modulation (PAM) schemes. They noted that polar coding and Imai–Hirakawa MLM are conceptually equivalent and developed optimum labeling methods in coded modulation.

In [101], Chen et al. studied BICM together with polar coding. They reported that, with CRC-aided list-decoding, their methods achieve better performance than the turbo-coded modulation scheme used in WCDMA (wideband code division multiple access) wireless communication systems by up to 0.5 dB.

In [102], Chen et al. studied a 2^m-ary polar coded modulation (PCM) scheme, in which they regarded the bit-to-symbol mapping as part of the polar transform. They found that the performance can be significantly improved by optimizing the labeling rule. They gave simulation results for 8-ary PAM over AWGN channels that showed that their proposed method can outperform the bit-interleaved turbo coded modulation scheme used in the WCDMA mobile communication systems by up to 1.5 dB.

In [103], Mahdavifar et al. studied polar coding with BICM. Instead of encoding the multiple channels created by BICM separately, they exploited the recursive structure of polar codes to construct a unified compound polar code for the multiple channels with a single encoder and decoder. They proved that this scheme achieves the multichannel capacity with the same error decay rate as in standard polar coding. They noted that the use of a single polar code (as opposed to multiple shorter polar codes) improves the performance.

In [104], Afser et al. considered a polar coding scheme for BICM in which they used list decoding with CRC. They evaluated their designs with various mapping schemes for 16-QAM modulation. They reported that their method provides significant advantages over the BICM schemes implemented with other well-known codes for moderate blocklengths.

In [105], Böcherer et al. presented a low-complexity algorithm for constructing polar codes for higher-order modulation based on information theoretic principles. They reported simulation results showing that polar codes constructed with their method can provide 1 dB better performance compared to the state-of-the-art AR4JA low-density parity-check codes.

8.8.2   Lattices

In [106], Liu et al. proposed a new class of lattices constructed from polar codes – which they call polar lattices – that achieve the capacity $\frac{1}{2}\log(1+\text{SNR})$ of the AWGN channel. Their construction retains the $O(N \log N)$ encoding and decoding complexity of polar codes. As an extension of [106], Liu and Ling [107] showed that polar lattices achieve the rate-distortion bound of a memoryless Gaussian source. In [108], Yan et al. used a polar lattice construction that achieves the secrecy capacity under the strong secrecy criterion over a Gaussian wiretap channel. In [109], Liu and Ling considered polar codes and polar lattices for independent fading channels when the channel state information is only available to the receiver. For the binary input case, they proposed a new design of polar codes through single-stage polarization to achieve the ergodic capacity. For the non-binary input case, they showed how to achieve the ergodic Poltyrev capacity when there is no power limit and the ergodic capacity when there is a power limit.

8.8.3   Parallel Channels, Multicarrier Systems

In [110], Hof et al. introduced a capacity-achieving polar coding scheme for communication over a set of parallel channels, where the channels are symmetric, binary-input, memoryless, and form a chain under degradation order, but are assumed to be arbitrarily permuted. In [111], this result was generalized to the case where the channels are not necessarily degraded with respect to each other.

In [112], Tse et al. proposed a polar coding scheme for parallel Gaussian channels, where the encoder knows the sum-rate of the parallel channels but does not know the capacity of any channel. By using the nesting property of polar codes, they designed a coding/decoding scheme that achieves the sum-rate.

In [113], Wasserman et al. studied polar coding for an orthogonal frequency division multiplexing (OFDM) scheme over a quasi-static multipath fading channel. They investigated the effect of interleaving of polar codeword bits on FER performance.

In [114], Mei et al. considered design of polar codes on impulsive noise channels for single-carrier and multicarrier systems. They used density-evolution to obtain tight bounds on FER for single-carrier systems. For multicarrier OFDM systems, they obtained a lower bound on FER by assuming that the noise on each subcarrier is Gaussian.

8.8.4 Optical Channels

As pointed out above, the lack of an error floor makes polar codes attractive for optical transport applications where extreme reliability is required. In order to overcome the throughput limitations of sequential decoders, BP decoding and other heuristic methods for speeding up the decoder have been tried.

In an early work [115] on the potential of polar coding for real-world applications, Eslami and Pishro-Nik provided a stopping-set analysis of polar codes under BP decoding and concluded that polar codes show superior error floor performance compared to the conventional capacity-approaching coding techniques. They considered concatenation schemes with polar codes as inner codes for optical transport networks (OTNs) as a potential application. They reported favorable results for polar codes in terms of rate, reliability, and complexity as compared to conventional schemes for OTNs.

In [116], Wu and Lankl reported that polar codes can compete with existing ITU G.975.1 standard forward error correction (FEC) schemes. In follow-up work [117, 118], they considered q-ary polar codes in combination with M-ary pulse-amplitude modulation, where their goal was to achieve simplified mapping and a high degree of flexibility at low to moderate complexity.

In [119], Ahmad performed extensive simulation studies and showed that polar codes under the simple SC decoding outdo all G.975.1 recommended FEC codes in terms of error performance with the same overhead and relatively shorter blocklengths.

In [120], Koike-Akino et al. studied bit-interleaved polar-coded modulation for low-latency short-block communications and showed that polar codes with list+CRC decoding can outperform state-of-the-art LDPC codes.

In [121] and [122], Koike-Akino et al. considered a polar coding scheme for high-throughput optical interfaces. They presented a polar turbo product code (TPC) that is suitable for parallel and pipelined implementation and reported a coding gain of 0.5 dB over BCH-constituent TPC.

8.8.5 Massive MIMO

In [123], Cao et al. considered irregular polar codes for wireless massive multiple-input multiple-output (MIMO) communication channels. They reported simulation results that show significant complexity reduction compared to regular polar codes while also yielding a marginal improvement in error rate performance.

8.8.6 Polar Codes in 3GPP NR

As part of Release 15 of the Third Generation Partnership Project New Radio (3GPP-NR) standardization process, polar codes have become part of the Fifth Generation New Radio (5G-NR) standard [7]. Polar codes have been designated as the channel coding scheme for the control channel of the enhanced mobile broadband (eMBB) service. More specifically, polar codes are going to protect the downlink control information

(DCI), uplink control information (UCI), and the broadcast channel (BCH). The data channels in the eMBB service, namely, uplink shared channel (UL-SCH) and downlink shared channel (DL-SCH), will be protected by LDPC codes. The control channels carry less data but require better protection than the data channels.

Two new services are planned for the future releases of 5G, which are called ultra-reliable low-latency communications (URLLC) and massive machine-type communications (mMTC). Polar coding is likely to play a role in these future releases as well.

A full description of the 5G eMBB polar code can be found in [7]. The polar codes in the 5G standard are designed so that they provide sufficient performance under a CRC-aided list-of-eight decoder. The list size eight is a compromise between performance and complexity. The frozen bits are chosen using a table that is designed to serve multiple code lengths and rates simultaneously. Shortening and puncturing methods are defined so as to adjust the code length to values that are not powers of two. For a tutorial description of the 5G polar code and its design philosophy, we refer to the article by Bioglio et al. [8].

8.8.7   Rateless and Rate-Compatible Codes

A rateless code incrementally sends packets over a channel until the data are correctly decoded. Rateless codes are closely related to rate-compatible codes used in constructing incremental redundancy (IR) hybrid automatic-repeat-request (HARQ) schemes. Such schemes are important in wireless communications where the channel state is unknown at the transmitter.

In [124], Li et al. gave a code based on polar coding that was capacity-achieving in the sense that the achievable rate was as good as the best code specifically designed for the unknown channel. While previous rateless coding schemes were designed for specific classes of channels, the rateless scheme here was capacity-achieving for broad classes of channels as long as they were ordered via degradation.

In [125], Hong et al. gave a method for constructing rate-compatible polar codes that are capacity-achieving at multiple code rates with low-complexity sequential decoders. Their proposed code consisted of parallel concatenation of multiple polar codes, which allowed the construction of lower-rate polar codes by adding more constituent polar codes, and which enabled incremental retransmissions at different rates in order to adapt to channel conditions. They showed that the proposed construction was capacity-achieving for an arbitrary sequence of rates and for any class of degraded channels. In [126], Mondelli et al. extended this construction to the case where no degradation assumption is made.

8.8.8   Joint Source-Channel Coding

Polar coding has attracted significant interest for potential uses as a joint-source channel coding (JSCC) scheme.

In [127] and [128] Wang et al. studied a JSCC scheme that exploits the redundancy of language-based sources during polar decoding. They showed that, by effective use of this redundancy, joint decoding schemes outperform stand-alone polar codes with CRC-aided successive cancellation list decoding by over 0.6 dB.

There have also been studies of JSCC with polar codes for potential Internet of Things (IoT) applications, where the main motivation is to conserve energy and reduce complexity at sensors. Furthermore, in case there are many sensors collecting correlated data, it is important to make use of Slepian–Wolf coding to reduce the traffic in the network. (As already mentioned above, [76] already demonstrated the practical potential of polar codes for Slepian–Wolf coding.)

In [129], Yaacoub and Sarkis proposed a JSCC method in which the source data is systematically encoded using a polar channel code. The transmitter sends only the parity bits (the systematic data bits are not sent). The decoder uses the channel outputs for the parity bits and side information (if available) for the systematic data bits and carries out decoding in the usual manner. A variant of this same method was studied also in [130] by Jin and Yang. The distributed version of the above problems was studied in [131] and [132].

Finally, we mention that JSCC with polar codes has been considered in [133] by Xie et al. for NAND-based solid-state drives (SSDs) to protect flash data. A novelty of their method is to use the non-host data (metadata), which is present for system management, as side information to assist the decoder. The paper reports competitive advantages of polar codes over LDPC codes on flash memories.

8.9 Conclusions

We have given a survey of polarization and polar coding covering both the theoretical and practical aspects of the subject. We believe that several trends are at work that will enhance the role of polar coding in future communication systems. First, there is an ever-present demand for higher data rates while advances in VLSI technology are no longer able to keep up with that demand. Furthermore, for battery powered communication terminals, the slow rate of progress in battery technology necessitates the use of algorithms that consume less energy per bit as the desired data rates keep increasing. Thus, the demand for higher data rates is likely to bring complexity of implementation – as measured by area efficiency (throughput per unit chip area) and energy efficiency (energy per bit) – to the forefront, making it as important as the traditional performance metric of bit error rate vs. SNR. Second, with the densification of mobile devices and base stations, interference will emerge as a major problem. In order to make better use of the congested spectral resources, it is likely that at some point it will become economically viable to implement sophisticated coding ideas for relay and interference channels. From this perspective, communication systems for 5G and beyond offer many challenges and opportunities. Polar codes, with their versatile and low-complexity nature, are likely to be one of the technology alternatives for meeting these challenges.

References

[1] E. Arıkan, "Channel polarization: A method for constructing capacity- achieving codes for symmetric binary-input memoryless channels," *IEEE Trans. Inform. Theory*, vol. 55, no. 7, pp. 3051–3073, 2009.

[2] M. Plotkin, "Binary codes with specified minimum distance," *IRE Trans. Inform. Theory*, vol. 6, no. 4, pp. 445–450, 1960.

[3] I. Reed, "A class of multiple-error-correcting codes and the decoding scheme," *Trans. IRE Prof. Group Inform. Theory*, vol. 4, no. 4, pp. 38–49, 1954.

[4] D. E. Muller, "Application of Boolean algebra to switching circuit design and to error detection," *Trans. IRE Prof. Group Electron. Computers*, vol. EC-3, no. 3, pp. 6–12, 1954.

[5] J. Błasiok, V. Guruswami, P. Nakkiran, A. Rudra, and M. Sudan, "General strong polarization," in *Proc. 50th Annual ACM SIGACT Symp. Theory Comput. (STOC 2018)*, Los Angeles, CA, 2018, pp. 485–492.

[6] J. Błasiok, V. Guruswami, P. Nakkiran, A. Rudra, and M. Sudan, "General strong polarization," *arXiv:1802.02718*, 2018.

[7] 3GPP, "3GPP TS 38.212: Multiplexing and channel coding (Release 15)," 2018.

[8] V. Bioglio, C. Condo, and I. Land, "Design of polar codes in 5G new radio," *arXiv: 1804.04389*, 2018.

[9] E. Arıkan, "Source polarization," in *IEEE Int. Symp. Inform. Theory (ISIT)*, Austin, TX, 2010, pp. 899–903.

[10] E. Arıkan and E. Telatar, "On the rate of channel polarization," in *IEEE Int. Symp. Inform. Theory, 2009 (ISIT 2009)*, Seoul, June 2009, pp. 1493–1495.

[11] S. H. Hassani, R. Mori, T. Tanaka, and R. Urbanke, "Rate-dependent analysis of the asymptotic behavior of channel polarization," *arXiv:1110.0194*, 2011.

[12] E. Şaşoğlu, "Polarization and polar codes," *Found. Trends Commun. Inform. Theory*, vol. 8, no. 4, pp. 259–381, 2012.

[13] M. Alsan and E. Telatar, "A simple proof of polarization and polarization for non-stationary memoryless channels," *IEEE Trans. Inform. Theory*, vol. 62, no. 9, pp. 4873–4878, 2016.

[14] I. Tal, "A simple proof of fast polarization," *arXiv: 1704.07179*, 2017.

[15] N. Hussami, S. B. Korada, and R. Urbanke, "Performance of polar codes for channel and source coding," in *2009 IEEE Int. Symp. Inform. Theory*, Seoul, June 2009, pp. 1488–1492.

[16] S. B. Korada, "Polar codes for channel and source coding," PhD dissertation, École Polytechnique Fédérale de Lausanne, 2009.

[17] H. S. Cronie and S. B. Korada, "Lossless source coding with polar codes," in *Proc. 2010 IEEE Int. Symp. Inform. Theory (ISIT)*, Austin, TX, June 2010, pp. 904–908.

[18] R. Mori and T. Tanaka, "Performance and construction of polar codes on symmetric binary-input memoryless channels," in *2009 IEEE Int. Symp. Inform. Theory*, Seoul, 2009, pp. 1496–1500.

[19] I. Tal and A. Vardy, "How to construct polar codes," *arXiv:1105.6164*, 2011.

[20] R. Pedarsani, S. H. Hassani, I. Tal, and I. E. Telatar, "On the construction of polar codes," in *Proc. 2011 IEEE Int. Symp. Inform. Theory (ISIT)*, Cambridge, MA, 2011, pp. 11–15.

[21] R. Pedarsani, "Polar codes: Construction and performance analysis," PhD dissertation, École Polytechnique Fédérale de Lausanne, 2011.

[22] V. Guruswami and P. Xia, "Polar codes: Speed of polarization and polynomial gap to capacity," *IEEE Trans. Inform. Theory*, vol. 61, no. 1, pp. 3–16, 2015.

[23] B. Geiger and B. C. Geiger, "The fractality of polar and Reed–Muller codes," *Entropy*, vol. 20, no. 1, p. 70, 2018.

[24] M. Bardet, V. Dragoi, A. Otmani, and J.-P. Tillich, "Algebraic properties of polar codes from a new polynomial formalism," in *2016 IEEE Int. Symp. Inform. Theory (ISIT)*, Barcelona, July 2016, pp. 230–234.

[25] C. Schurch, "A partial order for the synthesized channels of a polar code," in *2016 IEEE Int. Symp. Inform. Theory (ISIT)*, Barcelona, July 2016, pp. 220–224.

[26] M. Mondelli, S. H. Hassani, and R. Urbanke, "Construction of polar codes with sublinear complexity," in *2017 IEEE Int. Symp. Inform. Theory (ISIT)*, Aachen, June 2017, pp. 1853–1857.

[27] E. Arikan, "A performance comparison of polar codes and Reed–Muller codes," *IEEE Commun. Lett.*, vol. 12, no. 6, pp. 447–449, 2008.

[28] P. Trifonov, "Efficient design and decoding of polar codes," *IEEE Trans. Commun.*, vol. 60, no. 11, pp. 3221–3227, 2012.

[29] M. Qin, J. Guo, A. Bhatia, A. G. I. Fàbregas, and P. H. Siegel, "Polar code constructions based on LLR evolution," *IEEE Commun. Lett.*, vol. 21, no. 6, pp. 1221–1224, 2017.

[30] S. H. Hassani, K. Alishahi, and R. Urbanke, "Finite-length scaling of polar codes," *arXiv:1304.4778*, 2013.

[31] D. Goldin and D. Burshtein, "Improved bounds on the finite length scaling of polar codes," *IEEE Trans. Inform. Theory*, vol. 60, no. 11, pp. 6966–6978, 2014.

[32] H. D. Pfister and R. Urbanke, "Near-optimal finite-length scaling for polar codes over large alphabets," *arXiv:1605.01997*, 2016.

[33] I. Tal and A. Vardy, "List decoding of polar codes," *IEEE Trans. Inform. Theory*, vol. 61, no. 5, pp. 2213–2226, 2015.

[34] K. Chen, K. Niu, and J. Lin, "List successive cancellation decoding of polar codes," *Electron. Lett.*, vol. 48, no. 9, pp. 500–501, 2012.

[35] K. Niu and K. Chen, "CRC-aided decoding of polar codes," *IEEE Commun. Lett.*, vol. 16, no. 10, pp. 1668–1671, 2012.

[36] K. Niu and K. Chen, "Stack decoding of polar codes," *Electron. Lett.*, vol. 48, no. 12, pp. 695–697, 2012.

[37] I. Dumer, "On decoding algorithms for polar codes," *arXiv:1703.05307*, 2017.

[38] I. Dumer and K. Shabunov, "Recursive list decoding for Reed–Muller codes," *arXiv:1703.05304*, 2017.

[39] E. Arıkan, "A survey of Reed–Muller codes from polar coding perspective," in *2010 IEEE Information Theory Workshop (ITW)*, Cairo, January 2010, pp. 1–5.

[40] E. Arıkan, "Serially concatenated polar codes," *IEEE Access*, 2018. DOI: 10.1109/ACCESS.2018.2877720.

[41] S. B. Korada, E. Sasoglu, and R. Urbanke, "Polar codes: Characterization of exponent, bounds, and constructions," in *2009 IEEE Int. Symp. Inform. Theory*, Seoul, June 2009, pp. 1483–1487.

[42] N. Presman, O. Shapira, S. Litsyn, T. Etzion, and A. Vardy, "Binary polarization kernels from code decompositions," *IEEE Trans. Inform. Theory*, vol. 61, no. 5, pp. 2227–2239, 2015.

[43] H. Lin, S. Lin, and K. A. S. Abdel-Ghaffar, "Linear and nonlinear binary kernels of polar codes of small dimensions with maximum exponents," *IEEE Trans. Inform. Theory*, vol. 61, no. 10, pp. 5253–5270, 2015.

[44] R. Mori and T. Tanaka, "Channel polarization on q-ary discrete memoryless channels by arbitrary kernels," in *2010 IEEE Int. Symp. Inform. Theory*, Austin, TX, June 2010, pp. 894–898.

[45] E. Şaşoğlu, E. Telatar, and E. Arıkan, "Polarization for arbitrary discrete memoryless channels," in *2009 IEEE Inform. Theory Workshop*, Taormina, October 2009, pp. 144–148.

[46] R. Mori and T. Tanaka, "Source and channel polarization over finite fields and Reed–Solomon matrices," *IEEE Trans. Inform. Theory*, vol. 60, no. 5, pp. 2720–2736, 2014.

[47] R. Mori and T. Tanaka, "Non-binary polar codes using Reed–Solomon codes and algebraic geometry codes," *arXiv:1007.3661*, 2010.

[48] R. Mori and T. Tanaka, "Source and channel polarization over finite fields and Reed–Solomon matrix," *arXiv:1211.5264*, 2012.

[49] W. Park and A. Barg, "Multilevel polarization for nonbinary codes and parallel channels," *arXiv:1107.4965*, 2011.

[50] W. Park and A. Barg, "Polar codes for q-ary channels, q=2^r," *arXiv:1107.4965*, 2011.

[51] W. Park and A. Barg, "Polar codes for q-ary channels, q^2r," in *2012 IEEE Int. Symp. Inform. Theory (ISIT)*, Cambridge, MA, July 2012, pp. 2142–2146.

[52] W. Park and A. Barg, "polar codes for q-ary channels," *IEEE Trans. Inform. Theory*, vol. 59, no. 2, pp. 955–969, 2013.

[53] A. G. Sahebi and S. S. Pradhan, "Multilevel polarization of polar codes over arbitrary discrete memoryless channels," *arXiv:1107.1535*, 2011.

[54] N. Presman, O. Shapira, and S. Litsyn, "Polar codes with mixed kernels," in *2011 IEEE Int. Symp. Inform. Theory*, St Petersburg, 2011, pp. 6–10.

[55] F. Gabry, V. Bioglio, I. Land, and J. Belfiore, "Multi-kernel construction of polar codes," in *2017 IEEE Int. Conf. Commun. Workshops (ICC Workshops)*, Paris, 2017, pp. 761–765.

[56] S. B. Korada and R. Urbanke, "Polar codes for Slepian–Wolf, Wyner–Ziv, and Gelfand–Pinsker," in *2010 IEEE Inform. Theory Workshop*, Cairo, 2010, pp. 1–5.

[57] M. Karzand and E. Telatar, "Polar codes for q-ary source coding," in *2010 IEEE Int. Symp. Inform. Theory (ISIT)*, Austin, TX, 2010, pp. 909–912.

[58] M. Ye and A. Barg, "Polar codes for distributed hierarchical source coding," *arXiv:1404.5501*, 2014.

[59] R. G. Gallager, *Information Theory and Reliable Communication*. New York: Wiley, 1968.

[60] J. Honda and H. Yamamoto, "Polar coding without alphabet extension for asymmetric channels," in *2012 IEEE Int. Symp. Inform. Theory (ISIT)*, Cambridge, MA, 2012, pp. 2147–2151.

[61] E. Şaşoğlu, "Polar coding theorems for discrete systems," PhD dissertation, École Polytechnique Fédérale de Lausanne, 2011.

[62] S. H. Hassani and R. Urbanke, "Universal polar codes," in *2014 IEEE Int. Symp. Inform. Theory (ISIT)*, pp. 1451–1455, Honolulu, HI, June 2014.

[63] E. Şaşoğlu and L. Wang, "Universal polarization," *IEEE Trans. Inform. Theory*, vol. 62, no. 6, pp. 2937–2946, 2016.

[64] M. Ye and A. Barg, "Universal source polarization and an application to a multi-user problem," *arXiv:1408.6824*, 2014.

[65] M. Alsan and E. Telatar, "Polarization as a novel architecture to boost the classical mismatched capacity of B-DMCs," *arXiv:1401.6097*, 2014.

[66] M. Alsan, "Channel capacity of polar coding with a given polar mismatched successive cancellation decoder," *arXiv:1610.07297*, 2016.

[67] A. El Gamal and Y.-H. Kim, *Network Information Theory*. Cambridge: Cambridge University Press, 2011.

[68] N. Goela, "Modern low-complexity capacity-achieving codes for network communication," PhD dissertation, University of California, Berkeley, 2013.

[69] R. Blasco-Serrano, "On compression and coordination in networks," PhD dissertation, KTH Royal Institute of Technology, 2013.

[70] A. G. Sahebi and S. S. Pradhan, "Polar codes for some multi-terminal communications problems," in *2014 IEEE Int. Symp. Inform. Theory*, Honolulu, HI, June 2014, pp. 316–320.

[71] L. Wang, "Channel coding techniques for network communication," PhD dissertation, University of California, San Diego, 2015.

[72] T. C. Gulcu, "Design and analysis of communication schemes via polar coding," PhD Thesis, University of Maryland, 2015.

[73] M. Mondelli, "From polar to Reed–Muller Codes: Unified scaling, non-standard channels, and a proven conjecture," PhD dissertation, École Polytechnique Fédérale de Lausanne, 2016.

[74] D. Slepian and J. Wolf, "Noiseless coding of correlated information sources," *IEEE Trans. Inform. Theory*, vol. 19, no. 4, pp. 471–480, 1973.

[75] E. Arıkan, "Polar coding for the Slepian–Wolf problem based on monotone chain rules," in *2012 Proc. IEEE Int. Symp. Inform. Theory*, Cambridge, MA, 2012, pp. 566–570.

[76] S. Onay, "Successive cancellation decoding of polar codes for the two-user binary-input MAC," in *2013 IEEE Int. Symp. Inform. Theory*, Istanbul, July 2013, pp. 1122–1126.

[77] A. Bhatt, N. Ghaddar, and L. Wang, "Polar coding for multiple descriptions using monotone chain rules," in *2017 55th Ann. Allerton Conf. Commun. Control Comput. (Allerton)*, Monticello, IL, October 2017, pp. 565–571.

[78] S. Salamatian, M. Médard, and E. Telatar, "A successive description property of monotone-chain polar codes for Slepian–Wolf coding," in *2015 IEEE Int. Symp. Inform. Theory (ISIT)*, Hong Kong, June 2015, pp. 1522–1526.

[79] E. Sasoglu, E. Telatar, and E. Yeh, "Polar codes for the two-user multiple-access channel," *arXiv:1006.4255*, 2010.

[80] E. Abbe and E. Telatar, "MAC polar codes and matroids," in *Inform.Theory Appl. Workshop (ITA), 2010*, pp. 1–8, San Diego, CA, January 2010.

[81] R. Nasser and E. Telatar, "Fourier analysis of MAC polarization," *arXiv:1501.06076* , 2015.

[82] N. Goela, E. Abbe, and M. Gastpar, "Polar codes for broadcast channels," *IEEE Trans. Inform. Theory*, vol. 61, no. 2, pp. 758–782, 2015.

[83] M. Mondelli, S. H. Hassani, I. Sason, and R. Urbanke, "Achieving the superposition and binning regions for broadcast channels using polar codes," *arXiv:1401.6060*, 2014.

[84] M. Mondelli, S. H. Hassani, I. Sason, and R. L. Urbanke, "Achieving Marton's region for broadcast channels using polar codes," *IEEE Trans. Inform. Theory*, vol. 61, no. 2, pp. 783–800, 2015.

[85] M. Andersson, V. Rathi, R. Thobaben, J. Kliewer, and M. Skoglund, "Nested polar codes for wiretap and relay channels," *IEEE Commun. Lett.*, vol. 14, no. 8, pp. 752–754, 2010.

[86] R. Blasco-Serrano, R. Thobaben, V. Rathi, and M. Skoglund, "Polar codes for compress-and-forward in binary relay channels," in *2010 Conf. Rec. 44th Asilomar Conf. Sig. Syst. Computers (ASILO- MAR)*, Pacific Grove, CA, 2010, pp. 1743–1747.

[87] R. Blasco-Serrano, R. Thobaben, M. Andersson, V. Rathi, and M. Skoglund, "Polar codes for cooperative relaying," *IEEE Trans. Commun.*, vol. 60, no. 11, pp. 3263–3273, 2012.

[88] M. Karzand, "Polar codes for degraded relay channels," in *International Zurich Sem. Commun.*, Zurich, 2012, p. 59.

[89] L. Wang, "Polar coding for relay channels," in *2015 IEEE Int. Symp. Inform. Theory (ISIT)*, Hong Kong, 2015, pp. 1532–1536.

[90] L. Wang and E. Sasoglu, "Polar coding for interference networks," *arXiv:1401.7293*, 2014.

[91] M. Mondelli, S. H. Hassani, and R. Urbanke, "A new coding paradigm for the primitive relay channel," *arXiv:1801.03153*, 2018.

[92] Y. P. Wei and S. Ulukus, "Polar Coding for the general wiretap channel with extensions to multiuser scenarios," *IEEE J. Sel. Areas Commun.*, vol. 34, no. 2, pp. 278–291, 2016.

[93] O. O. Koyluoglu and H. El Gamal, "Polar coding for secure transmission and key agreement," in *2010 IEEE 21st Int. Symp. Personal Indoor and Mobile Radio Communications (PIMRC)*, Istanbul, September 2010, pp. 2698–2703.

[94] E. Hof and S. Shamai, "Secret and private rates on degraded wire-tap channels via polar coding," in *2010 IEEE 26th Conv. Electric. Electron. Eng. Israel (IEEEI)*, Eilat, November 2010, pp. 94–96.

[95] H. Mahdavifar and A. Vardy, "Achieving the secrecy capacity of wiretap channels using polar codes," *IEEE Trans. Inform. Theory*, vol. 57, no. 10, pp. 6428–6443, 2011.

[96] R. A. Chou and A. Yener, "Polar coding for the multiple access wiretap channel via rate-splitting and cooperative jamming," in *2016 IEEE Int. Symp. Inform. Theory (ISIT)*, Barcelona, July 2016, pp. 983–987.

[97] T. C. Gulcu and A. Barg, "Interactive function computation via polar coding," *arXiv:1405.0894*, 2014.

[98] H. Imai and S. Hirakawa, "A new multilevel coding method using error-correcting codes," *IEEE Trans. Inform. Theory*, vol. 23, no. 3, pp. 371–377, 1977.

[99] M. Seidl, A. Schenk, C. Stierstorfer, and J. B. Huber, "Polar-coded modulation," *IEEE Trans. Commun.*, vol. 61, no. 10, pp. 4108–4119, 2013.

[100] M. Seidl, A. Schenk, C. Stierstorfer, and J. B. Huber, "Aspects of polar-coded modulation," in *SCC 2013: 9th Int. ITG Conf. Syst. Commun. Coding*, Munich, January 2013, pp. 1–6.

[101] Kai Chen, Kai Niu, and Jia-Ru Lin, "An efficient design of bit-interleaved polar coded modulation," in *2013 IEEE 24th Ann. Int. Symp. Personal, Indoor, and Mobile Radio Communications (PIMRC)*, London, September 2013, pp. 693–697.

[102] Jiaru Lin, Kai Niu, and Kai Chen, "Polar coded modulation with optimal constellation labeling," in *National Doctoral Academic Forum on Information and Communications Technology 2013*, Institution of Engineering and Technology, 2013, pp. 25–25.

[103] H. Mahdavifar, M. El-Khamy, J. Lee, and I. Kang, "Polar coding for bit-interleaved coded modulation," *IEEE Trans. Veh. Technol.*, vol. 65, no. 5, pp. 3115–3127, 2016.

[104] H. Afser, N. Tirpan, H. Delic, and M. Koca, "Bit-interleaved polar-coded modulation," in *2014 IEEE Wireless Commun. Netw. Conf. (WCNC)*, Istanbul, April 2014, pp. 480–484.

[105] G. Bocherer, T. Prinz, P. Yuan, and F. Steiner, "Efficient polar code construction for higher-order modulation," in *2017 IEEE Wireless Commun. Netw. Conf. Workshops (WCNCW)*, San Francisco, CA, March 2017., pp. 1–6

[106] L. Liu, Y. Yan, C. Ling, and X. Wu, "Construction of capacity-achieving lattice codes: polar lattices," *arXiv:1411.0187*, 2014.

[107] L. Liu and C. Ling, "Polar lattices are good for lossy compression," in *2015 IEEE Inform. Theory Workshop – Fall (ITW)*, Jeju Island, October 2015, pp. 342–346.

[108] Y. Yan, L. Liu, and C. Ling, "Polar lattices for strong secrecy over the mod-Lambda; Gaussian wiretap channel," in *2014 IEEE Int. Symp. Inform. Theory*, Honolulu, HI, June 2014, pp. 961–965.

[109] L. Liu and C. Ling, "Polar codes and polar lattices for independent fading channels," *arXiv:1601.04967*, 2016.

[110] E. Hof, I. Sason, and S. Shamai, "Polar coding for reliable communications over parallel channels," in *2010 IEEE Inform. Theory Workshop*, Dublin, August 2010, pp. 1–5.

[111] E. Hof, I. Sason, S. Shamai, and C. Tian, "Capacity-achieving polar codes for arbitrarily permuted parallel channels," *IEEE Trans. Inform. Theory*, vol. 59, no. 3, pp. 1505–1516, 2013.

[112] D. Tse, B. Li, and K. Chen, "Polar coding for parallel Gaussian channel," *arXiv:1705.07275*, 2017.

[113] D. R. Wasserman, A. U. Ahmed, and D. W. Chi, "BER performance of polar Coded OFDM in multipath fading," *arXiv:1610.00057*, 2016.

[114] Z. Mei, B. Dai, M. Johnston, and R. Carrasco, "Design of polar codes with single and multi-carrier modulation on impulsive noise channels using density evolution," *arXiv:1712.00983*, 2017.

[115] A. Eslami and H. Pishro-Nik, "On finite-length performance of polar codes: Stopping sets, error floor, and concatenated design," *arXiv:1211.2187*, 2012.

[116] Z. Wu and B. Lankl, "Polar codes for low-complexity forward error correction in optical access networks," in *ITG Symp. Proc. Photon. Netw. 15*, Leipzig, June 2014, pp. 1–8.

[117] Z. Wu and B. Lankl, "Probabilistic shaping and polar-coded pulse-amplitude modulation for optical communications," in *Adv. Photon. 2016 (IPR, NOMA, Sensors, Networks, SPPCom, SOF) (2016)*, July 2016. DOI: 10.1364/SPPCOM.2016.SpTu1F.2.

[118] Z. Wu and B. Lankl, "Coded pulse-amplitude-modulation for intensity- modulated optical communications," in *SCC 2017: 11th Int. ITG Conf. Syst. Commun. Coding*, Hamburg, February 2017, pp. 1–6.

[119] T. Ahmad, "Polar codes for optical communications," PhD dissertation, Bilkent University, 2016.

[120] T. Koike-Akino, Y. Wang, S. C. Draper, et al., "Bit-interleaved polar-coded modulation for low- latency short-block transmission," in *Optic. Fiber Commun. Conf. (2017)*, March 2017. DOI: 10.1364/OFC.2017.W1J.6.

[121] T. Koike-Akino, C. Cao, Y. Wang, et al., "Irregular polar turbo product coding for high-throughput optical interface," in *Optic. Fiber Commun. Conf.*, San Diego, CA, March 2018, p. Tu3C.5.

[122] T. Koike-Akino, Y. Wang, D. S. Millar, K. Kojima, and K. Parsons, "Polar coding for multilevel shaped constellations," in *Adv. Photon. 2018 (BGPP, IPR, NP, NOMA, Sensors, Networks, SPPCom, SOF) (2018)*, Zürich, July 2018., p. SpW1G.1.

[123] C. Cao, T. Koike-Akino, Y. Wang, and S. C. Draper, "Irregular polar coding for massive MIMO channels," in *2017 IEEE Global Commun. Conf.*, Singapore, December 2017, pp. 1–7.

[124] B. Li, D. Tse, K. Chen, and H. Shen, "Capacity-achieving rateless polar codes," in *2016 IEEE Int. Symp. Inform. Theory (ISIT)*, Barcelona, July 2016, pp. 46–50.

[125] S. N. Hong, D. Hui, and I. Marić, "Capacity-achieving rate-compatible polar codes," *IEEE Trans. Inform. Theory*, vol. 99, p. 1, 2017.

[126] M. Mondelli, S. H. Hassani, I. Maric, D. Hui, and S. N. Hong, "Capacity- achieving rate-compatible polar codes for general channels," in *2017 IEEE Wireless Commun. Netw. Conf. Workshops (WCNCW)*, Barcelona, March 2017, pp. 1–6.

[127] Y. Wang, M. Qin, K. R. Narayanan, A. Jiang, and Z. Bandic, "Joint source-channel decoding of polar codes for language-based sources," in *2016 IEEE Global Commun. Conf. (GLOBECOM)*, Washington, DC, December 2016, pp. 1–6.

[128] Y. Wang, K. R. Narayanan, and A. A. Jiang, "Exploiting source redundancy to improve the rate of polar codes," in *2017 IEEE Int. Symp. Inform. Theory (ISIT)*, Aachen, June 2017, pp. 864–868.

[129] C. Yaacoub and M. Sarkis, "Systematic polar codes for joint source-channel coding in wireless sensor networks and the Internet of Things," *Procedia Computer Sci.*, vol. 110, pp. 266–273, 2017.

[130] L. Jin and H. Yang, "Joint source-channel polarization with side information," *IEEE Access*, vol. 6, pp. 7340–7349, 2018.

[131] C. Yaacoub and M. Sarkis, "Distributed compression of correlated sources using systematic polar codes," in *2016 9th Int. Symp. Turbo Codes Iterat. Inform. Proc. (ISTC)*, Brest, September 2016, pp. 96–100.

[132] L. Jin, P. Yang, and H. Yang, "Distributed joint source-channel decoding using systematic polar codes," *IEEE Commun. Lett.*, vol. 22, no. 1, pp. 49–52, 2018.

[133] T. Xie, Y. Y. Tai, and J. Zhu, "Polar codes for NAND-based SSD systems: A joint source channel coding perspective," in *2017 IEEE Inform. Theory Workshop (ITW)*, Kaohsiung, November 2017, pp. 196–200.

9 Massive MIMO and Beyond

Thomas L. Marzetta, Erik G. Larsson, and Thorkild B. Hansen

Massive MIMO is now a commercial reality, and in five years or so it should be widely utilized in its preferred, large-scale, time division duplex (TDD) form. Together with millimeter wave, Massive MIMO dominates the fifth generation of wireless technology (5G). Particularly in the scarce and costly sub-5 GHz bands, Massive MIMO promises to deliver area spectral efficiency (bits/second/hertz/square-kilometer) improvements over 4G ranging from 10 to over 1000, depending on the mobility of terminals. Other benefits include energy efficiency (bits/joule) gains in excess of 1000, and simple and effective power control that yields uniformly great service throughout the cell. This chapter addresses an existential question confronting the wireless industry and the academic research community: Is Massive MIMO the ultimate physical layer technology, or is it possible that there may be something much better? The implications of this question are huge. Should Massive MIMO represent the end of the line for wireless communication theory and practice, then all further progress would entail the adaptation of existing principles to ever-shorter wavelengths. On the other hand, if some entirely new physical layer principles could be identified and exploited, then the wireless industry would experience still another rebirth, and wireless researchers would have a brand new set of problems to solve.

The chapter is organized in three sections. The first section summarizes the state of the art in Massive MIMO and the ultimate scalability limit of the technology, which is directly related to the fundamental need for acquisition of channel state information (CSI). The second section is a critique of present-day wireless information theory: Essentially there has been little meaningful progress since the theory of the MIMO broadcast channel was developed, and there are no signs of further progress. The third section presents elements of electromagnetic theory that, when fused with communication theory, could possibly result in breakthroughs in wireless technology. This theory is presented in a compact form that should be readily understandable by communication theorists.

Section 9.1 delineates the essential features of Massive MIMO: large numbers of individually controlled service antennas, two-way communication with a smaller number of autonomous terminals via aggressive multiplexing over the same time/frequency resources, and the acquisition of CSI by the service antennas through direct measurements. Cellular deployments using concentrated antenna arrays are now well understood, and emphasis has shifted to decentralized (cell-free) Massive MIMO which promises greater shadow-fading diversity and multiplexing ability under line-of-sight conditions. As always, ultimate scalability is limited by the need for acquiring CSI.

The main point of Section 9.2 is that all of the powerful Shannon theory for wireless communications assumes that perfect CSI is available at no cost, wherever needed. But the absence of a-priori CSI is the fundamental feature of dynamic wireless systems. To better direct future research efforts, we present a clean, well-defined model for communication between the service antennas and the terminals: block-fading i.i.d. Rayleigh propagation, time-division duplex operation, and no prior small-scale CSI. The known information theory falls considerably short of calculating the simultaneous uplink/downlink achievable rate region for the terminals. Massive MIMO provides achievable lower bounds, but there is currently no tight upper bound. All current insight suggests that a tight upper bound would be much closer to the lower bound than to the perfect-CSI upper bound. As it stands, Shannon theory is unlikely to yield any new breakthroughs in wireless technology without a drastic reformulation of the wireless system model.

All of today's practical wireless systems rely on exceedingly elementary models for the function of antennas and the propagation of signals, and Maxwell's equations play only a peripheral role within the theory that underlies these systems. In Section 9.3, we present rigorous physical modeling of the wireless system through electromagnetic theory, which, in turn, will necessitate the formulation and development of commensurate communication theory. A full unification of electromagnetic theory and communication theory has never been achieved, it presents fascinating new research problems, and it has a chance of yielding breakthroughs in wireless technology. To underscore this point, we discuss two highly non-intuitive macrophenomena that rely on electromagnetic principles which are unknown to communication theorists: resonant evanescent wave coupling and super-directivity.

9.1   Massive MIMO

9.1.1   What Is Massive MIMO?

The essence of Massive MIMO is a base station comprising a numerically large array of low-power, individually controlled, physically small service antennas. The key characteristics and benefits of Massive MIMO technology are:

1. The use of *measured, rather than assumed, channels*. Operation in time-division duplex mode, and the exploitation of uplink–downlink reciprocity, enables the measurement of channel responses on the uplink for the subsequent use of these estimated responses in the downlink precoding. The absence of a-priori assumptions concerning sparsity or covariance structure implies seamless adaptation to propagation conditions, ranging from line-of-sight to rich scattering. Approaches based on grid-of-beams or feedback of quantized CSI (in frequency-division duplexing mode) can never compete in performance with the use of measured channels – as shown both theoretically [1, #3.12] and experimentally [2].

2. *Spatial multiplexing of many terminals* in the same time–frequency resource. This is possible as long as the channels to different terminals are sufficiently distinct, which has been shown to hold in diverse environments both theoretically [3, chapter 7] and experimentally [4–6]. Another benefit of extreme spatial multiplexing is increased macro-diversity, which permits the terminals to find the base station that offers the best signal plus interference to noise ratio (SINR). Furthermore, the many spatial degrees of freedom facilitate efficient algorithms for the resolution of collisions that may appear during initial access [7–9].

3. *Array gain*, resulting, in principle, in a coverage enhancement in closed-loop operation of $10 \log_{10}(M)$ dB, with M being the number of base station antennas. The array gain stems from coherent superposition of the signals transmitted by the service antennas – these signals add up constructively on the spots where the terminals are located. A similar array gain occurs on the uplink.

4. *Channel hardening*, effectively removing the effects of small-scale fading, and rendering the effective channel seen by each terminal deterministic and frequency-(subcarrier-) independent. This greatly improves reliability, simplifies resource allocation tasks, and cuts down latency. The hardening results from the law of large numbers, as the effective channel gain associated with each terminal comprises a sum of many fading coefficients.

5. A *reduction in requirements on hardware accuracy and resolution of analog–digital converters*, facilitating the use of low-complexity transceivers. Effects from nonlinearities and other impairments can in many cases aggregate non-coherently, while the useful signals aggregate coherently [10–12].

9.1.2 Models for Massive MIMO

A canonical sampled, complex-baseband model for radio propagation is shown in Figure 9.1. An antenna transmits the sample, x, and another antenna receives the signal,

$$y = \sqrt{\rho}gx + w, \tag{9.1}$$

where ρ represents the signal to noise ratio (SNR) and w is noise. The gain, g, factors as follows:

$$g = \sqrt{\beta}h, \tag{9.2}$$

where β is a deterministic constant (assumed known to all parties) that represents path loss including shadow fading, while h is random (assumed unknown a priori to everyone) and represents small-scale fading. The small-scale fading, in turn, is modeled as Rayleigh fading, $h \sim CN(0, 1)$. We assume that two or more small-scale fading

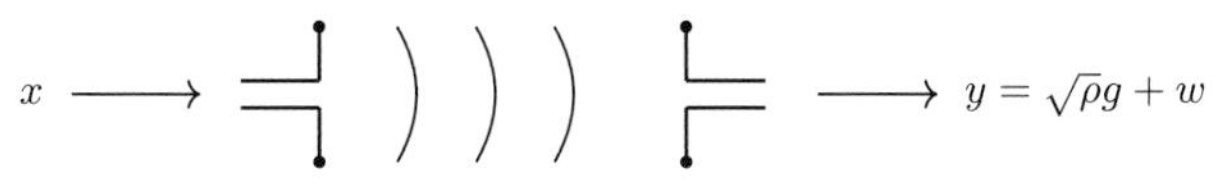

Figure 9.1 Basic wireless communications model.

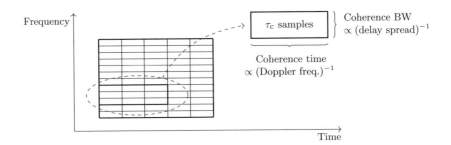

Figure 9.2 Channel coherence: Small-scale fading is piece-wise constant over time–frequency blocks comprising τ_c samples.

coefficients are jointly independent. The noise is Gaussian, $w \sim CN(0, 1)$, and different realizations are jointly independent as well as independent of all other random variables that enter the model. The deterministic constants, ρ_{ul} and ρ_{dl}, denote expected SNRs on uplink and downlink respectively, when full power is applied to a single transmitter for a large-scale fading coefficient of unity.

As illustrated in Figure 9.2, the time–frequency plane is divided into coherence intervals, each of duration τ_c samples. Within a coherence interval, the small-scale fading is constant, but independent from one coherence interval to another. This constitutes a block-fading model. In turn, τ_c is equal to the dimension of the channel coherence interval; that is, the coherence time in seconds multiplied by the coherence bandwidth in hertz [3, section 2.1].

9.1.3   Cellular Massive MIMO

The cellular Massive MIMO concept was originated in [13], and is described in refined forms in [3, 14, 15]. Figure 9.3 illustrates the concept: A multiplicity of cells cover a designated area, and in each cell an array of M co-located antennas serves K single-antenna terminals. The service antennas in each cell are close together, so the path loss between terminal k in cell l' and the base station in cell l, denoted $\beta^l_{l'k}$ here, does not depend on the antenna index, m. Different base stations do not cooperate, other than for pilot sequence assignment and power control – tasks that are only performed when the large-scale fading changes.

Cellular Massive MIMO in its very basic form is well understood, and system simulations based on capacity lower bounds have demonstrated its potential in various scenarios [3, chapter 6]. The main results include rigorous lower bounds on ergodic capacity, valid for any M and K, for multiple-cell systems – taking into account interference, channel estimation errors, power control, and the effects of pilot reuse. Initial field trials have lent credibility to these simulations [16].

In brief, the ergodic capacity for the kth terminal is lower bounded by

$$\alpha \log_2(1 + SINR), \tag{9.3}$$

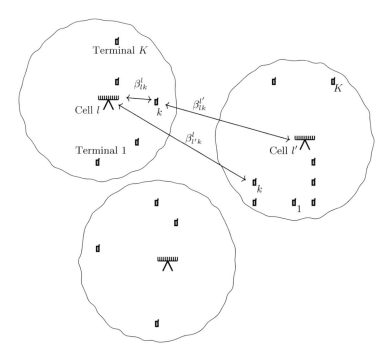

Figure 9.3 Cellular Massive MIMO: Each cell is served by a concentrated array of antennas; large-scale fading between every base station array and every terminal is assumed known.

where α represents the effective fraction of the channel coherence time spent on payload transmission (see also Section 9.1.5) and SINR is an *effective SINR*, given by

$$\text{SINR} = \frac{\mathcal{G}\rho_{\text{ul}}\gamma_{lk}^l\eta_{lk}}{1 + \rho_{\text{ul}}\sum_{l'\in\mathcal{P}_l}\sum_{k'=1}^K \zeta_{l'k'}^l\eta_{l'k'} + \rho_{\text{ul}}\sum_{l'\notin\mathcal{P}_l}\sum_{k'=1}^K \beta_{l'k'}^l\eta_{l'k'} + \mathcal{G}\rho_{\text{ul}}\sum_{l'\in\mathcal{P}_l\setminus\{l\}}\gamma_{l'k}^l\eta_{l'k}} \tag{9.4}$$

on the uplink, and

$$\text{SINR} = \frac{\mathcal{G}\rho_{\text{dl}}\gamma_{lk}^l\eta_{lk}}{1 + \rho_{\text{dl}}\sum_{l'\in\mathcal{P}_l}\zeta_{lk}^{l'}\sum_{k'=1}^K \eta_{l'k'} + \rho_{\text{dl}}\sum_{l'\notin\mathcal{P}_l}\beta_{lk}^{l'}\sum_{k'=1}^K \eta_{l'k'} + \mathcal{G}\rho_{\text{dl}}\sum_{l'\in\mathcal{P}_l\setminus\{l\}}\gamma_{lk}^{l'}\eta_{l'k}} \tag{9.5}$$

on the downlink. In (9.4) and (9.5), the following quantities are introduced:

- $\mathcal{P}_l$ is a set comprising the indices of the cells that use the same pilot sequences as the *l*th cell – including cell *l* itself.
- The gain $\mathcal{G}$ is given by

$$\mathcal{G} = \begin{cases} M, & \text{for maximum-ratio processing,} \\ M - K, & \text{for zero-forcing processing.} \end{cases} \tag{9.6}$$

- The mean-square of the channel estimate (not to be confused with the mean-square channel estimation error) is

$$\gamma_{l'k}^l = \frac{\tau_p \rho_{ul} \left(\beta_{l'k}^l\right)^2}{1 + \tau_p \rho_{ul} \sum\limits_{l'' \in \mathcal{P}_l} \beta_{l''k}^l}, \qquad l' \in \mathcal{P}_l, \tag{9.7}$$

where τ_p is the number of samples per coherence interval spent on pilots.
- The variable $\zeta_{lk}^{l'}$ is given by

$$\zeta_{lk}^{l'} = \begin{cases} \beta_{lk}^{l'}, & \text{for maximum-ratio processing,} \\ \beta_{lk}^{l'} - \gamma_{lk}^{l'}, & \text{for zero-forcing processing.} \end{cases} \tag{9.8}$$

- The variables $\{\eta_{lk}\}$ are nonnegative power control coefficients that satisfy

$$\begin{cases} \eta_{lk} \leq 1, & \text{on uplink} \\ \sum\limits_{k=1}^{K} \eta_{lk} \leq 1, & \text{on downlink.} \end{cases} \tag{9.9}$$

The numerators of (9.4) and (9.5) have the interpretation of *coherent beamforming gain*. This gain scales with M for maximum-ratio and $M-K$ for zero-forcing processing, reflecting the loss of degrees of freedom associated with the interference nulling performed by zero-forcing processing. The denominators of (9.4) and (9.5) comprise four terms, which are, in order from left to right:

- The first term "1" corresponds to additive receiver noise.
- The second term comprises *noncoherent interference* from cells that use the same pilots as cell l. This term furthermore includes *gain uncertainty*, a term that accounts for the fact that the terminals lack side information on their effective (scalar) channel. This noncoherent interference term is appreciably smaller for zero-forcing processing than for maximum-ratio processing; in fact, if zero-forcing had access to perfect CSI, it would disappear (except for the contribution from gain uncertainty).
- The third term comprises noncoherent interference from cells that use pilots different from those used in the home cell. Importantly, the magnitude of this term is the same regardless of whether maximum-ratio or zero-forcing processing is used. In practical multicell deployments, this term often dominates [3, section 6.3].
- The fourth term comprises *coherent interference*, which is beamformed – and therefore amplified by G – owing to pilot reuse.

Significantly, pilot reuse results in *pilot contamination*, which has two consequences: a decrease of the coherent gain (SINR numerator) resulting from increased channel estimation error, as seen in (9.7) and the appearance of coherent interference; for more detail, see [3, chapter 4]. While (9.4) and (9.5) are obtained under the assumption of synchronized cells, the magnitudes of the four terms (including the coherent interference) are essentially the same under nonsynchronous operation [3, section 4.4.3].

Importantly, the power control coefficients η_{lk} enter linearly into the numerators and denominators of (9.4) and (9.5). This facilitates effective, closed-form solutions for power control [3, chapter 5, and section 6.2.5] – particularly, max–min fairness power control, where every terminal in a cell receives the same quality of service. Obviously, in a single-cell system, the sum over $l' \in \mathcal{P}_l$ reduces to a single term ($l' = l$) and the sums over $l' \notin \mathcal{P}_l$ disappear; in this case there is no interference from other cells.

The "pre-log" factor α that weighs the spectral efficiency is equal to

$$\alpha = \frac{1}{2} \frac{\tau_c - n_{\text{reuse}} \tau_p}{\tau_c}, \tag{9.10}$$

where n_{reuse} is the reuse factor for the pilots. In (9.15), the factor $1/2$ results if we additionally assume that half of each coherence interval is spent on uplink and half on downlink.

Ultimately, since τ_c scales inverse proportionally to terminal velocity (Figure 9.2), the greater the mobility, the smaller α. A useful rule of thumb is that no more than half of the coherence interval should be spent on pilots; $\alpha \geq 1/4$.

We next summarize a few more important facts known about cellular Massive MIMO, which are not treated in [3]:

1. While careful power control is exceedingly important [3, chapter 6], judicious assignment of pilot sequences can also improve performance, especially when worst-case performance (max–min quality of service) guarantees are of concern. Optimal assignment entails the solution of combinatorial problems, for which only heuristics are known [17, 18].

2. The basic form of Massive MIMO assumes no beamformed pilots in the downlink. In independent Rayleigh fading, as assumed here, the effective gain of the (scalar) channel seen by each terminal fluctuates marginally around its expected value; the gain uncertainty contribution to the SINR denominator is then negligible. Under other fading models, this effective gain may vary significantly. In either case, each terminal may use a simple blind gain estimation algorithm that requires no downlink pilots to function and which yields performance close to a genie-aided terminal that knows the effective gain [19].

3. The use of different pilot sequences in neighboring cells (pilot reuse factor, n_{reuse}, greater than 1) enables the lth base station to estimate the channels not only to terminals in the lth cell but also to terminals in other cells that use different pilot sequences. The so-obtained estimates may be used to suppress part of the nonco-herent intercell interference through the use of more advanced signal processing at each base station [20–23]. With zero-forcing processing, the coherent gain reduces from $M - K$ to $M - n_{\text{reuse}} K$, rendering the resulting algorithm ineffective unless M is exorbitantly large [22]. Greatly improved performance may be achieved with minimum mean squared error (MMSE) processing [23], although rigorous capacity bound expressions are unavailable in closed form.

9.1.4   Cell-Free Massive MIMO

Cell-free Massive MIMO [24–26] – (Figure 9.4) departs from the cellular Massive MIMO paradigm in three fundamental ways:

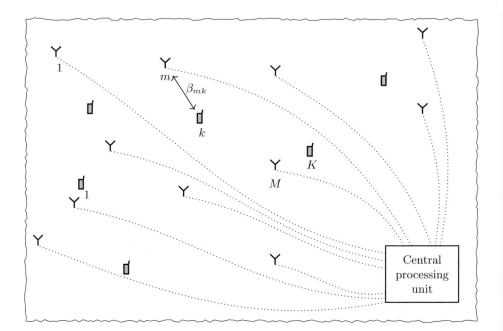

Figure 9.4 Cell-free Massive MIMO: A single distributed antenna array serves all of the terminals over a large area.

1. Each base station (called access point in the cell-free context) has a single antenna, although multiple antenna access points can be treated as well. These access points are spread out over the served area.
2. All access points are connected via backhaul to a central processing unit, and cooperate through coherent signal processing. However, while payload data are shared between the access points, no CSI is shared.
3. In cell-free Massive MIMO, there is no concept of cells; thus all resources, including pilots, are reused universally.

The distinction between cell-free Massive MIMO and the concept of "small cells" [27] is important. Small cells do not cooperate coherently, whereas the access points in cell-free Massive MIMO do.

One main advantage of cell-free Massive MIMO over cellular Massive MIMO is the improved macro-diversity against path loss and shadow fading. Every terminal is likely to be surrounded by many access points to which it has different path loss and shadow fading.

Another advantage of cell-free Massive MIMO over cellular is its superior ability to exploit multiple-antenna terminals under line-of-sight conditions. For a particular terminal, the significant access points occupy a longer baseline, so channels to the different antennas of the terminal can be sufficiently different such that transmission of several independent streams of data to the terminal is feasible under all propagation conditions.

An obvious disadvantage of the cell-free concept is the requirement for back-haul connections, and the possible difficulty of the radio-frequency phase reference distribution.

While cell-free Massive MIMO is not nearly as well understood as cellular Massive MIMO, a number of basic facts are known [25]. We state some of these facts here. In the model given here, β_{mk} represents the path loss and shadow fading between the mth access point and the kth terminal, and as before, ρ_{ul} and ρ_{dl} represent normalized SNRs.

With maximum-ratio processing, a lower bound on ergodic capacity for the kth terminal is given by $\alpha \log_2(1 + \text{SINR})$, where SINR is equal to

$$
\text{SINR} = \frac{\rho_{ul}\eta_k \left(\sum_{m=1}^{M} \gamma_{mk}\right)^2}{\rho_{ul}\sum_{k'=1,k'\neq k}^{K} \eta_{k'} \left(\sum_{m=1}^{M} \gamma_{mk}\frac{\beta_{mk'}}{\beta_{mk}}\right)^2 \phi_{kk'} + \rho_{ul}\sum_{k'=1}^{K}\eta_{k'}\sum_{m=1}^{M}\gamma_{mk}\beta_{mk'} + \sum_{m=1}^{M}\gamma_{mk}} \tag{9.11}
$$

on the uplink, and

$$
\text{SINR} = \frac{\rho_{dl} \left(\sum_{m=1}^{M} \sqrt{\eta_{mk}}\gamma_{mk}\right)^2}{\rho_{dl}\sum_{k'=1,k'\neq k}^{K} \left(\sum_{m=1}^{M} \sqrt{\eta_{mk'}}\gamma_{mk'}\frac{\beta_{mk}}{\beta_{mk'}}\right)^2 \phi_{kk'} + \rho_{dl}\sum_{k'=1}^{K}\sum_{m=1}^{M}\eta_{mk'}\gamma_{mk'}\beta_{mk} + 1} \tag{9.12}
$$

on the downlink. In these equations,

- γ_{mk} is the mean-square channel estimate,

$$
\gamma_{mk} = \frac{\tau_p \rho_{ul} \beta_{mk}^2}{\tau_p \rho_{ul} \sum_{k'=1}^{K} \phi_{kk'}\beta_{mk'} + 1}. \tag{9.13}
$$

- $\phi_{kk'}$ is the magnitude-squared of the inner product between the pilot sequences used by terminals k and k'.
- η_{mk} are nonnegative power control coefficients that satisfy

$$
\begin{cases} 0 \leq \eta_{mk} \leq 1, & \text{on uplink} \\ \sum_{k=1}^{K} \eta_{mk}\gamma_{mk} \leq 1, & \text{on downlink.} \end{cases} \tag{9.14}
$$

- the factor α is equal to

$$
\alpha = \frac{1}{2}\frac{\tau_c - \tau_p}{\tau_c}. \tag{9.15}
$$

The numerators in (9.11) and (9.12) have the interpretation of coherent beamforming gain, similarly to the cellular case. The denominators comprise three terms that originate from coherent interference due to pilot contamination (via the factors $\phi_{kk'}$), noncoherent interference, and thermal noise. Equations (9.11) and (9.12) are strikingly different from their counterparts in cellular Massive MIMO, and more difficult to interpret intuitively. They of course reduce to the formulas for single-cell cellular Massive MIMO in case all access points are co-located ($\beta_{mk} = \beta_k$) and all pilot sequences are orthogonal ($\phi_{kk'} = 0$).

A main distinction from the cellular case is that for the downlink, (9.12), the power control coefficients appear *nonlinearly* in the numerator and the denominator – requiring the use of nonlinear optimization for power control. Fortunately, the max–min fairness power control problem can be equivalently recast as a convex problem and solved to global optimality by using second-order cone programming [25].

With zero-forcing processing, no closed-form lower bound on capacity is known. A rigorous lower bound on capacity, which involves an indefinite expectation, is given in [26].

9.1.5   Ultimate Scalability of Massive MIMO

In the following we speculate on the maximum size of practical Massive MIMO systems. Our discussion assumes only basic Massive MIMO, so our conclusions are very conservative. Other researchers are making strenuous efforts to extend the scalability of Massive MIMO, variously by considering more sophisticated algorithms, or by postulating detailed propagation models to support CSI acquisition. Undoubtedly such measures will help, but in no case have they been proven to provide unlimited extensions of scale.

Scalability of Cellular Massive MIMO

Cellular Massive MIMO is scalable, in two senses. First, nothing precludes a multicell system from growing indefinitely – while additional interference does appear, the new interfering cells are further and further away. Second, although there are physics-based limits on the number of degrees of freedom that can be created in a finite volume (see Section 9.3), the practical number of antennas that can be usefully employed is quite large. In fact, performance monotonically improves with increased number of service antennas, M, and the required processing scales at most proportionally to M.

How many antennas can a cellular Massive MIMO base station usefully deploy? The answer depends on the terminal mobility.

- Current thinking for mobile access in macrocellular environments is that 100–200 antennas would be suitable.[1] This is unlikely to change in the future, because in macrocellular at 3 GHz, with highway-speed mobility, the coherence interval is on the order of 200 kHz × 1 millisecond, that is $\tau_c = 200$ samples. With pilot reuse of $n_{\text{reuse}} = 3$ (that practically does away with pilot contamination), each base station could then ultimately learn the channel to some 30 simultaneously served terminals – assuming mutually orthogonal pilots. Once the number of base station antennas M reaches beyond twice this number, with some margin – say $M = 100$, the spectral efficiency grows only logarithmically with M. Doubling M then yields only a 3 dB

---

[1] It is worth noting that fully digital 64-antenna arrays, such as the Nokia AirScale array, are in commercial service.

effective SINR increase, that is a single extra bit per second/Hz per terminal. Beyond $M = 100$ or $M = 200$, the increased performance may not be worth the cost of the additional antennas. Multiple antennas are only truly useful if they are used to multiplex, and mobility limits the amount of multiplexing we can perform.

Could one quadruple the number of antennas for additional coverage? This may not be worth the effort either. Increasing M from 200 to 2000 gives a 10 dB improvement in array gain. To put this increase of 10 dB in context, note that it is sufficient to achieve a range extension of some 75%. Alternatively, it could be used to overcome one-tenth of the propagation loss incurred by an energy-saving coated window glass (the latter loss is on the order of 20 dB).

- In stationary environments, the situation is different. As a thought experiment, suppose extra antennas and RF chains came at no material cost. How large an array could eventually be useful, and would power consumption eventually render Massive MIMO infeasible?

  The answer is somewhat speculative. One case study in [3, section 6.1] establishes the feasibility of providing (fixed) wireless broadband service to 3000 homes, using a single isolated base station with 3200 antennas (zero-forcing processing and max–min power control). The power consumption of the associated digital signal processing is estimated in [1, #6.6] to less than 500 watts. The service of this many terminals is enabled by the long channel coherence (50 ms in the example).

  Is this as massive as MIMO could ever get? Perhaps not. Conceivably, there will be environments with even larger channel coherence. Consider, for example, an outdoor city square with no cars or other traffic – hence no significant mobility. Eventually only measurements can determine the channel coherence, but assuming for the sake of argument 200 ms by 400 kHz gives room for training of 40,000 terminals (assuming no more than half of resources are spent on training). Multiplexing these terminals would require at least 40,000 antennas, which would, at 3 GHz and half wavelength-spacing, occupy a 10×10-meter area, say with a rectangular array for example – easily integrated onto the face of a large building.

  What gross rate would the base station offer? Assuming, conservatively, 1 bit/s/Hz spectral efficiency, with the uniformly great service design (max–min power control) described in [3], the gross rate in a 25 MHz bandwidth would amount to 1 Tbit/s. How much power would the digital processing require? A back-of-the-envelope calculation along the lines of the homework cited above suggests some tens of kW – the equivalent of a few domestic space heaters.

  How much transmit power is required? The exact value will depend on the coverage area, but to appreciate the order of magnitude, observe that doubling the number of antennas doubles the array gain. If, simultaneously, the number of terminals is doubled, then the total radiated power will be independent of the array size. Hence, the transmitted power is small compared to the power required for processing.

  Machine-to-machine, Internet of Things (IoT), or perhaps virtual-reality-type applications may eventually create the desirability, or need, to build the extreme version of Massive MIMO conceptualized here.

Scalability of Cell-Free Massive MIMO

Cell-free Massive MIMO may be ultimately limited by the scalability of the required signal processing. In its basic form, cell-free Massive MIMO requires all baseband data to be processed in one place, and the amount of data to be collected there grows in proportion to M, resulting in a fundamentally unscalable processing problem.

In addition, scalability of power control is poorly understood. While reasonably efficient algorithms exist that can find optimal power control coefficients in the max–min fairness sense, it is not known how the resulting max–min fairness optimal rate behaves when M and K increase – whether this rate approaches zero or whether it converges to a strictly positive limit. The answer may depend on the precise model applied to generate the locations of the terminals and access points, and the models used for path loss and shadow fading.

Other open questions include the need for frequency and time synchronization. The block fading model assumed here implicitly requires the impulse response of all terminal-to-access-point channels to be shorter than the reciprocal of the coherence bandwidth. In an orthogonal frequency division multiplexing (OFDM) implementation, this means that the impulse responses of channels even to distant access points must fit within the cyclic prefix. That may not hold in practice and refined models may be required.

9.2 Information Theory of Multiuser MIMO

A fundamental question of considerable interest is the following: Subject to the conventional Massive MIMO signal model, are significant improvements possible over the usual pilot-based training and linear precoding? We describe a particular single-cell TDD multiuser MIMO scenario which, in principle, is amenable to a complete analysis through Shannon theory. Because of the limitations of Shannon theory, we must resort to lower and upper bounds on performance.

9.2.1 System Model

The single-cell multiuser MIMO problem is concerned with two-way communication between a base station comprising a concentrated array of M antennas, and K single-antenna autonomous terminals.

We assume the propagation model of Sections 9.1.1 and 9.1.2, specified to a single cell. The propagation channel between the mth base station antenna and the kth terminal is $g_k^m = \sqrt{\beta_k} h_k^m$. The large-scale fading, $\{\beta_k\}$, is constant over an arbitrarily long coding interval, and is known a priori to both the base station and the terminals. The small-scale fading, $\{h_k^m\}$, constitutes independent zero-mean, circularly symmetric, unit-variance complex Gaussian random variables, it is a-priori unknown to everyone and it is piecewise-constant over coherence intervals of τ_c samples. We assume TDD half-duplex operation under which any combination of alternate uplink and downlink transmissions is permitted within the coherence interval.

Within a coherence interval the respective signal model for the uplink and downlink channels is

$$\mathbf{y} = \sqrt{\rho_{\mathrm{ul}}}\mathbf{G}\mathbf{x} + \mathbf{w}, \; \mathrm{E}|x_k|^2 \leq 1, \; k = 1, \cdots, K, \tag{9.16}$$

$$\mathbf{y} = \sqrt{\rho_{\mathrm{dl}}}\mathbf{G}^{\mathrm{T}}\mathbf{x} + \mathbf{w}, \; \mathrm{E}\mathbf{x}^{\mathrm{H}}\mathbf{x} \leq 1, \tag{9.17}$$

where $\mathbf{G} = \mathbf{H}\mathbf{D}_\beta^{1/2}$ is the $M \times K$ propagation matrix, $\mathbf{H}$ is the small-scale fading matrix, $\mathbf{D}_\beta$ is a diagonal matrix with $\{\beta_k\}$ on its diagonal, and $\mathbf{w}$ represents receiver noise consisting of i.i.d., $\mathrm{CN}(0,1)$ random variables. The quantities ρ_{ul} and ρ_{dl} are the expected SNRs that would be experienced at any receiver if full transmit power were applied to exactly one transmit antenna, with the corresponding large-scale fading coefficient equal to 1.

9.2.2   Most General Shannon-Theoretic Question

Channel coding is performed over an arbitrarily large number of coherence intervals, and therefore over many independent realizations of the small-scale fading and the receiver noise. Within a coherence interval, any combination of uplink and downlink transmissions is permitted, in any desired order, subject to the stipulated power constraints. In principle, therefore, Shannon theory is directly applicable, and for a given set of large-scale fading coefficients, the system performance is completely specified by a convex $2K$-dimensional achievable rate region, $\{\mathbf{r}_{\mathrm{ul}}, \mathbf{r}_{\mathrm{ul}}\}$, that specifies the simultaneous net uplink and downlink achievable rates for the K terminals. The achievable rate region is an explicit function of the large-scale fading coefficients.

There is a nontrivial interplay between uplink and downlink transmissions. For a given operating point in the achievable rate region, the coherence interval might well constitute a multiplicity of alternating up- and downlink transmissions, because any transmission, whether data-bearing or an explicit pilot, confers some information about the channel to the receiver. Consider the extreme cases, where either the uplink rates are equal to zero, or the downlink rates are equal to zero. In the former case, it is clear that some fraction of the coherence interval should be expended on uplink pilots, because otherwise the base station would never learn anything about the channel, and therefore could not transmit information selectively to the terminals. In the latter case, it seems unlikely that any downlink transmission would be of assistance, with the possible exception of data feedback. Apart from pilot signals, there is also the possibility that feedback on the reverse channel concerning received data signals could help as well.

At present, there is little hope for the explicit elucidation of the achievable rate region, the coding strategy for attaining capacity, and its scaling with M and K. In the absence of a theoretical breakthrough leading to a grand solution, one must rely on upper and lower bounds on performance.

9.2.3   Capacity Lower Bounds via Massive MIMO Theory

A tractable lower bound on the uplink or downlink achievable rate for a particular terminal is given by (9.3), with the respective SINRs given by (9.4) and (9.5) when specified to the single cell scenario. The achievable rates are functions of the K uplink power control coefficients, the K downlink power control coefficients, and the fractional allocation of the coherence interval to the three transmission activities of uplink pilots, uplink data, and downlink data.

In principle, the individual rates can be translated into achievable rate regions. Given the fractional allocation of the coherence interval for the three transmission activities, the derivation of the uplink and the downlink achievable rate regions occurs independently.

Because of the reliance on orthogonal pilots, the maximum optimized number of terminals, K, is less than half the sample duration of the coherence interval, which itself is inversely proportional to the mobility (e.g., speed) of the terminals. Performance always improves with an increasing number of service antennas, M, but eventually this growth occurs only logarithmically.

9.2.4 Capacity Upper Bounds

Existing capacity upper bounds result from pretending either that perfect CSI is available with no expenditure of resources, or that the terminals can exchange information as desired via some hidden no-cost auxiliary channels.

Perfect CSI Bound

Given perfect CSI (possession by the base station for the uplink, possession by both the base station and the terminals for the downlink), the respective uplink and downlink achievable rate regions are known [28, 29]. The perfect CSI system scalability is in sharp contrast to that of Massive MIMO, in that sum-throughput grows in proportion to $\min(M, K)$, for unlimited M and K. What is needed is a more realistic bound that directly incorporates the a-priori absence of CSI.

Terminal Cooperation Bound

The assumption that the terminals can freely exchange information transforms the multiuser MIMO system into a point-to-point noncoherent MIMO system, where CSI is totally unknown, a priori. We initially investigate the case where the coherence interval is dedicated entirely either to downlink or to uplink transmissions.

Consider downlink transmissions, where, in each coherence interval, we transmit a $M \times \tau_c$ matrix, $\mathbf{X}$, and we receive a $K \times \tau_c$ matrix, $\mathbf{Y}$, where, by assumption, $M > \tau_c$,

$$\mathbf{Y} = \sqrt{\rho_{\mathrm{dl}}} \mathbf{G}^{\mathrm{T}} \mathbf{X} + \mathbf{W}$$
$$= \sqrt{\rho_{\mathrm{dl}}} \mathbf{D}_\beta^{1/2} \mathbf{H}^{\mathrm{T}} \mathbf{X} + \mathbf{W}. \tag{9.18}$$

Following the treatment in [30], we replace $\mathbf{X}$ by its singular value decomposition, to obtain

$$\mathbf{Y} = \sqrt{\rho_{\mathrm{dl}}} \mathbf{D}_\beta^{1/2} \mathbf{H}^{\mathrm{T}} \Phi \mathbf{D}_\mathbf{v} \Psi^{\mathrm{H}} + \mathbf{W}$$
$$= \sqrt{\rho_{\mathrm{dl}}} \mathbf{D}_\beta^{1/2} \acute{\mathbf{H}}^{\mathrm{T}} \mathbf{D}_\mathbf{v} \Psi^{\mathrm{H}} + \mathbf{W}, \tag{9.19}$$

where $\mathbf{\Phi}$ and $\mathbf{\Psi}$ are unitary matrices of dimension $M \times M$ and $\tau_c \times \tau_c$ respectively, and $\mathbf{D_v}$ is a $M \times \tau_c$ positive-real diagonal matrix,

$$\mathbf{D_v} = \begin{bmatrix} \bar{\mathbf{D}}_\mathbf{v} \\ \mathbf{0} \end{bmatrix}. \tag{9.20}$$

In effect, the channel matrix, $\mathbf{H}$ is replaced by a statistically identical "virtual" channel matrix $\acute{\mathbf{H}} = \mathbf{\Phi}^T\mathbf{H}$, which is statistically independent of $\mathbf{D_v}$ and of $\mathbf{\Psi}$. The structure of $\mathbf{D_v}$ implies that only τ_c of the virtual antennas are powered. Thus a statistically identical received signal could be created by an array of only τ_c antennas which is fed by signals which are mutually orthogonal over the coherence interval, $\bar{\mathbf{D}}_\mathbf{v}\mathbf{\Psi}^H$. We conclude that the capacity for $M > \tau_c$ is equal to the capacity for $M = \tau_c$ antennas, so there is no point in using more base station transmit antennas than the sample duration of the coherence interval.

Consider, now, uplink transmission. We further upper bound the performance of this channel by making all of the large-scale fading coefficients equal to their maximum value, $\mathbf{D}_\beta \rightarrow \beta_{\max}\mathbf{I}_K$, and by replacing the power constraint in (9.15) by the less restrictive constraint $\mathrm{E}\mathbf{x}^H\mathbf{x} \leq K$. (It is necessary to equalize the large-scale fading coefficients so that the propagation matrix can absorb the $K \times K$ unitary factor of the transmitted signal.) We can now use the same argument as on the downlink to prove that the capacity for $K > \tau_c$ is equal to the capacity for $K = \tau_c$.

The exact capacity of the noncoherent point-to-point MIMO channel has eluded computation. The asymptotic (high SNR) capacity is known, however, for the case where the large-scale fading coefficients are equal [31]. The asymptotic capacity scales as $L(1 - L/\tau_c)\log_2 \rho$, where $L = \min(M, K, \tau_c/2)$. This expression is substantially in line with the sum-capacity lower bound attained by Massive MIMO, and it falls short drastically from what is promised by the perfect CSI bound.

In the preceding discussion, we assumed that the coherence interval is spent entirely on one-way transmission, either downlink or uplink. More generally a mixture of downlink and uplink transmissions could help things somewhat, particularly if explicit pilots were transmitted in the opposite direction. Consequently, the noncoherent MIMO bounds obtained above are not, strictly speaking, upper bounds. However, if orthogonal pilots are utilized, improvements to the bounds are limited by the constraint that $M + K \leq \tau_c$. There are possible advantages also in the employment of data feedback on the reverse link [32].

9.2.5   Discussion

For the single-cell scenario analyzed in this section, the only absolute upper bound is the perfect CSI bound. We have argued that a tight upper bound, if it could be computed, would be much closer to the lower bound provided by Massive MIMO theory and protocol. We cannot, of course, prove the impossibility of some unforeseen breakthrough that would negate this tentative conclusion.

A number of researchers are making strenuous efforts to improve upon basic orthogonal-pilot Massive MIMO performance. In what follows, we briefly discuss some representative approaches.

Pilot contamination precoding (PCP) mitigates coherent intercell interference and restores logarithmic improvements in throughput with large M, which otherwise would saturate, through coordinated activity over multiple cells based on large-scale fading [33]. For a finite number of cells, PCP enables continued throughput increase with M; however, this has not been proven for an infinite number of cells. (The order in which limits are taken may make a difference.)

Other researchers adopt a more structured model for small-scale fading than i.i.d. Rayleigh. Spatial correlations are exploited to reduce the burden of CSI acquisition [34], in what could be regarded as a generalization of cell sectorization.

A radically different approach is taken in [35], where it is shown that the coherent interference due to pilot contamination does not necessarily grow linearly with M, provided spatial correlations of the channels to the users are sufficiently different. Thus, as in PCP, throughput in a multiple-cell system continues to grow with M. Again, however, this result has not been established for an infinite-cell system.

While these extensions to basic Massive MIMO are undoubtedly useful, the practical scalability that they confer has, so far, been limited.

9.3   Physics-Based Models for Propagation

The previous section suggests that further breakthroughs beyond Massive MIMO may be unobtainable through purely mathematical theories of communication. The present section reviews the physics that may give wireless communication theory a new impetus. We review two solution methods for wave propagation: the plane-wave representation in Cartesian coordinates, and the spherical-wave solution in terms of orthogonal eigenfunctions. We use these tools to explain some non-obvious wireless phenomena: the very old concept of the super-directive array, and the more recently discovered resonant evanescent wave coupling for wireless transfer of power.

9.3.1   Scalar Wave Equation with Spatially Distributed Source

Wave propagation phenomena will be illustrated in this section through sound waves, which qualitatively behave as electromagnetic waves without polarization. Indeed, the formulas we derive all bear close resemblance to their electromagnetic analogs. A general point in space with rectangular coordinates (x, y, z) is given by the vector $\mathbf{r} = x\hat{\mathbf{x}} + y\hat{\mathbf{y}} + z\hat{\mathbf{z}}$ where $\hat{\mathbf{x}}$, $\hat{\mathbf{y}}$, and $\hat{\mathbf{z}}$ are the three orthogonal unit vectors. The length of the vector $\mathbf{r}$ is $r = |\mathbf{r}| = \sqrt{x^2 + y^2 + z^2}$, and we introduce the unit vector $\hat{\mathbf{r}} = \mathbf{r}/r$ which is parallel to $\mathbf{r}$. The results of this section are all given in terms of these vectors. However, in a few places we employ the angles (θ, ϕ) of the spherical coordinate system to express the unit vector as $\hat{\mathbf{r}} = \hat{\mathbf{x}} \sin\theta \cos\phi + \hat{\mathbf{y}} \sin\theta \sin\phi + \hat{\mathbf{z}} \cos\theta$.

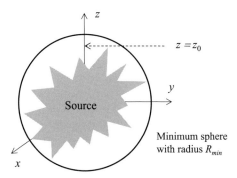

Figure 9.5 The source resides in the half-space $z < z_0$ inside the minimum sphere centered at the origin with radius R_{min}.

The source of the sound wave resides in the half-space $z < z_0$ inside the minimum sphere centered at the origin with radius R_{min}, as shown in Figure 9.5. From first principles, we now show that the pressure of the sound wave satisfies the wave equation. Specifically, we consider a compressible, non-viscous fluid (air) in equilibrium with no shear strength in which the total time-dependent pressure can be expressed as [36, 37]

$$P_T(\mathbf{r}, t) = P_0 + P(\mathbf{r}, t), \tag{9.21}$$

where P_0 is the equilibrium pressure and $P(\mathbf{r}, t)$ the excess pressure caused by the source. The excess pressure is orders of magnitude smaller than the equilibrium pressure for normal sound. Similarly, the total air density is given by $\rho_T(\mathbf{r}, t) = \rho_0 + \rho(\mathbf{r}, t)$, where ρ_0 is the equilibrium density and $\rho(\mathbf{r}, t)$ is the excess density with $\rho_0 \gg |\rho(\mathbf{r}, t)|$. Hence, the total pressure and density are positive, whereas the excess pressure and density can be both positive and negative.

Consider now a small portion of undisturbed air with volume V that is located at $\mathbf{r}$ and then displaced by the disturbance caused by the source, so that at time t it is located at $\mathbf{r} + \boldsymbol{\chi}(\mathbf{r}, t)$ where $\boldsymbol{\chi}(\mathbf{r}, t)$ is the displacement. The volume change of this portion of fluid is denoted by dV and given in terms of the divergence of $\boldsymbol{\chi}(\mathbf{r}, t)$ as $dV = \nabla \cdot \boldsymbol{\chi}(\mathbf{r}, t)\, V$. According to Hooke's law for acoustics, the associated excess pressure is proportional to the relative volume change

$$P(\mathbf{r}, t) = -\kappa \frac{dV}{V} = -\kappa \nabla \cdot \boldsymbol{\chi}(\mathbf{r}, t), \tag{9.22}$$

where the constant κ is the bulk modulus. We can add a term $\kappa S(\mathbf{r}, t)$ to (9.22) that accounts for a source that injects air to get

$$P(\mathbf{r}, t) = -\kappa \nabla \cdot \boldsymbol{\chi}(\mathbf{r}, t) + \kappa S(\mathbf{r}, t), \tag{9.23}$$

which is the first relation needed to derive the wave equation. To get the second relation, we note that the acceleration and mass of the small portion of air are $\frac{\partial^2}{\partial t^2} \boldsymbol{\chi}(\mathbf{r}, t)$ and $V\rho_0$, respectively. Here, the mass due to the excess density $\rho(\mathbf{r}, t)$ is neglected. The force applied to achieve this acceleration can be expressed in terms of the gradient of the excess pressure as $-\nabla \nabla P(\mathbf{r}, t)$, where the minus sign is due to the fact that air moves in the direction of lower pressure. Newton's second law then gives

$$\nabla P(\mathbf{r}, t) = -\rho_0 \frac{\partial^2}{\partial t^2} \chi(\mathbf{r}, t). \tag{9.24}$$

Combining (9.23) and (9.24) produces the inhomogeneous wave equation

$$\nabla^2 P(\mathbf{r}, t) - \frac{1}{c^2} \frac{\partial^2}{\partial t^2} P(\mathbf{r}, t) = -Q(\mathbf{r}, t), \tag{9.25}$$

where $c = \sqrt{\kappa/\rho_0}$ is the wavespeed and

$$Q(\mathbf{r}, t) = \frac{\kappa}{c^2} \frac{\partial^2}{\partial t^2} S(\mathbf{r}, t) \tag{9.26}$$

is the source function. As indicated in Figure 9.5, the source function is zero everywhere in the half-space $z \geq z_0$ and outside the minimum sphere $r = R_{min}$. The operator ∇^2 is given in rectangular coordinates by

$$\nabla^2 = \frac{\partial^2}{\partial x^2} + \frac{\partial^2}{\partial y^2} + \frac{\partial^2}{\partial z^2}, \tag{9.27}$$

and we refer to $P(\mathbf{r}, t)$ as simply the time-domain pressure or time-domain field.

The time-domain pressure $P(\mathbf{r}, t)$ can be expressed in terms of its frequency-domain analog $P_\omega(\mathbf{r})$ through the Fourier-transform relations

$$P(\mathbf{r}, t) = \frac{1}{2\pi} \int_{-\infty}^{\infty} d\omega \, P_\omega(\mathbf{r}) e^{-i\omega t}, \qquad P_\omega(\mathbf{r}) = \int_{-\infty}^{\infty} dt \, P(\mathbf{r}, t) e^{i\omega t}. \tag{9.28}$$

Therefore, if the frequency-domain pressure $P_\omega(\mathbf{r})$ is known, we can determine the pressure $P(\mathbf{r}, t)$ with general time dependence. Inserting the first part of (9.28) into the wave equation (9.25), we obtain the equation that must be satisfied by the frequency-domain pressure

$$\nabla^2 P_\omega(\mathbf{r}) + k^2 P_\omega(\mathbf{r}) = -Q_\omega(\mathbf{r}), \tag{9.29}$$

where $Q_\omega(\mathbf{r})$ is the frequency-domain analog of $Q(\mathbf{r}, t)$, and $k = \omega/c = 2\pi/\lambda$ is the propagation constant with λ being the wavelength. The frequency-domain wave equation (9.29) is also referred to as the inhomogeneous Helmholtz equation. The following sections deal only with frequency-domain fields and therefore omit the subscript ω.

9.3.2   Plane-Wave Solution

We shall now employ the Fourier transform with respect to the spatial coordinates (x, y, z) to obtain a general solution to the Helmholtz equation (9.29). The pressure and source function are first expressed in terms of their spatial Fourier transforms $\tilde{P}(k_x, k_y, k_z)$ and $\tilde{Q}(k_x, k_y, k_z)$ as

$$P(\mathbf{r}) = \frac{1}{(2\pi)^3} \int_{-\infty}^{\infty} \int_{-\infty}^{\infty} \int_{-\infty}^{\infty} dk_x \, dk_y \, dk_z \, \tilde{P}(k_x, k_y, k_z) e^{ik_x x + ik_y y + ik_z z} \tag{9.30}$$

and

$$Q(\mathbf{r}) = \frac{1}{(2\pi)^3} \int\limits_{-\infty}^{\infty} \int\limits_{-\infty}^{\infty} \int\limits_{-\infty}^{\infty} dk_x \, dk_y \, dk_z \, \tilde{Q}(k_x, k_y, k_z) \, e^{ik_xx+ik_yy+ik_zz}. \tag{9.31}$$

Inserting these expressions into the Helmholtz equation (9.29) shows that

$$\left[k^2 - k_x^2 - k_y^2 - k_z^2 \right] \tilde{P}(k_x, k_y, k_z) = -\tilde{Q}(k_x, k_y, k_z), \tag{9.32}$$

which in turn produces the Fourier representation of the pressure

$$P(\mathbf{r}) = -\frac{1}{(2\pi)^3} \int\limits_{-\infty}^{\infty} \int\limits_{-\infty}^{\infty} \int\limits_{-\infty}^{\infty} dk_x \, dk_y \, dk_z \, \tilde{Q}(k_x, k_y, k_z) \frac{e^{ik_xx+ik_yy+ik_zz}}{k^2 - k_x^2 - k_y^2 - k_z^2}. \tag{9.33}$$

To make the Fourier transforms well-behaved, one usually inserts a small loss into the medium so that k is replaced by $k + ik'$, where the loss is accounted for by $k' > 0$ [38]. This loss will ensure exponential decay of outgoing waves ($|e^{i(k+ik')r}|/r = e^{-k'r}/r$) so that integrals over infinite volumes and planes become absolutely convergent. After the derivation is complete, the loss is set to zero to obtain the plane-wave expansion for lossless media. Below we shall briefly discuss some of the difficulties (which are manageable) with spatial Fourier transforms in lossless media.

The k_z-integral in (9.33) can be computed in closed form using residue calculus. First we show that the contribution along the half-circle contour $k_z = \mathcal{K}e^{i\phi}$ with $0 \le \phi \le \pi$ vanishes as $\mathcal{K} \to \infty$. The inverse Fourier integral

$$\tilde{Q}(k_x, k_y, k_z) = \int dV \, Q(\mathbf{r}) \, e^{-ik_xx-ik_yy-ik_zz} \tag{9.34}$$

combined with the fact that $Q(\mathbf{r}) = 0$ for $z > z_0$ proves that $|\tilde{Q}(k_x, k_y, \mathcal{K}e^{i\phi})| \le C_1 e^{\mathcal{K}z_0 \sin\phi}$ where C_1 is a constant. Consequently, the integrand in (9.33) is bounded by $C_2 e^{-\mathcal{K}(z-z_0)\sin\phi}/\mathcal{K}^2$ where C_2 is another constant. These considerations show that the contribution from the half-circle contour in the upper complex k_z-plane vanishes as $\mathcal{K} \to \infty$ when $z > z_0$. Hence, the k_z-integral in (9.33) can be computed by adding the residues due to poles in the upper half of the complex k_z-plane.

The integrand of (9.33) has only one pole in the upper half of the complex k_z-plane given by $k_z = \sqrt{(k + ik')^2 - k_x^2 - k_y^2}$, and the associated residue is readily found to get

$$P(\mathbf{r}) = \frac{i}{2(2\pi)^2} \int\limits_{-\infty}^{\infty} \int\limits_{-\infty}^{\infty} dk_x \, dk_y \, \tilde{Q}(k_x, k_y, \gamma) \frac{e^{ik_xx+ik_yy+i\gamma z}}{\gamma}, \quad z > z_0, \tag{9.35}$$

where the propagation constant in the z-direction is given by

$$\gamma = \begin{cases} |\sqrt{k^2 - k_x^2 - k_y^2}|, & k_x^2 + k_y^2 \le k^2 \\ i|\sqrt{k^2 - k_x^2 - k_y^2}|, & k_x^2 + k_y^2 > k^2 \end{cases} \tag{9.36}$$

after the loss has been set to zero ($k' = 0$). With $k_z = \gamma$, the imaginary part of k_z is always nonnegative for observation points in the half-space $z > z_0$. Inserting the inverse Fourier expression (9.34) into (9.35) gives the final plane-wave expansion for the pressure [39]

$$P(\mathbf{r}) = \frac{1}{2\pi} \int_{-\infty}^{\infty} \int_{-\infty}^{\infty} dk_x \, dk_y \, T(k_x, k_y) \, e^{ik_x x + ik_y y + i\gamma z}, \quad z > z_0, \tag{9.37}$$

where $T(k_x, k_y)$ is the plane-wave spectrum,

$$T(k_x, k_y) = \frac{i}{4\pi\gamma} \int dV \, Q(\mathbf{r}) \, e^{-ik_x x - ik_y y - i\gamma z}. \tag{9.38}$$

The implication of (9.37) is that the entire external effect of the distributed source is the creation of a linear superposition of plane-waves, each of which satisfies the homogeneous Helmholtz equation $(\nabla^2 + k^2)e^{i(k_x x + k_y y + \gamma z)} = 0$. For $k_x^2 + k_y^2 < k^2$, the plane-waves are *propagating*, and for $k_x^2 + k_y^2 > k^2$, the plane-waves are *evanescent*, decaying exponentially fast with increasing z. The evanescent plane waves are needed to capture the rapid variation of the pressure near the source. Also, we note that the spectrum has a $1/\gamma$ singularity where $k_x^2 + k_y^2 = k^2$. This singularity is integrable and can be removed through a change of integration variables. It originates from the fact that the pressure in a plane $z = \text{constant} > z_0$ belongs to neither of the Lebesgue spaces $\mathcal{L}_1$ or $\mathcal{L}_2$ when the medium is lossless; see [40] for a detailed discussion.

The plane-wave spectrum $T(k_x, k_y)$ is determined through (9.38) in terms of the volume-source function $Q(\mathbf{r})$ that resides in the half-space $z < z_0$. An equally important expression is obtained by employing the Fourier transform in two dimensions to (9.37) to get

$$T(k_x, k_y) = \frac{e^{-i\gamma z}}{2\pi} \int_{-\infty}^{\infty} \int_{-\infty}^{\infty} dx \, dy \, P(\mathbf{r}) \, e^{-ik_x x - ik_y y}, \quad z > z_0, \tag{9.39}$$

which determines the plane-wave spectrum in terms of the pressure on any plane perpendicular to the z-axis in the source-free half-space $z > z_0$. Hence, the pressure on a plane acts as an "equivalent planar source" that can replace the original volume source $Q(\mathbf{r})$. This equivalent source is also referred to as a Huygens source. Equations (9.48) and (9.63) represent other examples of field equivalence principles. We note that the spectrum $T(k_x, k_y)$ is independent of z; the z-dependence of the integral over $P(\mathbf{r})$ and of the factor in front cancel each other.

The complex time-average power $\mathcal{P}$ transmitted across a plane perpendicular to the z-axis in the half-space $z > z_0$ can be expressed in terms of the pressure $P(\mathbf{r})$ and the particle velocity $\mathbf{u}(\mathbf{r}) = \nabla P(\mathbf{r})/(i\omega\rho_0)$ as[2]

$$\mathcal{P}(z) = \frac{1}{2} \int_{-\infty}^{\infty} \int_{-\infty}^{\infty} dx \, dy P(\mathbf{r}) \, \hat{\mathbf{z}} \cdot \mathbf{u}^*(\mathbf{r}) = -\frac{1}{2i\omega\rho_0} \int_{-\infty}^{\infty} \int_{-\infty}^{\infty} dx \, dy P(\mathbf{r}) \frac{\partial}{\partial z} P^*(\mathbf{r}), \tag{9.40}$$

$z > z_0$.

---

[2] The particle velocity $\mathbf{u}(\mathbf{r}, t)$ in the time domain is given in terms of the displacement $\boldsymbol{\chi}(\mathbf{r}, t)$ as $\mathbf{u}(\mathbf{r}, t) = \frac{\partial}{\partial t}\boldsymbol{\chi}(\mathbf{r}, t)$, so the frequency-domain analog of Newton's law (9.24) proves $\nabla P(\mathbf{r}) = i\omega\rho_0\mathbf{u}(\mathbf{r})$.

Inserting the plane-wave expansion (9.37) and applying Parseval's theorem gives

$$\mathcal{P}(z) = -\frac{1}{2i\omega\rho_0} \int\limits_{-\infty}^{\infty}\int\limits_{-\infty}^{\infty} dk_x\, dk_y |T(k_x, k_y)|^2 (-i\gamma^*)\, e^{i(\gamma-\gamma^*)z}, \qquad z > z_0. \qquad (9.41)$$

Invoking the expression (9.36) for γ shows that the real and imaginary parts of $\mathcal{P}(z)$ are

$$\mathrm{Re}\,\mathcal{P}(z) = \frac{1}{2\omega\rho_0} \int\limits_{k_x^2+k_y^2\leq k^2}\int dk_x\, dk_y |T(k_x, k_y)|^2 \gamma, \qquad z > z_0 \qquad (9.42)$$

and

$$\mathrm{Im}\,\mathcal{P}(z) = -\frac{1}{2\omega\rho_0} \int\limits_{k_x^2+k_y^2> k^2}\int dk_x\, dk_y |T(k_x, k_y)|^2 |\gamma|\, e^{-2|\gamma|z}, \qquad z > z_0. \qquad (9.43)$$

These expressions prove that the real power is independent of z and carried solely by the propagating plane-waves. The imaginary (reactive) power is carried solely by the evanescent plane-waves and decays as $z \to \infty$.

The formulas of this section were derived for sources in the half-space $z < z_0$ and observation points in the half-space $z > z_0$. If the sources had been in $z > z_0$ and the observation point in $z < z_0$, the formulas should be changed as follows: $e^{i\gamma z}$ is replaced by $e^{-i\gamma z}$ in (9.37), and $e^{-i\gamma z}$ is replaced by $e^{i\gamma z}$ in (9.38) and (9.39). Hence, the plane-wave spectrum depends not only on the source but also on the choice of half-space in which the pressure is computed. Equation (9.42) for the real power remains unchanged (except for the fact that $T(k_x, k_y)$ is different), whereas the formula (9.43) for the imaginary power has $e^{-2z|\gamma|z}$ replaced by $e^{2z|\gamma|z}$. With these results, it should be apparent that the external field produced by the volume source can be exactly duplicated by two planar sources, one on each side of the volume source. For example, the field of the volume source in Figure 9.5 can be computed in both the upper and lower half spaces from the pressure on the two planes $z = \pm R_{min}$. Time-domain analogs of the plane-wave expansions can be found in [40], and plane-wave expansions that involve only propagating waves can be found in [41, 42].

9.3.3 Green's Function Solution

The most common Green's function solution to (9.29) entails the free-space response, $G(\mathbf{r}, \mathbf{r}')$, to the point source $Q(\mathbf{r}) = \delta(\mathbf{r} - \mathbf{r}')$. From (9.29) we see that this Green's function satisfies the inhomogeneous Helmholtz equation

$$\nabla^2 G(\mathbf{r}, \mathbf{r}') + k^2\, G(\mathbf{r}, \mathbf{r}') = -\delta(\mathbf{r} - \mathbf{r}'). \qquad (9.44)$$

The expression for $G(\mathbf{r}, \mathbf{r}')$ is usually determined by solving (9.44) directly in the spatial domain. Here we shall take a different approach and determine $G(\mathbf{r}, \mathbf{r}')$ from the plane-wave solutions derived above. By inserting $Q(\mathbf{r}) = \delta(\mathbf{r} - \mathbf{r}')$ into (9.37) and (9.38), we find that $G(\mathbf{r}, \mathbf{r}')$ has the plane-wave representation (called the Weyl identity)

$$G(\mathbf{r}, \mathbf{r}') = \frac{i}{8\pi^2} \int\limits_{-\infty}^{\infty} \int\limits_{-\infty}^{\infty} dk_x \, dk_y \, \frac{e^{ik_x(x-x')+ik_y(y-y')+i\gamma|z-z'|}}{\gamma}. \tag{9.45}$$

Spherical symmetry implies that the solution is unchanged if $x - x' = 0$, $y - y' = 0$, and $|z - z'| = |\mathbf{r} - \mathbf{r}'|$. After transforming (k_x, k_y) to polar coordinates, the expression can be evaluated analytically to give

$$G(\mathbf{r}, \mathbf{r}') = \frac{e^{ik|\mathbf{r}-\mathbf{r}'|}}{4\pi|\mathbf{r} - \mathbf{r}'|}, \tag{9.46}$$

which is known as the free-space Green's function. Applying the principle of super-position, the frequency-domain pressure for the general source $Q(\mathbf{r}')$ is expressed as

$$P(\mathbf{r}) = \frac{1}{4\pi} \int dV' \, \frac{Q(\mathbf{r}') \, e^{ik|\mathbf{r}-\mathbf{r}'|}}{|\mathbf{r} - \mathbf{r}'|}, \tag{9.47}$$

which is the desired expression. One can show that (9.47) is a solution to the Helmholtz equation (9.29) both inside and outside the source region. Moreover, the time-domain analog of (9.47) is a causal outgoing wave that propagates away from the source region.

We shall also present an expression that is obtainable from Green's second identity [43, p. 806], which gives the pressure at the observation point $\mathbf{r}$ in terms of the pressure $P(\mathbf{r}')$ and its normal derivative $\hat{\mathbf{n}}' \cdot \nabla' P(\mathbf{r}')$ on a surface S with surface normal $\hat{\mathbf{n}}'$ that encloses the source (for example, S could be the minimum sphere in Figure 9.5):

$$P(\mathbf{r}) = \frac{1}{4\pi} \int\limits_{S} dS' \left[P(\mathbf{r}') \hat{\mathbf{n}}' \cdot \nabla' \left(\frac{e^{ik|\mathbf{r}-\mathbf{r}'|}}{|\mathbf{r} - \mathbf{r}'|} \right) - \frac{e^{ik|\mathbf{r}-\mathbf{r}'|}}{|\mathbf{r} - \mathbf{r}'|} \hat{\mathbf{n}}' \cdot \nabla' P(\mathbf{r}') \right]. \tag{9.48}$$

Specifically, one applies Green's second identity to the infinite volume that lies outside S (if S is a sphere, this volume is an infinite spherical shell) where the pressure satisfies the homogeneous Helmholtz equation. The integration variable is $\mathbf{r}'$ and the observation point is $\mathbf{r}$. By use of radiation conditions for the pressure and the Green's function, one proves that the contribution from the spherical surface at infinity vanishes. The volume integral reduces to $\int P(\mathbf{r}')\delta(\mathbf{r} - \mathbf{r}')dV'$, which equals $P(\mathbf{r})$ if $\mathbf{r}$ is in the volume of integration and zero if $\mathbf{r}$ is outside the volume of integration. Hence, (9.48) holds when the observation point $\mathbf{r}$ is outside the surface S. Moreover, it follows that if the observation point $\mathbf{r}$ is placed inside S, the right-hand side of (9.48) equals zero. Finally, one can use Neumann or Dirichlet Green's functions instead of the free-space Green's function used here to eliminate either the $P(\mathbf{r}')$ term or the $\hat{\mathbf{n}}' \cdot \nabla' P(\mathbf{r}')$ term of the integrand in (9.48). The plane-wave formula (9.39) and the spherical-wave formula (9.63) are examples of expressions that involve only the pressure (not its normal derivative).

The quantities $P(\mathbf{r}')$ and $\hat{\mathbf{n}}' \cdot \nabla' P(\mathbf{r}')$ are the Huygens sources that replace the original source $Q(\mathbf{r}')$. Since the pressure satisfies the homogeneous Helmholtz equation outside the source region, both $P(\mathbf{r}')$ and $\hat{\mathbf{n}}' \cdot \nabla' P(\mathbf{r}')$ are smooth functions. In contrast, the original source $Q(\mathbf{r}')$ may contain singularities.

Far from the source, the pressure can be approximated by a simple expression that involves an outgoing spherical wave with amplitude determined by a far-field pattern. To see this, we employ the far-zone approximation $|\mathbf{r} - \mathbf{r}'| \sim r - \hat{\mathbf{r}} \cdot \mathbf{r}'$ in (9.47) to get

$$P(\mathbf{r}) \sim \frac{e^{ikr}}{4\pi r} \int dV' \, Q(\mathbf{r}') e^{-ik\hat{\mathbf{r}} \cdot \mathbf{r}'}, \quad r \to \infty, \tag{9.49}$$

which holds only for sources of finite extent since the condition $r \gg r'$ must hold for all integration points. One can show that the first term neglected in (9.49) is of order r^{-2} so that

$$P(\mathbf{r}) = \frac{\mathcal{F}(\hat{\mathbf{r}}) \, e^{ikr}}{r} + O(r^{-2}), \quad r \to \infty, \tag{9.50}$$

where $\mathcal{F}(\hat{\mathbf{r}})$ is the far-field pattern

$$\mathcal{F}(\hat{\mathbf{r}}) = \frac{1}{4\pi} \int dV' \, Q(\mathbf{r}') e^{-ik\hat{\mathbf{r}} \cdot \mathbf{r}'}, \tag{9.51}$$

which has the precise definition

$$\mathcal{F}(\hat{\mathbf{r}}) = \lim_{r \to \infty} r \, e^{-ikr} P(r\hat{\mathbf{r}}). \tag{9.52}$$

Here, $O(r^{-2})$ represents a function with properties such that $r^2 O(r^{-2})$ is bounded as $r \to \infty$. We have now shown that any source of finite extent radiates a far field of the form $\mathcal{F}(\hat{\mathbf{r}}) \, e^{ikr}/r$. Sources of infinite extent, however, do not exhibit this behavior. From (9.48) and (9.52) we find that the far-field pattern can also be expressed in terms of the Huygens sources

$$\mathcal{F}(\hat{\mathbf{r}}) = -\frac{1}{4\pi} \int dS' \left[ik\hat{\mathbf{r}} \cdot \hat{\mathbf{n}}' P(\mathbf{r}') + \hat{\mathbf{n}}' \cdot \nabla' P(\mathbf{r}') \right] e^{-ik\hat{\mathbf{r}} \cdot \mathbf{r}'}. \tag{9.53}$$

For a point source at $\mathbf{r}_0$ with $Q(\mathbf{r}) = \delta(\mathbf{r} - \mathbf{r}_0)$, the far-field pattern is $\mathcal{F}(\hat{\mathbf{r}}) = e^{-ik\hat{\mathbf{r}} \cdot \mathbf{r}_0}/(4\pi)$. Since the distance r depends on the location of the origin of the coordinate system, so does the far-field pattern.

An alternative way to obtain (9.50) is to apply the method of stationary phase to (9.37). For large r, the exponent oscillates violently with k_x and k_y, and there is only a significant net contribution to the integral in the vicinity of the points where the derivatives of the phase vanish, where the phase is approximated by a quadratic function, and the factor $1/\gamma$ is approximated by a constant. This approximate integral can be evaluated analytically to give (9.50) and (9.51). The quantity $\mathcal{F}(\hat{\mathbf{r}})$ can be obtained directly from the spectrum $T(k_x, k_y)$ by noting the similarities between (9.51) and (9.38),

$$\mathcal{F}(\hat{\mathbf{r}}) = -ik \, \hat{\mathbf{z}} \cdot \hat{\mathbf{r}} \, T(k \hat{\mathbf{x}} \cdot \hat{\mathbf{r}}, k \hat{\mathbf{y}} \cdot \hat{\mathbf{r}}), \quad \hat{\mathbf{z}} \cdot \hat{\mathbf{r}} > 0, \tag{9.54}$$

which can be expressed in spherical coordinates as

$$\mathcal{F}(\theta, \phi) = -ik \cos \theta \, T(k \cos \phi \, \sin \theta, k \sin \phi \, \sin \theta), \quad \theta < \pi/2. \tag{9.55}$$

The far-field pattern in the other hemisphere $\hat{\mathbf{z}} \cdot \hat{\mathbf{r}} < 0$ is given by (9.54) with $-ik$ replaced by ik (as discussed at the end of Section 9.3.2, the plane-wave spectrum $T(k_x, k_y)$ also depends on which half-space is under consideration). The total power radiated into the far zone can be expressed conveniently in terms of the far-field pattern. To see this, we write the integral in (9.42) for the power radiated into the hemisphere $\hat{\mathbf{r}} \cdot \hat{\mathbf{z}} > 0$ in spherical coordinates, employ the expression (9.54), and add the contribution from the hemisphere $\hat{\mathbf{r}} \cdot \hat{\mathbf{z}} < 0$ to get

$$P_{rad} = \frac{1}{2c\rho_0} \int_{4\pi} d\Omega \, |\mathcal{F}(\hat{\mathbf{r}})|^2, \tag{9.56}$$

where the integral is over the unit sphere.

9.3.4   Spherical-Wave Solution

The Fourier/plane-wave solution to the Helmholtz equation derived in Section 9.3.2 is exceedingly powerful. For certain purposes, however, solutions in spherical coordinates are needed. The approach (not treated in detail here) starts with the Helmholtz equation in spherical coordinates, for which there are a countable set of outgoing orthogonal eigenfunction solutions, $h_n^{(1)}(kr)Y_{mn}(\theta, \phi)$, $n = 0, 1, 2, \ldots$, $m = -n, \ldots, n$, obtainable by the method of separation of variables, where $h_n^{(1)}(kr)$ are spherical Hankel functions [44] given by

$$h_n^{(1)}(Z) = (-i)^{n+1} \frac{e^{iZ}}{Z} \sum_{q=0}^{n} \frac{i^q}{q!(2Z)^q} \frac{(n+q)!}{(n-q)!}, \tag{9.57}$$

and $Y_{nm}(\hat{\mathbf{r}})$ are spherical harmonics [44] given in terms of the associated Legendre functions $P_n^m(\cos \theta)$ and complex exponentials by

$$Y_{nm}(\hat{\mathbf{r}}) = \sqrt{\frac{(2n+1)}{4\pi} \frac{(n-m)!}{(n+m)!}} \, P_n^m(\cos \theta) \, e^{im\phi} = \sum_{q=-n}^{n} \mathcal{Y}_{nmq} \, e^{iq\theta} e^{im\phi}, \tag{9.58}$$

where $\mathcal{Y}_{nmq}$ are Fourier coefficients that are obtainable from recursion relations [45, pp. 319–320]. The summations in both the expression (9.57) for the spherical Hankel function and in the expression (9.58) for the spherical harmonics contain a finite number of terms. Hence, these functions are well-behaved and easy to compute. The spherical Hankel function $h_n^{(1)}(Z)$ has precisely n zeros located in the lower half of the complex Z-plane. The functions $h_n^{(1)}(kr)Y_{nm}(\hat{\mathbf{r}})$, which are also called spherical modes, form a complete set of functions for expanding any outgoing solution to the Helmholtz equation in the region $r \geq R_{min}$. In particular, one can show that [38]

$$\frac{e^{ik|\mathbf{r}-\mathbf{r}'|}}{4\pi|\mathbf{r}-\mathbf{r}'|} = ik \sum_{n=0}^{\infty} \sum_{m=-n}^{n} j_n(kr')h_n^{(1)}(kr)Y_{nm}(\hat{\mathbf{r}}) \, Y_{nm}^*(\hat{\mathbf{r}}'), \qquad r > r', \tag{9.59}$$

where $\mathbf{r} = r\hat{\mathbf{r}}$, $\mathbf{r}' = r'\hat{\mathbf{r}}'$, $*$ indicates complex conjugation, and $j_n(Z)$ are spherical Bessel functions, which equal the real part of $h_n^{(1)}(Z)$ when Z is real. The spherical-wave expansion of the pressure is obtained by inserting (9.59) into (9.47) to get

$$P(\mathbf{r}) = \sum_{n=0}^{\infty} \sum_{m=-n}^{n} \Lambda_{nm} h_n^{(1)}(kr)Y_{nm}(\hat{\mathbf{r}}), \qquad r \geq R_{min}, \tag{9.60}$$

where the spherical expansion coefficients Λ_{nm} are determined in terms of the source function

$$\Lambda_{nm} = ik \int dV' \, j_n(kr') \, Y_{nm}^*(\hat{\mathbf{r}}')Q(\mathbf{r}'). \tag{9.61}$$

The spherical harmonics satisfy the orthogonality relation

$$\int_{4\pi} d\Omega \; Y_{nm}(\hat{\mathbf{r}}) \, Y_{n'm'}^*(\hat{\mathbf{r}}) = \delta_{nn'}\delta_{mm'}, \tag{9.62}$$

where the integral is over the unit sphere. This relation can be used with (9.60) and the fact that the spherical Hankel function has no zeros on the real axis to express the expansion coefficients in terms of the pressure on a sphere of radius r, thereby enabling the inference of external pressure everywhere, as

$$\Lambda_{nm} = \frac{1}{h_n^{(1)}(kr)} \int_{4\pi} d\Omega \, P(r\hat{\mathbf{r}}) \, Y_{nm}^*(\hat{\mathbf{r}}), \quad r \geq R_{min}. \tag{9.63}$$

Both the integral in (9.63) and the spherical Hankel function in front of the integral depend on the radius r of the sphere where the pressure is given. However, the expansion coefficients Λ_{nm} are independent of r. The formula (9.63) is similar to the Huygens expressions (9.39) and (9.48) since it determines the field from an equivalent source (the pressure on a sphere surrounding the original source) rather than from the original source function $Q(\mathbf{r})$.

The spherical-wave expansion of the far-field pattern can be obtained by combining (9.52), (9.60), and (9.57) to get

$$\mathcal{F}(\hat{\mathbf{r}}) = \frac{1}{k} \sum_{n=0}^{\infty} \sum_{m=-n}^{n} (-i)^{n+1} \Lambda_{nm} Y_{nm}(\hat{\mathbf{r}}), \tag{9.64}$$

and (9.62) shows that the spherical-wave expansion coefficients can be determined from the far-field pattern as

$$\Lambda_{nm} = k \, i^{n+1} \int_{4\pi} d\Omega \, \mathcal{F}(\hat{\mathbf{r}}) \, Y_{nm}^*(\hat{\mathbf{r}}). \tag{9.65}$$

Hence, we can compute the field everywhere in the region $r > R_{min}$ from the far-field pattern.

We shall now discuss the convergence properties of the spherical-wave expansion. One can show from (9.61) that for typical non-resonant sources[3] the behavior of the expansion coefficients Λ_{nm} for large n are determined by the behavior of $j_n(kR_{min})$, which is shown in Figure 9.6 for three values of kR_{min}. We see that the expansion coefficients start to decay when n reaches the value kR_{min} and exhibit exponential decay

---

[3] The term "typical non-resonant source" refers to most multi-wavelength sources encountered in practice. These sources have reactive fields that extend at most a few wavelengths beyond the surface of the source. Section 9.3.6 discusses properties of super-directive multi-wavelength resonant sources, and Section 9.3.7 considers sub-wavelength resonant sources for power transfer.

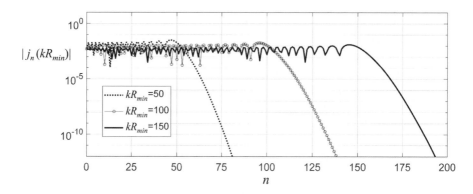

Figure 9.6 The magnitude of the spherical Bessel function $j_n(kR_{min})$ for varying n and three values of kR_{min}.

for $n > kR_{min}$. It can be shown that to achieve a relative accuracy of $\mathcal{E}$ for typical non-resonant sources, the expansions (9.60) and (9.64) should be truncated at [46],[4]

$$N = kR_{min} + \frac{(-3\ln \mathcal{E})^{2/3}}{2}(kR_{min})^{1/3}. \tag{9.66}$$

Assume that we have a source that can excite spherical modes up to $n = N$. Such a source has $\sum_{n=0}^{N}\sum_{m=-n}^{n} = N^2 + 2N + 1$ expansion coefficients Λ_{nm} and thus the number of degrees of freedom is also $N^2 + 2N + 1$. Moreover, one can show that $N^2 + 2N$ is the maximum directivity achievable with a spherical-wave expansion truncated at $n = N$ [47]. The directivity $D \geq 1$ is given in terms of the far-field pattern through

$$D = \max_{\hat{\mathbf{r}}}\left(\frac{|\mathcal{F}(\hat{\mathbf{r}})|^2}{\frac{1}{4\pi}\int_{4\pi} d\Omega'\, |\mathcal{F}(\hat{\mathbf{r}}')|^2}\right), \tag{9.67}$$

and is a measure for how directive a source is. Neglecting the factor $1/2c\rho_0$ from (9.56), the numerator $|\mathcal{F}(\hat{\mathbf{r}})|^2$ determines the radiation intensity in the direction $\hat{\mathbf{r}}$ and the denominator represents the average radiation intensity over the unit sphere (the total radiated power is 4π times the average radiation intensity). An isotropic source (like a point source) has directivity $D = 1$. One often characterizes a radiator by its gain G, which is given in terms of the directivity through $G = \eta D$, where η is the efficiency. Specifically, if the input power is P_{in} and the radiated power is P_{rad}, the efficiency is $\eta = P_{rad}/P_{in}$.

We can relate the upper limits on the directivity to the physical size of the source for the situation where (9.66) holds. For simplicity we set the truncation to $N = kR_{min}$ and find that the maximum number of degrees of freedom and the maximum achievable

---

[4] When the truncation formula (9.66) is applied to (9.60), the observation point should be outside the reactive zone, which extends a couple of wavelengths beyond the minimum sphere for typical non-resonant sources.

directivity are both roughly *three times the area of the minimum sphere in square wavelengths*. Hence, the number of degrees of freedom is proportional to the surface area of the source and not to the volume occupied by the source. This is in agreement with Franceschetti [48], who states that: "A key insight of our analysis is that the amount of information, in terms of degrees of freedom, scales with the surface boundary, rather than with the volume of the space." An analysis on the degrees of freedom contained in scattered fields can be found in [49].

Moreover, a far-field pattern produced by a typical non-resonant source can approximately be recreated by placing array elements on the minimum sphere at a spacing that is about half a wavelength. In Section 9.3.6 we shall discuss the properties of sources for which the truncation number is much larger than the value in (9.66). Time-domain analogs of the spherical-wave expansions can be found in [50].

9.3.5 Array Sources

An array source uses a collection of smaller sources (called the array elements) whose combined action produces a desired total field [51, 52]. In many applications the phase and amplitudes of the excitation signals that feed the individual array elements are adjusted so that the array far-field pattern has peaks and nulls in prescribed directions. In other applications one may want to ensure that the sidelobes are below certain thresholds outside a given main beam. Massive MIMO drives the antenna elements using knowledge gained by direct measurement of the propagation channels to the intended receivers. In the following, we confine the discussion to open-loop (e.g., not Massive MIMO) operation.

With most arrays the amplitude and phase of the excitation signals can be varied so that the same array can produce a wide range of fields. For example, one can steer the main beam into different directions dependent on requirements. We first present the most basic theory of arrays and then discuss some advanced topics.

Properties of a wide range of array geometries can be illustrated by using the planar array in Figure 9.7 as an example. Here the array elements reside in the $x-y$ plane

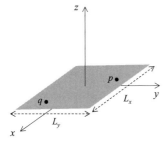

Figure 9.7 A planar array with elements residing in the $x-y$ plane inside the rectangle given by $-L_x/2 \leq x \leq L_x/2$ and $-L_y/2 \leq y \leq L_y/2$. The elements labeled p and q are shown. The spacing between adjacent elements is Δ, so that the total number of array elements is $N_a = N_x N_y = L_x L_y/(\Delta^2)$.

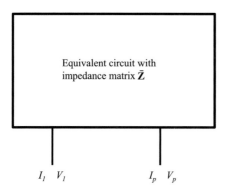

$I_1 \quad V_1$ $I_p \quad V_p$

Figure 9.8 The array of electroacoustic transducers represented by an equivalent circuit with impedance matrix $\bar{\mathbf{Z}}$. The voltage and current for element p are denoted by V_p and I_p.

inside the rectangle $-L_x/2 \leq x \leq L_x/2$ and $-L_y/2 \leq y \leq L_y/2$. The spacing between adjacent elements is Δ, so that the total number of array elements is $N_a = N_x N_y = L_x L_y/(\Delta^2)$ with $N_x = L_x/\Delta$ and $N_y = L_y/\Delta$. The array elements are identical, with far-field patterns given by $\mathcal{F}^e(\hat{\mathbf{r}})$ referenced to the origin and assuming unit-amplitude excitation. The element locations are $\mathbf{r} = x_p\hat{\mathbf{x}} + y_p\hat{\mathbf{y}}$, where p is a single index that covers all elements in the planar array.

Each array element is assumed to be an electroacoustic transducer that can be represented by an equivalent impedance. The voltages and currents of the transducers in the array are given by the voltage column vector $\mathbf{V}$ that contains the voltages V_p and the current column vector $\mathbf{I}$ that contains the currents I_p. As illustrated in Figure 9.8, the impedance matrix $\bar{\mathbf{Z}}$ relates these voltages and currents through [53]

$$\mathbf{V} = \bar{\mathbf{Z}}\mathbf{I}. \tag{9.68}$$

This formulation is quite general, and it can rigorously describe any system of transmit and receive transducers, whether local or remote, provided the propagation is linear and time-invariant. In theory, a transmit transducer is driven by an ideal zero-impedance current source, while a receive transducer is connected to an ideal infinite-impedance voltmeter. A finite impedance transmitter or receiver is handled by solving associated circuit equations.

The far-field pattern of array element p situated at $x_p\hat{\mathbf{x}} + y_p\hat{\mathbf{y}}$ is

$$\mathcal{F}(\hat{\mathbf{r}}) = I_p\, e^{-ik\hat{\mathbf{r}}\cdot(x_p\hat{\mathbf{x}}+y_p\hat{\mathbf{y}})}\, \mathcal{F}^e(\hat{\mathbf{r}}), \tag{9.69}$$

so the array far-field pattern $\mathcal{F}^a(\hat{\mathbf{r}})$ can be expressed in terms of the element pattern $\mathcal{F}^e(\hat{\mathbf{r}})$ and the array factor

$$\mathcal{A}(\hat{\mathbf{r}}) = \sum_{p=1}^{N_a} I_p\, e^{-ik\hat{\mathbf{r}}\cdot(x_p\hat{\mathbf{x}}+y_p\hat{\mathbf{y}})} \tag{9.70}$$

as

$$\mathcal{F}^a(\hat{\mathbf{r}}) = \mathcal{A}(\hat{\mathbf{r}})\,\mathcal{F}^e(\hat{\mathbf{r}}). \tag{9.71}$$

The array factor depends solely on the geometry of the array and equals the array far-field pattern when the array elements are point sources. All terms of the summation (9.70) for the array factor add up in phase in the direction $\hat{\mathbf{r}} = \hat{\mathbf{r}}_0$ if we choose the excitation currents to be $I_p = e^{ik\hat{\mathbf{r}}_0 \cdot (x_p \hat{\mathbf{x}} + y_p \hat{\mathbf{y}})}$. For this choice of I_p, the (normalized) array factor can be computed in closed form:

$$\mathcal{A}(\hat{\mathbf{r}}) = \left[\frac{\sin\left(\dfrac{N_x k \Delta \hat{\mathbf{x}} \cdot [\hat{\mathbf{r}} - \hat{\mathbf{r}}_0]}{2}\right)}{N_x \sin\left(\dfrac{k \Delta \hat{\mathbf{x}} \cdot [\hat{\mathbf{r}} - \hat{\mathbf{r}}_0]}{2}\right)} \right] \left[\frac{\sin\left(\dfrac{N_y k \Delta \hat{\mathbf{y}} \cdot [\hat{\mathbf{r}} - \hat{\mathbf{r}}_0]}{2}\right)}{N_y \sin\left(\dfrac{k \Delta \hat{\mathbf{y}} \cdot [\hat{\mathbf{r}} - \hat{\mathbf{r}}_0]}{2}\right)} \right]. \tag{9.72}$$

This array factor steers the main beam in the direction $\hat{\mathbf{r}} = \hat{\mathbf{r}}_0$, where $\hat{\mathbf{r}}_0$ is often referred to as the scan angle. Numerous analytical and numerical methods have been developed to determine excitations that achieve desired array factors [51, 52].

The element spacing Δ must be chosen appropriately to achieve the desired array factor. To illustrate, consider the planar array set to point its main beam in the direction $\hat{\mathbf{r}} = \hat{\mathbf{r}}_0 = \hat{\mathbf{z}}$, so that $\mathcal{A}(\hat{\mathbf{z}}) = 1$. In the observation direction $\hat{\mathbf{r}} = \hat{\mathbf{x}}$ the array factor is

$$\mathcal{A}(\hat{\mathbf{x}}) = \frac{\sin\left(\dfrac{N_x k \Delta}{2}\right)}{N_x \sin\left(\dfrac{k \Delta}{2}\right)}, \tag{9.73}$$

which also equals 1 if $k\Delta = 2\pi$ corresponding to $\Delta = \lambda$. Hence, with $\Delta = \lambda$ the array factor equals 1 in both the desired direction $\hat{\mathbf{r}} = \hat{\mathbf{z}}$ and in the undesired direction $\hat{\mathbf{r}} = \hat{\mathbf{x}}$. Such maxima in undesired directions are called grating lobes. To avoid grating lobes in general, one sets the element spacing to roughly $\Delta = \lambda/2$ [51, p. 34], which for (9.73) would result in $\mathcal{A}(\hat{\mathbf{x}}) = \sin(N_x \pi / 2)/N_x$.

Since each array element is in close proximity of other array elements, significant coupling can occur. Therefore, the impedance matrix must in general be determined by numerically solving a boundary value problem to ensure that the boundary values are satisfied on the physical body of the array. This is achieved through the use of integro-differential equations that cannot be solved analytically except for simple configurations. Additional complications arise from the fact that the array elements near the edges of the array experience different environments than the array elements near the center of the array. Hence, the array element patterns for elements near the edges of the array are different from the array element patterns for elements near the center of the array.

If a particular transducer were perfectly decoupled from the other transducers, then the ratio of its voltage to its current would be a constant. In general, however, this ratio does depend on scan angle. For some arrays one can encounter so-called array blindness, where the reflected signals cancel the incident signals for certain scan angles. Hence, for these "blind" scan angles it is difficult to feed power into the array.

9.3.6   Superdirective Arrays

The performance of an array improves as the overall array size and number of elements increase. Important performance metrics are directivity and number of available degrees of freedom. However, sometimes the array must fit into a limited physical space and thus only a certain maximum size can be tolerated. For standard half-wavelength spacing designs, these spatial constraints can often be incompatible with the required array performance. Hence, it would be advantageous to design a smaller array that has the performance of a larger array. We shall first investigate the properties of such smaller arrays using the spherical-wave solution of Section 9.3.4 and then create a directivity-optimized linear array.

The maximum possible directivity of a typical non-resonant source with minimum sphere radius R_{min} is roughly $D = (kR_{min})^2$ for large kR_{min}.[5] Sources that significantly exceed this directivity level are called superdirective. Section 9.3.4 also showed that to achieve a directivity D requires that the source is capable of exciting spherical modes up to roughly index $n = N = \sqrt{D}$ for large D.

Let us assume that someone has built an array with N_a array elements that can be excited by a set of excitation currents I_p, with $p = 1, 2, 3, \ldots, N_a$. Apparently we thus have N_a degrees of freedom. However, if the array can only excite spherical waves up to index $n = N$, the true number of degrees of freedom cannot exceed N^2, regardless of the number of array elements. Therefore, reducing the element spacing by adding more elements while keeping the overall array size constant does not necessarily add degrees of freedom to the array.

Assume that the array is capable of producing a directivity $D = N^2$. Without making any assumption about the array, we know that it must be capable of exciting the spherical wave $h_N^{(1)}(kr)Y_{Nm}(\hat{\mathbf{r}})$. In other words, the expansion coefficient Λ_{Nm} must have a magnitude that is comparable to the expansion coefficients Λ_{nm} with $n < N$. Figure 9.9 shows the magnitude of the spherical Hankel function $h_N^{(1)}(kr)$ as a function of r for $N = 25, N = 50$, and $N = 75$. For decreasing r, this magnitude increases dramatically beyond the point $r = N/k$. Therefore, since the truncated expansion

$$P(\mathbf{r}) = \sum_{n=0}^{N} \sum_{m=-n}^{n} \Lambda_{nm} h_n^{(1)}(kr) Y_{nm}(\hat{\mathbf{r}}) \tag{9.74}$$

holds everywhere is the region $r \geq R_{min}$, the magnitude of the total field of the array grows dramatically for decreasing r in the region $r < N/k$.

For example, if the array fits inside a minimum sphere with radius $R_{min} = N/(3k)$, the array size is about one-third of the size of a typical non-resonant array with the same

---

[5] For non-resonant sources, N is given in terms of kR_{min} by (9.66), which is approximately $N = kR_{min}$ for large R_{min}. Similarly, Section 9.3.4 showed that the maximum directivity is $N^2 + 2N$ and the number of degrees of freedom is at most $N^2 + 2N + 1$. Here we assume that N is so large that these two quantities can be approximated by N^2.

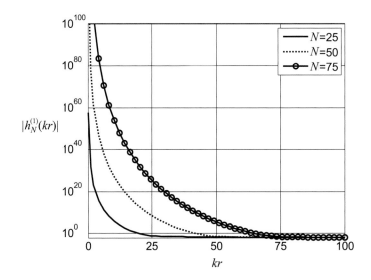

Figure 9.9 The magnitude of the spherical Hankel function $h_N^{(1)}(kr)$ of order $N = 25$, $N = 50$, and $N = 75$. For decreasing r, the magnitude increases dramatically beyond the point $r = N/k$.

directivity. When $N = 75$ we see from Figure 9.9 that the spherical modes in (9.74) with $n = N$ grow by a factor of about 10^{25} (corresponding to 500 dB) as we move the observation point from $r = 3R_{min}$ (corresponding to $kr = 75$ in Figure 9.9) to $r = R_{min}$ (corresponding to $kr = 25$ in Figure 9.9). In other words, near this array (regardless of how it is constructed) the field will attain enormous values, generating a tremendous amount of heat through i^2R losses. In fact, the reactive near field of a source that can excite spherical modes up to $n = N$ extends out to a radius of approximately N/k [54]. This is the main reason why multi-wavelength superdirective sources are not considered practical.

We shall now see how this plays out for a directivity-optimized linear array consisting of point sources along the z-axis at $\mathbf{r}_q = \hat{\mathbf{z}}q\Delta$ with $q = -Q, \dots, Q$. Here, the element spacing is Δ and the total number of elements is $2Q + 1$. With I_q denoting the array excitation currents, the array far-field pattern is

$$\mathcal{F}(\hat{\mathbf{r}}) = \sum_{q=-Q}^{Q} I_q\, e^{-ik\hat{\mathbf{r}}\cdot\hat{\mathbf{z}}q\Delta} = \mathbf{h}^H\mathbf{I}, \tag{9.75}$$

where $\mathbf{I}$ is the current column vector, $\mathbf{h}$ is the steering column vector with elements $e^{ik\hat{\mathbf{r}}\cdot\hat{\mathbf{z}}q\Delta}$, and superscript H denotes the Hermitian transpose. This pattern is omnidirectional with respect to the z-axis and can be expressed in terms of the spherical angle θ through use of $\cos\theta = \hat{\mathbf{r}}\cdot\hat{\mathbf{z}}$. To compute the directivity (9.67), we note that the radiation intensity $|\mathcal{F}(\hat{\mathbf{r}})|^2$ can be expressed as

$$|\mathcal{F}(\hat{\mathbf{r}})|^2 = \sum_{q=-Q}^{Q}\sum_{p=-Q}^{Q} I_q I_p^*\, e^{-ik\hat{\mathbf{r}}\cdot\hat{\mathbf{z}}(q-p)\Delta} = \mathbf{I}^H\mathbf{h}\mathbf{h}^H\mathbf{I} \tag{9.76}$$

and to compute the average radiation intensity over the unit sphere, we note that

$$\frac{1}{4\pi} \int\limits_{4\pi} d\Omega \, e^{-ik\hat{\mathbf{r}}\cdot\hat{\mathbf{z}}(q-p)\Delta} = \frac{1}{2} \int\limits_0^\pi d\theta \, \sin\theta \, e^{-ik[q-p]\Delta\cos\theta} = \text{sinc}\,([p-q]k\Delta), \quad (9.77)$$

where $\text{sinc}(x) = \sin(x)/x$ is the sinc function. With the real part of the impedance matrix given by

$$Z_{pq} = \text{sinc}\,([p-q]k\Delta), \tag{9.78}$$

the average radiation intensity can be expressed as

$$\frac{1}{4\pi} \int\limits_{4\pi} d\Omega \, |\mathcal{F}(\hat{\mathbf{r}})|^2 = \sum_{q=-Q}^{Q} \sum_{p=-Q}^{Q} I_q I_p^* Z_{pq} = \mathbf{I}^H \bar{\mathbf{Z}} \mathbf{I}. \tag{9.79}$$

For $\Delta = n\lambda/2$, with n being an integer, $\bar{\mathbf{Z}}$ is the identity matrix. The directivity in the direction $\hat{\mathbf{r}}$ is found from (9.67) to be

$$D(\hat{\mathbf{r}}) = \frac{\mathbf{I}^H \mathbf{h}\mathbf{h}^H \mathbf{I}}{\mathbf{I}^H \bar{\mathbf{Z}} \mathbf{I}}. \tag{9.80}$$

Using the Schwartz inequality, it can be shown that [55, section 5.2.2]

$$\mathbf{I}_{opt} = \frac{\bar{\mathbf{Z}}^{-1}\mathbf{h}}{\sqrt{\mathbf{h}^H \bar{\mathbf{Z}}^{-1}\mathbf{h}}} \tag{9.81}$$

is the unique (up to a factor) current vector that maximizes the directivity. The denominator in (9.81) ensures that the average radiation intensity $\mathbf{I}_{opt}^H \bar{\mathbf{Z}} \mathbf{I}_{opt}$ equals 1, and we used the fact that $(\bar{\mathbf{Z}}^{-1})^H = \bar{\mathbf{Z}}^{-1}$ to get (9.81). The corresponding optimum value of the directivity is

$$D_{opt}(\hat{\mathbf{r}}) = \mathbf{h}^H \bar{\mathbf{Z}}^{-1}\mathbf{h} \tag{9.82}$$

and the array quality factor, which is a measure of the ratio between the power stored in the reactive field and the total radiated power, is set to [56]

$$Q_{fac} = \mathbf{I}_{opt}^H \mathbf{I}_{opt} = \frac{\mathbf{h}^H \bar{\mathbf{Z}}^{-1} \bar{\mathbf{Z}}^{-1}\mathbf{h}}{\mathbf{h}^H \bar{\mathbf{Z}}^{-1}\mathbf{h}}. \tag{9.83}$$

The theoretical maximum directivity of a linear array with N_a equally spaced point sources approaches N_a^2 as the element spacing approaches zero [57]. However, due to the finite accuracy of computer codes, this upper limit is not even achievable in numerical simulations, as we shall see.

The discussion above based on the properties of the spherical Hankel function demonstrated that the field values near a multi-wavelength superdirective source would be enormous. Therefore, we should expect the quality factor of such a source to be large as well. The quality factor is also a measure of the operational bandwidth of the array, and a large quality factor results in a narrow-band array.

Table 9.1 Directivity-optimized linear arrays of equally spaced point sources along the z-axis. All arrays have length 10λ and are optimized for the end-fire direction $\hat{\mathbf{r}} = \hat{\mathbf{z}}$. The average radiation intensity $\mathbf{I}_{opt}^H \bar{\mathbf{Z}} \mathbf{I}_{opt}$ equals 1, so that the total radiated power is 4π. Theoretical $D_{opt}(\hat{\mathbf{z}})$ equals the square of the number of elements.

No. of elements	Δ/λ	Theoretical $D_{opt}(\hat{\mathbf{z}})$	Computed $D_{opt}(\hat{\mathbf{z}})$	Q_{fac}
21	1/2	441	21.00	1
26	2/5	676	276.99	$1.6 \cdot 10^5$
31	1/3	961	581.60	$1.7 \cdot 10^{12}$
41	1/4	1681	1320.45	$7.0 \cdot 10^{14}$

Table 9.1 shows results of a simulation based on (9.82) and (9.83) for a linear array with length 10λ optimized in the end-fire direction $\hat{\mathbf{r}} = \hat{\mathbf{z}}$. With element spacing of $\Delta = \lambda/2$ the directivity is $D = 21$ and the quality factor is $Q_{fac} = 1$. This array with standard $\lambda/2$ spacing is a conventional non-resonant source. The fact that the directivity is equal to the number of array elements follows from a general theorem on linear arrays with $\lambda/2$ spacing [58]. As we increase the number of elements by decreasing the element spacing Δ, the directivity increases by considerable amounts while the quality factor increases dramatically. For $\Delta = 1/4$ the quality factor (sum of the excitation current magnitudes squared) is $7.0 \cdot 10^{14}$, which results from the average magnitude of the excitation currents being equal to $2.7 \cdot 10^6$. These enormous excitation currents produce very high field value near the array but a moderate far-field pattern, which carries exactly the same total power as the pattern for the well-behaved $\lambda/2$ array. Moreover, for $\Delta = \lambda/4$ the condition number of $\bar{\mathbf{Z}}$ is $3.4 \cdot 10^{16}$, so the solution is extremely unstable.

These problems associated with achieving supergain do not imply that dense arrays (see for example [59, 60]) cannot be beneficial in some instances. However, if a dense array is significantly smaller than an original half-wavelength spacing array and has the same performance, the fields in the vicinity of the denser array will possess the unfortunate properties explained here. Interestingly, the appendix of one of the original MIMO publications [61] discusses the benefits and practical problems related to supergain arrays.

Reference [62] computes the quality factor of the field outside a sphere of radius R_{min} of each spherical mode. Due to the behavior of the spherical Hankel function illustrated in Figure 9.9, the quality factor grows with increasing n for $n > kR_{min}$. Hence, the modes with $n > kR_{min}$ generate large reactive fields, and [62] concludes that: "For a very narrowband system, superdirectivity might be exploited to marginally increase the number of spatial degrees of freedom." Interesting parallels between superdirectivity and singular detection have been pointed out in [63].

The discussion in this section was restricted to sources and arrays that are several wavelengths across. For electrically small sources with dimensions smaller than or comparable to the wavelength, supergain directivity that far exceeds the limit $D = (kR_{min})^2$ is easily realized in practice by use of resonant structures.

9.3.7   Resonant Evanescent Wave Coupling

We shall now examine the problem of wireless power transfer [64] from the point of view of plane-wave expansions. Wireless power transfer is the transmission of electrical energy from a power source to an electrical load (such as the battery of an electric car) without the use of wires or cables. The power is carried by electromagnetic fields typically in the frequency range from 10 kHz to 25 MHz. We show that this transfer of power can be accomplished through evanescent plane-waves.

Consider the primary and secondary coils in Figure 9.10 with the equivalent circuit model in Figure 9.11. The primary coil transmits electromagnetic energy that induces a current in the secondary coil. The primary coil circuit has inductance L_p, capacitance C_p, and resistance R_p. The corresponding parameters for the secondary coil circuit are L_s, C_s, and R_s. The load attached to the secondary coil has resistance R_o, and we assume that the secondary coil operates at resonance, so that $\omega^2 L_s C_s = 1$. The coils constitute an air-core transformer having a very low coupling coefficient. Notwithstanding, an application of elementary circuit theory – which itself constitutes the quasi-static approximation – shows that, if the coils have sufficiently high Q, and if the secondary circuit is tuned to resonance, power can be transmitted with relatively high efficiency.

The entire action of the primary coil is the creation of a spectrum of plane-waves, dominated by evanescent waves. Paradoxically, real power can still be transmitted despite the fact that, in the ordinary course of things, evanescent waves carry only reactive power. The electric field $\mathbf{E}_p(\mathbf{r})$ generated by the primary coil satisfies Maxwell's equations and can be expressed in terms of plane-waves as [39, 40]

$$\mathbf{E}_p(\mathbf{r}) = \frac{1}{2\pi} \int\limits_{-\infty}^{\infty} \int\limits_{-\infty}^{\infty} dk_x \, dk_y \, \mathbf{T}_p(k_x, k_y) \, e^{ik_x x + ik_y y + i\gamma z}, \tag{9.84}$$

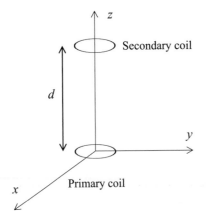

Figure 9.10 The primary coil resides in the $z = 0$ plane with its center at the origin. The secondary coil resides in the $z = d$ plane with its center at $(x, y, z) = (0, 0, d)$. Both coils are circular and have radius a. The primary coil transmits electromagnetic energy, a part of which is dissipated in the load attached to the secondary coil.

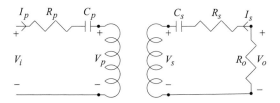

Figure 9.11 Equivalent circuit model for the coil system in Figure 9.10. The primary coil circuit has inductance L_p, capacitance C_p, and resistance R_p. The corresponding parameters for the secondary coil circuit are L_s, C_s, and R_s. The secondary coil operates at resonance so that $\omega^2 L_s C_s = 1$, and its load has resistance R_o.

where the spectrum $\mathbf{T}_p(k_x, k_y)$ is given in terms of the current density $\mathbf{J}_p(\mathbf{r})$ of the primary coil as

$$\mathbf{T}_p(k_x, k_y) = \frac{1}{4\pi k \gamma} \sqrt{\frac{\mu}{\epsilon}}\, \mathbf{k} \times \mathbf{k} \times \int dV\, \mathbf{J}_p(\mathbf{r})\, e^{-i\mathbf{k}\cdot\mathbf{r}}, \tag{9.85}$$

with propagation vector $\mathbf{k} = k_x \hat{\mathbf{x}} + k_y \hat{\mathbf{y}} + \gamma \hat{\mathbf{z}}$ and γ given by (9.36). Here, ϵ is the free-space permittivity, and μ is the free-space permeability. Equations (9.84) and (9.85) are the electromagnetic analogs of the acoustic formulas (9.37) and (9.38).

The primary coil in Figure 9.10 occupies the circle $\mathbf{r} = a\cos\phi\hat{\mathbf{x}} + a\sin\phi\hat{\mathbf{y}}$, with $0 \leq \phi < 2\pi$, so that the current I_p flows in the direction $-\sin\phi\hat{\mathbf{x}} + \cos\phi\hat{\mathbf{y}}$. For this source geometry, we find from (9.85) that the spectrum is

$$\mathbf{T}_p(k_x, k_y) = I_p \frac{i\omega\mu a J_1\left(a\sqrt{k_x^2 + k_y^2}\right)}{2\gamma\sqrt{k_x^2 + k_y^2}}\left[-k_y\hat{\mathbf{x}} + k_x\hat{\mathbf{y}}\right], \tag{9.86}$$

where $J_1(Z)$ is the cylindrical Bessel function of order one. To compute the electromotive force V_s' that is induced in the secondary coil, we introduce the plane-wave receiving characteristic $\mathbf{R}_s(k_x, k_y)$. Specifically, the induced electromotive force V_s' in the secondary coil due to a single incident plane-wave $\mathbf{E}_o e^{ik_x x + ik_y y + i\gamma z}$ is

$$V_s' = \mathbf{R}_s(k_x, k_y) \cdot \mathbf{E}_o e^{i\gamma d}, \tag{9.87}$$

where the second factor is simply the incident plane-wave evaluated at the location of the center (reference point) of the secondary coil. Using the Lorentz reciprocity theorem [65, section 5.1], one can show that

$$\mathbf{R}_s(k_x, k_y) = 4\pi i\gamma \frac{\mathbf{T}_s(-k_x, -k_y)}{i\omega\mu I_s}, \tag{9.88}$$

where $\mathbf{T}_s(k_x, k_y)$ is the plane-wave spectrum of the field radiated by the secondary coil when its reference point is at the origin and its current is I_s. Hence, $\mathbf{T}_s(k_x, k_y)$ is readily obtained from (9.86) to get

$$\mathbf{R}_s(k_x, k_y) = -\frac{2\pi i a J_1\left(a\sqrt{k_x^2 + k_y^2}\right)}{\sqrt{k_x^2 + k_y^2}}\left[-k_y\hat{\mathbf{x}} + k_x\hat{\mathbf{y}}\right]. \tag{9.89}$$

The field that impinges on the secondary coil is expressed in (9.84) as a superposition of plane-waves $\frac{1}{2\pi} \mathbf{T}_p(k_x, k_y) e^{ik_x x + ik_y y + i\gamma z}$, so the total induced electromotive force is

$$V_s' = \frac{1}{2\pi} \int_{-\infty}^{\infty} \int_{-\infty}^{\infty} dk_x \, dk_y \, \mathbf{R}_s(k_x, k_y) \cdot \mathbf{T}_p(k_x, k_y) \, e^{i\gamma d}, \tag{9.90}$$

which, through the use of a new integration variable $k_\rho = \sqrt{k_x^2 + k_y^2}$, can be expressed as

$$V_s' = -i\omega M I_p, \tag{9.91}$$

where

$$M = i\pi a^2 \mu \int_0^{\infty} dk_\rho \, k_\rho \, J_1^2(k_\rho a) \, \frac{e^{i\gamma d}}{\gamma}. \tag{9.92}$$

In the low-frequency limit ($k \to 0$), the expression (9.92) reduces to

$$M = \pi a^2 \mu \int_0^{\infty} dk_\rho \, J_1^2(k_\rho a) \, e^{-k_\rho d}, \tag{9.93}$$

which is the well-known formula for the mutual inductance of two coaxial coils [66, p. 263].[6] An analytical solution based on (9.93) for $a \ll d$ was shown in [64] to agree with a numerical solution to Maxwell's equations.

To determine the power dissipated in the load of the secondary coil, we need the current I_s. Since the voltage V_s can be expressed as $V_s = -i\omega M I_p - i\omega L_s I_s$ and as $V_s = -I_s([-i\omega C_s]^{-1} + R_s + R_o)$, we find

$$I_s = \frac{i\omega M}{R_s + R_o} I_p, \tag{9.94}$$

where the fact that the secondary coil operates at resonance has been used. Hence, the power dissipated in the load is

$$P_o = \frac{1}{2} R_o |I_s|^2 = \frac{R_o \, \omega^2 |M|^2}{2(R_s + R_o)^2} |I_p|^2. \tag{9.95}$$

The induced electromotive force in the secondary coil, which enables the transfer for power, is expressed through (9.92) in terms of propagating and evanescent plane-waves. For $k_\rho \leq k$ the waves are propagating since $\gamma = \sqrt{k^2 - k_\rho^2}$ is real. For $k_\rho > k$ the waves are evanescent since $\gamma = i\sqrt{k_\rho^2 - k^2}$ is purely imaginary. In the limit $k \to 0$, only evanescent plane-waves are in play and we obtain (9.93). Hence, in the low-frequency limit, the power is carried solely by evanescent plane-waves. How can this be true when it was shown in Section 9.3.2 that the real power is carried solely by propagating plane-waves? The answer to this apparent paradox lies in the fact that the field of the evanescent waves induce currents in conductors, which in turn dissipate

---

[6] Jackson uses Gaussian units so conversion to MKSA units is required; see [66, p. 66].

heat. This is exactly what happens in the load of the secondary coil. Naturally, the power absorbed by the load decreases with coil separation distance, as can be seen from the $e^{-k_\rho d}$ factor of the integrand in (9.93). The effect is analogous to driving an inductor with an AC voltage source, which produces only reactive power. Nevertheless, a resistor that is placed in series with the inductor will dissipate real power, and maximum power is dissipated if a series capacitance is added, tuned to resonance.

A similar phenomena can be observed in a waveguide that operates below its cutoff frequency. To illustrate, place a source and a probe in this waveguide. The source will excite evanescent waveguide modes which decay exponentially away from the source. Nevertheless, their fields can be picked up by the probe and a resistor attached to the probe can produce heat.

9.3.8 Discussion

The details of how antennas produce their radiated fields are complicated, and can only be calculated numerically. However, the entire external effect of the antennas has a relatively simple description in terms of a spectrum of plane-waves, or equivalently, a series comprising orthogonal eigenfunctions. We anticipate that these physics-based tools will serve to supplement the familiar mathematical tools of the wireless communication theorist.

References

[1] E. G. Larsson, H. Q. Ngo, T. L. Marzetta, and H. Yang, "Homework problems to go with 'Fundamentals of Massive MIMO'." [Online]. Available: www.cambridge.org/Marzetta, 2016.

[2] J. Flordelis, F. Rusek, F. Tufvesson, E. G. Larsson, and O. Edfors, "FDD massive MIMO versus TDD massive MIMO performance: What do the data say?" *IEEE Trans. Wireless Commun.*, vol. 17, no. 4, pp. 2247–2261, 2018.

[3] T. L. Marzetta, E. G. Larsson, H. Yang, and H. Q. Ngo, *Fundamentals of Massive MIMO*. Cambridge: Cambridge University Press, 2016.

[4] X. Gao, O. Edfors, F. Rusek, and F. Tufvesson, "Massive MIMO performance evaluation based on measured propagation data," *IEEE Trans. Wireless Commun.*, vol. 14, no. 7, pp. 3899–3911, July 2015.

[5] J. Hoydis, C. Hoek, T. Wild, and S. ten Brink, "Channel measurements for large antenna arrays," in *Proc. Int. Symp. Wireless Commun. Syst. (ISWCS)*, Paris, Aug. 2012.

[6] A. O. Martinez, E. D. Carvalho, and J. Ø. Nielsen, "Towards very large aperture massive MIMO: A measurement based study," in *Proc. IEEE Global Telecommun. Conf. (GLOBECOM)*, Austin, TX, Dec. 2014.

[7] E. Björnson, E. De Carvalho, J. H. Sorensen, E. G. Larsson, and P. Popovski, "A random access protocol for pilot allocation in crowded massive MIMO systems," *IEEE Trans. Wireless Commun.*, vol. 16, pp. 2220–2234, Apr. 2017.

[8] E. de Carvalho, E. Björnson, E. G. Larsson, and P. Popovski, "Random access for massive MIMO systems with intra-cell pilot contamination," in *Proc. IEEE Int. Conf. Acoust. Speech Signal Process. (ICASSP)*, Shanghai, Mar. 2016.

[9] L. Liu and W. Yu, "Massive connectivity with massive MIMO – Part I: Device activity detection and channel estimation," *IEEE Trans. Signal Process.*, vol. 66, pp. 2933–2946, 2018.

[10] E. Björnson, J. Hoydis, M. Kountouris, and M. Debbah, "Massive MIMO systems with non-ideal hardware: Energy efficiency, estimation, and capacity limits," *IEEE Trans. Inf. Theory*, vol. 60, no. 11, pp. 7112–7139, Nov. 2014.

[11] U. Gustavsson, C. Sanchéz-Perez, T. Eriksson, et al., "On the impact of hardware impairments on massive MIMO," in *Proc. IEEE Global Telecommun. Conf. (GLOBECOM)*, Austin, TX, Dec. 2014.

[12] C. Mollén, J. Choi, E. G. Larsson, and J. R. W. Heath, "Uplink performance of wideband massive MIMO with one-bit ADCs," *IEEE Trans. Wireless Commun.*, vol. 16, pp. 87–100, Jan. 2017.

[13] T. L. Marzetta, "Noncooperative cellular wireless with unlimited numbers of base station antennas," *IEEE Trans. Wireless Commun.*, vol. 9, no. 11, pp. 3590–3600, Nov. 2010.

[14] E. G. Larsson, F. Tufvesson, O. Edfors, and T. L. Marzetta, "Massive MIMO for next generation wireless systems," *IEEE Commun. Mag.*, vol. 52, no. 2, pp. 186–195, Feb. 2014.

[15] E. Björnson, E. G. Larsson, and T. L. Marzetta, "Massive MIMO: 10 myths and one critical question," *IEEE Commun. Mag.*, vol. 54, no. 2, pp. 114–123, Feb. 2016.

[16] P. Harris, S. Malkowsky, J. Vieira, et al., "Performance characterization of a real-time massive MIMO system with LOS mobile channels," *IEEE J. Sel. Areas Commun.*, vol. 35, no. 6, pp. 1244–1253, 2017.

[17] X. Zhu, Z. Wang, L. Dai, and C. Qian, "Smart pilot assignment for massive MIMO," *IEEE Commun. Letters*, vol. 19, no. 9, pp. 1644–1647, 2015.

[18] I. Atzeni, J. Arnau, and M. Debbah, "Fractional pilot reuse in massive MIMO systems," in *Proc. IEEE ICCW*, 2015, pp. 1030–1035.

[19] H. Q. Ngo and E. G. Larsson, "No downlink pilots are needed in TDD massive MIMO," *IEEE Trans. Wireless Commun.*, vol. 16, pp. 2921–2935, May 2017.

[20] K. F. Guo, Y. Guo, G. Fodor, and G. Ascheid, "Uplink power control with MMSE receiver in multi-cell MU-massive-MIMO systems," in *Proc. IEEE ICC*, Sydney, Jun. 2014, pp. 5184–5190.

[21] N. Krishnan, R. D. Yates, and N. B. Mandayam, "Uplink linear receivers for multi-cell multiuser MIMO with pilot contamination: Large system analysis," *IEEE Trans. Wireless Commun.*, vol. 13, no. 8, pp. 4360–4373, Aug. 2014.

[22] E. Björnson, E. G. Larsson, and M. Debbah, "Massive MIMO for maximal spectral efficiency: How many users and pilots should be allocated?" *IEEE Trans. Wireless Commun.*, vol. 15, no. 2, pp. 1293–1308, Feb. 2016.

[23] X. Li, E. Björnson, E. G. Larsson, S. Zhou, and J. Wang, "Massive MIMO with multi-cell MMSE processing: Exploiting all pilots for interference suppression," *EURASIP J. Wireless Comm. and Netw.*, 2017. DOI: 10.1186/s13638-017-0879-2.

[24] H. Q. Ngo, A. Ashikhmin, H. Yang, E. G. Larsson, and T. L. Marzetta, "Cell-free massive MIMO: Uniformly great service for everyone," in *Proc. IEEE Workshop Signal Process. Adv. Wireless Commun. (SPAWC)*, Stockholm 2015 pp. 201–205.

[25] H. Q. Ngo, A. Ashikhmin, H. Yang, E. G. Larsson, and T. L. Marzetta, "Cell-free massive MIMO versus small cells," *IEEE Trans. Wireless Commun.*, vol. 16, pp. 1834–1850, Mar. 2017.

[26] E. Nayebi, A. Ashikhmin, T. L. Marzetta, H. Yang, and B. D. Rao, "Precoding and power optimization in cell-free massive MIMO systems," *IEEE Trans. Wireless Commun.*, vol. 16, no. 7, pp. 4445–4459, 2017.

[27] J. G. Andrews, S. Buzzi, W. Choi, et al., "What will 5G be?" *IEEE J. Sel. Areas Commun.*, vol. 32, no. 6, pp. 1065–1082, 2014.

[28] S. Verdú, *Multiuser Detection*. Cambridge: Cambridge University Press, 1998.

[29] H. Weingarten, Y. Steinberg, and S. Shamai (Shitz), "The capacity region of the Gaussian multiple-input multiple-output broadcast channel," *IEEE Trans. Inf. Theory*, vol. 52, no. 9, pp. 3936–3964, Sept. 2006.

[30] T. L. Marzetta and B. M. Hochwald, "Capacity of a mobile multiple-antenna communication link in Rayleigh flat fading," *IEEE Trans. Inf. Theory*, vol. 45, no. 1, pp. 139–157, Jan. 1999.

[31] L. Zheng and D. N. C. Tse, "Communication on the Grassmann manifold: A geometric approach to the noncoherent multiple-antenna channel," *IEEE Trans. Inf. Theory*, vol. 48, no. 2, Feb. 2002.

[32] G. Kramer, "Information networks with in-block memory," *IEEE Trans. Inf. Theory*, vol. 60, no. 4, pp. 2105–2120, 2014.

[33] A. Adhikary, A. Ashikhmin, and T. L. Marzetta, "Uplink interference reduction in large-scale antenna systems," *IEEE Trans. Commun.*, vol. 65, no. 5, May 2017.

[34] A. Adhikary, J. Nam, J.-Y. Ahn, and G. Caire, "Joint spatial division and multiplexing: The large-scale array regime," *IEEE Trans. Inf. Theory*, vol. 9, no. 10, Oct. 2013.

[35] E. Björnson, J. Hoydis, and L. Sanguinetti, "Massive MIMO has unlimited capacity," *IEEE Trans. Wireless Commun.*, vol. 17, no. 1, pp. 574–590, 2018.

[36] R. P. Feynman, R. B. Leighton, and M. Sands, *The Feynman Lectures on Physics*. Boston MA: Addison-Wesley, 1963.

[37] J. E. White, *Seismic Waves: Radiation, Transmission, and Attenuation*. New York: McGraw-Hill, 1965.

[38] W. C. Chew, *Waves and Fields in Inhomogeneous Media*. New York IEEE Press, 1995.

[39] D. M. Kerns, "Plane-wave scattering-matrix theory of antennas and antenna-antenna inter-actions," NBS Monograph 162, US Government Printing Office, Washington, DC, 1981.

[40] T. B. Hansen and A. D. Yaghjian, *Plane-Wave Theory of Time-Domain Fields: Near-Field Scanning Applications*. New York IEEE Press, 1999.

[41] R. Coifman, V. Rokhlin, and S. Wandzura, "The fast multipole method for the wave equation: A pedestrian prescription," *IEEE Antenn. Propag. Mag.*, vol. 35, no. 3, pp. 7–12, 1993.

[42] T. B. Hansen, "Translation operator based on Gaussian beams for the fast multipole method in three dimensions," *Wave Motion*, vol. 50, no. 5, pp. 940–954, 2013.

[43] P. M. Morse and H. Feshbach, *Methods of Theoretical Physics, Part I*. New York: McGraw-Hill, 1953.

[44] F. W. Olver, D. W. Lozier, R. F. Boisvert, and C. W. Clark, *NIST Handbook of Mathematical Functions*. Cambridge: Cambridge University Press, 2010.

[45] J. E. Hansen, Ed., *Spherical Near-Field Antenna Measurements*. London: Institution Of Engineering and Technology, 1988.

[46] W. C. Chew, E. Michielssen, J. Song, and J.-M. Jin, *Fast and Efficient Algorithms in Computational Electromagnetics*. London: Artech House, 2006.

[47] R. Harrington, "On the gain and beamwidth of directional antennas," *IRE Trans. Antenn. Propag.*, vol. 6, no. 3, pp. 219–225, 1958.

[48] M. Franceschetti, "On Landau's eigenvalue theorem and information cut-sets," *IEEE Trans. Inf. Theory*, vol. 61, no. 9, pp. 5042–5051, 2015.

[49] O. M. Bucci and G. Franceschetti, "On the degrees of freedom of scattered fields," *IEEE Trans. Antenn. Propag.*, vol. 37, no. 7, pp. 918–926, 1989.

[50] T. B. Hansen, "Formulation of spherical near-field scanning for time-domain acoustic fields," *J. Acoust. Soc. Am.*, vol. 98, pp. 1204–1215, 1995.

[51] R. J. Mailloux, *Phased Array Antenna Handbook*. London: Artech House, 1994.

[52] R. C. Hansen, *Phased Array Antennas*. Chichester: Wiley, 1998.

[53] M. T. Ivrlac and J. A. Nossek, "Toward a circuit theory of communication," *IEEE Trans. Circuits Syst. I: Regular Papers*, vol. 57, no. 7, pp. 1663–1683, 2010.

[54] A. D. Yaghjian, "Sampling criteria for resonant antennas and scatterers," *J. Appl. Phys.*, vol. 79, pp. 7474–7482, 1996, erratum 80, 2547.

[55] M. A. Richards, *Fundamentals of Radar Signal Processing* 2n ed. New York: McGraw-Hill, 2014.

[56] Y. Lo, S. Lee, and Q. Lee, "Optimization of directivity and signal-to-noise ratio of an arbitrary antenna array," *Proc. IEEE*, vol. 54, no. 8, pp. 1033–1045, 1966.

[57] A. I. Uzkov, "An approach to the problem of optimum directive antennae design," *Comptes Rendus (Doklady) de lAcademie des Sciences de lURSS*, vol. 53, pp. 35–38, 1946.

[58] E. E. Altshuler, T. H. O'Donnell, A. D. Yaghjian, and S. R. Best, "A monopole superdirective array," *IEEE Trans. Antenn. Propag.*, vol. 53, no. 8, pp. 2653–2661, 2005.

[59] C. Masouros, M. Sellathurai, and T. Ratnarajah, "Large-scale MIMO transmitters in fixed physical spaces: The effect of transmit correlation and mutual coupling," *IEEE Trans. Commun.*, vol. 61, no. 7, pp. 2794–2804, 2013.

[60] M. T. Ivrlac and J. A. Nossek, "High-efficiency super-gain antenna arrays," in *Int. ITG Workshop Smart Antenn. (WSA)*, Bremen, 2010, pp. 369–374.

[61] G. J. Foschini and M. J. Gans, "On limits of wireless communications in a fading environment when using multiple antennas," AT&T Bell Laboratories Technical Memorandum, Order Number ITD-95-27994C, September, 1995.

[62] A. S. Poon and D. N. C. Tse, "Does superdirectivity increase the degrees of freedom in wireless channels?" in *Proc. IEEE Int. Symp. Inf. Theory (ISIT)*, Hong Kong 2015, pp. 1232–1236.

[63] A. B. Baggeroer, "Space/time random processes and optimum array processing," Naval Undersea Center, San Diego, CA, Tech. Rep., 1976.

[64] A. Karalis, J. D. Joannopoulos, and M. Soljačić, "Efficient wireless non-radiative mid-range energy transfer," *Ann. Phys.*, vol. 323, no. 1, pp. 34–48, 2008.

[65] R. E. Collin, *Antennas and Radiowave Propagation*. New York McGraw-Hill, 1985.

[66] J. D. Jackson, *Classical Electrodynamics* 2nd ed. Chichester: Wiley, 1975.

10 Short-Packet Transmission

Giuseppe Durisi, Gianluigi Liva, and Yury Polyanskiy

10.1 Introduction

10.1.1 The Objective of this Chapter

Wireless communication will be a key enabler of future autonomous systems, be they connected vehicles, smart meters, automated factories, or remote healthcare infrastructures. Indeed, most future wireless communications will be originated by autonomous devices [1] – a typology of traffic often referred to as machine-type communication (MTC). For an important class of MTC applications, the wireless traffic generated by these autonomous devices will be drastically different from the human-generated one [2]: a low data volume will be sporadically transmitted in the form of *short data packets*. Due to the very large number of such devices, the total traffic produced by them is expected to constitute a large part or even the dominant part of the overall transmitted data. Furthermore, the system requirements in terms of *reliability* (packet error probability) and *end-to-end latency* may be an order of magnitude more stringent than those of current systems.

For example, in MTC for factory automation one may need to transfer packets of 100 bits (carrying, e.g., sensor readings or commands for actuators) within 100 μs and with error probability not exceeding 10^{-9} [3, 4]. This makes current wireless broadband systems, which are (1) are optimized for the transmission of long data packets (1000 bits or more), and (2) operate at moderate packet error probability (around 10^{-2} before high-layer retransmission protocols, which cause a further increase in latency), unsuitable to provide MTC connectivity. The need to support MTC in next-generation wireless systems (e.g., 5G and beyond) is widely acknowledged [5, 6]. However, most current approaches to design wireless systems, which are based on asymptotic information theoretic principles, cease to be useful for MTC systems. In addition, short-packet transmission is also crucial for the exchange of control information in conventional cellular and satellite communication systems [2, 7].

In this chapter, we shall review recent advances in *finite-blocklength* information theory, that provide the theoretical principles governing the transmission of short packets. Specifically, by focusing on channel models of interest in wireless communication, we shall discuss how to rigorously characterize the inherent tradeoff between packet size (blocklength), error probability, and transmission rate. Finally, we shall discuss the design of channel codes that are specifically tailored to short-packet transmissions.

10.1.2   Short Packets and Latency

Latency in wireless communication systems is a notoriously difficult quantity to define. Take as latency definition the end-to-end delay in the wireless access network, i.e., the time it takes for a data packet at a given source to be decoded correctly at the intended destination. This quantity involves many components: encoding and decoding delays, queuing delays, and transmission delays. In this chapter, we shall use the packet size, i.e., the blocklength, as a proxy for the transmission delay – an assumption that is sensible for short-distance wireless communication. Specifically, for a given packet size n measured in coded symbols, and a given system bandwidth W, we shall approximate the delay as n/W. We will not touch upon the other delay components listed above, with the exception of decoding complexity (a proxy for the decoding delay), which we will briefly discuss in Section 10.5.

10.1.3   A Motivating Example: Communication Over an additive Gaussian Noise Channel

Consider the problem of transmitting information over the following real-valued (additive white-Gaussian noise (AWGN)) channel:

$$Y_k = X_k + W_k, \quad k = 1, \ldots, n. \tag{10.1}$$

Here, n denotes the blocklength, i.e., the number of discrete-time channel uses that are available for the transmission of a given packet. We shall assume that the additive noise process $\{W_k\}$ is stationary and memoryless with marginal distribution $\mathcal{N}(0, 1)$ and that it does not depend on the input symbols $\{X_k\}$, which are subject to a power constraint ρ, which we will formally introduce in (10.3).

In the asymptotic limit of large blocklength n, the largest rate at which reliable communication over the channel (10.1) is possible is given by the Shannon capacity [8]

$$C(\rho) = \frac{1}{2} \log(1 + \rho). \tag{10.2}$$

This quantity is asymptotic in the following sense: For a given rate below $C(\rho)$, arbitrarily low error probability can be achieved provided that one uses a channel code with sufficiently long blocklength. As we shall demonstrate in this chapter, Shannon's capacity is of limited relevance when the blocklength n is small (due, for example, to a latency constraint); in this scenario, a more refined performance metric needs to be used. To be concrete, let us consider the following question: *What is the minimum blocklength n^* needed for a channel code that operates at a signal-to-noise ratio (SNR) of 0 dB (i.e., $\rho = 1$ in (10.2)) and at a rate equal to 70% of Shannon's capacity (i.e., 0.35 bit per channel use), to yield a packet error probability not exceeding 10^{-4}?* As we shall see, finite-blocklength information theory provides us with an accurate answer to this question.

As a first step towards answering this question, we need to formalize the notion of a channel code. An (M, n, ϵ, ρ) code for the channel (10.1) consists of the following two elements:

1. An encoder, which maps each message J – a random variable uniformly distributed on the message set $\{1, \ldots, M\}$ – to a codeword, i.e., a vector of n channel inputs $x^n = [x_1, \ldots, x_n]$. We shall assume that each codeword satisfies the power constraint

$$\sum_{k=1}^{n} x_k^2 \leq n\rho. \tag{10.3}$$

2. A decoder, which maps each vector of n channel outputs $y^n = [y_1, \ldots, y_n]$ into a message estimate $\hat{J}$ so that the average error probability $\mathbb{P}[\hat{J} \neq J]$ does not exceed ϵ.

Note that the rate of this code is $(\log_2 M)/n$ bit per channel use.

The fundamental quantity of interest in this chapter is the *maximum coding rate* $R^*(n, \epsilon, \rho)$, which is the largest rate among all channel codes with blocklength n, error probability no larger than ϵ, and power ρ:

$$R^*(n, \epsilon, \rho) = \max \left\{ \frac{\log M}{n} : \exists(M, n, \epsilon, \rho) \text{ code} \right\}. \tag{10.4}$$

Shannon's result (10.2) can be restated in terms of $R^*(n, \epsilon, \rho)$ as follows [9]:

$$C(\rho) = \lim_{n \to \infty} R^*(n, \epsilon, \rho). \tag{10.5}$$

Note that (10.5) holds for all $\epsilon \in (0, 1)$.

Differently from the Shannon capacity $C(\rho)$, which is given by the simple formula (10.2), the maximum coding rate $R^*(n, \epsilon, \rho)$ does not admit a closed-form expression. Indeed, the computation of $R^*(n, \epsilon, \rho)$ is in general an NP-hard problem [10]. Hence, one has to resort to the evaluation of upper and lower bounds on $R^*(n, \epsilon, \rho)$. Obtaining such bounds is a key problem in information theory, with a long history [9, 11–15].

Following the line of work initiated by [16], in this chapter we shall discuss nonasymptotic bounds on $R^*(n, \epsilon, \rho)$ that are provably tight for small values of the blocklength n. As we shall see, such bounds are by now available for a large class of communication channels of practical interest for next-generation wireless systems (5G and beyond). Furthermore, these bounds are numerically tight and provide useful engineering insights on the design of the physical layer of such systems.

Returning to our example, some of the best known bounds on $R^*(n, \epsilon, \rho)$ for the AWGN channel (10.1) are illustrated in Figure 10.1. The upper bound is based on the converse theorem [16, theorem 27], which we shall review in Section 10.3. The lower bound is Shannon's achievability bound [9].[1] The bounds are tight enough to provide a fairly accurate answer to our question: The minimum blocklength n^* needed by a coding scheme of rate $R = 0.35$ bit per channel use, operating over the AWGN channel (10.1) with signal-to-noise ratio (SNR) 0 dB, to achieve a packet error probability not exceeding 10^{-4} is between 321 and 399. In Figure 10.1, we also plot the follow-

---

[1] The numerical routines used to generate the figures in this chapter are contained in the SPECTRE toolbox [17] and can be downloaded at `http://github.com/yp-mit/spectre`.

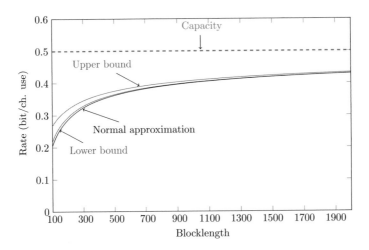

Figure 10.1 Upper bound, lower bound, and normal approximation on $R^*(n, \epsilon, \rho)$ for the AWGN channel (10.1) as a function of the blocklength n. Here, SNR $= 0\,$dB and $\epsilon = 10^{-4}$.

ing simple asymptotic approximation of $R^*(n, \epsilon, \rho)$, which is commonly referred to as *normal approximation* [16, eq. (296)], [18]:

$$R^*(n, \epsilon, \rho) \approx C(\rho) - \sqrt{\frac{V(\rho)}{n}} Q^{-1}(\epsilon) + \frac{1}{2n} \log n. \tag{10.6}$$

Here, $C(\rho)$ is the Shannon capacity defined in (10.2),

$$V(\rho) = \frac{\rho}{2} \frac{\rho + 2}{(\rho + 1)^2} \log^2 e \tag{10.7}$$

is the so-called *channel dispersion* [16, eq. (221)], and $Q(\cdot)$ denotes the Gaussian Q-function. This approximation, which is obtained by analyzing the converse and the achievability bound in the large n limit using the Berry–Esseén central-limit theorem [19, chapter XVI.5], is much easier to evaluate numerically than the nonasymptotic bounds, and yields the fairly accurate prediction $n^* \approx 420$. It is worth stressing that the normal approximation tends to be accurate only when the transmission rate is close to capacity, which rules out its application in scenarios when low latency is combined with an ultra-high reliability requirement and low SNR.

The normal approximation (10.6) provides a natural way to benchmark the performance of practical coding schemes, which is more accurate than the classic ϵ vs. E_b/N_0 curves. Specifically, one fixes a desired packet error probability ϵ and determines, for a given code with rate R, the minimum SNR $\rho_{\min}(\epsilon)$ needed to achieve ϵ. It is then natural to define the normalized rate $R_{\text{norm}}(\epsilon)$ as

$$R_{\text{norm}}(\epsilon) = \frac{R}{R^*(n, \epsilon, \rho_{\min}(\epsilon))}, \tag{10.8}$$

where the denominator in (10.8) can be evaluated using the normal approximation (10.6).

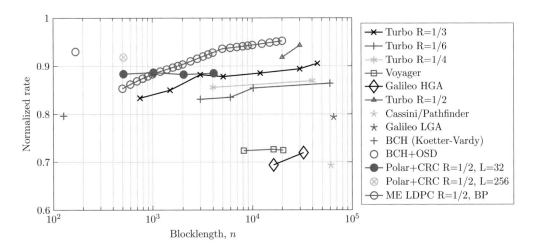

Figure 10.2 Normalized rate $R_{\mathrm{norm}}(\epsilon)$ of some practical coding schemes over a binary-input AWGN channel; $\epsilon = 10^{-4}$.

In Figure 10.2, we present an overview of the performance of some practical coding schemes for the binary-input AWGN channel from 1980 to the present, in terms of the normalized rate $R_{\mathrm{norm}}(\epsilon)$ for $\epsilon = 10^{-4}$. An earlier version of this figure appeared first in [20] and was later updated in [21] and [2]. The top eight codes in the legend are taken from the NASA JPL report [22]. The Bose–Chaudhuri–Hocquenghem (BCH) code with Koetter–Vardy decoder is taken from [23, figure 2]. The normalized rate reported here corresponds to the use of soft-decision list decoding.

Ordered statistics decoding (OSD) [24] yields a further improvement in performance. OSD decoding of nonbinary low-density parity-check (LDPC) codes and of convolutional codes with large constraint length turn out to yield similar performance as Bose–Chaudhuri–Hocquenghem (BCH)–OSD. Indeed, as we shall discuss in more detail in Section 10.5, this decoding technique yields state-of-the-art performance for short packets (between 100 and 200). For larger packet size, list decoding of polar codes combined with cyclic redundancy check (CRC) [21] and multi-edge (ME) type low-density parity-check (LDPC) codes [25] are a competitive benchmark. Indeed, the last two coding schemes have been chosen for 5G control and data channels, respectively [26]. The rest of this chapter is organized as follows: In Section 10.2, we introduce the system model model we shall focus on in most of the chapter, i.e., a memoryless block fading multiple-antenna channel. As we shall see, this model yields a good compromise between accuracy and analytical tractability. We shall also review the fundamental performance limits for such a channel in the asymptotic limits of large blocklength, under various assumptions on the availability of channel-state information (CSI). In Section 10.3, we introduce the theoretical tools needed to characterize the maximum coding rate $R^*(n, \epsilon, \rho)$ and describe general techniques to bound it. In Section 10.4, we present tight bounds on $R^*(n, \epsilon, \rho)$ for the multiantenna block-fading channel model introduced in Section 10.2. Finally, in Section 10.5, we discuss the design of practical coding schemes for short-packet transmissions.

10.2   Multiantenna Fading Channels

10.2.1   The Memoryless Block-Fading Model

We shall focus on a point-to-point multiple-input multiple-output (MIMO) communication link where the transmitter is equipped with m_t antennas and the receiver with m_r antennas. The channels connecting each antenna pairs is assumed to be affected by both additive (thermal) noise and multiplicative (fading) noise. We assume that the additive noise is independent and identically distributed (i.i.d.) and that the fading process evolves according to a block-fading model. Specifically, the fading coefficients stay constant over a block of n_c channel uses (coherence interval) and change independently across blocks. We also assume that each packet spans ℓ blocks, i.e., that the blocklength n is equal to ℓn_c (see Figure 10.3).

We next point out how such a channel model is relevant to the 5G physical layer. Like its predecessor, 5G will rely on orthogonal frequency-division multiplexing (OFDM). Specifically, each packet spans ℓ resource blocks, each consisting of n_o OFDM symbols of n_s (usually contiguous) subcarriers. To obtain frequency diversity, which is of tantamount importance in ultra-reliable communications, one needs to space the resource blocks in frequency so that they are separated by more than the coherence bandwidth of the channel. This yields independently fading resource blocks. Finally, one obtains a block-fading channel if each resource block lies within the coherence time and the coherence bandwidth of the channel so that the fading coefficients remain constant. An example of packet structure in 5G is depicted in Figure 10.4.

In the remainder of the chapter, we shall focus on a multiple-antenna extension of the block-fading model illustrated in Figure 10.3. It will be convenient to describe the channel's input–output relation within the kth coherence interval ($k = 1, \ldots, \ell$) in matrix form as

$$\mathbb{Y}_k = \mathsf{X}_k \mathbb{H}_k + \mathbb{W}_k. \tag{10.9}$$

Here, $\mathsf{X}_k \in \mathbb{C}^{n_c \times m_t}$ and $\mathbb{Y}_k \in \mathbb{C}^{n_c \times m_r}$ contain the transmitted and received symbols within the kth coherence interval and over all antennas, respectively; the entries of the matrix $\mathbb{H}_k \in \mathbb{C}^{m_t \times m_r}$ are the fading coefficients, which stay constant within the coherence block; finally, $\mathbb{W}_k \in \mathbb{C}^{n_c \times m_r}$ denotes the additive thermal noise at the receiver, which is independent of X_k and $\mathbb{H}_k$, and has i.i.d. $\mathcal{CN}(0, 1)$ entries. Both the noise process $\{\mathbb{W}_k\}$ and the fading process $\{\mathbb{H}_k\}$ are assumed to be stationary and memoryless. Throughout most of this chapter, we shall assume for simplicity that the entries of $\mathbb{H}_k$

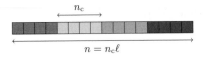

Figure 10.3 Each codeword of length n spans ℓ coherence blocks, each one involving n_c channel uses. The fading coefficient stays constant within each coherence block and takes independent values across coherence blocks.

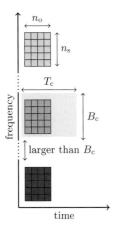

Figure 10.4 $\ell = 3$ independently fading resource blocks, each consisting of $n_o \times n_s$ channel uses in the time–frequency plane. Each resource block lies within the coherence bandwidth B_c and the coherence time T_c of the channel. Furthermore, the resource blocks are spaced in frequency by more than B_c. This yields a memoryless block-fading model.

are independent $\mathcal{CN}(0, 1)$ random variables. This corresponds to a spatially uncorrelated Rayleigh fading. As we shall point out, however, some of the results reviewed in this chapter hold for more general fading models.

In the remainder of this section, we shall first define the notion of maximum coding rate for the system model just described and illustrate the dependence of this quantity on the available channel state information (CSI). Then, we shall obtain classic asymptotic performance metrics such as ergodic and outage capacity as properly chosen asymptotic limits of the maximum coding rate. Finally, we will illustrate the limitations of these asymptotic performance metrics.

10.2.2   Maximum Coding Rate and CSI

The maximum coding rate achievable on the channel (10.9) depends crucially on the availability of CSI at the encoder and/or the decoder.

Four scenarios are possible (see Figure 10.5):

1. no-CSI: neither the encoder nor the decoder is aware of the realizations of the fading process $\{\mathbb{H}_k\}$.
2. CSIT: the encoder knows $\{\mathbb{H}_k\}$.
3. CSIR: the decoder knows $\{\mathbb{H}_k\}$.
4. CSIRT: both encoder and decoder know $\{\mathbb{H}_k\}$.

To keep the notation compact, we shall abbreviate in mathematical expressions the acronyms no-CSI, CSIT, CSIR, and CSIRT as no, tx, rx, and rt, respectively.

The no-CSI case is perhaps the most relevant in sporadic short-packet transmissions under stringent latency constraints. Indeed, in such a scenario, the fading coefficients are typically not known a priori to the encoder and the decoder. They may be estimated,

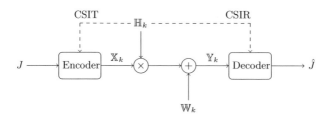

Figure 10.5 A pictorial representation of the input–output relation (10.9). The dashed lines represent the availability of CSI at encoder and decoder.

for example through the transmission of pilot symbols, but this yields a rate loss, which is automatically captured in the no-CSI setup [27]. Furthermore, stringent latency constraints may prevent the transmission of the CSI acquired at the decoder to the encoder via a feedback channel.

Next, we introduce the notion of channel code for each of the four settings.

DEFINITION 10.1 (no-CSI)   *An* $(\ell, n_c, M, \epsilon, \rho)_{no}$ *code for the channel (10.9) entails:*

1. *an encoder* f_{no}: $\{1, \ldots, M\} \mapsto \mathbb{C}^{n_c \times m_t \ell}$ *that maps the message* $J \in \{1, \ldots, M\}$ *to a codeword in the set* $\{C_1, \ldots, C_M\}$. *As each codeword* C_m, $m = 1, \ldots, M$, *spans* ℓ *coherence intervals; it is convenient to express it as the concatenation of* ℓ *subcodewords*

$$C_m = [C_{m,1}, \cdots, C_{m,\ell}]. \tag{10.10}$$

We assume that each subcodeword $C_{m,k} \in \mathbb{C}^{n_c \times m_t}$ *satisfies*[2]

$$\mathrm{tr}\{C_{m,k}^H C_{m,k}\} = n_c \rho, \quad m = 1, \ldots, M, \quad k = 1, \ldots, \ell. \tag{10.11}$$

2. *A decoder* g_{no} : $\mathbb{C}^{n_c \times m_r \ell} \to \{1, \ldots, M\}$ *satisfying the maximum error probability constraint*

$$\max_{1 \le j \le M} \mathbb{P}\left[g_{no}(\mathbb{Y}^\ell) \neq J \mid J = j\right] \le \epsilon \tag{10.12}$$

where

$$\mathbb{Y}^\ell = [\mathbb{Y}_1, \cdots, \mathbb{Y}_\ell] \tag{10.13}$$

is the channel output induced by the transmitted codeword

$$X^\ell = [X_1, \cdots, X_\ell] = f_{no}(j) \tag{10.14}$$

according to (10.9).

DEFINITION 10.2 (CSIR)   *An* $(\ell, n_c, M, \epsilon, \rho)_{rx}$ *code for the channel (10.9) entails:*

1. *an encoder* f_{no}: $\{1, \ldots, M\} \mapsto \mathbb{C}^{n_c \times m_t \ell}$ *that maps the message* $J \in \{1, \ldots, M\}$ *to a codeword satisfying the power constraint (10.11).*

---

[2] The constraint (10.11) is stronger than the (analog of) (10.3). Indeed, (10.11) implies that each codeword satisfies $\mathrm{tr}\{C_m^H C_m\} = \ell n_c \rho$, but the reverse implication does not hold. Because of the noise-variance normalization, ρ can be thought of as the SNR.

2. *A decoder* $g_{rx}: \mathbb{C}^{n_c \times m_r \ell} \times \mathbb{C}^{m_t \times m_r \ell} \mapsto \{1, \ldots, M\}$ *satisfying*

$$\max_{1 \leq j \leq M} \mathbb{P}[g_{rx}(\mathbb{Y}^\ell, \mathbb{H}^\ell) \neq J \,|\, J = j] \leq \epsilon. \qquad (10.15)$$

DEFINITION 10.3 (CSIT)   *An* $(\ell, n_c, M, \epsilon, \rho)_{tx}$ *code for the channel* (10.9) *entails:*

1. *an encoder* $f_{tx}: \{1, \ldots, M\} \times \mathbb{C}^{m_t \times m_r \ell} \mapsto \mathbb{C}^{n_c \times m_t \ell}$ *that maps the message* $J \in \{1, \ldots, M\}$ *and the channel* $\mathbb{H}^\ell$ *to a codeword* $C_j = f_{tx}(j, \mathbb{H}^\ell)$ *satisfying the power constraint* (10.11) *for all* $j \in \{1, \ldots, M\}$ *and all* $\mathbb{H}^\ell \in \mathbb{C}^{m_t \times m_r \ell}$.
2. *A decoder* $g_{no}: \mathbb{C}^{n_c \times m_r \ell} \mapsto \{1, \ldots, M\}$ *satisfying* (10.12).

DEFINITION 10.4 (CSIRT)   *An* $(\ell, n_c, M, \epsilon, \rho)_{rt}$ *code for the channel* (10.9) *entails:*

1. *an encoder* $f_{tx}: \{1, \ldots, M\} \times \mathbb{C}^{m_t \times m_r \ell} \mapsto \mathbb{C}^{n_c \times m_t \ell}$ *that maps the message* $J \in \{1, \ldots, M\}$ *and the channel* $\mathbb{H}^\ell$ *to a codeword* $C_j = f_{tx}(j, \mathbb{H}^\ell)$ *satisfying the power constraint* (10.11) *for all* $j \in \{1, \ldots, M\}$ *and all* $\mathbb{H}^\ell \in \mathbb{C}^{m_t \times m_r \ell}$.
2. *A decoder* $g_{rx}: \mathbb{C}^{n_c \times m_r \ell} \times \mathbb{C}^{m_t \times m_r \ell} \mapsto \{1, \ldots, M\}$ *satisfying* (10.15).

The maximum coding rate for the four cases listed above is

$$R_p^*(\ell, n_c, \epsilon, \rho) = \sup \left\{ \frac{\log M}{n} : \exists (\ell, n_c, M, \epsilon, \rho)_p \text{ code} \right\}, \quad p \in \{no, rx, tx, rt\}. \quad (10.16)$$

By definition, we have

$$R_{csi}^* \leq R_{rx}^* \leq R_{rt}^* \qquad (10.17)$$

and

$$R_{no}^* \leq R_{tx}^* \leq R_{rt}^*. \qquad (10.18)$$

10.2.3   Asymptotic Behavior of R^*: Ergodic and Outage Capacity

As we shall discuss next, classic results on the capacity of fading channels (see e.g., [28, 29] for a comprehensive literature review on this topic) can be obtained by analyzing the asymptotic behavior of R^*.

Ergodic Capacity

Analyzing $R_{rx}^*(\ell, n_c, \epsilon, \rho)$ in the asymptotic limit $\ell \to \infty$ (codewords spanning an arbitrarily large number of coherence intervals) for fixed n_c, ϵ, and ρ, we recover the ergodic-capacity formula [30][3]

$$\lim_{\ell \to \infty} R_{rx}^*(\ell, n_c, \epsilon, \rho) = C_{rx}^{erg} = \mathbb{E}\left[\log \det \left(I_{m_r} + \frac{\rho}{m_t} \mathbb{H}^H \mathbb{H} \right) \right]. \qquad (10.19)$$

Here, $\mathbb{H}$ is a $m_t \times m_r$ complex matrix with i.i.d. $\mathcal{CN}(0, 1)$ entries. An important observation is that the right-hand side (RHS) of (10.19) does not depend on ϵ and n_c.

---

[3]   To keep the notation consistent with (10.9), we shall gather the fading coefficients in a matrix of dimension $m_t \times m_r$ rather than of dimension $m_r \times m_t$, as more commonly done.

If CSIRT is available, adapting the input covariance matrix to the channel realizations allows one to improve the achievable rates. Specifically,

$$\lim_{\ell \to \infty} R_{\mathrm{rt}}^*(\ell, n_{\mathrm{c}}, \epsilon, \rho) = C_{\mathrm{rt}}^{\mathrm{erg}} = \sup_{\mathsf{Q}(\mathbb{H}) : \mathbb{E}\,[\mathrm{tr}\,\mathsf{Q}] = \rho} \mathbb{E}\left[\log \det\!\left(\mathsf{I}_{m_{\mathrm{r}}} + \mathbb{H}^H \mathsf{Q}\mathbb{H}\right)\right]. \tag{10.20}$$

Here, Q is a positive-defined matrix whose eigenvectors are the same as that of $\mathsf{H}\mathsf{H}^H$ and whose eigenvalues are obtained through water-filling on the eigenvalues of $\mathsf{H}\mathsf{H}^H$.

In the no-CSI case, no closed-form expression for the ergodic capacity, which we shall denote by $C_{\mathrm{no}}^{\mathrm{erg}}$, is available. Its high-SNR behavior is, however, known [31, 32]; specifically, when $n_{\mathrm{c}} > 1$,

$$\lim_{\ell \to \infty} R_{\mathrm{no}}^*(\ell, n_{\mathrm{c}}, \epsilon, \rho) = C_{\mathrm{no}}^{\mathrm{erg}} = m^*\left(1 - \frac{m^*}{n_{\mathrm{c}}}\right)\log \rho + c + o_\rho(1), \quad \rho \to \infty \tag{10.21}$$

where

$$m^* = \min\{m_{\mathrm{t}}, m_{\mathrm{r}}, \lfloor n_{\mathrm{c}}/2 \rfloor\}, \tag{10.22}$$

where c is a constant that depends on m_{t}, m_{r}, and n_{c} (see [32, Eq. (9)]) and where $o_\rho(1)$ denotes terms that vanish when $\rho \to \infty$. Tight bounds on the ergodic capacity for the no-CSI case are also available [33, 34]. For the case $n_{\mathrm{c}} = 1$, capacity grows double-logarithmically in ρ; similarly as in (10.21), the second-order term in the asymptotic expansion of capacity (the so-called *fading number*) depends on the fading distribution and on the number of available antennas [35–37].

Outage Capacity
The outage capacity [38] can be recovered from the maximum coding rate by taking the limit $n_{\mathrm{c}} \to \infty$ (arbitrarily large coherence interval) for fixed ϵ, ℓ, and ρ. Specifically,

$$\lim_{n_{\mathrm{c}} \to \infty} R_{\mathrm{rx}}^*(\ell, n_{\mathrm{c}}, \epsilon, \rho) = \lim_{n_{\mathrm{c}} \to \infty} R_{\mathrm{no}}^*(\ell, n_{\mathrm{c}}, \epsilon, \rho) = C_{\mathrm{rx}}^{\mathrm{out}, \epsilon} \tag{10.23}$$

$$= \sup\left\{R : \inf_{\{\mathsf{Q}_k\}_{k=1}^{\ell}} P_{\mathrm{rx}}^{\mathrm{out}}\!\left(\{\mathsf{Q}_k\}_{k=1}^{\ell}, R\right) \leq \epsilon\right\}, \tag{10.24}$$

where the $\{\mathsf{Q}_k\}$ are diagonal matrices with nonnegative entries satisfying $\mathrm{tr}\{\mathsf{Q}_k\} = \rho$ (recall that we assumed Rayleigh fading) and where $P_{\mathrm{rx}}^{\mathrm{out}}(\cdot, \cdot)$ is the outage probability

$$P_{\mathrm{rx}}^{\mathrm{out}}\!\left(\{\mathsf{Q}_k\}_{k=1}^{\ell}, R\right) = \mathbb{P}\left\{\frac{1}{\ell}\sum_{k=1}^{\ell} \log \det(\mathsf{I}_{m_{\mathrm{r}}} + \mathbb{H}_k^H \mathsf{Q}_k \mathbb{H}_k) \leq R\right\}. \tag{10.25}$$

For the case $\ell = 1$, Telatar [30] conjectured that the optimal diagonal matrix Q_1 is of the form

$$\mathsf{Q}_1 = \frac{\rho}{m}\, \mathrm{diag}\{\underbrace{1, \ldots, 1}_{m}, \underbrace{0, \ldots, 0}_{m_{\mathrm{t}} - m}\} \tag{10.26}$$

for some $m \in \{1, \ldots, m_{\mathrm{t}}\}$. This conjecture was proved in [39] for the multiple-input single-output case.

An important observation is that (10.24) holds irrespective of whether CSI is available at the receiver. Indeed, when n_{c} gets large, the cost of learning the channel at the receiver, which determines the gap in achievable rate between the CSIR and the no-CSI cases, vanishes.

As in the ergodic case, the availability of CSIRT allows for an optimization of the input covariance matrix, which yields

$$\lim_{n_c \to \infty} R_{rt}^*(\ell, n_c, \epsilon, \rho) = C_{rt}^{out,\epsilon} = \sup\left\{R : P_{rt}^{out}(R) < \epsilon\right\}, \tag{10.27}$$

where

$$P_{rt}^{out}(R) = \mathbb{P}\left\{\sup_{\{Q_k\}_{k=1}^{\ell}} \frac{1}{\ell} \sum_{k=1}^{\ell} \log\det(I_{m_r} + \mathbb{H}_k^H Q_k \mathbb{H}_k) \leq R\right\}. \tag{10.28}$$

As in the ergodic case, the supremum in (10.28) is achieved by water-filling.

Diversity-Multiplexing Tradeoff

The last asymptotic regime of interest is when $\rho \to \infty$, ℓ and n_c are fixed, and ϵ is the following function of ρ:

$$\epsilon(\rho) = \rho^{-d\ell}. \tag{10.29}$$

Here, $d \in \{0, 1, \ldots, m_t m_r\}$ is the so-called *spatial diversity gain*. When CSIR is available, the multiplexing gain $r(d)$, as defined by Zheng and Tse [40], can be obtained by evaluating the following limit:

$$\lim_{\rho \to \infty} \frac{R_{rx}^*(\ell, n_c, \epsilon(\rho), \rho)}{\log \rho} = r(d). \tag{10.30}$$

As proven in [40], $r(d)$ is the piece-wise linear function connecting the points

$$r\big((m_t - k)(m_r - k)\big) = k, \quad k = 0, \ldots, \min\{m_t, m_r\}. \tag{10.31}$$

In the no-CSI case, we have that [41]

$$\lim_{\rho \to \infty} \frac{R_{no}^*(n_c, \ell, \epsilon(\rho), \rho)}{\log \rho} = \left(1 - \frac{m^*}{n_c}\right) r(d), \tag{10.32}$$

with m^* given in (10.22).

Limitations of Asymptotic Analyses

As we shall see, the asymptotic performance metrics just reviewed, i.e., ergodic and outage capacity, and diversity-multiplexing tradeoffs (DMTs), have limited relevance for short-packet communications. The ergodic capacity, for example, requires the availability of a large number of independently fading coherence blocks, and does not depend on the targeted packet error probability ϵ and the size of the coherence interval n_c. The outage capacity is unable to capture the overhead due to channel estimation, which is often significant for short packets. Finally, the DMT is asymptotic in the SNR whereas typical applications for which short packets are needed, require low SNR.

Motivated by these limitations, we shall present in Section 10.4 tight nonasymptotic bounds on R^* and discuss the novel engineering insights on the design of short-packet communication systems enabled by these bounds. But first we shall present in the next section the fundamental tools that allow one to bound R^* tightly.

10.3   Finite-Blocklength Bounds: Theoretical Foundations

A communication channel is simply a system with an input and an output. Most commonly, we think of a channel as acting on n-letter inputs and producing n-letter outputs. In such cases, the end-to-end conditional probability $P_{Y^n|X^n}$ comprises the complete description of the channel. A natural question is how many bits of data one can transmit per letter. Shannon's formalization of this question leads to a rule of thumb, that the maximal data rate should be given by

$$\frac{1}{n} \max_{P_{X^n}} I(X^n; Y^n) \, . \tag{10.33}$$

This answer is precise when the blocklength n is very large, and becomes somewhat meaningless for short-to-moderate blocklengths. One way to see this is that the above expression does not even depend on the required end-to-end probability of error ϵ. In this section[4] we will present some basic tools for bounding the maximal achievable rate over any abstract channel. The overall summary is that instead of (10.33) one should use the $(1 - \epsilon)$-th quantile of

$$\log \frac{dP_{Y^n|X^n}}{dP_{Y^n}} \, , \tag{10.34}$$

where the pair (X^n, Y^n) is distributed according to the joint distribution induced through $P_{Y^n|X^n}$ by the maximizer of (10.33). Note that the mean of (10.34) is precisely $I(X^n; Y^n)$.

10.3.1   Preliminary Definitions

A transition probability kernel acting between two alphabets $T : \mathsf{A} \to \mathsf{B}$ assigns to each $x \in \mathsf{A}$ a probability measure $T(\cdot|x)$ on B, such that for every $E \in \sigma\mathsf{B}$ the function $x \mapsto T(E|x)$ is measurable. Here, $\sigma\mathsf{A}$ and $\sigma\mathsf{B}$ are σ-fields. Transition probability kernels will also be called "randomized functions" since a deterministic function $f : \mathsf{A} \to \mathsf{B}$ corresponds to the kernel

$$T_f(E|x) = 1\{f(x) \in E\}. \tag{10.35}$$

Similar to maps, transition probability kernels $T : \mathsf{A} \to \mathsf{B}$ and $S : \mathsf{B} \to \mathsf{W}$ can be composed to give a kernel $S \circ T : \mathsf{A} \to \mathsf{W}$ by

$$S \circ T(E|x) = \int_{\mathsf{B}} S(E|w) T(dw|x) \, , \tag{10.36}$$

where integration is over the conditional measure $T(\cdot|x)$ on W. A special case of this composition is when we write

$$P_{Y|X} \circ P_X(E) = \int_{\mathsf{B}} P_{Y|X}(E|w) P_X(dx) \, ,$$

---

[4] This material is primarily based on [16, 20, 42].

for a marginal distribution P_Y induced by $P_{Y|X}$ acting on P_X. Given a pair of random transformations $P_{Y_i|X_i} : A_i \to B_i$, $i = 1, 2$, we define their product as a random transformation $P_{Y^2|X^2} : A_1 \times A_2 \to B_1 \times B_2$ with

$$P_{Y^2|X^2}(\cdot|x_1, x_2) = P_{Y_1|X_1}(\cdot|x_1) \times P_{Y_2|X_2}(\cdot|x_2).$$

Following [43], we shall typically take A and B to be n-fold Cartesian products of some single-letter alphabets $\mathcal{A}$ and $\mathcal{B}$. For memoryless channel kernels, $P_{Y^n|X^n}$ is given as the n-fold product of a single-letter kernel $P_{Y|X}$. Thus, the elements of A and B (and the values of random variables X and Y) throughout the subsequent sections can be thought of as vectors of fixed dimension equal to the blocklength n. Throughout this subsection, the dependence of the channel kernel on n will not be made explicit, to keep the notation compact.

We next generalize the notion of channel code provided in Section 10.1.3 for the AWGN channel to general channel kernels.

DEFINITION 10.5 (Channel code) *An M-code for the random transformation $P_{Y|X}$:* A → B *is defined by an (encoder) map $f : \{1, \ldots, M\} \to$ A and a transition probability kernel (decoder) $g :$ B → $\{1, \ldots, M\}$. The elements of the image of f are called codewords. The pair (f, g) is an (M, ϵ)-code if*

$$\frac{1}{M} \sum_{j=1}^{M} \left(1 - P_{Y|X}(g^{-1}(j)|f(j))\right) \leq \epsilon, \tag{10.37}$$

where the left-hand side is known as the (average) probability of (block/frame) error.

We are frequently interested in codes further satisfying an *input constraint*, such as the power constraint (10.3). In such cases, the input space A should be understood as the collection of all possible n-dimensional vectors satisfying the input constraint. For example, as we have already seen, for the AWGN channel we have $A = \{x \in \mathbb{R}^n : \sum_{i=1}^n |x_i|^2 \leq n\rho\}$ where $\rho > 0$ is the power constraint.

For every output distribution P_Y, we define the *information density* as

$$i(x; y) = i_{P_{Y|X} \| P_Y}(x; y) = \log \frac{dP_{Y|X=x}}{dP_Y}(y), \tag{10.38}$$

where $\frac{dP}{dQ}$ is the Radon–Nikodym derivative of P with respect to Q. For the applications of interest in this book, this derivative can always be taken to be the ratio of corresponding probability density functions (PDFs).

10.3.2 Binary Hypothesis Testing

Many of the finite-blocklength tools we shall review in this section entail the evaluation of the optimal performance of a suitably designed binary hypothesis test. Consider a random variable W on W that can take one of the two distributions P and Q. A randomized test between these two distributions is a random transformation (a transition probability kernel) $P_{Z|W} : W \to \{0, 1\}$, where 0 indicates that the test chooses Q. The optimal performance among all such transformations is denoted by[5]

$$\beta_\alpha(P, Q) = \inf_{\substack{P_{Z|W} : \\ \sum_{w \in W} P_{Z|W}(1|w)P(w) \geq \alpha}} \left[\sum_{w \in W} P_{Z|W}(1|w)Q(w) \right]. \qquad (10.39)$$

Thus, $\beta_\alpha(P, Q)$ gives the minimum probability of error under hypothesis Q if the probability of success under hypothesis P is at least α. Next, we recall some basic properties of this function (see [44] for a more complete survey).

The function $\alpha \mapsto \beta_\alpha$ is nondecreasing and convex in $\alpha \in [0, 1]$. Furthermore, the infimum in (10.39) is achieved by the randomized test specified in the following lemma due to Neyman and Pearson (see, e.g., [45]).

LEMMA 10.6 (Neyman–Pearson)   *Consider a space W and two probability measures P and Q. Then for every $\alpha \in [0, 1]$ there exists a $\gamma > 0$ and a $\tau \in [0, 1)$ such that*

$$\beta_\alpha(P, Q) = Q[Z_\alpha^* = 1], \qquad (10.40)$$

and where[6] the conditional probability $P_{Z^*|W}$ is defined via

$$Z_\alpha^*(W) = 1\left\{\frac{dP}{dQ} > \gamma\right\} + Z_\tau 1\left\{\frac{dP}{dQ} = \gamma\right\}, \qquad (10.41)$$

where $Z_\tau \in \{0, 1\}$ equals 1 with probability τ and is independent of W. The constants γ and τ are uniquely determined by the constraint

$$P[Z_\alpha^* = 1] = \alpha. \qquad (10.42)$$

Moreover, any other test Z satisfying $P[Z = 1] \geq \alpha$ either differs from Z_α^* only on the set $\left\{\frac{dP}{dQ} = \gamma\right\}$ or results in $Q[Z = 1] > \beta_\alpha(P, Q)$.

In other words, β_α is a piecewise linear function, joining the points

$$\begin{cases} \beta_\alpha = Q\left[\frac{dP}{dQ} \geq \gamma\right], \\ \alpha = P\left[\frac{dP}{dQ} \geq \gamma\right] \end{cases} \qquad (10.43)$$

for all $\gamma > 0$.

The following bounds are easy to show ([20, Eq. (2.59)]):

$$\beta_\alpha(P, Q) \geq \frac{1}{\gamma}\left(\alpha - P\left[\frac{dP}{dQ} \geq \gamma\right]\right) \qquad (10.44)$$

$$\beta_\alpha(P, Q) \leq \frac{1}{\gamma_0} P\left[\frac{dP}{dQ} \geq \gamma_0\right] \qquad (10.45)$$

---

[5] Here and in the remainder of the section, we write summations over alphabets instead of integrals, whenever this does not cause confusion. However, all of the results in this section hold also for nondiscrete measures and uncountable alphabets.

[6] In the case in which P is not absolutely continuous with respect to Q, we can define $\frac{dP}{dQ}$ to be equal to $+\infty$ on the singular set and, hence, to be automatically included in every optimal test.

$$\leq \frac{1}{\gamma_0}, \tag{10.46}$$

where $\gamma > 0$ is arbitrary and γ_0 satisfies

$$P\left[\frac{dP}{dQ} \geq \gamma_0\right] \geq \alpha. \tag{10.47}$$

The bounds (10.44) and (10.46) imply that β_α behaves approximately as the exponent of the negative of the α-th quantile of $\log\frac{dP}{dQ}$ under P. In this chapter, we will mostly deal with distributions that are n-fold products of a fixed distribution. In this case, $\log\frac{dP}{dQ}$ is a sum of i.i.d. random variables and the quantile behavior is governed by the central-limit theorem (CLT), or, more precisely, by the Berry–Esseen theorem (see, e.g., [19, theorem 2, chapter XVI.5]). In particular, we have [16, app. A]:

LEMMA 10.7 *Let $\mathcal{A}$ be a measurable space with measures $\{P_i\}$ and $\{Q_i\}$, with $P_i \ll Q_i$ defined on it for $i = 1, \ldots, n$. Define two measures on $\mathcal{A}^n$: $P = \prod_{i=1}^{n} P_i$ and $Q = \prod_{i=1}^{n} Q_i$, and let*

$$D_n = \frac{1}{n}\sum_{i=1}^{n} D(P_i\|Q_i), \tag{10.48}$$

$$V_n = \frac{1}{n}\sum_{i=1}^{n} V(P_i\|Q_i) = \frac{1}{n}\sum_{i=1}^{n}\int\left(\log\frac{dP_i}{dQ_i}\right)^2 dP_i - D(P_i\|Q_i)^2, \tag{10.49}$$

$$T_n = \frac{1}{n}\sum_{i=1}^{n}\int\left|\log\frac{dP_i}{dQ_i} - D(P_i\|Q_i)\right|^3 dP_i, \tag{10.50}$$

$$B_n = 6\frac{T_n}{V_n^{3/2}}. \tag{10.51}$$

Assume that all these quantities are finite and that $V_n > 0$. Then, for every $\Delta > 0$

$$\log\beta_\alpha(P,Q) \geq -nD_n - \sqrt{nV_n}Q^{-1}\left(\alpha - \frac{B_n + \Delta}{\sqrt{n}}\right) - \frac{1}{2}\log n + \log\Delta, \tag{10.52}$$

$$\log\beta_\alpha(P,Q) \leq -nD_n - \sqrt{nV_n}Q^{-1}\left(\alpha + \frac{B_n}{\sqrt{n}}\right) - \frac{1}{2}\log n$$

$$+ \log\left(\frac{2\log 2}{\sqrt{2\pi V_n}} + 4B_n\right). \tag{10.53}$$

Each bound holds provided that the argument of Q^{-1} lies in the interval $(0, 1)$.

In particular, when $P_i = P$ and $Q_i = Q$, $i = 1, \ldots, n$, $V(P\|Q) > 0$, and the third moment of $\log\frac{dP}{dQ}$ is finite, we have

$$\log\beta_\alpha\left(P^n, Q^n\right) = -nD(P\|Q) - \sqrt{nV(P\|Q)}Q^{-1}(\alpha) - \frac{1}{2}\log n + O(1). \tag{10.54}$$

If $V(P\|Q) = 0$, then we trivially have

$$\log\beta_\alpha\left(P^n, Q^n\right) = -nD(P\|Q) + \log\alpha. \tag{10.55}$$

The asymptotic expression (10.53) is one of the key steps to establish the normal approximation (10.6).

We shall next introduce the nonasymptotic bounds on the maximum coding rate we will use in Section 10.4 to characterize the performance achievable over fading channels under different assumptions on the available CSI. We start with two achievability bounds, namely the dependence-testing (DT) bound in Section 10.3.3 and the $\kappa\beta$ bound in Section 10.3.4, and then present a converse bound in Section 10.3.5.

10.3.3   Achievability I: Dependence Testing Bound

THEOREM 10.8 (DT)   *For every distribution P_X on A, there exists a code with M codewords and average probability of error not exceeding*

$$\epsilon \leq \mathbb{E}\left[\exp\left\{-\left|i(X;Y) - \log\frac{M-1}{2}\right|^+\right\}\right] \tag{10.56}$$

where $P_{XY}(a,b) = P_X(a)P_{Y|X}(b|a)$.

Before proving the theorem, we formulate and prove a useful lemma.

LEMMA 10.9   *Fix a P_X on A and a measurable function $\gamma : \mathsf{A} \mapsto [0,\infty]$; furthermore, let $P_Y(y) = \sum_{x\in\mathsf{A}} P_{Y|X}(y|x)P_X(x)$. There exists an (M,ϵ) code (average probability of error) satisfying*

$$\epsilon \leq \mathbb{P}[i(X;Y) \leq \log\gamma(X)] + \frac{M-1}{2}\mathbb{P}[i(X;\bar{Y}) > \log\gamma(X)], \tag{10.57}$$

where $P_{XY\bar{Y}}(a,b,c) = P_X(a)P_{Y|X}(b|a)P_Y(c)$.

Proof of Lemma 10.9.   The idea of the proof is to average the probability of error over random codebooks generated using the distribution P_X; the decoder runs M binary hypothesis tests in parallel, the jth of which is between the true distribution $P_{Y|X=c_j}$ and the "average noise" P_Y.

Let $\{Z_x\}_{x\in\mathsf{A}}$ be a collection of deterministic functions over B defined as

$$Z_x(y) = 1\{i(x;y) > \log\gamma(x)\}.$$

First we describe the operation of the decoder given the codebook $\{c_i\}_{i=1}^M$; the decoder computes the values $Z_{c_j}(y)$ for the received channel output y and returns the first index j for which $Z = 1$ (or 0 if all of them are 0). In this way, the average probability of error is given as

$$\epsilon(c_1,\ldots c_M) = \frac{1}{M}\sum_{i=1}^M \lambda_i, \tag{10.58}$$

where

$$\lambda_j = \mathbb{P}\left[\{Z_{c_j}(Y) = 0\}\bigcup_{i<j}\{Z_{c_i}(Y) = 1\}\,\middle|\, X = c_j\right], \tag{10.59}$$

or, using the union bound and the definition of $Z_x(y)$,

$$\lambda_j \leq \mathbb{P}[i(c_j, Y) \leq \log \gamma(c_j) \mid X = c_j] + \sum_{i<j} \mathbb{P}[i(c_i, Y) > \log \gamma(c_i) \mid X = c_j]. \quad (10.60)$$

We will now average each expression in (10.60) over codebooks $\{c_i\}$ that are generated as (pairwise) independent random variables with distribution P_X:

$$\mathbb{E}[\lambda_j] \leq \mathbb{P}[i(X; Y) \leq \log \gamma(X)] + (j-1)\mathbb{P}[i(X; \bar{Y}) > \log \gamma(X)].$$

Then, from (10.58) we find that the ensemble average of ϵ satisfies

$$\mathbb{E}[\epsilon(c_1, \ldots c_M)] \leq \mathbb{P}[i(X; Y) \leq \log \gamma(X)] + \frac{M-1}{2}\mathbb{P}[i(X; \bar{Y}) > \log \gamma(X)]. \quad \blacksquare$$

Proof of Theorem 10.8. Notice that by taking a conditional expectation given $X = x$ in (10.57) we obtain

$$P_{Y|X=x}[i(x; Y) \leq \log \gamma(x)] + \frac{M-1}{2}P_Y[i(x; Y) > \gamma(x)], \quad (10.61)$$

which is a weighted sum of two types of errors. This thus corresponds to the average error probability in a Bayesian hypothesis testing problem for which the optimal solution is the likelihood ratio test with threshold $\gamma(x) = (M-1)/2$. Simple algebra shows that with this choice of $\gamma(x)$, the bound (10.57) becomes (10.56). $\quad \blacksquare$

10.3.4   Achievability II: $\kappa\beta$ Bound

The DT bound is difficult to compute numerically or to analyze analytically when the output distribution P_Y in the information density (10.38) is not a product distribution. In such cases, it is convenient to replace P_Y by an easier to analyze output distribution Q_Y. This is exactly the idea behind the $\kappa\beta$ bound described in this section.

Let $F \subset A$ be the set of permissible inputs. For an arbitrary $F \subset A$, we define a related measure of performance for the composite hypothesis test between Q_Y and the collection $\{P_{Y|X=x}\}_{x \in F}$:

$$\kappa_\tau(F, Q_Y) = \inf_{\substack{P_{Z|Y}: \\ \inf_{x \in F} P_{Z|X}(1|x) \geq \tau}} \sum_{y \in B} Q_Y(y)P_{Z|Y}(1|y). \quad (10.62)$$

As long as Q_Y is the output distribution induced by an input distribution Q_X, the quantity (10.62) satisfies the bound

$$\tau Q_X[F] \leq \kappa_\tau(F, Q_Y) \leq \tau. \quad (10.63)$$

The RHS bound is achieved by choosing the test Z that is equal to 1 with probability τ regardless of Y. To prove the left-hand bound, note that for any $P_{Z|Y}$ that satisfies the condition in (10.62), we have

$$\sum_{y \in B} Q_Y(y)P_{Z|Y}(1|y)$$

$$= \sum_{x\in A} \sum_{y\in B} Q_X(x) P_{Y|X}(y|x) P_{Z|Y}(1|y) \tag{10.64}$$

$$\geq \sum_{x\in F} Q_X(x) \sum_{y\in B} P_{Y|X}(y|x) P_{Z|Y}(1|y) \tag{10.65}$$

$$\geq \sum_{x\in F} Q_X(x) \left\{ \inf_{x\in F} \sum_{y\in B} P_{Y|X}(y|x) P_{Z|Y}(1|y) \right\} \tag{10.66}$$

$$\geq \tau Q_X[F]. \tag{10.67}$$

THEOREM 10.10 ($\kappa\beta$ bound) *For every $0 < \epsilon < 1$, every $0 < \tau < \epsilon$ and every distribution Q_Y on B, there exists an (M,ϵ) code with codewords chosen from $F \subset A$, satisfying*

$$M \geq \frac{\kappa_\tau(F, Q_Y)}{\sup_{x\in F} \beta_{1-\epsilon+\tau}(x, Q_Y)}. \tag{10.68}$$

Note: It is possible[7] that (10.68) will be of the form $M \geq \alpha/0$ with $\alpha > 0$. In this case, the statement of the theorem should be understood as "(M,ϵ) codes with arbitrarily high M exist."

Proof.   We first describe the operation of the decoder given a codebook $\{c_i\}_{i=1}^M$. Upon reception of $y \in B$ the decoder sequentially tests whether codeword c_i was sent, where i runs from 1 to M. The test for c_i is performed as a binary hypothesis test discriminating $P_{Y|X=c_i}$ (hypothesis H_0) against "average noise" Q_Y (hypothesis H_1). We would like to select each such test as an optimal one within the constraint $P(\text{decide } H_0|H_0) \geq 1 - \epsilon + \tau$. To do this we define a collection of random variables $Z(x), x \in F$ conditionally independent given Y and with $P_{Z(x)|Y}$ chosen so that it achieves $\beta_{1-\epsilon+\tau}(x, Q_Y)$ in (10.39). In other words,

$$P[Z(x) = 1|X = x] \geq 1 - \epsilon + \tau, \tag{10.69}$$

$$Q[Z(x) = 1] = \beta_{1-\epsilon+\tau}(x, Q_Y), \tag{10.70}$$

where we denoted

$$Q[Z(x) = 1] = \int_B P_{Z(x)|Y}(1|y) Q(dy).$$

The decoder applies the M independent random transformations $P_{Z(c_1)|Y}, \dots, P_{Z(c_M)|Y}$ to the channel output Y and produces as output the first index j such that $Z(c_j) = 1$, or 1 if all Z are zero.

Having specified the decoder operation, we proceed to generate the codebook $\{c_i\}_{i=1}^M$. This will be done in a manner similar to the maximal coding idea of Feinstein.

As a first step, we choose an arbitrary $c_1 \in F$. Then, by (10.69) we know that the described decoder will decode c_1 correctly with probability of at least $1 - \epsilon + \tau$, which is better than $1 - \epsilon$. So c_1 does not violate the probability of error criterion. Next,

---

[7] For an example of such a case, take $A = B = [0, 1]$ with the Borel σ-algebra. Define $P_{Y|X=x}(y) = \delta_x(y)$, i.e., a point measure at $y = x$, and take Q_Y to be Lebesgue measure. Then, $\beta_\alpha(x, Q_Y) = 0$ for every x and α, and $\kappa_\tau(Q_Y) = 1$ for every $\tau > 0$.

suppose that j codewords have already been selected. We can select some $x \in \mathsf{F}$ as the next codeword c_{j+1} only provided that

$$P\left[Z(x) = 1, Z(c_1) = \cdots = Z(c_j) = 0 | X = x\right] \geq 1 - \epsilon. \tag{10.71}$$

If we cannot find any such x then we stop; otherwise we choose any x satisfying (10.71).

There are two cases. Either the process continues indefinitely, in which case there is nothing to prove, or it stops after a finite number of steps M. In the latter case, we have found an (M, ϵ) code and we need only to show that M satisfies the bound in (10.68). Note that there is a large amount of freedom in the process: Each random variable $Z(c_i)$ is perhaps not uniquely defined by c_i, the choice of c_{j+1} is not unique, etc. However, the lower bound on M will be independent of all those choices.

Let

$$V_M = \max\{Z(c_j), j = 1, \ldots, M\}.$$

If the process stops after M steps, this implies that for every $x \in \mathsf{F}$ we have

$$P[Z(x) = 1, V_M = 0 | X = x] < 1 - \epsilon. \tag{10.72}$$

But by the definition of $Z(x)$ and by (10.69), it follows that

$$1 - \epsilon + \tau \leq P[Z(x) = 1 | X = x] \tag{10.73}$$
$$\leq P[Z(x) = 1, V_M = 0 | X = x] + P[V_M = 1 | X = x] \tag{10.74}$$
$$< 1 - \epsilon + P[V_M = 1 | X = x]. \tag{10.75}$$

Thus, V_M is a random variable taking values in $\{0, 1\}$ and such that, for every $x \in \mathsf{F}$

$$P[V_M = 1 | X = x] \geq \tau. \tag{10.76}$$

But then, V_M defines a composite hypothesis test and, by the definition of κ_τ in (10.62), we have

$$Q[V_M = 1] \geq \kappa_\tau(\mathsf{F}, Q_Y). \tag{10.77}$$

Note also that

$$Q[V_M = 1] = Q\left[\bigcup_{j=1}^{M} \{Z(c_j) = 1\}\right] \tag{10.78}$$

$$\leq \sum_{j=1}^{M} Q[Z(c_j) = 1] \tag{10.79}$$

$$= \sum_{j=1}^{M} \beta_{1-\epsilon+\tau}(c_j, Q_Y) \tag{10.80}$$

$$\leq M \sup_{x \in \mathsf{F}} \beta_{1-\epsilon+\tau}(x, Q_Y), \tag{10.81}$$

where (10.80) follows by (10.70). Finally, (10.81) and (10.77) imply (10.68). ∎

Using (10.63) in Theorem 10.10 we obtain a weakened but useful bound:

$$M \geq \sup_{0<\tau<\epsilon} \sup_{Q_X} \frac{\tau Q_X[\mathsf{F}]}{\sup_{x\in\mathsf{F}} \beta_{1-\epsilon+\tau}(x, Q_Y)}, \tag{10.82}$$

where the supremum is over all input distributions, and Q_Y denotes the distribution induced by Q_X on the output. An interesting connection between the weakened form of the $\kappa\beta$ bound in (10.82) and the DT bound comes from the following observation. By a judicious choice of $\gamma(x)$ in Lemma 10.9 we could have obtained the bound (10.82) for average probability error with supremum in the denominator replaced by the average over Q_X.

In (10.39) and (10.62) we have defined β_α and κ_τ using randomized tests. Then, in Theorem 10.10 we have constructed the coding scheme with a randomized decoder. Correspondingly, if we define β_α and κ_τ using nonrandomized tests, then the analog of Theorem 10.10 for a nonrandomized decoder can be proved.

10.3.5   Converse Bound: Meta-Converse Method

In this section, we develop a method for proving converse, i.e., "impossibility," results. The central idea can be summarized as follows: We take an arbitrary code for the channel $P_{Y|X}$; we prove that, if used on a different channel $Q_{Y|X}$, this code must have large probability of error with a guaranteed lower bound; we then show that there is a link between the probability of error on the Q-channel and the probability of error on the P-channel; since the former is lower-bounded, so is the latter.

THEOREM 10.11   *Consider two random transformations* $(\mathsf{A}, \mathsf{B}, P_{Y|X})$ *and* $(\mathsf{A}, \mathsf{B}, Q_{Y|X})$. *Fix a code* (f, g) *(with possibly randomized encoder and decoder) and let* ϵ *and* ϵ' *be its average probability of error under the P-transformation and the Q-transformation, respectively. Also denote by* $P_X = Q_X$ *the probability distribution induced by the encoder* f *on the input alphabet* A. *Then we have*

$$\beta_{1-\epsilon}(P_{XY}, Q_{XY}) \leq 1 - \epsilon'. \tag{10.83}$$

Proof.   Denote by W and $\hat{W}$ the random variable representing the input to the encoder (i.e., the message) and the output of the decoder (i.e., the message estimate), respectively. Then we have two joint distributions $P_{WXY\hat{W}}$ and $Q_{WXY\hat{W}}$ defined as follows:

$$P_{WXY\hat{W}}(w, x, y, \hat{w}) = \frac{1}{M} f(x|w) P_{Y|X}(y|x) g(\hat{w}|y), \tag{10.84}$$

$$Q_{WXY\hat{W}}(w, x, y, \hat{w}) = \frac{1}{M} f(x|w) Q_{Y|X}(y|x) g(\hat{w}|y), \tag{10.85}$$

where $\frac{1}{M}$ represents the fact that W is equiprobable on $\{1, \ldots, M\}$. We define the following random variable

$$Z = 1\{W = \hat{W}\}.$$

The crucial observation is that the conditional distribution of Z given (X, Y) is the same for both P and Q; namely, we have

$$P_{Z|XY} = Q_{Z|XY}. \tag{10.86}$$

Indeed,

$$\mathbb{P}[Z = 1|X, Y] = \sum_{j=1}^{M} \mathbb{P}[W = j, \hat{W} = j|X, Y] \tag{10.87}$$

$$= \sum_{j=1}^{M} \mathbb{P}[W = j|X, Y]\mathbb{P}[\hat{W} = j|X, Y] \tag{10.88}$$

$$= \sum_{j=1}^{M} \mathbb{P}[W = j|X]\mathbb{P}[\hat{W} = j|Y] \tag{10.89}$$

$$= \sum_{j=1}^{M} \mathbb{P}[W = j|X]g(j|Y), \tag{10.90}$$

where (10.87) follows from the definition of Z, (10.88) follows since under both P and Q we have a Markov chain: $W - X - Y - \hat{W}$ and therefore, W and $\hat{W}$ are conditionally independent given (X, Y); and (10.89) is also a consequence of the Markov chain property. Finally, (10.89) implies (10.86) since $\mathbb{P}[W = j|X]$ in each term of the sum depends only on the joint distribution of X and W, while by construction we have that $P_{W,X} = Q_{W,X}$.

Overall, $P_{Z|XY}$ defines a transition probability kernel $A \times B \to \{0, 1\}$ and therefore constitutes a binary hypothesis test between P_{XY} and Q_{XY} satisfying

$$\sum_{x \in A} \sum_{y \in B} P_{Z|XY}(1|xy)P_{XY}(x, y) = 1 - \epsilon \tag{10.91}$$

$$\sum_{x \in A} \sum_{y \in B} P_{Z|XY}(1|xy)Q_{XY}(x, y) = 1 - \epsilon'. \tag{10.92}$$

Therefore, by the definition of β_α in (10.39) we have

$$\beta_{1-\epsilon}(P_{XY}, Q_{XY}) \le 1 - \epsilon'. \qquad \blacksquare$$

Theorem 10.11 allows one to use any converse for the channel $Q_{Y|X}$ to prove a converse for channel $P_{Y|X}$. It has many interesting generalizations (for example, to list decoding and channels with feedback) and applications, which we are not able to survey here.

A simple application of Theorem 10.11 yields the following result.

THEOREM 10.12 (Max–min converse)   *Every (M, ϵ) code with codewords belonging to* F *satisfies*

$$M \le \sup_{P_X} \inf_{Q_Y} \frac{1}{\beta_{1-\epsilon}(P_{XY}, P_X \times Q_Y)}, \tag{10.93}$$

where P_X ranges over all distributions on F, *and Q_Y ranges over all distributions on* B.

Proof. Denote the distribution of the encoder output by $\bar{P}_X$ and particularize Theorem 10.11 by choosing $Q_{Y|X} = Q_Y$ for an arbitrary Q_Y, in which case we obtain $\epsilon' = 1 - \frac{1}{M}$. Therefore, from (10.83) we obtain

$$\frac{1}{M} \geq \sup_{Q_Y} \beta_{1-\epsilon}(\bar{P}_X P_{Y|X}, \bar{P}_X \times Q_Y) \tag{10.94}$$

$$\geq \inf_{P_X} \sup_{Q_Y} \beta_{1-\epsilon}(P_{XY}, P_X \times Q_Y). \tag{10.95}$$

∎

As we will see shortly, $\beta_\alpha(x, Q_Y)$ is constant on F in important special cases. In those cases the following converse is particularly useful.

THEOREM 10.13   *Fix a probability measure Q_Y on B. Suppose that $\beta_\alpha(P_{Y|X=x}, Q_Y)$ does not depend on $x \in$ F. Then every (M, ϵ) code satisfies*

$$M \leq \frac{1}{\beta_{1-\epsilon}(P_{Y|X=x}, Q_Y)}. \tag{10.96}$$

Proof.   The result follows from Theorem 10.12 and the following auxiliary result [16, lemma 31]: Suppose that $\beta_\alpha(P_{Y|X=x}, Q_{Y|X=x}) = \beta_\alpha$ is independent of $x \in$ F. Then, for every P_X supported on F we have

$$\beta_\alpha(P_X P_{Y|X}, P_X Q_{Y|X}) = \beta_\alpha(P_{Y|X=x}, Q_{Y|X=x}). \tag{10.97}$$

∎

10.4   Finite-Blocklength Bounds for Multiple-Antenna Fading Channels

In this section, we shall apply the finite-blocklength bounds described in Section 10.3 to characterize the maximum coding rate R^* achievable on the multiple-antenna fading channel (10.9) under various assumption on the availability of CSI. We will first consider a quasi-static fading scenario, where each codeword spans exactly one coherence interval (i.e., $\ell = 1$). For this setup, we will present nonasymptotic bounds and a normal approximation on $R^*(\ell = 1, n_c, \epsilon, \rho)$ that hold for a large class of fading distributions (including the Rayleigh distribution we focused on in Section 10.3). Our main finding is that, in quasi-static fading channels, the outage capacity, which is the asymptotic performance metric of relevance, is approached much faster in the blocklength than in the case of the AWGN channel, because the channel dispersion is zero.

We will then consider the general block-fading setup, and characterize $R^*(\ell, n_c, \epsilon, \rho)$ for the Rayleigh-fading case. As we shall show in Section 10.4.3, these bounds allow one to optimally select, for a given blocklength, the number of diversity branches and of transmit antennas to use. The material contained in this section is primarily based on [46–48], to which we shall refer for detailed proofs of the results. Here, we shall provide only the intuition behind the results. Throughout this section, we shall denote the minimum between the number of transmit and receive antennas by $m = \min\{m_t, m_r\}$. We

warn the reader that the material contained in the following subsections is rather technical. Readers interested in the engineering insights obtainable from the finite-blocklength
bounds are invited to proceed directly to Section 10.4.3.

10.4.1 Quasi-Static Fading Channels

Achievability Bounds

Since $\ell = 1$, we shall omit ℓ in the parametrization of coding schemes and of the
maximum coding rate. We first present an achievability bound for the no-CSI case. The
bound is based on the $\kappa\beta$-bound (10.10).

THEOREM 10.14 (achievability–no-CSI)   *Assume that for a given $0 < \epsilon < 1$ there
exists a $m_t \times m_t$ positive semidefinite matrix Q^* such that*

$$\inf_Q P_{rx}^{out}\left(Q, C_{rx}^{out,\epsilon}\right) = P_{rx}^{out}\left(Q^*, C_{rx}^{out,\epsilon}\right), \tag{10.98}$$

*where the infimum is over all positive-semidefinite matrices Q satisfying the power con-
straint $tr\{Q\} = \rho$. Then, for every $0 < \tau < \epsilon$ there exists an $(n_c, M, \epsilon, \rho)_{no}$ code with rate
$R_{no}^*(n_c, \epsilon, \rho) = (\log M)/n_c$ that satisfies*

$$\frac{\log M}{n_c} \geq \frac{1}{n_c} \log \frac{\tau}{\mathbb{P}\left[\prod_{j=1}^{m_r} B_j \leq \gamma_{n_c}\right]}. \tag{10.99}$$

*Here, $B_j \sim \text{Beta}(n_c - t^* - j + 1, t^*)$, $j = 1, \ldots, m_r$, are independent beta-distributed
random variables, $t^* = \text{rank}(Q^*)$, and $\gamma_{n_c} \in [0, 1]$ is chosen so that*[8]

$$\mathbb{P}\left[\sin^2\{I_{n_c,t^*}, \sqrt{n_c}I_{n_c,t^*}U\mathbb{H} + \mathbb{W}\} \leq \gamma_{n_c}\right] \geq 1 - \epsilon + \tau, \tag{10.100}$$

*with $U \in \mathbb{C}^{t^* \times m_t}$ satisfying $U^H U = Q^*$. Here, $\sin^2\{I_{n_c,t^*}, \sqrt{n_c}I_{n_c,t^*}U\mathbb{H} + \mathbb{W}\}$ denotes
the product of the sine of the principle angles between the subspaces spanned by the
columns of I_{n_c,t^*} and of $\sqrt{n_c}I_{n_c,t^*}U\mathbb{H} + \mathbb{W}$, respectively.*

Sketch of the proof. We consider a decoder that computes the sine of the principle angles
between the subspace spanned by the received matrix Y and the one spanned by each
codeword. The decoder chooses the first codeword for which the sine of the angles is
below the threshold γ_{n_c}. Note that this "angle" decoder requires neither CSIR nor the
knowledge of the fading distribution. The achievability bound then follows from the $\kappa\beta$
bound in Theorem 10.10 applied to a physically degraded channel whose output is the
subspace spanned by the columns of $\mathbb{Y}$. See [46, app. II] for a complete proof.

For the case when CSIT (but not CSIR) is available, the achievability bound in (10.99)
can be improved through water-filling as follows:

THEOREM 10.15 (achievability–CSIT)   *Let $\Lambda_1 \geq \cdots \geq \Lambda_m$ be the m largest eigen-
values of $\mathbb{H}\mathbb{H}^H$. For every $0 < \epsilon < 1$ and every $0 < \tau < \epsilon$, there exists an $(n_c, M, \epsilon)_{tx}$ code
that satisfies*

$$\frac{\log M}{n_c} \geq \frac{1}{n_c} \log \frac{\tau}{\mathbb{P}\left[\prod_{j=1}^{m_r} B_j \leq \gamma_{n_c}\right]}. \tag{10.101}$$

---

[8]   We denote by $I_{a,b}$ ($a > b$) the $a \times b$ matrix containing the first b columns of the identity matrix I_a.

Here, $B_j \sim \text{Beta}(n_c - m_t - j + 1, m_t), j = 1, \ldots, m_r$, are independent beta-distributed random variables, and $\gamma_{n_c} \in [0, 1]$ is chosen so that

$$\mathbb{P}\left[\sin^2\left\{I_{n_c, m_t}, \sqrt{n_c}I_{n_c, m_t} \text{ diag}\left\{\sqrt{v_1^* \Lambda_1}, \ldots, \sqrt{v_m^* \Lambda_m}, \underbrace{0, \ldots, 0}_{m_t - m}\right\} + \mathbb{W}\right\} \leq \gamma_{n_c}\right]$$
$$\geq 1 - \epsilon + \tau, \quad (10.102)$$

where

$$v_j^* = [\bar{\gamma} - 1/\Lambda_j]^+, \quad j = 1, \ldots, m_r \quad (10.103)$$

are the water-filling power gains and $\bar{\gamma}$ is chosen so that the power constraint is satisfied.

Converse Bounds

Next, we present two converse bounds that are based on the meta-converse method (Theorem 10.12). The first bound holds for the CSIR case.

THEOREM 10.16 (converse CSIR)   *Let Q be an arbitrary $m_t \times m_t$ positive-semidefinite matrix satisfying the power constraint $\text{tr}\{Q\} = \rho$. Let*

$$L_{n_c}^{rx}(Q) = \sum_{i=1}^{n_c} \sum_{j=1}^{m} \left(\log(1 + \Lambda_j) + 1 - \left|\sqrt{\Lambda_j} Z_{ij} - \sqrt{1 + \Lambda_j}\right|^2\right) \quad (10.104)$$

and

$$S_{n_c}^{rx}(Q) = \sum_{i=1}^{n_c} \sum_{j=1}^{m} \left(\log(1 + \Lambda_j) + 1 - \frac{\left|\sqrt{\Lambda_j} Z_{ij} - 1\right|^2}{1 + \Lambda_j}\right), \quad (10.105)$$

where $Z_{ij}, i = 1, \ldots, n_c, j = 1, \ldots, m$, are i.i.d. $\mathcal{CN}(0, 1)$ distributed. Then, for every $n_c \geq m_r$ and every $0 < \epsilon < 1$, the maximum coding rate $R_{rx}^*(n_c, \epsilon, \rho)$ is upper bounded by

$$R_{rx}^*(n_c, \epsilon, \rho) \leq \frac{1}{n_c} \log \frac{c_{rx}(n_c)}{\inf_Q \mathbb{P}[L_{n_c}^{rx}(Q) \geq n_c \gamma_{n_c}(Q)]}. \quad (10.106)$$

Here,

$$c_{rx}(n_c) = \frac{\pi^{m_r(m_r-1)}}{\Gamma_{m_r}(n_c)\Gamma_{m_r}(m_r)} \mathbb{E}\left[\left(1 + \rho \|\mathbb{H}\|_F^2\right)^{\lfloor(m_r+1)^2/4\rfloor}\right]$$
$$\times \prod_{i=1}^{m_r}\left[(n_c + m_r - 2i)^{n_c+m_r-2i+1} e^{-(n_c+m_r-2i)}\right.$$
$$\left. + \Gamma(n_c + m_r - 2i + 1, n_c + r - 2i)\right], \quad (10.107)$$

with $\Gamma_{(\cdot)}(\cdot)$ denoting the complex multivariate gamma function [49, eq. (83)], and $\gamma_{n_c}(Q)$ is the solution of

$$\mathbb{P}[S_{n_c}^{rx}(Q) \leq n_c \gamma_{n_c}(Q)] = \epsilon. \quad (10.108)$$

Sketch of the proof. We use the meta-converse method (Theorem 10.11) with a Gaussian auxiliary channel that depends on the input codeword X through its Gramian matrix $X^H X$. The bound (10.106) is established by using the Neyman–Pearson lemma to evaluate the β-function and by showing that the probability of error over the auxiliary channel is lower bounded by $1 - c_{rx}/M$. See [46, app. VII] for more details.

When CSIRT is available, the following converse bound holds.

THEOREM 10.17 (converse CSIRT) *Let $\Lambda_1 \geq \cdots \geq \Lambda_m$ be the m largest eigenvalues of $\mathbb{H}\mathbb{H}^H$, and let $\boldsymbol{\Lambda} = [\Lambda_1, \ldots, \Lambda_m]^T$. Consider an arbitrary power-allocation function $\boldsymbol{v} : \mathbb{R}_+^m \mapsto \mathcal{V}_m$, where*

$$\mathcal{V}_m = \left\{ [p_1, \ldots, p_m] \in \mathbb{R}_+^m : \sum_{j=1}^m p_j \leq \rho \right\}. \tag{10.109}$$

Let

$$L_{n_c}^{rt}(\boldsymbol{v}, \boldsymbol{\Lambda}) = \sum_{i=1}^{n_c} \sum_{j=1}^m \left(\log\left(1 + \Lambda_j v_j(\boldsymbol{\Lambda})\right) + 1 \right.$$
$$\left. - \left| \sqrt{\Lambda_j v_j(\boldsymbol{\Lambda})} Z_{i,j} - \sqrt{1 + \Lambda_j v_j(\boldsymbol{\Lambda})} \right|^2 \right) \tag{10.110}$$

and

$$S_{n_c}^{rt}(\boldsymbol{v}, \boldsymbol{\Lambda}) = \sum_{i=1}^{n_c} \sum_{j=1}^m \left(\log\left(1 + \Lambda_j v_j(\boldsymbol{\Lambda})\right) + 1 - \frac{\left| \sqrt{\Lambda_j v_j(\boldsymbol{\Lambda})} Z_{ij} - 1 \right|^2}{1 + \Lambda_j v_j(\boldsymbol{\Lambda})} \right), \tag{10.111}$$

where $v_j(\cdot)$ is the jth coordinate of $\boldsymbol{v}(\cdot)$, and Z_{ij}, $i = 1, \ldots, n_c$, $j = 1, \ldots, m$, are i.i.d. $\mathcal{CN}(0, 1)$ distributed random variables. For every n_c and every $0 < \epsilon < 1$, the maximum coding rate $R_{rt}^*(n_c, \epsilon)$ is upper bounded by

$$R_{rt}^*(n_c, \epsilon) \leq \frac{1}{n_c} \log \frac{c_{rt}(n_c)}{\inf_{\boldsymbol{v}(\cdot)} \mathbb{P}[L_{n_c}^{rt}(\boldsymbol{v}, \boldsymbol{\Lambda}) \geq n_c \gamma_{n_c}(\boldsymbol{v})]}, \tag{10.112}$$

where

$$c_{rt}(n_c) = \left(\frac{(n_c - 1)^{n_c} e^{-(n_c - 1)}}{\Gamma(n_c)} + \frac{\Gamma(n_c, n_c - 1)}{\Gamma(n_c)} \right)^m$$
$$\times \mathbb{E}\left[\det(I_{m_t} + \rho \mathbb{H}\mathbb{H}^H) \right] \tag{10.113}$$

and the scalar $\gamma_{n_c}(\boldsymbol{v})$ is the solution of

$$\mathbb{P}[S_{n_c}^{rt}(\boldsymbol{v}, \boldsymbol{\Lambda}) \leq n_c \gamma_{n_c}(\boldsymbol{v})] = \epsilon. \tag{10.114}$$

The infimum on the RHS of (10.112) is taken over all power allocation functions $\boldsymbol{v} : \mathbb{R}_+^m \mapsto \mathcal{V}_m$.

Sketch of the proof. Since CSI is available at both the encoder and the decoder, we can transform the MIMO quasi-static channel into a set of parallel quasi-static channels. The proof then builds on [20, section 4.5], which characterizes the nonasymptotic coding rate of parallel AWGN channels.

The converse bounds in Theorems 10.16 and 10.17 are difficult to evaluate numerically because the minimizers in (10.106) and (10.112) are not known in closed form. We next focus on two scenarios where these minimizations are not needed, namely the single-input multiple-output (SIMO) case and the case of isotropic codewords.

SIMO Quasi-Static Fading Channel

In the SIMO case, CSIT is not beneficial and the converse bounds in Theorems 10.16 and 10.17 can be tightened as follows.

THEOREM 10.18 (converse SIMO)   *Let*

$$L_{n_c} = n_c \log(1 + \rho G) + \sum_{i=1}^{n_c} \left(1 - \left|\sqrt{\rho G} Z_i - \sqrt{1 + \rho G}\right|^2\right) \qquad (10.115)$$

and

$$S_{n_c} = n_c \log(1 + \rho G) + \sum_{i=1}^{n_c} \left(1 - \frac{\left|\sqrt{\rho G} Z_i - 1\right|^2}{1 + \rho G}\right) \qquad (10.116)$$

with $G = \|\boldsymbol{H}\|^2$ and Z_i, $i = 1, \dots, n_c$, i.i.d. $\mathcal{CN}(0, 1)$ distributed. For every n_c and every $0 < \epsilon < 1$, the maximum coding rate for the SIMO case with CSIR (with or without CSIT) is upper bounded by

$$R_{rx}^*(n_c, \epsilon, \rho) = R_{rt}^*(n_c, \epsilon, \rho) \le \frac{1}{n_c} \log \frac{1}{\mathbb{P}[L_{n_c} \ge n_c \gamma_{n_c}]}, \qquad (10.117)$$

where γ_{n_c} is the solution of

$$\mathbb{P}[S_{n_c} \le n_c \gamma_{n_c}] = \epsilon. \qquad (10.118)$$

Isotropic Codewords

We next consider the case of codes in which all codewords X are isotropic in the sense that they satisfy $\mathsf{X}^H \mathsf{X} = (n_c \rho / m_t) \mathsf{I}_{m_t}$. For this class of codes, we shall denote by $R_{no,iso}^*$ and $R_{rx,iso}^*$ the maximum coding rate for the case of no-CSI and CSIR, respectively. Note that since the code consists of isotropic codewords, CSIT is not beneficial also in this setup. In the asymptotic limit $n_c \to \infty$, we have that

$$\lim_{n_c \to \infty} R_{no,iso}^*(n_c, \epsilon, \rho) = \lim_{n_c \to \infty} R_{rx,iso}^*(n_c, \epsilon, \rho) = C_{rx,iso}^{out,\epsilon}$$

$$= \sup \left\{ R : \mathbb{P}\left[\log \det\left(\mathsf{I}_{m_r} + \frac{\rho}{m_t} \mathbb{H}^H \mathbb{H}\right) < R\right] \le \epsilon \right\}. \qquad (10.119)$$

In the next theorem, we specialize Theorem 10.16 to the case of isotropic codes.

THEOREM 10.19 (converse ISO)   *Let $L_{n_c}^{rx}(\cdot)$ and $S_{n_c}^{rx}(\cdot)$ be as in (10.104) and (10.105), respectively. Then, for every n_c and every $0 < \epsilon < 1$, the maximum coding rate $R_{rx,iso}^*(n_c, \epsilon, \rho)$ is upper bounded by*

$$R_{rx,iso}^*(n_c, \epsilon, \rho) = R_{rt,iso}^*(n_c, \epsilon, \rho) \le \frac{1}{n_c} \log \frac{1}{\mathbb{P}[L_{n_c}^{rx}((\rho/m_t)\mathsf{I}_{m_t}) \ge n_c \gamma_{n_c}]}, \qquad (10.120)$$

where γ_{n_c} is the solution of

$$\mathbb{P}[S_{n_c}^{\mathrm{rx}}((\rho/m_t)I_{m_t}) \leq n_c \gamma_{n_c}] = \epsilon. \tag{10.121}$$

Second-Order Asymptotic Analysis

The bounds reported in the previous section allows one to study the rate at which the maximum coding rate approaches the outage capacity as the length of the coherence interval n_c increases. Following [20, definition 2], we define the ϵ-dispersion V^ϵ of the MIMO quasi-static channel as:

$$V_p^\epsilon = \limsup_{n_c \to \infty} n_c \left(\frac{C_p^{\mathrm{out},\epsilon} - R_p^*(n_c, \epsilon, \rho)}{Q^{-1}(\epsilon)} \right)^2$$

$$\epsilon \in (0, 1)\backslash\{1/2\}, \mathrm{p} = \{\mathrm{no, rx, tx, rt}\}. \tag{10.122}$$

The following result holds for the case when CSIT is available.

THEOREM 10.20 (zero-dispersion CSIT) *Assume that the fading channel $\mathbb{H}$ satisfies the following conditions:*

1. *the expectation $\mathbb{E}\left[\det(I_{m_t} + \rho\mathbb{H}\mathbb{H}^H)\right]$ is finite;*
2. *the joint PDF of the ordered nonzero eigenvalues of $\mathbb{H}^H\mathbb{H}$ exists and is continuously differentiable;*
3. *$C_{\mathrm{rt}}^{\mathrm{out},\epsilon}$ is a point of growth of the outage probability function, i.e.,[9]*

$$\left(P_{\mathrm{rt}}^{\mathrm{out}}\right)'\left(C_{\mathrm{rt}}^{\mathrm{out},\epsilon}\right) > 0. \tag{10.123}$$

Then

$$\left\{R_{\mathrm{rx}}^*(n_c, \epsilon, \rho), R_{\mathrm{rt}}^*(n_c, \epsilon, \rho)\right\} = C_{\mathrm{rt}}^{\mathrm{out},\epsilon} + \mathcal{O}\left(\frac{\log n_c}{n_c}\right). \tag{10.124}$$

Hence, the ϵ-dispersion is zero for both the CSIRT and the CSIT case:

$$V_{\mathrm{tx}}^\epsilon = V_{\mathrm{rt}}^\epsilon = 0, \quad \epsilon \in (0, 1)\backslash\{1/2\}. \tag{10.125}$$

Sketch of the proof. The result is established by analyzing the achievability bound (10.101) and the converse bound (10.112) in the large n_c limit using a Cramer–Esseen-type central-limit theorem. For details, see [46, app. IV and V].

The assumptions on the fading-matrix distribution in Theorem 10.20 are satisfied by most probability distributions used to model MIMO fading channels, including i.i.d. or correlated Rayleigh, Rician, and Nakagami distributions. The conditions in the theorem do not hold, however, for the nonfading AWGN channel (a quasi-static channel in which the cumulative distribution of the fading coefficient is a step function), which has positive dispersion, as shown in (10.6).

A similar result holds for the case when CSIT is not available, under more stringent conditions on the fading distribution. This is due to the analytical intractability of the minimization in the converse bound (10.106).

---

[9] Note that this condition implies that $C_{\mathrm{rt}}^{\mathrm{out},\epsilon}$ is a continuous function of ϵ.

THEOREM 10.21 (zero-dispersion no-CSIT) *Let $f_{\mathbb{H}}$ be the PDF of the fading matrix $\mathbb{H}$. Assume that $\mathbb{H}$ satisfies the following conditions:*

1. *$f_{\mathbb{H}}$ is a smooth function, i.e., it has derivatives of all orders;*
2. *there exists a positive constant a such that*

$$f_{\mathbb{H}}(\mathsf{H}) \leq a\|\mathsf{H}\|_F^{-2m_t m_r - \lfloor (m_r+1)^2/2 \rfloor - 1} \tag{10.126}$$

$$\|\nabla f_{\mathbb{H}}(\mathsf{H})\|_F \leq a\|\mathsf{H}\|_F^{-2m_t m_r - 5}; \tag{10.127}$$

3. *the function $P_{\mathrm{rx}}^{\mathrm{out}}(\cdot, \cdot)$ satisfies*

$$\liminf_{\delta \to 0} \frac{\inf_{\mathsf{Q}}\{P_{\mathrm{rx}}^{\mathrm{out}}(\mathsf{Q}, C_{\mathrm{rx}}^{\mathrm{out},\epsilon} + \delta) - P_{\mathrm{rx}}^{\mathrm{out}}(\mathsf{Q}, C_{\mathrm{rx}}^{\mathrm{out},\epsilon})\}}{\delta} > 0. \tag{10.128}$$

Then,

$$\{R_{\mathrm{no}}^*(n_c, \epsilon, \rho), R_{\mathrm{rx}}^*(n_c, \epsilon, \rho)\} = C_{\mathrm{rx}}^{\mathrm{out},\epsilon} + \mathcal{O}\left(\frac{\log n_c}{n_c}\right). \tag{10.129}$$

Hence, the ϵ-dispersion is zero for both the CSIR and the no-CSI case:

$$V_{\mathrm{no}}^\epsilon = V_{\mathrm{rx}}^\epsilon = 0, \quad \epsilon \in (0,1)\backslash\{1/2\}. \tag{10.130}$$

Sketch of the proof. The proof is significantly more involved than the proof of Theorem 10.20. Indeed, since the outage-capacity-achieving covariance matrix is not known, one has to prove in the converse part that the $\mathcal{O}(\cdot)$ terms in the asymptotic expansion are uniform in Q. This is achieved in [46] through the use of tools from Riemmanian geometry, including a version of Stoke's theorem on oriented manifolds. See [46, app. VIII and IX] for details.

The conditions on $f_{\mathbb{H}}$ in Theorem 10.21 are mild and hold for most commonly used fading distributions. In the SIMO case and in the case of isotropic codewords, these condition can be relaxed significantly [46].

The zero-dispersion result of Theorems 10.20 and 10.21 suggests that the maximum coding rate in quasi-static fading channels converges faster to the outage capacity than in the AWGN case, where the dispersion is positive and the rate of convergence is $\mathcal{O}(1/\sqrt{n})$ (see (10.6)). This implies that the outage capacity is a relevant performance metric for delay-constrained communication over quasi-static fading channels. The reason for this is that, in quasi-static fading channels, packet errors are mainly caused by deep fades and channel codes are ineffective against them. Numerical evidence for these claims is provided in Section 10.4.3

To reduce the negative impact of deep fades, which is critical in ultra-reliable communication systems, one needs to exploit time–frequency diversity. This yields the block-memoryless fading model we shall analyze in Section 10.4.2.

Normal Approximation

Motivated by (10.6) and by the second-order asymptotic analyses in Theorems 10.20 and 10.21, we next propose normal approximations for the maximum coding rate that are more accurate than the outage capacity.

For the CSIRT case, following [46], we define the normal approximation $R_{\text{rt,n}}^*$ of R_{rt}^* as the solution of

$$\epsilon = \mathbb{E}\left[Q\left(\frac{C(\mathbb{H}) - R_{\text{rt,n}}^*}{\sqrt{V(\mathbb{H})/n_{\text{c}}}}\right)\right]. \tag{10.131}$$

Here,

$$C(\mathsf{H}) = \sum_{j=1}^{m} \log(1 + v_j^* \lambda_j) \tag{10.132}$$

is the capacity of the quasi-static channel when $\mathbb{H} = \mathsf{H}$; furthermore, $\{v_j^*\}$ are the water-filling power allocation values and $\{\lambda_j\}$ are the eigenvalues of $\mathsf{H}^H\mathsf{H}$). Finally,

$$V(\mathsf{H}) = m - \sum_{j=1}^{m} \frac{1}{(1 + v_j^* \lambda_j)^2} \tag{10.133}$$

is the dispersion when $\mathbb{H} = \mathsf{H}$ (cf. (10.7)).

When only CSIR is available, not even the outage capacity is known in closed form in the general MIMO case. This prevents us from obtaining a normal approximation for R_{rx}^*. Such approximation can, however, be readily derived in the SIMO case (just set $m = 1$, $v_1^* = \rho$ in (10.131)–(10.133) and note that $\lambda_1 = \|\boldsymbol{h}\|^2$).

For the case of isotropic codes, a normal approximation $R_{\text{rx,iso,n}}^*$ to $R_{\text{rx,iso}}^*$ can be obtained as the solution of

$$\epsilon = \mathbb{E}\left[Q\left(\frac{C_{\text{iso}}(\mathbb{H}) - R_{\text{rx,iso,n}}^*(n_{\text{c}}, \epsilon, \rho)}{\sqrt{V_{\text{iso}}(\mathbb{H})/n_{\text{c}}}}\right)\right]. \tag{10.134}$$

Here,

$$C_{\text{iso}}(\mathsf{H}) = \sum_{j=1}^{m} \log(1 + \rho\lambda_j/m_{\text{t}}) \tag{10.135}$$

and

$$V_{\text{iso}}(\mathsf{H}) = m - \sum_{j=1}^{m} \frac{1}{(1 + \rho\lambda_j/m_{\text{t}})^2}, \tag{10.136}$$

where $\{\lambda_j\}$ are the eigenvalues of $\mathsf{H}^H\mathsf{H}$.

10.4.2   Block-Memoryless Fading Channels

We next consider the case $\ell > 1$, but we shall focus exclusively on the Rayleigh-fading case.

The No-CSI Case

Let $\mathbb{A}$ be an $n \times m$ $(n > m)$ random matrix. We say that $\mathbb{A}$ is isotropically distributed if, for every deterministic $n \times n$ unitary matrix V, the matrix $\mathsf{V}\mathbb{A}$ has the same probability distribution as $\mathbb{A}$. A key ingredient of the nonasymptotic bounds on R_{no}^* described in this section is a closed-form expression for the PDF induced on the channel output $\mathbb{Y}_k$

in (10.9) when $\mathbb{X}_k$ is a scaled isotropically distributed matrix with orthonormal columns. Such an input distribution, which is commonly referred to as unitary space-time modulation (USTM), achieves the high-SNR ergodic-capacity expansion in (10.21) when $n_c \geq m_t + m_r$.[10] In the following theorem, we consider a minor modification of the USTM distribution, in which only $\tilde{m}_t$ out of the available m_t transmit antennas are used.

THEOREM 10.22 (USTM-induced output distribution)   *Assume that* $n_c \geq m_t + m_r$. *Let* $q = \min\{\tilde{m}_t, m_r\}$ *and* $p = \max\{\tilde{m}_t, m_r\}$. *Let also* $\mathbb{X} = \sqrt{\rho n_c/\tilde{m}_t} \mathbb{U}$ *where* $\mathbb{U} \in \mathbb{C}^{n_c \times \tilde{m}_t}$ $(1 \leq \tilde{m}_t \leq m_t)$ *satisfies* $\mathbb{U}^H \mathbb{U} = \mathbb{I}_{\tilde{m}_t}$ *and is isotropically distributed. Further, let* $\mathbb{Y} = \mathbb{X}\mathbb{H} + \mathbb{W}$ *where* $\mathbb{H} \in \mathbb{C}^{\tilde{m}_t \times m_r}$ *and* $\mathbb{W} \in \mathbb{C}^{n_c \times m_r}$ *are defined as in* (10.9). *The PDF of* $\mathbb{Y}$ *is given by*

$$f_{\mathbb{Y}}(\mathbb{Y}) = \frac{\prod\limits_{u=n_c-q+1}^{n_c} \Gamma(u)}{\pi^{m_r n_c} \prod\limits_{u=1}^{\tilde{m}_t} \Gamma(u)} \frac{(1+\mu)^{\tilde{m}_t(n_c-\tilde{m}_t-m_r)}}{\mu^{\tilde{m}_t(n_c-\tilde{m}_t)}} \psi_{\tilde{m}_t}(\sigma_1^2, \ldots, \sigma_{m_r}^2). \tag{10.137}$$

Here, $\sigma_1 > \cdots > \sigma_{m_r}$ *denote the* m_r *nonzero singular values of* $\mathbb{Y}$, *which are positive and distinct almost surely [50],* $\mu = \rho n_c/\tilde{m}_t$, *and*

$$\psi_{\tilde{m}_t}(\sigma_1^2, \ldots, \sigma_{m_r}^2) = \frac{\det(\mathsf{M})}{\prod\limits_{i<j}^{m_r}(\sigma_i^2 - \sigma_j^2)} \prod_{k=1}^{m_r} \frac{e^{-\sigma_k^2/(1+\mu)}}{\sigma_k^{2(n_c-m_r)}}. \tag{10.138}$$

The entries of the $p \times p$ *real matrix* M *are given by*

$$[\mathsf{M}]_{ij} = \begin{cases} \sigma_i^{2(\tilde{m}_t-j)} \tilde{\gamma}\left(n_c+j-p-\tilde{m}_t, \sigma_i^2\mu/(1+\mu)\right), & \begin{array}{l} 1 \leq i \leq m_r \\ 1 \leq j \leq \tilde{m}_t \end{array} \\[12pt] \exp\left(-\sigma_i^2\mu/(1+\mu)\right) \left[\dfrac{\partial^{\tilde{m}_t-j}}{\partial\delta^{\tilde{m}_t-j}} \delta^{n_c-i} \bigg|_{\delta=\mu/(1+\mu)} \right], & \begin{array}{l} m_r < i \leq p \\ 1 \leq j \leq \tilde{m}_t \end{array} \\[12pt] \sigma_i^{2(n_c-j)} \exp\left(-\sigma_i^2\mu/(1+\mu)\right), & \begin{array}{l} 1 \leq i \leq m_r \\ \tilde{m}_t < j \leq p \end{array} \end{cases} \tag{10.139}$$

where

$$\tilde{\gamma}(n, x) \triangleq \frac{1}{\Gamma(n)} \int_0^x t^{n-1} e^{-t} dt \tag{10.140}$$

denotes the regularized incomplete gamma function.

We are now ready to state an achievability bound, which is a simple application of the DT bound in Theorem 10.8 (see [47, app. A]).

---

[10] A different distribution, described in [32], is required when $n_c < m_t + m_r$.

THEOREM 10.23 (achievability no-CSI) *Let* $\Lambda_{k,\widetilde{m}_t,1} > \cdots > \Lambda_{k,\widetilde{m}_t,m_r}$ *be the ordered eigenvalues of* $\mathbb{Z}_k^H D_{\widetilde{m}_t} \mathbb{Z}_k$ *where* $\{\mathbb{Z}_k\}_{k=1}^{\ell}$ *are independent complex Gaussian* $n_c \times m_r$ *matrices with i.i.d.* $\mathcal{CN}(0,1)$ *entries, and*

$$D_{\widetilde{m}_t} = \text{diag}\Big\{ \underbrace{1 + \rho n_c/\widetilde{m}_t, \ldots, 1 + \rho n_c/\widetilde{m}_t}_{\widetilde{m}_t}, \underbrace{1, \ldots, 1}_{n_c - \widetilde{m}_t} \Big\} \tag{10.141}$$

for $\widetilde{m}_t \in \{1, \ldots, m_t\}$. *It can be shown that the eigenvalues are positive and distinct almost surely. Let*

$$S_{k,\widetilde{m}_t} = \widetilde{m}_t(n_c - \widetilde{m}_t) \log \frac{\rho n_c}{\widetilde{m}_t + \rho n_c} - \sum_{u=n_c-q+1}^{n_c} \log \Gamma(u) + \sum_{u=1}^{\widetilde{m}_t} \log \Gamma(u)$$
$$- \text{tr}\{\mathbb{Z}_k^H \mathbb{Z}_k\} - \log \psi_{\widetilde{m}_t}(\Lambda_{k,\widetilde{m}_t,1}, \ldots, \Lambda_{k,\widetilde{m}_t,m_r}) \tag{10.142}$$

where $q = \min\{\widetilde{m}_t, m_r\}$ *and the function* $\psi_{\widetilde{m}} : \mathbb{R}_+^{m_r} \to \mathbb{R}$ *was defined in* (10.138). *Finally, let*

$$\epsilon_{ub}(M) = \min_{1 \le \widetilde{m}_t \le m_t} \mathbb{E}\left[\exp\left\{ -\left[\sum_{k=1}^{\ell} S_{k,\widetilde{m}_t} - \log(M-1) \right]^+ \right\} \right]. \tag{10.143}$$

Then

$$R_{no}^*(\ell, n_c, \epsilon, \rho) \ge \max\left\{ \frac{\log M}{n_c \ell} : \epsilon_{ub}(M) \le \epsilon \right\}. \tag{10.144}$$

We next provide a converse bound, which is based on the meta-converse technique (Theorem 10.12) together with the converse bound on the β-function in (10.44), and uses the output distribution induced by USTM as auxiliary distribution.

THEOREM 10.24 (converse no-CSI) *Fix a* $\widetilde{m}_t \in [1, \ldots, m_t]$. *Let* $\{\Sigma_k\}_{k=1}^{\ell}$ *be* $m_t \times m_t$ *diagonal matrices with nonnegative diagonal entries, satisfying* $\text{tr}\{\Sigma_k\} = n_c \rho$, $k = 1, \ldots, \ell$, *and let*

$$\widetilde{\Sigma}_k = \begin{bmatrix} I_{m_t} + \Sigma_k & \mathbf{0} \\ \mathbf{0} & I_{n_c-m_t} \end{bmatrix}. \tag{10.145}$$

Furthermore, let $\{\bar{\mathbb{Z}}_k\}_{k=1}^{\ell}$ *be independent complex Gaussian* $n_c \times m_r$ *matrices with i.i.d.* $\mathcal{CN}(0,1)$ *entries. Finally, let*

$$\bar{c}_{\widetilde{m}_t}(\Sigma_k) = \widetilde{m}_t(n_c - \widetilde{m}_t) \log \frac{\rho n_c}{\widetilde{m}_t} - \widetilde{m}_t(n_c - \widetilde{m}_t - m_r) \log\left(1 + \frac{\rho n_c}{\widetilde{m}_t}\right)$$
$$- m_r \log \det \widetilde{\Sigma}_k - \sum_{u=n_c-p+1}^{n_c} \log \Gamma(u) + \sum_{u=1}^{\widetilde{m}_t} \log \Gamma(u) \tag{10.146}$$

and

$$\bar{S}_{k,\widetilde{m}_t}(\Sigma_k) = \bar{c}_{\widetilde{m}_t}(\Sigma_k) - \text{tr}\{\bar{\mathbb{Z}}_k^H \bar{\mathbb{Z}}_k\} - \log \psi_{\widetilde{m}_t}(\bar{\Lambda}_{k,\widetilde{m}_t,1}, \ldots, \bar{\Lambda}_{k,\widetilde{m}_t,m_r}). \tag{10.147}$$

Here, $\bar{\Lambda}_{k,\tilde{m}_t,1} > \cdots > \bar{\Lambda}_{k,\tilde{m}_t,m_r}$ are the ordered eigenvalues of $\mathbb{Z}_k^H \tilde{\Sigma}_k \mathbb{Z}_k$ (which are positive and distinct almost surely), $p = \max\{\tilde{m}_t, m_r\}$, and $\psi_{\tilde{m}_t}$ is defined in (10.138). Then, the maximum coding rate $R_{no}^*(\ell, n_c, \epsilon, \rho)$ is upper bounded by

$$R_{no}^*(\ell, n_c, \epsilon, \rho) \leq \min_{1 \leq \tilde{m}_t \leq m_t} \sup_{\{\Sigma_k\}_{k=1}^\ell} \inf_{\lambda > 0}$$

$$\frac{1}{n_c \ell} \left[\lambda - \log \left(\left[\mathbb{P}\left\{ \sum_{k=1}^\ell \bar{S}_{k,\tilde{m}_t}(\Sigma_k) \leq \lambda \right\} - \epsilon \right]^+ \right) \right]. \quad (10.148)$$

The bound (10.148) is easier to evaluate numerically than the tighter bound provided in [47]. See [47] for more details on the computation of (10.148). Note also that, differently from the quasi-static case, no normal approximation is currently available for the block-fading case.[11] This makes the problem of characterizing the maximum coding rate involved, because the evaluation of the achievability and converse bounds in Theorems 10.23 and 10.24 is time-consuming. A high-SNR normal approximation for the single-input single-output (SISO) case has been recently proposed in [51]. Again for the SISO case, an extension of the bounds in Theorems 10.23 and 10.24 to Rician-fading channels was reported in [52], where the optimality of pilot-based transmission schemes is also investigated.

The CSIR Case

For the CSIR case, the following normal approximation for the asymptotic regime $\ell \to \infty$ is established in [48] for general block-fading distributions (which include Rayleigh-fading as a special case), under some technical condition on the growth of the peak power of each codeword with the number of coherence intervals.

THEOREM 10.25   *Consider a general block-fading memoryless MIMO channel of the form (10.9). Assume that $\{\mathbb{H}_k\}_{k=1}^\ell$ are i.i.d. isotropic matrices and also assume that*

$$\mathbb{P}[\mathbb{H}_1 \neq 0] > 0. \quad (10.149)$$

For every $0 < \epsilon < 1/2$, there exists a $(\ell, n_c, M, \epsilon, \rho)_{rx}$ code satisfying

$$R_{rx}^*(\ell, n_c, \epsilon, \rho) \geq \frac{\log M}{\ell n_c} \geq C_{rx}^{erg}(\rho) - \sqrt{\frac{V_{rx}(P)}{\ell n_c}} Q^{-1}(\epsilon) + o\left(\frac{1}{\sqrt{\ell}}\right). \quad (10.150)$$

Here, C_{rx}^{erg} is given in (10.19) and V_{rx} is the conditional variance of the information density, minimized over all capacity-achieving distributions:

$$V_{rx} = \inf_{P_\mathbb{X} : I(\mathbb{X};\mathbb{Y}\,|\,\mathbb{H})=C_{rx}^{erg}} \text{Var}\left[\log \frac{dP_{\mathbb{Y}\mathbb{H}\,|\,\mathbb{X}}}{dP_{\mathbb{Y}\mathbb{H}}^*} \,\Big|\, \mathbb{X} \right]. \quad (10.151)$$

Here, $P_{\mathbb{Y}\mathbb{H}}^*$ is the (unique) capacity-achieving output distribution.

---

[11] A high-SNR normal approximation was recently proposed in [51].

Furthermore, for every sequence $\delta_\ell \to 0$, there exists a sequence $\delta'_\ell \to 0$ so that every $(\ell, n_c, M, \epsilon, \rho)_{rx}$ code satisfying the extra constraint $\max_k \|X_k\|_F \leq \delta_\ell \ell^{1/4}$, $k = 1, \ldots, \ell$, must also satisfy

$$R^*_{rx}(\ell, n_c, \epsilon, \rho) \leq C^{erg}_{rx}(\rho) - \sqrt{\frac{V_{rx}(P)}{\ell n_c}} Q^{-1}(\epsilon) + \frac{\delta'_\ell}{\sqrt{\ell}}. \tag{10.152}$$

Under the additional assumption that $\mathbb{P}[\text{rank}(\mathbb{H}) > 1] > 0$, the conditional dispersion V_{rx} in (10.151) admits the following closed-form expression:

$$V_{rx}(\rho) = n_c \, \text{Var}\left(\sum_{i=1}^{m} \log\left(1 + \frac{\rho}{m_t} \Lambda_i \right) \right)$$

$$+ (\log e)^2 \sum_{i=1}^{m} \mathbb{E}\left[1 - \frac{1}{\left(1 + \rho \Lambda_i / m_t \right)^2} \right] + \left(\frac{P}{m_t} \right)^2 \left(\eta_1 - \frac{\eta_2}{m_t} \right). \tag{10.153}$$

Here, $\{\Lambda_i\}_{i=1}^{m}$ are the nonzero eigenvalues of $\mathbb{H}\mathbb{H}^H$,

$$c(\sigma) = \frac{\sigma}{1 + \sigma \rho / m_t} \tag{10.154}$$

$$\left(\log^2 e \right) \sum_{i=1}^{m} \mathbb{E}\left[c^2(\Lambda_i^2) \right] \tag{10.155}$$

and

$$\left(\log^2 e \right) \left(\sum_{i=1}^{m} \mathbb{E}\left[c(\Lambda_i^2) \right] \right)^2. \tag{10.156}$$

10.4.3 Numerical Results

Quasi-Static Fading Channels

We first consider a quasi-static 1×2 SIMO channel. The subchannels between the transmit antenna and each of the receive antennas are independent and follow a Rician distribution with K factor equal to $20 \, \text{dB}$. We set $\epsilon = 10^{-3}$ and choose $\rho = -1.55 \, \text{dB}$. This yields $C^{out,\epsilon}_{rx} = 1$ bit/channel use. In Figure 10.6 we plot the achievability bound (10.99), the converse bound (10.117) and the normal approximation (10.131). We also plot a lower bound on R^*_{rx} obtained by using the $\kappa\beta$ bound in Theorem 10.10 and assuming CSIR. The blocklength required to achieve 90% of the outage capacity $C^{out,\epsilon}_{rx}$ is in the range $[120, 320]$ for the CSIRT case and in the range $[120, 480]$ for the no-CSI case. For an AWGN channel with the same capacity (and, hence, $\rho = -3 \, \text{dB}$), this number is approximately 1420. Hence, for the parameters chosen in Figure 10.6, the prediction (based on zero dispersion) of fast convergence to capacity is validated. The gap between the normal approximation $R^*_{rt,n}$ defined implicitly in (10.131) and both the achievability (CSIR) and the converse bounds is less than 0.02 bit/(ch. use) for blocklengths larger than 400.

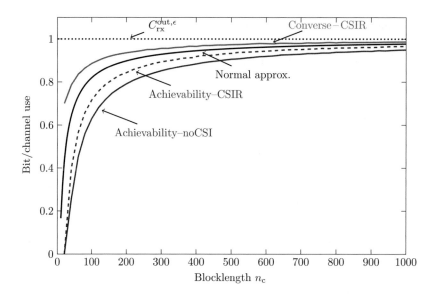

Figure 10.6 Achievability and converse bounds for a quasi-static SIMO Rician-fading channel with K-factor equal to 20 dB, two receive antennas, SNR $= -1.55$ dB, and $\epsilon = 10^{-3}$. Note that in the SIMO case $C_{\mathrm{rt}}^{\mathrm{out},\epsilon} = C_{\mathrm{rx}}^{\mathrm{out},\epsilon}$.

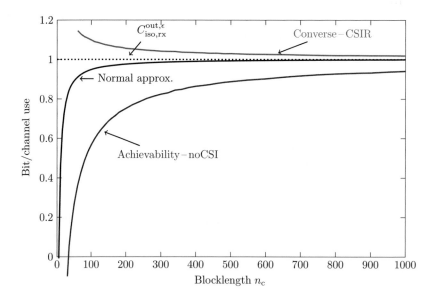

Figure 10.7 Achievability and converse bounds for codes with isotropic codewords over a quasi-static MIMO Rayleigh-fading channel with two transmit and three receive antennas, SNR $= 2.12$ dB, and $\epsilon = 10^{-3}$.

In Figure 10.7 we compare the normal approximation $R_{\mathrm{rx,iso,n}}^{*}$ defined by (10.134) with the achievability bound (10.99) and the converse bound (10.120) on the maximum coding rate achievable with isotropic codes over a 2×3 MIMO quasi-static, spatially white, Rayleigh-fading channel. We set $\epsilon = 10^{-3}$ and choose $\rho = 2.12$ dB so that $C_{\mathrm{rx,iso}}^{\mathrm{out},\epsilon} = 1$ bit/(ch. use). For this scenario, the blocklength required to achieve 90% of $C_{\mathrm{rx,iso}}^{\mathrm{out},\epsilon}$ is less than 500, which again demonstrates fast convergence to $C_{\mathrm{rx,iso}}^{\mathrm{out},\epsilon}$.

Block-Memoryless Fading Channel

We consider the transmission of a packet of size $n = 164$ channel uses over a 2×2 MIMO Rayleigh block-fading channel. We assume $\rho = 6$ dB and consider both $\epsilon = 10^{-3}$ and $\epsilon = 10^{-5}$ and study how the maximum coding rate R_{no}^* depends on the number of time–frequency branches ℓ or, equivalently, the size of the coherence interval n_{c}. As shown in Figure 10.3, these quantities can be controlled by sizing and spacing the resource blocks in the time–frequency plane appropriately. Intuitively, for a fixed $n = \ell n_{\mathrm{c}}$, the larger ℓ the lower the probability of deep fades; but also the larger the penalty due lack of CSI. Indeed the smaller n_{c}, the more difficult it is to track channel variations at the receiver.

In Figure 10.8, we plot the DT lower bound (10.144) and the MC upper bound (10.148). The achievability and the converse bounds delimit R_{no}^* tightly and demonstrate that R_{no}^* is not monotonic in the coherence interval n_{c}, but that there exists an optimal value n_{c}^*, or, equivalently, an optimal number $\ell^* = n/n_{\mathrm{c}}^*$ of time–frequency diversity branches, that maximizes R_{no}^*. For $n_{\mathrm{c}} < n_{\mathrm{c}}^*$, the cost of estimating the channel dominates. For $n_{\mathrm{c}} > n_{\mathrm{c}}^*$, the bottleneck is the limited number of time–frequency diversity branches offered by the channel. For the parameters considered in Figure 10.8, the optimal coherence interval length is $n_{\mathrm{c}}^* \approx 24$, which corresponds to about seven time–frequency diversity branches.

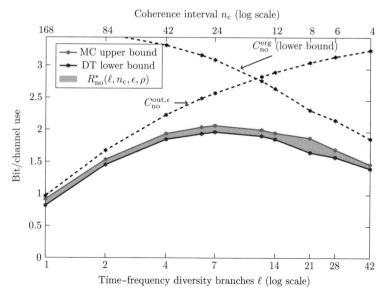

Figure 10.8 $m_{\mathrm{t}} = m_{\mathrm{r}} = 2$, $n = 168$, $\epsilon = 10^{-3}$, $\rho = 6$ dB.

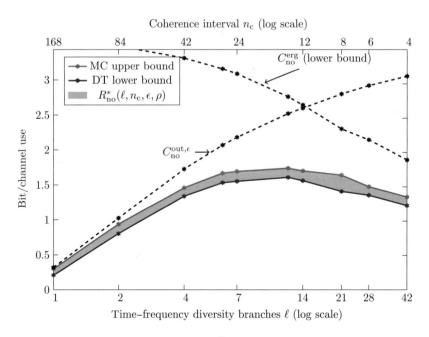

Figure 10.9 $m_t = m_r = 2, n = 168, \epsilon = 10^{-5}, \rho = 6\,\text{dB}$.

For comparison, we plot also the outage capacity $C_{rx}^{out,\epsilon}$ in (10.24) and a lower bound on the ergodic capacity C_{no}^{erg} in (10.19) based on USTM distribution [34]. As shown in the figure, $C_{no}^{out,\epsilon}$ provides a good approximation for R_{no}^* only when ℓ is small ($n_c \approx 168$), i.e., when the fading channel is essentially constant over the duration of the packet (quasi-static scenario). Furthermore, $C_{no}^{out,\epsilon}$ fails to capture the loss in throughput due to the channel estimation overhead, which is relevant for small n_c. For example, for $n_c = 4$, the outage capacity overestimates R_{no}^* by a factor of two.

The lower bound on C_{no}^{erg} plotted in the figure approximates R_{no}^* poorly when n_c is large. For example, it overestimates R_{no}^* by a factor of four when $n_c = 168$. As expected, the approximation gets better as n_c becomes smaller. The number of active transmit antennas $\tilde{m}_t$ that maximizes the DT achievability bound is $\tilde{m}_t = 2$ (both antennas active) for $1 \le \ell \le 21$, and it is $\tilde{m}_t = 1$ (only one antenna active) for $\ell > 21$. The lower bound on C_{no}^{erg}, which also involves a maximization over the number of active antennas, exhibits the same behavior. We also note that the intersection between $C_{no}^{out,\epsilon}$ and C_{no}^{erg} predicts coarsely the optimal number ℓ^* of time–frequency diversity branches.

In Figure 10.9, we consider the case $\epsilon = 10^{-5}$. A similar behavior can be observed.

10.5   Channel Codes for Short-Packet Transmissions

While for large blocklengths the design of capacity-approaching channels codes is a problem that finds a simple and effective solution by resorting to turbo-like codes (e.g., turbo codes or LDPC codes) with iterative decoding, in the short blocklength regime the code design task unfolds in a number of nontrivial tradeoffs. The main reasons that render the problem "richer" from both research and engineering viewpoints are (1) the strongly suboptimal performance attained by turbo-like codes under belief propagation (BP) decoding in the short blocklength regime; and (2) the possibility to employ near-optimum (i.e., near maximum-likelihood) decoding algorithms, which are complexity-wise feasible thanks to the limited dimension of the codes (the same algorithms being not practical for long blocks, due to an unfavorable complexity scaling with the blocklength). Due to these reasons, solutions to the short code design problem are many: besides the nowadays omnipresent turbo and LDPC codes, "classical" error correcting schemes based on strong algebraic construction or on exhaustive searches of powerful codes with moderate-complexity trellis representations become viable candidates, often outperforming their turbo-like counterparts.

In the following, we provide an overview of some of the most promising classes of channel codes (and decoding algorithms) for short blocklengths. We do so by introducing a classification that has its roots into the history of channel coding. We will begin with so-called "classical" channels codes. Here, we will first introduce a decoding algorithm that enables near-maximum-likelihood (ML) performance for codesof length up to few hundred bits, irrespective to the code structure. The lack of any requirement (besides linearly) on the code structure opens the way to the adoption of well-known algebraic codes (such as BCH or Reed–Muller [RM] codes) with good distanceproperties.[12] We include, among classical channel codes, convolutional codes possessing a tail-biting trellis with state complexity profile that enables near-optimum decoding with moderate–low complexity. We will then discuss "modern" channel codes, which include turbo, LDPC, and polar codes. These classes of codes were introduced (or gained popularity, in the case of LDPC codes) after the "turbo revolution" that took place in the early 1990s, hence deserving to be introduced in the group of modern solutions to the coding problem.

In the following, we will focus for simplicity on the binary-input AWGN channel. We will often differentiate between *short* and *moderate* blocklength regimes. Although a clear distinction between the two regimes is hard to define, we may consider short codes as channel codes whose blocklength extends up to $n \approx 400$ bits, while by moderate blocklengths we mean lengths up to $n \approx 1000$. Whenever possible, we will try to compare the various coding schemes not only in terms of error rate performance, but also in terms of their decoding complexity and of other interesting features they may offer to the communication-protocol design (for example, their capability to include an error detection step in the decoding process).

---

[12] We shall see that the same class of algorithms can be applied to turbo-like codes as well, producing remarkable coding gains with respect to BP decoding. Note though that the knowledge of some important code parameters (such as the code minimum distance), which is available for many algebraic constructions, allows one to reduce the complexity of some near-ML decoding algorithms. These parameters are not known for most turbo-like constructions.

10.5.1   Classical Channel Codes

Ordered Statistics Decoding of Binary Linear Block Codes

Consider a (n, k) binary linear block code $\mathcal{C}$ (here, n is the blocklength in bits, and k is the code dimension, i.e., $M = 2^k$ is the size of the codebook). An optimum (maximum-likelihood (ML)) decoder would provide at its output

$$\hat{x}^n = \arg\max_{x^n \in \mathcal{C}} P_{Y^n|X^n}\left(y^n|x^n\right). \tag{10.157}$$

The evaluation of (10.157) requires (in the absence of a code structure that enables savings) the computation of 2^k likelihood values, hindering any feasible implementation of an ML decoder already for fairly small codes (i.e., k in the order of a few tens of bits). An intuitive way to reduce the complexity of (10.157) is by limiting the search of the codeword maximizing the likelihood $P_{Y^n|X^n}$ to a subset of the code. We shall refer to such a subset as "list." Let us denote the list by $\mathcal{L}$ (with $\mathcal{L} \subset \mathcal{C}$); then decoding reduces to

$$\hat{x}^n = \arg\max_{x^n \in \mathcal{L}} P_{Y^n|X^n}\left(y^n|x^n\right). \tag{10.158}$$

The evaluation of (10.158) may require considerably less effort than (10.157), provided that the size of the list $L = |\mathcal{L}|$ is much smaller that the size of the code. The saving does not come, however, for free: the solution of (10.157) may not always lie in the list. As a result, a decoder based on (10.158) is guaranteed to be always outperformed (in terms of error probability) by the ML rule of (10.157). A key ingredient for the application of the low-complexity decoding rule (10.158) is to construct, from an observation y^n, a list that contains a set of candidate codewords with a large likelihood $P_{Y^n|X^n}$. For the binary-input AWGN channel, this corresponds to constructing a list whose codewords lie close to y^n in Euclidean norm. Obviously, the larger the list size, the larger the probability of correct detection, but also the larger the complexity of the decoding operation.

A particularly effective approach to the construction of lists that trade optimally between performance and complexity is to use algorithms based on *ordered statistics* [24, 53]. In one of its simplest forms, the list construction in an OSD algorithm works as follows. The channel output samples are ordered in decreasing reliability, yielding, in the binary-input AWGN case, the vector $\mathring{y}^n$ with entries

$$\left|\mathring{y}_1\right| \geq \left|\mathring{y}_2\right| \geq \ldots \geq \left|\mathring{y}_n\right|. \tag{10.159}$$

We denote by π the permutation applied to y^n to get $\mathring{y}^n$, i.e., $\mathring{y}^n = \pi\left(y^n\right)$. The same permutation is applied to the n columns of the code generator matrix G leading to the permuted generator matrix $\mathring{G} = \pi\left(G\right)$. Let us assume that the k leftmost columns of $\mathring{G}$ are linearly independent.[13] By performing row operations, we can then put the permuted generator matrix in the systematic form

$$\mathring{G}_{\text{sys}} = (I|P), \tag{10.160}$$

---

[13] If this assumption does not hold, simple strategies can be put in place by applying a few additional permutations to the columns of the generator matrix (we refer the reader to [24] for more details).

where I is the $k \times k$ identity matrix. A hard decision is then performed on the first (most reliable) k elements of $\mathring{y}^n$, yielding a k-bit vector

$$\left(u_1^{(1)}, u_2^{(1)}, \ldots, u_k^{(1)} \right). \tag{10.161}$$

For a given value of the OSD parameter T, a number of $\sum_{t=1}^{T} \binom{k}{t}$ additional k-bit vectors is obtained by adding all the possible binary vectors with Hamming weight equal to or smaller than T to (10.161). Let

$$L = 1 + \sum_{t=1}^{T} \binom{k}{t}. \tag{10.162}$$

The set of obtained $L - 1$ vectors together with the hard decision vector (10.161),

$$\left(u_1^{(1)}, u_2^{(1)}, \ldots, u_k^{(1)} \right)$$
$$\left(u_1^{(2)}, u_2^{(2)}, \ldots, u_k^{(2)} \right)$$
$$\vdots$$
$$\left(u_1^{(L)}, u_2^{(L)}, \ldots, u_k^{(L)} \right)$$

is then encoded via the generator matrix (10.160) to produce the list $\mathcal{L}$ of L codewords, i.e.,

$$\left(x_1^{(1)}, x_2^{(1)}, \ldots, x_n^{(1)} \right) = \left(u_1^{(1)}, u_2^{(1)}, \ldots, u_k^{(1)} \right) \mathring{G}_{\mathrm{sys}}$$
$$\left(x_1^{(2)}, x_2^{(2)}, \ldots, x_n^{(2)} \right) = \left(u_1^{(2)}, u_2^{(2)}, \ldots, u_k^{(2)} \right) \mathring{G}_{\mathrm{sys}}$$
$$\cdots$$
$$\left(x_1^{(L)}, x_2^{(L)}, \ldots, x_n^{(L)} \right) = \left(u_1^{(L)}, u_2^{(L)}, \ldots, u_k^{(L)} \right) \mathring{G}_{\mathrm{sys}}.$$

The list is then used, upon inverse permutation π^{-1} of the code coordinates, in (10.158) to obtain the final decision.

The intuition on why OSD performs well is that hard decisions on the k most reliable channel outputs yields a only few erroneous bits: Most of the errors are confined to the $n - k$ hard decisions associated with the least reliable channel observations. The list is hence built by choosing codewords which "disagree" with the hard decision on y^n, in the k coordinates associated to the most reliable channel outputs, in T or fewer positions.

The parameter T plays a crucial role in the tradeoff between error probability and complexity. A large value of T results in a large list size, implying a near-ML performance at the expense of a high decoding complexity. A small value of T yields a small list size, and, hence, to simple decoders accompanied by an unavoidable performance degradation. It turns out that the tradeoff depends largely on the blocklength: The larger n, the larger should be T in order to approach the ML code performance. For example, for a $(24, 12)$ Golay code, $T = 2$ provides already an error probability that tightly follows that of an ML decoder. For a $(128, 64)$ extended BCH code, a value of T as high as 3 or 4 may be required to approach the performance of an ML decoder.

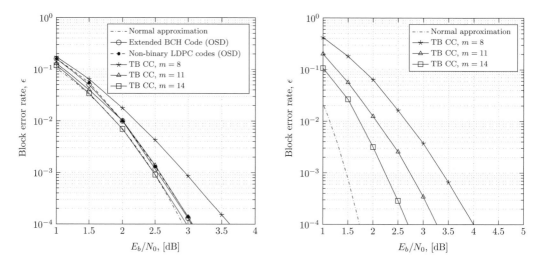

Figure 10.10 Block error rate vs. SNR for $(128, 64)$ codes (left) and $(512, 256)$ codes (right), compared with the normal approximation (10.6) for the binary-input AWGN channel. For the shortest blocklength, results for tail-biting convolutional codes as well as for a $(128, 64)$ extended BCH code are provided. For the longest blocklength, the results are provided only for tail-biting convolutional codes.

The OSD algorithm described above can be modified in several ways to reduce complexity [53]. A simple yet effective means to limit the average number of operations in an OSD is by introducing an early stopping criterion. On the binary-input AWGN channel, for example, if the code minimum distance $d_{\min}$ is known, the information can be used to stop the generation of codeword candidates. In fact, assuming binary signaling $\mathcal{X} = \{-\sqrt{\rho}, \sqrt{\rho}\}$, if a codeword x^n is found that lies at an Euclidean distance from the channel output y^n that is less than $\sqrt{\rho d_{\min}}$, then list construction can be halted by returning the decision x^n, without incurring any performance loss. Indeed, there cannot be any closer codeword, in the Euclidean space, to the channel output. The savings can be dramatic for codes that possess a large minimum distance and at medium–high SNRs. Figure 10.10 depicts the performance of a $(128, 64)$ extended BCH code with minimum distance of 22 with OSD where the parameter T is set to 4 [24]. On the same chart, the normal approximation (10.6) for the binary-input AWGN channel is provided. The chart, as the other charts included in this section, is in terms of block error probability vs. E_b/N_0, where E_b is the energy per information bit and N_0 is the single-sided noise power spectral density. This quantity is related to ρ via

$$\frac{E_b}{N_0} = \frac{\rho}{2R}. \tag{10.163}$$

The block error probability of the BCH code turns out to be very close to the normal approximation down to a block error rate $\epsilon = 10^{-4}$. OSD can be applied to any binary linear block code. For example, one may decode a (n, k) LDPC code with OSD [54]. In Figure 10.10 the performance of a $(128, 64)$ LDPC constructed over an order-256 field is provided, under OSD with $T = 4$. Down to the simulated error rates, its block error rate is almost indistinguishable from that of the extended BCH code. On the one hand, the advantage of using a code with a strong algebraic structure lies in the possibility of selecting codes with large (and known) minimum distance, with possible savings in the decoding complexity. On the other hand, one may consider using an LDPC (or a turbo) code with a hybrid decoding approach [54]: A first attempt is performed with iterative decoding, resorting to OSD only if iterative decoding fails.

Linear Block Codes Based on Tail-Biting Trellises

Some short codes with good distance spectra admit a reasonably compact tail-biting representation [55–59]. Thanks to this, they are amenable of near-optimum decoding by means of variants of the Viterbi algorithm tailored to operate over tail-biting trellises. The main difficulty in performing ML decoding over a tail-biting trellis lies in the uncertainty on the initial encoder state. To cope with this lack of information, one may resort to a list decoding approach, where a Viterbi decoding attempt is performed for each hypothesis on the initial (and thus, final) state. This would result in running L Viterbi decoders, where L is the number of possible initial (final) states. The L codewords produced at the output of each Viterbi decoder would form a list, out of which the most likely codeword is picked as the final decision. Observe that, differently from the OSD case, here it is guaranteed that the ML decision is always in the list, at the cost of an almost L-fold increase in complexity with respect to a single Viterbi decoder.

A simple suboptimum approach, which typically yields an error probability close to that of an ML decoder, is to use the wrap-around Viterbi algorithm (WAVA) [60]. The algorithm is very similar to the Viterbi algorithm with equally probable initial states. In its log domain implementation, at the beginning of the decoding process all initial state metrics are set to zero. A first iteration of the Viterbi algorithm is performed over the code trellis, producing for each of the L final states one path survivor. We assume next that the path metric is given by the squared Euclidean distance between the (modulated) branch labels and the corresponding channel observations. Among the survivors, the most likely one (i.e., the one at minimum Euclidean distance from the received vector) is selected. If this most likely path is also a tail-biting one, then decoding stops and the most likely tail-biting path is produced as output. If the most likely path is not tail-biting, then the initial state metrics of the trellis are updated with the metrics computed at the corresponding final states of the previous iteration. A new iteration of the Viterbi algorithm is performed. The process continues until either the termination condition is satisfied, i.e., the path with minimum final state metric is a tail-biting one, or a maximum number of iterations is reached (it was shown in [60] that four iterations are typically sufficient to obtain near-ML performance). In the former case, the algorithm outputs the path with minimum final state metric. In the latter case, the decoder outputs again the path corresponding to the path with minimum final state metric even if the path does not satisfy the tail-biting constraint, or alternatively it declares a decoding failure.

Table 10.1 Example of good tail-biting convolutional codes.

R	Generators	m	(n, k)	$A(x)$
	[515, 677]	8		$1 + 576x^{12} + 1152x^{13} + \dots$
	[5537, 6131]	11	(128, 64)	$1 + 64x^{14} + 960x^{15} + \dots$
	[75063, 56711]	14		$1 + 8x^{16} + 1856x^{18} + \dots$
1/2				
	[515, 677]	8		$1 + 1152x^{12} + 2304x^{13} + \dots$
	[5537, 6131]	11	(256, 128)	$1 + 128x^{14} + 1920x^{15} + \dots$
	[75063, 56711]	14		$1 + 3328x^{18} + 21120x^{20} + \dots$
	[435, 526, 717]	8		$1 + 64x^{17} + 128x^{18} + \dots$
	[4653, 5435, 6257]	11	(192, 64)	$1 + 192x^{22} + 576x^{24} + \dots$
	[47671, 55245, 63217]	14		$1 + 384x^{27} + 256x^{28} + \dots$
1/3				
	[435, 526, 717]	8		$1 + 128x^{17} + 256x^{18} + \dots$
	[4653, 5435, 6257]	11	(384, 128)	$1 + 384x^{22} + 1152x^{24} + \dots$
	[47671, 55245, 63217]	14		$1 + 768x^{27} + 512x^{28} + \dots$

Good codes can be obtained by applying a tail-biting termination to properly chosen convolutional code [57–59, 61].[14] Examples of some tail-biting convolutional codes are provided in Table 10.1. The table includes the description of the convolutional codes generator polynomials in octal form, as well as the lowest part of their distance spectra via the weight enumerator function $A(x) = \sum_{i=0}^{n} A_i x^i$, with A_i being the multiplicity of codeword with Hamming weight i. The weight enumerator functions are truncated to the first three nonzero terms. The code descriptions are provided for two rates ($R = 1/2$ and $R = 1/3$) and for two code dimensions ($k = 64$ and $k = 128$). Observe that, in some cases, an increase in the blocklength does not result in improved minimum distance. For example, for rate 1/2 codes with memory 8, the minimum distance is 12 with both $k = 64$ and $k = 128$, with the multiplicity of minimum-weight codewords that is double in the $k = 128$ case. The larger blocklength can be in fact exploited only by increasing the convolutional code memory.

In Figure 10.10 the performance of some tail-biting convolutional codes with rate 1/2 and generators from Table 10.1 is compared to the normal approximation. For the smallest blocklength case ($n = 128$) all codes perform remarkably close to the normal approximation, with the memory 14 code block error rate nearly on top of the finite length benchmark. The memory 11 convolutional code shows a slight loss, and provides performance that matches that of the (128, 64) extended BCH code with OSD. The memory 8 code suffers a visible performance loss, which is around 0.5 dB at a

---

[14] The performance of convolutional codes with respect to decoding latency has been addressed also in [62–64], where the back-tracking approach to provide early bit estimated with Viterbi decoding was analyzed.

block error rate of 10^{-4}. In the same figure, a second chart provides the performance for the three tail-biting convolutional codes with the same generator polynomials, at a blocklength of $n = 512$. As predicted by the considerations on the codes distance spectra, for a given generator polynomial pair the increased blocklength may result in an actual performance loss (this is the case for both the memory 8 and 11 codes, whereas for the memory 14 code the effect will be visible at a larger blocklength). The weakness of moderate-memory convolutional codes at increasing blocklengths is well-known [64] and poses a limitation to their use at blocklengths of several hundred bits.

10.5.2　Modern Channel Codes

While performing exceptionally well in the short blocklength regime, the two approaches discussed in the previous section can hardly be extended to moderate blocklength. The complexity of OSD, in fact, does not scale favorably with the blocklength. Tail-biting convolutional codes based on a given set of generators suffer from an increased block error probability when the blocklength grows beyond a given value, and larger memories are required to approach the theoretical limits for increasing n. Turbo and LDPC codes (under iterative decoding) fall short of matching the performance of short algebraic codes under OSD and of short tail-biting convolutional codes under WAVA. Nevertheless, turbo and LDPC codes become competitive in the moderate blocklength regime thanks to their linear (in the blocklength, per iteration) decoding complexity.

Polar codes under (successive cancellation) list decoding in concatenation with high-rate outer codes perform remarkably well both in the short blocklength regime (where they approach finite-length benchmarks by a few tenths of a dB) and in the moderate blocklength regime (where they outperform binary turbo/LDPC codes, with the latter capable of challenging polar codes only when constructed on high-order finite fields).

We discuss these three classes of codes in more details in following subsections.

Binary Turbo Codes

Binary turbo codes[15] have been successfully employed as channel codes in the 3G/4G cellular standards. Turbo codes are known to provide excellent coding gains in the moderate blocklength regime and (if carefully designed) at short blocklengths as well. If low error rates are required (block error rate $\epsilon < 10^{-4}$), a convenient design choice is to adopt 16-state component codes, i.e., to use memory 4 convolutional codes in the parallel concatenation. Less complex, memory 3 convolutional codes can be adopted when the target error rates are not too demanding. Visible gains can also be attained when a tail-biting termination is employed for the component codes (the gain is especially visible at short blocklengths, where the rate loss due to the terminations can be simply too large) [67, 68]. The interleaver design plays also an important role if low error floors are demanded. In this case, a small size of the information word (resulting in a small dimension of the interleaving matrix) allows for a careful optimization with respect to

---

[15] Here, by *turbo codes*, we refer to parallel concatenated convolutional codes [65]. We do not address the case of serial concatenated convolutional codes [66].

the code distance properties. Such code-matched interleaver designs [69], which are essential to attain low error floors in the short blocklength regime, are optimized over each blocklength and choice of convolutional codes. This, however, sacrifices flexibility in the description of the interleaver for large sets of input block sizes. The block error rates of turbo codes with parameters $(128, 64)$ and $(512, 256)$, based on memory 4 tail-biting convolutional codes, are provided in Figure 10.14, and compared to the normal approximation. Iterative (turbo) decoding with 10 iterations has been used for the simulations. The two codes perform almost 1 dB away from the normal approximation, down to a block error rate $\epsilon = 10^{-4}$.

Binary Low-Density Parity-Check Codes

LDPC codes [70] attracted considerable attention in the past two decades thanks to their capacity-approaching performance and the possibility of developing highly parallel iterative decodes (which are typically required in high-data-rate applications). An (n, k) LDPC code is defined by an $m \times n$ *sparse* parity-check matrix H (with $m \geq n - k$). Denote by $w_{c,i}$ the Hamming weight (i.e., number of 1s) of the ith column in H, and by $w_{r,j}$ the Hamming weight (i.e., number of 1s) of the jth column in H. An LDPC code is said to be *regular* if the Hamming weight of each column in H is fixed, $w_{c,i} = w_c$ for all i, and the Hamming weight of each row in H is fixed, $w_{r,j} = w_r$ for all j. A regular LDPC code is often referred to as a (w_c, w_r) LDPC code. An LDPC code is said to be *irregular* [71] if the Hamming weight of the columns and/or rows of H varies. LDPC codes can be well described through a bipartite (Tanner) graph $\mathscr{G}$. The graph is composed by

- a set $V = \{V_1, V_2, \ldots, V_n\}$ of n variable nodes (VNs), one for each codeword bit (i.e., one for each column of H)
- a set $C = \{C_1, C_2, \ldots, C_m\}$ of m check nodes (CNs), one for each parity-check equation (i.e., row) in H
- a set $\mathcal{E} = \{e_{i,j}\}$ of edges.

An edge $e_{j,i}$ connects the VN V_i to the CN C_j if and only if $h_{j,i} = 1$. It follows that the Tanner graph and the parity-check matrix provide an equivalent representation of the code.

From a design viewpoint, it is convenient to categorize LDPC codes into unstructured and structured LDPC codes. An unstructured LDPC code is typically constructed according to a given degree distribution profile, i.e., according to a target distribution of its parity-check matrix row and column weights. The construction often follows a (computer-based) pseudorandom algorithm that places the nonzero entries in the code parity-check matrix trying to maximize the *girth* (defined as the length of the shortest cycle) of the corresponding Tanner graph. Unstructured LDPC codes can be designed to achieve near-capacity performance for large block lengths [71], but can be hardly implemented in practice due to their lack of structure. *Structured* LDPC codes are hence a preferred option in practice. A class of structured LDPC codes that is rather popular in modern communication standards is the class of protograph-based codes [72]. Protograph codes belong to the larger class of multi-edge type LDPC codes [25]. A protograph is a relatively small graph from which a larger Tanner graph can be obtained by

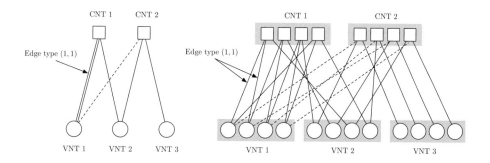

Figure 10.11 Protograph for a rate-1/3 LDPC (left) and its fourfold expansion (right).

a copy-and-permute procedure: The protograph is copied Q times, and then the edges of the individual replicas are permuted among the replicas (under restrictions described below) to obtain a single, large graph. The parameter Q is often referred to as a *lifting factor*. An example is presented in Figure 10.11. The nodes in the protograph identify variable node types (VNTs) and check node types (CNTs). The nodes of the protograph are labeled so that if VNT i is connected to check node CNT j in the protograph, then a variable node V of type i in the expanded graph has to be connected to one of the Q replicated check nodes of type j (see example in Figure 10.11). A protograph can possess parallel edges, i.e., two nodes can be connected by more than one edge. To obtain proper Tanner graph codes, the copy-and-permute procedure must eliminate such parallel connections.

As a Tanner graph provides a description of a code parity-check matrix, a protograph can be put in correspondence of a so-called base matrix. In particular, if the protograph possesses N VNTs and M CNTs, then its base matrix B is a $M \times N$ matrix with $b_{j,i}$ equal to number of edges connecting the VNT i with the CNT j. As an example, the base matrix for the protograph in Figure 10.11 is given by

$$B = \begin{pmatrix} 2 & 1 & 0 \\ 1 & 1 & 1 \end{pmatrix},$$

while the parity-check matrix associated to the protograph expansion of Figure 10.11 is

$$H = \left(\begin{array}{cccc|cccc|cccc} 1 & 1 & 0 & 0 & 0 & 0 & 1 & 0 & 0 & 0 & 0 & 0 \\ 0 & 1 & 1 & 0 & 0 & 1 & 0 & 0 & 0 & 0 & 0 & 0 \\ 0 & 0 & 1 & 1 & 1 & 0 & 0 & 0 & 0 & 0 & 0 & 0 \\ 1 & 0 & 0 & 1 & 0 & 0 & 0 & 1 & 0 & 0 & 0 & 0 \\ \hline 1 & 0 & 0 & 0 & 1 & 0 & 0 & 0 & 1 & 0 & 0 & 0 \\ 0 & 0 & 1 & 0 & 0 & 0 & 0 & 1 & 0 & 1 & 0 & 0 \\ 0 & 1 & 0 & 0 & 0 & 1 & 0 & 0 & 0 & 0 & 1 & 0 \\ 0 & 0 & 0 & 1 & 0 & 0 & 1 & 0 & 0 & 0 & 0 & 1 \end{array} \right).$$

It turns that the protograph expansion can be described in terms of the base matrix expansion, where each entry in the base matrix is replaced by a $Q \times Q$ binary matrix whose row/column weights equal the value of the corresponding entry in B. Very often, the base matrix expansion is performed by means of $Q \times Q$ cyclic permutation matrices (also referred to as circulants), or by the superposition of $Q \times Q$ cyclic permutation matrices. The so-obtained LDPC codes are *quasi-cyclic*, a property that facilitates the implementation of efficient encoders and decoders [73]. Powerful protograph LDPC codes have been designed during the past decade. A class of protograph LDPC codes that performs remarkably well down to short blocklengths is accumulate-repeat-accumulate (ARA) codes [74]. For rate-1/2 codes, an ARA protograph that provides excellent performance at moderate–short blocklengths has base matrix

$$
B = \begin{pmatrix} 2 & 1 & 1 & 1 & 0 \\ 1 & 2 & 1 & 1 & 0 \\ 2 & 1 & 0 & 0 & 1 \end{pmatrix},
$$

with the first column (i.e., first VNT) associated to bits that are punctured (thus, not transmitted over the channel).[16] The performance of accumulate–repeat–accumulate (ARA) codes with parameters $(128, 64)$ and $(512, 256)$ is shown in Figure 10.14. A maximum of 200 iterations has been assumed in the simulations. While for the smallest blocklength the ARA code loses some tenths of a dB with respect to the turbo code, at $n = 512$ the ARA and the turbo code attain very similar performance. It is interesting to compare the performance of both codes with those provided in Figure 10.10. While at the smallest blocklength short algebraic codes (with OSD decoding) and tail-biting convolutional codes visibly outperform both ARA and turbo codes, at a blocklength of $n = 512$ bits ARA and turbo codes turn out to be very competitive.

A further class of protograph LDPC codes with excellent performance is that proposed in [77], which relies on the concatenation of an outer (high-rate) LDPC code with an inner (potentially low-rate) LDPC code. The inner LDPC code construction resembles an LT code [78], yielding an overall LDPC code structure that closely mimics that of a Raptor code [79] (the main difference being that here the bits at the input of the inner LT encoder are, with the exception of the punctured ones, sent over the channel). A protograph proposed in [80] (and dubbed *lowest family*) is depicted in Figure 10.12. The protograph consists of 12 CNTs and 22 VNTs with two punctured VNTs yielding a code rate $R = 1/2$. The protograph, which describes the outer code, can be concatenated

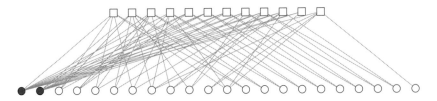

Figure 10.12 *Lowest family* protograph proposed in [80] as a channel coding option in 5G. Darkened circles represent punctured variable node types, i.e., variable node types associated to bits that are not sent through the channel.

---

[16] The use of punctured bits can help to obtain codes with capacity-approaching performance [75, 76].

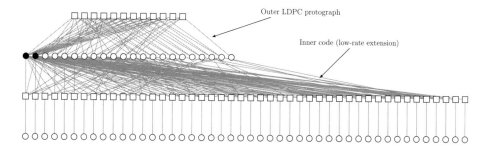

Figure 10.13 Low-rate extension of the *lowest family* protograph proposed in [80] as a channel coding option in 5G. Darkened circles represent punctured variable node types, i.e., variable node types associated to bits that are not sent through the channel.

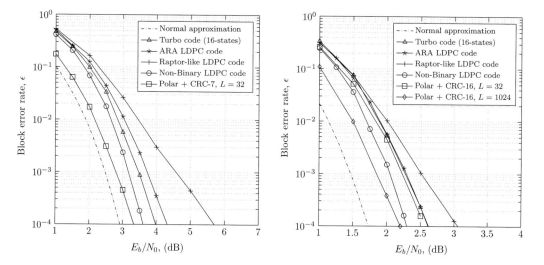

Figure 10.14 Block error rate vs. SNR for $(128, 64)$ codes (left) and $(512, 256)$ codes (right), compared with the normal approximation (10.6) for the binary-input AWGN channel. Results are provided for various turbo, LDPC, and polar code constructions.

with an additional protograph to produce a protograph extension toward low rates (see Figure 10.13).

The block error rate for two codes constructed from the protograph in Figure 10.12 is shown, for completeness, in Figure 10.14. The two codes, due to the protograph dimensions, have parameters that slightly differ from the ones of the other codes on the same charts. In particular, the shortest code has blocklength 120 and dimension 60, while the longest one has blocklength 520 and dimension 260. The two codes, decoded with 200 iterations, do not perform as well as their ARA counterparts. The reason for this can be found in the original design target of the codes from [80], which is enhanced mobile broadband links, where larger blocklengths are typically considered. The adaptation of the codes to the short blocklength regime is impaired by the large protograph size, which yields very limited degrees of freedom in the protograph expansion. In fact, the lifting factor Q turns out to be particularly small, reducing the effectiveness of girth optimization techniques.

Non-Binary Turbo and Low-Density Parity-Check Codes

Among turbo-like codes, non-binary LDPC [81, 82] and turbo [83, 84] constructed on large-order finite fields provide the best performance under iterative decoding in the short blocklength regime [84–86]. Non-binary LDPC codes based on ultra-sparse parity-check matrices (i.e., parity-check matrices with constant column weight equal to 2), in particular, are capable of providing a performance within 0.5 dB from finite-length benchmarks down to very low error rates when constructed on finite fields of order larger than or equal to 64. While their decoding can be largely simplified by employing a fast Fourier transform at the check nodes (with probability-domain decoding), efficient (low-complexity) implementations in the log domain are still an area of active research [87]. In Figure 10.14, $(128, 64)$ and $(512, 256)$ LDPC codes constructed over a finite field of order 256 are compared with binary LDPC and turbo codes, and with polar codes. In both cases, the codes (which possess a constance parity-check matrix column weight equal to 2, with row weight equal to 4) have been decoded with 200 iterations. The codes outperform binary turbo and LDPC codes, attaining a block error rate of 10^{-4} at roughly 0.5 dB from the normal approximation.

Polar Codes

Polar codes [88, 89] have attracted a lot of attention during the past decade thanks to their provable capacity-achieving properties under low-complexity successive cancellation (SC) decoding. Remarkably, their structure was revealed to be an asset in the short-blocklength regime as well, enabling the implementation of efficient list decoders [90]. The list decoder of [90] relies on the successive cancellation decoder, by bifurcating the decoding process at each bit decision: Two instances of the successive cancellation decoder are run, by assuming two opposite decisions (i.e., 0 and 1) for the last decided bit value. The number of active decoders thus grows as the decoding process proceeds from the least reliable information bit decisions to the most reliable ones. A limit to the number of active decoders at a given time is set to constrain the decoding complexity. The maximum number of active decoders, which we denote by L, determines how many final decisions will be provided at the end by the parallel instances of SC decoders. Suppose the list decoder to be at the ith step of the decoding process, i.e., for each active SC decoder instance, two instances are created by taking the ith information bit to be either 0 or 1. Assume furthermore that, prior to the bifurcation, there are L active SC decoders. After splitting the decoding processes, $2L$ SC decoders are active. The set of decoders is pruned at this point by stopping the L decoders that yield the lowest (partial) likelihood. After processing the last information bit, the set of L active decoders delivers the L list elements (i.e., L information words), out of which the most likely one is selected as the final decision. In [90] it was observed that, with a nonnegligible probability, the transmitted information word may be present in the list, but the maximum likelihood search within the list returns a different word. The result is due to the relatively poor minimum distance of polar codes. In [90] it was suggested to overcome the problem by introducing a different criterion for the selection of the information word within the list: By precoding the information message with a high-rate binary linear block code (e.g., CRC code), the word in the list that satisfies the outer code constraints is picked as the final decision. If more than one word satisfies the outer code constraints, then the one with the largest likelihood is picked. The introduction of the outer code yields in fact a concatenation of two binary linear block codes, resulting in a new binary linear block code with improved distance properties with respect to the original polar code. The resulting coding scheme turns out to be remarkably powerful at both short- and moderate-length blocks. Figure 10.14 depicts the performance of two polar codes in concatenation with an outer CRC code. A first code is obtained by concatenating a $(128, 71)$ polar code with a $(71, 64)$ shortened cyclic code. The list size is here limited to $L = 32$. The code performs remarkably close to the normal approximation and to the best codes in Figure 10.10. For the longer blocklength case, a $(512, 272)$ polar code is concatenated with a $(272, 256)$ shortened cyclic code. With a list size of 32, the code performance is in line with that of turbo and ARA codes, while by increasing the list size to 1024 the code outperforms even the non-binary LDPC code.

10.5.3 Remarks on Short Code Designs

Besides coding gain, some important figures of merit have been neglected in the comparison provided in the previous subsections. Decoding complexity plays a crucial role in the choice of an appropriate coding scheme for a given application. Non-binary LDPC codes, OSD with large values of the parameter T, tail-biting convolutional codes with large memory are all characterized by a complexity that may be unacceptable in many applications where the decoder has either to operate at high data rates, or it has to rely on very few computational resources. Tail-biting convolutional codes with small memory, binary LDPC codes, and turbo codes feature less complex decoders, which may be well-tailored to implementation in devices with stringent energy and computational power constraints, at the expense of some performance loss (at medium–large block lengths for convolutional codes, and in the short blocklength regime for binary turbo and LDPC codes). Polar codes allow for a fine performance–complexity tradeoff thanks to the possibility of adjusting the list size (though, to attain near-optimum performance at moderate block sizes, large lists may be required). Another important aspect that has not been discussed is the possibility to embed, at the end of the decoding process, an error-detection stage. Some decoders (referred to as *incomplete decoders*) provide a natural error-detection mechanism by producing (with very high probability) words that do not satisfy the code constraints when the decoder fails, thus allowing the detection of an erroneous output. This feature is particularly appealing in applications where low *undetected* error rates are required (for example, telecommand links [91]). LDPC codes under iterative decoding can be designed to target very low undetected error rates, as proposed in [92, 93]. For (tail-biting) convolutional codes, post-decoding verification can be put in place based on the general metric defined by Forney in [94], yielding a fine control on the undetected error probability [95, 96]. When a polar code is concatenated with an outer code, the list size can be used to trade coding gain with low undetected error rates.

10.6   Conclusions

As shown in this chapter, finite-blocklength tools are extremely useful for the design of short-packet communication systems: They provide a natural benchmark for the performance of actual codes (see Figure 10.2), they give insights on the design of the communication systems, such as the optimal choice of the number of transmit antennas and of diversity branches to code across (see Figure 10.8), and they yield simple approximations on the maximum coding rate, such as (10.6), (10.131), and (10.134), that can be used in place of the Shannon capacity in the design of latency-aware higher-layer algorithms such as user scheduling and resource allocation.

Many important extensions of this theory and application areas have been omitted due to space constraints. More insights on how to use finite-blocklength bounds to design actual short-packet communication systems for 5G scenarios, including the impact of pilot transmission and design of short codes approaching nonasymptotic bounds for the fading case, can be found in [52, 97]. Finite-blocklength bounds for the case in which variable-length codes are used and feedback (either full or just decision feedback, as in hybrid automatic repeat request (HARQ) strategies) can be found in [98]. Actual coding schemes for the variable-length decision-feedback setup are discussed in [99, 100]. Finite-blocklength results have been recently used in [101] to perform a joint queuing–coding analysis of the delay–violation probability in communication systems with random packet arrivals. Another relevant line of work uses finite-blocklength tools to analyze the minimum energy per bit required to transmit a small number of bits at a target reliability [102, 103] and to design optimal strategies for massive random multiple access [104].

References

[1] A. Osseiran, F. Boccardi, V. Braun, et al., "Scenarios for 5G mobile and wireless communications: The vision of the METIS project," *IEEE Commun. Mag.*, vol. 52, no. 5, pp. 26–35, May 2014.

[2] G. Durisi, T. Koch, and P. Popovski, "Towards massive, ultra-reliable, and low-latency wireless communication with short packets," *Proc. IEEE*, vol. 104, no. 9, pp. 1711–1726, Sep. 2016.

[3] N. A. Johansson, Y.-P. E. Wang, E. Eriksson, and M. Hessler, "Radio access for ultra-reliable and low-latency 5G communications," in *Proc. IEEE Int. Conf. Commun. (ICC)*, London, Jun. 2015.

[4] O. N. C. Yilmaz, Y.-P. E. Wang, N. A. Johansson, et al., "Analysis of ultra-reliable and low-latency 5G communication for a factory automation use case," in *Proc. IEEE Int. Conf. Commun. (ICC)*, London, Jun. 2015.

[5] ITU-R, "Recommendation ITU-R M.2083-0: IMT vision – framework and overall objectives of the future development of IMT for 2020 and beyond," Tech. Rep., Sep. 2015.

[6] "3GPP TS 22.261 3rd generation partnership project: Technical specification group services and system aspects; service requirements for the 5G systems; stage 1; (release 16)," Tech. Rep. V16.2.0, Dec. 2017.

[7] T. de Cola, E. Paolini, G. Liva, and G. P. Calzolari, "Reliability options for data communications in the future deep-space missions," *Proc. IEEE*, vol. 99, no. 11, pp. 2056–2074, Nov. 2011.

[8] C. E. Shannon, "Communication in the presence of noise," *Proc. IRE*, vol. 37, pp. 10–21, Jan. 1949.

[9] C. E. Shannon, "Probability of error for optimal codes in a Gaussian channel," *Bell Syst. Tech. J.*, vol. 38, pp. 611–656, 1959.

[10] R. Costa, M. Langberg, and J. Barros, "One-shot capacity of discrete channels," in *Proc. IEEE Int. Symp. Inf. Theory (ISIT)*, Austin, TX, Jun. 2010, pp. 211–215.

[11] A. Feinstein, "A new basic theorem of information theory," *IRE Trans. Info. Theory*, vol. 4, no. 4, pp. 2–22, May 1954.

[12] R. L. Dobrushin, "Mathematical problems in the Shannon theory of optimal coding of information," in *Proc. Berkeley Symp. Math, Statist. Prob.*, vol. 1, 1961, pp. 211–252.

[13] V. Strassen, "Asymptotische Abschätzungen in Shannon's Informationstheorie," in *Prague Conf. Inf. Theory*, Prague, 1962, pp. 689–723.

[14] C. E. Shannon, R. G. Gallager, and E. R. Berlekamp, "Lower bounds to error probability for coding on discrete memoryless channels: Part I," *Inf. Contr.*, vol. 10, pp. 65–103, 1967.

[15] R. Gallager, "A simple derivation of the coding theorem and some applications," *IEEE Trans. Inf. Theory*, vol. 11, no. 1, pp. 3–18, Jan. 1965.

[16] Y. Polyanskiy, H. V. Poor, and S. Verdú, "Channel coding rate in the finite blocklength regime," *IEEE Trans. Inf. Theory*, vol. 56, no. 5, pp. 2307–2359, May 2010.

[17] A. Collins, G. Durisi, T. Erseghe, et al., *SPECTRE: Short-packet communication toolbox*, v2.0, Sep. 2016. [Online]. Available: https://github.com/yp-mit/spectre.

[18] V. Y. F. Tan and M. Tomamichel, "The third-order term in the normal approximation for the AWGN channel," *IEEE Trans. Inf. Theory*, vol. 61, no. 5, pp. 2430–2438, May 2015.

[19] W. Feller, *An Introduction to Probability Theory and Its Applications*, 3rd ed. Chichester: Wiley, 1971, vol. II.

[20] Y. Polyanskiy, "Channel coding: Non-asymptotic fundamental limits," PhD dissertation, Princeton University, 2010.

[21] I. Tal and A. Vardy, "List decoding of polar codes," *IEEE Trans. Inf. Theory*, vol. 61, no. 5, pp. 2213–2226, May 2015.

[22] S. Dolinar, D. Divsalar, and F. Pollara, "Code performance as a function of the block size," Jet Propulsion Laboratory, Pasadena, CA, JPL TDA Progress Report 133, 1998.

[23] R. Koetter and A. Vardy, "Algebraic soft-decision decoding of Reed–Solomon codes," U.S. Patent WO2 001 035 536 A1, 2001.

[24] M. Fossorier and S. Lin, "Soft-decision decoding of linear block codes based on ordered statistics," *IEEE Trans. Inf. Theory*, vol. 41, no. 5, pp. 1379–1396, Sep. 1995.

[25] T. Richardson and R. Urbanke, "Multi-edge type LDPC codes," 2004. [Online]. Available: http://citeseerx.ist.psu.edu/viewdoc/summary?doi=10.1.1.106.7310.

[26] "3GPP TS 38.212 V15.0.0: Multiplexing and channel coding," Dec. 2017.

[27] A. Lapidoth, "On the asymptotic capacity of stationary Gaussian fading channels," *IEEE Trans. Inf. Theory*, vol. 51, no. 2, pp. 437–446, Feb. 2005.

[28] E. Biglieri, J. G. Proakis, and S. Shamai (Shitz), "Fading channels: Information-theoretic and communications aspects," *IEEE Trans. Inf. Theory*, vol. 44, no. 6, pp. 2619–2692, Oct. 1998.

[29] A. Lozano, A. M. Tulino, and S. Verdú, "Multiantenna capacity: Myths and realities," in *Space-Time Wireless Systems: From Array Processing to MIMO Communications*, H. Bölcskei, D. Gesbert, C. Papadias, and A. J. van der Veen, Eds. Cambridge: Cambridge University Press, 2006, pp. 87–107.

[30] İ. E. Telatar, "Capacity of multi-antenna Gaussian channels," *Eur. Trans. Telecommun.*, vol. 10, pp. 585–595, Nov. 1999.

[31] L. Zheng and D. N. C. Tse, "Communication on the Grassmann manifold: A geometric approach to the noncoherent multiple-antenna channel," *IEEE Trans. Inf. Theory*, vol. 48, no. 2, pp. 359–383, Feb. 2002.

[32] W. Yang, G. Durisi, and E. Riegler, "On the capacity of large-MIMO block-fading channels," *IEEE J. Sel. Areas Commun.*, vol. 31, no. 2, pp. 117–132, Feb. 2013.

[33] F. Rusek, A. Lozano, and N. Jindal, "Mutual information of IID complex Gaussian signals on block Rayleigh-faded channels," *IEEE Trans. Inf. Theory*, vol. 58, no. 1, pp. 331–340, Jan. 2012.

[34] R. Devassy, G. Durisi, J. Östman, et al., "Finite-SNR bounds on the sum-rate capacity of Rayleigh block-fading multiple-access channels with no a priori CSI," *IEEE Trans. Commun.*, vol. 63, no. 10, pp. 3621–3632, Oct. 2015.

[35] A. Lapidoth and S. M. Moser, "Capacity bounds via duality with applications to multiple-antenna systems on flat-fading channels," *IEEE Trans. Inf. Theory*, vol. 49, no. 10, pp. 2426–2467, Oct. 2003.

[36] S. M. Moser, "The fading number of multiple-input multiple-output fading channels with memory," *IEEE Trans. Inf. Theory*, vol. 55, no. 6, pp. 2716–2755, Jun. 2009.

[37] T. Koch, *On Heating Up and Fading in Communication Channels*. Konstanz: Hartung-Gorre Verlag, 2009.

[38] L. H. Ozarow, S. Shamai (Shitz), and A. D. Wyner, "Information theoretic considerations for cellular mobile radio," *IEEE Trans. Veh. Technol.*, vol. 43, no. 2, pp. 359–378, May 1994.

[39] E. Abbe, E. Telatar, and S. Huang, "Proof of the outage probability conjecture for MISO channels," *IEEE Trans. Inf. Theory*, vol. 59, no. 5, pp. 2596–2602, May 2013.

[40] L. Zheng and D. Tse, "Diversity and multiplexing: A fundamental tradeoff in multiple-antenna channels," *IEEE Trans. Inf. Theory*, vol. 49, no. 5, pp. 1073–1096, May 2003.

[41] L. Zheng, "Diversity-multiplexing tradeoff: A comprehensive view of multiple antenna systems," PhD dissertation, University of California at Berkeley, Nov. 2002.

[42] Y. Polyanskiy, H. V. Poor, and S. Verdú, "New channel coding achievability bounds," in *Proc. IEEE Int. Symp. Inf. Theory (ISIT)*, Toronto, Jul. 2008.

[43] S. Verdú and T. S. Han, "A general formula for channel capacity," *IEEE Trans. Inf. Theory*, vol. 40, no. 4, pp. 1147–1157, Jul. 1994.

[44] Y. Polyanskiy and S. Verdú, "Empirical distribution of good channel codes with nonvanishing error probability," *IEEE Trans. Inf. Theory*, vol. 60, no. 1, pp. 5–21, Jan. 2014.

[45] H. V. Poor, *An Introduction to Signal Detection and Estimation*, 2nd ed. New York: Springer-Verlag, 1994.

[46] W. Yang, G. Durisi, T. Koch, and Y. Polyanskiy, "Quasi-static multiple-antenna fading channels at finite blocklength," *IEEE Trans. Inf. Theory*, vol. 60, no. 7, pp. 4232–4265, Jul. 2014.

[47] G. Durisi, T. Koch, J. Östman, Y. Polyanskiy, and W. Yang, "Short-packet communications over multiple-antenna Rayleigh-fading channels," *IEEE Trans. Commun.*, vol. 64, no. 2, pp. 618–629, Feb. 2016.

[48] A. Collins and Y. Polyanskiy, "Dispersion of the coherent MIMO block-fading channel," in *Proc. IEEE Int. Symp. Inf. Theory (ISIT)*, Barcelona, Jul. 2016.

[49] A. T. James, "Distributions of matrix variates and latent roots derived from normal samples," *Ann. Math. Statist.*, vol. 35, no. 2, pp. 475–501, 1964.

[50] A. M. Tulino and S. Verdú, "Random matrix theory and wireless communications," in *Found. Trends Commun. Inform. Theory*, vol. 1, no. 1, pp. 1–182, 2004.

[51] A. Lancho, T. Koch, and G. Durisi, "A high-SNR normal approximation for single-antenna Rayleigh block-fading channels," in *Proc. IEEE Int. Symp. Inf. Theory (ISIT)*, Aachen, Jun. 2017.

[52] J. Östman, G. Durisi, and E. G. Ström, "Finite-blocklength bounds on the maximum coding rate of Rician fading channels with applications to pilot-assisted transmission," in *Proc. IEEE Int. Workshop Signal Process. Advances Wireless Commun. (SPAWC)*, Sapporo, Jul. 2017.

[53] S. Lin and D. Costello, Jr., *Error Control Coding*. Englewood Cliffs, NJ: Prentice Hall, 2004.

[54] M. P. C. Fossorier, "Iterative reliability-based decoding of low-density parity check codes," *IEEE J. Sel. Areas Commun.*, vol. 19, no. 5, pp. 908–917, May 2001.

[55] S. Lin, T. Kasami, T. Fujiwara, and M. Fossorier, *Trellises and Trellis-Based Decoding Algorithms for Linear Block Codes*. New York: Springer Science & Business Media, 1998.

[56] A. R. Calderbank, G. D. Forney, and A. Vardy, "Minimal tail-biting trellises: The Golay code and more," *IEEE Trans. Inf. Theory*, vol. 45, no. 5, pp. 1435–1455, Jul. 1999.

[57] P. Stahl, J. B. Anderson, and R. Johannesson, "Optimal and near-optimal encoders for short and moderate-length tail-biting trellises," *IEEE Trans. Inf. Theory*, vol. 45, no. 7, pp. 2562–2571, Nov. 1999.

[58] I. E. Bocharova, R. Johannesson, B. D. Kudryashov, and P. Stahl, "Tailbiting codes: Bounds and search results," *IEEE Trans. Inf. Theory*, vol. 48, no. 1, pp. 137–148, Jan. 2002.

[59] R. Johannesson and K. S. Zigangirov, *Fundamentals of Convolutional Coding*, 2nd ed. Chichester: Wiley, 2015.

[60] R. Y. Shao, S. Lin, and M. P. Fossorier, "Two decoding algorithms for tailbiting codes," *IEEE Trans. Commun.*, vol. 51, no. 10, pp. 1658–1665, Oct. 2003.

[61] G. Solomon and H. Van Tilborg, "A connection between block and convolutional codes," *SIAM J. Appl. Math.*, vol. 37, no. 2, pp. 358–369, 1979.

[62] T. Hehn and J. B. Huber, "LDPC codes and convolutional codes with equal structural delay: A comparison," *IEEE Trans. Commun.*, vol. 57, no. 6, pp. 1683–1692, Jun. 2009.

[63] S. V. Maiya, D. J. Costello, and T. E. Fuja, "Low latency coding: Convolutional codes vs. LDPC codes," *IEEE Trans. Commun.*, vol. 60, no. 5, pp. 1215–1225, May 2012.

[64] C. Rachinger, J. B. Huber, and R. R. Müller, "Comparison of convolutional and block codes for low structural delay," *IEEE Trans. Commun.*, vol. 63, no. 12, pp. 4629–4638, Dec. 2015.

[65] C. Berrou, A. Glavieux, and P. Thitimajshima, "Near Shannon limit error-correcting coding and decoding: Turbo-codes," in *Proc. IEEE Int. Conf. Commun. (ICC)*, Geneva, May 1993.

[66] S. Benedetto, D. Divsalar, G. Montorsi, and F. Pollara, "Serial concatenation of interleaved codes: Performance analysis, design, and iterative decoding," *IEEE Trans. Inf. Theory*, vol. 44, no. 3, pp. 909–926, May 1998.

[67] S. Crozier and P. Guinand, "High-performance low-memory interleaver banks for turbo-codes," in *IEEE VTC*, Atlantic City, NJ, Oct. 2001 vol. 4, pp. 2394–2398.

[68] T. Jerkovits and B. Matuz, "Turbo code design for short blocks," in *Proc. 7th Adv. Satellite Mobile Syst. Conf.*, Majorca (Spain), Sept. 2016.

[69] W. Feng, J. Yuan, and B. S. Vucetic, "A code-matched interleaver design for turbo codes," *IEEE Trans. Commun.*, vol. 50, no. 6, pp. 926–937, Jun. 2002.

[70] R. G. Gallager, *Low-Density Parity-Check Codes*. Cambridge, MA: MIT Press, 1963.

[71] T. Richardson and R. Urbanke, *Modern Coding Theory*. Cambridge: Cambridge University Press, 2008.

[72] J. Thorpe, "Low-density parity-check (LDPC) codes constructed from protographs," JPL INP, Tech. Rep., Aug. 2003.

[73] W. E. Ryan and S. Lin, *Channel Codes Classical and Modern*. Cambridge: Cambridge University Press, 2009.

[74] A. Abbasfar, D. Divsalar, and K. Yao, "Accumulate-repeat-accumulate codes," *IEEE Trans. Commun.*, vol. 55, no. 4, pp. 692–702, Apr. 2007.

[75] G. Liva and M. Chiani, "Protograph LDPC codes design based on EXIT analysis," in *Proc. IEEE Global Telecommun. Conf. (GLOBECOM)*, Washington, DC, Nov. 2007, pp. 3250–3254.

[76] D. Divsalar, S. Dolinar, C. R. Jones, and K. Andrews, "Capacity-approaching protograph codes," *IEEE J. Sel. Areas Commun.*, vol. 27, no. 6, pp. 876–888, Aug. 2009.

[77] T. Y. Chen, K. Vakilinia, D. Divsalar, and R. D. Wesel, "Protograph-based raptor-like LDPC codes," *IEEE Trans. Commun.*, vol. 63, no. 5, pp. 1522–1532, May 2015.

[78] M. Luby, "LT codes," in *Proc. 43rd IEEE Symp. Found. Computer Sci.*, Vancouver, Nov. 2002, pp. 271–282.

[79] M. Shokrollahi, "Raptor codes," *IEEE Trans. Inf. Theory*, vol. 52, no. 6, pp. 2551–2567, Jun. 2006.

[80] Qualcomm Inc., "LDPC rate compatible design overview," 3GPP TSG-RAN WG1, Oct. 2016.

[81] M. C. Davey and D. MacKay, "Low density parity-check codes over GF(q)," *IEEE Commun. Lett.*, vol. 2, no. 6, pp. 165–167, Jun. 1998.

[82] C. Poulliat, M. Fossorier, and D. Declercq, "Design of regular $(2, d_c)$-LDPC codes over GF(q) using their binary images," *IEEE Trans. Commun.*, vol. 56, no. 10, pp. 1626–1635, 2008.

[83] J. Berkmann, *"Iterative decoding of nonbinary codes,"* PhD dissertation, Technical University München, 2000.

[84] G. Liva, E. Paolini, B. Matuz, S. Scalise, and M. Chiani, "Short turbo codes over high order fields," *IEEE Trans. Commun.*, vol. 61, no. 6, pp. 2201–2211, Jun. 2013.

[85] G. Liva, E. Paolini, T. D. Cola, and M. Chiani, "Codes on high-order fields for the CCSDS next generation uplink," in *2012 6th Adv. Satellite Multimedia Syst. Conf. (ASMS) and 12th Signal Proces. Space Commun. Workshop (SPSC)*, Sept. 2012, pp. 44–48.

[86] B. Y. Chang, D. Divsalar, and L. Dolecek, "Non-binary protograph-based LDPC codes for short block-lengths," in *Proc. IEEE Inf. Theory Workshop (ITW)*, Lausanne, Sep. 2012.

[87] D. Declercq and M. Fossorier, "Decoding algorithms for nonbinary LDPC codes over gf(q)," *IEEE Trans. Commun.*, vol. 55, no. 4, pp. 633–643, Apr. 2007.

[88] N. Stolte, *"Rekursive Codes mit der Plotkin-Konstruktion und ihre Decodierung,"* PhD dissertation, TU Darmstadt, 2002.

[89] E. Arikan, "Channel polarization: A method for constructing capacity-achieving codes for symmetric binary-input memoryless channels," *IEEE Trans. Inf. Theory*, vol. 55, no. 7, pp. 3051–3073, Jul. 2009.

[90] I. Tal and A. Vardy, "List decoding of polar codes," *IEEE Trans. Inf. Theory*, vol. 61, no. 5, pp. 2213–2226, May 2015.

[91] *Next Generation Uplink*, Green Book, Issue 1, Consultative Committee for Space Data Systems (CCSDS) Report Concerning Space Data System Standards 230.2-G-1, Jul. 2014.

[92] D. Divsalar, S. Dolinar, and C. Jones, "Short protograph-based LDPC codes," in *Proc. IEEE Milcom*, Orlando, FL, 2007, pp. 1–6.

[93] *Short Block Length LDPC Codes for TC Synchronization and Channel Coding*, Orange Book, Consultative Committee for Space Data Systems (CCSDS) Experimental Specification 231.1-O-1, Apr. 2015.

[94] G. Forney, "Exponential error bounds for erasure, list, and decision feedback schemes," *IEEE Trans. Inf. Theory*, vol. 14, no. 2, pp. 206–220, Mar. 1968.

[95] E. Hof, I. Sason, and S. Shamai, "On optimal erasure and list decoding schemes of convolutional codes," in *Proc. Tenth Int. Symp. Commun. Theory and Applications (ISCTA)*, Ambleside, Jul. 2009, pp. 6–10.

[96] A. R. Williamson, M. J. Marshall, and R. D. Wesel, "Reliability-output decoding of tail-biting convolutional codes," *IEEE Trans. Commun.*, vol. 62, no. 6, pp. 1768–1778, Jun. 2014.

[97] G. C. Ferrante, J. Östman, G. Durisi, and K. Kittichokechai, "Pilot-assisted short-packet transmission over multiantenna fading channels: A 5G case study," in *Proc. Conf. Inf. Sci. Sys. (CISS)*, Princeton, NJ, Mar. 2018.

[98] Y. Polyanskiy, H. V. Poor, and S. Verdú, "Feedback in the non-asymptotic regime," *IEEE Trans. Inf. Theory*, vol. 57, no. 8, pp. 4903–4925, Aug. 2011.

[99] T.-Y. Chen, A. R. Williamson, N. Seshadri, and R. D. Wesel, "Feedback communication systems with limitations on incremental redundancy," Sep. 2013. [Online]. Available: http://arxiv.org/abs/1309.0707.

[100] A. R. Williamson, T.-Y. Chen, and R. D. Wesel, "Variable-length convolutional coding for short blocklengths with decision feedback," *IEEE Trans. Commun.*, vol. 63, no. 7, pp. 2389–2403, Jul. 2015.

[101] R. Devassy, G. Durisi, G. C. Ferrante, O. Simeone, and E. Uysal-Biyikoglu, "Delay and peak-age violation probability in short-packet transmission," in *Proc. IEEE Int. Symp. Inf. Theory (ISIT)*, Vail, CO, Jun. 2018. [Online]. Available: https://arxiv.org/abs/1805.03271.

[102] Y. Polyanskiy, H. Poor, and S. Verdú, "Minimum energy to send k bits through the Gaussian channel with and without feedback," *IEEE Trans. Inf. Theory*, vol. 57, no. 8, pp. 4880–4902, Aug. 2011.

[103] W. Yang, G. Durisi, and Y. Polyanskiy, "Minimum energy to send k bits over Rayleigh-fading channels," in *Proc. IEEE Int. Symp. Inf. Theory (ISIT)*, Hong Kong, Jun. 2015.

[104] Y. Polyanskiy, "A perspective on massive random access," in *IEEE Int. Symp. Inf. Theory (ISIT)*, Aachen Jan. 2017.

11 Information Theoretic Perspectives on Nonorthogonal Multiple Access (NOMA)

Peng Xu, Zhiguo Ding, and H. Vincent Poor

11.1 Introduction

Because of its promise of superior spectral efficiency, nonorthogonal multiple access (NOMA) has been recognized as a promising technique to be used in the fifth generation (5G) of wireless networks [1–8]. Unlike conventional orthogonal multiple access (OMA) techniques, such as time division multiple access (TDMA) and orthogonal frequency division multiple access (OFDMA), which serve a single user in each orthogonal resource block, NOMA aims to serve more than one user in each orthogonal resource block, e.g., a time slot, a frequency channel, or a spreading code. Several 5G multiple access techniques have been proposed by academia and industry, where power-domain NOMA (PD-NOMA) is one of the most well-known examples.

The basic concept of PD-NOMA is to exploit the power domain at the same time slot, OFDMA subcarrier, or spreading code, for achieving multiple access (MA), i.e., different users are served at different power levels. Unlike conventional orthogonal MA techniques, such as TDMA, NOMA faces strong co-channel interference between different users, and successive interference cancellation (SIC) is used by PD-NOMA users having better channel conditions for interference management. From an information theoretic perspective, downlink PD-NOMA can achieve the capacity region of the single-input single-output (SISO) Gaussian broadcast channel (BC). Note that the SISO Gaussian BC is a type of the degraded discrete memoryless (DM) BC, whose capacity region was established in [9] by using superposition coding proposed by Cover for the degraded DM BC [10]. Moreover, the capacity region of the multiple-input multiple-output (MIMO) Gaussian BC was found in [11] by applying dirty paper coding (DPC) instead of superposition coding. Compared to the design of SISO-NOMA, the design of MIMO-NOMA is more challenging, since the optimal user ordering scheme has not been revealed and it is not clear whether the use of MIMO-NOMA can achieve the optimal system performance. However, various practical MIMO-NOMA schemes have been developed and shown to outperform MIMO-OMA [12, 13].

Another well-known form of NOMA, code-domain NOMA (CD-NOMA), has also been proposed. It is worth pointing out that CD-NOMA belongs to a more general framework, termed hybrid NOMA, since CD-NOMA is based on the concept that more than one user is served on the same orthogonal bandwidth resource block, such as an OFDMA subcarrier or a time slot, and each user could occupy multiple subcarriers or time slots [14–19]. Practical forms of CD-NOMA include low-density spreading (LDS), sparse code multiple access (SCMA), pattern division multiple access (PDMA), and more. For LDS and SCMA, the number of subcarriers assigned to each user is much smaller than the total number of subcarriers, and this low spreading (sparse) feature ensures that the number of users utilizing the same subcarrier is not too large, so that the system complexity remains manageable [15, 16]. However, for PDMA, the LDS (sparse) feature of LDS and SCMA is no longer strictly present, i.e., the number of subcarriers occupied by one user is not necessarily much smaller than the total number of subcarriers [18]. As a result, PDMA can potentially yield better performance than LDS and SCMA, but at the price of higher complexity.

Because of its compatibility with other communication technologies, NOMA can be straightforwardly integrated into existing and future wireless systems. Motivated by this, multiuser superposition transmission (MUST), which refers to a downlink PD-NOMA scheme with two users, has been recently included in Third-Generation Partnership Project long-term evolution advanced (3GPP-LTE-A) networks [20]. In particular, the use of the PD-NOMA principle can guarantee that two users can be served simultaneously on the same OFDMA subcarrier which does not require any changes to the LTE resource blocks (i.e., OFDMA subcarriers). Furthermore, NOMA has also been recently included in the forthcoming digital TV standard (ATSC 3.0), where it is referred to as layered division multiplexing (LDM) [21]. In particular, by superimposing multiple data streams according to the NOMA principle, the spectral efficiency of TV broadcasting can be improved. Based on the above examples, it is clear that NOMA has substantial potential not only for 5G networks but also for other existing and emerging communication networks.

In this chapter, PD-NOMA will be taken as a representative example to illustrate the general principle of NOMA from an information theoretic perspective. In particular, the single-antenna PD-NOMA system is the main focus, which is described in the following.

11.1.1   System Model

Consider a downlink NOMA scenario which consists of a base station and K users. Each of the K users and the base station is equipped with a single antenna. The channels from the base station to all users are assumed to undergo quasi-static block fading, i.e., the channel gains are constant during one fading block consisting of T channel uses, but change independently from block to block. All channel gains are also assumed to undergo frequency-flat quasi-static Rayleigh fading. The base station sends a superposition coded signal, denoted by $x(t, b) = \sum_{k=1}^{K} s_k(t, b)$, at time instant t within fading block b. Note that the signal $s_k(t, b)$ carries the message for user k, and the signals sent to different users are independent. The signal received at user k can be written as follows:

$$y_k(t, b) = h_k(b) \sum_{i=1}^{K} s_i(t, b) + n_k(t, b), \ t \in [1 : T], \tag{11.1}$$

at time instant t within fading block b. Here, $h_k(b)$ denotes the channel gain from the base station to user k in block b, which is independent of all the other channel gains, and

$n_k(t, b)$ denotes the noise at time t within block b. The noise is assumed to be independent and identically distributed (i.i.d.) with respect to both t and b, and is complex Gaussian with zero mean and unit variance. For the sake of brevity, the fading block index b will be omitted in the rest of this chapter whenever this does not cause any confusion.

User Ordering

User ordering is determined by the base station according to the obtained channel state information at the transmitter (CSIT). Note that the base station could have either perfect or imperfect CSIT. Denote user ordering by $\{|h_{\pi_1}|^2, |h_{\pi_2}|^2, \cdots, |h_{\pi_K}|^2\}$, which means that user π_k has a stronger channel than users $\pi_1, \cdots, \pi_{k-1}$, and thus can first decode the messages for users $\pi_1, \cdots, \pi_{k-1}$ successfully, before decoding its own message. For perfect CSIT, the base station can distinguish all the channel gains such that $|h_{\pi_1}|^2 \leq |h_{\pi_2}|^2 \leq \cdots \leq |h_{\pi_K}|^2$. However, for the scenario with imperfect CSIT, the base station may not distinguish the users whose gains are close to each other, so the user ordering determined by the base station may be different from that obtained with perfect CSIT.

Successive Interference Cancellation

Assume that the base station broadcasts the user ordering information to each user correctly at the beginning of each fading block. Based on the user ordering, SIC is carried out in the NOMA system at each user to decode messages. In the SIC process, the messages of users π_l, $l \in [1 : k]$, will be sequentially decoded by user π_k. In particular, user π_k can cancel the interference generated by users π_1 to π_l when detecting the message of user π_{l+1}.

Power Constraints

Two different types of power constraint will be considered: *short-term* and *long-term* power constraints. Denote $\mathcal{H}(b) \triangleq \{h_1(b), \cdots, h_K(b)\}$. For block b, the power allocated for each user is denoted by a function of $\mathcal{H}(b)$. Specifically, the power allocated to user π_k, whose ordering index in the SIC process is k, is denoted by $P_k(\mathcal{H}(b))$ within block b. The short-term power constraint means that the sum power allocated to all the users within any block is constrained, i.e.,

$$\sum_{k=1}^{K} P_k(\mathcal{H}(b)) \leq P, \quad \forall b, \tag{11.2}$$

where P is the constraint. Alternatively, the power constraint in (11.2) can be relaxed such that the sum transmit power averaged over a large number of fading blocks does not exceed P. Based on the strong law of large numbers, this long-term power constraint can be expressed as follows:

$$\lim_{B \to \infty} \frac{1}{B} \sum_{b=1}^{B} \sum_{k=1}^{K} P_k(\mathcal{H}(b)) = \mathbb{E}_{\mathcal{H}} \left[\sum_{k=1}^{K} P_k(\mathcal{H}) \right] \leq P. \tag{11.3}$$

11.1.2 Instantaneous NOMA Region

Consider a NOMA system with two users and a short-term power constraint. Denote the two channel gains as h_w and h_b, and, without loss of generality, assume that $h_w = h_{\pi_1}$ and $h_b = h_{\pi_2}$, i.e., $|h_w|^2 < |h_b|^2$. For a given instantaneous channel pair (h_w, h_b), the corresponding instantaneous BC capacity region is given by [9]

$$\mathcal{C}^{\mathrm{BC}} \triangleq \bigcup_{a_1 + a_2 = 1, a_1, a_2 \geq 0} \left\{ (R_1, R_2) : R_1, R_2 \geq 0, \right.$$

$$\left. R_1 \leq \log\left(1 + \frac{a_1 x}{1 + a_2 x}\right), R_2 \leq \log\left(1 + a_2 y\right) \right\}, \tag{11.4}$$

where a_i denotes the fraction of total power allocated to user i, $x = |h_w|^2 P$, and $y = |h_b|^2 P$. However, the NOMA rate region, denoted by $\mathcal{R}^{\mathrm{N}}$, imposes an additional power allocation constraint on $\mathcal{C}^{\mathrm{BC}}$, that is $a_1 \geq a_2$. Such a constraint is used to guarantee fairness for the user with a weaker channel gain.

On the other hand, the TDMA rate region is given by

$$\mathcal{R}^{\mathrm{T}} \triangleq \left\{ (R_1, R_2) : R_1, R_2 \geq 0, \frac{R_1}{R_1^*} + \frac{R_2}{R_2^*} \leq 1 \right\}, \tag{11.5}$$

where $R_1^* \triangleq \log(1 + x)$ and $R_2^* \triangleq \log(1 + y)$.

Figure 11.1 illustrates the three regions, where point B is located at

$$\left(\log\left(1 + \frac{y}{2}\right), \log\left(1 + \frac{x}{2 + x}\right) \right).$$

Obviously, $\log(1 + \frac{y}{2}) > \log(1 + \frac{x}{2}) > \log\left(1 + \frac{x}{2+x}\right)$ holds since $x < y$. Note that the curve A–B and the segment A–C represent the optimal rate pairs achieved by NOMA and TDMA, respectively, where point B corresponds to equal power sharing among the two users.

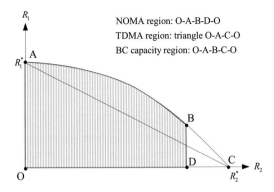

NOMA region: O-A-B-D-O
TDMA region: triangle O-A-C-O
BC capacity region: O-A-B-C-O

Figure 11.1 The BC capacity region, and NOMA and TDMA rate regions for an instantaneous pair of channel gains (h_w, h_b), where the point B is located at $\left(\log(1 + \frac{y}{2}), \log\left(1 + \frac{x}{2+x}\right)\right)$.

11.1.3   NOMA with Perfect or Imperfect CSIT

The performance of the fading BC has been extensively investigated from an information theoretic perspective. For the fading BC with perfect CSIT, the ergodic capacity and the outage capacity/probability were studied in [22] and [23], respectively. To achieve ergodic capacity, user rates are adapted to instantaneous channel state information (CSI). Compared to ergodic capacity, the concept of outage assumes transmission with a predefined rate, which is more appropriate for applications with strict delay constraints. Furthermore, since multiple users coexist in a fading BC, two types of outage probabilities can be defined, namely the common outage probability (COP) and the individual outage probability (IOP) [23]. The COP is the probability that any of the users is in outage, whereas the IOP is the probability that an individual user is in outage. Note that the COP captures short-term user fairness, which refers to fairness within a single fading block. The work in [23] considered outage with CSIT, in which the base station can adapt the power allocation strategy according to the instantaneous CSI. For the case where CSIT is not available, the outage performance of the fading BC was established in [24], where the power allocation strategy at the base station remains unchanged under any channel conditions.

Many recent works have investigated the block fading BC from the new perspective of NOMA (e.g., [25–33]). The works in [25–28, 32, 33] assume perfect user ordering at the base station, which requires perfect CSIT at the beginning of each block. With perfect CSIT, the user's data rates can be adapted to the channel conditions, which can achieve the optimal performance in terms of both ergodic rates and outage probabilities. However, perfect CSIT requires a large amount of CSI feedback or requires the base station to estimate each channel gain based on pilot symbols broadcast by the users, which reduces spectral efficiency. Compared to the works on NOMA assuming the availability of perfect CSIT (e.g., [25–28, 32, 33]), the works in [29–31] propose NOMA schemes with imperfect CSIT, which require less channel feedback and lower overhead. In [29, 30], the outage performance of NOMA with only statistical CSIT knowledge is investigated. In addition, the work in [31] investigates NOMA with imperfect CSIT realized by one-bit feedback, where each user feeds the one-bit of channel information back to the base station at the beginning of each fading block. Note that this one-bit feedback scheme is an indication of how NOMA can be implemented in practice. For example, in the MUST scheme described in [20], the quantized CSI feedback scheme has been particularly highlighted as a potential enhancement to assist the base station in performing user ordering and power allocation, and the one-bit feedback scheme is the simplest form of the quantized feedback scheme.

This chapter takes the works in [27] and [31] as examples to investigate the performance of NOMA with perfect CSIT and limited channel feedback, in terms of ergodic rates and outage probabilities, respectively. Then, some promising open problems in this area are discussed.

Throughout this chapter, $\mathbb{P}(\cdot)$ and $\mathbb{E}(\cdot)$ denote the probability of an event and the expectation of a random variable, respectively. $\{x_i\}$ denotes the sequence formed by all the possible x_i, and $[1 : K]$ denotes the set $\{1, \cdots, K\}$. Moreover, $\log(\cdot)$ or $\ln(\cdot)$ denote the base 2 logarithm and the natural logarithm, respectively; $C_K^n \triangleq \frac{K!}{n!(K-n)!}$, for $n \leq K$; and $[x]^+ \triangleq \max\{x, 0\}$. Finally, "$\doteq$" denotes exponential equality, i.e., $f(P) \doteq P^x$ means that $\lim_{P \to \infty} \frac{\log f(P)}{\log P} = x$.

11.2   NOMA with Perfect CSIT: A New Evaluation Criterion

In this section, the ergodic rate performance of NOMA with two users will be investigated based on a new evaluation criterion that treats TDMA as a benchmark, under the short-term power constraint. For the considered NOMA, the weaker and stronger gains within each fading block are denoted respectively by h_w and h_b within each fading block for simplicity, i.e., $|h_w|^2 < |h_b|^2$.

11.2.1   Preliminary Results

Some preliminary results are provided in this subsection, which will be used in the next subsection. Specifically, define the following two functions as:

$$f^N(z) = \log\left(\frac{(1+x)y}{y + (2^z - 1)x}\right), \quad 0 \leq z \leq R_2^*, \tag{11.6}$$

$$f^T(z) = \left(1 - \frac{z}{R_2^*}\right) R_1^*, \quad 0 \leq z \leq R_2^*, \tag{11.7}$$

where R_1^* and R_2^* are defined in Section 11.1.2.

For a given $z_0 \in (0, R_2^*)$, we have the following two propositions.

PROPOSITION 11.1   ([27, proposition 1]): If $z > z_0$, then $f^N(z) + z > f^T(z_0) + z_0$.

PROPOSITION 11.2   ([27, proposition 2]): If $z < z_0$, then $f^N(z) > f^T(z_0)$.

Proofs of these propositions can be found in [27].

11.2.2   New Evaluation Criterion

In this subsection, a new evaluation criterion will be discussed, in which NOMA will be compared with TDMA in terms of users' individual rates and the sum-rate.

Consider sets of points N and T in the two-dimensional plane of paired R_2 and R_1 values, where the point N is located at

$$(R_2^N, R_1^N) = \left(\log(1 + a_2 y), \log\left(1 + \frac{a_1 x}{1 + a_2 x}\right)\right), \tag{11.8}$$

with $a_1 + a_2 = 1$ and $0 \leq a_2 \leq a_1$ In addition, point T is located at

$$(R_2^T, R_1^T) = \left(b_2 R_2^*, b_1 R_1^*\right), \tag{11.9}$$

with $b_1 + b_2 = 1$, and $b_1, b_2 \geq 0$.

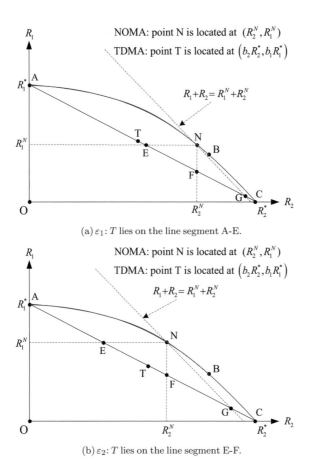

(a) ε_1: T lies on the line segment A-E.

(b) ε_2: T lies on the line segment E-F.

Figure 11.2 Comparison of the achievable rate pairs of NOMA and TDMA for a given instantaneous channel pair (h_w, h_b), where (R_2^N, R_1^N) is defined in (11.8).

Figure 11.2 shows two cases of the relative locations of points N and T for a given instantaneous channel pair (h_w, h_b), where N and T lie on the curve A–B (corresponding to a NOMA rate pair) and the segment A–C (corresponding to a TDMA rate pair), respectively. Moreover, consider three important lines: $R_1 = R_1^N$, $R_2 = R_2^N$ and $R_1 + R_2 = R_1^N + R_2^N$, which represent individual rates and the sum-rate of NOMA, respectively. It is easy to prove that $R_1^N + R_2^N < R_2^*$, and these three lines will divide the segment A–C into four subsegments with intersection points E, F, and G, as depicted in Figure 11.2. In the rest of this section, $b_1 = b_2 = 1/2$ will be considered for simplicity. Such a time allocation scheme is called "naive TDMA" in [34], where each user is allocated an equal-length time slot.

With respect to the parameters (h_b, h_w, a_2, P), four different events can be defined according to the location of the point T as follows:

$$\varepsilon_1 \triangleq \{\text{point } T \text{ lies on subsegment A-E}\}, \tag{11.10}$$

$$\varepsilon_2 \triangleq \{\text{point } T \text{ lies on subsegment E-F}\}, \tag{11.11}$$

$$\varepsilon_3 \triangleq \{\text{point T lies on subsegment F-G}\}, \tag{11.12}$$

$$\varepsilon_4 \triangleq \{\text{point T lies on subsegment G-C}\}. \tag{11.13}$$

Therefore, Figure 11.2 precisely shows the two events ε_1 and ε_2, respectively. Actually, these events comprehensively reflect the relationship between NOMA and TDMA in terms of individual rates and the sum-rate, i.e.,

$$\varepsilon_1 = \{R_1^N < R_1^T, R_2^N > R_2^T, R_1^N + R_2^N > R_1^T + R_2^T\}, \tag{11.14}$$

$$\varepsilon_2 = \{R_1^N > R_1^T, R_2^N > R_2^T, R_1^N + R_2^N > R_1^T + R_2^T\}, \tag{11.15}$$

$$\varepsilon_3 = \{R_1^N > R_1^T, R_2^N < R_2^T, R_1^N + R_2^N > R_1^T + R_2^T\}, \tag{11.16}$$

$$\varepsilon_4 = \{R_1^N > R_1^T, R_2^N < R_2^T, R_1^N + R_2^N < R_1^T + R_2^T\}. \tag{11.17}$$

Note that $(R_1^N, R_2^N, R_1^T, R_2^T)$ satisfies $R_1^N = f^N(R_2^N)$ and $R_1^T = f^T(R_2^T)$ shown in (11.6) and (11.7). Thus, by replacing (f^N, z, f^T, z_0) with $(R_1^N, R_2^N, R_1^T, R_2^T)$ for Propositions 11.1 and 11.2, some redundant conditions for each event can be removed, i.e.,

$$\varepsilon_1 \overset{(a)}{=} \{R_1^N < R_1^T, R_2^N > R_2^T\}, \tag{11.18}$$

$$\varepsilon_2 \overset{(b)}{=} \{R_1^N > R_1^T, R_2^N > R_2^T\}, \tag{11.19}$$

$$\varepsilon_3 \overset{(c)}{=} \{R_2^N < R_2^T, R_1^N + R_2^N > R_1^T + R_2^T\}, \tag{11.20}$$

$$\varepsilon_4 \overset{(d)}{=} \{R_1^N + R_2^N < R_1^T + R_2^T\}, \tag{11.21}$$

where (a) and (b) are based on Proposition 11.1; (c) is based on Proposition 11.2; and (d) follows by the converse-negative property of Proposition 11.1 (i.e., $\{R_1^N + R_2^N < R_1^T + R_2^T\} \Rightarrow \{R_2^N < R_2^T\}$) and Proposition 11.2.

REMARK 11.3   *Among these four events, ε_2 is particularly important since it represents the event that NOMA outperforms TDMA in terms of not only the sum-rate but also each individual rate.*

Now, four events have been defined for the instantaneous CSI (h_b, h_w). Since h_w and h_b are random variables, $\varepsilon_1, \varepsilon_2, \varepsilon_3,$ and ε_4 are also random events for a fixed power allocation policy, a_i. This means that the point T will be probabilistically located on any of the four segments shown in Figure 11.2. Thus, the ergodic rate performance of NOMA can be statistically characterized by analyzing the probability of each event ε_i by treating TDMA as a benchmark, in terms of both the sum-rate and individual rates. This summarizes the basic idea of this evaluation criterion.

REMARK 11.4   *This section considers fixed power allocation, i.e., a_1 and a_2 are constant for any channel realization, the same as in [1] and [25]. The use of dynamic power allocation schemes, e.g., the power allocation is dynamically optimized according to instantaneous CSI, obviously can achieve better performance. However, dynamic power allocation suffers a high overhead cost since each user has to access the power parameters set at the base station.*

REMARK 11.5   *Existing works mainly focus on the ergodic sum-rate achieved by NOMA, (i.e., $\sum_{i=1}^{2} \mathbb{E}[R_i^N]$) [25, 29], and compare it with the ergodic sum-rate achieved by TDMA (i.e., $\sum_{i=1}^{2} \mathbb{E}[R_i^T]$). However, the considered new evaluation criterion characterizes NOMA from a different viewpoint. Particularly, the considered criterion focuses on instantaneous CSI, and compares the sum-rate and the individual rates achieved by NOMA and TDMA statistically, i.e., by analyzing $\mathbb{P}(\sum_{i=1}^{2} R_i^T > \sum_{i=1}^{2} R_i^N, R_i^T > R_i^N, i \in \{1, 2\}$. The considered criterion reveals more details of the statistical characteristics of the instantaneous sum-rate and the instantaneous individual rates.*

The performance of NOMA will be evaluated by using the considered criterion in comparison with TDMA in the next section.

11.2.3   Performance Evaluation of NOMA

Assume that the channel pair (h_w, h_b) is chosen from a set of K channel gains, $(h_1, \cdots, h_K)$, in a wireless downlink BC, where $|h_1|^2 \leq \cdots \leq |h_K|^2$, and $(h_w, h_b) = (h_m, h_n)$, $m < n$, $m, n \in [1 : K]$. This means that user m is paired with user n to perform NOMA, as motivated in [35]. Also, define $x \triangleq P|h_w|^2$ and $y \triangleq P|h_b|^2$, which are random variables with a joint probability density function (PDF) as follows [36]:

$$f_{X,Y}(x, y) = w_1 f(x)f(y)[F(x)]^{m-1}[1 - F(y)]^{K-n}$$
$$\times [F(y) - F(x)]^{n-1-m}, \quad 0 < x < y, \tag{11.22}$$

where $w_1 \triangleq \frac{K!}{(m-1)!(n-1-m)!(K-n)!}$. Considering Rayleigh fading, $f(x) = \frac{1}{P}e^{-\frac{x}{P}}$ and $F(x) = 1 - e^{-\frac{x}{P}}$ can be obtained.

Using the joint PDF of x and y, the probability of each event defined in the previous section can be calculated in order to evaluate the performance of NOMA. The probability of each event ε_i is given in the following lemma, where $b_2 = 1/2$ (i.e., the naive TDMA) is set for simplicity.

LEMMA 11.6   *([27, lemma 1]): Given (K, m, n, P, a_2) and $b_2 = 1/2$, the probability of the event ε_1 defined in the previous subsection can be expressed as*

$$\mathbb{P}(\varepsilon_1) = 1 - w_3 \sum_{i=0}^{n-1} \frac{(-1)^i C_{n-1}^i}{K - n + i + 1}\left(1 - d^{K-n+i+1}\right) - \mathbb{P}(\varepsilon_2), \tag{11.23}$$

where $w_3 = \frac{K!}{(m-1)!(K-m)!}$ and $\mathbb{P}(\varepsilon_2)$ can be expressed as

$$\mathbb{P}(\varepsilon_2) = w_1 \sum_{k=0}^{m-1}(-1)^{m-1-k}C_{m-1}^k\left[\sum_{i=0}^{n-1}\frac{(-1)^{n-1-i}C_{n-1}^i d^{K-i}}{K-i}\right.$$
$$\left. - \sum_{i=0}^{k}\sum_{j=0}^{n-1-k}\frac{(-1)^{n-1-i-j}C_k^i C_{n-1-k}^j d^{K-i-j}}{K-i-j}\right], \tag{11.24}$$

with $w_2 \triangleq \frac{1-2a_2}{a_2^2}$, and $d \triangleq e^{-\frac{w_2}{P}}$. In addition,

$$\mathbb{P}(\varepsilon_3) = 1 - \mathbb{P}(\varepsilon_1) - \mathbb{P}(\varepsilon_2) - \mathbb{P}(\varepsilon_4), \tag{11.25}$$

where

$$P(\varepsilon_4) = 1 - w_1 \sum_{i=0}^{n-1-m} \frac{(-1)^i C_{n-1-m}^i}{m+i} \int_{\sqrt{w_2+1}-1}^{w_2} f(y)[F(y)]^{n-1-m-i}$$

$$\times [1 - F(y)]^{K-n} \left([F(y)]^{m+i} - \left[F\left(\frac{w_2 - y}{1+y}\right) \right]^{m+i} \right) dy$$

$$- w_3 \sum_{j=0}^{n-1} \frac{(-1)^j C_{n-1}^j}{K-n+j+1} d^{-(K-n+j+1)}. \tag{11.26}$$

Special case: As noted previously, we are most interested in $P(\epsilon_2)$. The expression for $P(\varepsilon_2)$ in Lemma 11.6 is too complicated to provide significant insight. However, $P(\varepsilon_2)$ can be simplified for a special pairing case, i.e., $m = 1$, $n = K$. In this case, $k = 0$, and $P(\varepsilon_2)$ can be simplified and greater insight can be obtained as

$$P(\varepsilon_2) = 1 - (1-d)^K - d^K. \tag{11.27}$$

Now, based on (11.27), the optimal a_2 can be easily obtained in the following lemma.

LEMMA 11.7   *([27, corollary 1]): To maximize $P(\varepsilon_2)$ in (11.27), the optimal power allocation in the case $(m,n) = (1,K)$ is $a_2 = \frac{-1+\sqrt{1+P \ln 2}}{P \ln 2} < \frac{1}{2}$ and $a_1 = 1 - a_2$, and the maximum value of $P(\varepsilon_2)$ is $P(\varepsilon_2) = 1 - \frac{1}{2^{K-1}}$.*

REMARK 11.8   *Lemma 11.7 shows that $P(\varepsilon_2) \to 1$ if K is sufficiently large. This means that, for almost any instantaneous channel states, NOMA yields larger individual rates than naive TDMA for both users m and n as long as the difference between their channel gains is sufficiently large. This phenomenon sheds light on the user pairing principle for NOMA. In addition, as verified via some numerical examples in the next subsection, this phenomenon is also valid for some other pairing cases (i.e., $(m,n) \neq (1,K)$).*

11.2.4   Numerical Results

Some numerical results are provided to evaluate the performance of NOMA in comparison with TDMA. Rayleigh fading channels are assumed with i.i.d. channel gains. In addition, since the noise power has a unit value, the power P and the transmit signal to noise ratio (SNR) have the same value.

Figure 11.3 depicts the probabilities of the four events defined in Section 11.2.2 via column diagrams. Here the total number of users is set as $K = 10$, and different choices of (m,n) are considered. Specifically, the probabilities of the events that point T lies on subsegments A–E, E–F, F–G, and G–C in Figure 11.2 are displayed, where SNR = 30 dB, and a_2 is set as the optimal value which is obtained based on a one-dimensional search for maximizing $P(\varepsilon_2)$ in Lemma 11.6. Four different user pairs (m,n) are considered in this figure. From this figure, one can observe that the analytical results match the simulation results perfectly. In addition, the probability that point T lies on the subsegment E–F (i.e., $P(\varepsilon_2)$) increases with the value of $(n-m)$, as discussed in Remark 11.8. When $(m,n) = (2,10)$, $P(\varepsilon_2) \to 1$, which implies that NOMA outperforms TDMA in

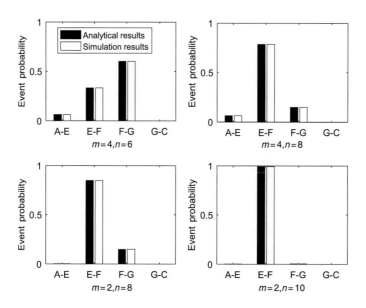

Figure 11.3 Probability that point T lies on each of the four segments described in Figure 11.2, where $K = 10$, and SNR = 30 dB.

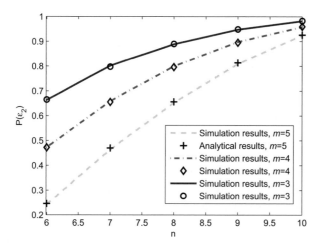

Figure 11.4 Probability of the event ε_2, where $K = 10$, and SNR = 30 dB.

terms of not only the sum rate but also each individual rate almost for any instantaneous channel states.

Figure 11.4 provides additional numerical results to show $\mathbb{P}(\varepsilon_2)$ as a function of n, where SNR = 30 dB and the optimal choice of a_2 is also obtained numerically according to Lemma 11.6. Here, the total number of users is also set as $K = 10$. From this figure, one can observe that $\mathbb{P}(\varepsilon_2)$ can decrease significantly with m. Take the case $n = 6$ for example, the probability that point T lies on the segment E–F is 0.25 if $m = 3$,

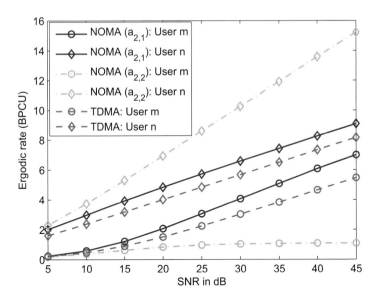

Figure 11.5 Ergodic individual rates, $a_{2,1} = \frac{\sqrt{P\ln(2)+1}-1}{P\ln(2)}$, $a_{2,2} = \frac{K-1}{2K-1}$, and $m = 1, n = K = 9$.

whereas the value of this probability becomes 0.66 if $m = 5$. In addition, $\mathbb{P}(\varepsilon_2)$ increases significantly with n. Take the case $m = 4$ for example, $\mathbb{P}(\varepsilon_2)$ increases from 0.48 to 0.8, as n increases from 6 to 8. In summary, this figure demonstrates that $\mathbb{P}(\varepsilon_2)$ increases with $(n - m)$, which means that the performance gain between NOMA and TDMA increases as the difference between the two channel gains increases.

Figure 11.5 depicts individual ergodic rates of NOMA and TDMA averaged over the fading channels as functions of SNR, where $K = 9$, $(m, n) = (1, K)$, and $a_{2,1} = \frac{\sqrt{P\ln(2)+1}-1}{P\ln(2)}$ which is set according to Lemma 11.7. Moreover, the performance of NOMA with the power allocation strategy $a_{2,2} = \frac{K-1}{2K-1}$ used in [25] has also been shown for comparison. Note that the choice of $a_{2,2}$ in [25] can ensure that $a_{2,2} < 1/2$, but makes $a_{2,2}$ approach $1/2$ when K is large. As shown in this figure, NOMA with power allocation $a_{2,2}$ yields a larger individual ergodic rate for user n, which is the user with better channel conditions, but suffers a large individual ergodic rate loss for the user with poorer channel conditions. Thus, although the ergodic sum rate achieved by NOMA with this power allocation is large, it suffers poor fairness between the two users, in comparison with naive TDMA. However, NOMA with power allocation parameter $a_{2,1}$ (obtained from the considered evaluation criterion) enjoys an almost constant performance gain over TDMA for each user's rate, when the SNR is larger than 30 dB. Furthermore, when SNR = 45 dB, the performance gains between the two schemes for users m and n are about 1.5 bits per channel use (BPCU) and 1 BPCU, respectively. Thus, the considered evaluation criterion provides a new analytical framework to enable a better tradeoff between throughput and user fairness achieved by NOMA. Specifically, a suitable fixed power allocation strategy can be obtained using the considered evaluation criterion such that NOMA can outperform conventional orthogonal MA schemes in terms of not only the sum-rate, but also the individual rates.

11.3   NOMA with One-Bit Feedback: Short-Term Power Constraint

In this section, the outage performance of PD-NOMA with one-bit feedback will be investigated under the short-term power constraint. Note that the short-term power constraint is particularly important for applications with a stringent power consumption requirement at the base station. In particular, a downlink NOMA system with one-bit feedback as well as the corresponding power allocation scheme is investigated for delay-sensitive applications. The outage probability is used as the performance metric, and a closed-form expression for the COP is analyzed. Furthermore, in order to minimize the COP, a dynamic power allocation policy is studied according to instantaneous CSI feedback. The formulated power allocation problem is challenging, since the objective function for minimizing the COP is nonconvex. To make this problem tractable, the properties of the optimal power allocation solution are characterized first and then used to transform the problem into a series of convex problems.

11.3.1   One-Bit Feedback System Model

The wireless channels from the base station to the users are assumed to be i.i.d. Rayleigh fading with mean zero and variance one. In each fading block, assume that the base station transmits one message to each user with the same fixed rate r_0 BPCU to guarantee fairness [26]. Moreover, assume that all users have perfect CSI and compare their fading gains to a predefined threshold, denoted by α. Particularly, given h_k, user k feeds a single bit back to the base station in each fading block via a zero-delay reliable link. Note that the one-bit feedback scheme considered in this section is the simplest form of a quantized feedback scheme, whose overhead is negligible when the length of each fading block is moderate or large. Denote the one-bit feedback symbol for user k by $Q(h_k)$, where $Q(h_k) = 1$ if $|h_k|^2 \geq \alpha$, and $Q(h_k) = 0$, otherwise.

Consider the channel feedback sequence $\{Q(h_k)\} \triangleq \{Q(h_1), \cdots, Q(h_K)\}$. Obviously, $\{Q(h_k)\}$ is a 0–1 sequence, which has 2^K possible realizations. Power allocation is performed by the base station based on this feedback, where only $(K+1)$ categories for the realizations are considered by the base station. Define a corresponding random variable as follows.

DEFINITION 11.9   *Define a random variable N with respect to the K-dimensional random binary feedback sequence $\{Q(h_k)\}$ as $N \triangleq K - \sum_{k=1}^{K} Q(h_k)$. Obviously, N has $(K + 1)$ possible realizations, and the event $N = n$ represents the case in which n users send "0" and the other $(K - n)$ users send "1", $n \in [0 : K]$.*

Three steps will be used by the base station for the event $N = n$ to determine the user ordering. Firstly, the users are divided into two groups corresponding to feedback symbol "0" and "1," which are denoted by $\mathcal{G}_{0|n}$ and $\mathcal{G}_{1|n}$, respectively. Second the users in

$\mathcal{G}_{0|n}$ and $\mathcal{G}_{1|n}$ are allocated the ordering index sequences $\{1, \cdots, n\}$ and $\{n + 1, \cdots, K\}$, respectively. Finally, the users in the same group are *randomly* indexed (ordered) since their fading gains cannot be distinguished by the base station.

The channel gains for the ordered users are denoted by $\{|h_{\pi_1}|^2, |h_{\pi_2}|^2, \cdots, |h_{\pi_K}|^2\}$, where $\pi_k \in [1 : K]$, and $\pi_i \neq \pi_j$ if $i \neq j$. Now, for the event $N = n$, $Q(h_{\pi_k}) = 0$ if $1 \leq k \leq n$, and $Q(h_{\pi_k}) = 1$ if $n + 1 \leq k \leq K$. Then, the superimposed message $\sum_{k=1}^{K} s_{\pi_k}(t)$ is transmitted by the base station using the power allocation policy discussed in the next subsection, where $s_{\pi_k}(t)$ denotes the signal of user π_k in the tth channel use during a fading block.

Now, based on one-bit feedback, the power constraint in (11.2) is rewritten as follows:

$$\sum_{k=1}^{K} P_{k,n} \leq P, \forall n \in [0 : K], \tag{11.28}$$

where $P_{k,n}$ denotes the power allocated to user π_k for the event $N = n$.

11.3.2   Outage Probability

The outage probability of the NOMA system with the short-term power constraint will be discussed in this subsection. Some useful preliminary results are provided in the following.

Preliminary Results

The conditional probability $\mathbb{P}(|h_{\pi_k}|^2 < x_k|N = n)$ is crucial for the derivation of the outage probability, where $x_k > 0$, $k \in [1 : K]$. According to the user ordering, one can observe that, for the event $N = n$, $|h_{\pi_k}|^2 < \alpha$ if $k \in [1 : n]$, and $|h_{\pi_k}|^2 \geq \alpha$ otherwise. Moreover, all the channels h_{π_k} are mutually independent according to the applied user ordering principle, if conditioned on the event $N = n$, since $\mathcal{G}_{0|n}$ and $\mathcal{G}_{1|n}$ are determined by the event $N = n$ and all the users in each group are randomly ordered. Thus, $\mathbb{P}(|h_{\pi_k}|^2 < x_k|N = n)$, $k \in [1 : n]$, can be obtained as follows:

$$
\begin{aligned}
\mathbb{P}(|h_{\pi_k}|^2 < x_k \mid N = n) &= \mathbb{P}\left(|h_{\pi_k}|^2 < x_k \mid |h_{\pi_k}|^2 < \alpha\right) \\
&= \frac{\mathbb{P}\left(|h_{\pi_k}|^2 \leq x_k, |h_{\pi_k}|^2 < \alpha\right)}{\mathbb{P}\left(|h_{\pi_k}|^2 < \alpha\right)} \\
&= \min\left\{\frac{1 - e^{-x_k}}{1 - e^{-\alpha}}, 1\right\}, \ x_k \geq 0.
\end{aligned}
\tag{11.29}
$$

Similarly, $\mathbb{P}(|h_{\pi_k}|^2 < x_k \mid N = n)$, $k \in [n + 1 : K]$ can be obtained as follows:

$$\mathbb{P}(|h_{\pi_k}|^2 < x_k \mid N = n) = \left[1 - e^{-(x_k - \alpha)}\right]^+, \ x_k \geq 0. \tag{11.30}$$

In addition, the expressions for the signal-to-interference-plus-noise ratios (SINRs) at the receivers have to be developed. Since SIC is adopted in the decoding process with the channel ordering of $\{|h_{\pi_1}|^2, |h_{\pi_2}|^2, \cdots, |h_{\pi_K}|^2\}$, the SINR when user π_k decodes the message of user π_l is given by [10]

$$\text{SINR}_{l \to k} = \frac{P_{l,n}|h_{\pi_k}|^2}{|h_{\pi_k}|^2 \sum_{m=l+1}^{K} P_{m,n} + 1}, \quad l \in [1:k]. \tag{11.31}$$

Outage Probability

The COP [23] is adopted as the performance criterion for the considered NOMA system to guarantee short-term fairness within each fading block. The COP of NOMA with one-bit feedback is provided in the following theorem.

THEOREM 11.10   *([31, theorem 1]): The COP of the considered one-bit NOMA scheme can be expressed as follows:*

$$\mathbb{P}^{Common}(\alpha, \{P_{k,n}\}) = \sum_{n=0}^{K} \mathbb{P}_n(\alpha) \left[1 - \prod_{k=1}^{K}(1 - \mathbb{P}_{k,n}^{Indiv}(\alpha, \mathbf{P}_n)) \right], \tag{11.32}$$

where $\mathbf{P}_n \triangleq \{P_{1,n}, \cdots, P_{K,n}\}$ is the power allocation sequence for the event $N = n$; and $\mathbb{P}_n(\alpha)$ and $\mathbb{P}_{k,n}^{Indiv}(\alpha, \mathbf{P}_n)$ are defined as follows:

$$\mathbb{P}_n(\alpha) \triangleq C_K^n (1 - e^{-\alpha})^n e^{-\alpha(K-n)}, \tag{11.33}$$

$$\mathbb{P}_{k,n}^{Indiv}(\alpha, \mathbf{P}_n) \triangleq \begin{cases} \min\left\{\frac{1-e^{-\hat{\zeta}_{k,n}}}{1-e^{-\alpha}}, 1\right\}, & k \in [1:n], \\ \left[1 - e^{-(\hat{\zeta}_{k,n}-\alpha)}\right]^+, & k \in [n+1:K], \end{cases} \tag{11.34}$$

with the definition $\hat{\zeta}_{k,n} \triangleq \max\{\zeta_{1,n}, \cdots, \zeta_{k,n}\}$, and

$$\zeta_{k,n} = \frac{\hat{r}_0}{P_{k,n} - \hat{r}_0 \sum_{m=k+1}^{K} P_{m,n}}, \forall k \in [1:K], \text{ where } \hat{r}_0 = 2^{r_0} - 1. \tag{11.35}$$

Note that in (11.35), $\zeta_{k,n}$ decreases with the allocated power for user k, i.e., $P_{k,n}$, and it is implicitly assumed that $\zeta_{k,n} \geq 0$, i.e.,

$$P_{k,n} \geq \hat{r}_0 \sum_{m=k+1}^{K} P_{m,n}, \forall k \in [1:K-1], n \in [0:K]. \tag{11.36}$$

Such a power constraint is typical for NOMA systems [25, 26, 28], where a user with a weaker channel gain is allocated more power to guarantee fairness. Furthermore, $\{P_{k,n}\}$ can be expressed as a function of $\{\zeta_{k,n}\}$ as follows:

$$P_{k,n} = \frac{\hat{r}_0}{\zeta_{k,n}} + \hat{r}_0 \sum_{m=k+1}^{K} (\hat{r}_0 + 1)^{m-k-1} \frac{\hat{r}_0}{\zeta_{m,n}}, \forall k \in [1:K], n \in [0:K], \tag{11.37}$$

which can be obtained from (11.35) based on mathematical induction. Thus, the sum power with respect to the event $N = n$ can be expressed as follows:

$$\sum_{k=1}^{K} P_{k,n} = \sum_{k=1}^{K} \left(\frac{\hat{r}_0}{\zeta_{k,n}} + \hat{r}_0 \sum_{m=k+1}^{K} (\hat{r}_0 + 1)^{m-k-1} \frac{\hat{r}_0}{\zeta_{m,n}} \right)$$

$$= \sum_{k=1}^{K} \left(\frac{\hat{r}_0}{\zeta_{k,n}} + \frac{\hat{r}_0^2}{\zeta_{k,n}} \sum_{i=0}^{k-2} (\hat{r}_0 + 1)^{i-2} \right)$$

$$= \sum_{k=1}^{K} \frac{(\hat{r}_0 + 1)^{k-1} \hat{r}_0}{\zeta_{k,n}}. \tag{11.38}$$

As can be observed in (11.38), the sum power can be expressed as the sum of the weighted inverses of $\zeta_{k,n}$, where these weighting factors form a geometric sequence.

Diversity Gain

DEFINITION 11.11   *The diversity gain based on the COP is defined as follows:*

$$d = -\lim_{P \to \infty} \frac{\log \mathbb{P}^{Common}}{\log P}. \tag{11.39}$$

In addition, the diversity gain in (11.39) can also be expressed as $\mathbb{P}^{Common} \doteq P^{-d}$.

To provide greater insight into the outage performance, the diversity gain of the COP in (11.32) is given in the following lemma under the short-term constraint.

LEMMA 11.12   *([31, lemma 1]): Under the short-term power constraint in (11.28), the maximum achievable diversity gain of the considered NOMA scheme is 1.*

REMARK 11.13   *As can be observed from Lemma 11.12, the COP-based diversity achieved by NOMA is the same as the IOP-based diversity achieved by the NOMA user with the weakest channel [25]. This is expected since the COP of NOMA is mainly determined by the user with the weakest channel.*

11.3.3   Power Allocation

Power allocation has been demonstrated to have a significant impact on the outage performance in many multiple access scenarios [23, 37, 38]. Motivated by this, a power allocation problem will be formulated to minimize the COP $\mathbb{P}^{Common}$ in (11.32) under the short-term power constraint, which is given as follows:

$$\min_{\alpha, \{P_{k,n}\}} \sum_{n=0}^{K} \mathbb{P}_n(\alpha) \left[1 - \prod_{k=1}^{K} (1 - \mathbb{P}_{k,n}^{Indiv}(\alpha, \mathbf{P}_n)) \right] \tag{11.40a}$$

s.t.   (11.28) and (11.36), $P_{k,n} \geq 0, \forall k \in [1 : K], n \in [0 : K]. \tag{11.40b}$

The above problem can be simplified by applying the transformation of variables in (11.35). In particular, an equivalent form for the problem in (11.40) is the following:

$$\text{(P1)} \quad \min_{\alpha, \{\zeta_{k,n}\}} \sum_{n=0}^{K} \mathbb{P}_n(\alpha) \left[1 - \prod_{k=1}^{K} (1 - \mathbb{P}_{k,n}^{\text{Indiv}}(\alpha, \boldsymbol{\zeta}_n)) \right] \tag{11.41a}$$

$$\text{s.t.} \quad \sum_{k=1}^{K} \frac{(\hat{r}_0 + 1)^{k-1} \hat{r}_0}{\zeta_{k,n}} \leq P, \, n \in [0 : K]; \tag{11.41b}$$

$$\zeta_{k,n} \geq 0, \, \forall k \in [1 : K], n \in [0 : K], \tag{11.41c}$$

where $\boldsymbol{\zeta}_n = \{\zeta_{1,n}, \cdots, \zeta_{K,n}\}$ and $\mathbb{P}_{k,n}^{\text{Indiv}}$ becomes a function of $\boldsymbol{\zeta}_n$; (11.41b) is based on (11.28) and (11.38). From (11.37), one can observe that the optimal power allocation scheme can be obtained as long as the optimal values of $\{\zeta_{k,n}\}$ are found.

Note that the number of constraints has been reduced when using the transformed problem in (11.41). However, problem (P1) still involves a nonconvex objective function that makes it difficult to solve. Problem (P1) involves $(K(K + 1) + 1)$ optimization variables, consisting of $K(K + 1)$ power variables $\zeta_{k,n}$ and one threshold variable α. In the following, the power allocation problem for a fixed threshold α is first addressed, and then a one-dimensional search will be used to find the optimal α.

When α is fixed, $\mathbb{P}_n(\alpha)$ is also fixed, so the objective in (11.41a) is additive of the subfunctions $\mathbb{P}_n(\alpha) \left[1 - \prod_{k=1}^{K} (1 - \mathbb{P}_{k,n}^{\text{Indiv}}(\alpha, \boldsymbol{\zeta}_n)) \right]$, where the nth subfunction is with respect to the variable vector $\boldsymbol{\zeta}_n$, $0 \leq n \leq K$. Now, one can observe that the constraints in (11.41b) and (11.41c) are uncoupled for the $(K + 1)$ variable vectors $\boldsymbol{\zeta}_n$, $0 \leq n \leq K$. This means that $(K + 1)$ decoupled subproblems can be obtained by decomposing the joint optimization problem (P1) *without loss of optimality*, where the nth subproblem can be formulated as follows:

$$\max_{\boldsymbol{\zeta}_n} f_{1,n}(\alpha, \boldsymbol{\zeta}_n) \triangleq \prod_{k=1}^{K} (1 - \mathbb{P}_{k,n}^{\text{Indiv}}(\alpha, \boldsymbol{\zeta}_n)) \tag{11.42a}$$

$$\text{s.t.} \quad \sum_{k=1}^{K} \frac{(\hat{r}_0 + 1)^{k-1} \hat{r}_0}{\zeta_{k,n}} \leq P, \, \zeta_{k,n} \geq 0, \, \forall k \in [1 : K]. \tag{11.42b}$$

Note that $\mathbb{P}_{k,n}^{\text{Indiv}}$ is a nonconvex function shown in (11.34). The following proposition simplifies $\mathbb{P}_{k,n}^{\text{Indiv}}$ for $k \in [n + 1 : K]$.

PROPOSITION 11.14 *([31, proposition 1]): The optimal solution of problem (11.42) satisfies $\hat{\zeta}_{k,n} \geq \alpha$, $\forall k \in [n + 1 : K]$, $n \in [0 : K - 1]$.*

The functions $\mathbb{P}_{k,n}^{\text{Indiv}}$ for $k \in [1 : n]$ can also be simplified by considering $\hat{\zeta}_{k,n} \leq \alpha$ only. Specifically, as shown in (11.34), when $\hat{\zeta}_{k,n} > \alpha$, $\forall k \in [1 : n]$, $\mathbb{P}_{k,n}^{\text{Indiv}} = 1$ can be obtained; and the objective function in (11.42a) becomes zero (i.e., $f_{1,n} = 0$). Therefore, the problem in (11.42) can be simplified as follows:

$$\max_{\boldsymbol{\zeta}_n} f_{1,n}(\alpha, \boldsymbol{\zeta}_n) = \prod_{k=1}^{n} \frac{e^{-\hat{\zeta}_{k,n}} - e^{-\alpha}}{1 - e^{-\alpha}} \prod_{k=n+1}^{K} e^{-(\hat{\zeta}_{k,n} - \alpha)} \tag{11.43a}$$

$$\text{s.t. (11.42b); and } \hat{\zeta}_{k,n} \leq \alpha, \forall k \in [1 : n]; \, \hat{\zeta}_{k,n} \geq \alpha, \forall k \in [n + 1 : K]. \tag{11.43b}$$

REMARK 11.15   *The constraint in (11.43b) requires $P \geq \frac{(\hat{r}_0+1)^n-1}{\alpha}$ to satisfy $\hat{\zeta}_{k,n} \leq \alpha$, $\forall k \in [1 : n]$, which is obtained from (11.37). If this power requirement is not satisfied, i.e., $P < \frac{(\hat{r}_0+1)^n-1}{\alpha}$, $\mathbb{P}_n^{Common} = 1 - \prod_{k=1}^{K}(1 - \mathbb{P}_{k,n}^{Indiv}) = 1$ for any power allocation, the COP for the event $N = n$ becomes 1 in this case.*

The following proposition further simplifies this problem by allowing the elimination of $\hat{\zeta}_{k,n}$.

PROPOSITION 11.16   *([31, proposition 2]): The optimal solution of problem (11.43) satisfies $\zeta_{k,n} \leq \zeta_{k+1,n}$, $\forall k \in [1 : K - 1]$.*

Based on Proposition 11.16, the problem in (11.43) can be transformed into the following:

$$\max_{\boldsymbol{\zeta}_n} f_{2,n}(\alpha, \boldsymbol{\zeta}_n) \triangleq \prod_{k=1}^{n} \left(e^{-\zeta_{k,n}} - e^{-\alpha}\right) \prod_{k=n+1}^{K} e^{-\zeta_{k,n}} \tag{11.44a}$$

$$\text{s.t. (11.42b); } \zeta_{k,n} \leq \alpha, \forall k \in [1 : n]; \; \zeta_{k,n} \geq \alpha, \forall k \in [n + 1 : K]; \tag{11.44b}$$

$$\zeta_{k,n} \leq \zeta_{k+1,n}, \; \forall k \in [1 : K - 1]. \tag{11.44c}$$

Now, by taking the natural logarithm of $f_{2,n}$, the problem in (11.43) can be transformed into the following equivalent *convex* problem:

$$(P1.n) \; \max_{\boldsymbol{\zeta}_n} \sum_{k=1}^{n} \ln\left(e^{-\zeta_{k,n}} - e^{-\alpha}\right) - \sum_{k=n+1}^{K} \zeta_{k,n} \tag{11.45a}$$

$$\text{s.t. } \sum_{k=1}^{K} \frac{(\hat{r}_0 + 1)^{k-1}\hat{r}_0}{\zeta_{k,n}} \leq P; \tag{11.45b}$$

$$\zeta_{1,n} \geq 0; \; \zeta_{n,n} \leq \alpha; \; \zeta_{n+1,n} \geq \alpha; \tag{11.45c}$$

$$\zeta_{k,n} \leq \zeta_{k+1,n}, \; \forall k \in [1 : K - 1]. \tag{11.45d}$$

This problem can be verified to be convex by calculating the Hessian matrix of the objective function and the constraint in (11.45b). Some corresponding numerical solvers will be used to solve this convex optimization problem in Section 11.3.4.

Furthermore, a one-dimensional search is needed to find the optimal value of α in problem (P1) in (11.41). Note that if α is sufficiently large the probability that all users feed back the message "0" goes to 1 (i.e., $\mathbb{P}_K(\alpha) \to 1$), and hence the optimal α always has a finite value.

11.3.4   Numerical Results

In this subsection, the outage performance of the considered NOMA scheme with one-bit feedback under the short-term power constraint will be evaluated via computer simulation results.

Benchmark Schemes

(1) *TDMA*: TDMA transmission with one-bit feedback is taken as one benchmark scheme since it is equivalent to any other OMA scheme [39, section 6.1.3]. Specifically, assume that each fading block is divided into K equal time slots, where user k is served in the kth time slot. $P_{k,n}^T$ is used to denote the power allocated to user k for the event $N = n$. For TDMA transmission, the short-term power constraint is $\frac{1}{K} \sum_{k=1}^{K} P_{k,n}^T \leq P$. Furthermore, redefine $\{\zeta_{k,n}\}$ in (11.35) as $\zeta_{k,n} = \frac{2^{Kr_0}-1}{P}$. Now, the short-term power constraint can be rewritten as follows:

$$\frac{2^{Kr_0} - 1}{K} \sum_{k=1}^{K} \frac{1}{\zeta_{k,n}} \leq P, \ \forall n \in [0 : K]. \tag{11.46}$$

Then, similar to problem (P1) in (11.41), the power allocation problem can be formulated for TDMA transmission under the short-term power constraint. This problem can be solved using an approach similar to that developed in Section 11.3.3.

(2) *Fixed NOMA*: To show the benefits of the optimal power allocation scheme, NOMA with one-bit feedback and fixed power allocation is taken as the second benchmark scheme. Note that fixed NOMA has been adopted in many existing works (e.g., [25, 29]) for simplicity. Specifically, the NOMA transmission scheme in Section 11.3.1 can also be used, but the power allocation scheme is fixed as $\zeta_{k,n} = \frac{(\hat{r}_0+1)^K-1}{P}$, $\forall k \in [1 : K], n \in [0 : K]$. Then, the optimal α is also obtained via a one-dimensional search.

(3) *NOMA without feedback*: To show the benefits of one-bit feedback, NOMA without CSI feedback is taken as the third benchmark scheme, in which the base station has only statistical CSI, but does not have access to instantaneous CSI nor the ordering information [29]. In this case, user ordering is randomly determined by the base station. Obviously, NOMA without CSI is a special case of NOMA with one-bit feedback when $\alpha = 0$ or $\alpha = \infty$.

(4) *NOMA with perfect CSIT*: Finally, the COP of NOMA with perfect CSIT is considered as a lower bound on the COP of NOMA with one-bit feedback. With perfect CSIT, the base station knows the exact user ordering and informs the users of this ordering. In addition, the required power threshold for the users within any block for correct decoding is also known by the base station. In this case, an outage event occurs only when the required power threshold for the users is larger than P [40].

Outage Performance

The outage performance of NOMA with one-bit feedback is shown under the short-term power constraint in (11.40). In Figures 11.6–11.8, the outage performance of NOMA employing the optimal power allocation scheme discussed in this section is compared with the benchmark schemes defined previously as a function of the SNR, the transmission rate r_0, and the number of users K, respectively. From these figures, one can observe that NOMA with optimal power allocation outperforms the benchmark schemes, i.e., TDMA, fixed NOMA, and NOMA without feedback. As shown in Figure 11.6, all the curves experience almost the same slope at high SNR, but there exists a constant gap

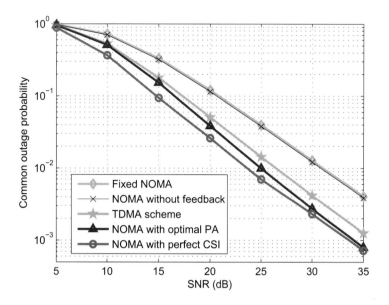

Figure 11.6 COP vs. SNR for different transmission schemes, where $K = 3$, the target rate is set as $r_0 = 0.8$ BPCU, and "PA" denotes "power allocation."

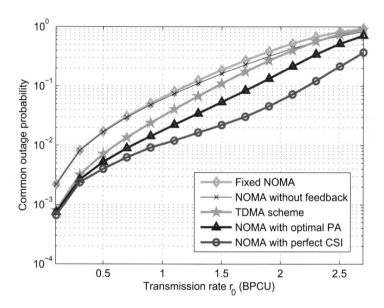

Figure 11.7 COP vs. r_0 for different transmission schemes, where $K = 3$, and the SNR is set as 25 dB.

between the considered NOMA scheme and each benchmark scheme, which is due to the fact that all the schemes achieve the same diversity gain of 1 (Lemma 11.12) under the short-term power constraint. Moreover, the COP of the considered NOMA scheme with one-bit feedback approaches the COP of NOMA with perfect CSIT at high SNR,

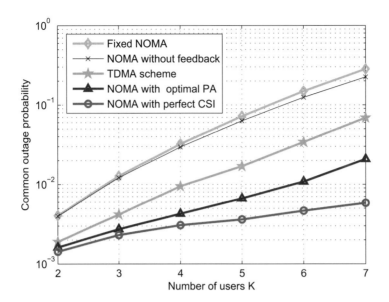

Figure 11.8 COP vs. K for different transmission schemes, where the target transmission rate is set as $r_0 = 0.8$ BPCU, and the SNR is set as 30 dB.

so one-bit feedback is useful to the considered NOMA scheme for improving the outage performance. From Figure 11.7, NOMA with the optimal power allocation and TDMA have almost the same COP when $r_0 = 0.1$. However, NOMA outperforms TDMA as r_0 increases; e.g., for the case $r_0 = 1.9$, the two schemes have COPs of approximately 0.14 and 0.28, respectively. In this case, the successful transmission probability can be improved from 0.72 to 0.86 if we change the MA technique from TDMA to NOMA. Finally, Figure 11.8 shows that the COPs of all schemes increase with K (the number of users). Particularly, the performance gap between NOMA and TDMA with one-bit feedback increases as K increases. This can be explained by the fact that NOMA is more spectrally efficient in the sense that it serves all users simultaneously, compared to TDMA. In particular, when the number of users increases, each user will be allocated less power but can still use the same transmission time period when PD-NOMA is employed; however, when TDMA is employed, each user will be allocated less power as well as less time, which significantly affects the transmission efficiency.

11.4   NOMA with One-Bit Feedback: Long-Term Power Constraint

In this section, the outage performance of NOMA with one-bit feedback will be investigated under the long-term power constraint. Note that the long-term power constraint is appropriate for applications with a slack power consumption requirement at the base station. Specifically, a power allocation problem is first formulated for minimizing the COP, under the long-term power constraint, which is challenging since the objective function is nonconvex. To make this problem tractable, we apply a high SNR approximation and show that the approximated problem is convex. Our analysis shows that, for each feedback event, the optimal solution is in the form of two increasing geometric progressions. An efficient iterative search algorithm is proposed to determine the length of each geometric progression.

11.4.1 System Model

The system model is almost the same as that in Section 11.3, except that the short-term power constraint in (11.28) is replaced by the long-term power constraint defined in the following.

The long-term power constraint in (11.3) should be further elaborated for NOMA with one-bit feedback. Specifically, the considered long-term power constraint ensures that the average sum transmission power for all the users is constrained, which can be expressed as follows [31]:

$$
\sum_{k=1}^{K} \mathbb{E}\left[\mathcal{P}_k(\{Q(h_k)\})\right] = \sum_{k=1}^{K}\sum_{n=0}^{K}\left[P_{k,n}\mathbb{P}(N=n)\right]
$$

$$
= \sum_{n=0}^{K}\left[\mathbb{P}(N=n)\sum_{k=1}^{K}P_{k,n}\right] \le P, \tag{11.47}
$$

where the random variable N is defined in Definition 11.9.

The outage probability has been given in Theorem 11.10 for a fixed α and fixed power allocation parameters. Now, the diversity gain of NOMA with one-bit feedback under the long-term power constraint is given in the follow lemma.

LEMMA 11.17 *([31, Lemma 2]): Under the long-term power constraint in (11.47), the maximum achievable diversity gain of the considered NOMA scheme is 2, which is achieved only if α satisfies $\alpha \doteq P^{-1}$.*

REMARK 11.18 *As can be observed from Lemma 11.17, the diversity gain under the long-term power constraint is twice as large as that under the short-term power constraint. This is because the long-term power constraint is more relaxed compared to the short-term power constraint, which is beneficial to achieve a lower COP as well as a larger diversity gain.*

11.4.2 Power Allocation

The power allocation optimization problem under the long-term power constraint can be formulated as follows:

$$
\min_{\alpha,\{P_{k,n}\}} \sum_{n=0}^{K}\mathbb{P}_n(\alpha)\left[1 - \prod_{k=1}^{K}(1 - \mathbb{P}_{k,n}^{\text{Indiv}}(\alpha,\mathbf{P}_n))\right] \tag{11.48a}
$$

s.t. (11.47) and (11.36), $P_{k,n} \ge 0, \forall k \in [1:K], n \in [0:K]. \tag{11.48b}$

Similar to the problem in (11.40), the transformation of variables in (11.35) can be applied, and the problem in (11.48) can be transformed into the following equivalent form:

$$(P2) \quad \min_{\alpha,\{\zeta_{k,n}\}} \sum_{n=0}^{K} \mathbb{P}_n(\alpha) \left[1 - \prod_{k=1}^{K} (1 - \mathbb{P}_{k,n}^{\text{Indiv}}(\alpha, \zeta_n)) \right] \tag{11.49a}$$

$$\text{s.t.} \quad \sum_{n=0}^{K} \mathbb{P}(\alpha) \sum_{k=1}^{K} \frac{(\hat{r}_0 + 1)^{k-1} \hat{r}_0}{\zeta_{k,n}} \leq P; \tag{11.49b}$$

$$\zeta_{k,n} \geq 0, \forall k \in [1 : K], n \in [0 : K]. \tag{11.49c}$$

Approximation for High SNR

Problem (P2) in (11.49) is more challenging than problem (P1) because the power constraint is coupled and the decoupling approach used to solve problem (P1) is not applicable. Thus, a high SNR approximation of the objective function (i.e., $\mathbb{P}^{\text{Common}}$) will be used in order to simplify the problem. Specifically, an approximation of the objective function is first obtained at high SNR; then, for a fixed α, the optimal solution of this approximated problem is found; and finally, the optimal value for α is obtained using a one-dimensional search.

Using Propositions 11.14 and 11.16, problem (P2) can be simplified as follows:

$$(P3) \quad \min_{\{\zeta_{k,n}\}} \sum_{n=0}^{K} \mathbb{P}_n(\alpha) f_{3,n}(\alpha, \zeta_n) \tag{11.50a}$$

$$\text{s.t. (11.49b) and } \zeta_{k,n} \geq 0, k \in [1 : n], n \in [1 : K]; \tag{11.50b}$$

$$\zeta_{k,n} \geq \alpha, \forall k \in [n+1 : K], \ n \in [0 : K-1]; \tag{11.50c}$$

$$\zeta_{k,n} \leq \zeta_{k+1,n}, \forall k \in [1 : K-1], n \in [0 : K], \tag{11.50d}$$

where

$$f_{3,n}(\alpha, \zeta_n) \triangleq 1 - \prod_{k=1}^{n} \frac{\left[e^{-\zeta_{k,n}} - e^{-\alpha} \right]^+}{1 - e^{-\alpha}} \prod_{k=n+1}^{K} e^{-(\zeta_{k,n} - \alpha)}.$$

Problem (P3) can be approximately transformed into a convex problem at high SNR as shown in the following proposition.

PROPOSITION 11.19   *([31, proposition 3]): At high SNR, problem (P3) in (11.50) can be approximately transformed into convex problem (P4), which is defined as follows:*

$$(P4) \quad \min_{\{\zeta_{k,n}\}} \sum_{n=0}^{K} \mathbb{P}_n(\alpha) \left[\sum_{k=1}^{n} \frac{\zeta_{k,n}}{1 - e^{-\alpha}} + \sum_{k=n+1}^{K} (\zeta_{k,n} - \alpha) \right] \tag{11.51a}$$

$$\text{s.t. (11.49b) and } \zeta_{k,n} \geq 0, k \in [1 : n], n \in [1 : K]; \tag{11.51b}$$

$$\zeta_{k,n} \geq \alpha, k \in [n+1 : K], \ n \in [0 : K-1]; \tag{11.51c}$$

$$\zeta_{k,n} \leq \zeta_{k+1,n}, \ k \in [1 : K-1], n \in [0 : K]. \tag{11.51d}$$

REMARK 11.20   *Proposition 11.19 is a high SNR approximation for problem (P3). However, such a suboptimal solution can still provide a significant performance gain compared to benchmark schemes even in the moderate SNR regime, as shown later in the numerical result section.*

Optimal Solution of Problem (P4)

Problem (P4) can be further simplified. To do so, as new problem is first defined as follows.

DEFINITION 11.21   *Define a new optimization problem, denoted by (P5), obtained by removing the last constraint in (11.51d) of problem (P4).*

Problems (P4) and (P5) will be shown to be equivalent in Proposition 11.22, i.e., the optimal solution of problem (P5) satisfies constraint (11.51d). Note that problem (P4) is a convex optimization problem for a given α. The Lagrangian function of problem (P5) is given by

$$\mathcal{L}(\{\zeta_{k,n}\}, w, \{\lambda_{k,n}\})$$

$$\triangleq \mathbb{P}_n(\alpha) \left[\sum_{k=1}^{n} \frac{\zeta_{k,n}}{1 - e^{-\alpha}} + \sum_{k=n+1}^{K} (\zeta_{k,n} - \alpha) \right]$$

$$+ \omega \left(\sum_{n=0}^{K} \mathbb{P}_n(\alpha) \sum_{k=1}^{K} \frac{(\hat{r}_0 + 1)^{k-1} \hat{r}_0}{\zeta_{k,n}} - P \right)$$

$$- \sum_{n=1}^{K} \sum_{k=1}^{n} \lambda_{k,n} \zeta_{k,n} - \sum_{n=0}^{K-1} \sum_{k=n+1}^{K} \lambda_{k,n} (\zeta_{k,n} - \alpha), \tag{11.52}$$

where $\lambda_{k,n}, \omega \geq 0$ are Lagrange multipliers. The Karush–Kuhn–Tucker (KKT) conditions of problem (P5) are given by

$$\frac{\partial \mathcal{L}}{\partial \zeta_{k,n}} = \begin{cases} \frac{\mathbb{P}(\alpha)}{1 - e^{-\alpha}} - \frac{\omega \mathbb{P}(\alpha)(\hat{r}_0 + 1)^{k-1} \hat{r}_0}{\zeta_{k,n}^2} - \lambda_{k,n} = 0, \\ \quad \text{if } k \in [1 : n], n \in [1 : K]; \\ \mathbb{P}(\alpha) - \frac{\omega \mathbb{P}(\alpha)(\hat{r}_0 + 1)^{k-1} \hat{r}_0}{\zeta_{k,n}^2} - \lambda_{k,n} = 0, \\ \quad \text{if } k \in [n + 1 : K], n \in [0 : K - 1]. \end{cases} \tag{11.53}$$

The complementary slackness conditions are given as follows:

$$\omega \left(\sum_{n=0}^{K} \mathbb{P}_n(\alpha) \sum_{k=1}^{K} \frac{(\hat{r}_0 + 1)^{k-1} \hat{r}_0}{\zeta_{k,n}} - P \right) = 0 \tag{11.54a}$$

$$\lambda_{k,n} \zeta_{k,n} = 0 \text{ if } k \in [1 : n], n \in [1 : K]; \tag{11.54b}$$

$$\lambda_{k,n} (\zeta_{k,n} - \alpha) = 0 \text{ if } k \in [n + 1 : K], n \in [0 : K - 1]. \tag{11.54c}$$

From (11.53) and (11.54a)–(11.54c), it can be obtained that $\omega > 0$, $\lambda_{k,n} = 0$, for $k \in [1:n]$, $n \in [1:K]$, and the optimal $\zeta_{k,n}$ can be expressed as

$$\zeta_{k,n} = \begin{cases} \sqrt{\omega(\hat{r}_0 + 1)^{k-1}\hat{r}_0(1 - e^{-\alpha})}, \\ \qquad \text{if } k \in [1:n], n \in [1:K]; \\ \sqrt{\frac{\omega\mathbb{P}(\alpha)(\hat{r}_0+1)^{k-1}\hat{r}_0}{\mathbb{P}(\alpha)-\lambda_{k,n}}}, \\ \qquad \text{if } k \in [n+1:K], n \in [0:K-1]. \end{cases} \tag{11.55}$$

It is difficult to obtain the Lagrange multipliers directly. Alternatively, the properties of the optimal power allocation are first studied. The following proposition asserts that the constraint in (11.51d) is satisfied.

PROPOSITION 11.22 *([31, proposition 4]): The optimal solution of problem (P5) in (11.55) satisfies $\zeta_{k,n} \leq \zeta_{k+1,n}$, $\forall k \in [1:K-1]$, $n \in [0:K]$, i.e., problems (P4) and (P5) are equivalent.*

One can observe from Proposition 11.22 and constraint (11.51c) that, if $\zeta_{k,n} = \alpha$ for a given $k \in [n+1:K]$ and $n \in [0:K-1]$, $\zeta_{l,n} = \alpha$ also holds $\forall l \in [n+1:k]$. Motivated by this, a series of integers representing the number of $\zeta_{k,n}$ that are equal to α is defined as follows.

DEFINITION 11.23 *For each n, let $i_n \in [0:K-n]$ denote the number of $\zeta_{k,n}$ whose values are equal to α, i.e., $\zeta_{k,n} = \alpha$, for $k \in [n+1:n+i_n]$ and $\zeta_{k,n} > \alpha$ for $k \in [n+i_n+1:K]$.*

In summary, for given i_n's, the optimal solution of the $\zeta_{k,n}$ is given by the following theorem.

THEOREM 11.24 *([31, Theorem 2]): If all integers $i_n \in [0:K-n]$ defined in Definition 11.23 are known, the optimal solution of problems (P4) and (P5) can be expressed as follows:*

$$\zeta_{k,n} = \begin{cases} \sqrt{\omega(\hat{r}_0 + 1)^{k-1}\hat{r}_0(1 - e^{-\alpha})}, & \text{if } k \in [1:n], \\ \alpha, & \text{if } k \in [n+1:n+i_n], \\ \sqrt{\omega(\hat{r}_0 + 1)^{k-1}\hat{r}_0}, & \text{if } k \in [n+i_n+1:K], \end{cases} \tag{11.56}$$

for each $n \in [0:K]$, where

$$\sqrt{w} = \frac{\sum_{n=0}^{K}\mathbb{P}_n(\alpha)A_n(i_n)}{P - \sum_{n=0}^{K}\mathbb{P}_n(\alpha)B_n(i_n)} \tag{11.57}$$

and

$$A_n(i_n) \triangleq \sum_{k=1}^{n}\sqrt{\frac{(\hat{r}_0 + 1)^{k-1}\hat{r}_0}{1 - e^{-\alpha}}} + \sum_{k=n+i_n+1}^{K}\sqrt{(\hat{r}_0 + 1)^{k-1}\hat{r}_0}, \tag{11.58}$$

$$B_n(i_n) \triangleq \sum_{k=n+1}^{n+i_n}\frac{(\hat{r}_0 + 1)^{k-1}\hat{r}_0}{\alpha}. \tag{11.59}$$

Note that $A_0(i_0) \triangleq 0$ if $i_0 = K - n$, and $B_n(i_n) \triangleq 0$ if $i_n = 0$, $\forall n \in [0:K]$.

REMARK 11.25 *From Theorem 11.24, the optimal solution $\{\zeta_{1,n}, \cdots, \zeta_{K,n}\}$ has the structure of two increasing geometric progressions and with several constant values of α between them. Interestingly, the feedback parameter n affects only the lengths of the two geometric progressions, but does not affect their values.*

Search Algorithm for $\{i_n^*\}$

The remaining work is to find the unique integer sequence, denoted by $\{i_n^*\}$, such that all complementary slackness conditions are satisfied. Since $\lambda_{k,n} = 0$ for $k \in [1 : n]$, it is only needed to choose $\{i_n^*\}$ such that

$$\lambda_{k,n} \geq 0 \text{ for } k \in [n+1 : n+i_n^*]$$

$$\text{and } \zeta_{k,n} > \alpha \text{ for } k \in [n+i_n^*+1 : K]. \tag{11.60}$$

Note that, given $\{i_n\}$, since $\zeta_{k,n}^{(t)} = \alpha$ for $k \in [n+1 : n+i_n]$ in (11.55), $\lambda_{k,n}$ can be obtained as

$$\lambda_{k,n} = \mathbb{P}_n(\alpha) \left(1 - \frac{\omega(\hat{r}_0 + 1)^{k-1} \hat{r}_0}{\alpha^2} \right), \; k \in [n+1 : n+i_n]. \tag{11.61}$$

A closed-form solution for the i_n^* cannot be found. Alternatively, an efficient iterative algorithm will be designed to find $\{i_n^*\}$, which is summarized in the following algorithm.

---

1. Initialize $t = 1$, $i_n^{(1)} = 0$ for $n \in [0 : K]$, and $\lambda_{k,n}^{(1)} = 0$ for $k \in [n+1 : K], n \in [0 : K]$.
2. At the tth iteration:
   (a) Update $\omega^{(t)}$, $\lambda_{k,n}^{(t)}$, and $\zeta_{k,n}^{(t)}$ in (11.57), (11.61), and (11.56), respectively.
   (b) If $i_n^{(t)} = K - n$ or $\zeta_{n+i_n^{(t)}+1,n}^{(t)} > \alpha$, $\forall n \in [0 : K]$, break the loop and the algorithm ends.
   (c) Else, for each n satisfying $\zeta_{n+i_n^{(t)}+1,n}^{(t)} \leq \alpha$, set $i_n^{(t+1)}$ as

   $$i_n^{(t+1)} = \arg \max_{i \in \left[i_n^{(t)}+1:K-n \right]} \{i : \zeta_{n+i,n}^{(t)} \leq \alpha\},$$

   whereas, for each n satisfying $\zeta_{n+i_n^{(t)}+1}^{(t)} > \alpha$, set $i_n^{(t+1)}$ as $i_n^{(t+1)} = i_n^{(t)}$.
3. Update $t = t + 1$ and repeat Step 2 until $\{i_n^*\}$ is found.

The unique sequence $\{i_n^*\}$ can be found by the above algorithm as asserted in the following theorem, which assures that the proposed algorithm converges.

THEOREM 11.26 *[31, theorem 3] The strategy involved in the proposed algorithm, i.e., updating each $i_n^{(t)}$ satisfying $\zeta_{n+i_n^{(t)}+1,n}^{(t)} \leq \alpha$ as $i_n^{(t+1)} = \arg\max_{i \in \left[i_n^{(t)}:K-n \right]} \{i : \zeta_{n+i}^{(t)} \leq \alpha\}$, guarantees that $\{i_n^*\}$ will be found.*

From (11.56), $\zeta_{k,0}^{(t)} = \zeta_{k,n}^{(t)}$, $\forall k \in [n+1], n \in [0 : K-1]$. Thus, according to Step 2(c) in the proposed algorithm, $i_n^{(t)} = i_0^{(t)} - n$ can be obtained if $n \in [0 : i_0^{(t)} - 1]$ and $i_n^{(t)} = 0$, otherwise. Since $i_0^{(t)} \in [0 : K]$, the proposed algorithm requires at most $K + 1$ iterations to find $\{i_n^*\}$. This means that the proposed algorithm enjoys much lower complexity compared to an exhaustive search having complexity $O((K + 1)!)$.

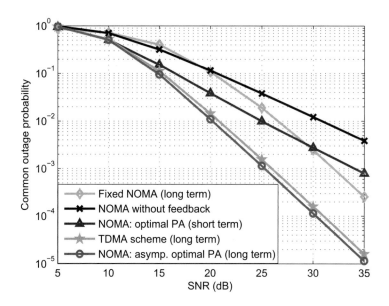

Figure 11.9 COP vs. SNR for different transmission schemes, where the number of users is set as $K = 3$, and the target transmission rate is set as $r_0 = 0.8$ BPCU for each user.

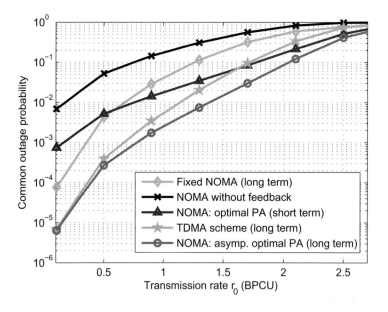

Figure 11.10 COP vs. r_0 for different transmission schemes, where the number of users is set as $K = 3$, and the SNR is set as 25 dB.

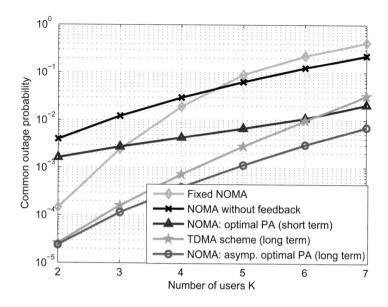

Figure 11.11 COP vs K for different transmission schemes, where the target transmission rate is set as $r_0 = 0.8$ BPCU for each user, and the SNR is set as 30 dB.

11.4.3   Numerical Results

The outage performance of NOMA with one-bit feedback is determined under the long-term power constraint in (11.48). In Figures 11.9–11.11, the outage performance of NOMA with the asymptotically optimal power allocation scheme is compared with the benchmark schemes in Section 11.3.4 and NOMA under the short-term power constraint, as a function of the SNR, the transmission rate r_0, and the number of users K, respectively. As shown in Figure 11.9, the same slope is achieved by the COPs of NOMA with the considered power allocation, TDMA, and fixed NOMA have at high SNR, under the long-term power constraint. This is because a diversity gain of 2 can be achieved by all these schemes (Lemma 11.17). However, among these three schemes, fixed NOMA realizes the poorest performance, especially at high SNR. This implies that the asymptotically optimal power allocation scheme plays an important role in improving the outage performance. Although the considered asymptotically optimal power allocation scheme is based on the high-SNR approximation, it also has better performance at low SNR than NOMA under the short-term power constraint. As shown in Figure 11.10, NOMA with the asymptotically optimal long-term power allocation scheme achieves the lowest outage probability among the considered schemes. Moreover, the fixed NOMA scheme does not perform well especially when the transmission rate r_0 is large. When $r_0 = 2.1$, NOMA with the asymptotically optimal power allocation and TDMA achieve COPs of approximately 0.12 and 0.35, respectively. In this case, the successful transmission probability can be improved from 0.65 to 0.88 if we change the MA technique from TDMA to NOMA. Finally, as can be observed in

Figure 11.11, the COP gap between NOMA with the asymptotically optimal power allocation and TDMA increases as K increases. TDMA with the long-term power constraint achieves a COP even higher than the NOMA scheme with the short-term power constraint. This means that TDMA is not appropriate to implement for downlink scenarios with large numbers of users due to its poor spectral efficiency.

11.5   Conclusions

This chapter has considered several typical system models to investigate the ergodic rate and outage performance of downlink NOMA with perfect CSIT and limited channel feedback, respectively, from an information theoretic perspective. The ergodic rate performance of NOMA with perfect CSIT has been first investigated based on a new evaluation criterion proposed in [27]. The basic idea is to describe the relationships among the BC capacity region, the NOMA rate region, and the TDMA rate region. Then, this new evaluation criterion for NOMA was used to evaluate the ergodic rate performance of NOMA in wireless fading scenarios. Specifically, NOMA and TDMA have been compared statistically in terms of not only the sum rate but also the individual rates. Based on this evaluation criterion, the ergodic rate performance of NOMA with user pairing has been evaluated in terms of both the sum-rate and users' individual rates, by considering TDMA as a benchmark. Numerical results have shown that NOMA outperforms TDMA in terms of both the sum-rate and users' individual rates.

The outage performance of downlink NOMA with one-bit CSI feedback has also been investigated. A closed-form expression for the COP and its optimal diversity gain have been analyzed under short-term and long-term power constraints. The diversity gain has been demonstrated to be one and two under the short-term and long-term power constraints, respectively. Then, dynamic power allocation policies according to the feedback state have also been discussed to minimize the COP. Under the short-term power constraint, the original nonconvex problem has been transformed into a series of convex problems. Under the long-term power constraint, a high-SNR approximation has been applied to obtain an asymptotically optimal solution. The considered NOMA schemes with one-bit feedback have been demonstrated to outperform various existing MA schemes in terms of the outage probability via numerical results.

Future work of interest includes extending the considered evaluation criterion to NOMA with more than two users and MIMO-NOMA to evaluate the ergodic rate performance. In addition, the ergodic rate performance of NOMA with different power allocation strategies could also be studied. With limited channel feedback, an important topic for future research is to investigate the outage performance of NOMA with multi-bit feedback, where the one-bit feedback scheme for NOMA can be extended to multi-bit feedback. Moreover, the extension of the analysis of the ergodic rate and the outage probability to asymmetric scenarios with different distances and different rates for different users is also of interest. Finally, user fairness is an important topic for NOMA, and many other fairness criteria could be adopted for NOMA to ensure user fairness, such as max–min fairness, proportional fairness and α fairness [32, 41].

Acknowledgments

The work of H. V. Poor was supported by the U.S. National Science Foundation under grants CCF-0939370, CCF-1513915, and CCF-190830.

References

[1] Y. Saito, A. Benjebbour, Y. Kishiyama, and T. Nakamura, "System-level performance evaluation of downlink non-orthogonal multiple access (NOMA)," in *Proc. IEEE Int. Symp. Personal, Indoor and Mobile Radio Commun. (PIMRC)*, London, Sep. 2013.

[2] Y. Saito, Y. Kishiyama, A. Benjebbour, et al., "Nonorthogonal multiple access (NOMA) for cellular future radio access," in *Proc. IEEE 77th Veh. Techn. Conf. (VTC Spring)*, Dresden, Jun. 2013.

[3] Y. Tao, L. Liu, S. Liu, and Z. Zhang, "A survey: Several technologies of non-orthogonal transmission for 5G," *China Commun.*, vol. 12, no. 10, pp. 1–15, Oct. 2015.

[4] Z. Wei, J. Yuan, D. W. K. Ng, M. Elkashlan, and Z. Ding, "A survey of downlink nonorthogonal multiple access for 5G wireless communication networks," *ZTE Commun.*, vol. 14, no. 4, pp. 17–25, Oct. 2016.

[5] Q. Li, H. Niu, A. Papathanassiou, and G. Wu, "5G network capacity: Key elements and technologies," *IEEE Veh. Technol. Mag.*, vol. 9, no. 1, pp. 71–78, Mar. 2014.

[6] Z. Ding, Y. Liu, J. Choi, et al., "Application of non-orthogonal multiple access in LTE and 5G networks," *IEEE Commun. Mag.*, vol. 55, no. 2, pp. 185–191, Feb. 2017.

[7] Z. Ding, X. Lei, G. K. Karagiannidis, et al., "A survey on non-orthogonal multiple access for 5G networks: Research challenges and future trends," *IEEE J. Sel. Areas Commun.*, vol. 35, no. 10, pp. 2181–2195, Oct. 2017.

[8] Z. Ding, M. Xu, Y.Chen, M. Peng, and H. V. Poor, "Embracing non-orthogonal multiple access in future wireless networks," *Frontiers Inf. Technol. Electronic Eng.*, vol. 19, no. 3, pp. 322–339, Mar. 2018.

[9] P. Bergmans, "A simple converse for broadcast channels with additive white Gaussian noise (corresp.)," *IEEE Trans. Inf. Theory*, vol. 20, no. 2, pp. 279–280, Mar. 1974.

[10] T. Cover, "Broadcast channels," *IEEE Trans. Inf. Theory*, vol. 18, no. 1, pp. 2–14, Jan. 1972.

[11] H. Weingarten, Y. Steinberg, and S. Shamai, "The capacity region of the Gaussian multiple-input multiple-output broadcast channel," *IEEE Trans. Inf. Theory*, vol. 52, no. 9, pp. 3936–3964, Sep. 2006.

[12] Z. Ding, F. Adachi, and H. V. Poor, "The application of MIMO to nonorthogonal multiple access," *IEEE Trans. Wireless Commun.*, vol. 15, no. 1, pp. 537–552, Jan. 2016.

[13] Z. Ding, R. Schober, and H. V. Poor, "A general MIMO framework for NOMA downlink and uplink transmission based on signal alignment," *IEEE Trans. Wireless Commun.*, vol. 15, no. 6, pp. 4438–4454, Jun. 2016.

[14] M. Yoshida and T. Tanaka, "Analysis of sparsely-spread CDMA via statistical mechanics," in *Proc. IEEE Int. Symp. Inf. Theory*, Seattle, WA, Jul. 2006.

[15] M. Al-Imari, M. A. Imran, R. Tafazolli, and D. Chen, "Performance evaluation of low density spreading multiple access," in *Proc. 8th Int. Wireless Commun. Mobile Comput. Conf.*, Limassol, Aug. 2012.

[16] H. Nikopour and H. Baligh, "Sparse code multiple access," in *Proc. IEEE Int. Symp. Personal Indoor Mobile Radio Commun.*, London, Sep. 2013.

[17] G. C. Ferrante and M.-G. Di Benedetto, "Spectral efficiency of random time-hopping CDMA," *IEEE Trans. Inf. Theory*, vol. 61, no. 12, pp. 6643–6662, Dec. 2015.

[18] S. Chen, B. Ren, Q. Gao, et al., "Pattern division multiple access: A novel nonorthogonal multiple access for fifth-generation radio networks," *IEEE Trans. Veh. Technol.*, vol. 66, no. 4, pp. 3185–3196, Apr. 2017.

[19] B. M. Zaidel, O. Shental, and S. Shamai (Shitz), "Sparse NOMA: A closed-form characterization," in *Proc. IEEE Int. Symp. Inf. Theory*, Vail, CO, Jun. 2018.

[20] 3GPP TR 36.859: "Study on downlink multiuser superposition transmission (MUST) for LTE," Nov. 2015.

[21] L. Zhang W. Li, Y. Wu, et al., "Layered-division-multiplexing: Theory and practice," *IEEE Trans. Broadcast*, vol. 62, no. 1, pp. 216–232, Mar. 2016.

[22] L. Li and A. J. Goldsmith, "Capacity and optimal resource allocation for fading broadcast channels – Part I: Ergodic capacity," *IEEE Trans. Inf. Theory*, vol. 47, no. 3, pp. 1083–1102, Mar. 2001.

[23] L. Li and A. J. Goldsmith, "Capacity and optimal resource allocation for fading broadcast channels – Part II: Outage capacity," *IEEE Trans. Inf. Theory*, vol. 47, no. 3, pp. 1103–1127, Mar. 2001.

[24] W. Zhang, S. P. Kotagiri, and J. N. Laneman, "On downlink transmission without transmit channel state information and with outage constraints," *IEEE Trans. Inf. Theory*, vol. 55, no. 9, pp. 4240–4248, Sep. 2009.

[25] Z. Ding, Z. Yang, P. Fan, and H. V. Poor, "On the performance of non-orthogonal multiple access in 5G systems with randomly deployed users," *IEEE Signal Process. Lett.*, vol. 21, no. 12, pp. 1501–1505, Sep. 2014.

[26] S. Timotheou and I. Krikidis, "Fairness for non-orthogonal multiple access in 5G systems," *IEEE Signal Process. Lett.*, vol. 22, no. 10, pp. 1647–1651, Oct. 2015.

[27] P. Xu, Z. Ding, X. Dai, and H. V. Poor, "A new evaluation criterion for non-orthogonal multiple access in 5G software defined networks," *IEEE Access*, vol. 3, pp. 1633–1639, Sep. 2015.

[28] Z. Ding, M. Peng, and H. Poor, "Cooperative non-orthogonal multiple access in 5G systems," *IEEE Commun. Lett.*, vol. 19, no. 8, pp. 1462–1465, Aug. 2015.

[29] Z. Yang, Z. Ding, P. Fan, and G. Karagiannidis, "On the performance of 5G non-orthogonal multiple access systems with partial channel information," *IEEE Trans. Commun.*, vol. 64, no. 2, pp. 654–667, Feb. 2016.

[30] S. Shi, L. Yang, and H. Zhu, "Outage balancing in downlink non-orthogonal multiple access with statistical channel state information," *IEEE Trans. Wireless Commun.*, vol. 15, no. 7, pp. 4718–4731, July 2016.

[31] P. Xu, Y. Yuan, Z. Ding, X. Dai, and R. Schober, "On the outage performance of non-orthogonal multiple access with 1-bit feedback," *IEEE Trans. Wireless Commun.*, vol. 15, no. 10, pp. 6716–6730, Oct. 2016.

[32] P. Xu and K. Cumanan, "Optimal power allocation scheme for non-orthogonal multiple access with α-fairness," *IEEE J. Sel. Areas Commun.*, vol. 35, no. 10, pp. 2357–2369, Oct. 2017.

[33] P. Xu, Z. Yang, Z. Ding, and Z. Zhang, "Optimal relay selection scheme for cooperative NOMA," *IEEE Trans. Veh. Technol.* DOI: 10.1109/TVT.2018.2821900.

[34] T. Cover and J. Thomas, *Elements of Information Theory*, 2nd ed. Hoboken, NJ: Wiley-Interscience 2006.

[35] Z. Ding, P. Fan, and H. V. Poor, "Impact of user pairing on 5G non-orthogonal multiple access," *IEEE Trans. Veh. Tech.*, vol. 65, no. 8, pp. 6010–6023, Aug. 2016.

[36] H. A. David and H. N. Nagaraja, *Order Statistics*, 3rd ed. New York: Wiley, 2003.

[37] L. Li, N. Jindal, and A. Goldsmith, "Outage capacities and optimal power allocation for fading multiple-access channels," *IEEE Trans. Inf. Theory*, vol. 51, no. 4, pp. 1326–1347, Apr. 2005.

[38] J. Luo, R. Yates, and P. Spasojevic, "Service outage based power and rate allocation for parallel fading channels," *IEEE Trans. Inf. Theory*, vol. 51, no. 7, pp. 2594–2611, Jul 2005.

[39] D. Tse and P. Viswanath, *Fundamentals of Wireless Communication*. New York: Cambridge University Press, 2005.

[40] G. Caire, G. Taricco, and E. Biglieri, "Optimum power control over fading channels," *IEEE Trans. Inf. Theory*, vol. 45, no. 5, pp. 1468–1489, Jul 1999.

[41] H. Shi, R. V. Prasad, E. Onur, and I. Niemegeers, "Fairness in wireless networks: Issues, measures and challenges," *IEEE Commun. Surv. Tuts.*, vol. 16, no. 1, pp. 5–24, 2014.

12 Compute–Forward Strategies for Next-Generation Wireless Systems

Bobak Nazer, Michael Gastpar, and Sung Hoon Lim

12.1 Introduction

The demand for mobile, high-data-rate service continues to grow at a rapid pace. To meet this demand, networks are becoming denser and employing increasingly sophisticated communication strategies, often leading to more interference between users. From the lens of (classical) network information theory, this interference can be handled either by treating it as noise or decoding and canceling it. This chapter examines another possibility: A receiver can decode a linear combination of simultaneously transmitted codewords, so long as they are all drawn from the same, linear codebook. This *compute–forward* strategy can act as a building block for network communication architectures that can sometimes outperform conventional approaches [1].

Error-correcting codes with algebraic structure have a long history [2]. First and foremost, algebraic structure can enable low-complexity encoding and decoding and is thus crucial for engineering practice. Several other properties of algebraic codes have been important in the past, including the worst-case performance guarantees they carry. In parallel, information theoretic techniques were developed to understand the fundamental performance limits of error-correcting codes for data transmission over noisy channels [3, 4]. These techniques make no reference to the underlying algebraic structure of the codes. In particular, random coding proofs for establishing the achievability side of capacity theorems are non-constructive and indifferent toward complexity.

More recently, it was discovered that error-correcting codes with algebraic structure have another important property: They can serve to shape interference between multiple users into a useful form. Perhaps the simplest proxy to understand this behavior is the problem of decoding the sum of codewords. Specifically, consider an uplink scenario with two users and a decoder wishing only to recover the sum of the two codewords, rather than the codewords themselves. For this scenario in isolation, it can be shown that the data rates of the users can be substantially increased if linear codes are used, as first discovered in [5–7]. It is perhaps initially unclear what use an individual decoder would get out of learning the sum of two codewords. Yet, if we consider a larger network setting where, say, another decoder learns an independent linear combination of these two codewords, then these two decoders can work together to solve for the original codewords. This overall strategy is often referred to as compute–forward [1].

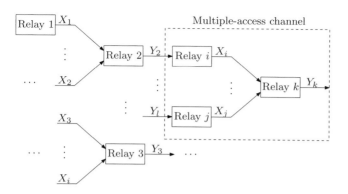

Figure 12.1 Multiple-access channel as a canonical component of a larger network.

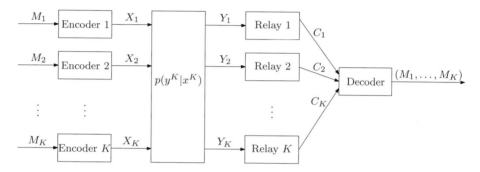

Figure 12.2 Cloud radio access network. The relays have finite capacity links to the decoder.

Figure 12.1 illustrates a basic setting where compute–forward will play a role. It shows a larger network, composed of many relay terminals. Inside, one can find component networks referred to as *multiple-access channels* (MAC; the dashed box in the figure). From the perspective of compute–forward, these constitute the canonical basis elements, and thus they receive substantial attention throughout the remainder of this chapter. Figure 12.2 shows a network that takes this idea one step further: Here, we have K simultaneous MACs (the fan-in of the K relays). Each relay is connected via a rate-constrained link to a central decoder. This could model a so-called cloud radio access network (C-RAN) [8–10]. From the perspective of compute–forward, the goal here is for each relay to decode a differently weighted sum of the messages in such a way that together, these weighted sums constitute an invertible system, allowing the central decoder to recover all messages individually.

A third example of a different flavor is illustrated in Figure 12.3. This figure illustrates a multiple-input multiple-output (MIMO) system rather than a network setting. Yet, compute–forward can play an interesting role here, too, when we study the implementation of the receiver. Specifically, instead of using a joint decoder at the receiver that simultaneously retrieves all K messages, we can reduce complexity substantially by using a bank of separate decoders. If we force each of these to decode only one of the

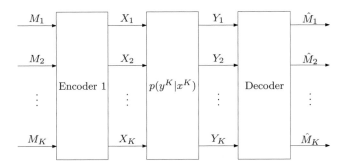

Figure 12.3 A MIMO transceiver.

messages, then the performance will be disappointing due to the possibly severe interference between the K transmitted signals. If instead we let each of the separate decoders extract a linear combination of the codewords, then this interference can be leveraged and a satisfying near-optimal performance is attained at a much reduced complexity. This architecture is known as an *integer-forcing receiver* [11].

As these examples illustrate, compute–forward is an important technique for networks with interference between simultaneously transmitted signals. This occurs very prominently in wireless networks, which are often modeled as Gaussian fading channels. For these channels, algebraic structure is most commonly exploited via the concept of lattice codes [12], which have been studied extensively over the past several decades. Consequently, the vast majority of prior work on compute–forward has relied on nested lattice codebooks as a building block.

By contrast, the present chapter mostly focuses on an alternative approach based on joint typicality techniques combined with nested linear codes. We will demonstrate that this approach captures the performance of lattice-based compute–forward for Gaussian networks as a special case. An important additional feature of this approach is that it permits bridging the gap toward classical results in network information theory. That is, most of the foundations of network information theory (in terms of achievability proofs) rely on independent and identically distributed (i.i.d.) random codes and joint typicality encoding and decoding techniques. As discussed above, early work on compute–forward relied on specialized lattice coding techniques, and thus stood apart from the bulk of network information theory. In this chapter, we describe a set of techniques that can be used to characterize the performance of nested linear codes under joint typicality encoding and decoding, and thus bring compute–forward closer to the core of network information theory as described in [4].

12.2   Lattice–Based Compute–Forward

We now provide a brief overview of the main achievability result from lattice-based compute–forward. Recall that a lattice Λ is a discrete, additive subgroup of $\mathbb{R}^n$ that is closed under addition and reflection. Thus, any integer-linear combination is itself a lattice point, $a_1\lambda_1 + a_2\lambda_2 \in \Lambda$ for any $a_1, a_2 \in \mathbb{Z}$ and $\lambda_1, \lambda_2 \in \Lambda$. The compute–forward strategy exploits this property to enable receivers to recover integer-linear combinations of lattice codewords, often at higher rates than are available for decoding the individual codewords. Of course, a lattice has infinite extent and a bit more care is needed to construct good lattice codebooks that satisfy an average power constraint. One approach is to employ a fine lattice Λ_F in conjunction with a coarse lattice Λ_C that is nested within the fine lattice $\Lambda_C \subset \Lambda_F$. Then, the nested lattice codebook $\mathcal{L}$ consists of all fine lattice points that are closer to the zero vector than to any other coarse lattice point (i.e., they fall within fundamental Voronoi region of the coarse lattice). It can be argued that sequences of good nested lattice codebooks of rate $1/2\log(P/\sigma^2)$ exist such that all codewords satisfy an average power constraint P and can tolerate noise with variance up to σ^2.

Consider a K-user Gaussian MAC with channel inputs $x_k^n \in \mathbb{R}^n$ for $k \in [1 : K]$ and channel output

$$Y^n = \sum_{k=1}^{K} h_k X_k^n + Z^n,$$

where h_k are real-valued channel coefficients and Z^n is i.i.d. $\mathcal{N}(0, 1)$. The goal is for the receiver to recover an integer-linear combination of the codewords, $\sum_{k=1}^{K} a_k X_k^n$ for some integer coefficients a_k. See Figure 12.4 for an illustration.

The encoding and decoding steps are as follows. Encoder k maps its message to a lattice codeword X_k^n. The receiver, upon observing Y^n, scales it by $\alpha \in \mathbb{R}$ and quantizes the result to the nearest fine lattice codeword. See Figure 12.5 for an illustration. The effective channel can be written as

$$\alpha Y^n = \sum_{k=1}^{K} a_k X_k^n + Z_{\text{eff}}^n$$

$$Z_{\text{eff}}^n = \alpha Z^n + \sum_{k=1}^{K} (\alpha h_k - a_k) X_k^n .$$

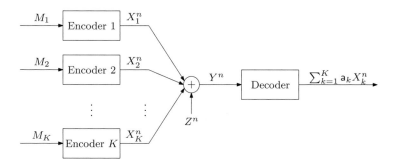

Figure 12.4 The K-user Gaussian multiple-access channel.

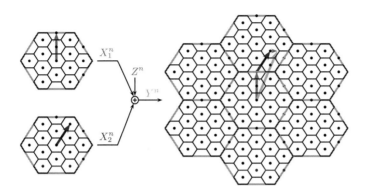

Figure 12.5 Illustration of lattice-based compute–forward for the $K = 2$ user scenario where the receiver wants to decode the sum of codewords $X_1^n + X_2^n$ from $Y^n = X_1^n + X_2^n + Z^n$. ©2011 IEEE. Reprinted, with permission, from [13].

The effective noise variance is

$$\sigma_{\text{eff}}^2 = \frac{1}{n} \mathsf{E} \|Z_{\text{eff}}^n\|^2 = \alpha^2 + P \sum_{k=1}^{K} (\alpha h_k - a_k)^2 .$$

Since $\sum_{k=1}^{K} a_k X_k^n$ belongs to the fine lattice, it can be decoded successfully so long as $\sigma_{\text{eff}}^2 < \sigma^2$. Equivalently, decoding is successful if the rate of the nested lattice codebook satisfies

$$R < \max_{\alpha \in \mathbb{R}} \frac{1}{2} \log^+ \left(\frac{P}{\alpha^2 + P \sum_{k=1}^{K} (\alpha h_k - a_k)^2} \right) \tag{12.1}$$

$$= \frac{1}{2} \log^+ \left(\frac{P}{\sigma_{\text{eff}}^2(\mathbf{h}, \mathbf{a})} \right), \tag{12.2}$$

where

$$\sigma_{\text{eff}}^2(\mathbf{h}, \mathbf{a}) := \mathbf{a}(P^{-1}\mathbf{I} + \mathbf{h}\mathbf{h}^\mathsf{T})^{-1}\mathbf{a}^\mathsf{T}, \tag{12.3}$$

$\log^+(\beta) = \max\{0, \log(\beta)\}$, $\mathbf{a} = [a_1 \cdots a_K]$ and $\mathbf{h} = [h_1 \cdots h_K]$ are the row vectors of integer coefficients and channel coefficients, respectively, and $\mathbf{I}$ is the identity matrix. (Note that the last step follows by plugging in the optimal minimum mean squared error (MMSE) coefficient for α and then applying the Woodbury matrix identity.)

This achievability strategy offers two advantages. First, the receiver is free to decode any integer-linear combination satisfying (12.2). Second, the rates can exceed those available for decoding the messages individually.

EXAMPLE 12.1   Consider the two-user channel $Y^n = X_1^n + X_2^n + Z^n$ and a receiver that wishes to recover the sum $X_1^n + X_2^n$. According to (12.2), this is possible for any rate up to $\frac{1}{2} \log^+(\frac{1}{2} + P)$ whereas the symmetric capacity for recovering both messages is $\frac{1}{4} \log(1 + 2P)$. This example is illustrated in Figure 12.5.

Below, we develop an alternative approach to compute–forward based on joint typicality encoding and decoding. This will allow us to derive compute–forward rate regions for discrete memoryless (DM) channels, recover the Gaussian results above as a special case, and evaluate rate regions for more sophisticated decoding techniques.

12.3 Computation Over Discrete Memoryless Multiple-Access Channels

We give a formal problem statement for computing linear combinations over a K-user DM-MAC. A DM-MAC

$$(\mathcal{X}_1 \times \cdots \times \mathcal{X}_K, p_{Y|X_1,\dots,X_K}(y|x_1,\dots,x_K), \mathcal{Y})$$

consists of K sender alphabets $\mathcal{X}_k$, $k \in [1:K]$, one receiver alphabet $\mathcal{Y}$, and a collection of conditional probability mass functions (pmf) $p_{Y|X_1,\dots,X_K}(y|x_1,\dots,x_K)$.

We begin by selecting a finite field $\mathbb{F}$ over which linear combinations will be taken. The goal of the receiver is to recover L linear combinations according to the coefficient vectors $\mathbf{a}_1,\dots,\mathbf{a}_L \in \mathbb{F}^K$. Let

$$A = \begin{bmatrix} \mathbf{a}_1 \\ \vdots \\ \mathbf{a}_K \end{bmatrix} \in \mathbb{F}^{L \times K} \tag{12.4}$$

denote the coefficient matrix.

A $(2^{nR_1},\dots,2^{nR_K},n)$ computation code consists of

- K message sets $[2^{nR_k}] := \{0,1,2,\dots,2^{nR_k}-1\}$, $k \in [1:K]$;[1]
- K encoders, where encoder k maps each message $m_k \in [2^{nR_k}]$ to a pair of sequences $(u_k^n, x_k^n)(m_k) \in \mathbb{F}^n \times \mathcal{X}_k^n$ such that $u_k^n(m_k)$ is *injective*;
- L linear combinations for each message tuple $(m_1,\dots,m_K)$

$$w_A^n(m_1,\dots,m_K) := \begin{bmatrix} w_{\mathbf{a}_1}^n(m_1,\dots,m_K) \\ \vdots \\ w_{\mathbf{a}_L}^n(m_1,\dots,m_K) \end{bmatrix} := A \begin{bmatrix} u_1^n(m_1) \\ \vdots \\ u_K^n(m_K) \end{bmatrix},$$

where the linear combinations are defined over the vector space $\mathbb{F}^n$; and

- a decoder that assigns estimates $\hat{w}_{\mathbf{a}_1}^n,\dots,\hat{w}_{\mathbf{a}_L}^n \in \mathbb{F}^n \times \cdots \times \mathbb{F}^n$ to each received sequence $y^n \in \mathcal{Y}^n$.

We assume that each message M_k is independently and uniformly drawn from $[2^{nR_k}]$. The average probability of error is defined as

$$P_e^{(n)} = P\left\{(\hat{W}_{\mathbf{a}_1}^n,\dots,\hat{W}_{\mathbf{a}_L}^n) \neq (W_{\mathbf{a}_1}^n,\dots,W_{\mathbf{a}_L}^n)\right\}.$$

We say that a rate tuple $(R_1,\dots,R_K)$ is achievable for computing (recovering) the linear combinations with coefficient matrix A (or simply computing the A linear combinations) if there exists a sequence of $(2^{nR_1},\dots,2^{nR_K},n)$ codes such that $\lim_{n\to\infty} P_e^{(n)} = 0$.

---

[1] We index our messages starting at 0 to create a more natural correspondence with the finite field.

The capacity region $\mathcal{C}_A$ for computing A linear combinations over the DM-MAC is the closure of the set of achievable rate tuples $(R_1, \ldots, R_K)$.

The role of the mappings $u_k^n(m_k)$ is to embed the messages into the vector space $\mathbb{F}^n$ such that the coding strategy is compatible with a properly defined algebra. The restriction to injective mappings ensures that once the decoder has enough linearly independent combinations, the original messages can be recovered.

Notation. For a matrix A, we denote the jth row vector by a_j and denote the submatrix formed by the rows $k \in S$ of A by $A(S)$. We denote the standard basis vector (unit vector) with 1 in the jth position by e_j.

12.4   Multiple–Access for Computation

12.4.1   Recovering the Individual Messages

Recall that the capacity region for multiple access (i.e., recovering all transmitted messages) $\mathcal{C}_{\text{MAC}}$ is the set of rate tuples $(R_1, \ldots, R_K)$ such that

$$\sum_{k \in S} R_k < I(X(S); Y | X(S^c), Q), \text{ for every } S \subseteq [1 : K] \tag{12.5}$$

for some pmf $p(q) \prod_{k=1}^K p(x_k | q)$ with $|Q| \le K$ where $X(S) = (X_k : k \in S)$.

Since the decoder can recover the linear combinations explicitly once it has all the individual messages separately, any achievable rate region for multiple access is an achievable strategy for computing any $A \in \mathbb{F}^{L \times K}$. Accordingly, we have the relation $\mathcal{C}_{\text{MAC}} \subseteq \mathcal{C}_A$. This observation can be generalized to the following remark.

REMARK 12.2   *Consider a matrix $\tilde{A}$ such that span$(A) \subseteq$ span$(\tilde{A})$ where span(A) is the row-span of a matrix A. Since the decoder can recover the A linear combinations explicitly assuming that it has the $\tilde{A}$ linear combinations, any achievable rate region for computing $\tilde{A}$ linear combinations such that span$(A) \subseteq$ span$(\tilde{A})$ is an achievable rate region for computing $A \in \mathbb{F}^{L \times K}$ linear combinations. This results in the relation $\mathcal{C}_{\tilde{A}} \subseteq \mathcal{C}_A$.*

Gaussian MAC

The capacity region of the Gaussian MAC (for recovering the individual messages) consists of the set of rate tuples $(R_1, \ldots, R_K)$ such that

$$\sum_{k \in S} R_k < C \left(\sum_{k \in S} h_k^2 P \right), \quad S \subseteq [1 : K], \tag{12.6}$$

where $C(x) = \dfrac{1}{2} \log(1 + x)$.

12.5   Nested Linear Codes

12.5.1   Random Linear Code Ensembles

In this section, we present the ensemble of nested linear codes that we will use throughout the chapter. First, let $\mathbb{F}_q$ denote a finite field with q elements. For compatibility with linear codes, we define the q-ary expansion of the messages $m \in [2^{nR}]$ by $\boldsymbol{m} \in \mathbb{F}_q^\kappa$, where $\kappa = nR/\log(q)$, and, for simplicity, we assume that R and q are such that κ is an integer. A random linear code is generated by multiplying the message with a generator matrix G and adding a dither vector $\boldsymbol{d}^n$:

$$u^n(m) = \boldsymbol{m}G \oplus \boldsymbol{d}^n, \tag{12.7}$$

where each element of G and $\boldsymbol{d}^n$ is generated independently according to the uniform distribution over $\mathbb{F}_q$.

One important property of linear codes used in this chapter is that a linear code is closed under addition. More specifically, a linear code of length n and rank κ is a linear subspace $\mathcal{C}$ with dimension κ of the vector space $\mathbb{F}_q^n$.

The marginal distribution of each codeword is

$$P\{U^n(m) = u^n\} = \prod_{i=1}^{n} p_q(u_i), \tag{12.8}$$

where $p_q = \text{Unif}(\mathbb{F}_q)$. The codewords are also pairwise independent, i.e.,

$$P\{U^n(m) = u^n, U^n(\hat{m}) = \hat{u}^n\} = \prod_{i=1}^{n} p_q(u_i)p_q(\hat{u}_i), \tag{12.9}$$

for $m \neq \hat{m}$; however, a tuple of codewords (more than two) are not independent due to their linear structure. For example, consider field size $q = 5$, and the codeword triple $(\boldsymbol{m}, 2\boldsymbol{m}, 3\boldsymbol{m})$ for any nonzero message vector $\boldsymbol{m}$. The general joint distribution of the codewords resulting from this construction can be found in [14, Theorem 1]. See Figure 12.6 for an illustration of the difference between random i.i.d. and random linear code books.

Now, consider a discrete memoryless channel (DMC) $p(y|u)$ where the inputs are from a finite field, $u \in \mathbb{F}_q$. Then, by using the linear codes (12.7), and by the pairwise independence property, it follows from standard typicality arguments that any rate R such that

$$R < I(U; Y) \tag{12.10}$$

is achievable, where $U \sim \text{Unif}(\mathbb{F}_q)$. Note that, due to the fact that each linear codeword is uniformly distributed, the input distribution in (12.10) is fixed to a uniform distribution.

12.5.2   Shaping via Joint Typicality Encoding

Channel Transformation

For general DMCs, we can introduce a mapping $x(u) \in \mathcal{X}$ to map the symbols of the linear code which are elements of $\mathbb{F}_q$ to the channel input alphabet $\mathcal{X}$. Equivalently, the

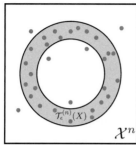 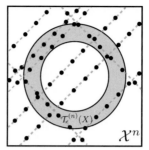

Random i.i.d. Codes   Random Linear Codes

Figure 12.6 An illustration of the typicality of random i.i.d. (gray) and random linear (black) codewords. Due to the weak law of large numbers, most random i.i.d. codewords are typical for large n. In contrast, since random linear codewords are uniformly distributed, exponentially many codewords will be atypical with respect to nonuniform distributions. This issue is resolved via *multicoding*, i.e., we generate exponentially more linear codewords than needed and use an auxiliary index to select the typical ones. ©2018 IEEE. Reprinted, with permission, from [18].

$$
\begin{array}{lllll}
p_U & U & x(u) & X & p_X
\end{array}
$$

p_U	U		X	p_X
$\frac{1}{5}$	0			
$\frac{1}{5}$	1		0	$\frac{3}{5}$
$\frac{1}{5}$	2			
$\frac{1}{5}$	3		1	$\frac{2}{5}$
$\frac{1}{5}$	4			

Figure 12.7 Shaping via channel transformation: Uniformly distributed in $\mathbb{F}_5$ is mapped to a binary alphabet $\mathcal{X}$ with probability distribution $p(x_1) = \dfrac{3}{5}$ and $p(x_2) = \dfrac{2}{5}$.

channel $p(y|x)$ with the mapping $x(u)$ is "transformed" to a DMC with channel transition probability $p(y|u) = p(y|x(u))$. The mapping function can also be used to "shape" the codewords such that the codewords are typical with respect to a desired distribution (see, for example, Figure 12.7). By choosing the field size to be arbitrarily large, it can be shown that any rate R such that

$$
R < \max_{p(x)} I(X; Y) \tag{12.11}
$$

is achievable.

Multicoding

The drawback of the channel transformation shaping is that the field size q can grow arbitrarily large. Alternatively, we can shape the input distribution for a fixed field size via multicoding. Multicoding was originally introduced for state-dependent channels with noncausal state information at the transmitters and broadcast channels [15–17]. As in Section 12.5.1, we employ q-ary random linear codes, but will generate many possible codewords for each message, and then search through these to find one that lies in the desired typical set. Specifically, we generate multiple codewords for each message using a *nested linear code*. As before, we write the q-ary expansion of the message $m \in [2^{nR}]$ as $\boldsymbol{m} \in \mathbb{F}_q^\kappa$, where $\kappa = nR/\log(q)$. Additionally, we introduce an auxiliary index $l \in [2^{n\hat{R}}]$ and its q-ary expansion $\boldsymbol{l} \in \mathbb{F}_q^{\hat{\kappa}}$, where $\hat{\kappa} = n\hat{R}/\log(q)$. Let $\tilde{R} = R + \hat{R}$ and $\tilde{\kappa} = \kappa + \hat{\kappa}$. Then, a random nested linear code is generated by multiplying $[\boldsymbol{m}, \boldsymbol{l}]$ with a generator matrix $G \in \mathbb{F}_q^{\tilde{\kappa} \times n}$ and adding a dither vector d^n:

$$u^n(m, l) = [\boldsymbol{m}, \boldsymbol{l}]G \oplus \mathsf{d}^n, \tag{12.12}$$

where

$$G = \left[\begin{array}{cccc}
g_{11} & g_{12} & \cdots & g_{1n} \\
g_{21} & g_{22} & \cdots & g_{2n} \\
\vdots & \vdots & \ddots & \vdots \\
g_{\kappa,1} & g_{\kappa,2} & \cdots & g_{\kappa,n} \\
\hline
g_{\kappa+1,1} & g_{\kappa+1,2} & \cdots & g_{\kappa+1,n} \\
g_{\kappa+2,1} & g_{\kappa+2,2} & \cdots & g_{\kappa+2,n} \\
\vdots & \vdots & \ddots & \vdots \\
g_{\kappa+\hat{\kappa},1} & g_{\kappa+\hat{\kappa},2} & \cdots & g_{\kappa+\hat{\kappa},n}
\end{array}\right],$$

and the elements of G and d^n are randomly generated uniformly and independently from $\mathbb{F}_q$. Using the nested linear codebook, we find for each message $m \in [2^{nR}]$, an index $l \in [2^{n\hat{R}}]$ such that the codeword $u^n(m, l)$ is typical with respect to p_U, i.e.,

$$u^n(m, l) \in \mathcal{T}_{\epsilon'}^{(n)}(U), \tag{12.13}$$

where the typical set is with respect to some desired distribution $p_U(u)$. Recall that by the construction given in (12.12), the codewords are uniformly distributed. Thus, we are looking for a codeword that is typical with respect to some generic distribution p_U while the codewords are actually generated from a uniform distribution. Nevertheless, by the covering lemma for linear codes [18], the probability that there exists such a codeword for each message tends to 1 as $n \to \infty$ if

$$\hat{R} > D(p_U \| p_\mathsf{q}), \tag{12.14}$$

where $D(p_U \| p_\mathsf{q})$ is the Kullback–Leibler divergence (aka relative entropy) and p_q is the uniform distribution. The end-to-end nested linear coding architecture is shown in Figure 12.8.

12.5.3   Nested Linear Coding Architecture for the MAC

In this section, we introduce the nested linear coding architecture for computing linear combinations over the MAC (see Figure 12.9 for a black diagram). In the previous

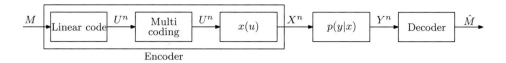

Figure 12.8 Nested linear coding architecture: Linear encoding followed by multicoding for shaping and channel transformation.

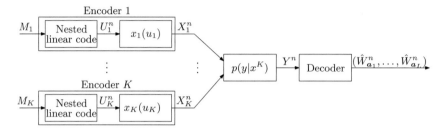

Figure 12.9 Nested linear coding architecture for computing the A linear combinations over a K-user DM-MAC. Each user selects, via multicoding, a linear codeword U_k^n of the desired type, maps it into the channel input alphabet via the function $x_k(u_k)$, and transmits it as X_k^n. The receiver observes Y^n over the DM-MAC specified by $p(y|x_1,\ldots,x_K)$ and outputs an estimate $\hat{W}_A$. Decoding is successful if $\hat{W}_A = W_A$.

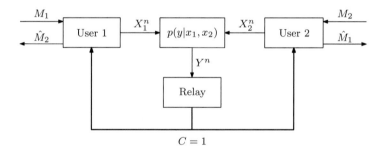

Figure 12.10 A two-way relay channel example where user 1 and 2 exchange messages through a relay node. The link to the relay node is a MAC and the relay has a noiseless link with unit capacity.

point-to-point communication case, we used a nested linear coding structure for shaping. In the MAC, we introduce an additional nesting structure between user codebooks by using the same generator matrix for all users, i.e., every codebook is a subcode of the same linear codebook. To gain some intuition on why nested linear codes can be useful for communication over networks, consider the two-way relay channel example in Figure 12.10. In this example network, we assume that users 1 and 2 exchange messages through a relay node. The link to the relay node is a MAC and the relay has a noiseless link with unit capacity to the users. Note that in this example, if the relay sends the sum of the codewords sent by user 1 and user 2, each user can subtract the message it sent from the sum to recover the other user's message. Thus, one would like to design a code such that the relay can compute the sum efficiently using a set of distinct codewords (such that the messages can be ultimately recovered from the codeword). This observation suggests the use of nested linear codes since any sum of user 1's and user 2's codewords is included in the larger codebook, which is the smallest possible sum set. This property reduces the number of competing (sum) codewords when decoding at the relay.

We continue by specifying some definitions. For $k \in [1 : K]$, user k has a message $m_k \in [2^{nR_k}]$ with q-ary expansion $\boldsymbol{m}_k \in \mathbb{F}_{\mathsf{q}}^{\kappa_k}$ where $\kappa_k = nR_k/\log(\mathsf{q})$. We will also require auxiliary indices $l_k \in [2^{n\hat{R}_k}]$, $k = 1, \ldots, K$, and similarly define their q-ary expansion by $\boldsymbol{l}_k \in \mathbb{F}_{\mathsf{q}}^{\hat{\kappa}_k}$ where $\hat{\kappa}_k = n\hat{R}_k/\log(\mathsf{q})$. Let $\tilde{R}_k := R_k + \hat{R}_k$, $R_{\max} := \max\{R_1, R_2, \ldots, R_K\}$ and $\tilde{R}_{\max} := \max\{\tilde{R}_1, \tilde{R}_2, \ldots, \tilde{R}_K\}$. For simplicity, we assume that $nR_k/\log(\mathsf{q})$ and $n\hat{R}_k/\log(\mathsf{q})$ are integers for all rates in the sequel. Further define

$$\mathsf{m}_k(m_k, l_k) = [\boldsymbol{m}_k, \boldsymbol{l}_k, \mathbf{0}], \quad k \in \mathcal{K}, \tag{12.15}$$

where $\mathsf{m}_k(m_k, l_k) \in \mathbb{F}_{\mathsf{q}}^{\kappa}$, $\kappa = n\tilde{R}_{\max}/\log(\mathsf{q})$, and $\mathbf{0}$ is a vector of zeros with length $n(\tilde{R}_{\max} - \tilde{R}_k)/\log(\mathsf{q})$. Note that all $\mathsf{m}_k(m_k, l_k)$ have the same length due to zero padding. When it is clear from the context, we will simply write the zero-padded index pair of user k as m_k. Moreover, with some abuse of notation and for simplicity, we define the set

$$[2^{n\tilde{R}_k}] := \{\mathsf{m}_k(m_k, l_k) : m_k \in [2^{nR_k}], l_k \in [2^{n\hat{R}_k}]\}. \tag{12.16}$$

We define a $(2^{nR_1}, \ldots, 2^{nR_K}, 2^{n\hat{R}_1}, \ldots, 2^{n\hat{R}_K}, n)$ *nested linear code* as the collection of K codebooks generated by the following procedure.

Codebook Generation
Fix a finite field $\mathbb{F}_{\mathsf{q}}$ and a parameter $\epsilon' \in (0, 1)$. Randomly generate a $\kappa \times n$ matrix, $\mathsf{G} \in \mathbb{F}_{\mathsf{q}}^{\kappa \times n}$, and sequences $\mathsf{d}_k^n \in \mathbb{F}_{\mathsf{q}}^n$, $k = 1, \ldots, K$ where each element of G and d_k^n is independently and randomly generated according to $\mathrm{Unif}(\mathbb{F}_{\mathsf{q}})$, and $\kappa = n\tilde{R}_{\max}/\log(\mathsf{q})$.
For each $k \in \mathcal{K}$, generate a linear code $\mathcal{C}_k$ with parameters $(R_k, \hat{R}_k, n, \mathsf{q})$ by

$$u_k^n(m_k, l_k) = \mathsf{m}_k(m_k, l_k)\mathsf{G} \oplus \mathsf{d}_k^n, \tag{12.17}$$

for $m_k \in [2^{nR_k}]$, $l_k \in [2^{n\hat{R}_k}]$. The generator matrix G and dithers d_k^n, $k \in [1 : K]$ are known at all parties prior to transmission.

As in the point-to-point case, the auxiliary indices are used for multicoding to find codewords that are typical with respect to a desired distribution.

12.6 Compute–Forward with Nested Linear Codes

We now present our compute–forward strategy for DM channels. At a high level, the encoders simply map their messages to linear codewords (using the auxiliary indices to select codewords of the desired type) and the decoder searches for linear combinations that are jointly typical with the channel output.

12.6.1   Encoding Strategy

Fix a pmf $\prod_{k=1}^{K} p(u_k)$, and functions $x_k(u_k)$, $k \in \mathcal{K}$. For $k \in \mathcal{K}$, given $m_k \in [2^{nR_k}]$, find an index $l_k \in [2^{n\hat{R}_k}]$ such that $u_k^n(m_k, l_k) \in \mathcal{T}_{\epsilon'}^{(n)}(U_k)$. If there is more than one, select one randomly and uniformly. If there is none, randomly choose an index from $[2^{n\hat{R}_k}]$. Node k transmits $x_{ki}(u_{ki})$, $i = 1, \ldots, n$.

As in the point-to-point case, the probability of successful encoding tends to 1 as $n \to \infty$ if

$$\hat{R}_k > D(p_{U_k} \| p_q) + \delta(\epsilon'), \quad k \in [1:K], \tag{12.18}$$

where $\delta(\epsilon')$ tends to zero as ϵ' tends to zero and p_q is the uniform distribution.

We denote the set of index tuples $(m_1, \ldots, m_K)$ in matrix form by

$$\mathsf{M} = \begin{bmatrix} \mathsf{m}_1 \\ \vdots \\ \mathsf{m}_K \end{bmatrix}, \quad \mathsf{m}_k \in [2^{n\tilde{R}_k}], k \in \mathcal{K} \tag{12.19}$$

and define the collections of such M matrices by the set $\mathcal{I}$. We also define the sumset of $\mathsf{M} \in \mathcal{I}$ with respect to the coefficient matrix A by

$$\mathcal{I}_{\text{sumset}}(\mathsf{A}) = \{\mathsf{M}_\mathsf{A} : \mathsf{M}_\mathsf{A} = \mathsf{A}\mathsf{M}, \mathsf{M} \in \mathcal{I}\}. \tag{12.20}$$

We incrementally develop several decoding strategies.

12.6.2   "Single-User" Decoding

In this section, we introduce a joint typicality decoder for computing a single linear combination, i.e., $\mathsf{A} = \boldsymbol{a} \in \mathbb{F}_q^{1 \times K}$. Recall the property that a linear combination of linear codes is a codeword. Thus, we can define a sum codebook

$$\mathcal{C}_{\boldsymbol{a}} = \{w_{\boldsymbol{a}}^n(\mathsf{M}_{\boldsymbol{a}}) \in \mathbb{F}_q^n : w_{\boldsymbol{a}}^n(\mathsf{M}_{\boldsymbol{a}}) = \mathsf{M}_{\boldsymbol{a}}\mathsf{G} \oplus \mathsf{d}_{\boldsymbol{a}}^n \text{ for some } \mathsf{M}_{\boldsymbol{a}} \in \mathcal{I}_{\text{sumset}}(\boldsymbol{a})\}, \tag{12.21}$$

where $\mathsf{d}_{\boldsymbol{a}}^n = \bigoplus_{k=1}^{K} a_k \mathsf{d}_k^n$ and a_k is the kth element of vector $\boldsymbol{a}$. Our decoding strategy treats the channel output as if it came from a single user with codebook $\mathcal{C}_{\boldsymbol{a}}$.

Decoding

Let $\epsilon' < \epsilon$. Upon receiving y^n, the decoder finds a unique index tuple $\mathsf{M}_{\boldsymbol{a}} \in \mathcal{I}_{\text{sumset}}(\boldsymbol{a})$ such that

$$(w_{\boldsymbol{a}}^n(\mathsf{M}_{\boldsymbol{a}}), y^n) \in \mathcal{T}_{\epsilon}^{(n)}. \tag{12.22}$$

By the Markov lemma and the joint typicality lemma [18], the probability of error tends to zero as $n \to \infty$ if

$$R_k + \hat{R}_k < I(W_{\boldsymbol{a}}; Y) + D(p_{W_{\boldsymbol{a}}} \| p_q) + \tilde{D} - \delta(\epsilon), \quad k \in \mathcal{K}(\boldsymbol{a}), \tag{12.23}$$

where $\mathcal{K}(\boldsymbol{a}) = \{k : a_k \neq 0\}$, $\tilde{D} = \sum_{k=1}^{K} D(p_{U_k} \| p_q) - \hat{R}_k$, and $\delta(\epsilon) \to 0$ as $\epsilon \to 0$. Finally, by eliminating the auxiliary rates in the conditions (12.18) and (12.23), we have the following theorem.

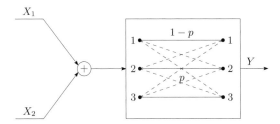

Figure 12.11 A two-user binary adder channel with symmetric noise.

THEOREM 12.3 *Consider the case of computing a single linear combination, i.e.,* $\mathsf{A} = \boldsymbol{a} \in \mathbb{F}_q^{1 \times K}$. *Then, a rate tuple* $(R_1, \ldots, R_K)$ *is achievable if*

$$R_k < H(U_k) - H(W_{\boldsymbol{a}}|Y), \quad k \in \mathcal{K}(\boldsymbol{a}) \tag{12.24}$$

for some $\prod_{k=1}^{K} p(u_k)$ *and* $x_k(u_k)$, $k \in [1 : K]$ *where* $W_{\boldsymbol{a}} = \bigoplus_{k=1}^{K} a_k U_k$.

EXAMPLE 12.4 *Consider a two-user binary MAC with* $\mathcal{X}_1 = \mathcal{X}_2 = \{0, 1\}$, $\mathcal{Y} = \{0, 1, 2\}$, *and*

$$Y = [X_1 + X_2 + Z] \bmod 3, \tag{12.25}$$

where $[x] \bmod 3$ *is the modulo-3 operation,* $p_Z(0) = 1 - p$, *and* $p_Z(1) = p_Z(2) = p/2$. *The channel can be equivalently seen as a channel with output as the sum of the channel inputs* $X_1 + X_2$ *followed by a ternary symmetric channel with crossover probabilities* $p/2$, *as shown in Figure 12.11. For computing* $\mathsf{A} = [1, 1]$ *with crossover probability* $p = 0.1$, *the rate region (12.24) is compared with the MAC capacity in Figure 12.12.*

12.6.3 Sequential Decoding

Now consider the problem of computing L linearly independent combinations, i.e., $\mathsf{A} \in \mathbb{F}_q^{L \times K}$. We can directly extend the single-user decoding strategy by using the single-user decoder L times sequentially. In the jth step, since the decoder has $W_{\mathsf{A}^{j-1}}^n$, where $\mathsf{A}^{j-1} = \mathsf{A}([1 : j - 1])$, the decoder recovers $\hat{W}_{\boldsymbol{a}_j}^n$ by treating $(Y, \hat{W}_{\mathsf{A}^{j-1}}^n)$ as the channel output (see Figure 12.13).

Sequential Decoder

The decoding is done in L steps. In the first step, the decoder finds a unique index tuple $\hat{\mathsf{M}}_{\boldsymbol{a}_1} \in \mathcal{I}_{\text{sumset}}(\boldsymbol{a}_1)$ such that

$$(w_{\boldsymbol{a}_1}^n(\hat{\mathsf{M}}_{\boldsymbol{a}_1}), y^n) \in \mathcal{T}_\epsilon^{(n)}. \tag{12.26}$$

At step $j = 2, \ldots, L$, upon having the estimates $w_{\mathsf{A}^{j-1}}^n(\hat{\mathsf{M}}_{\mathsf{A}^{j-1}})$, the decoder finds a unique index tuple $\hat{\mathsf{M}}_{\boldsymbol{a}_j} \in \mathcal{I}_{\text{sumset}}(\boldsymbol{a}_j)$ such that

$$(w_{\boldsymbol{a}_j}^n(\hat{\mathsf{M}}_{\boldsymbol{a}_j}), y^n, w_{\mathsf{A}^{j-1}}^n(\hat{\mathsf{M}}_{\mathsf{A}^{j-1}})) \in \mathcal{T}_\epsilon^{(n)}. \tag{12.27}$$

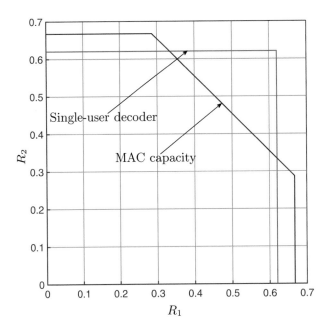

Figure 12.12 Comparison of the MAC capacity vs. computation rate (12.24) for the channel in Figure 12.11 with crossover probability $p = 0.1$.

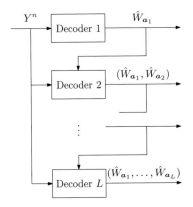

Figure 12.13 Sequential decoder for recovering multiple linear combinations.

The performance analysis follows the same steps of the single-user decoding case by treating $(y^n, w^n_{A^{j-1}})$ as the channel output. Thus, the probability of decoding error tends to zero as $n \to \infty$ if

$$R_k + \hat{R}_k < I(W_{a_j}; Y, W_{A^{j-1}}) + D(p_{W_{a_j}} \| p_q) + \tilde{D} - \delta(\epsilon), \quad k \in \mathcal{K}(a_j), \quad (12.28)$$

where $W_{a_j} = \sum_{k=1}^{K} a_{jk} U_k$, $W_{A^{j-1}} = (W_{a_1}, \ldots, W_{a_{j-1}})$, $\mathcal{K}(A) = \{k : a_{jk} \neq 0\}$ and $\tilde{D} = \sum_{k=1}^{K} D(p_{U_k} \| p_q) - \hat{R}_k$. By eliminating the auxiliary rates along with the

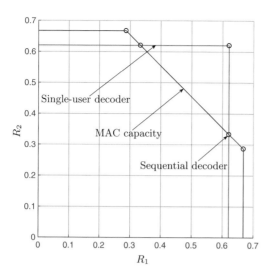

Figure 12.14 Comparison of sequential decoding points, the MAC capacity, and the computation rate (12.24) for the channel in Figure 12.11 with crossover probability $p = 0.1$.

condition (12.18), and using the sequential decoding strategy, we establish the following theorem.

THEOREM 12.5 (Sequential decoding)   *A rate tuple* $(R_1, \ldots, R_K)$ *is achievable for computing the* A-*linear combinations if for all* $1 \leq j \leq L, k \in \mathcal{K}(A_j)$,

$$R_k < H(U_k) - H(W_{a_j} | Y, W_{A^{j-1}}),\qquad(12.29)$$

for some $\prod_{k=1}^{K} p(u_k)$, $x_k(u_k)$, $k \in \mathcal{K}$ *where* a_j *is the jth row of* A, $A^j = A([1 : j])$, $\mathcal{K}(A_j) = \{k \in \mathcal{K} : A_{jk} \neq 0\}$, $W_{a_j} = \sum_{k=1}^{K} a_{jk} U_k$, *and* $W_{A^{j-1}} = (W_{a_1}, \ldots, W_{a_{j-1}})$.

REMARK 12.6   *In light of Remark 12.2, the rate region in Theorem 12.5 can be further improved by taking the union over all rate regions* (12.29) *with* $\tilde{A} \in \mathbb{F}_q^{\mathrm{rank}(\tilde{A}) \times K}$ *in place of* A *such that* $\mathrm{span}(A) \subseteq \mathrm{span}(\tilde{A})$.

EXAMPLE 12.7   We revisit Example 12.4. For computing $A = [1, 1]$ with crossover probability $p = 0.1$, the rate region in Theorem 12.5 is compared with rate region (12.24) and the MAC capacity in Figure 12.11. For simplicity, for all cases we plot the rate regions with input distributions fixed to uniform distributions.

12.6.4   Simultaneous Joint Decoding

In this subsection, we introduce a joint decoding strategy to compute the A linear combinations. As the name suggests, the decoder outputs estimates of all of the linear combinations simultaneously using a joint typicality decoding strategy.

Simultaneous Joint Typicality Decoder

Upon receiving y^n, the decoder finds a unique index tuple $\mathsf{M_A} \in \mathcal{I}_{\text{sumset}}(\mathsf{A})$ such that $\mathsf{M_A} = \mathsf{AM}$ and

$$(u_1^n(m_1, l_1), \ldots, u_K^n(m_K, l_K), y^n) \in \mathcal{T}_\epsilon^{(n)}, \qquad (12.30)$$

for some $\mathsf{M} \in \mathcal{I}$. If there is no such index tuple, or more than one, the decoder declares an error.

For simplicity, we focus on the two-user case. For the two-user case, there are two subproblems, namely, computing two linear combinations and computing a single linear combination. The following theorem establishes a joint decoding rate region for computing two linear combinations for two users using nested linear codes. We note that computing A linear combinations where $\text{rank}(\mathsf{A}) = 2$ (for the two-user case) is equivalent to recovering the individual messages. For the probability of error analysis, we refer the reader to [18].

THEOREM 12.8   *A rate pair is achievable for the DM-MAC via nested linear codes if* $(R_1, R_2) \in \mathscr{R}_{\text{LMAC}} := (\mathcal{R}_1 \cup \mathcal{R}_2)$ *for some* $p(u_1)p(u_2)$ *and* $x_1(u_1), x_2(u_2)$, *where*

$$\mathcal{R}_k = \{(R_1, R_2) : R_1 < I(X_1; Y|X_2)$$

$$R_2 < I(X_2; Y|X_1)$$

$$R_1 + R_2 < I(X_1, X_2; Y)$$

$$R_k < \min_{\mathsf{B} \in \hat{\mathbb{F}}_q^{1 \times 2}} I(U_k; Y, W_\mathsf{B})\}, \qquad (12.31)$$

and $\hat{\mathbb{F}}_q = \mathbb{F}_q \setminus \{0\}$.

For computing a single linear combination, i.e., $\text{rank}(\mathsf{A}) = 1$, the joint decoding rate region results in the following theorem.

THEOREM 12.9   *A rate pair is achievable for computing an* $\mathsf{A} \in \mathbb{F}_q^{1 \times 2}$ *linear combination if* $(R_1, R_2) \in (\mathscr{R}_{\text{LMAC}} \cup \mathscr{R}_{\text{CF}})$ *for some* $p(u_1)p(u_2)$ *and* $x_1(u_1), x_2(u_2)$, *where*

$$\mathscr{R}_{\text{CF}} = \{(R_1, R_2) : R_1 < H(U_1) - H(W_\mathsf{A}|Y),$$

$$R_2 < H(U_2) - H(W_\mathsf{A}|Y)\}, \qquad (12.32)$$

and the rate region $\mathscr{R}_{\text{LMAC}}$ *is given in Theorem 12.8.*

EXAMPLE 12.10   We revisit Example 12.4. For computing $\mathsf{A} = [1, 1]$ with crossover probability $p = 0.1$, the rate region in Theorem 12.9 is compared with the sequential decoding rate region and the MAC capacity in Figure 12.15. For simplicity, for all cases we plot the rate regions with input distributions fixed to uniform distributions.

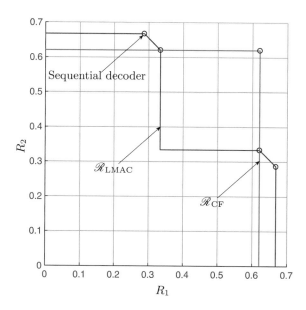

Figure 12.15 The rate region $\mathscr{R}_{\mathrm{LMAC}} \cup \mathscr{R}_{\mathrm{CF}}$ and the sequential decoding rate points for the channel in Figure 12.11 with crossover probability $p = 0.1$. The rate region $\mathscr{R}_{\mathrm{CF}}$ is the same as the single-user decoding rate region for the two-user case.

12.6.5   Case Study: Gaussian Channel

We now demonstrate that our DM compute–forward results can be extended to Gaussian networks. Consider the following Gaussian MAC:

$$Y^n = \sum_{k=1}^{K} h_k x_k^n + Z^n,$$

where the h_k are real-valued channel coefficients, the x_k^n are codewords satisfying power constraint $\sum_{i=1}^{n} (x_{ki})^2 \leq nP$, and Z^n is i.i.d. $\mathcal{N}(0, 1)$. One approach to proving an achievability result is to first establish a result for a DM network and then use a discretization argument to extend this result to Gaussian input distributions and channels. That is, we use a quantized Gaussian as our input distribution and quantize the channel output as well. Finally, we take a limit that sends the quantization bin size to zero. For instance, this approach can be used to establish the achievability of the Gaussian MAC capacity region (12.6) starting from the DM-MAC capacity region (12.5).

In a similar fashion, we can derive a Gaussian compute–forward result by combining Theorem 12.3 with a discretization argument.

THEOREM 12.11   *Consider the case of computing a single linear combination, i.e.,* $\mathsf{A} = \boldsymbol{a} = [a_1 \ \cdots \ a_K] \in \mathbb{F}_{\mathsf{q}}^{1 \times K}$ *over a Gaussian MAC with channel vector* $\mathbf{h} = [h_1 \ \cdots \ h_K]$. *Then, for* q *large enough, a rate tuple* $(R_1, \ldots, R_K)$ *is achievable if, for some*[2] $\mathbf{a} = [\mathsf{a}_1 \ \cdots \ \mathsf{a}_K] \in \mathbb{Z}^K$ *such that* $[\mathsf{a}_k] \bmod \mathsf{q} = a_k$, *we have that*

---

[2] Note that we use "script" lowercase a to refer to finite field coefficients and "Roman" lowercase a to refer to integers.

$$R_k < h(X_k) - h(V_{\mathbf{a}}|Y) + \log \gcd(\mathbf{a}), \quad k \in \mathcal{K}(\mathbf{a}) \tag{12.33}$$

$$= \frac{1}{2} \log^+ \left(\frac{P}{\sigma_{\text{eff}}^2(\mathbf{h}, \mathbf{a})} \right), \quad k \in \mathcal{K}(\mathbf{a}), \tag{12.34}$$

where $\gcd(\mathbf{a})$ is the greatest common divisor of the elements in $\mathbf{a}$, X_k are i.i.d. $\mathcal{N}(0, P)$, $V_{\mathbf{a}} = \sum_{k=1}^{K} a_k X_k$, and the effective noise variance is

$$\sigma_{\text{eff}}^2(\mathbf{h}, \mathbf{a}) := \mathbf{a}(P^{-1}\mathbf{I} + \mathbf{h}\mathbf{h}^{\mathsf{T}})^{-1}\mathbf{a}^{\mathsf{T}}. \tag{12.35}$$

We refer interested readers to [18] for the proof. The same approach can be used to obtain Gaussian versions of any of the DM compute–forward theorems. For instance, we can derive the Gaussian version of the two-user multiple-access region from Theorem 12.8.

THEOREM 12.12 *A rate pair is achievable for recovering both messages over the two-user Gaussian MAC via nested linear codes if $(R_1, R_2) \in \mathcal{R}_{\text{GLMAC}} := (\mathcal{R}_1 \cup \mathcal{R}_2)$ where*

$$\mathcal{R}_k = \{(R_1, R_2) : R_1 < I(X_1; Y|X_2)$$
$$R_2 < I(X_2; Y|X_1)$$
$$R_1 + R_2 < I(X_1, X_2; Y)$$
$$R_k < \min_{b_k \in \mathbb{Z}\setminus\{0\}} I(X_k; Y, b_1 X_1 + b_2 X_2)\}, \tag{12.36}$$

where X_1, X_2 are i.i.d. $\mathcal{N}(0, P)$.

The proof of Theorem 12.12 can be found in [18].

Building on this result and Theorem 12.9, we can derive a Gaussian compute–forward rate region for recovering a linear combination via simultaneous decoding.

THEOREM 12.13 *Consider the case of computing a single linear combination, i.e., $\mathsf{A} = \boldsymbol{a} = [a_1\ a_2] \in \mathbb{F}_{\mathsf{q}}^{1 \times 2}$ over a two-user Gaussian MAC with channel vector $\mathbf{h} = [h_1\ h_2]$. Define*

$$\mathcal{R}_{\text{GCF}} = \left\{ (R_1, R_2) : R_k < \frac{1}{2} \log^+ \left(\frac{P}{\mathbf{a}(P^{-1}\mathbf{I} + \mathbf{h}\mathbf{h}^{\mathsf{T}})^{-1}\mathbf{a}^{\mathsf{T}}} \right) \text{ for } k \in \mathcal{K}(\bar{\mathbf{a}}) \right\}, \tag{12.37}$$

where $\mathbf{a} \in \mathbb{Z}^2$ is an integer vector satisfying $\mathbf{a} = \arg\min_{\bar{\mathbf{a}} \in \mathbb{Z}^2\setminus\{0\}} \bar{\mathbf{a}}(P^{-1}\mathbf{I} + \mathbf{h}\mathbf{h}^{\mathsf{T}})^{-1}\bar{\mathbf{a}}^T$. A rate pair is achievable for computing an $\boldsymbol{a} = [\mathbf{a}] \bmod q$ linear combination over $\mathbb{F}_{\mathsf{q}}$ for q large enough if $(R_1, R_2) \in (\mathcal{R}_{\text{GLMAC}} \cup \mathcal{R}_{\text{GCF}})$.

In the next section, we will demonstrate applications of these Gaussian compute–forward theorems to wireless networks.

12.7   Applications

As discussed above, compute–forward is a framework for communicating linear combinations of messages from transmitters to receivers. However, in the vast majority

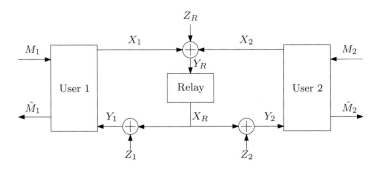

Figure 12.16 A Gaussian two-way relay channel.

of communication applications, each receiver only wishes to obtain a subset of the transmitters' messages, rather than linear combinations of them. We now explore some network topologies where the compute–forward framework provides significant advantages over conventional approaches, even though the overarching goal is to communicate messages. In particular, we consider a two-way relay channel, MAC, MIMO channel, and C-RAN.

12.7.1   Two-Way Relay Channel

The *Gaussian two-way relay channel* is a canonical example for demonstrating the performance advantages of compute–forward. In this scenario, $K = 2$ users wish to exchange messages via a relay as illustrated in Figure 12.16. We focus on the important special case of symmetric rates.

For $k = 1, 2$, user k has a message $m_k \in [2^{nR}]$ that it maps to a codeword $x_k^n(m_k) \in \mathbb{R}^n$ that satisfies an average power constraint $\sum_{i=1}^n x_{ki}^2 \leq nP$. The relay observes

$$Y_R^n = x_1^n + x_2^n + Z_R^n,$$

where Z_R^n is i.i.d. $\mathcal{N}(0, 1)$. Based on this observation, it emits its own channel input $X_R^n(Y_R^n)$, which is again subject to an average power constraint $\sum_{i=1}^n X_{Ri}^2 \leq nP$. User 1 sees a noisy version of the relay signal,

$$Y_1^n = X_R^n + Z_1^n,$$

where Z_1^n is i.i.d. $\mathcal{N}(0, 1)$ to which it assigns an estimate $\hat{m}_2$ of its desired message m_2. Similarly, user 2 observes

$$Y_2^n = X_R^n + Z_2^n,$$

where Z_2^n is i.i.d. $\mathcal{N}(0, 1)$ and produces an estimate $\hat{m}_1$ of its desired message m_1.

i.i.d. Random Coding

The best-known i.i.d. random coding strategy consists of two stages. First, the transmitters use a good Gaussian multiple-access code to communicate their messages to the relay. This succeeds with high probability so long as the (symmetric) rate satisfies

$R < \frac{1}{4} \log(1 + 2P)$. Now, assuming the relay has correctly received m_1 and m_2, it takes the sum of their q-ary expansions $\boldsymbol{m}_1 \oplus \boldsymbol{m}_2$ over some finite field $\mathbb{F}_q$ and maps the resulting message to a good point-to-point channel code. Both users can recover the relay's broadcasted message if $R < \frac{1}{2} \log(1 + P)$. Overall, any rate satisfying

$$R < \frac{1}{4} \log(1 + 2P)$$

is achievable under this strategy.

Compute–Forward

Note that, given the entirety of the messages, an optimal strategy for the relay is to compute the modulo-sum and transmit it. Therefore, it is natural to ask whether higher end-to-end rates are available by instead decoding the sum directly via compute–forward. In all of the compute–forward regions evaluated below, we implicitly assume that the relay can successfully broadcast the sum back to the users if $R < \frac{1}{2} \log(1 + P)$. Note that a simple cut-set bound shows that the rate at which the sum can be decoded at the relay is upper bounded by $\frac{1}{2} \log(1 + P)$. Therefore, the achievable rate region is entirely determined by the performance of the compute–forward scheme.

To begin, we evaluate the performance of "single-user" Gaussian compute–forward to directly decode the sum. Applying Theorem 12.11 with $K = 2$ and $\mathbf{h} = [1 \ \ 1]$, and $\mathbf{a} = [1 \ \ 1]$, the relay can recover the sum if

$$R < \frac{1}{2} \log^+ \left(\frac{1}{2} + P \right). \tag{12.38}$$

We can further improve this performance by employing simultaneous decoding. According to Theorem 12.13 with $\mathbf{h} = [1 \ \ 1]$ and $\mathbf{a} = [1 \ \ 1]$, the following rate is achievable:

$$R < \max \left\{ \frac{1}{2} \log^+ \left(\frac{1}{2} + P \right), \ \frac{1}{4} \log(1 + 2P) \right\}. \tag{12.39}$$

That is, we can attain the maximum between the i.i.d. random coding and "single-user" compute–forward rates.

The DM results can be used to evaluate the compute–forward performance for specific constellations. For instance, taking the field size to be $q = 2$, $U_k \sim \text{Bernoulli}(\frac{1}{2})$, and

$$x_k = \begin{cases} +\sqrt{P} & u_k = 0, \\ -\sqrt{P} & u_k = 1, \end{cases}$$

in Theorem 12.3, we find that the following rate is achievable for binary phase shift keying (BPSK) compute–forward:

$$R < 1 - H(U_1 \oplus U_2 | Y). \tag{12.40}$$

These achievable rate expressions are plotted in Figure 12.17 along with the cut-set upper bound of $\frac{1}{2} \log(1 + P)$.

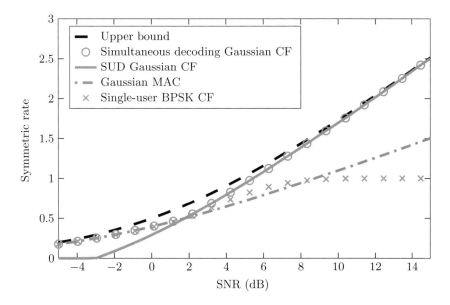

Figure 12.17 Symmetric rates for the Gaussian two-way relay channel.

12.7.2   Compute–Forward Multiple–Access

Compute–forward can be used as a building block for multiple-access communication. That is, if the transmitters employ nested linear (or lattice) codebooks, the receiver can first recover K linearly independent linear combinations, and then solve these for the K desired messages. If the linear combinations are recovered via "single-user" decoding, then this approach can be realized with modest implementation complexity as compared to simultaneous decoding.

In general, this approach may not yield the full MAC capacity region (without the aid of time-sharing). For instance, in Figure 12.11, the $\mathscr{R}_{\mathrm{LMAC}}$ region is the multiple-access performance of compute–forward with simultaneous decoding and uniform input distributions for the channel from Example 12.4. Clearly, this region has a "hole" as compared to the multiple-access rate region available to i.i.d. random coding with uniform inputs. Interestingly, for the two-user case, it can be shown that this hole corresponds exactly to a reflection of the maximum "single-user" computation rate across the multiple-access sum-rate line. That is, if a MAC is well-suited for computation, then it is poorly suited for multiple access via nested linear codes.

Additionally, it should be noted that sequential compute–forward can attain the sum capacity. These include the usual corner points associated with successive interference cancellation (corresponding to taking A as the identity matrix) and often also include points on the interior of the sum-capacity face. For example, in Figure 12.11, there are four distinct rate pairs on the sum-rate boundary that can be attained by sequential compute–forward.

The linear structure of the Gaussian MAC makes it possible for compute–forward multiple-access to always operate within a constant gap of the sum-rate.

LEMMA 12.14   *Consider a Gaussian MAC with channel vector* $\mathbf{h} = [h_1 \cdots h_K]$. *There always exists a choice of full-rank integer matrix* $\mathbf{A} \in \mathbb{Z}^{K \times K}$ *with rows* $\mathbf{a}_1, \ldots, \mathbf{a}_K$ *such that*

$$\sum_{k=1}^{K} \frac{1}{2} \log^+ \left(\frac{P}{\sigma_{\text{eff}}^2(\mathbf{h}, \mathbf{a}_k)} \right) \geq \frac{1}{2} \log \left(1 + P \|\mathbf{h}\|^2 \right) - \frac{K}{2} \log K . \tag{12.41}$$

Note that the effective noise variance formula (12.35) can be written as the squared Euclidean norm of an element of the K-dimensional lattice $\Lambda_{\text{channel}} = \left(P^{-1}\mathbf{I} + \mathbf{h}\mathbf{h}^{\mathsf{T}} \right)^{-1/2} \mathbb{Z}^K$. It can be shown that the problem of finding K linearly independent integer vectors to maximize the sum-rate is equivalent to finding the successive minima of Λ_{channel}. The result follows by applying Minkowski's second theorem on successive minima.

On its own, Lemma 12.14 does not have an operational meaning, i.e., a decoding technique is needed to link each user's rate to a unique effective noise variance. Clearly, "single-user" decoding does not suffice on its own, since each user's rate would be set by the minimum, i.e., $R_k < \min_k \frac{1}{2} \log^+ \left(P/\sigma_{\text{eff}}^2(\mathbf{h}, \mathbf{a}_k) \right)$. Instead, it can be shown via sequential decoding that $R_{\pi(k)} < \frac{1}{2} \log^+ \left(P/\sigma_{\text{eff}}^2(\mathbf{h}, \mathbf{a}_k) \right)$ is achievable for some permutation π. Therefore, sequential compute–forward can always operate within a constant gap of the Gaussian MAC sum capacity. (This performance can also be attained via a lattice-based compute–forward scheme that only requires digital, rather than analog, successive cancellation. See [19] for more on this scheme.)

In fact, sequential compute–forward can often attain a higher sum-rate, since recovered linear combinations can be used to further reduce the effective noise variance. Under certain technical conditions, it can be shown that it attains the exact sum capacity. See [20] for further details.

In Figure 12.18, we have evaluated the performance of a compute–forward multiple-access strategy over the two-user Gaussian MAC

$$Y^n = x_1^n + h x_2^n + Z^n,$$

where h is a real-valued channel gain and Z^n is i.i.d. $\mathcal{N}(0, 1)$. In particular, the first and second computation rates are

$$R_{\text{comp},1} = \max_{\substack{\bar{\mathbf{a}}_1 \in \mathbb{Z}^2 \\ \bar{\mathbf{a}}_1 \neq \mathbf{0}}} \frac{1}{2} \log^+ \left(\frac{P}{\sigma_{\text{eff}}^2([1 \; h], \bar{\mathbf{a}}_1)} \right) \tag{12.42}$$

$$R_{\text{comp},2} = \max_{\substack{\bar{\mathbf{a}}_2 \in \mathbb{Z}^2 \\ \text{rank}(\mathbf{A})=2}} \frac{1}{2} \log^+ \left(\frac{P}{\sigma_{\text{eff}}^2([1 \; h], \bar{\mathbf{a}}_2)} \right) \tag{12.43}$$

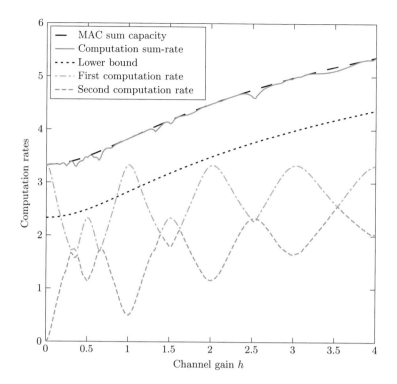

Figure 12.18 Computation rates for the two-user Gaussian MAC $Y^n = x_1^n + h x_2^n + Z^n$, the sum capacity, and the achievable compute–forward multiple-access sum-rate as well as its lower bound from Lemma 12.14.

where $\mathbf{A} = \begin{bmatrix} \mathbf{a}_1 \\ \mathbf{a}_2 \end{bmatrix}$ and $\mathbf{a}_1$ is the integer vector chosen to maximize $R_{\mathrm{comp},1}$. The computation sum-rate is $R_{\mathrm{comp},1} + R_{\mathrm{comp},2}$, the MAC sum capacity is $\frac{1}{2} \log(1 + h^2 P)$, and the lower bound is $\frac{1}{2} \log(1 + h^2) - 1$ (from Lemma 12.14). From the figure, we can observe that although the individual computation rates vary significantly with respect to the channel gain h, the sum-rate is relatively stable, and lies quite close to the MAC sum capacity.

An alternative perspective on compute–forward multiple-access is developed in [21], where it is assumed that the channel gains of the Gaussian MAC are fixed and known throughout. In that scenario, one can fine-tune the lattice code construction to the channel parameters, essentially by scaling them appropriately, separately at each encoder. In that case, it can be shown [21] for the two-user Gaussian MAC that compute–forward multiple-access can attain any point in the full capacity region as soon as the per-user power P exceeds the threshold characterized by

$$\frac{h_1 h_2 P}{\sqrt{1 + h_1^2 P + h_2^2 P}} \geq 1. \tag{12.44}$$

12.7.3   Integer-Forcing

Modern wireless systems often utilize multiple antennas to attain higher throughputs. One resulting challenge is the design of architectures that can approach the capacity of MIMO channels at manageable implementation complexity. That is, while it is now known how to design efficient encoding and decoding algorithms to approach the Gaussian point-to-point capacity in practice (e.g., low-density parity check [LDPC], polar codes), there is still much room to improve upon conventional MIMO transceiver architectures.

One approach is to presume the availability of good "single-user" encoding and decoding algorithms, and design architectures that reduce the MIMO channel into several effective single-user channels. For instance, conventional linear architectures employ linear equalization to mitigate the interference between codewords followed by single-user decoding that treats interference as noise. Below, we use "single-user" compute–forward to develop *integer-forcing* architectures for uplink and downlink scenarios.

Uplink Channels

Consider K single-antenna users that wish to communicate with a base station that is equipped with N antennas, i.e., a MIMO MAC. The channel output observed at the base station can be expressed as

$$\mathbf{Y} = \mathbf{HX} + \mathbf{Z}, \tag{12.45}$$

where $\mathbf{H} \in \mathbb{R}^{N \times K}$ is the channel matrix, the kth row of $\mathbf{X} \in \mathbb{R}^{K \times n}$ is the codeword x_k^n emitted by the kth user, and $\mathbf{Z} \in \mathbb{R}^{N \times n}$ is elementwise i.i.d. $\mathcal{N}(0, 1)$. Each codeword is subject to an average power constraint $\sum_{i=1}^{n} x_{ki}^2 \leq nP$ and we assume the rates to be symmetric, i.e., each user's message takes values in $[2^{nR}]$.

REMARK 12.15   *Note that the real-valued setup above can be applied to a complex-valued channel* $\mathbf{Y}_C = \mathbf{H}_C \mathbf{X}_C + \mathbf{Z}_C$ *by using the real-valued decomposition,*

$$\begin{bmatrix} \mathrm{Re}(\mathbf{Y}_C) \\ \mathrm{Im}(\mathbf{Y}_C) \end{bmatrix} = \begin{bmatrix} \mathrm{Re}(\mathbf{H}_C) & -\mathrm{Im}(\mathbf{H}_C) \\ \mathrm{Im}(\mathbf{H}_C) & \mathrm{Re}(\mathbf{H}_C) \end{bmatrix} \begin{bmatrix} \mathrm{Re}(\mathbf{X}_C) \\ \mathrm{Im}(\mathbf{X}_C) \end{bmatrix} + \begin{bmatrix} \mathrm{Re}(\mathbf{Z}_C) \\ \mathrm{Im}(\mathbf{Z}_C) \end{bmatrix},$$

to obtain an effective real-valued channel.

We focus on the important special case where channel state information (CSI) is only available at the receiver, meaning that we need to allow for some probability of outage. Specifically, let $R_{\mathrm{scheme}}(\mathbf{H})$ be the symmetric rate that is achievable for a coding scheme under channel matrix $\mathbf{H}$. The outage probability is simply the probability that the targeted rate exceeds the available rate, $p_{\mathrm{outage}} = \mathrm{P}\{R_{\mathrm{scheme}}(\mathbf{H}) < R\}$. For a fixed outage probability p, we define the symmetric outage rate as $R_{\mathrm{outage}}(p) = \sup\{R : p_{\mathrm{outage}}(R) \leq p\}$.

The symmetric outage capacity is given by

$$R_{\mathrm{MIMO\text{-}MAC}}(\mathbf{H}) = \min_{\mathcal{S} \subseteq [1:K]} \frac{1}{2|\mathcal{S}|} \log \det \left(\mathbf{I} + P\mathbf{H}_{\mathcal{S}}\mathbf{H}_{\mathcal{S}}^{\mathsf{T}}\right), \tag{12.46}$$

where $\mathbf{H}_{\mathcal{S}}$ is the submatrix of $\mathbf{H}$ consisting of the columns indexed by $\mathcal{S}$. This can be derived using i.i.d. Gaussian encoding and simultaneous joint typicality decoding.

As mentioned above, we would like to reduce the MIMO channel into a collection of effective single-user channels. A natural solution is to reduce the interference between individual codewords via linear equalization. Specifically, we form the effective channels

$$\tilde{y}_k^n = \mathbf{b}_k \mathbf{Y} = \underbrace{\mathbf{b}_k \mathbf{h}_k^{\mathsf{T}} x_k^n}_{\text{signal}} + \underbrace{\sum_{j \neq k} \mathbf{b}_k \mathbf{h}_j^{\mathsf{T}} x_j^n}_{\text{interference}} + \underbrace{\mathbf{b}_k \mathbf{Z}}_{\text{noise}}, \quad (12.47)$$

where $\mathbf{b}_k \in \mathbb{R}^N$ is the kth equalization vector and $\mathbf{h}_k^{\mathsf{T}}$ is the kth column of the channel matrix $\mathbf{H}$. A single-user decoder is then employed to recover x_k^n from $\tilde{y}_k^n$. The resulting symmetric rate for this linear receiver is

$$R_{\text{linear}}(\mathbf{H}) = \min_{k \in [1:K]} \max_{\mathbf{b}_k} \frac{1}{2} \log \left(\frac{P(\mathbf{b}_k \mathbf{h}_k^{\mathsf{T}})^2}{\|\mathbf{b}_k\|^2 + P \sum_{j \neq k} (\mathbf{b}_k \mathbf{h}_j^{\mathsf{T}})^2} \right). \quad (12.48)$$

The optimal equalization vectors are simply the MMSE projections

$$\mathbf{b}_{\text{MMSE},k} = P \mathbf{h}_k (\mathbf{I} + P \mathbf{H} \mathbf{H}^{\mathsf{T}})^{-1}, \quad (12.49)$$

and plugging these in we obtain the performance of the MMSE linear receiver.

The performance of a linear receiver can be further improved via sequential decoding, which is often referred to as successive interference cancellation (SIC) in the MIMO literature. The receiver recovers the codewords according to some permutation π of $[1:K]$. After each decoding step, it removes the contribution of the recovered codeword from the subsequent effective channels. Thus, the symmetric rate is

$$R_{\text{SIC}}(\mathbf{H}) = \max_{\pi} \min_{k \in [1:K]} \max_{\mathbf{b}_k} \frac{1}{2} \log \left(\frac{P(\mathbf{b}_k \mathbf{h}_{\pi(k)}^{\mathsf{T}})^2}{\|\mathbf{b}_k\|^2 + P \sum_{j > k} (\mathbf{b}_k \mathbf{h}_{\pi(j)}^{\mathsf{T}})^2} \right). \quad (12.50)$$

The optimal equalization vectors are again given by MMSE projections

$$\mathbf{b}_{\text{MMSE-SIC},k} = P \mathbf{h}_{\pi(k)} (\mathbf{I} + P \tilde{\mathbf{H}}_k \tilde{\mathbf{H}}_k^{\mathsf{T}})^{-1}, \quad (12.51)$$

where $\tilde{H}_k$ consists of the columns $\pi(1), \ldots, \pi(k)$ of $\mathbf{H}$. The performance of the MMSE-SIC linear receiver follows from plugging in these vectors.

The integer-forcing linear receiver uses the linear equalization step for a different purpose: to create favorable effective channels for compute–forward. That is, for a given full-rank integer matrix $\mathbf{A} \in \mathbb{Z}^{K \times K}$, we form the effective channels

$$\tilde{y}_k^n = \mathbf{b}_k \mathbf{Y} = \underbrace{\mathbf{a}_k \mathbf{X}}_{\text{integer combination}} + \underbrace{(\mathbf{b}_k \mathbf{H} - \mathbf{a}_k) \mathbf{X} + \mathbf{b}_k \mathbf{Z}}_{\text{effective noise}} \quad (12.52)$$

where $\mathbf{a}_k$ is the kth row of $\mathbf{A}$. The goal of the receiver is to recover the integer-linear combinations $\mathbf{A}\mathbf{X}$ and then solve these for the desired codewords as shown in Figure 12.19. Specifically, each integer-linear combination $\mathbf{a}_k \mathbf{X}$ corresponds to a linear combination

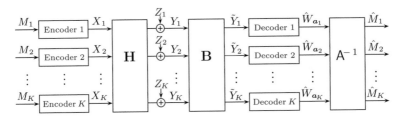

Figure 12.19 Integer-forcing receiver for the MIMO uplink channel.

W_{a_k}, and the inverse matrix $\mathbb{A}^{-1} \in \mathbb{F}_q^{K \times K}$ for $A = [\mathbf{A}] \bmod q$ can be applied to these linear combinations to recover the q-ary representation of the messages.

Optimizing over all full-rank integer matrices $\mathbf{A} \in \mathbb{Z}^{K \times K}$ and equalization vectors, it follows from our single-user compute–forward result (Theorem 12.11) that the integer-forcing symmetric rate is

$$R_{\mathrm{IF}}(\mathbf{H}) = \max_{\mathbf{A}} \min_{k \in [1:K]} \max_{\mathbf{b}_k} \frac{1}{2} \log\left(\frac{P}{\|\mathbf{b}_k\|^2 + P\|\mathbf{b}_k\mathbf{H} - \mathbf{a}_k\|^2}\right). \tag{12.53}$$

Plugging in the optimal MMSE equalization vectors

$$\mathbf{b}_{\mathrm{IF},k} = P\,\mathbf{a}_k\mathbf{H}^{\mathsf{T}}\left(\mathbf{I} + P\mathbf{H}\mathbf{H}^{\mathsf{T}}\right)^{-1}, \tag{12.54}$$

and applying Woodbury's matrix identity, we can express (12.53) as

$$R_{\mathrm{IF}}(\mathbf{H}) = \max_{\mathbf{A}} \min_{k \in [1:K]} \frac{1}{2} \log\left(\frac{P}{\sigma_{\mathrm{eff}}^2(\mathbf{H}, \mathbf{a}_k)}\right) \tag{12.55}$$

$$\sigma_{\mathrm{eff}}^2(\mathbf{H}, \mathbf{a}_k) = \mathbf{a}_k\left(P^{-1}\mathbf{I} + \mathbf{H}^{\mathsf{T}}\mathbf{H}\right)^{-1}\mathbf{a}_k^{\mathsf{T}}. \tag{12.56}$$

The problem of finding the optimal $\mathbf{A}$ is equivalent to finding the shortest basis vectors for the lattice generated by $\left(P^{-1}\mathbf{I} + \mathbf{H}^{\mathsf{T}}\mathbf{H}\right)^{-1/2}$. While this is a challenging optimization problem in general, the LLL algorithm serves as a good approximation and runs in polynomial time in K [22].

In Figure 12.20, we have plotted the symmetric outage rates for the MMSE, MMSE-SIC, and integer-forcing, as well as the symmetric outage capacity for a MIMO uplink channel with $K = 3$ users and $N = 2$ receive antennas under i.i.d. Rayleigh fading at $p_{\mathrm{outage}} = 0.05$. The MMSE receiver performs poorly since $K > N$, i.e., it is impossible to fully separate the users via linear equalization. The MMSE-SIC receiver performs quite well at the lower SNR range, but falls away from the capacity at higher SNR values. In contrast, the integer-forcing receiver is able to operate near the capacity for the entire plotted SNR range.

Integer-forcing also scales well as the number of users increases. In Figure 12.21, we have plotted the symmetric outage rates of integer-forcing and the MMSE linear receiver for $N = K$ with respect to increasing K at SNR $= 20$ dB and $p_{\mathrm{outage}} = 0.01$. We have also plotted the upper bound $\frac{1}{2K} \log \det \left(\mathbf{I} + P\mathbf{H}\mathbf{H}^{\mathsf{T}}\right)$ on the symmetric outage capacity.

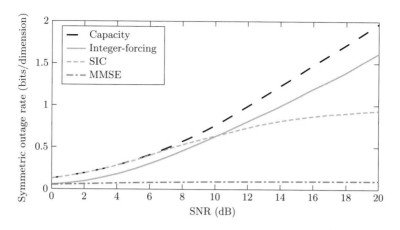

Figure 12.20 Five percent symmetric outage rate for an uplink MIMO channel with $K = 3$ single-antenna users and $N = 2$ receive antennas under i.i.d. Rayleigh fading.

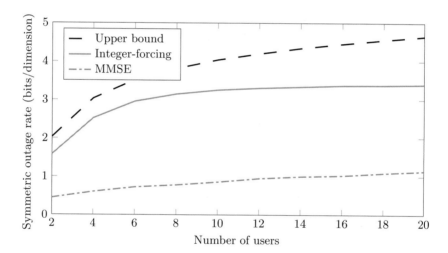

Figure 12.21 Five percent symmetric outage rate for a MIMO uplink channel with K single-antenna users communicating and a K-antenna receiver under i.i.d. Rayleigh fading.

Downlink Channels

Now we turn to the downlink channel where an N-antenna transmitter communicates with K single-antenna users, i.e., a MIMO broadcast channel. The channel output seen at the kth user is

$$Y_k^n = \mathbf{h}_k \mathbf{X} + Z_k^n, \tag{12.57}$$

where $\mathbf{h}_k \in \mathbb{R}^N$ is the channel vector, $\mathbf{X} \in \mathbb{R}^{N \times n}$ is the channel input subject to the total power constraint $\mathrm{Tr}(\mathbf{X}\mathbf{X}^\mathsf{T}) \leq P$, and Z_k^n is i.i.d. $\mathcal{N}(0, 1)$ noise. The channel is known at the transmitter and receivers (i.e., full CSI). The transmitter's goal is to transmit the message $m_k \in [2^{nR_k}]$ for $k \in [1 : K]$

The downlink capacity region is attained via dirty-paper coding and joint typicality decoding and can be found in [4, Section 9.6.4]. An alternative, suboptimal approach is to employ a single-user encoder to generate a codeword $\bar{x}_k^n$ to carry the message m_k for $k \in [1 : K]$. Let $\bar{\mathbf{X}}$ be the matrix whose kth row is $\bar{x}_k^n$. The transmitted signal is generated by multiplying this matrix of codewords by a beamforming matrix $\mathbf{B} \in \mathbb{R}^{N \times K}$,

$$\mathbf{X} = \mathbf{B}\bar{\mathbf{X}}. \tag{12.58}$$

The kth user thus observes

$$Y_k^n = \mathbf{h}_k \mathbf{B}\bar{\mathbf{X}} + Z_k^n, \tag{12.59}$$

from which it attempts to decode $\bar{x}_k^n$ while treating the other codewords as noise. Note that the role of $\mathbf{B}$ is to attempt to orthogonalize the users. Overall, this strategy is often referred to as zero-forcing beamforming.

Another approach is to use the beamforming matrix to steer the channel towards integer values and have each user recover a linear combination of codewords with overall coefficient matrix $\mathsf{A} \in \mathbb{F}_q^{K \times K}$. While this will produce effective channels with better SNRs, the goal of each receiver is to recover a specific message, rather than a linear combination. Thus, we need the additional step of pre-inverting the linear combinations at the transmitter to complete this integer-forcing beamforming strategy [23]. Specifically, as illustrated in Figure 12.22, the transmitter applies the inverse matrix A^{-1} (with the inverse taken over $\mathbb{F}_q$) to the q-ary representations of the messages, encodes the resulting linear combinations to obtain the codewords $\bar{x}_k^n$ and then applies linear equalization. The effective channel at the kth receiver can be expressed as

$$Y_k^n = \mathbf{a}_k \bar{\mathbf{X}} + (\mathbf{h}_k \mathbf{B} - \mathbf{a}_k)\bar{\mathbf{X}} + Z_k^n \tag{12.60}$$

for some $\mathbf{a}_k \in \mathbb{Z}^K$ such that $[\mathbf{a}_k] \bmod q = \mathbf{a}_k$. From here, the receiver can apply a single-user compute–forward decoder to obtain a linear combination, which, due to the pre-inversion step, corresponds to its desired message.

In Figure 12.23, we have plotted the average capacity under i.i.d. Rayleigh fading as well as the performance of zero-forcing and integer-forcing receivers for $K = 4$ users and $N = 2$ transmit antennas. The beamforming and equalization matrices are optimized via an iterative algorithm that relies on uplink–downlink duality [24]. Since the number of users exceeds the number of transmit antennas, zero-forcing cannot invert the channel and thus performs poorly, whereas integer-forcing closely tracks the sum capacity.

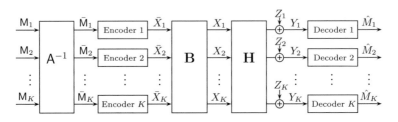

Figure 12.22 Integer-forcing beamforming for the MIMO downlink channel.

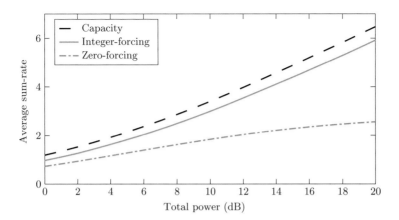

Figure 12.23 Average sum-rate for a downlink MIMO channel with $N = 2$ transmit antennas and $K = 4$ single-antenna users under i.i.d. Rayleigh fading.

12.7.4 Cloud-Radio Access Networks

Cloud radio access networks are an emerging architecture for cellular networks. For instance, in an uplink C-RAN network, neighboring base stations cooperate to decode by forwarding some version of their received signals to a central processor. For example, a natural compress–forward strategy is for each base station to quantize their received signal and forward it to the central processor, which reconstructs an effective MIMO MAC channel and decodes. The compute–forward approach is for each base station to decode a linear combination of the codewords and send it to the central processor, which then solves for the original messages.

As demonstrated in Figure 12.24, compress–forward and compute–forward both offer advantages in different regimes. Specifically, this figure considers an uplink scenario with two users that communicate to two base stations,

$$Y_1^n = h_{1,1}x_1^n + h_{1,2}x_2^n + Z_1^n \tag{12.61}$$
$$Y_2^n = h_{2,1}x_1^n + h_{2,2}x_2^n + Z_2^n, \tag{12.62}$$

where the $h_{j,k}$ are real-valued channel gains, x_k^n is the codeword emitted by the kth user (subject to power constraint P), and Z_j^n is i.i.d. $\mathcal{N}(0, 1)$ noise. Each base station is connected to the central processor with a bit pipe with capacity $C = 2$ bits/channel use. We focus on the special case of symmetric rates and assume the channel gains are known at all terminals.

The strategies plotted include the cut-set upper bound (which is taken to be the minimum of C and (12.46)), single-user Gaussian compute–forward, simultaneous decode–forward, and treating interference as noise (i.e., single-user decode–forward). In the compute–forward strategy, we optimize over all linearly independent integer vectors. The single-user decode–forward strategy has each base station recover a single message while treating the other as noise. The simultaneous decode–forward strategy includes the additional possibility of joint "multiple-access" decoding of both messages

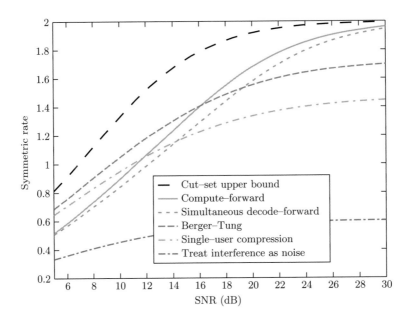

Figure 12.24 Average symmetric rates for an uplink C-RAN with two users, two base stations, and $C = 2$ bits/channel use/base station for the backhaul rate.

at each base station. For both decode–forward strategies, we optimize over which base station recovers which messages. The Berger–Tung compress–forward strategy employs a quantize-bin strategy (here with a symmetric distortion target) whereas single-user compression simply treats the base station's observation as an i.i.d. Gaussian source, ignoring any potential correlations between the base stations.

12.8   Practical Codes

In this section, we briefly introduce practical code design examples for compute–forward.

12.8.1   Compute–Forward Multiple-Access via Low-Density Parity-Check Codes

Historically, theoretical results based on i.i.d. random coding techniques have inspired many modern communication systems. These information theoretic limits are approached in practice by employing linear codes as an efficient coding strategy. In this sense, the compute–forward results discussed in this chapter with nested linear codes are already one step closer to practice.

Many state-of-the-art communication standards utilize a combination of linear codes and quadrature amplitude modulation (QAM). In particular, LDPCs are a class of linear codes with near-optimal performance for point-to-point Gaussian channels [25].

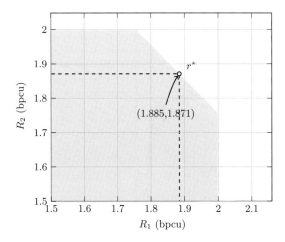

Figure 12.25 Target rate pair $r^\star$ for the 4-QAM modulation case. The achievable rate pair $r^\star$ of the CFMA strategy achieves a non-corner point on the dominant face of $\mathcal{R}_{\mathsf{MAC-UI}}$. ©2019 IEEE. Reprinted, with permission, from [26].

In [26], the authors introduce a low-complexity compute–forward multiple-access strategy based on nested LDPC codes. The strategy uses binary off-the-shelf LDPC codes designed for point-to-point Gaussian channels and 4-QAM modulation over a complex Gaussian MAC. The encoder does not use multicoding, which results in shaping loss, although this is often a reasonable compromise for low complexity in practice.

As an example of this scheme, we consider a MAC with channel gains $h_1 = 1, h_2 = 1$, and the rate region (12.5) with discrete uniform QAM inputs which we will denote by $\mathcal{R}_{\mathsf{MAC-UI}}$. The simplified nested LDPC compute–forward strategy is evaluated for the reference point $r^\star$ in Figure 12.25, corresponding to a target rate pair of $(1.885, 1.871)$. Decoding consists of two stages using a modified version of the sum-product algorithm. The first stage recovers the sum of the codewords, and the second stage recovers one of the messages based on the observation Y^n and the sum of codewords recovered in the previous stage. Due to the use of sum-product algorithms and sequential decoding, the complexity of the strategy is roughly the same order as that of a point-to-point LDPC decoder. Figure 12.26 depicts the performance evaluation of the practical codes compared with the theoretical bound. For reference, the base LDPC code performance over a point-to-point Gaussian channel is also given in the figure. Overall, this architecture can serve as a low-complexity multiple-access strategy that can closely approach a point on the dominant face of the rate region $\mathcal{R}_{\mathsf{MAC-UI}}$.

12.8.2   Software-Defined Radio

An important step toward practical deployment of compute–forward strategies is to evaluate their performance in realistic settings. To this end, a series of software-defined radio experiments were performed in [27] using WARP boards to compare the performance

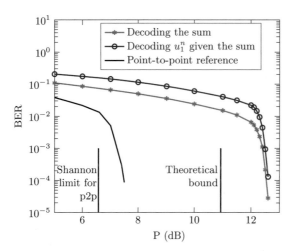

Figure 12.26 Bit error rate (BER) simulation results for each decoding step for 4-QAM modulation and target rate pair $(R_1, R_2) = (1.885, 1.871)$. The base code performance over the AWGN point-to-point channel is included for reference. ©2019 IEEE. Reprinted, with permission, from [26].

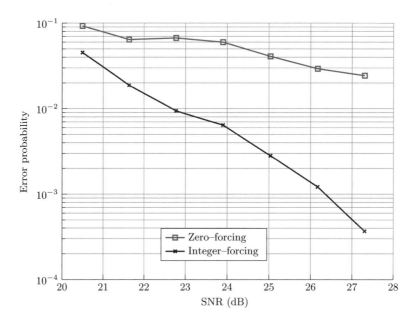

Figure 12.27 Error probability (averaged over codewords) for two-user uplink integer-forcing on a WARP software-defined radio.

of integer-forcing with zero-forcing. In Figure 12.27, we can see that, in experiments, integer-forcing indeed has a significant advantage over zero-forcing.

The details of this experiment are as follows. The transmitter and receiver are 6 m apart and each have two antennas, which are spaced 40 cm apart. They use orthogonal frequency division multiplexing (OFDM) signaling with 48 data subcarriers at carrier frequency 2.4 GHz. Each subcarrier has an independent codeword of length 500 generated from a regular $(3, 6)$ LDPC code over $\mathbb{F}_3$. Modulation is via a 9-QAM constellation and 3900 frames are sent in total for the experiment. The error probability is the number of codewords that are incorrect across all frames and subcarriers.

12.8.3   Historical Notes

The significance and usefulness of codes with algebraic structure for problems in network information theory was first observed by Körner and Marton in [28] for a source coding problem. The extension of this insight to channel coding problems appeared in [29], and to Gaussian channels through the lens of lattice codes in [5–7]. Since then, many other applications of algebraic network information theory have been uncovered, beyond the discussion in this chapter, including distributed source coding [30–36], dirty-paper MACs [37–40], broadcast channels [41], interference channels [19, 42–47], and physical-layer secrecy [48–50].

This book chapter specifically focuses on compute–forward, which was first studied using lattice codes [1, 20]. The focus of the present chapter is on a different way of implementing compute–forward, namely, via linear codes over finite fields. A key ingredient is the multicoding for nested linear codes, which was studied in [51, 52]. Compute–forward for discrete memoryless networks was developed in [18, 52–54].

Among the many applications of compute–forward that have been studied, there are integer-forcing strategies for MIMO transceivers in the uplink [11] and downlink [23], as well as a corresponding duality result [24]; there are also integer-forcing strategies for C-RANs [55, 56]; there is the Gaussian two-way relay channel [57, 58]; and more recently, the development of compute–forward multiple-access [19, 21, 59, 60]. Finally, practical overall communication strategies for compute–forward have been studied extensively, most recently in [26, 27].

References

[1] B. Nazer and M. Gastpar, "Compute-and-forward: Harnessing interference through structured codes," *IEEE Trans. Inf. Theory*, vol. 57, no. 10, pp. 6463–6486, Oct. 2011.

[2] D. J. Costello and G. D. Forney, "Channel coding: The road to channel capacity," *Proceedings of the IEEE*, vol. 95, no. 6, pp. 1150–1177, 2007.

[3] T. M. Cover and J. A. Thomas, *Elements of Information Theory*, 2nd ed. New York: Wiley, 2006.

[4] A. El Gamal and Y.-H. Kim, *Network Information Theory*. Cambridge: Cambridge University Press, 2011.

[5] K. Narayanan, M. P. Wilson, and A. Sprintson, "Joint physical layer coding and network coding for bi-directional relaying," in *Proc. 45th Ann. Allerton Conf. Comm. Control Comput.*, Monticello, IL, Sep. 2007, pp. 5641–5654.

[6] B. Nazer and M. Gastpar, "Lattice coding increases multicast rates for Gaussian multiple-access networks," in *Proc. 45th Ann. Allerton Conf. Comm. Control Comput.*, Monticello, IL, Sep. 2007, pp. 1089–1096.

[7] W. Nam, S.-Y. Chung, and Y. H. Lee, "Capacity bounds for two-way relay channels," in *Proc. Int. Zurich Seminar on Comm.*, Zurich, Mar. 2008.

[8] China Mobile, "C-RAN: The road towards green RAN," White paper, ver. 2.5, China Mobile Research Institute, 2011.

[9] Intel Corp., "Intel heterogeneous network solution brief," Solution brief, Intel core processor, Telecommunications industry.

[10] S.-H. Park, O. Simeone, O. Sahin, and S. Shamai, "Robust and efficient distributed compression for cloud radio access networks," *IEEE. Trans. Veh. Tech.*, vol. 62, no. 2, pp. 692–703, Feb. 2013.

[11] J. Zhan, B. Nazer, U. Erez, and M. Gastpar, "Integer-forcing linear receivers," *IEEE Trans. Inf. Theory*, vol. 55, no. 12, pp. 7661–7685, Dec. 2014.

[12] R. Zamir, *Lattice Coding for Signals and Networks*. Cambridge: Cambridge University Press, 2014.

[13] B. Nazer and M. Gastpar, "Reliable physical layer network coding," *Proc. IEEE*, vol. 99, no. 3, pp. 438–460, Mar. 2011.

[14] Y. Domb, R. Zamir, and M. Feder, "The random coding bound is tight for the average linear code or lattice," *IEEE Trans. Inf. Theory*, vol. 62, no. 1, pp. 121–130, Jan. 2016.

[15] S. I. Gelfand and M. S. Pinsker, "Coding for channel with random parameters," *Probl. Control Inf. Theory*, vol. 9, no. 1, pp. 19–31, 1980.

[16] S. I. Gelfand and M. S. Pinsker, "Capacity of a broadcast channel with one deterministic component," *Probl. Inf. Transm.*, vol. 16, no. 1, pp. 24–34, 1980.

[17] K. Marton, "A coding theorem for the discrete memoryless broadcast channel," *IEEE Trans. Inf. Theory*, vol. 25, no. 3, pp. 306–311, 1979.

[18] S. H. Lim, C. Feng, A. Pastore, B. Nazer, and M. Gastpar, "A joint typicality approach to compute–forward," *IEEE Trans. Inf. Theory*, vol. 64, no. 12, pp. 7657–7685, 2018.

[19] O. Ordentlich, U. Erez, and B. Nazer, "The approximate sum capacity of the symmetric Gaussian-user interference channel," *IEEE Trans. Inf. Theory*, vol. 60, no. 6, pp. 3450–3482, Jun. 2014.

[20] B. Nazer, V. Cadambe, V. Ntranos, and G. Caire, "Expanding the compute-and-forward framework: Unequal powers, signal levels, and multiple linear combinations," *IEEE Trans. Inf. Theory*, vol. 62, no. 9, pp. 4879–4909, Sep. 2016.

[21] J. Zhu and M. Gastpar, "Gaussian multiple access via compute-and-forward," *IEEE Trans. Inf. Theory*, vol. 63, no. 5, pp. 2678–2695, May 2017.

[22] D. Micciancio and S. Goldwasser, *Complexity of Lattice Problems: A Cryptographic Perspective*. Cambridge: Kluwer Academic Publishers, 2002.

[23] S. N. Hong and G. Caire, "Compute-and-forward strategies for cooperative distributed antenna systems," *IEEE Trans. Inf. Theory*, vol. 59, no. 9, pp. 5227–5243, Sep. 2013.

[24] W. He, B. Nazer, and S. S. Shitz, "Uplink–downlink duality for integer-forcing," *IEEE Trans. Inf. Theory*, vol. 64, no. 3, pp. 1992–2011, 2018.

[25] S.-Y. Chung, D. Forney, T. Richardson, and R. Urbanke, "On the design of low-density parity-check codes with 0.0045 dB of the Shannon limit," *IEEE Commun. Lett.*, vol. 5, no. 2, pp. 58–60, Feb. 2001.

[26] E. Sula, J. Zhu, A. Pastore, S. H. Lim, and M. Gastpar, "Compute–forward multiple access (CFMA): Practical implementations," *IEEE Trans. Commun.*, vol. 67, no. 2, pp. 1133–1147, 2019.

[27] C. I. Ionita, B. Nazer, C. Feng, and B. Aazhang, "An experimental study on the robustness of integer-forcing linear receivers," in *IEEE Int. Conf. Commun. (ICC)*, Glasgow, Jun. 2017.

[28] J. Körner and K. Marton, "How to encode the modulo-two sum of binary sources," *IEEE Trans. Inf. Theory*, vol. 25, no. 2, pp. 219–221, 1979.

[29] B. Nazer and M. Gastpar, "Reliable computation over multiple access channels," in *Proc. 43rd Ann. Allerton Conf. Comm. Control Comput.*, Monticello, IL, Sep. 2005.

[30] D. Krithivasan and S. S. Pradhan, "Lattices for distributed source coding: Jointly Gaussian sources and reconstruction of a linear function," *IEEE Trans. Inf. Theory*, vol. 55, no. 12, pp. 5628–5651, Dec. 2009.

[31] D. Krithivasan and S. S. Pradhan, "Distributed source coding using Abelian group codes," *IEEE Trans. Inf. Theory*, vol. 57, no. 3, pp. 1495–1519, Mar. 2011.

[32] A. B. Wagner, "On distributed compression of linear functions," *IEEE Trans. Inf. Theory*, vol. 57, no. 1, pp. 79–94, Jan. 2011.

[33] D. N. C. Tse and M. A. Maddah-Ali, "Interference neutralization in distributed lossy source coding," in *Proc. IEEE Int. Symp. Inf. Theory*, Austin, TX, Jun. 2010.

[34] V. Lalitha, N. Prakash, K. Vinodh, P. V. Kumar, and S. S. Pradhan, "Linear coding schemes for the distributed computation of subspaces," *IEEE J. on Sel. Areas Commun.*, vol. 31, no. 4, pp. 678–690, 2013.

[35] Y. Yang and Z. Xiong, "Distributed compression of linear functions: Partial sum-rate tightness and gap to optimal sum-rate," *IEEE Trans. Inf. Theory*, vol. 60, no. 5, pp. 2835–2855, May 2014.

[36] O. Ordentlich and U. Erez, "Integer-forcing source coding," *IEEE Trans. Inf. Theory*, vol. 63, no. 2, pp. 1253–1269, Feb. 2017.

[37] T. Philosof and R. Zamir, "On the loss of single-letter characterization: The dirty multiple access channel," *IEEE Trans. Inf. Theory*, vol. 55, no. 6, pp. 2442–2454, Jun. 2009.

[38] T. Philosof, R. Zamir, U. Erez, and A. J. Khisti, "Lattice strategies for the dirty multiple access channel," *IEEE Trans. Inf. Theory*, vol. 57, no. 8, pp. 5006–5035, Aug. 2011.

[39] I.-H. Wang, "Approximate capacity of the dirty multiple-access channel with partial state information at the encoders," *IEEE Trans. Inf. Theory*, vol. 58, no. 5, pp. 2781–2787, May 2012.

[40] A. Padakandla and S. S. Pradhan, "An achievable rate region based on coset codes for multiple access channel with states," *IEEE Trans. Inf. Theory*, vol. 63, no. 10, pp. 6393–6415, 2017.

[41] A. Padakandla and S. S. Pradhan, "Achievable rate region for three user discrete broadcast channel based on coset codes," *IEEE Trans. Inf. Theory*, vol. 64, no. 4, pp. 2267–2297, 2018.

[42] G. Bresler, A. Parekh, and D. N. C. Tse, "The approximate capacity of the many-to-one and one-to-many Gaussian interference channel," *IEEE Trans. Inf. Theory*, vol. 56, no. 9, pp. 4566–4592, Sep. 2010.

[43] A. S. Motahari, S. Oveis-Gharan, M.-A. Maddah-Ali, and A. K. Khandani, "Real interference alignment: Exploiting the potential of single antenna systems," *IEEE Trans. Inf. Theory*, vol. 60, no. 8, pp. 4799–4810, Aug. 2014.

[44] U. Niesen and M. A. Maddah-Ali, "Interference alignment: From degrees-of-freedom to constant-gap capacity approximations," *IEEE Trans. Inf. Theory*, vol. 59, no. 8, pp. 4855–4888, Aug. 2013.

[45] I. Shomorony and S. Avestimehr, "Degrees of freedom of two-hop wireless networks: Everyone gets the entire cake," *IEEE Trans. Inf. Theory*, vol. 60, no. 5, pp. 2417–2431, May 2014.

[46] V. Ntranos, V. R. Cadambe, B. Nazer, and G. Caire, "Integer-forcing interference alignment," in *Proc. IEEE Int. Symp. Inf. Theory*, Istanbul, Jul. 2013.

[47] A. Padakandla, A. G. Sahebi, and S. S. Pradhan, "An achievable rate region for the three-user interference channel based on coset codes," *IEEE Trans. Inf. Theory*, vol. 62, no. 3, pp. 1250–1279, Mar. 2016.

[48] X. He and A. Yener, "Providing secrecy with structured codes: Tools and applications to two-user Gaussian channels," *IEEE Trans. Inf. Theory*, vol. 60, no. 4, pp. 2121–2138, Apr. 2014.

[49] S. Vatedka, N. Kashyap, and A. Thangaraj, "Secure compute-and-forward in a bidirectional relay," *IEEE Trans. Inf. Theory*, vol. 61, no. 5, pp. 2531–2556, May 2015.

[50] J. Xie and S. Ulukus, "Secure degrees of freedom of one-hop wireless networks," *IEEE Trans. Inf. Theory*, vol. 60, no. 6, pp. 3359–3378, Jun. 2014.

[51] S. Miyake, "Coding theorems for point-to-point communication systems using sparse matrix codes." PhD Thesis, University of Tokyo, 2010.

[52] A. Padakandla and S. S. Pradhan, "Computing the sum of sources over an arbitrary multiple access channel," in *Proc. IEEE Int. Symp. Inf. Theory*, Istanbul, 2013.

[53] P. Sen, S. H. Lim, and Y.-H. Kim, "Optimal achievable rates for computation with random homologous codes," *arXiv:1805.03338*, 2018.

[54] S. H. Lim, C. Feng, A. Pastore, B. Nazer, and M. Gastpar, "Towards an algebraic network information theory: Simultaneous joint typicality decoding," *arXiv:1901.03274*, 2019.

[55] I. El Bakoury and B. Nazer, "Integer-forcing architectures for uplink cloud radio access networks," in *Allerton Conf. Comm. Control Comput.* Monticello, IL, 2017.

[56] I. El Bakoury and B. Nazer, "Uplink-downlink duality for integer-forcing in cloud radio access networks," in 2018.

[57] M. P. Wilson, K. Narayanan, H. D. Pfister, and A. Sprintson, "Joint physical layer coding and network coding for bidirectional relaying," *IEEE Trans. Inf. Theory*, vol. 56, no. 11, pp. 5641–5654, Nov. 2010.

[58] W. Nam, S.-Y. Chung, and Y. H. Lee, "Capacity of the Gaussian two-way relay channel to within $\frac{1}{2}$ bit," *IEEE Trans. Inf. Theory*, vol. 56, no. 11, pp. 5488–5494, Nov. 2010.

[59] J. Zhu and M. Gastpar, "Asymmetric compute-and-forward with CSIT," in *Int. Zurich Sem. Commun.*, Zurich, 2014.

[60] P. Sen and Y.-H. Kim, "Homologous codes for multiple access channels," *arXiv:1801.07695*, 2018.

13 Waveform Design

Paolo Banelli, Giulio Colavolpe, Luca Rugini, and Alessandro Ugolini

13.1 Introduction

The need for higher data rates has been the main driver that has led from 2G systems to long-term evolution (LTE) systems, with data-rates evolving from tens of kbit/s up to the current state of the art of tens of Mbit/s. Focusing on the adopted modulation schemes, the transition has been from single-carrier modulations with binary constellations, such as the Gaussian minimum-shift keying (GMSK) used in the 2G GSM systems, to multicarrier modulations with multilevel constellations, such as the quadrature-amplitude modulation (QAM) with adaptively chosen cardinality, currently used in LTE 4G systems [1]. 5G cellular communications [2, 3] promise to deliver the gigabit experience to mobile users, with a capacity increase of up to three orders of magnitude with respect to current LTE systems. Such an ambitious goal will be realized through a combination of innovative techniques involving different network layers. At the physical layer, the modulation format to be adopted is the subject of intense research activity (see [4] and references therein). The issue has also been extensively addressed in EU-funded research projects such as 5GNOW [5, 6] and METIS 2020 [7].

Orthogonal frequency-division multiplexing (OFDM) and orthogonal frequency-division multiple access (OFDMA) are the modulation technique and the multiple-access strategy adopted in LTE cellular network standards, respectively [1, 8]. OFDM and OFDMA have been selected for several reasons such as, to cite a few, the ease of implementation of both transmitter and receiver thanks to the use of fast Fourier transform (FFT) and inverse fast Fourier transform (IFFT) blocks, the ability to counteract multipath distortion, the orthogonality of subcarriers which eliminates intercell interference, the possibility of adapting the transmitted power and the modulation cardinality, and the ease of integration with multiantenna hardware, both at the transmitter and receiver.

Despite these positive aspects, OFDM and OFDMA present several drawbacks that call into question their adoption in 5G networks. First of all, the spectral efficiency of OFDM is limited by the need for a cyclic prefix and by its large sidelobes (which require some null guard tones at the spectrum edges). In addition, OFDM signals may exhibit large peak-to-average-power ratio values [9], thus suffering from nonlinear distortions possibly introduced by the high-power amplifier or by analog-to-digital (A/D) and digital-to-analog (D/A) converters. Furthermore, the impossibility of having strict frequency synchronization among subcarriers makes OFDM and OFDMA not really orthogonal techniques [10, 11]. In particular, synchronization is a key issue in the uplink of a cellular network wherein different mobile terminals transmit separately [11], and, also, in the downlink when base station coordination is used [12, 13].

Contrary to previous generations of cellular networks, 5G systems will feature several innovative strategies with respect to existing LTE systems, including, among others, extensive adoption of small cells, use of millimeter wave communications for short-range links [14–17], large-scale antenna arrays installed on macro base stations [18], cloud-based radio access network [19, 20], and, possibly, opportunistic exploitation of spectrum holes through a cognitive approach [21]. In addition, 5G systems have to accommodate a variety of services and emerging new applications. In particular, according to the classification in [22], the main reference scenarios currently envisioned for 5G networks are the following:

- **Very large data-rate wireless connectivity**. Users will be able to quickly download large amounts of data (e.g., high-definition video streaming). In this case, a modulation scheme with large spectral and energy efficiency is required.
- **Internet of Things (IoT)** [23]. Up to one trillion devices are expected to be connected through the 5G network. These connected things will have quite limited processing capabilities and will have to sporadically transmit small amounts of data, thus requiring a modulation scheme robust to time synchronizations errors and good performance for short communications.
- **Tactile internet** [24]. This scenario requires a communication service that must be reliable and with small latency, on the order of 1 ms, more than one order of magnitude smaller than the latency of current 4G systems. Other kinds of applications such as online gaming and car-to-car and car-to-infrastructure communications, although not directly related to the concept of tactile internet, can also take advantage of the low latency requirements [25].
- **Wireless regional area networks (WRANs)**. The increased throughput of 5G networks will make it suited for granting internet broadband access in sparsely populated areas that are not yet covered by wired technologies such as ADSL and fiber optics. In this scenario, network devices will have a very low mobility, so Doppler effects will be negligible, and also latency will not be a key requirement. In this case, the use of the so-called white spaces, i.e., frequency bands licensed to other services but actually not used, seems unavoidable. The modulation format of future 5G system should thus be able to efficiently exploit the available fragmented and heterogeneous spectrum.

These considerations are at the base of an intense debate on what modulation format and multiple-access strategy should be in next-generation cellular networks. This chapter provides a review of some of the most interesting waveforms for 5G networks [4, 5, 26–29]. We will limit ourselves to a critical mutual comparison of them in terms of spectral efficiency by employing a general signal processing framework coupled with an information theoretic approach, which permits evaluating the practical information rate associated with a specific signal format. The aim is not to be exhaustive with respect to all possible implementation issues and scenarios, but to highlight possible research directions and approaches that deserve to further investigation. However, the proposed signal processing is general enough to enable further comparison among different waveforms with respect to other design criteria, such as sensitivity to synchronization errors, nonlinear distortions, fast-fading channels, sidelobes' spectral mask, and so on.

13.2   System Model

This section starts with a single-antenna single-user system model for filterbank multicarrier (FBMC) [27, 30]. Then, the obtained system model is modified to consider OFDM [12, 31], generalized frequency-division multiplexing (GFDM) [22], and universal filtered multicarrier (UFMC) [32, 33]. Successively, we introduce a unified model valid for all the above techniques. Finally, the single-antenna single-user unified model is extended to the multiantenna multiuser case.

13.2.1   Continuous-Time Model

We first focus on FBMC [27, 30]. FBMC is a multicarrier technique with some key differences with respect to OFDM, which is the most popular among multicarrier techniques. A distinctive feature of FBMC is that successive data blocks overlap in time. This overlap is obtained by enlarging, with respect to OFDM, the length of the prototype filter applied to each data block. This way, a longer prototype filter produces a signal with larger duration and allows for reduced sidelobes in the frequency domain. As a direct consequence of the reduced sidelobes, in case frequency offset or time synchronization errors, FBMC gathers a reduced amount of intercarrier interference (ICI), with respect to OFDM. Because of the reduced spectral sidelobes, FBMC generates less adjacent channel interference (ACI) than OFDM, thereby simplifying the coexistence with other signals allocated in nearby bands. Therefore, FBMC is particularly suitable for multiple-access and cognitive radio applications. On the other hand, in FBMC, the time-domain overlap of successive data blocks requires more advanced data detection techniques with respect to OFDM. Indeed, FBMC intentionally introduces intersymbol interference (ISI) between successive data blocks, and ICI also in the absence of frequency offsets. Instead, OFDM, which uses a prototype filter with length equal to the data block, avoids the ISI between successive data blocks, by temporal separation of the waveforms of different data blocks, and avoids ICI in the absence of frequency offsets, thanks to its orthogonal design. The advanced detection methods required by FBMC may be interpreted as ISI mitigation (or cancellation) techniques, which require additional complexity with respect to single-block detection methods used in OFDM.

Herein we introduce a continuous-time signal model for FBMC [27, 30]. We assume that N is the number of subcarriers and T_c is the sampling period. We also define the sampling frequency $F_c = 1/T_c$, the frequency separation index δ_f, and the time separation index δ_t. Note that a time separation index δ_t corresponds to signals staggered in time by $N\delta_t T_c$; $\delta_t = 1$ corresponds to absence of a cyclic prefix or a guard time. In addition, a frequency separation index δ_f corresponds to a frequency spacing of $\delta_f/(NT_c)$, among the data; $\delta_f = 1$ corresponds to orthogonal frequency spacing $1/(NT_c)$, like in OFDM. We denote with $d_{i,k}$ the ith data symbol on the kth subcarrier, and with $p_0(t)$ the pulse shape of duration $D = MN\delta_t T_c$, where M, a positive integer, is the overlapping factor between consecutive data blocks. The use of a larger value of M implies longer pulse shapes and hence reduced spectral sidelobes; however, a larger M also implies an increased ISI, due to the larger number M of data blocks that overlap in the time domain.

In FBMC transmissions, the data $\{d_{i,k}\}$ are transmitted in multiple signal blocks. The time-domain FBMC signal $x_i(t)$ generated by the ith data block of symbols can be expressed by

$$x_i(t) = \frac{1}{\sqrt{N}} \sum_{k=0}^{N-1} d_{i,k} p_0(t) e^{j2\pi k \frac{\delta_f}{NT_c}(t+iN\delta_t T_c)}, \tag{13.1}$$

for $0 \leq t < D$. This FBMC signal $x_i(t)$ generated by the ith data block may be overlapped with other signals $\{x_l(t)\}$ generated by other data blocks, which are staggered in time by integer multiples of the data block duration $N\delta_t T_c$. Let us denote with $x_{t,i}(t)$ the superposition of all those signals $\{x_l(t)\}$ whose time support overlaps with $x_i(t)$. Since the time support of $x_i(t)$, expressed by $D = MN\delta_t T_c$, is an integer multiple of the data block duration $N\delta_t T_c$, and considering that two consecutive signals $x_l(t)$ and $x_{l+1}(t)$ are staggered in time by the data block duration $N\delta_t T_c$, the transmitted signal $x_{t,i}(t)$ includes all those signals $\{x_l(t)\}$ with index $i - M + 1 \leq l \leq i + M - 1$. However, since a multipath channel usually widens the time support of transmitted signals, for convenience let us extend the definition of the transmitted FBMC signal $x_{t,i}(t)$ to include also the two signals $x_{i-M}(t)$ and $x_{i+M}(t)$. Therefore, the transmitted signal can be expressed by

$$x_{t,i}(t) = \sum_{l=i-M}^{i+M} x_l(t - lN\delta_t T_c), \tag{13.2}$$

which is valid for $-N\delta_t T_c \leq t - iN\delta_t T_c < (M+1)N\delta_t T_c$. After convolution with a linear and time-invariant multipath channel $h(t)$ with time support $D_h = LT_c$, where the channel order L is assumed integer and lower than the number N of subcarriers, the received FBMC signal $y(t)$, for $(-N\delta_t + L)T_c \leq t - iN\delta_t T_c < (M+1)N\delta_t T_c$, can be expressed by

$$y(t) = x_{t,i}(t) * h(t) + n(t), \tag{13.3}$$

where $*$ denotes convolution and $n(t)$ is an additive white Gaussian noise (AWGN) term. We now define

$$y_i(t) = y(t - iN\delta_t T_c), \tag{13.4}$$

which allows us to work with a simplified time axis $(-N\delta_t + L)T_c \leq t < (M+1)N\delta_t T_c$ that includes the interval of interest $0 \leq t < (MN\delta_t + L)T_c$. Note that the mathematical model of (13.2)–(13.4) is exact for the time interval of interest (related to the ith data), because (13.2) includes the nearby blocks of interest (with index from $i - M$ to $i + M$);

in order to be exact also outside the time interval of interest, (13.2) should include all the transmitted blocks.

In the following, we will use a frequency separation index $\delta_f = 1$. For transmissions without guard time or cyclic prefix, we will use a time separation index $\delta_t = 1$, whereas, for transmissions with guard time or cyclic prefix of duration GT_c, we will use $\delta_t = (N + G)/N$. We now introduce the corresponding discrete-time signal models, starting from FBMC. Then, we will extend the FBMC discrete model to other transmission schemes like OFDM, GFDM, and UFMC.

13.2.2 Discrete-Time Models

FBMC

For classical FBMC [27, 30], we define $\mathbf{d}_i = [d_{i,0}, \ldots, d_{i,k}, \ldots, d_{i,N-1}]^T$ as the $N \times 1$ ith data block, and

$$\mathbf{x}_i = [x_i(0), \ldots, x_i(nT_c), \ldots, x_i((MN - 1)T_c)]^T = \mathbf{P}_t \mathbf{d}_i$$

$$= \mathbf{F}_{MN}^H \mathbf{P}_f \mathbf{d}_i = \mathbf{F}_{MN}^H \sum_{k=0}^{N-1} d_{i,k} \mathbf{p}_{f_k} = \sum_{k=0}^{N-1} d_{i,k} \mathbf{p}_{t_k} = \sum_{k=0}^{N-1} d_{i,k} \mathbf{D}^k \mathbf{p}_{t_0} \tag{13.5}$$

as the $MN \times 1$ ith signal block in the time domain, where $\mathbf{p}_{t_0}$ is the vector containing the samples $p_0(nT_c)$ of the prototype pulse-shaping filter $p_0(t)$ with duration $D = MNT_c$, $\mathbf{F}_N$ is the unitary FFT matrix of size N, $\mathbf{D}^k = \text{diag}\left(\left[1, \ldots, e^{j\frac{2\pi}{MN}kMn}, \ldots, e^{j\frac{2\pi}{MN}kM(N-1)}\right]\right)$ is the diagonal matrix that performs modulation on the kth subcarrier, $\mathbf{P}_t = [\mathbf{p}_{t_0}, \ldots, \mathbf{p}_{t_k}, \ldots, \mathbf{p}_{t_{N-1}}]$ is the $MN \times N$ pulse-shaping matrix in the critically sampled time domain, and $\mathbf{P}_f = [\mathbf{p}_{f_0}, \ldots, \mathbf{p}_{f_k}, \ldots, \mathbf{p}_{f_{N-1}}]$ is the $MN \times N$ pulse-shaping matrix, which contains the pulse shaper associated to each subcarrier, in the M-oversampled frequency domain, where

$$\mathbf{p}_{f_0} = \left[P_0^{(a)}(0), \ldots, P_0^{(a)}\left(\frac{mF_c}{MN}\right), \ldots, P_0^{(a)}\left(\frac{(MN-1)F_c}{MN}\right)\right]^T \tag{13.6}$$

is the $MN \times 1$ frequency-domain oversampled prototype pulse shaper (on the MN FFT grid) applied to the data on the zeroth subcarrier, where $P_0^{(a)}(f)$ is the (periodic) continuous Fourier transform of the sampled version of $p_0(t)$, i.e., the aliased version of the pulse-shaper spectrum $P_0(f) = \mathcal{F}\{p_0(t)\}$, as expressed by $P_0^{(a)}(f) = \sum_{i=-\infty}^{+\infty} P_0(f - i/T_c)$.

The structure of the FBMC signal block (13.5) is shown in Figure 13.1.

Consequently, by denoting with $\langle i \rangle_M$ the remainder of i after integer division by M,

$$\mathbf{p}_{f_k} = \left[P_0^{(a)}\left(\frac{\langle -kM \rangle_{MN} F_c}{MN}\right), \ldots, P_0^{(a)}\left(\frac{\langle m - kM \rangle_{MN} F_c}{MN}\right), \ldots, \right.$$

$$\left. P_0^{(a)}\left(\frac{\langle MN - 1 - kM \rangle_{MN} F_c}{MN}\right)\right]^T = \mathbf{Z}_c^{kM} \mathbf{p}_{f_0}, \tag{13.7}$$

is the $MN \times 1$ kth filter in the frequency domain obtained by (cyclically) shifting of kM frequency bins the (periodic) prototype filter of the zeroth subcarrier,

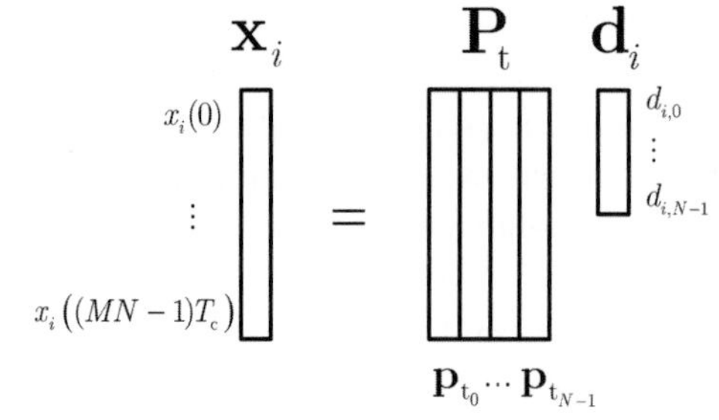

Figure 13.1 Vector structure for $\mathbf{x}_i$ of FBMC.

$\mathbf{Z}_\mathrm{c} = \mathrm{circ}([0, 1, 0, \ldots, 0]^T)$ is the circulant matrix that, when applied to a vector, cyclically downshifts the vector elements of one position, and $\mathbf{Z}_\mathrm{c}^k = (\mathbf{Z}_\mathrm{c})^k = \mathrm{circ}([\mathbf{0}_{1 \times k}, 1, 0, .., 0]^T)$ is the circulant matrix that, when applied to a vector, cyclically downshifts the vector elements of k positions. On the other hand,

$$\mathbf{p}_{\mathrm{t}_k} = \left[p_k(0), \ldots, p_k(nT_\mathrm{c}), \ldots, p_k((MN-1)\, T_\mathrm{c})\right]^T = \mathbf{F}_{MN}^H \mathbf{p}_{\mathrm{f}_k}$$
$$= \mathbf{F}_{MN}^H \mathbf{Z}_\mathrm{c}^{kM} \mathbf{p}_{\mathrm{f}_0} = \mathbf{F}_{MN}^H \mathbf{Z}_\mathrm{c}^{kM} \mathbf{F}_{MN} \mathbf{F}_{MN}^H \mathbf{p}_{\mathrm{f}_0} = \mathbf{D}^k \mathbf{p}_{\mathrm{t}_0} \tag{13.8}$$

is the $MN \times 1$ kth pulse shaper in the time domain, and $\mathbf{D}^k = \mathbf{F}_{MN}^H \mathbf{Z}_\mathrm{c}^{kM} \mathbf{F}_{MN} = \mathrm{diag}\left(\left[1, \ldots, e^{j\frac{2\pi}{MN}kMn}, \ldots, e^{j\frac{2\pi}{MN}kM(N-1)}\right]\right)$, where $\mathbf{D} = \mathbf{F}_{MN}^H \mathbf{Z}_\mathrm{c}^{M} \mathbf{F}_{MN}$.

Although the $N \times 1$ data block $\mathbf{d}_i$ is transmitted every NT_c seconds, we should bear in mind that the signal block $\mathbf{x}_i$ has a larger time duration $D = MNT_\mathrm{c}$, due to the pulse shaper. Therefore, each signal block produces ISI to the subsequent $M-1$ signal blocks and receives ISI from the previous $M-1$ signal blocks. In the case of multipath channels with maximum duration LT_c, where $L+1$ is the number of resolvable taps of the discrete-time channel at T_c sampling, i.e., $h[n] = h(nT_\mathrm{c})$, the signal block $\mathbf{x}_i$ related to the data block $\mathbf{d}_i$ will be spread over a duration $\tilde{D} = (MN + L)T_\mathrm{c}$. Thus, in order to recover the data block $\mathbf{d}_i$, the received signal should be observed over an extended time span of duration at least equal to $\tilde{D}$, to avoid the loss of signal energy related to the data block $\mathbf{d}_i$. (The observation window could be even larger, in the case of multiblock equalization or ISI estimation-cancellation techniques.)

Herein we want to model the transmitted signal $\mathbf{x}_{\mathrm{t},i}$ as the superposition of signal blocks $\mathbf{x}_i$ shifted by N samples in the time domain. In order to include the ISI caused by the previous signal block and the ISI that affects the next signal block, it is convenient to extend the length of a signal block from MN samples to $MN + 2L$ samples. Therefore, we use an $(MN + 2L) \times MN$ padding shift matrix $\mathbf{Z}$ that, when multiplied with a column vector of size MN, first adds L zeros on the top and on the bottom of the column vector, and then performs an acyclic downshift of one position of the element of the vector. The padding shift matrix $\mathbf{Z}$ is defined as the Toeplitz matrix with 1s in the diagonal with elements $(n + L + 1, n)$, for $n = 0, \ldots, MN - 1$, and all 0s elsewhere. Similarly, we use the multiple-shift matrix $\mathbf{Z}_s$ as the $(MN + 2L) \times MN$ matrix that performs an acyclic downshift of s positions. $\mathbf{Z}_s$ has 1s in the diagonal with elements $(n + L + s, n)$, for

$n = 0, \ldots, MN - 1$, and all 0s elsewhere. Obviously, the single-shift matrix is $\mathbf{Z}_1 = \mathbf{Z}$, and the zero-shift matrix $\mathbf{Z}_0 = [\mathbf{0}_{MN \times L}, \mathbf{I}_{MN}, \mathbf{0}_{MN \times L}]^T$ pads the vector with $2L$ zeros (L zeros at the top and L zeros at the bottom) but does not perform any shift. Using $\mathbf{Z}_s$ with $s < 0$ allows for an acyclic upshift of $|s|$ positions. Now, we can define the $(MN+2L) \times 1$ transmitted vector $\mathbf{x}_{t,i}$ associated with the ith observation. Since the channel order L is assumed lower than the number of subcarriers N, the multipath channel increases the ISI span from $M - 1$ to M signal blocks. Hence, the transmitted vector $\mathbf{x}_{t,i}$ is the shifted superposition of the $MN \times 1$ signal blocks $\{\mathbf{x}_{i-M}, \ldots, \mathbf{x}_i, \ldots, \mathbf{x}_{i+M}\}$, as expressed by

$$\mathbf{x}_{t,i} = \sum_{m=-M}^{M} \mathbf{Z}_{mN} \mathbf{x}_{i+m} = \sum_{m=-M}^{M} \mathbf{Z}_{mN} \mathbf{P}_t \mathbf{d}_{i+m}. \tag{13.9}$$

The structure of the FBMC transmitted vector (13.9) is shown in Figure 13.2.

Consequently, the $(MN+L) \times 1$ received vector $\mathbf{y}_i$ associated with the ith observation can be expressed by

$$\mathbf{y}_i = \mathbf{H}\mathbf{x}_{t,i} + \mathbf{w}_i = \mathbf{H} \sum_{m=-M}^{M} \mathbf{Z}_{mN} \mathbf{P}_t \mathbf{d}_{i+m} + \mathbf{w}_i, \tag{13.10}$$

where $\mathbf{H}$ is the $(MN+L) \times (MN+2L)$ Toeplitz channel matrix implementing the discrete-time linear convolution caused by the multipath channel $h[n] = h(nT_c)$. The structure of the FBMC received vector (13.10) is shown in Figure 13.3.

We now introduce an alternative model for the received vector $\mathbf{y}_i$. Indeed, for the ith observation, the $(MN + L) \times 1$ received signal block $\mathbf{y}_i$ can be expressed by

$$\mathbf{y}_i = \mathbf{H}_{tot} \mathbf{d}_{i,long} + \mathbf{w}_i = \sum_{m=-M}^{M} \mathbf{H}_m \mathbf{d}_{i+m} + \mathbf{w}_i, \tag{13.11}$$

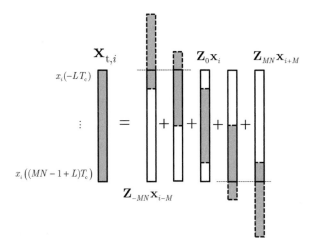

Figure 13.2 Vector structure for $\mathbf{x}_{t,i}$ of FBMC. Nonzero elements in gray.

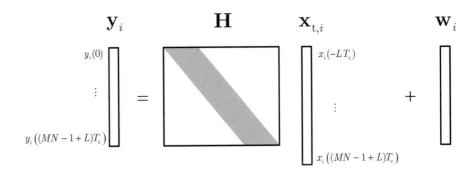

Figure 13.3 Vector structure for $\mathbf{y}_i$ of FBMC. Nonzero elements of $\mathbf{H}$ in gray.

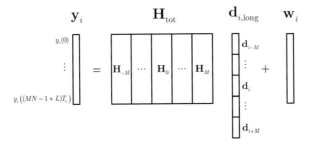

Figure 13.4 Alternative vector structure for $\mathbf{y}_i$ of FBMC.

where $\mathbf{H}_{\text{tot}} = [\mathbf{H}_{-M}, \ldots, \mathbf{H}_0, \ldots, \mathbf{H}_M]$ is the total $(MN + L) \times (2M + 1)N$ matrix accounting for both the multipath channel and the time-domain pulse shapers, $\mathbf{d}_{i,\text{long}} = [\mathbf{d}_{i-M}, \ldots, \mathbf{d}_i, \ldots, \mathbf{d}_{i+M}]$ collects all the symbol vectors interfering with $\mathbf{d}_i$, and $\mathbf{w}_i$ is the AWGN. The submatrices $\{\mathbf{H}_m\}$ in $\mathbf{H}_{\text{tot}}$ highlight the ISI received from the M previous signal blocks and the ISI generated to the following M signal blocks. By comparing (13.11) with (13.10), we have $\mathbf{H}_m = \mathbf{HZ}_{mN}\mathbf{P}_t$ for $m = -M, \ldots, 0, \ldots, M$. We can also exploit the relation $\mathbf{P}_t = \mathbf{F}_{MN}^H\mathbf{P}_f$ to rewrite $\mathbf{H}_m$ as $\mathbf{H}_m = \mathbf{HZ}_{mN}\mathbf{F}_{MN}^H\mathbf{P}_f$ to make explicit the IFFT operation. The alternative structure of the FBMC received vector (13.11) is shown in Figure 13.4.

If the FBMC receiver wants to apply an observation window with reduced size $W \leq MN + L$, the received vector can be modified as follows

$$\mathbf{z}_i = \mathbf{Ry}_i = \mathbf{RH} \sum_{m=-M}^{M} \mathbf{Z}_{mN}\mathbf{P}_t\mathbf{d}_{i+m} + \mathbf{Rw}_i, \tag{13.12}$$

where $\mathbf{z}_i$ is the $W \times 1$ reduced-size received vector and $\mathbf{R}$ is a $W \times (MN + L)$ selection matrix obtained by extracting the required W rows from the matrix $\mathbf{I}_{MN+L}$. For instance, to eliminate the ISI from $\mathbf{d}_{i-M}$, $\mathbf{z}_i$ can discard the first L samples of $\mathbf{y}_i$ by using a receiving matrix $\mathbf{R}$ that keeps only the last MN rows of $\mathbf{I}_{MN+L}$, thereby leading to (13.12) but with summation index starting from $m = -M + 1$. On the other hand, when all the

$W = MN + L$ samples of $\mathbf{y}_i$ need to be kept, $\mathbf{R} = \mathbf{I}_{MN+L}$ and therefore $\mathbf{z}_i = \mathbf{y}_i$. As a consequence, the received vector $\mathbf{z}_i$ is more general than $\mathbf{y}_i$.

The model described so far can be easily extended to include a cyclic prefix (CP) or a guard time interval. Please note that, in FBMC, often the CP is absent, since the block overlapping structure of the transmitted data produces ISI also in the case of CP. Anyway, the extension of the previous model to include CP-based transmissions can be useful for some special cases, like OFDM, that avoid the data block overlapping at the transmitter and eliminate the ISI at the receiver, using a CP. As a consequence, we introduce a $(MN + G) \times MN$ precoding matrix $\mathbf{T}$ that acts at the waveform level (after pulse shaping). We assume that $G \geq L$. Therefore, the extended signal block $\mathbf{x}_i$ becomes an $(MN + G) \times 1$ vector, as expressed by

$$\mathbf{x}_i = \mathbf{T}\mathbf{P}_t\mathbf{d}_i = \mathbf{T}\mathbf{F}_{MN}^H\mathbf{P}_f\mathbf{d}_i. \tag{13.13}$$

Similarly, the extended transmitted signal vector $\mathbf{x}_{t,i}$ associated with the ith observation becomes an $(MN + G + 2L) \times 1$ transmitted vector, as expressed by

$$\mathbf{x}_{t,i} = \sum_{m=-M}^{M} \mathbf{Z}_{mN}\mathbf{x}_{i+m} = \sum_{m=-M}^{M} \mathbf{Z}_{mN}\mathbf{T}\mathbf{F}_{MN}^H\mathbf{P}_f\mathbf{d}_{i+m}, \tag{13.14}$$

where now the padding shift matrix $\mathbf{Z}_s$ has size $(MN + G + 2L) \times (MN + G)$. The $(MN + G + L) \times 1$ received signal vector $\mathbf{y}_i$ over the extended observation window can be expressed by

$$\mathbf{y}_i = \mathbf{H}\mathbf{x}_{t,i} + \mathbf{w}_i = \mathbf{H} \sum_{m=-M}^{M} \mathbf{Z}_{mN}\mathbf{T}\mathbf{F}_{MN}^H\mathbf{P}_f\mathbf{d}_{i+m} + \mathbf{w}_i, \tag{13.15}$$

where $\mathbf{H}$ is the $(MN + G + L) \times (MN + G + 2L)$ Toeplitz channel matrix implementing the discrete-time linear convolution caused by the multipath channel (at sampling time T_c). Note that in classical FBMC the CP is absent: therefore, $G = 0$ and $\mathbf{T} = \mathbf{I}_{MN}$; consequently, for classical FBMC, (13.10) and (13.15) are identical.

Also in this case, we can use a $W \times (MN + G + L)$ receiving matrix $\mathbf{R}$, with $W \leq MN + G + L$, to discard some unwanted samples of $\mathbf{y}_i$ (for instance, in the presence of a CP, the first G samples could be removed). Therefore, the received vector can be expressed by

$$\mathbf{z}_i = \mathbf{R}\mathbf{y}_i = \mathbf{R}\mathbf{H} \sum_{m=-M}^{M} \mathbf{Z}_{mN}\mathbf{T}\mathbf{F}_{MN}^H\mathbf{P}_f\mathbf{d}_{i+m} + \mathbf{R}\mathbf{w}_i, \tag{13.16}$$

where $\mathbf{R}$ contains only the necessary W rows of $\mathbf{I}_{MN+G+L}$.

OFDM

Among the multicarrier techniques, OFDM is the simplest and the most popular [12, 31]. One of the key features of OFDM is the easy generation of the transmitted signal through simple IFFT processing, and the easy recovery of the received data through simple FFT processing. Differently from FBMC, the OFDM data blocks do not overlap in the time domain; therefore, the OFDM transmission is performed on a block-by-block basis, and also the OFDM reception is performed independently on each block. The use of CP in OFDM converts a frequency-selective multipath channel into a parallel set of frequency-flat single-path channels and enables a simple per-subcarrier equalization with scalar gain compensation only. This low-complexity equalization is one of the main reasons why OFDM has gained increased popularity for multipath channels. In addition, CP-OFDM well matches multiple-input multiple-output (MIMO) systems based on multiple antennas, because each subcarrier can be considered as a separate MIMO channel. One of the weaknesses of OFDM is the spectral efficiency loss caused by the use of the CP. Moreover, the relevant sidelobes of the transmitted signal makes OFDM sensitive to frequency offsets and Doppler effects, which break the frequency orthogonality and produce ICI. Another drawback of OFDM, and in general of multicarrier techniques, is related to the relevant nonlinear distortions caused by high-power amplification at the transmitter side, because the signal to be amplified has a large peak-to-average power ratio, caused by the IFFT combination of the signals of many subcarriers.

The mathematical model described in the previous section can be exploited for classical OFDM too [12, 31]. For OFDM with CP, the CP length is $G > 0$, and the overlapping factor is $M = 1$, because the signal blocks do not overlap at the transmitter. This corresponds to $\delta_t = (N + G)/N$ and to $\delta_f = 1$. In addition, for OFDM, the frequency-domain pulse shape $\mathbf{p}_{f_0} = \mathbf{e}_0 = [1, 0, \ldots, 0]$ is obtained by sampling of the digital $\mathrm{sinc}(\cdot)$ in the frequency domain; similarly, for the kth subcarrier, $\mathbf{p}_{f_k} = \mathbf{e}_k = [0, \ldots, 0, 1, 0, \ldots, 0]$, so that the frequency-domain pulse-shaping matrix is $\mathbf{P}_f = \mathbf{I}_N$. The precoding matrix is a CP inserting matrix, as expressed by $\mathbf{T} = \mathbf{T}_{\mathrm{CP},(N+G)\times N} = [\mathbf{I}_{\mathrm{CP},G\times N}^T, \mathbf{I}_N]^T$, where $\mathbf{I}_{\mathrm{CP},G\times N}$ contains the last G rows of $\mathbf{I}_N$. In addition, the time-shift should include the CP. Furthermore, usually OFDM detection discards both the CP and the last L samples (that would affect the CP of the following block), that is, $\mathbf{R}$ is an $N \times (N + G + L)$ matrix obtained from $\mathbf{I}_{N+G+L}$ by discarding the first G rows and the last L rows, defined as $\mathbf{R} = [\mathbf{0}_{N\times G}, \mathbf{I}_N, \mathbf{0}_{N\times L}]$. Therefore, for OFDM with CP, we can express the signal block $\mathbf{x}_i$, the transmitted block $\mathbf{x}_{t,i}$, and the received blocks $\mathbf{y}_i$ and $\mathbf{z}_i$ (before and after CP discarding, respectively), using the previous model introduced for FBMC, which for OFDM simplifies to

$$\mathbf{x}_i = \mathbf{T}\mathbf{P}_t\mathbf{d}_i = \mathbf{T}\mathbf{F}_N^H\mathbf{d}_i, \tag{13.17}$$

$$\mathbf{x}_{t,i} = \mathbf{Z}_{-N-G}\mathbf{T}\mathbf{F}_N^H\mathbf{d}_{i-1} + \mathbf{Z}_0\mathbf{T}\mathbf{F}_N^H\mathbf{d}_i + \mathbf{Z}_{N+G}\mathbf{T}\mathbf{F}_N^H\mathbf{d}_{i+1}, \tag{13.18}$$

$$\mathbf{y}_i = \mathbf{H}\mathbf{Z}_{-N-G}\mathbf{T}\mathbf{F}_N^H\mathbf{d}_{i-1} + \mathbf{H}\mathbf{Z}_0\mathbf{T}\mathbf{F}_N^H\mathbf{d}_i + \mathbf{H}\mathbf{Z}_{N+G}\mathbf{T}\mathbf{F}_N^H\mathbf{d}_{i+1} + \mathbf{w}_i, \tag{13.19}$$

$$\mathbf{z}_i = \mathbf{R}\mathbf{y}_i = \mathbf{R}\mathbf{H}\mathbf{Z}_0\mathbf{T}\mathbf{F}_N^H\mathbf{d}_i + \mathbf{R}\mathbf{w}_i. \tag{13.20}$$

Specifically, in OFDM, $\mathbf{R}\mathbf{H}\mathbf{Z}_{-N-G}\mathbf{T}\mathbf{F}_N^H\mathbf{d}_{i-1} = \mathbf{R}\mathbf{H}\mathbf{Z}_{N+G}\mathbf{T}\mathbf{F}_N^H\mathbf{d}_{i+1} = \mathbf{0}_{N\times 1}$, because the ISI from the previous signal block is eliminated by discarding the CP. Note that $\mathbf{R}\mathbf{H}\mathbf{Z}_0\mathbf{T}$ in (13.20) is an $N \times N$ circulant matrix, hence the FFT of the received vector $\mathbf{z}_i$ produces

$$\mathbf{F}_N\mathbf{z}_i = \mathbf{F}_N\mathbf{R}\mathbf{H}\mathbf{Z}_0\mathbf{T}\mathbf{F}_N^H\mathbf{d}_i + \mathbf{F}_N\mathbf{R}\mathbf{w}_i = \mathbf{\Lambda}\mathbf{d}_i + \mathbf{F}_N\mathbf{R}\mathbf{w}_i, \tag{13.21}$$

where $\mathbf{\Lambda}$ is $N \times N$ diagonal. This enables the well-known low-complexity per-subcarrier equalization of OFDM, which is one of the main advantages of OFDM with respect to many other multicarrier and single-carrier techniques.

GFDM

GFDM [22] is a multicarrier technique that tries to combine the advantages of both FBMC and OFDM. Indeed, similarly to FBMC, GFDM adopts prototype filters with length larger than the duration of a data block; like in FBMC, this choice reduces the spectral sidelobes with respect to OFDM. However, differently than FBMC, multiple data blocks of GFDM can be interpreted as a single superblock that does not interfere with other superblocks. Indeed, in GFDM, the prototype filters are chosen in such a way that the time-domain overlap among consecutive superblocks is avoided. In addition, similarly to OFDM, GFDM uses a CP to reduce the implementation complexity in multipath channels. Anyway, differently from OFDM, the CP of GFDM is inserted on a superblock basis; therefore, since the same CP is shared by multiple data blocks, the spectral efficiency loss caused by the CP is reduced in GFDM with respect to OFDM. The main drawback of GFDM is its increased complexity with respect to OFDM, since the signal processing has to be performed on a data superblock basis, rather than on a data block basis like in OFDM.

The FBMC and OFDM signal models described in the previous section can be modified to include GFDM [22]. The $N \times 1$ ith data block is defined in the same way as in FBMC, i.e., $\mathbf{d}_i = [d_{i,0}, \ldots, d_{i,k}, \ldots, d_{i,N-1}]^T$, and the corresponding $MN \times 1$ ith signal block is defined as

$$\mathbf{x}_i = \mathbf{P}_{t,i}\mathbf{d}_i = \sum_{k=0}^{N-1} d_{i,k}\mathbf{p}_{t_k,\langle i \rangle_M}. \tag{13.22}$$

Differently from FBMC in (13.13), GFDM employs a different pulse shape for each data block, which is obtained by cyclically shifting both in time (by iN) and in frequency (by k/N) the same prototype filter $g[n]$ with length MN samples. Therefore, the pulse-shaping matrix $\mathbf{P}_{t,i}$ and the pulse-shaping vector $\mathbf{p}_{t_k,\langle i \rangle_M}$ depend also on the time index i, as expressed by

$$\mathbf{P}_{t,i} = \left[\mathbf{p}_{t_0,\langle i \rangle_M}, \ldots, \mathbf{p}_{t_k,\langle i \rangle_M}, \ldots, \mathbf{p}_{t_{N-1},\langle i \rangle_M} \right], \tag{13.23}$$

$$\mathbf{p}_{t_k,i} = [g[\langle -iN \rangle_{MN}], g[\langle 1 - iN \rangle_{MN}]e^{-j2\pi \frac{k}{N}}, \ldots,$$
$$\ldots, g[\langle M - 1 - iN \rangle_{MN}]e^{-j2\pi \frac{k}{N}(MN-1)}]^T. \tag{13.24}$$

Alternatively, the $MN \times N$ matrix $\mathbf{P}_{t,i}$ can be expressed as

$$\mathbf{P}_{t,i} = (\mathbf{1}_M \otimes \mathbf{F}_N^H) \circ (\mathbf{g}_{\langle i \rangle_M} \otimes \mathbf{1}_N^T), \tag{13.25}$$

$$\mathbf{g}_i = [g[\langle -iN \rangle_{MN}], g[\langle 1 - iN \rangle_{MN}], \ldots, g[\langle M - 1 - iN \rangle_{MN}]]^T, \tag{13.26}$$

where $\otimes$ and $\circ$ represent the Kronecker and the Hadamard (element-wise) product, respectively.

Note that, in GFDM, the duration $D = MNT_c$ of the pulse-shaping filter is identical to FBMC; however, the shape is different, and the overlap is also different. Indeed, in FBMC, the pulse shapes related to consecutive data blocks are time-shifted by N samples, using a sliding window approach; conversely, in GFDM, due to the cyclic shifting philosophy, M consecutive data blocks have the same time span and form a superblock that is separated in time from the other superblocks.

Assuming that i is the index of the $MN \times 1$ signal block $\mathbf{x}_i$, we want to understand which are the indices of the other $M - 1$ signal blocks that overlap with $\mathbf{x}_i$: since M signal blocks share the same superblock, the M indices of the M signal blocks depend on the value of $\langle i \rangle_M$. In general, the M overlapping blocks have the following indices: $\{i - \langle i \rangle_M, i + 1 - \langle i \rangle_M, \dots, i + M - 1 - \langle i \rangle_M\}$. For instance, when $i = 0$, the indices of the blocks in the same superblock are $\{0, 1, \dots, M - 1\}$. If we focus on the next block with $i = 1$, the indices of the blocks in the superblock are again $\{0, 1, \dots, M - 1\}$, since the blocks with $i = 0$ and $i = 1$ share the same superblock. Similarly, the indices of the blocks in the superblock stay the same also for the blocks $2 \le i \le M - 1$. Then, when $i = M$, the indices of the blocks in the same superblock are $\{M, M + 1, \dots, 2M - 1\}$, and they stay the same also for $M + 1 \le i \le 2M - 1$.

In GFDM, usually, the observation window is equal to $MN + G$ samples, where G denotes the CP size, with $G \ge L$. As in OFDM, the CP is inserted at the transmitter and discarded at the receiver, in order to eliminate the ISI between consecutive superblocks. Therefore, the signal block for GFDM is redefined as

$$\mathbf{x}_i = \mathbf{T}\mathbf{P}_{\mathrm{t},i}\mathbf{d}_i, \tag{13.27}$$

where $\mathbf{T}$ is the $(MN + G) \times MN$ matrix that inserts the CP, defined as $\mathbf{T} = \mathbf{T}_{\mathrm{CP},(MN+G)\times MN} = [\mathbf{I}_{\mathrm{CP},G\times MN}^T, \mathbf{I}_{MN}]^T$, where $\mathbf{I}_{\mathrm{CP},G\times MN}$ contains the last G rows of $\mathbf{I}_{MN}$. The structure of the GFDM signal block (13.27) is shown in Figure 13.5.

To keep the compatibility with the previous FBMC model, we use an extended observation window with size $MN + G + L$ that includes L additional (future) samples. By denoting with $\mathbf{x}_{\mathrm{t},i}$ the $(MN + G + 2L) \times 1$ transmitted vector associated with the ith observation, the central $MN + G$ samples of $\mathbf{x}_{\mathrm{t},i}$ are determined by the superposition of M signal blocks that belong to the same superblock of data block i, while the first (last) L samples of $\mathbf{x}_{\mathrm{t},i}$ are determined by the superposition of M signal blocks that belong to the previous (next) superblock. Therefore, the transmitted signal $\mathbf{x}_{\mathrm{t},i}$ can be expressed by

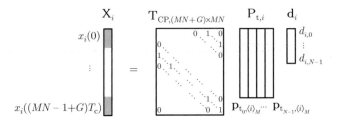

Figure 13.5 Vector structure for $\mathbf{x}_i$ of GFDM. Identical parts in $\mathbf{x}_i$ in gray.

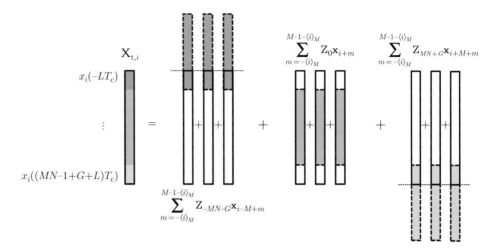

Figure 13.6 Vector structure for $\mathbf{x}_{t,i}$ of GFDM. Zero elements in white.

$$\mathbf{x}_{t,i} = \sum_{m=-\langle i\rangle_M}^{M-1-\langle i\rangle_M} (\mathbf{Z}_{-MN-G}\mathbf{x}_{i-M+m} + \mathbf{Z}_0\mathbf{x}_{i+m} + \mathbf{Z}_{MN+G}\mathbf{x}_{i+M+m})$$

$$= \sum_{m=-\langle i\rangle_M}^{M-1-\langle i\rangle_M} \mathbf{Z}_{-MN-G}\mathbf{TP}_{t,i-M+m}\mathbf{d}_{i-M+m} + \sum_{m=-\langle i\rangle_M}^{M-1-\langle i\rangle_M} \mathbf{Z}_0\mathbf{TP}_{t,i+m}\mathbf{d}_{i+m}$$

$$+ \sum_{m=-\langle i\rangle_M}^{M-1-\langle i\rangle_M} \mathbf{Z}_{MN+G}\mathbf{TP}_{t,i+M+m}\mathbf{d}_{i+M+m}. \tag{13.28}$$

The structure of the GFDM transmitted vector (13.28) is shown in Figure 13.6.

As an alternative to (13.28), we can define an extended data vector $\mathbf{d}_{i,\text{long}} = \left[\mathbf{d}_{i-\langle i\rangle_M}^T, \mathbf{d}_{i+1-\langle i\rangle_M}^T, \dots, \mathbf{d}_{i+M-1-\langle i\rangle_M}^T\right]^T$ and express $\mathbf{x}_{t,i}$ as a function of $\mathbf{d}_{i,\text{long}}$ and $\mathbf{d}_{i+M,\text{long}}$, as expressed by

$$\mathbf{x}_{t,i} = \mathbf{Z}_{-MN-G}\mathbf{TP}_{t,\text{big}}\mathbf{d}_{i-M,\text{long}} + \mathbf{Z}_0\mathbf{TP}_{t,\text{big}}\mathbf{d}_{i,\text{long}} + \mathbf{Z}_{MN+G}\mathbf{TP}_{t,\text{big}}\mathbf{d}_{i+M,\text{long}}, \tag{13.29}$$

where $\mathbf{P}_{t,\text{big}} = \left[\mathbf{P}_{t,0}, \dots, \mathbf{P}_{t,M-1}\right]$ is an $MN \times MN$ matrix that contains the pulse shapes of M data blocks belonging to the same superblock. This last expression of $\mathbf{x}_{t,i}$ eliminates the time dependence on the pulse-shaping matrix, since the pulse shape cyclically repeats every M data blocks.

The $(MN+G+L) \times 1$ received signal vector $\mathbf{y}_i$ over the extended observation window can be expressed by

$$\mathbf{y}_i = \mathbf{H}\mathbf{x}_{t,i} + \mathbf{w}_i$$

$$= \mathbf{H}\mathbf{Z}_{-MN-G}\mathbf{TP}_{t,\text{big}}\mathbf{d}_{i-M,\text{long}} + \mathbf{H}\mathbf{Z}_0\mathbf{TP}_{t,\text{big}}\mathbf{d}_{i,\text{long}}$$

$$+ \mathbf{H}\mathbf{Z}_{MN+G}\mathbf{TP}_{t,\text{big}}\mathbf{d}_{i+M,\text{long}} + \mathbf{w}_i, \tag{13.30}$$

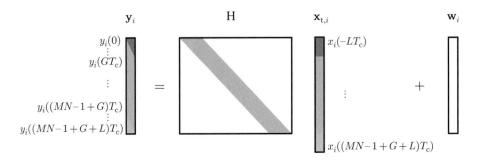

Figure 13.7 Vector structure for $\mathbf{y}_i$ of GFDM. Nonzero elements of $\mathbf{H}$ in gray.

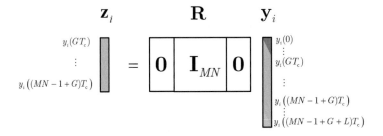

Figure 13.8 Vector structure for $\mathbf{z}_i$ of GFDM.

where $\mathbf{H}$ is the $(MN+G+L) \times (MN+G+2L)$ Toeplitz channel matrix implementing the discrete-time linear convolution caused by the multipath channel (at sampling time T_c). The structure of the GFDM received vector (13.30) is shown in Figure 13.7. Note that, as in OFDM, the ISI from the previous superblock can be eliminated by discarding the CP, and the data of the next superblock can be avoided by discarding the last L samples. Therefore, GFDM can use a receiving matrix $\mathbf{R}$ obtained from $\mathbf{I}_{MN+G+L}$ by discarding the first G rows and the last L rows, which leads to a reduced-size received vector

$$\mathbf{z}_i = \mathbf{R}\mathbf{y}_i = \mathbf{R}\mathbf{H}\mathbf{Z}_0\mathbf{T}\mathbf{P}_{t,\text{big}}\mathbf{d}_{i,\text{long}} + \mathbf{R}\mathbf{w}_i. \tag{13.31}$$

The structure of the GFDM reduced-size received vector (13.31) is shown in Figure 13.8. As in OFDM, $\mathbf{R}\mathbf{H}\mathbf{Z}_{-MN-G}\mathbf{T}\mathbf{P}_{t,\text{big}}\mathbf{d}_{i-M,\text{long}} = \mathbf{R}\mathbf{H}\mathbf{Z}_{MN+G}\mathbf{T}\mathbf{P}_{t,\text{big}}\mathbf{d}_{i+M,\text{long}} = \mathbf{0}_{MN \times 1}$, since the CP eliminates the ISI. Similarly to OFDM, $\mathbf{R}\mathbf{H}\mathbf{Z}_0\mathbf{T}$ is circulant: hence, GFDM can exploit FFT processing (of size MN) in order to diagonalize the channel [22].

UFMC

UFMC [32, 33] is a multicarrier multiuser scheme where each user occupies a subband with consecutive subcarriers. The main purpose of UFMC is to reduce the ICI with respect to OFDM and OFDMA. With this aim in mind, the key idea of UFMC is to use prototype filters that reduce the sidelobes on a subband basis, rather than on a subcarrier basis as done in FBMC. This choice of UFMC allows for prototype filters whose length is significantly reduced with respect to FBMC: Typically, the UFMC filter length is similar to the CP length used in OFDM, and hence by far lower than the data block length, while the FBMC filter length is an integer multiple of the data block length. Similarly to OFDM, the UFMC data blocks do not overlap in the time domain; however, since UFMC does not employ a CP, some ISI between consecutive data blocks may arise in multipath channels. With respect to OFDM, another drawback of UFMC is the increased complexity of equalization; anyway, since the UFMC signal detection is performed on a data block basis, the UFMC equalization complexity is lower than for FBMC. Note that each UFMC subband includes adjacent subcarriers; on the contrary, the optimal allocation for maximum frequency diversity would require maximally separated subcarriers, rather than adjacent subcarriers.

The model of UFMC, as expressed in [32, 33], is a multiuser model that contains the signals of multiple users, analogously to OFDMA. Therefore, in order to fairly compare with single-user techniques like FBMC, OFDM, and GFDM, we assume that all the UFMC data are transmitted by the same user. Thus, for UFMC, we assume that a single user occupies all the subbands. We define $\mathbf{d}_i = [d_{i,0}, \ldots, d_{i,k}, \ldots, d_{i,N-1}]^T$ as the $N \times 1$ ith data block, and

$$\mathbf{x}_i = \mathbf{T}\mathbf{F}_N^H \mathbf{d}_i \tag{13.32}$$

as the $(N+G) \times 1$ ith signal block, where $\mathbf{T}$ is an $(N+G) \times N$ filtering matrix that applies a filter (with the same order G) to each subband. By denoting with $\mathbf{x}_{t,i}$ the $(N+G+2L) \times 1$ transmitted vector associated with the ith observation, the central $N + G$ samples of $\mathbf{x}_{t,i}$ are determined by the current data block $\mathbf{d}_i$, while the first (last) L samples of $\mathbf{x}_{t,i}$ are determined by the previous (next) data block. Therefore, the transmitted signal $\mathbf{x}_{t,i}$ can be expressed by

$$\begin{aligned} \mathbf{x}_{t,i} &= \mathbf{Z}_{-N-G}\mathbf{x}_{i-1} + \mathbf{Z}_0\mathbf{x}_i + \mathbf{Z}_{N+G}\mathbf{x}_{i+1} \\ &= \mathbf{Z}_{-N-G}\mathbf{T}\mathbf{F}_N^H\mathbf{d}_{i-1} + \mathbf{Z}_0\mathbf{T}\mathbf{F}_N^H\mathbf{d}_i + \mathbf{Z}_{N+G}\mathbf{T}\mathbf{F}_N^H\mathbf{d}_{i+1}, \end{aligned} \tag{13.33}$$

which is formally identical to OFDM. However, in UFMC, the filtering matrix $\mathbf{T}$ is different from the CP insertion matrix used in OFDM. The $(N + G + L) \times 1$ received signal vector $\mathbf{y}_i$ over an extended observation window can be expressed by

$$\begin{aligned} \mathbf{y}_i &= \mathbf{H}\mathbf{x}_{t,i} + \mathbf{w}_i \\ &= \mathbf{H}\mathbf{Z}_{-N-G}\mathbf{T}\mathbf{F}_N^H\mathbf{d}_{i-1} + \mathbf{H}\mathbf{Z}_0\mathbf{T}\mathbf{F}_N^H\mathbf{d}_i + \mathbf{H}\mathbf{Z}_{N+G}\mathbf{T}\mathbf{F}_N^H\mathbf{d}_{i+1} + \mathbf{w}_i, \end{aligned} \tag{13.34}$$

which is also identical to its OFDM version (13.19), apart for the different matrix $\mathbf{T}$. Again, the ISI of the previous and next data blocks can be avoided by reducing the size of the observation window from $N + G + L$ to N. This can be done using the same $N \times (N + G + L)$ matrix $\mathbf{R}$ used for OFDM, defined as $\mathbf{R} = [\mathbf{0}_{N \times G}, \mathbf{I}_N, \mathbf{0}_{N \times L}]$. This leads to

$$\mathbf{z}_i = \mathbf{R}\mathbf{y}_i = \mathbf{R}\mathbf{H}\mathbf{Z}_0\mathbf{T}\mathbf{F}_N^H\mathbf{d}_i + \mathbf{R}\mathbf{w}_i, \tag{13.35}$$

which is again similar to the OFDM case. Note that, in UFMC, the additional samples G are due to filtering and may be useful for detection purposes. However, discarding these G samples eliminates the ISI, provided that the channel order L does not exceed the filter order G. However, in UFMC, in general $\mathbf{R}\mathbf{H}\mathbf{Z}_0\mathbf{T}$ is not circulant, so OFDM-like per-subcarrier equalization is not enabled. Anyway, with respect to FBMC and GFDM, the UFMC receiver processing acts on a vector of reduced size.

Table 13.1 Scalar parameters of the unified model.

Technique	M	G	$M_1(i)$	$M_2(i)$	$s(m,i)$
FBMC	> 1	0	$-M$	M	mN
OFDM	1	> 0	-1	1	$m(N+G)$
GFDM	> 1	> 0	$-M - \langle i \rangle_M$	$2M - 1 - \langle i \rangle_M$	$\lfloor \frac{m + \langle i \rangle_M}{M} \rfloor (MN+G)$
UFMC	1	> 0	-1	1	$m(N+G)$

Table 13.2 Vector-matrix parameters of the unified model: transmitter.

Technique	$\mathbf{T}$	$\mathbf{P}_{\mathrm{t},i}$	$\mathbf{x}_i$
FBMC	$\mathbf{I}_{MN}$	$\mathbf{F}_{MN}^H \mathbf{P}_{\mathrm{f}}$	$\mathbf{F}_{MN}^H \mathbf{P}_{\mathrm{f}} \mathbf{d}_i$
OFDM	$\mathbf{T}_{\mathrm{CP},(N+G)\times N}$	$\mathbf{F}_N^H$	$\mathbf{T}_{\mathrm{CP},(N+G)\times N} \mathbf{F}_N^H \mathbf{d}_i$
GFDM	$\mathbf{T}_{\mathrm{CP},(MN+G)\times MN}$	$(\mathbf{1}_M \otimes \mathbf{F}_N^H) \circ (\mathbf{g}_{\langle i \rangle_M} \otimes \mathbf{1}_N^T)$	$\mathbf{T}_{\mathrm{CP},(MN+G)\times MN} \mathbf{P}_{\mathrm{t},i} \mathbf{d}_i$
UFMC	$\mathbf{T}$	$\mathbf{F}_N^H$	$\mathbf{T}\mathbf{F}_N^H \mathbf{d}_i$

Table 13.3 Vector-matrix parameters of the unified model: receiver.

Technique	$\mathbf{R}$
FBMC	$\mathbf{I}_{MN+G+L}$
OFDM	$[\mathbf{0}_{N\times G}, \mathbf{I}_N, \mathbf{0}_{N\times L}]$
GFDM	$[\mathbf{0}_{MN\times G}, \mathbf{I}_{MN}, \mathbf{0}_{MN\times L}]$
UFMC	$[\mathbf{0}_{N\times G}, \mathbf{I}_N, \mathbf{0}_{N\times L}]$

13.2.3   Unified Discrete-Time Model

Note that the four signal models discussed above (FBMC, OFDM, GFDM and UFMC) have many similarities and can be seen as special cases of a unique unified model. In all cases, the $N \times 1$ ith data block is defined as $\mathbf{d}_i = [d_{i,0}, \ldots, d_{i,k}, \ldots, d_{i,N-1}]^T$, while the $(MN + G) \times 1$ ith signal block is defined as

$$\mathbf{x}_i = \mathbf{T}\mathbf{P}_{\mathrm{t},i}\mathbf{d}_i, \tag{13.36}$$

where $\mathbf{T}$ is an $(MN + G) \times MN$ precoding matrix, used for CP insertion or filtering, and $\mathbf{P}_{\mathrm{t},i}$ is the $MN \times N$ pulse shaping matrix. The unified model reduces to a specific model by selecting the overlapping factor M, the precoder length G, the precoding matrix $\mathbf{T}$, the pulse shaping matrix $\mathbf{P}_{\mathrm{t},i}$, and the receiving matrix $\mathbf{R}$. The specific parameters for each multicarrier technique discussed above (FBMC, OFDM, GFDM, and UFMC) are expressed in Tables 13.1–13.3.

In this unified model, the $(MN + G + 2L) \times 1$ transmitted vector $\mathbf{x}_{t,i}$ related to the ith observation can be expressed as

$$\mathbf{x}_{t,i} = \sum_{m=M_1(i)}^{M_2(i)} \mathbf{Z}_{s(m,i)} \mathbf{x}_{i+m}, \tag{13.37}$$

where $\mathbf{Z}_{s(m,i)}$ is the $(MN + G + 2L) \times (MN + G)$ padding shift matrix, where the shift $s(m, i)$ in general depends on both indices m and i. For each multicarrier technique, the specific values of the minimum block index $M_1(i)$ for ISI, of the maximum block index $M_2(i)$ for ISI, and of the shift $s(m, i)$, are expressed in Table 13.1, where $\lfloor x \rfloor$ represents the greatest integer less than or equal to x. Then, for all four models, the received signal is expressed by

$$\mathbf{y}_i = \mathbf{H}\mathbf{x}_{t,i} + \mathbf{w}_i, \tag{13.38}$$

where the channel matrix $\mathbf{H}$ is $(MN + G + L) \times (MN + G + 2L)$, and the received signal after discarding unwanted samples is expressed by

$$\mathbf{z}_i = \mathbf{R}\mathbf{y}_i, \tag{13.39}$$

where the receiving matrix is $\mathbf{R} = \mathbf{I}_{MN+G+L}$ for FBMC, $\mathbf{R} = [\mathbf{0}_{N \times G}, \mathbf{I}_N, \mathbf{0}_{N \times L}]$ for OFDM and UFMC, and $\mathbf{R} = [\mathbf{0}_{MN \times G}, \mathbf{I}_{MN}, \mathbf{0}_{MN \times L}]$ for GFDM, as shown in Table 13.3. Consequently, the unified expression that links the received vector $\mathbf{z}_i$ to the data vectors $\{\mathbf{d}_m\}$ is

$$\mathbf{z}_i = \mathbf{R}\mathbf{y}_i = \mathbf{R}\mathbf{H}\mathbf{x}_{t,i} + \mathbf{R}\mathbf{w}_i = \sum_{m=M_1(i)}^{M_2(i)} \mathbf{R}\mathbf{H}\mathbf{Z}_{s(m,i)} \mathbf{x}_{i+m} + \mathbf{R}\mathbf{w}_i$$

$$= \sum_{m=M_1(i)}^{M_2(i)} \mathbf{R}\mathbf{H}\mathbf{Z}_{s(m,i)} \mathbf{T}\mathbf{P}_{t,i+m} \mathbf{d}_{i+m} + \mathbf{R}\mathbf{w}_i$$

$$= \sum_{m=M_1(i)}^{M_2(i)} \mathbf{A}_{m,i} \mathbf{d}_{i+m} + \mathbf{R}\mathbf{w}_i, \tag{13.40}$$

where the matrix $\mathbf{A}_{m,i}$ that links the data vector $\mathbf{d}_{i+m}$ to the observed vector $\mathbf{z}_i$ is defined by

$$\mathbf{A}_{m,i} = \mathbf{R}\mathbf{H}\mathbf{Z}_{s(m,i)} \mathbf{T}\mathbf{P}_{t,i+m}. \tag{13.41}$$

Note that (13.40) can be rewritten as

$$\mathbf{z}_i = \mathbf{A}_{0,i}\mathbf{d}_i + \sum_{m=M_1(i), m \neq 0}^{M_2(i)} \mathbf{A}_{m,i} \mathbf{d}_{i+m} + \mathbf{R}\mathbf{w}_i$$

$$= \mathbf{A}_{0,i}\mathbf{d}_i + \mathbf{i}_i + \mathbf{v}_i, \tag{13.42}$$

where $\mathbf{i}_i = \sum_{m=M_1(i), m \neq 0}^{M_2(i)} \mathbf{A}_{m,i}\mathbf{d}_{i+m}$ is the total interference term due to data (i.e., a zero vector for OFDM), and $\mathbf{v}_i = \mathbf{R}\mathbf{w}_i$ is the noise term.

Alternatively, the unified model of (13.42) can be expressed as

$$\mathbf{z}_i = \mathbf{A}_{i,\text{tot}} \mathbf{d}_{i,\text{long}} + \mathbf{v}_i, \tag{13.43}$$

where $\mathbf{A}_{i,\text{tot}} = [\mathbf{A}_{M_1(i),i}, \cdots, \mathbf{A}_{M_2(i),i}]$ and $\mathbf{d}_{i,\text{long}} = [\mathbf{d}_{i+M_1(i)}^H, \cdots, \mathbf{d}_{i+M_2(i)}^H]^H$. The unified model of (13.43) is more appealing for FBMC and GFDM systems, since in these cases the data vectors $\{\mathbf{d}_m\}$ strongly interfere with one another.

13.2.4   Extension to Multiantenna and Multiuser Systems

The unified model of Section 13.2.3 can be extended to multiantenna and multiuser systems. We assume U users and we distinguish between downlink and uplink cases.

Downlink Model

We assume that the signal of user u uses a subset of $N^{(u)}$ subcarriers, out of N total subcarriers. The base station has N_{tx} transmit antennas. We denote with $\mathbf{d}_{i,j}^{(u)}$ the $N^{(u)} \times 1$ ith data vector of user u transmitted by the antenna j of the base station, and with $\mathbf{P}_{t,i,j}^{(u)}$ the $MN \times N^{(u)}$ pulse-shaping matrix related to the data vector $\mathbf{d}_{i,j}^{(u)}$ of user u. Hence, the ith signal vector $\mathbf{x}_{i,j}$ transmitted by the jth antenna can be expressed by

$$\mathbf{x}_{i,j} = \sum_{u=1}^{U} \mathbf{T}_j^{(u)} \mathbf{P}_{t,i,j}^{(u)} \mathbf{d}_{i,j}^{(u)}, \tag{13.44}$$

where $\mathbf{T}_j^{(u)}$ is the $(MN+G) \times MN$ precoding matrix for user u and antenna j. By defining $\mathbf{x}_i = [\mathbf{x}_{i,1}^T, \ldots, \mathbf{x}_{i,N_{\text{tx}}}^T]^T$, $\mathbf{d}_i^{(u)} = [(\mathbf{d}_{i,1}^{(u)})^T, \ldots, (\mathbf{d}_{i,N_{\text{tx}}}^{(u)})^T]^T$, $\mathbf{P}_{t,i}^{(u)} = \text{diag}\left(\mathbf{P}_{t,i,1}^{(u)}, \ldots, \mathbf{P}_{t,i,N_{\text{tx}}}^{(u)}\right)$, and $\mathbf{T}^{(u)} = \text{diag}\left(\mathbf{T}_1^{(u)}, \ldots, \mathbf{T}_{N_{\text{tx}}}^{(u)}\right)$, we obtain

$$\mathbf{x}_i = \sum_{u=1}^{U} \mathbf{T}^{(u)} \mathbf{P}_{t,i}^{(u)} \mathbf{d}_i^{(u)}. \tag{13.45}$$

Due to the time-domain superposition of consecutive vectors, we obtain

$$\mathbf{x}_{t,i,j} = \sum_{m=M_1(i)}^{M_2(i)} \mathbf{Z}_{s(m,i)} \mathbf{x}_{i+m,j}, \tag{13.46}$$

which, by defining $\mathbf{x}_{t,i} = [\mathbf{x}_{t,i,1}^T, \ldots, \mathbf{x}_{t,i,N_{\text{tx}}}^T]^T$ and $\mathbf{Z}_{m,i} = \left(\mathbf{I}_{N_{\text{tx}}} \otimes \mathbf{Z}_{s(m,i)}\right)$, can be rewritten as

$$\mathbf{x}_{t,i} = \sum_{m=M_1(i)}^{M_2(i)} \mathbf{Z}_{m,i} \mathbf{x}_{i+m}. \tag{13.47}$$

The vectors $\{\mathbf{x}_{t,i,j}\}$, or equivalently the aggregate vector $\mathbf{x}_{t,i}$, are transmitted by the base station and received by a generic user v, with $1 \leq v \leq U$, equipped with N_{rx} receive antennas. We denote with L the maximum channel order of all the $N_{\text{tx}} N_{\text{rx}}$ single-antenna channel links between the base station and the user v. By denoting with $\mathbf{H}_{l,j}^{(v)}$ the $(MN + G + L) \times (MN + G + 2L)$ channel matrix between the jth transmit antenna and the lth receive antenna of the vth user, and with $\mathbf{w}_{i,l}^{(v)}$ the AWGN gathered by the lth receive antenna, the signal $\mathbf{y}_{i,l}^{(v)}$ received by the lth receive antenna can be expressed by

$$\mathbf{y}_{i,l}^{(v)} = \sum_{j=1}^{N_{\text{tx}}} \mathbf{H}_{l,j}^{(v)} \mathbf{x}_{\text{t},i,j} + \mathbf{w}_{i,l}^{(v)}, \tag{13.48}$$

or equivalently by

$$\mathbf{y}_i^{(v)} = \mathbf{H}^{(v)} \mathbf{x}_{\text{t},i} + \mathbf{w}_i^{(v)}, \tag{13.49}$$

where $\mathbf{y}_i^{(v)} = [(\mathbf{y}_{i,1}^{(v)})^T, \ldots, (\mathbf{y}_{i,N_{\text{rx}}}^{(v)})^T]^T$, $\mathbf{w}_i^{(v)} = [(\mathbf{w}_{i,1}^{(v)})^T, \ldots, (\mathbf{w}_{i,N_{\text{rx}}}^{(v)})^T]^T$, and

$$\mathbf{H}^{(v)} = \begin{bmatrix} \mathbf{H}_{1,1}^{(v)} & \cdots & \mathbf{H}_{1,N_{\text{tx}}}^{(v)} \\ \vdots & & \vdots \\ \mathbf{H}_{N_{\text{rx}},1}^{(v)} & \cdots & \mathbf{H}_{N_{\text{rx}},N_{\text{tx}}}^{(v)} \end{bmatrix}. \tag{13.50}$$

The received signal $\mathbf{y}_{i,l}^{(v)}$ can further be processed using a receive matrix $\mathbf{R}_l^{(v)}$, which can also reduce the time support of the received vector, thereby obtaining

$$\mathbf{z}_{i,l}^{(v)} = \mathbf{R}_l^{(v)} \mathbf{y}_{i,l}^{(v)}, \tag{13.51}$$

or equivalently

$$\mathbf{z}_i^{(v)} = \mathbf{R}^{(v)} \mathbf{y}_i^{(v)}, \tag{13.52}$$

where $\mathbf{z}_i^{(v)} = [(\mathbf{z}_{i,1}^{(v)})^T, \ldots, (\mathbf{z}_{i,N_{\text{rx}}}^{(v)})^T]^T$ and $\mathbf{R}^{(v)} = \text{diag}\left(\mathbf{R}_1^{(v)}, \ldots, \mathbf{R}_{N_{\text{rx}}}^{(v)}\right)$.

By combining (13.45), (13.47), (13.49), and (13.52), we obtain

$$\mathbf{z}_i^{(v)} = \sum_{m=M_1(i)}^{M_2(i)} \mathbf{R}^{(v)} \mathbf{H}^{(v)} \mathbf{Z}_{m,i} \mathbf{x}_{i+m} + \mathbf{R}^{(v)} \mathbf{w}_i^{(v)} \tag{13.53}$$

$$= \sum_{u=1}^{U} \sum_{m=M_1(i)}^{M_2(i)} \mathbf{R}^{(v)} \mathbf{H}^{(v)} \mathbf{Z}_{m,i} \mathbf{T}^{(u)} \mathbf{P}_{\text{t},i+m}^{(u)} \mathbf{d}_{i+m}^{(u)} + \mathbf{R}^{(v)} \mathbf{w}_i^{(v)} \tag{13.54}$$

$$= \sum_{u=1}^{U} \sum_{m=M_1(i)}^{M_2(i)} \mathbf{A}_{m,i}^{(v,u)} \mathbf{d}_{i+m}^{(u)} + \mathbf{R}^{(v)} \mathbf{w}_i^{(v)}, \tag{13.55}$$

where

$$\mathbf{A}_{m,i}^{(v,u)} = \mathbf{R}^{(v)} \mathbf{H}^{(v)} \mathbf{Z}_{m,i} \mathbf{T}^{(u)} \mathbf{P}_{\text{t},i+m}^{(u)}. \tag{13.56}$$

Note that $\mathbf{A}_{0,i}^{(v,v)}$ is the useful matrix, $\{\mathbf{A}_{m,i}^{(v,v)}\}$ with $m \neq 0$ representing the self-user ISI matrices, while $\{\mathbf{A}_{m,i}^{(v,u)}\}$ with $u \neq v$ represents the interuser interference matrices.

Uplink Model

We assume that each user u has N_{tx} transmit antennas and uses a subset of $N^{(u)}$ subcarriers. We denote with $\mathbf{d}_{i,j}^{(u)}$ the $N^{(u)} \times 1$ ith data vector of user u transmitted by the antenna j, and with $\mathbf{P}_{\text{t},i,j}^{(u)}$ the $MN \times N^{(u)}$ pulse-shaping matrix related to the data vector $\mathbf{d}_{i,j}^{(u)}$ of user u. Hence, the ith signal vector $\mathbf{x}_{i,j}^{(u)}$ transmitted by the jth antenna of user u can be expressed by

$$\mathbf{x}_{i,j}^{(u)} = \mathbf{T}_j^{(u)}\mathbf{P}_{t,i,j}^{(u)}\mathbf{d}_{i,j}^{(u)}, \tag{13.57}$$

where $\mathbf{T}_j^{(u)}$ is the $(MN + G) \times MN$ precoding matrix for user u and antenna j. By defining $\mathbf{x}_i^{(u)} = [(\mathbf{x}_{i,1}^{(u)})^T, \dots, (\mathbf{x}_{i,N_{tx}}^{(u)})^T]^T$, $\mathbf{d}_i^{(u)} = [(\mathbf{d}_{i,1}^{(u)})^T, \dots, (\mathbf{d}_{i,N_{tx}}^{(u)})^T]^T$, $\mathbf{P}_{t,i}^{(u)} = \mathrm{diag}\left(\mathbf{P}_{t,i,1}^{(u)}, \dots, \mathbf{P}_{t,i,N_{tx}}^{(u)}\right)$, and $\mathbf{T}^{(u)} = \mathrm{diag}\left(\mathbf{T}_1^{(u)}, \dots, \mathbf{T}_{N_{tx}}^{(u)}\right)$, we obtain

$$\mathbf{x}_i^{(u)} = \mathbf{T}^{(u)}\mathbf{P}_{t,i}^{(u)}\mathbf{d}_i^{(u)}. \tag{13.58}$$

Due to the time-domain superposition of consecutive vectors, we have that

$$\mathbf{x}_{t,i,j}^{(u)} = \sum_{m=M_1(i)}^{M_2(i)} \mathbf{Z}_{s(m,i)}\mathbf{x}_{i+m,j}^{(u)}, \tag{13.59}$$

which, by defining $\mathbf{x}_{t,i}^{(u)} = [(\mathbf{x}_{t,i,1}^{(u)})^T, \dots, (\mathbf{x}_{t,i,N_{tx}}^{(u)})^T]^T$ and $\mathbf{Z}_{m,i} = \left(\mathbf{I}_{N_{tx}} \otimes \mathbf{Z}_{s(m,i)}\right)$, can be rewritten as

$$\mathbf{x}_{t,i}^{(u)} = \sum_{m=M_1(i)}^{M_2(i)} \mathbf{Z}_{m,i}\mathbf{x}_{i+m}^{(u)}. \tag{13.60}$$

The vectors $\{\mathbf{x}_{t,i,j}^{(u)}\}$, or equivalently the vectors $\{\mathbf{x}_{t,i}^{(u)}\}$, are transmitted by the U users and received by a base station equipped with N_{rx} receive antennas. We denote with L the maximum channel order of all the $UN_{tx}N_{rx}$ single-antenna channel links, including a possible time asynchronism among users. By denoting with $\mathbf{H}_{l,j}^{(u)}$ the $(MN + G + L) \times (MN + G + 2L)$ channel matrix between the jth transmit antenna and the lth receive antenna of the uth user, and with $\mathbf{w}_{i,l}$ the AWGN gathered by the lth receive antenna, the signal $\mathbf{y}_{i,l}$ received by the lth receive antenna of the base station can be expressed by

$$\mathbf{y}_{i,l} = \sum_{u=1}^{U}\sum_{j=1}^{N_{tx}} \mathbf{H}_{l,j}^{(u)}\mathbf{x}_{t,i,j}^{(u)} + \mathbf{w}_{i,l}, \tag{13.61}$$

and the total signal $\mathbf{y}_i = [\mathbf{y}_{i,1}^T, \dots, \mathbf{y}_{i,N_{rx}}^T]^T$ collected by all the receive antennas is expressed by

$$\mathbf{y}_i = \sum_{u=1}^{U} \mathbf{H}^{(u)}\mathbf{x}_{t,i}^{(u)} + \mathbf{w}_i, \tag{13.62}$$

where $\mathbf{w}_i = [\mathbf{w}_{i,1}^T, \dots, \mathbf{w}_{i,N_{rx}}^T]^T$ and

$$\mathbf{H}^{(u)} = \begin{bmatrix} \mathbf{H}_{1,1}^{(u)} & \cdots & \mathbf{H}_{1,N_{tx}}^{(u)} \\ \vdots & & \vdots \\ \mathbf{H}_{N_{rx},1}^{(u)} & \cdots & \mathbf{H}_{N_{rx},N_{tx}}^{(u)} \end{bmatrix}. \tag{13.63}$$

The received signal $\mathbf{y}_{i,l}$ can be processed using a receive matrix $\mathbf{R}_l^{(v)}$, where v represents the user to be detected, thus obtaining

$$\mathbf{z}_{i,l}^{(v)} = \mathbf{R}_l^{(v)} \mathbf{y}_{i,l}, \tag{13.64}$$

or equivalently

$$\mathbf{z}_i^{(v)} = \mathbf{R}^{(v)} \mathbf{y}_i, \tag{13.65}$$

where $\mathbf{z}_i^{(v)} = [(\mathbf{z}_{i,1}^{(v)})^T, \ldots, (\mathbf{z}_{i,N_{\mathrm{rx}}}^{(v)})^T]^T$ and $\mathbf{R}^{(v)} = \mathrm{diag}\left(\mathbf{R}_1^{(v)}, \ldots, \mathbf{R}_{N_{\mathrm{rx}}}^{(v)}\right)$.

By combining (13.58), (13.60), (13.62), and (13.65), we obtain

$$\mathbf{z}_i^{(v)} = \sum_{u=1}^{U} \sum_{m=M_1(i)}^{M_2(i)} \mathbf{R}^{(v)} \mathbf{H}^{(u)} \mathbf{Z}_{m,i} \mathbf{T}^{(u)} \mathbf{P}_{\mathrm{t},i+m}^{(u)} \mathbf{d}_{i+m}^{(u)} + \mathbf{R}^{(v)} \mathbf{w}_i \tag{13.66}$$

$$= \sum_{u=1}^{U} \sum_{m=M_1(i)}^{M_2(i)} \mathbf{A}_{m,i}^{(v,u)} \mathbf{d}_{i+m}^{(u)} + \mathbf{R}^{(v)} \mathbf{w}_i, \tag{13.67}$$

where

$$\mathbf{A}_{m,i}^{(v,u)} = \mathbf{R}^{(v)} \mathbf{H}^{(u)} \mathbf{Z}_{m,i} \mathbf{T}^{(u)} \mathbf{P}_{\mathrm{t},i+m}^{(u)}. \tag{13.68}$$

Similarly to the downlink case, $\mathbf{A}_{0,i}^{(v,v)}$ is the useful matrix, $\{\mathbf{A}_{m,i}^{(v,v)}\}$ with $m \neq 0$ represents the self-user ISI matrices, while $\{\mathbf{A}_{m,i}^{(v,u)}\}$ with $u \neq v$ represents the interuser interference matrices.

13.3   Equalization

Herein we discuss some equalization and detection strategies that are common to all the techniques mentioned in the previous section, such as linear equalization and interference cancellation. For the sake of simplicity, we focus on single-antenna single-user systems; the same equalization strategies can be applied also to multiantenna multiuser systems. For convenience, we rewrite the received observation vector in (13.42) as

$$\mathbf{z}_i = \mathbf{A}_{0,i} \mathbf{d}_i + \mathbf{i}_i + \mathbf{v}_i, \tag{13.69}$$

where $\mathbf{A}_{0,i}$ is the $W \times N$ system matrix, defined in (13.41), that includes transmitter pre-processing and channel (the value of W and the elements in $\mathbf{A}_{0,i}$ depend on the specific transmission scheme), $\mathbf{d}_i$ is the $N \times 1$ data vector of the ith symbol, $\mathbf{i}_i$ is the ISI, which is the interference term due to the data of the other symbols, and $\mathbf{v}_i$ is the noise term. Similarly, we rewrite the received observation vector in (13.43) as

$$\mathbf{z}_i = \mathbf{A}_{i,\mathrm{tot}} \mathbf{d}_{i,\mathrm{long}} + \mathbf{v}_i, \tag{13.70}$$

where $\mathbf{A}_{i,\mathrm{tot}} = \left[\mathbf{A}_{M_1(i),i}, \cdots, \mathbf{A}_{M_2(i),i}\right]$ is the $W \times N_{\mathrm{tot}}$ system matrix, with $N_{\mathrm{tot}} = (M_2(i) - M_1(i) + 1)N$ and $\mathbf{d}_{i,\mathrm{long}} = [\mathbf{d}_{i+M_1(i)}^H, \cdots, \mathbf{d}_{i+M_2(i)}^H]^H$ is the $N_{\mathrm{tot}} \times 1$ data vector that contains multiple symbols. Note that $\mathbf{A}_{0,i}$ is a submatrix of $\mathbf{A}_{i,\mathrm{tot}}$.

In the following, we assume that the matrices $\mathbf{A}_{0,i}$ and $\mathbf{A}_{i,\text{tot}}$ are known to the receiver (i.e., perfect channel estimation is assumed), and that $\mathbf{v}_i$ is AWGN. In addition, we assume that $\mathbf{A}_{0,i}$ is tall (i.e., $W \geq N$) and that $\mathbf{A}_{i,\text{tot}}$ is large (i.e., $W \leq (M_2(i) - M_1(i) + 1)N$).

13.3.1   Linear Equalization

A linear equalizer obtains the soft-estimated data $\hat{\mathbf{d}}_i$ by a linear combination of the elements of the received observation vector $\mathbf{z}_i$, as expressed by

$$\hat{\mathbf{d}}_i = \mathbf{E}_i \mathbf{z}_i, \tag{13.71}$$

where $\mathbf{E}_i$ is the $N \times W$ matrix that represents the equalizer. Different choices for the equalizer matrix $\mathbf{E}_i$ are possible. For instance, using the model (13.69), a matched filter (MF) equalizer can be expressed by

$$\mathbf{E}_i = \mathbf{A}_{0,i}^H, \tag{13.72}$$

which produces the soft output

$$\hat{\mathbf{d}}_i = \mathbf{A}_{0,i}^H \mathbf{A}_{0,i} \mathbf{d}_i + \mathbf{A}_{0,i}^H \boldsymbol{\eta}_i. \tag{13.73}$$

where

$$\boldsymbol{\eta}_i = \mathbf{i}_i + \mathbf{v}_i \tag{13.74}$$

is the ISI-plus-noise term before equalization. Note that the MF equalizer maximizes the signal-to-noise ratio (SNR) on the soft output. The MF produces relevant ICI among the data belonging the same symbol $\mathbf{d}_i$, because $\mathbf{A}_{0,i}^H \mathbf{A}_{0,i}$ is not diagonal. In addition, the MF treats the ISI $\mathbf{i}_i$ (produced by the different data symbols $\mathbf{d}_m$, with $m \neq i$) as a noise term.

An alternative approach is the use of a least-squares (LS) equalizer designed from (13.69), as expressed by

$$\mathbf{E}_i = \left(\mathbf{A}_{0,i}^H \mathbf{A}_{0,i} \right)^{-1} \mathbf{A}_{0,i}^H, \tag{13.75}$$

which produces the soft output

$$\hat{\mathbf{d}}_i = \mathbf{d}_i + \left(\mathbf{A}_{0,i}^H \mathbf{A}_{0,i} \right)^{-1} \mathbf{A}_{0,i}^H \boldsymbol{\eta}_i. \tag{13.76}$$

Note that $\mathbf{E}_i \mathbf{A}_{0,i} = \mathbf{I}_N$ is diagonal and hence the LS equalizer (13.75) completely removes the ICI produced by the data $\mathbf{d}_i$ on the other subcarriers. However, the ISI $\mathbf{i}_i$ is treated as noise and, due to the matrix inversion, the ISI-plus-noise term is often amplified. In order to take the ISI into account, an alternative LS equalizer can be designed from (13.70), as expressed by

$$\mathbf{E}_i = \mathbf{A}_{0,i}^H \left(\mathbf{A}_{i,\text{tot}} \mathbf{A}_{i,\text{tot}}^H \right)^{-1}. \tag{13.77}$$

This second LS equalizer tries to jointly remove both ICI and ISI. Note that the LS equalizer of (13.77) is more complex than that of (13.75), since $\mathbf{A}_{i,\text{tot}}$ is bigger than $\mathbf{A}_{0,i}^H$.

A third option is the linear minimum mean-squared error (MMSE) equalizer, as expressed by

$$\mathbf{E}_i = \left(\mathbf{\Sigma}_{\mathbf{d}_i}^{-1} + \mathbf{A}_{0,i}^H \mathbf{\Sigma}_{\boldsymbol{\eta}_i}^{-1} \mathbf{A}_{0,i} \right)^{-1} \mathbf{A}_{0,i}^H \mathbf{\Sigma}_{\boldsymbol{\eta}_i}^{-1}, \tag{13.78}$$

where $\mathbf{\Sigma}_{\boldsymbol{\eta}_i}$ stands for the covariance matrix of the ISI-plus-noise term before equalization. This MMSE equalizer leads to the soft output

$$\hat{\mathbf{d}}_i = \left(\mathbf{\Sigma}_{\mathbf{d}_i}^{-1} + \mathbf{A}_{0,i}^H \mathbf{\Sigma}_{\boldsymbol{\eta}_i}^{-1} \mathbf{A}_{0,i} \right)^{-1} \mathbf{A}_{0,i}^H \mathbf{\Sigma}_{\boldsymbol{\eta}_i}^{-1} \mathbf{A}_{0,i} \mathbf{d}_i \tag{13.79}$$

$$+ \left(\mathbf{\Sigma}_{\mathbf{d}_i}^{-1} + \mathbf{A}_{0,i}^H \mathbf{\Sigma}_{\boldsymbol{\eta}_i}^{-1} \mathbf{A}_{0,i} \right)^{-1} \mathbf{A}_{0,i}^H \mathbf{\Sigma}_{\boldsymbol{\eta}_i}^{-1} \boldsymbol{\eta}_i. \tag{13.80}$$

The MMSE equalizer tries to balance the residual ICI caused by the non-diagonal matrix $\mathbf{E}_i \mathbf{A}_{0,i}$ and the ISI-plus-noise amplification caused by the matrix inversion, aiming at minimizing the mean-squared error (MSE). Similarly to the LS approach of (13.77), another MMSE equalizer can be designed using (13.70); also in this MMSE case, the whole system matrix $\mathbf{A}_{i,\text{tot}}$ is required, as expressed by

$$\mathbf{E}_i = \mathbf{S}_i \left(\mathbf{\Sigma}_{\mathbf{d}_{i,\text{long}}}^{-1} + \mathbf{A}_{i,\text{tot}}^H \mathbf{\Sigma}_{\mathbf{v}_i}^{-1} \mathbf{A}_{i,\text{tot}} \right)^{-1} \mathbf{A}_{i,\text{tot}}^H \mathbf{\Sigma}_{\mathbf{v}_i}^{-1}, \tag{13.81}$$

where $\mathbf{S}_i = [\mathbf{0}_{N \times N}, \cdots, \mathbf{0}_{N \times N}, \mathbf{I}_N, \mathbf{0}_{N \times N}, \cdots, \mathbf{0}_{N \times N}]$ is the $N \times N_{\text{tot}}$ selection matrix that extracts the desired data estimate $\hat{\mathbf{d}}_i$ from $\hat{\mathbf{d}}_{i,\text{long}}$.

Many other linear equalizers can be designed. Some of them are obtained as convenient modifications of the three equalization approaches described above [34]. For instance, an LS equalizer can be modified by including the covariance of either the signal $\mathbf{d}_i$ or the ISI-plus-noise term $\boldsymbol{\eta}_i$, yielding a weighted LS equalizer. As another example, an MMSE equalizer can be modified by subsequent application of $\mathbf{D}_i^{-1}$, where $\mathbf{D}_i = \text{diag}(\mathbf{E}_i \mathbf{A}_{0,i})$ is diagonal; this way, since the bias is removed, we obtain an unbiased MMSE (UMMSE) equalizer, which gives the same signal-to-interference-plus-noise ratio (SINR) of its corresponding (biased) MMSE equalizer.

Since the equalizer $\mathbf{E}_i$ depends on the system matrix $\mathbf{A}_{0,i}$ or $\mathbf{A}_{i,\text{tot}}$, which in turn depend on the transmission technique, the mathematical expression of $\mathbf{E}_i$ is different for each transmission scheme. For instance, in OFDM with CP (assuming perfect synchronization), both ICI and ISI are absent. Therefore, in the OFDM case, the three equalizers MF, LS, and UMMSE coincide, leading to

$$\mathbf{E}_i = \mathbf{\Lambda}^{-1} \mathbf{F}_N, \tag{13.82}$$

where $\mathbf{\Lambda}$ is the diagonal matrix representing the frequency-domain channel matrix. This per-subcarrier equalization technique produces a soft output expressed by

$$\hat{\mathbf{d}}_i = \mathbf{d}_i + \mathbf{\Lambda}^{-1} \mathbf{F}_N \mathbf{v}_i. \tag{13.83}$$

On the contrary, in FBMC usually the ISI $\mathbf{i}_i$ is not zero, and hence the equalizers are different; in general, the MMSE approaches produce better performance, since the interference (or its covariance) is taken into account.

13.3.2   Nonlinear Equalization

A popular nonlinear equalization approach is to use interference cancellation techniques [35–38]. Interference cancellers usually perform four steps: (1) preliminary estimation of some data vectors, obtained by using a first linear equalizer; (2) reconstruction of the part of the interference, by using the output of the preliminary estimator; (3) hard or soft cancellation of the interference, performed by subtracting the reconstructed interference from the received vector; and (4) final equalization, by means of a second linear equalizer. If we denote with $\hat{\mathbf{i}}_i$ the interference reconstructed by the first linear equalizer, we can express the interference canceller output as

$$\hat{\mathbf{d}}_i = \mathbf{E}_i \left(\mathbf{z}_i - f_{\mathrm{NL}}(\hat{\mathbf{i}}_i) \right), \tag{13.84}$$

where $\mathbf{E}_i$ denotes the second linear equalizer and $f_{\mathrm{NL}}(\cdot)$ denotes a nonlinear increasing function that acts component-wise. The nonlinear function can be selected in order to increase the amount of cancellation for the well-estimated interference components. In case of hard cancellation, the nonlinear function is omitted (or is replaced by the identity function). Note that the interference cancellation can be done in parallel or successively. In parallel cancellation, many interference components are preliminarily estimated and subtracted, before the equalization of many data subcarriers. In successive cancellation, a single interference component is estimated and canceled, in order to equalize a single data subcarrier, and then the whole process is repeated for the other data subcarriers in a decision-feedback way. Turbo approaches are also possible, in order to iteratively refine the equalization of the already equalized data [39, 40].

Another nonlinear equalization approach is to use maximum-likelihood equalization techniques [34, 41]. The key idea of the ML equalization is to exploit the finite alphabet of the constellations of the data symbols. The hard-estimated data symbol $\hat{\mathbf{d}}_i$ is obtained as

$$\hat{\mathbf{d}}_i = \mathbf{S}_i \hat{\mathbf{d}}_{i,\mathrm{long}}, \tag{13.85}$$

$$\hat{\mathbf{d}}_{i,\mathrm{long}} = \arg \max_{\mathbf{d}_{i,\mathrm{long}}} \left\| \mathbf{z}_i - \mathbf{A}_{i,\mathrm{tot}} \mathbf{d}_{i,\mathrm{long}} \right\|^2, \tag{13.86}$$

where $\mathbf{S}_i$ is an $N \times N_{\mathrm{tot}}$ selection matrix that extracts the desired components.

From a performance viewpoint, ML equalization is a promising approach; indeed, for equiprobable data symbols, the ML detector maximizes the probability of correct detection of the vector $\mathbf{d}_i$. However, the main drawback of the ML detector is its computational complexity. The computational complexity of ML equalization depends on the chosen transmission scheme and on many system parameters: the number of subcarriers N that determines the size of the data vector $\mathbf{d}_i$; the number N_{tot} of interfering symbols $\{\mathbf{d}_m\}$ that affect the size of $\mathbf{A}_{i,\mathrm{tot}}$; and the constellation sizes of the data. The complexity of ML equalization can be reduced in two ways. First, a special structure of the transmitted signal can be exploited. For instance, if the ISI is absent (as in OFDM with CP removal), the ML equalizer can be simplified by using $\mathbf{A}_{0,i}$ instead of $\mathbf{A}_{i,\mathrm{tot}}$. If, in addition, also the ICI is absent, then the ML detector is equivalent to a linear equalizer; this way, the computational complexity can be reduced (to be linear with the data size N, instead of exponential). Second, low-complexity approximate ML detectors can reduce the search space in some ways, for instance by neglecting some improbable data vectors $\mathbf{d}_{i,\text{long}}$, as done by some sphere-decoding algorithms, or by including additional assumptions on the data vector $\mathbf{d}_{i,\text{long}}$, such as incorporating the knowledge of previously detected symbols [42]. In any case, the complexity of ML equalization is often not affordable by wireless communication systems. Therefore, in the following we will focus on linear equalization strategies only.

13.4   Channel Capacity and Spectral Efficiency

Rather than focusing on complexity issues, in this chapter we are more interested in comparing the different transceiver architectures from an information theoretic point of view. Our aim is to take into account that different nonorthogonal waveforms may have a different use of the time–frequency resource, may introduce and may be able to tolerate a different amount of interference, and also that suboptimal receivers could be adopted (the optimal ML receiver has clearly a computational complexity which is out of reach). As a figure of merit we will use the achievable spectral efficiency (ASE) with the constraint of arbitrarily small bit error rate (BER). The ASE is obtained by dividing the channel capacity (or an achievable lower bound of it) by the employed symbol time and frequency spacing (the time–frequency resource of every waveform).

As far as the computation of (an achievable lower bound of) the channel capacity is concerned, since we are considering simple receivers based on linear processing, as described in Section 13.3, followed by symbol-by-symbol detection, we are mainly interested in the achievable performance when using these suboptimal low-complexity detectors. We thus resort to the framework described in [43, 44]. This framework allows us to compute a proper lower bound on the channel capacity (and thus on the ASE) obtained by substituting the actual channel with an arbitrary auxiliary one with the same input and output alphabets as the original channel (mismatched detection [43, 44]) – the more accurate the auxiliary channel to approximate the actual one, the closer the bound. If the considered suboptimal detector is optimal for the adopted auxiliary channel, the obtained lower bound is *achievable* by that receiver, according to mismatched detection theory [44]. We thus say, with a slight abuse of terminology, that the computed lower bound is the ASE of the considered channel with the considered waveforms when that receiver is employed. When Gaussian inputs are considered, we will see that closed-form achievable lower bounds of the ASE can be provided. The same framework can also be used when finite constellations are employed at the channel input but, this time, no closed-form expressions for the lower bounds are available. However, they can be numerically computed by feeding the optimal detector for the auxiliary channel with the output of the real channel [44].

This section is divided into four subsections. The first three subsections present some channel capacity results using a vector input–output model and a scalar input–output model, where the output is taken before equalization and after equalization. The last subsection uses the results of the first three subsections to evaluate the ergodic channel capacity and the ASE. Throughout this section we will assume a single-user system with perfect synchronization and perfect channel knowledge at the receiver. We will also assume time-invariant multipath channels characterized by zero-mean complex Gaussian taps (Rayleigh fading).

13.4.1   Vector Input, Vector Output: Before Equalization

Herein we exploit some channel capacity results originally developed for multiantenna systems. Indeed, also the single-user single-antenna system described in Section 13.2 is a MIMO system, due to the multiple data symbols over multiple subcarriers. Let us consider the system equation (13.42) for the single-user case. The ith received block of size W, before equalization, can be written as

$$\mathbf{z}_i = \mathbf{A}_{0,i}\mathbf{d}_i + \sum_{m\neq 0} \mathbf{A}_{m,i}\mathbf{d}_{i+m} + \mathbf{v}_i \tag{13.87}$$

$$= \mathbf{A}_{0,i}\mathbf{d}_i + \mathbf{i}_i + \mathbf{v}_i, \tag{13.88}$$

where $\mathbf{d}_i$ is the $N \times 1$ data vector of the ith block containing uniformly and identically distributed complex random variables with Gaussian distribution having zero mean and covariance $\sigma_d^2 \mathbf{I}_N$, and $\mathbf{i}_i$ models the ISI from previous (and successive) data blocks.

If we consider the model (13.88), it is possible to obtain a lower bound on the mutual information for complex Gaussian alphabets and a given channel realization, by considering the following auxiliary channel model

$$\mathbf{z}_i = \mathbf{A}_{0,i}\mathbf{d}_i + \boldsymbol{\eta}_i, \tag{13.89}$$

and by modeling the ISI-plus-noise term $\boldsymbol{\eta}_i = \mathbf{i}_i + \mathbf{v}_i$ as an irreducible Gaussian noise. For this auxiliary channel model the mutual information can be expressed in closed form by

$$I^{(\mathrm{L})}(\mathbf{d}_i; \mathbf{z}_i|\mathbf{H}) = \log_2 \left| \mathbf{I}_W + \mathbf{A}_{0,i}\boldsymbol{\Sigma}_{\mathbf{d}_i}\mathbf{A}_{0,i}^H \boldsymbol{\Sigma}_{\boldsymbol{\eta}_i}^{-1} \right|, \tag{13.90}$$

where $\boldsymbol{\Sigma}_{\mathbf{d}_i} = E\left\{\mathbf{d}_i\mathbf{d}_i^H\right\}$ is the covariance matrix of the data signal and, similarly,

$$\boldsymbol{\Sigma}_{\boldsymbol{\eta}_i} = \sum_{m\neq 0} \mathbf{A}_{m,i+m}\boldsymbol{\Sigma}_{\mathbf{d}_{i+m}}\mathbf{A}_{m,i+m}^H + \sigma_v^2 \mathbf{I}_W \tag{13.91}$$

is the covariance matrix of the ISI-plus-noise term. The mutual information in (13.90) depends on the system matrix $\mathbf{A}_{0,i}$ defined in (13.41); therefore, the mutual information (13.90) is conditioned on the channel matrix $\mathbf{H}$ and depends on the pulse-shaping waveform. Assuming uniform power allocation $\boldsymbol{\Sigma}_{\mathbf{d}_i} = \sigma_d^2 \mathbf{I}_N$, (13.90) reduces to

$$I^{(\mathrm{L})}(\mathbf{d}_i; \mathbf{z}_i|\mathbf{H}) = \log_2 \left| \mathbf{I}_W + \sigma_d^2 \mathbf{A}_{0,i}\mathbf{A}_{0,i}^H \boldsymbol{\Sigma}_{\boldsymbol{\eta}_i}^{-1} \right|. \tag{13.92}$$

13.4.2   Vector Input, Vector Output: After Equalization

The use of a linear equalizer usually improves the uncoded symbol error rate (SER) performance of a suboptimal symbol-by-symbol equalizer. Basically, a linear equalizer should boost the useful component and possibly reduce the unwanted interference. If $\mathbf{E}_i$ represents the block linear equalizer (such as the LS and the linear MMSE), using the same assumptions of Subsection 13.4.1, we will come up with a soft estimate $\hat{\mathbf{d}}_i$ of the transmitted data $\mathbf{d}_i$ expressed by

$$\hat{\mathbf{d}}_i = \mathbf{E}_i \mathbf{z}_i \tag{13.93}$$

$$= \mathbf{E}_i \mathbf{A}_{0,i} \mathbf{d}_i + \mathbf{E}_i \mathbf{i}_i + \mathbf{E}_i \mathbf{v}_i \tag{13.94}$$

$$= \mathbf{E}_i \mathbf{A}_{0,i} \mathbf{d}_i + \boldsymbol{\eta}_i^{(e)}, \tag{13.95}$$

where $\mathbf{E}_i \mathbf{A}_{0,i} \mathbf{d}_i$ contains the useful information (and possibly the ICI for scalar symbol-by-symbol suboptimum detection), $\mathbf{E}_i \mathbf{i}_i$ contains only the residual ISI, and $\boldsymbol{\eta}_i^{(e)} = \mathbf{E}_i \mathbf{i}_i + \mathbf{E}_i \mathbf{v}_i$ is the ISI-plus-noise term after equalization. Thus, with a proper definition of the auxiliary channel, (13.95) leads to a lower bound for the equalized system, expressed by

$$I_e^{(\mathrm{L})}(\mathbf{d}_i; \mathbf{z}_i | \mathbf{H}) = \log_2 \left| \mathbf{I}_N + \sigma_d^2 \mathbf{E}_i \mathbf{A}_{0,i} \mathbf{A}_{0,i}^H \mathbf{E}_i^H \boldsymbol{\Sigma}_{\boldsymbol{\eta}_i^{(e)}}^{-1} \right|, \tag{13.96}$$

where

$$\boldsymbol{\Sigma}_{\boldsymbol{\eta}_i^{(e)}} = \sigma_d^2 \sum_{m \neq 0} \mathbf{A}_{m,i}^{(e)} \mathbf{A}_{m,i}^{(e)H} + \sigma_v^2 \mathbf{E}_i \mathbf{E}_i^H, \tag{13.97}$$

being $\mathbf{A}_{m,i}^{(e)} = \mathbf{E}_i \mathbf{A}_{m,i}$.

13.4.3 Scalar Input, Scalar Output: After Equalization

Now we consider a scalar-input scalar-output model, using the same assumption of the previous two subsections. We focus on the data symbol transmitted on subcarrier k at symbol time i, denoted with $d_{i,k}$, and on the corresponding equalized symbol $\hat{d}_{i,k}$.

After the linear equalizer $\mathbf{E}_i$, the soft symbol estimates in (13.95) can be expressed in the form

$$\hat{d}_{i,k} = \beta_{i,k} d_{i,k} + \sum_{(m,j) \neq (0,k)} \alpha_{(m,i,j,k)} d_{i+m,j} + v_{i,k}^{(e)}, \tag{13.98}$$

where $\alpha_{(m,i,j,k)}$ is the element on the kth row and on the jth column of the matrix $\mathbf{E}_i \mathbf{A}_{m,i}$, $\beta_{i,k} = \alpha_{(0,i,k,k)}$, and $v_{i,k}^{(e)}$ is the kth element of the colored noise $\mathbf{v}_i^{(e)} = \mathbf{E}_i \mathbf{v}_i$ after equalization. Equation (13.98) basically says that the soft equalized symbol is equal to a scaled version of the transmitted symbol $d_{i,k}$ plus a linear combination, with proper coefficients $\{\alpha_{(m,i,j,k)}\}$ of all remaining (interfering) symbols and the colored noise $\{v_{i,k}^{(e)}\}$. The coefficients $\{\alpha_{(m,i,j,k)}\}$ depend on the channel matrix and can be considered as Gaussian random variables with proper covariance matrix. As mentioned, we will assume that symbol-by-symbol detection is employed at the receiver after the linear equalizer.[1] This corresponds to the adoption of the following auxiliary channel:

$$\hat{d}_{i,k} = \beta_{i,k} d_{i,k} + \mu_{i,k}, \tag{13.99}$$

---

[1] Receivers based on interference cancellation cannot be considered within this framework because an auxiliary channel for which these receivers are optimal cannot be found.

where

$$\mu_{i,k} = \sum_{(m,j)\neq(0,k)} \alpha_{(m,i,j,k)} d_{i+m,j} + v_{i,k}^{(e)} \tag{13.100}$$

takes into account both thermal noise and interference and is modeled as a Gaussian random variable with mean zero and variance

$$\sigma_{\mu_{i,k}}^2 = \sigma_d^2 \sum_{(m,j)\neq(0,k)} |\alpha_{(m,i,j,k)}|^2 + \sigma_{v_{i,k}^{(e)}}^2, \tag{13.101}$$

where $\sigma_{v_{i,k}^{(e)}}^2$ is the variance of the thermal noise after equalization, i.e., σ_v^2 multiplied by the element (k,k) of $\mathbf{E}_i\mathbf{E}_i^H$. The samples $\mu_{i,k}$ are correlated. However, this correlation is not taken into account at the receiver, since the receiver acts on a symbol-by-symbol basis.

For a given realization of the channel matrix $\mathbf{H}$, the capacity for this auxiliary channel is known and given by

$$I_{i,k}^{(L)}(\mathbf{H}) = \log_2 \left(1 + \frac{|\beta_{i,k}|^2 \sigma_d^2}{\sigma_d^2 \sum_{(m,j)\neq(0,k)} |\alpha_{(m,i,j,k)}|^2 + \sigma_{v_{i,k}^{(e)}}^2} \right). \tag{13.102}$$

This will be a lower bound for the actual channel, achievable by the considered waveform with the considered linear equalizer.

The closed-form lower bound in (13.102) has been obtained under the assumption of Gaussian inputs. On the other hand, when the input symbols are independent and uniformly distributed (i.u.d.) random variables belonging to a given constellation, no closed form exists and we have to resort to numerical computation [45].

13.4.4 Ergodic Capacity

For any expression bound that we are going to consider, we are dealing with mutual information conditioned on a single channel realization. Assuming (as standard) a block fading channel model, it is possible to define an ergodic mutual information (or ergodic capacity) as the statistical average of the conditional mutual information over the channel. For instance, for the bound of (13.96), we obtain

$$I_e^{(L)}(\mathbf{d}_i; \mathbf{z}_i) = E_{\mathbf{H}} \left\{ \log_2 \left| \mathbf{I}_N + \sigma_d^2 \mathbf{E}_i \mathbf{A}_{0,i} \mathbf{A}_{0,i}^H \mathbf{E}_i^H \mathbf{\Sigma}_{\eta_i^{(e)}}^{-1} \right| \right\}, \tag{13.103}$$

since the system matrix $\mathbf{A}_{0,i}$, the equalizer matrix $\mathbf{E}_i$, and the interference-plus-noise covariance matrix $\mathbf{\Sigma}_{\eta_i^{(e)}}$ depend on the channel matrix $\mathbf{H}$.

In order to compute (13.103) or similar quantities, we can use a semi-analytical approach that first evaluates (13.96) for many different channel realizations $\mathbf{H}$ and then estimates the expected value (13.103) by a sample average of the values obtained using (13.96).

The ASE ρ is simply obtained by dividing the (ergodic or outage) achievable capacity lower bound by the product between the frequency spacing $\delta_f/(NT_c)$ and the symbol time of each subcarrier, i.e., $N\delta_t T_c$. Thus, considering for example the ergodic rate, the ergodic ASE can be defined as

$$\rho = \frac{E_{\mathbf{H}}\{I^{(L)}(\mathbf{H})\}}{\delta_f \delta_t}. \tag{13.104}$$

This allows a fair comparison between orthogonal and nonorthogonal waveforms. In fact, the interference that characterizes nonorthogonal waveforms will reduce the capacity lower bound – the denser the packing, the denser the interference. However, a denser packing will also reduce the denominator. This could result in an increase in the ASE value. In general, δ_f and δ_t have to be optimized, according to the *time–frequency packing principle* [46].

When the channel is slowly varying or when it is modeled as constant over the length of one or more data frames, we should consider the concept of outage capacity [47]. In fact, it is very likely that there will be blocks during which it is impossible to achieve a low error probability regardless of the signaling rate. Under such circumstances the channel will be considered to be in outage. As a result, we need to consider the channel capacity when there is a nonzero probability P_{out} that the channel is in outage, and therefore unusable in the sense that the desired data rate cannot be achieved with arbitrarily low error rate. In fact, under such conditions, $I^{(L)}(\mathbf{H})$ may take on arbitrarily small values with nonzero probability, so that arbitrarily low probability of error cannot be achieved regardless of the chosen codes [47]. The outage probability, P_{out}, thus specifies the probability of not achieving a given channel capacity. The maximum rate that can be supported by the channel with a given outage probability is the outage capacity.

Due to limits of space, we will limit our attention to ASE. This means that the transceiver models with different pulse shapes are compared without taking into account specific coding schemes, being understood that, with a properly designed channel code, the information theoretic performance can be closely approached.

13.5   Simulation Results and Comparison

This section is dedicated to the comparison of the different signal waveforms in terms of spectral efficiency, exploiting the unifying framework developed so far. As explicitly outlined by the theoretical formulation, the information rate, and thus the spectral efficiency, is highly influenced by the channel matrix and its statistical characteristics. A comprehensive and detailed channel propagation model for 5G systems is rather complex because, in addition to path-loss attenuation, frequency dependence, multipath and time variability, it should possibly take into account also very large bandwidths, a wide range of frequencies (from 500 MHz up to several GHz), pico-cell and large-cell scenarios, etc.; furthermore, 5G channel models are typically space-dependent and take into account the direction of arrival of each path and correct modeling of antenna-arrays and (Massive) MIMO configurations [48–50]. A detailed comparison of the different signal waveforms in all the possible scenarios is beyond the scope of this chapter, since a thorough comparison would depend on too many parameters. Thus, in this chapter, we limit the comparisons in terms of total capacity of single-antenna downlink systems, which is enough to give meaningful insights on the merits, drawbacks, and signal processing challenges that characterize each of the considered signal waveforms. To this end, rather than picking a specialized channel model among the several that are available, we stay with classical and general multipath fading channel models, such as those with exponentially decaying power–delay profiles (widely used for WLAN applications) and those with sparse power–delay profiles (widely used for cellular communications). The parameters of these channels, such as the power-decaying exponent and the number of paths, can be adjusted to roughly model the typical delay-spread and frequency selectivity of pico-cells, rather than micro-cells, or urban cells, which are foreseen to be simultaneously employed in 5G cellular systems. Furthermore, some parameters of the candidate signal waveforms, such as the number of subcarriers, the number of blocks in a frame, and the possible CP, are fixed in the simulations and comparisons, in order to be compliant both with the propagation environment (channel delay spread) and possibly also with a given maximum latency that the system wishes to guarantee.

In the following, we compare the ASE (measured as in (13.104)) of different signal waveforms and equalization techniques by semianalytically computing the lower bounds (LBs) derived in the previous section. The semianalytical computation is performed using three steps: First, we randomly generate several channel realizations, according to the statistics of each specific channel model; second, we analytically evaluate the LBs conditioned on the specific channel realization; and third, we average the conditional LBs over the generated channel realizations. We consider three types of LB:

1. the LB (13.92) of the vector input–output model (13.89) before equalization;
2. the LB (13.96) of the vector input–output model (13.95) after equalization;
3. the LB (13.102) of the scalar input–output model (13.99) after equalization, referred to as a *memoryless receiver*.

For a fair comparison, the third LB is summed over all the scalar symbols that constitute the symbol vector used for the vector bounds. For the first two cases, we also consider the corresponding upper bounds (UBs) obtained by neglecting the interference term within the input–output model. The bounds obtained after equalization make use of unbiased linear equalizers, such as LS and UMMSE; with a slight abuse of notation, we will refer to the UMMSE as the MMSE, since we have verified that both MMSE and UMMSE produce the same results for (13.96) (and also for (13.102)).

Since the maximum latency of 5G systems is expected to be in the order of 1 ms, we consider pulse-shaping waveforms with limited duration $D \lesssim 50\,\mu s$: for a multicarrier system with bandwidth $W = 20$ MHz; this duration approximately corresponds to $T_c = 50$ ns, $M \leq 4$, and $N = 256$ subcarriers. Constraining the duration D for latency purposes corresponds to fixing the subcarrier separation; thus, a higher number of subcarriers would correspond to a proportionally increased bandwidth. For OFDM, the CP length G is set equal to the channel order L. Similarly, for UFMC, the filter length is set to $G = L$, and, additionally, the N subcarriers are split into eight subbands with $N/8$ subcarriers each.

Sparse multipath channels are considered, with power–delay profile (PDP) selected according to LTE channel models [51]. In addition, multipath channels with exponentially decaying PDP are considered; in this case, the PDP is expressed by $E\{|h[n]|^2\} = C\exp(-\Delta n)$, where Δ is the PDP decay and C is a normalization constant. The PDP decay has been selected according to WLAN channel models [52]. Rayleigh fading is assumed for all the paths. In case of Rice fading, a line-of-sight (LOS) component is added to the first path of the channel; the Rice factor, defined as the power ratio between the LOS component and the sum of the non-LOS components, is set to $K = 10$.

For FBMC and UFMC, different pulse-shaping waveforms have been considered. For UFMC, Dolph–Chebyshev pulses [53] are assumed, and the capacity optimization has been performed over different values of the attenuation A, assuming eight subbands (each with 32 contiguous subcarriers) in a downlink transmission. For FBMC, PHYDYAS prototype pulses [54] are assumed, and the capacity optimization has been performed over different values of $M \leq 4$. In Figure 13.9 the PSD of FBMC PHYDYAS prototype pulses are compared with OFDM. The figure also reports the PSD of GFDM with $M = 4$ using a raised cosine (RC) pulse with rolloff factor $r_{\text{off}} = 1$. In the subsequent results, r_{off} will be optimized to maximize the capacity. In Figure 13.10, the PSD of the UFMC signal in a single subband is compared with the PSD of an OFDMA signal in the same subband. Figures 13.9 and 13.10 only show half the subband.

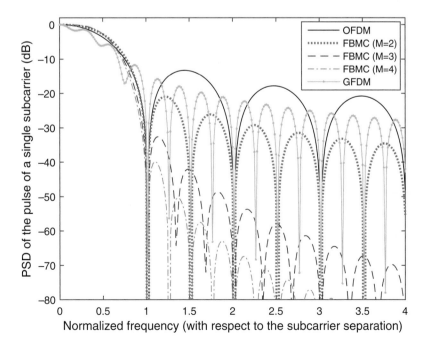

Figure 13.9 PSD of investigated pulses.

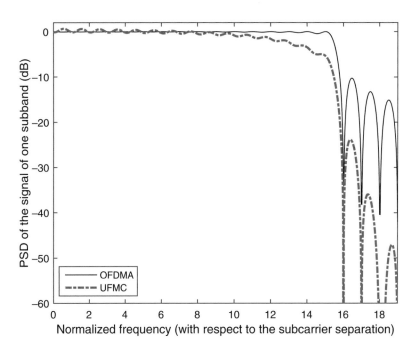

Figure 13.10 PSD in one subband. Only half of the subband is shown.

13.5.1 Comparison of Different Bounds and Equalizers

We first compare the different LBs (13.92), (13.96), (13.102). We remind the reader that the LB in (13.92) is obtained using a vector model before equalization, the LB in (13.96) is obtained using a vector model after equalization, and the LB in (13.102) is obtained using a scalar model after equalization. We also compare the vector UBs corresponding to LBs (13.92) and (13.96); the vector UBs are simply obtained by neglecting the interference contributions. We assume a multipath channel with a PDP decay that follows the Extended Pedestrian A (EPA) sparse model [51], but similar results would be obtained also with other channel models. Figure 13.11 shows the ASE ρ obtained for FBMC with PHYDYAS pulses with overlapping factor $M = 4$, that is, with pulse shape length four times larger with respect to the length that would be used in OFDM for the same number of subcarriers. We assume either MMSE or LS equalization at the receiver.

Figure 13.11 shows that, for both MMSE and LS equalization, the scalar LB (after equalization) of the ASE is below the corresponding vector LB (after equalization). This is an expected result; indeed, the scalar LB (13.102) neglects the correlation of the interference-plus-noise terms $\mu_{i,k}$, while the vector LB (13.96) exploits this correlation.

In general, we also expect that the vector LBs (UBs) after equalization cannot exceed the vector LB (UB) before equalization, because equalization can potentially discard some information. This is confirmed by Figure 13.11. In addition, we also observe that the vector LB after MMSE equalization tends to coincide with the vector LB before equalization, whereas the vector UB after LS equalization tends to be the same as the

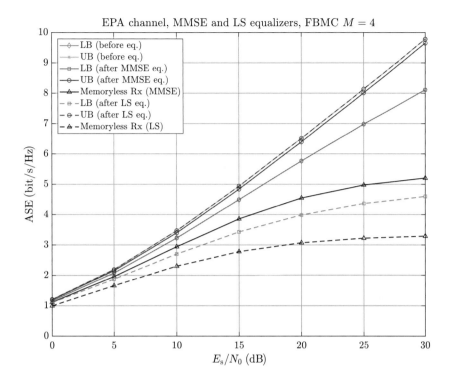

Figure 13.11 ASE versus SNR (FBMC with $M = 4$, EPA channel).

vector UB before equalization. These two behaviors can be explained as follows. MMSE equalization, which maximizes the SINR ratio, tries to achieve data separation without amplifying the interference-plus-noise component that worsens the LB; hence, MMSE equalization aims at the maximum value of the vector LB, which is the vector LB before equalization. On the other hand, LS equalization eliminates the interference term that is neglected in the UBs; hence, LS equalization aims at the maximum value of the vector UB, which is the vector UB before equalization. Since LBs are achievable, the two behaviors described above suggest employing MMSE rather than LS equalization. Similar advice is also given by the scalar equalizers, because the scalar LB after equalization (indicated as *memoryless* in Figure 13.11) for MMSE is higher than for LS equalization. Thus, also in the view of the vector UBs and LBs, in the following we will consider the MMSE bounds only.

For the other techniques, not shown in Figure 13.11 because of lack of space, we report that, in the OFDM case, all the bounds coincide. Indeed, OFDM with sufficient CP eliminates the ISI and the ICI by design; in this case, all the LBs and UBs coincide, since there is no interference at all. We also remind the reader that, for OFDM with CP, LS and MMSE equalization coincide, apart from a scalar constant that does not affect the ASE. Hence, for OFDM with CP, all the vector and scalar bounds (both LBs and UBs) for LS and MMSE equalization are identical. On the other hand, the results for the UFMC case can be considered an intermediate case between the results

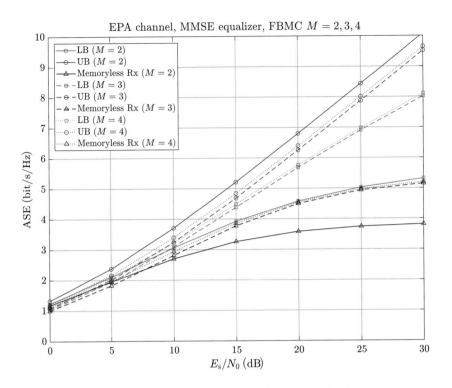

Figure 13.12 ASE versus SNR (FBMC, EPA channel, MMSE equalizer).

for FBMC shown in Figure 13.11 and the results for OFDM just described. Indeed, UFMC produces a reduced interference with respect to FBMC, but the interference is not zero as in OFDM, because UFMC does not use a CP or a guard time in multipath channels. Consequently, the ASE bounds for UFMC are conceptually similar to those of FBMC. Anyway, differently from FBMC, for UFMC both LBs and UBs are closer together, because of the reduced amount of interference; in some cases (i.e., when the amount of interference is small enough), the differences between the bounds are negligible, so that the results for UFMC would be practically similar to those for OFDM.

13.5.2   Comparison of Different Pulse-Shaping Waveforms

In Figures 13.12–13.14 we compare different pulse-shaping waveforms for FBMC, GFDM, and UFMC, assuming the EPA channel model and MMSE equalizers. The bounds shown in Figures 13.12–13.14 are those after equalization.

For the FBMC results in Figure 13.12, we consider the PHYDYAS pulses [30] with overlapping parameter $2 \leq M \leq 4$. The results of Figure 13.12 explain that, for FBMC, the best ASE UB is obtained for $M = 2$ and the best LB for $M = 4$. However, note that $M = 2$ produces also the worst ASE LBs (referred to as *Memoryless Rx* in the figure key), and also the widest gap between UB and LB. These results can be explained as

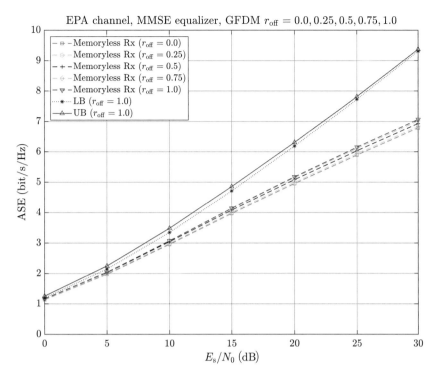

Figure 13.13 ASE versus SNR (GFDM, EPA channel, MMSE equalizer).

follows. With reference to Figure 13.9, when the overlapping parameter M increases, the sidelobes reduce; consequently, when M increases, the subcarrier spectra are *more separated* and the equalization task is somewhat easier, thereby producing less residual interference at the equalizer output. This reduced interference is the reason why, for $M = 4$, the gap between the LB and the UB is lower than the gap for $M = 2$. For the same reason, the reduced residual interference for $M = 4$ produces a larger LB than for $M = 2$, where the increased residual interference worsens the LB. On the other hand, the UBs neglect the interference; this partially explains why the UB for $M = 2$ is higher than the UB for $M = 4$. In addition, note that the bounds for $M = 3$ are in-between the bounds for $M = 2$ and $M = 4$; however, the bounds for $M = 3$ are closer to those for $M = 4$ rather than to those for $M = 2$. This result can be partially explained by the spectra of Figure 13.9, where the sidelobes for $M = 3$ are somewhat close to the sidelobes for $M = 4$ and somewhat distant from the sidelobes for $M = 2$. As a summary of the FBMC results, although $M = 2$ could be attractive for reduced complexity detection (not considered herein), in the following for FBMC we often select an overlapping parameter $M = 4$, which provides the largest LBs.

Figure 13.13 displays the ASE results for GFDM. Specifically, Figure 13.13 compares the ASE results for GFDM with $M = 4$ and RC pulses with different values of the rolloff factor r_{off}. We notice that there are no large ASE differences among the different rolloff values. The ASE LBs increase with increasing rolloff factor; the maximum value

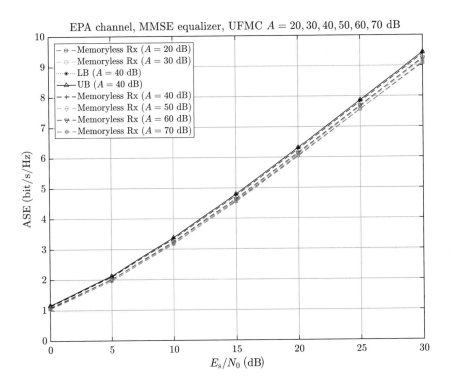

Figure 13.14 ASE versus SNR (UFMC, EPA channel, MMSE equalizer).

$r_{off} = 1$ ensures the highest ASE. Indeed, the waveform with $r_{off} = 1$ produces smaller sidelobes than $r_{off} < 1$, thereby reducing the interference. Note that the generated interference is not negligible; this is confirmed by the large gap between the ASE of the scalar LB (*memoryless*) and the ASE of the vector bounds (LB and UB), as shown in Figure 13.13 for $r_{off} = 1$.

Figure 13.14 illustrates the ASE results for UFMC. We remind the reader that the bounds in Figure 13.14 are those after equalization. We consider Dolph–Chebyshev pulse-shaping waveforms with different sidelobe attenuation A [53]. The ASE results shown in Figure 13.14 are obtained by averaging the ASE of all eight subbands. As a first result in Figure 13.14, we observe that all the ASE curves have a similar trend, with reduced gaps among the different curves. Indeed, in UFMC, the generated ISI is reduced, and therefore the residual interference after equalization is reduced as well, for all the values of sidelobe attenuation A. Figure 13.14 shows that, for increasing values of the sidelobe attenuation A (e.g., from $A = 20$ dB to $A = 40$ dB), the ASE LB increases; then, with a further increase of the sidelobe attenuation A (e.g., from $A = 40$ dB to $A = 70$ dB), the ASE LB reduces. Therefore, the best ASE performance is obtained for $A = 40$ dB. On the contrary, extreme pulse-shape choices with very low attenuation (e.g., $A = 20$ dB) or very large attenuation (e.g., $A = 70$ dB) produce slightly reduced ASE LBs. However, the ASE gap among different values of A is not critical. In addition, if we focus on the optimum sidelobe attenuation $A = 40$ dB, Figure 13.14 shows that the ASE distance between the scalar LB (*memoryless*) and the vector bounds (both LB and UB) is quite small; this is another consequence of the reduced residual interference in UFMC. This explains that a scalar MMSE equalizer for UFMC produces ASE results that are very close to the ASE of the vector UB.

The previous ASE results for GFDM and UFMC shown in Figures 13.13 and 13.14 highlight that the ASE slightly changes with the parameters r_{off} and A, respectively. In addition, the ASE results for FBMC in Figure 13.12 emphasize that the ASE difference between $M = 3$ and $M = 4$ is reduced as well. These results suggest that pulse shapes with a single parameter have a small margin for ASE maximization; the design of a pulse-shaping waveform to maximize the ASE would request a larger number of parameters.

13.5.3 Comparison of Different Techniques

In Figures 13.15–13.17, we compare the different techniques. In addition to OFDM with CP (CP-OFDM), we also include the zero-padding (ZP) version, which replaces the CP with zeros, denoted with ZP-OFDM. The vector LBs shown in Figures 13.15–13.17 are those after MMSE equalization. For FBMC, GFDM, and UFMC, we assume $M = 4$, $r_{off} = 1$, and $A = 40$ dB, respectively, which gave the best ASE in the previous comparisons. Several channel models are investigated, such as EPA and Extended Typical

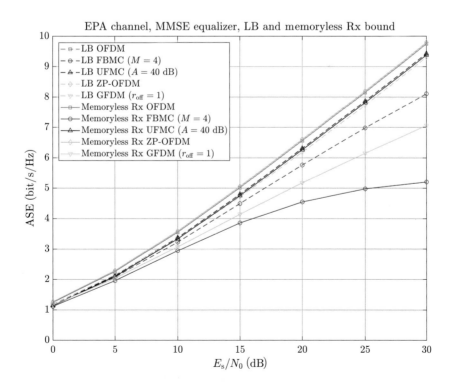

Figure 13.15 ASE versus SNR (EPA channel).

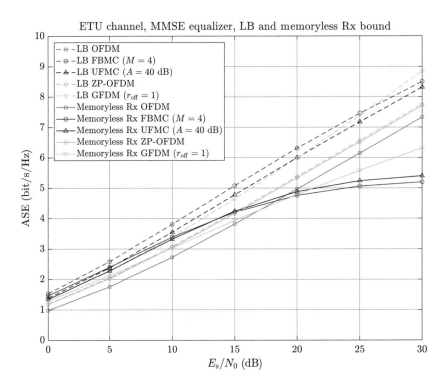

Figure 13.16 ASE versus SNR (ETU channel).

Urban (ETU) sparse models [51] and WLAN exponential models [52] with Rayleigh and Rice fading.

Figures 13.15–13.17, clearly show that the best technique depends on the channel model. In the EPA channel (Figure 13.15), the best performance is obtained by ZP-OFDM, followed by CP-OFDM, and then by UFMC. Therefore, from the ASE viewpoint, orthogonal techniques are preferable for EPA channels. Note that the loss of CP-OFDM (with respect to ZP-OFDM) is due to the additional energy spent for transmitting the CP; this loss is reduced, since the EPA channel is a pedestrian channel with reduced delay spread, for which a short CP is sufficient. In this EPA channel, CP-OFDM is second-best in terms of ASE. Also GFDM and UFMC give good ASE results; in the case of UFMC, the ASE is quite large also for the scalar LB (memoryless). On the other hand, FBMC and scalar LB of GFDM give inferior ASE results due to the presence of interference. Therefore, for the EPA channel, orthogonal and non-overlapping multicarrier techniques like OFDM and UFMC are preferable with respect to overlapping methods like GFDM and FBMC.

We consider the ETU channel in Figure 13.16. This figure displays that, at low–medium SNR, the bounds for FBMC, GFDM, and UFMC are larger than CP-OFDM and ZP-OFDM. Note that, at low–medium SNR, the difference between the vector LB and the scalar LB (*memoryless*) is small. Indeed, the ETU channel is noticeably longer than the EPA channel, and hence OFDM techniques suffer from some rate reduction due

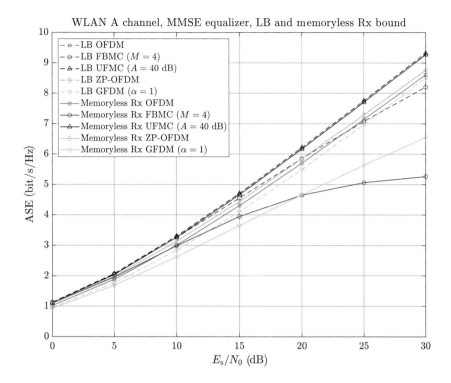

Figure 13.17 ASE versus SNR (WLAN channel A).

to the insertion of a long CP or ZP. Hence, in ETU channels at low SNR, nonorthogonal techniques are preferable to OFDM from the ASE viewpoint. However, Figure 13.16 shows that, at large SNR, the gap between the scalar LB of the three nonorthogonal techniques (FBMC, GFDM, and UFMC) and the corresponding vector LB increases: As the SNR increases, the vector LBs tend to increase more than the scalar LBs, which are affected by the larger interference power compared to the noise power. In this high-SNR case, the LBs for OFDM techniques are larger than the scalar (memoryless) LBs of the other techniques. Consequently, in ETU channels at high SNR, the nonorthogonal techniques (FBMC, GFDM, and UFMC) are preferable only when the receiver aims at the vector LB; if the receiver aims at the scalar LB, then ZP-OFDM and CP-OFDM are preferable, from the ASE viewpoint. Note that, also in this case, ZP-OFDM outperforms CP-OFDM, due to the power saving of ZP with respect to CP.

Figure 13.17 exhibits the ASE comparison in the WLAN channel A. In this scenario, the best ASE performance is achieved by UFMC; similar results are obtained for other WLAN channels not shown in Figure 13.17. The ASE performance of the orthogonal techniques (CP-OFDM and ZP-OFDM) is still acceptable. On the other hand, the ASE performance of the overlapping techniques (FBMC and GFDM) is quite good at low SNR: at high SNR, the FBMC or GFDM receiver should target the vector LB. In conclusion, among FBMC, OFDM, GFDM, and UFMC, there is no single winner in all the cases, at least from the ASE viewpoint. The choice among different techniques can

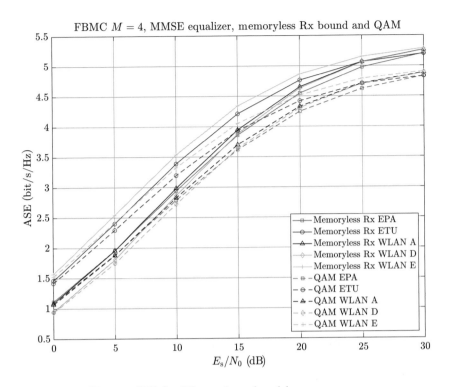

FBMC $M = 4$, MMSE equalizer, memoryless Rx bound and QAM

- —□— Memoryless Rx EPA
- —○— Memoryless Rx ETU
- —△— Memoryless Rx WLAN A
- —◇— Memoryless Rx WLAN D
- —+— Memoryless Rx WLAN E
- – □ – QAM EPA
- – ○ – QAM ETU
- – △ – QAM WLAN A
- – ◇ – QAM WLAN D
- – + – QAM WLAN E

Figure 13.18 ASE versus SNR for different channel models.

be driven by the specific channel scenario. In addition, the complexity of the receiver should be taken into account, to grant such a capacity promise. Orthogonal methods (CP-OFDM and ZP-OFDM) produce good results in all scenarios, except when the length of the CP or ZP is excessive. Overlapping techniques (FBMC and GFDM) promise larger ASE if the receiver can achieve the vector LB: however, in many scenarios, the scalar LB for overlapping techniques is reduced with respect to OFDM. UFMC is able to outperform OFDM in some scenarios. Summarizing, no specific technique is able to largely outperform its competitors in all scenarios.

13.5.4   Comparison of Different Channels and Finite Constellations

In Figures 13.18 and 13.19, we compare the effect of different channels and finite constellations. We assume FBMC with overlapping factor $M = 4$ and MMSE equalizer. Figure 13.18 shows that the best ASE performance is obtained in the WLAN channel E (exponential PDP, with Rayleigh fading) and in the ETU channel (sparse PDP, with large delay spread). Conversely, the worst ASE performance is obtained in the WLAN channel D (exponential PDP, with Rice fading) at low SNR and in the EPA channel (sparse PDP, with low delay spread) at large SNR. Indeed, both WLAN channel E and ETU channel have several relevant paths, which provide large frequency diversity; the WLAN channel D and the EPA channel have a single path or few relevant paths only,

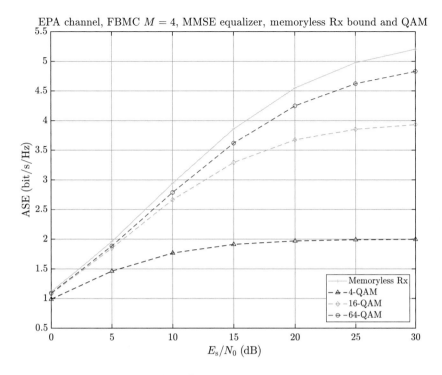

Figure 13.19 ASE versus SNR for different QAM constellations.

hence the diversity gain is reduced. The same considerations are valid for both the scalar LB (memoryless) and for the finite constellation results. The finite constellation results have been obtained by selecting the largest information rate among different QAM constellations; the information rate has been evaluated as explained in [4]. In addition, Figure 13.19 compares the information rate results for different QAM constellations: Clearly, 64-QAM is able to gather a significant percentage of the information rate predicted by the scalar (memoryless) bound.

13.6 Concluding Remarks

In this chapter, we have introduced a general signal processing framework that is capable of capturing in a unified view the several signal waveforms proposed for 5G cellular systems and beyond. In the light of the information theoretic view of this book, we have shown how the proposed framework can be exploited to easily compare different signal waveforms in terms of achievable spectral efficiency and we have discussed results for the single-antenna case. The overall system model can be extended to also take into account multiantenna scenarios, as well as typical impairments of communication systems such as phase noise, nonlinear distortions, synchronization errors, Doppler spread, and the associated signal processing challenges. In any case, even in the simpler scenario investigated in this chapter, it is clear that the choice of a specific signal waveform may depend on environmental constraints, as well as on the system requirements in terms of bandwidth and latency. Furthermore, the intrinsic implementation complexity associated with each specific waveform may also drive the choice. Thus, research efforts can be directed to exploit the specific structure of each signal waveform to derive low-complexity signal processing algorithms that enable practical implementation in critical multiuser multiantenna systems. Besides, future studies could exploit the proposed framework for a parametric design of the signal processing waveform, for fast and practical adaptation of the radio interface to the specific communication scenario, according to a cognitive radio paradigm.

References

[1] ETSI, "LTE; Evolved Universal Terrestrial Radio Access (E-UTRA); Base Station (BS) radio transmission and reception (3GPP TS 36.104 version 11.6.0 Release 11)," 2013.

[2] J. G. Andrews, S. Buzzi, W. Choi, et al., "What will 5G be?," *IEEE J. Select. Areas Commun.*, vol. 32, no. 6, pp. 1065–1082, 2014.

[3] 3GPP, "New Radio (NR): Physical channels and modulation (3GPP TS 38.211 version 15.1.0 Release 15)," 2018.

[4] P. Banelli, S. Buzzi, C. Colavolpe, et al., "Modulation formats and waveforms for 5G networks: Who will be the heir of OFDM?," *IEEE Signal Processing Mag.*, vol. 31, no. 6, pp. 80–93, 2014.

[5] G. Wunder, P. Jung, M. Kasparick, et al., "5GNOW: Non-orthogonal, asynchronous waveforms for future mobile applications," *IEEE Commun. Mag,.* vol. 52, no. 2, pp. 97–105, 2014.

[6] M. Kasparick, et al., "5G waveform candidate selection," Tech. Rep. D3.1 of 5G-Now, FP7 European Research Project, 2013.

[7] A. Osseiran, F. Boccardi, V. Braun et al., "Scenarios for 5G mobile and wireless communications: The vision of the METIS project," *IEEE Commun. Mag.*, vol. 52, no. 5, pp. 26–35, 2014.

[8] A. Ghosh, J. Zhang, J. G. Andrews, and R. Muhamed, *Fundamentals of LTE.* Harlow: Pearson Education, 2010.

[9] H. Ochiai and H. Imai, "On the distribution of the peak-to-average power ratio in OFDM signals," *IEEE Trans. Commun.*, vol. 49, no. 2, pp. 282–289, 2001.

[10] L. Rugini and P. Banelli, "BER of OFDM systems impaired by carrier frequency offset in multipath fading channels," *IEEE Trans. Wireless Commun.*, vol. 4, no. 5, pp. 2279–2288, 2005.

[11] M. Morelli, "Timing and frequency synchronization for the uplink of an OFDMA system," *IEEE Trans. Commun.*, vol. 52, no. 2, pp. 296–306, 2004.

[12] T. Hwang, C. Yang, G. Wu, S. Li, and G. Ye Li, "OFDM and its wireless applications: a survey," *IEEE Trans. Veh. Technol.*, vol. 58, no. 4, pp. 1673–1694, 2009.

[13] R. Irmer, H. Droste, P. Marsch, et al., "Coordinated multipoint: Concepts, performance, and field trial results," *IEEE Commun. Mag.*, vol. 49, no. 2, pp. 102–111, 2011.

[14] M. Cudak, A. Ghosh, T. Kovarik, et al., Moving towards mmwave-based beyond-4G (b-4G) technology," in *Proc. IEEE 77th Veh. Technol. Conf. (VTC)*, Dresden, 2013.

[15] A. Ghosh, T. A. Thomas, M. C. Cudak, et al., "Millimeter wave enhanced local area systems: A high data rate approach for future wireless networks," *IEEE J. Select. Areas Commun.*, vol. 32, no. 6, pp. 1152–1163, 2014.

[16] Z. Pi and F. Khan, "An introduction to millimeter-wave mobile broadband systems," *IEEE Commun. Mag.*, vol. 49, no. 6, pp. 101–107, 2011.

[17] T. S. Rappaport, S. Sun, R. Mayzus, et al., "Millimeter wave mobile communications for 5G cellular: It will work!" *IEEE Access*, vol. 1, pp. 335–349, 2013.

[18] F. Rusek, D. Persson, B. K. Lau, et al., "Scaling up MIMO: Opportunities and challenges with very large arrays," *IEEE Signal Processing Mag.,* vol. 30, no. 1, pp. 40–60, 2013.

[19] P. Rost, C. J. Bernardos, A. De Domenico, et al., "Cloud technologies for flexible 5G radio access networks," *IEEE Commun. Mag.*, vol. 52, no. 5, pp. 68–76, 2014.

[20] Z. Zhu, P. Gupta, Q. Wang, et al., "Virtual base station pool: Towards a wireless network cloud for radio access networks," in *Proc. ACM Int. Conf. on Computing Frontiers*, Ischia, 2011, pp. 1–10.

[21] S. Haykin, "Cognitive radio: Brain-empowered wireless communications," *IEEE J. Select. Areas Commun.,* vol. 23, no. 2, pp. 201–220, 2005.

[22] N. Michailow, M. Matthé, I. Simões Gaspar "Generalized frequency division multiplexing for 5th generation cellular networks," *IEEE Trans. Commun.*, vol. 62, no. 9, pp. 3045–3061, 2014.

[23] L. Atzori, A. Iera, and G. Morabito, "The Internet of Things: A survey," *Elsevier Computer Netw.*, vol. 54, no. 15, pp. 2787–2805, 2010.

[24] G. Fettweis, "The tactile Internet: Applications and challenges," *IEEE Veh. Technol. Mag.,* vol. 9, no. 1, pp. 64–70, 2014.

[25] F. Schaich, T. Wild, and Y. Chen, "Waveform contenders for 5G: Suitability for short packet and low latency transmissions," in *Proc. IEEE 79th Vehicular Technology Conf. (VTC),*" Seoul, 2014, pp. 1–5.

[26] R. Gerzaguet, N. Bartzoudis, and L. Gomes Baltar, et al., "The 5G candidate waveform race: A comparison of complexity and performance," *EURASIP J. Wireless Commun. Networking,* vol. 17, no. 1, p. 13, 2017.

[27] B. Farhang-Boroujeny, "OFDM versus filter bank multicarrier," *IEEE Signal Processing Mag.*, vol. 28, no. 3, pp. 92–112, 2011.

[28] W. Gerstacker, F. Adachi, H. Myung, and R. Dinis, "Broadband single-carrier transmission techniques," *Elsevier Physical Commun.*, vol. 8, pp. 1–4, 2013.

[29] A. Sahin, I. Guvenc, and H. Arslan, "A survey on multicarrier communications: Prototype filters, lattice structures, and implementation aspects," *IEEE Commun. Surveys Tuts.*, vol. 16, no. 3, pp. 1312–1338, 2014.

[30] M. Bellanger, "FBMC physical layer: A primer," Tech. Rep. FP7-ICT, Future Networks, 2010.

[31] Z. Wang and G. B. Giannakis, "Wireless multicarrier communications: Where Fourier meets Shannon," *IEEE Signal Processing Mag,*. vol. 17, no. 3, pp. 29–48, 2000.

[32] F. Schaich and T. Wild, "Waveform contenders for 5G: OFDM vs. FBMC vs. UFMC," in *Proc. IEEE 6th Int. Symp. Commun., Control Signal Processing (ISCCSP),*" Athens, 2014, pp. 457–460.

[33] V. Vakilian, T. Wild, F. Schaich, et al., "Universal-filtered multi-carrier technique for wireless systems beyond LTE," in *Proc. IEEE GLOBECOM Workshops*, Atlanta, GA, 2013, pp. 223–228.

[34] L. G. Baltar, A. Mezghani, J. Nossek, "MLSE and MMSE subchannel equalization for filter bank based multicarrier systems: Coded and uncoded results," in *Proc. 18th Eur. Signal Processing Conf. (EUSIPCO),*" Rhodes, 2010, pp. 2186–2190.

[35] A. Stamoulis, S. N. Diggavi, and N. Al-Dhahir, "Intercarrier interference in MIMO OFDM," *IEEE Trans. Signal Processing,*vol. 50, no. 10, pp. 2451–2464, 2002.

[36] N. Benvenuto, R. Dinis, D. Falconer, and S. Tomasin, "Single carrier modulation with non-linear frequency domain equalization: An idea whose time has come again," *Proc. IEEE,* vol. 98, no. 1, pp. 69–96, 2010.

[37] X. Cai and G. B. Giannakis, "Bounding performance and suppressing intercarrier interference in wireless mobile OFDM," *IEEE Trans. Commun,* vol. 51, no. 12, pp. 2047–2056, 2003.

[38] L. Rugini, P. Banelli, and G. Leus, "Low-complexity banded equalizers for OFDM systems in Doppler spread channels," *EURASIP J. Appl. Signal Processing*, 2006. DOI: 10.1155/ASP/2006/67404.

[39] K. Fang, L. Rugini, and G. Leus., "Low-complexity block turbo equalization for OFDM systems in time-varying channels," *IEEE Trans. Signal Processing,*vol. 56, no. 11, pp. 5555–5566, 2008.

[40] P. Schniter, "Low-complexity equalization of OFDM in doubly selective channels," *IEEE Trans. Signal Processing*, vol. 52, no. 4, pp. 1002–1011, 2004.

[41] R. Zakaria and D. Le Ruyet, "On maximum likelihood MIMO detection in QAM-FBMC systems," in *Proc. IEEE 21st Int. Symposium on Personal Indoor and Mobile Radio Commun. (PIMRC)*, Istanbul, 2010, pp. 183–187.

[42] R. Zakaria, D. Le Ruyet, and Y. Medjahdi, "On ISI cancellation in MIMO-ML detection using FBMC/QAM modulation," in *Proc. IEEE Int. Symposium Wireless Commun. Syst.(ISWCS)*, Paris, 2012, pp. 949–953.

[43] N. Merhav, G. Kaplan, A. Lapidoth, and S. Shamai Shitz, "On information rates for mismatched decoders," *IEEE Trans. Inform. Theory*, vol. 40, no. 6, pp. 1953–1967, 1994.

[44] D. M. Arnold, H.-A. Loeliger, P. O. Vontobel, et al., "Simulation-based computation of information rates for channels with memory," *IEEE Trans. Inform. Theory*, vol. 52, no. 8, pp. 3498–3508, 2006.

[45] R. E. Blahut, *Principles and Practice of Information Theory*. Reading, MA: Addison-Wesley, 1987.

[46] A. Barbieri, D. Fertonani, G. Colavolpe, et al., "Time–frequency packing for linear modulations: Spectral efficiency and practical detection schemes," *IEEE Trans. Commun.*, vol. 57, pp. 2951–2959, 2009.

[47] E. Biglieri, J. Proakis, and S. Shamai, "Fading channels: Information-theoretic and communications aspects," *IEEE Trans. Inform. Theory*, vol. 44, no. 6, pp. 2619–2692, 1998.

[48] H. Q. Ngo, A. Ashikhmin, H. Yang et al., "Cell-free massive MIMO: Uniformly great service for everyone," in *Proc. IEEE 16th Int. Workshop Signal Processing Adv. Wireless Commun. (SPAWC)*, Stockholm, 2015, pp. 201–205.

[49] S. Buzzi and C. D'Andrea, "On clustered statistical MIMO millimeter wave channel simulation," *arXiv:1604.00648*, 2016.

[50] S. Buzzi and C. D'Andrea, "User-centric communications versus cell-free massive MIMO for 5G cellular networks," in *Proc. 21st Int. ITG Workshop on Smart Antennas (WSA)*, Munich, 2017, pp. 1–6.

[51] J. A. del Peral-Rosado, J. Lopez-Salcedo, G. Seco-Granados, et al., "Evaluation of the LTE positioning capabilities under typical multipath channels," in *Proc. IEEE 6th Adv. Satellite*

Multimedia Syst. Conf. (ASMS) and 12th Signal Processing for Space Commun. Workshop (SPSC), Baiona, 2012, pp. 139–146.

[52] A. Doufexi, S. Armour ; M. Butler, et al., "A comparison of the HIPERLAN/2 and IEEE 802.11a wireless LAN standards," *IEEE Commun. Mag.*, vol. 40, no. 5, pp. 172–180, 2002.

[53] C. Dolph, "A current distribution for broadside arrays which optimizes the relationship between beam width and side-lobe level," *Proc. IRE*, vol. 34, no. 6, pp. 335–348, 1946.

[54] A. Viholainen, et al., "Prototype filter and structure optimization," Tech. Rep. D3.1 of PHYsical layer for DYnamic AccesS and cognitive radio (PHYDYAS), FP7-ICT Future Networks, 2008.

[55] B. Muquet, Z. Wang, G. B. Giannakis, et al., "Cyclic prefixing or zero padding for wireless multicarrier transmissions?," *IEEE Trans. Commun.*, vol. 50, no. 12, pp. 2136–2148, 2002.

Part III

Protocols

14 Information Theoretic Aspects of 5G Protocols

Čedomir Stefanović, Kasper F. Trillingsgaard, and Petar Popovski

14.1 The Gap between Information Theory and Communication Protocols

The gap that has existed between information theory and networking has been documented in several well-known works, such as [1, 2]. It stems mainly from the different types of questions asked by information theorists versus the questions asked by a designer of a communication protocol. The question asked by an information theorist is mathematical: *What is the best one can do in terms of, e.g., data rate or probability of error given a certain model of a communication system?* On the other hand, the question asked by a communication protocol designer is often a heuristic one: *How to make a protocol/transmission technique that solves a certain practical problem and, actually, works?*

The advantage of the information theoretic question is that, once the assumptions and the model are fixed, it attempts to find the ultimate limits of communication, without putting restrictions on the, e.g., codebooks that can be used, as long as they are conforming with the model. For example, let us assume that node 1 wants to communicate with node 2 over a binary symmetric channel (BSC). Information theory asks, e.g., what is the highest data rate at which node 1 can transmit to node 2 by assuming that it sends coded information over N channel uses and the maximal probability of error is ϵ? However, information theory usually does not ask how did node 1 and node 2 establish a BSC and how have they agreed that from a certain channel use node 1 starts to transmit its coded data block and node 2 starts to receive it? These are the questions asked by a designer of a communication protocol, who, in contrast, typically does not seek the ultimate limits, but rather tries to attain an acceptable performance.

In this chapter we will make an attempt to shed light on the bridges that exist between two types of questions and the associated research problems.

14.2 Communication Model with Protocol Information

Figure 14.1(a) shows the communication model proposed by C. E. Shannon in his groundbreaking work on reliable communication over unreliable channels. The central point of the communication model is the *uncertainty*, represented through the conditional probability $p(y|x)$ used to model the random disturbances (noise, interference) that affect the transmission of symbols. Other important elements are the input alphabet

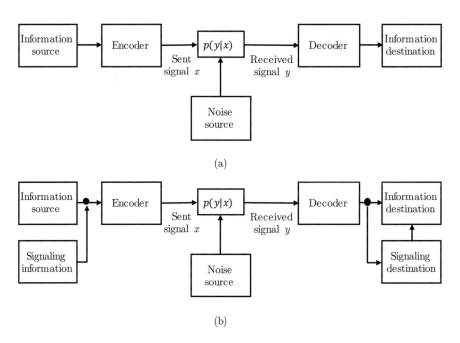

Figure 14.1 (a) Shannon's mathematical model of communication. (b) Extended communication model that takes into account the transmission of protocol information.

$\mathcal{X}$ and the output alphabet $\mathcal{Y}$ to the communication channel, as well as the restrictions on how to use those alphabets. For example, a standard restriction is the constraint on the transmit power for continuous-valued channels (e.g., Gaussian channels).

This is usually where the information theoretic model stops, keeping tacit assumptions on all the other elements that are necessary to bring the concepts toward a practical realization. One such assumption is the knowledge that the two communicating parties, node 1 and node 2, have already agreed to communicate and, furthermore, have agreed that the first upcoming channel use will be a start of a new data packet from node 1 to node 2. On the other hand, a communication protocol designer starts from a model that represents a certain physical situation and treats all the steps that can lead to successful exchange of data. As an example, node 1 and node 2 can be two wireless devices that come into each other's range, need to go through a rendezvous protocol in order to establish a connection, and then start to exchange data. Hence some of the communication resources need to be used to exchange protocol information[1] and make a transition to a state in which data are exchanged. An elementary example of this can be seen in the very structure of a communication packet or frame:[2] one part of the resources need to be used as a preamble to detect that the frame is there and to synchronize to it, one part to send metadata that describes what is contained in the data, one part for the data, and almost always a part for checking the packet integrity (checksum bits).

---

[1] In this chapter we will use the terms "protocol information," "metadata," and "control information" interchangeably.

[2] As often called when considered at the physical/MAC layer.

This is not taken into account in the model in Figure 14.1(a) and therefore we need to use a modified communication model, such as the one in Figure 14.1(b). Consider a situation in which there is a pre-established relation between the two communicating parties, e.g., node 1 is a base station (BS) and node 2 is a terminal. The model in Figure 14.1(b) still describes a situation with unidirectional communication, as does the model in Figure 14.1(a), but it explicitly accounts for the need to transmit control information. Specifically, in this model, node 2 is in the receiving state and waits for protocol information that describes the actual data that are being sent. The arrow from the signaling destination to the information destination reflects the fact that signaling information is used as metadata to interpret the data. When the size of the data that node 1 needs to send to node 2 is large, then the efficiency of the overall communication is dominated by the efficiency of the method for data transmission and the resources spent on the signaling (protocol) information are negligible. In other words, for asymptotically large data sizes, the model in Figure 14.1(b) is accurately captured by the simplified model in Figure 14.1(a). However, if the size of the data and the signaling information (i.e., metadata) are comparable, the simplified model loses its accuracy, and the overall optimization of the communication protocol has to be considered. This is the case in machine-type communications (MTCs), which typically involve sporadic exchanges of small data amounts, which can be in the order of tens of bytes.

It is therefore critical for systems that support MTC, as is envisioned for 5G wireless systems, to properly model the transmission of control information. In general, protocol information is essential to support the communication over multiuser channels, such as the multiple access channel MAC, the broadcast channel, or the relay channel. Nevertheless, in classical information theoretic models of these channels, the protocol information is usually neglected. In this chapter we shed light on the interplay between the protocol information and data for two basic multiuser channels relevant for 5G systems: uplink (multiple access) and downlink (broadcast) channel. We note that the models in Figure 14.1, as well as the models for uplink/downlink communication, assume unidirectional communication. The first step in generalizing the models toward bidirectional communication is the use of feedback, often idealized, from the receiver to the transmitter(s). This feedback is in fact a protocol information and its "honest" treatment would mean that the model in Figure 14.1, as well as the corresponding models for uplink/downlink, should be modified to reflect the bidirectional communication and the fact that the feedback is also sent through an unreliable channel. This shows only a tip of the complexity iceberg that one faces in trying to capture complete protocols with an information theoretic approach.

The reader will notice that the treatment of the uplink and the downlink case is, in a sense, unbalanced. The discussion of the uplink case is based on the rich history of access protocols and provides an overview of some of the relevant recent results. On the other hand, the case of protocol aspects in the downlink has been less studied and our discussion mainly relies on [3] as, to the best of our knowledge, one of the first works on the topic.

14.3   Uplink Communication Channel

Uplink communications in cellular technologies are commonly based on reservations. The communication session begins with the *connection-establishment* procedure, whose purpose is to remove the uncertainty related to which users are currently active in the cell and to reserve the uplink resources (i.e., channel uses) to the users detected as active. The result of the connection-establishment procedure is a collection of point-to-point communication links, modeled by the setup depicted in Figure 14.1, over which users exchange data with the BS. In the current mobile cellular standards, connection-establishment procedures are dominated by approaches based on slotted ALOHA [4], characterized by simplicity, but also by modest performance.

The focus of cellular technologies so far has been on human-oriented services, where the amount of resources used for data exchanges significantly exceeds the amount of resources spent on the connection establishment, justifying the use of robust but inefficient connection-establishment procedures. On the other hand, one of the goals of the 5G standardization is to efficiently support machine-type services, characterized by short data exchanges. In this respect, inefficient connection establishment is one of the main obstacles to the realization of MTC in cellular networks [5], since the associated overhead either dominates the communication session or requires an amount of resources that is unattainable. In the latter case, the connection establishment collapses and leaves most of the active users unserved. This problem is particularly significant for the massive MTC (mMTC) services, characterized by a huge number (orders of tens of thousands or more in a cell) of users that are sporadically active. Here we note that, despite the shortcomings, connection establishment and resource reservation are still integral parts of medium-access control in the new 3GPP cellular standards, i.e., in NB-IoT [6] and in the first set of Release 15 specifications [7], which is the release corresponding to 5G.

Connection establishment is part of the protocol information transmission and can be represented by the model depicted in the Figure 14.2, in which there are T active users out of K users in total, communicating over a MAC to a single BS. In its full extent, besides uncertainty related to synchronization, channel state information (CSI), and noise, the problem of connection establishment primarily involves uncertainty related

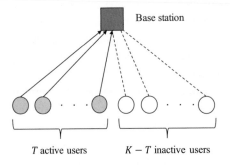

Figure 14.2 Model of connection establishment. In the general case, there are T active users out of K in total, where neither the value of T nor the identities of the active users are a priori known.

to the actual value of T as well as who these T active users are.[3] This is often referred to as activity detection and/or user identification. The latter inevitably results in a lack of coordination among users over the channel uses, i.e., in their mutual interference, which may prevent successful reception of their packets at the BS.

At the heart of the connection establishment in wireless networking solutions is the use of *random access* algorithms. Design of random access algorithms usually relaxes the problem of connection establishment by assuming that the synchronization among the users and the BS is in place. Further, the channel uses are typically partitioned into equal-length slots, which corresponds to the assumption that the active users have equal-length packets; in the case of connection establishment, the packets contain metadata that is used for identification of the active users. The active users *contend* by transmitting in randomly selected slots. The prototype model for contention is the *collision channel* model, in which it is assumed that a packet is always successfully decoded if it was the only one occurring in a slot (i.e., the slot was a singleton), and cannot be decoded otherwise. In other words, in the collision channel model, the interference among the active users is by definition destructive. Therefore, the goal of a random-access algorithm is to maximize the probability that there is just a single user transmitting in a slot. The basic performance parameter of random access algorithm is the efficiency of the slot use, denoted as *throughput* and expressed in the expected number of successfully received packets versus the number of used slots. In this respect, packets in random access are treated as atomic units of communication and the notion of throughput is decoupled from the usual measures of rate and capacity in the information theoretic sense [2, 8].

Slotted ALOHA [4] and tree-splitting [9] are two prominent examples of random-access algorithms, characterized by simplicity, but also with rather low throughput. In particular, slotted ALOHA asymptotically, when $T \rightarrow \infty$, achieves throughput of $1/e$ (packet/slot) on a collision channel. The best-performing tree-splitting scheme on a collision channel achieves throughput that is approximately equal to 0.4878 (packet/slot) [10].

An alternative to reservation-based access is *grant-free* access, in which the active users through the random-access protocol deliver both the protocol information (i.e., the information about their identities) and the data. Grant-free access may be a viable option for mMTC services, as they typically involve rather short messages sent intermittently (e.g., periodic measurements and reports), such that having a lasting connection between the BS and a user is not required. Indeed, recent research related to future wireless access networking considers grant-free access as one of the main topics [11, 12].

Note that slotted transmission means that there is a decision to group the channel uses from the models in Figure 14.1 into equally sized groups of a predefined size. This choice is not necessarily optimal in an information theoretic sense, while it determines the performance of any scheme that is based on the slotted assumption.

We proceed with an overview of information theoretic models of MACs that are relevant for the uplink cellular communications. We then provide a brief overview

---

[3] Note that K, or at least some upper bound on its value, is assumed to be known.

of recently proposed advanced random-access algorithms that by far outperform the standard schemes, showing promise to be used in 5G wireless access networking.

14.3.1 *T*-User Multiple Access Channel

In the classical information theoretic treatment of the T-user MAC, the objective is to find the capacity region, which is the largest achievable rate region in the large n limit when T is held fixed. The key assumption is that the receiver and the transmitters know who is active and can appropriately select the codebooks and the data rates. A well-known result states that the sum capacity of the T-user Gaussian MAC when all active users experience the same signal-to-noise ratio (SNR) per (real) channel use at the point of reception P is

$$C(T) = \frac{1}{2}\log\left(1 + TP\right), \tag{14.1}$$

which is known to be asymptotically achievable. In this case, the largest achievable per-user rate on the T-user Gaussian MAC is given by

$$R_u = \frac{1}{2T}\log\left(1 + TP\right), \tag{14.2}$$

which tends to zero as T increases.

The capacity-achieving strategies rely on coordination among the active users and the BS in terms of code rates, transmission powers, and/or resources used. This in turn implies that the BS needs to know the set of active users, i.e., that the connections between each of them and the BS have already been established. Achieving all the rate tuples from the capacity region with asymptotically large data sizes requires nonorthogonal transmissions. An example strategy is rate splitting, implemented through superposition coding at the transmitters and successive interference cancellation (SIC) at the receiver.

Traditionally, nonorthogonal access has been achieved by different variants of code division multiple access (CDMA). The emergence of 5G wireless systems has reignited the interest in nonorthogonal transmissions and there is a whole class of strategies based on nonorthogonal multiple access (NOMA) [13, 14]. As indicated in these works, the use of NOMA is motivated by its capacity-achieving features on the information theoretic model of a MAC. Our focus in this chapter are the protocol aspects, which means that we are interested not only in the rates that are achieved, but in how the system handles the intermittent arrivals and the uncertainty about the set of transmitting devices.

On one hand, CDMA (and NOMA) can help to deal with the intermittent aspects, as code-based nonorthogonal transmissions offer the inherent possibility to detect the set of active users. This will be indicated whenever suitable in the rest of the chapter. On the other hand, nonorthogonal multiplexing in the power domain, by the use of superposition coding, requires that the active users have knowledge of T (i.e., how many of them there are), such that they could tune their transmission powers. In practice, decoding of superposition codes is implemented via SIC and its performance becomes increasingly challenging as T grows. We note that, when the set of T active terminals is known, it is simpler in practice to use the connection-establishment procedure and orthogonal multiplexing strategies, sacrificing a fraction of the data rate. However, if the objective is to achieve the full rate region for the known set of T active terminals, one should use NOMA techniques.

14.3.2 T-out-of-K-User Multiple-Access Channel

The T-out-of-K-user MAC is better suited to modeling the problem of random access. In T-out-of-K-user MAC, a random and a-priori unknown subset of up to T out of K users is active. This model is also referred to as the model with partial activity. Every user is assigned a unique codeword, i.e., a *signature*, and all active users transmit their signatures in a slot.[4] A signature corresponds to metadata that can be used for user identification, which is the purpose of the connection establishment. The signatures are derived using the T-out-of-K coding scheme, such that if the channel output represents any combination of up to T signatures out of K, the BS will be able to successfully decode the corresponding signatures. There are several variants of the T-out-of-K-user MAC [15], all based on the additive nature of the wireless channel: T-out-of-K-user Gaussian MAC, T-out-of-K-user adder MAC, or T-out-of-K-user Euclidean (real-adder) MAC.

A simple but insightful model is T-out-of-K-user binary-input integer-adder MAC, given by

$$\mathbf{y} = \sum_{\ell \in \mathcal{L}} \mathbf{s}_\ell, \tag{14.3}$$

where $\mathcal{L}$ denotes the set of active users and $\mathbf{s}_\ell \in \{0, 1\}^L$ is a binary vector of length L that is the channel input of user ℓ. The channel output $\mathbf{y} \in \mathbb{N}^L$ is a vector of L integers, whose ith symbol is the integer sum of ith symbols of $\mathbf{s}_\ell, \forall \ell \in \mathcal{L}$. Clearly, this model approximates the scenario in which CSI is known and the active users apply power control such that the powers of the transmitted symbols at the point of reception are equal for every user, while the impact of noise is assumed to be negligible (e.g., through the application of a channel code). Achieving equal power for every user can be done by using channel reciprocity in combination with a suitable transmission of protocol information, such as a downlink beacon signal preceding the multiple access phase that can be used by the users to estimate the suitable transmit power resulting in the desired received power. The task of T-out-of-K signature coding is to ensure unique decodability of any sum of signatures on this channel, as long as $|\mathcal{L}| \leq T$. A construction of such a code was proposed by Lindström [15, 16], with the signature length of

$$L(T, K) = \lceil \log(K_P^T - 1) \rceil + 1 \leq T(1 + \log K) + 2, \tag{14.4}$$

where K_P is the smallest prime number for which $K_P \geq K$, and where each signature includes an indicator bit appended to it. In particular, the sum of the indicator bits informs the receiver whether $|\mathcal{L}| \leq T$, i.e., whether the observed sum of signatures can

---

[4] This can be seen as a form of NOMA based on code multiplexing.

be uniquely decoded or not. Equation (14.4) shows, in essence, that the length of the signature scales approximately as $T \log K$. In other words, the unique decodability of a sum of up to T-out-of-K signatures requires that the signature length scales approximately linearly with T, and is approximately T times longer compared to the case when a signature only serves as a unique identifier of one user (without T-out-of-K decodability), which is $\lceil \log K \rceil$.

The bounds on the minimal signature length $L_{\min}(T, K)$, when $1 \ll T \ll K$, are [15]

$$\left\lceil \frac{2T}{\log T} \log K \right\rceil + 1 \le L_{\min}(T, K) \le \left\lceil \frac{4T}{\log T} \log K \right\rceil + 1. \tag{14.5}$$

In terms of signature length, Lindström's construction is not optimal, but it is the best known so far.

T-out-of-K signature coding can be generalized to scenarios in which users are assigned multiple signatures. In this case, a signature can, besides the unique identification of the user, also represent data [17].

Finally, note that if T-out-of-K signature coding is applied in a random-access scheme where the active users contend with their signatures, then the contention is effectively performed on a *T-collision channel*. Specifically, the T-collision channel is the generalization in which all packets are successfully decoded from a slot as long as there are up to T of them occurring in the slot, and no packet can be decoded otherwise [18, 19].

14.3.3   Many-Access Channel

Recently, the model of the *many*-access channel was proposed and investigated in [11]. Distinguishing features of a many-access channel are: (1) the number of users grows with the blocklength n; (2) there is a single receiver (i.e., BS); and (3) an unknown subset of users transmits in a block. Moreover, the many-access channel models the grant-free access paradigm, in which the active users through the random-access protocol deliver both metadata (i.e., the information about their identities) and the data.

The results presented in [11] concern the Gaussian many-access channel, where the channel output is described by the relation:

$$\mathbf{y} = \sum_{\ell=1}^{K_n} \mathbf{s}_\ell(w_\ell) + \mathbf{z}. \tag{14.6}$$

Here, w_ℓ is the message of user ℓ, $\mathbf{s}_\ell(w_\ell) \in \mathbb{R}^n$ is the corresponding codeword consisting of n symbols, $\mathbf{z}$ is the Gaussian noise, and K_n is the number of users that scales with the blocklength n. User ℓ can be active when $w_\ell \ne 0$, or inactive when $w_\ell \ne 0$ and when the zero codeword $\mathbf{s}_\ell(w_\ell) = \mathbf{0}$ is transmitted. In the rest of the text, the focus is on the symmetric Gaussian many-access channel, where all the users experience the same channels and have the same activity patterns. The set of messages for all users is $\{0, 1, \ldots, M\}$, and every user is active with probability $0 < \alpha_n \le 1$, independently of all other users. When active, a user chooses one of M nonzero messages with probability α_n/M. Every user is assigned its own unique codebook, in which codewords that represent nonzero messages also carry the information related to the user identity; all codewords satisfy an average power constraint. The error probability at the point of reception is standardly defined as the probability that the set of decoded messages is not equal to the set of transmitted messages.

The performance of the Gaussian many-access channel is analyzed in terms of the bound on asymptotically achievable per-user rate with vanishingly small average error probability; note that the per-user rate is the same for all users in the symmetric case. It was shown in [11] that the per-user rate depends on the relation between number of users K_n, average number of active users $T_n = \alpha_n K_n$, and blocklength n, as elaborated next.

In the case when K_n is bounded, i.e., $K_n = K < \infty$ for large enough n, the per-user rate R_u tends to $\frac{1}{2K} \log(1 + KP)$, where P is the SNR per channel use at the point of reception. Clearly, this result bears direct analogy to (14.2).

In the case when K_n is unbounded and T_n is unbounded, where $T_n = O(n)$ and $K_n e^{-\delta T_n} \to 0$ and $\alpha_n \to \alpha \in [0, 1]$, then if $K_n H(\alpha_n) < \frac{n}{2} \log(1 + T_n P)$, where $H(p)$ is the binary entropy function, the achievable per-user rate is

$$R_u = \frac{1}{2T_n} \log(1 + T_n P) - \frac{K_n H(\alpha_n)}{n T_n}. \tag{14.7}$$

The first term on the right-hand side in (14.7) is also analogous to (14.2), and can be understood as the per-user rate when the user activities are known to the BS. The second term on the right-hand side in (14.7) can be understood as the penalty on the per-user rate due to the lack of knowledge of user activities, where $K_n H(\alpha_n)$ is total uncertainty related to the activity of all users. This term also explains the role of the condition $K_n H(\alpha_n) < \frac{n}{2} \log(1 + T_n P)$. Moreover, if $\alpha = 1$, then every user is always active and the per-user rate reduces again to an expression analogous to (14.2). It can also be shown that if K_n grows linearly with n, the penalty of user identification tends to 0. A strategy that achieves (14.7) is to concatenate user signatures and codewords that only represent messages [20], where the signature length is $K_n H(\alpha_n)/T_n$. In other words, the ratio of the signature length and the blocklength n matches the penalty term in (14.7).

Finally, in all other cases related to the growth of K_n and T_n with n, the per-user rate tends to 0. The reader is referred to [11] for details.

14.3.4   Massive Grant-Free Access without User Identification

An interesting perspective on grant-free access was provided in [19], where the users are assigned the same codebook. The number of active users T is fixed and there are M (nonzero) messages, i.e., codewords. An active user chooses one of the M messages uniformly at random and independently from any other user. Therefore, there is a nonzero probability that two or more users choose the same message, which is included in the error-event by default.[5] The decoding is done up to the permutation of messages, where the decoder declares a list of $J \leq T$ decoded messages. The error probability is defined as

---

[5] In scenarios where M is large and T is reasonably low, this probability is arguably small.

$$P_e = \frac{1}{T} \sum_{\ell=1}^{T} P_r \left\{ \{ W_\ell \notin D(\mathbf{y}) \} \cup \{ W_\ell = W_\kappa, \ell \neq \kappa \} \right\}, \qquad (14.8)$$

where W_ℓ is the message of user ℓ and $D(\mathbf{y})$ is the list of the decoded messages, based on the channel output $\mathbf{y}$.

A random coding bound for this model of grant-free access and Gaussian MAC proves the existence of a code under an average power constraint P satisfying $P_e \leq \epsilon$, with ϵ being a function of M, T, P, and n. The expression for ϵ is rather involved and thus omitted, while the interested reader is referred to [19] for details. It was also shown that when M, K, n, and ϵ are selected according to the characteristics of some low-power wide-area networking standards, most of the contribution to the error probability comes from the finite-blocklength effects if the number of active users T is small to moderate. As T grows, the impact of multiple-access interference begins to dominate.

In [21], the authors propose a low-complexity coding scheme for grant-free access without user identification on a Gaussian MAC. The scheme effectively results in T-out-of-K-user MAC experienced by the active users. It consists of an outer code, designed for the multiple-access binary-adder channel, and an inner compute-and-forward code that operates on the Gaussian MAC. Specifically, the compute-and-forward code turns the Gaussian MAC into a multiple-access binary-adder channel, while the outer code enables T-out-of-K decodability of the users' messages.

14.3.5   Coded Slotted ALOHA

The inadequate performance of slotted ALOHA for mMTC prompted investigations of its potential enhancements, primarily focusing on the throughput increase. A notable approach in this regard is represented by the coded-slotted ALOHA (CSA) family of protocols [22]. In the basic version of CSA [23], the slots are organized in frames. An active user contends by transmitting multiple replicas of its packet in several randomly selected slots of the frame; each replica also embeds information about the slots where the other replicas were transmitted. The BS is able to decode replicas from singleton slots, as in the standard setup. However, once a replica becomes decoded, the BS is also able to implement interference cancellation (IC) and remove all other replicas from the slots in which they occurred. This way, some of the collision slots become singletons, enabling decoding of replicas of other packets, and propeling new iterations of IC (Figure 14.3).

Application of SIC in CSA is analogous to iterative belief-propagation decoding of erasure-correcting codes, motivating use of theory and tools from codes-on-graphs to design and analyze CSA schemes. In this respect, a variety of erasure-correcting code schemes have been exploited so far, where the common goal is to design probability mass functions (pmfs) that govern the number of transmitted replicas per user (referred to as degree distributions); the interested reader is referred to [22] for an overview of CSA schemes. The outstanding result is that asymptotically, when number of users proportionally scales with number of slots and they both tend to infinity, the throughput of

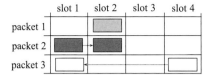

	slot 1	slot 2	slot 3	slot 4
packet 1				
packet 2				
packet 3				

Figure 14.3 Slotted ALOHA with successive interference cancellation: Replica of of packet 3 is decoded in slot 4, enabling removal of the other replica of packet 3 from slot 1 via interference cancellation. Slot 1 becomes singleton, replica of packet 2 becomes decoded from it, and the other replica of packet 2 is removed from slot 2 via interference cancellation. Slot 2 becomes singleton and packet 1 becomes decoded.

CSA tends to 1 (packet/slot) for the collision channel model. This is the ultimate performance limit for the collision channel model, equaling the performance of scheduled access.

In the general version of CSA, inspired by doubly generalized LDPC codes [24], the active users contend by transmitting encoded segments of their packets. Specifically, an active user segments its packet into N_f equal-length segments, encodes the segments using a segment-oriented linear block code that is chosen randomly and independently of any other user from a set of available codes, and contends by transmitting encoded segments instead of replicas. In this case, the design problem involves both the design of the set of the available codes, as well as the design of the pmf governing their selection.

The performance of CSA was also investigated on a T-collision channel model, which can be assumed as an abstraction of T-out-of-K-user MAC (see Section 14.3.2). Denote by $G = N_u/N_s$ the load of the CSA scheme, where N_u is the number of active users[6] and N_s is the number of slots. Since every user has a single packet to report to the BS, the throughput of the scheme is equal to $R_{\text{CSA}} = (1 - P_e)G$ (packet/slot), where P_e is the packet error probability due to the unresolvable collisions. Further, denote by $\rho = N_f/\bar{N}_h$ the ratio between the number of packet segments N_f and the average number of encoded segments $\bar{N}_h$ transmitted per user.[7] In other words, ρ is inversely proportional to the redundant transmissions within the random-access scheme. It was shown that the converse bound $\Gamma(\rho, T)$ on the asymptotically achievable load of any CSA scheme, when $N_s \propto N_u$ and $N_u, N_s \to \infty$, for which $P_e \to 0$ is given as the solution of the equation [25]

$$\frac{\Gamma(\rho, T)}{T} = 1 - \frac{1}{T} \exp\left(-\frac{\Gamma(\rho, T)}{\rho}\right) \sum_{t=0}^{T-1} \frac{T-t}{t!} \left(\frac{\Gamma(\rho, T)}{\rho}\right)^t. \tag{14.9}$$

Here, $\Gamma(\rho, T)$ can be understood as the bound on the asymptotically achievable throughput, i.e., $R_{\text{CSA}} < \Gamma(\rho, T)$. Investigation of (14.9) shows that the ratio $\Gamma(\rho, T)/T$

---

[6] Note that N_u should not be confused with T or K from Section 14.3.2. T and K are design parameters of the signature code, while N_u is the number of active users of the random-access algorithm. Number of active users in a slot is a function of N_u and, in general, can be lower, equal, or greater than T. On the other hand, if the unique identifiability of the active users in random access is required, then $N_u \leq K$.

[7] In the basic variant of CSA, where active users contend with replicas, $N_f = 1$ and $\bar{N}_h$ is the average number of replicas transmitted per user.

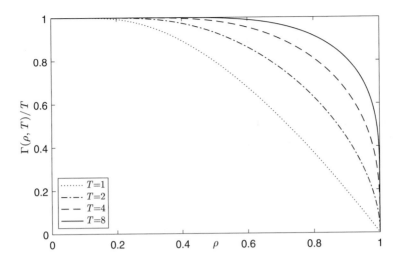

Figure 14.4 Converse bound on the normalized throughput $\Gamma(\rho, T)/T$ that is valid for any CSA scheme, as a function of ρ.

increases with T for *any* $\rho \in (0, 1]$, as illustrated in Figure 14.4. Putting this result in the context of Lindström's construction (see (14.4)) where the signature length linearly scales with T, it can be stated that increasing T pays off in terms of the bound on the normalized throughput $\Gamma(\rho, T)/T$. It was also shown that, asymptotically, normalized throughput R_{CSA}/T of spatially coupled CSA (a variant of the CSA scheme) indeed benefits from increasing T, tending to the converse bound [25].

14.3.6   Capture Effect

The collision channel model is a useful abstraction in the cases when the powers of the active users' signals at the point of reception are close to each other, power of the noise is negligible, and the receiver can perform only single packet reception. However, the collision channel model does not describe adequately the wireless communication scenarios with significant power variations among the interfering signals (e.g., due to fading) and nonnegligible noise powers. Such scenarios are conveniently described by information theoretic models with, e.g., block fading and Gaussian noise, and one can use various multiuser decoding strategies to extract signals from multiple collided users. On the other hand, in the protocol design community, the collision model is extended to describe such scenarios with the *capture effect*, where the strongest signal is successfully received in the presence of interference and/or noise if a certain condition is satisfied.[8]

Investigations of the performance of slotted ALOHA under various models of capture effect have been the subject of numerous works [4, 8, 26, 27]. A popular model is the threshold-based one, in which the packet of some user u is successfully decoded if its signal-to-interference-and-noise ratio (SINR) is higher than or equal to a certain

---

[8] We note that NOMA schemes based on multiplexing in the power domain essentially rely on the capture effect, i.e., decoding in the presence of interference of other users.

threshold b; otherwise, the signal cannot be decoded. Denoting by P_ℓ the power of the transmission of user ℓ at the point of reception, by $\mathcal{L}$ the set of active users in a slot, and by P_N the noise power, the threshold-based criterion can be expressed as

$$\frac{P_u}{\sum_{\ell \in \mathcal{L} \setminus \{u\}} P_\ell + P_N} \geq b, \, u \in \mathcal{L}. \tag{14.10}$$

Clearly, in (14.10) it is implicitly assumed that the sum of interference contributions behaves as a Gaussian noise process, which is only an approximation. Nevertheless, the threshold-based model is more successful than the collision channel model in representing operation of the physical layer when it comes to design and analysis of random-access protocols.

In scenarios with fading when there is no dominant component (e.g., Rayleigh fading), the collision channel model is too conservative and provides throughput values that are lower than those obtained with the threshold-based criterion. Consider the example of Rayleigh block-fading in which the received powers of the active users' transmissions in a slot are independent and identically distributed (i.i.d.) as

$$f_P(p_\ell) = \frac{1}{\bar{P}} e^{-\frac{p_\ell}{\bar{P}}}, \, \forall \ell \in \mathcal{L}, \tag{14.11}$$

where $\bar{P}$ is the expected received power. It can be shown that in the case when $b \geq 1$,[9] the probability that a capture effect occurs, i.e., that the transmission of some user becomes successfully received in a slot in which there are $|\mathcal{L}|$ interfering users, is

$$D(|\mathcal{L}|) = \frac{1}{(|\mathcal{L}| - 1)!} \frac{e^{-b\frac{P_N}{\bar{P}}}}{(1 + b)^{|\mathcal{L}| - 1}}. \tag{14.12}$$

The throughput of the conventional slotted ALOHA [4] in this case is simply

$$S = \sum_{i=1}^{\infty} D(i) \Pr\{|\mathcal{L}| = i\}. \tag{14.13}$$

In the case of Poisson arrivals with infinite population, the above expression becomes

$$S = \sum_{i=1}^{\infty} \frac{1}{(i - 1)!} \frac{e^{-b\frac{P_N}{\bar{P}}}}{(1 + b)^{i-1}} \cdot \frac{\lambda^i}{i!} e^{-\lambda}, \tag{14.14}$$

where λ is the average arrival rate per slot. Assuming that the expected signal-to-noise ratio (SNR) is $10 \log_{10}(\bar{P}/P_N) = 20 \, \text{dB}$ and that the value of the capture threshold is $b = 2$, the maximal throughput of the conventional slotted ALOHA is $S_{\max} = 0.43 \, (\text{packet/slot})$ when $\lambda \approx 1.19 \, (\text{packet/slot})$. In contrast, on the collision channel the throughput is $S = \Pr\{|\mathcal{L}| = 1\}$, whose maximum in the case of Poisson arrivals is $S_{\max} = 1/e \, (\text{packet/slot})$ when $\lambda = 1 \, (\text{packet/slot})$.

In the case of CSA with Rayleigh block-fading, where the received powers of the active users' transmissions are i.i.d. over both users and slots, with a perfect SIC and the same values of SNR and b as in the previous example, there are degree distributions

---

[9] When the capture threshold b is equal to or greater than 1, only one of the colliding users can capture a slot.

that asymptotically achieve throughputs as high as $S \approx 1.84$ (packet/slot) [28] when the load of the scheme $G \approx 1.86$ (packet/slot).[10] In other words, due to the combination of the capture effect and SIC, close to two packets are successfully received per slot on average.

14.3.7   The Impact of Finite Blocklength

The usual layering abstraction used for design of random-access algorithms can also be demonstrated to be suboptimal when finite-blocklength effects are taken into account. Specifically, consider a simple example of framed slotted ALOHA [29], in which there are N_u users, and every user contends with a packet that is N_b bits long. In the case when the number of slots N_s in the frame is decided solely on the basis of N_u, the classical result states that the value of N_s that maximizes the throughput of slotted ALOHA on the collision channel should be equal to N_u. Specifically, the probability that a slot is a singleton (which determines the throughput) when the impact of the physical layer is neglected is

$$\binom{N_u}{1} \frac{1}{N_s} \left(1 - \frac{1}{N_s}\right)^{N_u-1}. \tag{14.15}$$

However, if the collision channel model is extended to include effects of the finite blocklength, the probability that a slot is a singleton *and* that the transmission occurring in it is actually decoded can be approximated as

$$\binom{N_u}{1} \frac{1}{N_s} \left(1 - \frac{1}{N_s}\right)^{N_u-1} \left(1 - Q\left(\frac{n_s C - n_b + \log(n_s)/2}{\sqrt{n_s V}}\right)\right), \tag{14.16}$$

where n_s is the number of channel uses in a slot, n_b is the packet size in bits, and C and V denote the channel capacity and channel dispersion for the AWGN channel, respectively [30].

If the total number of channel uses in the frame is limited (i.e., the value of the product $N_s \cdot n_s$ is fixed), the optimal number of slots that maximizes throughput that is determined by (14.15) does not match the one determined by (14.16) [31].

14.4   Downlink Communication Channel

In a typical downlink communication channel, there is a single BS which broadcasts distinct messages to a subset of the available users. Assuming that CSI is available at the BS, this setup is conventionally modeled as a Gaussian broadcast channel in the classical information theoretic framework [32, chapter 5]. In this setup, the optimal coding scheme turns out to be based on superposition coding, which, unless all channel gains are equal, achieves a rate region strictly larger than orthogonal coding schemes such as time division multiple access (TDMA) and frequency division multiple access

---

[10]   The load of CSA scheme G (see Section 14.3.5) can be understood as the average number of innovative packets transmitted per slot, and, as such, bears analogy with λ.

(FDMA) [33]. We remark that downlink broadcast for superposition coding is the main inspiration behind the use of NOMA in the downlink [13].

Although superposition coding is optimal from an information theoretic perspective, it is rarely used for serving several users in practical wireless systems. First, TDMA achieves a significant fraction of the capacity region in the low-power regime and in scenarios where the channel gains are similar [33]. Second, the practical performance of the conventional decoding algorithm for superposition-coded transmissions, successive cancellation decoding, suffers as the number of encoded layers increases. Finally, superposition-coded transmissions require each user to observe the full downlink frame and to decode all messages to users encoded with stronger signals. This final disadvantage inevitably leads to more complex decoding algorithms at the users. In addition, a large number of receiving devices today are battery-powered and often operate in one of two states: an "active" state, where the receiver circuitry is active, and a "sleep" state, where the receiver circuitry is turned off. The receiver circuitry typically consumes a certain amount of power proportional to the time it is in the "active" state [34]. The use of superposition-coded transmissions implies that all users need to be in the "active" state in the whole frame, and as a consequence it has been shown by [34] that superposition coding is not necessarily optimal when the power consumption at the users is taken into account.

A second important aspect of downlink transmission is the need to communicate metadata. Metadata is required to inform users about which users the transmitted messages are destined for, sizes of the messages, security information, and packet structure. In modern communication systems, including 4G/LTE [35, chapter 10], metadata and messages are conventionally delivered using a conceptually simple protocol that we depict in Figure 14.5(a). Specifically, the messages are encoded separately using error-correcting codes and delivered orthogonally using TDMA and/or FDMA. The messages are preceded in time by a header containing metadata. The metadata include pointers to locations in a time–frequency grid that enable users to locate and decode their messages.

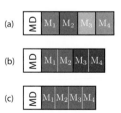

Figure 14.5 Three examples of frames for downlink broadcasting. *MD* denotes metadata, which is encoded in a separate packet and communicated in the beginning of the frame. There are four messages $\{M_k\}_{k=1}^4$ which are grouped in three different ways. In (a), each message is encoded separately, which means that a user only needs to observe the packet encoding metadata and the packet encoding that intended message. In (b), the messages are grouped and encoded in pairs, and hence the users need to observe and decode a packet encoding two messages. Finally, in (c), the four messages are jointly encoded in one packet. In this case, all users need to observe the whole frame.

The design of downlink communication systems is further complicated by the fact that mMTC services often involve short messages, which implies that the size of the metadata is comparable to the sizes of the messages. In the remaining part of this section, we consider some aspects of designing a downlink communication protocol for the transmission of short messages.

14.4.1   Transmission of Short Messages

We consider a generalization of the conventional approach depicted in Figure 14.5(a). Assume that there are K users, whose channels are Gaussian with equal and deterministic channel gains. In order to capture the need for communicating metadata, the sizes of the messages to the users are random variables, which are independently empty (i.e., size of 0 bits) with probability q. This implies that the transmitter needs to communicate metadata about the sizes of the messages and the structure of the transmission. The messages need to be successfully received at the users with an error probability not exceeding ϵ. This also implies that the *frame duration*, the total number of channel uses spent by the transmitter in order to communicate all messages, can be a random variable that depends on the realizations of message sizes. Finally, there is a power consumption associated to each channel use the users observe. In particular, in each channel use, each user is in one of two states (ON/OFF). If it is in the ON state, it consumes one unit of power and observes the corresponding channel output, while if it is in the OFF state, it consumes zero units of power and does not observe the corresponding channel output. The power consumption at a user is defined by the number of channel uses in which the user is in the ON state. We are interested in the tradeoff between the average frame duration and average power consumption at the users when the messages are short.

Throughout the section, assume that the maximum coding rate $R^*(n, \epsilon)$, for fixed blocklength n and error probability not exceeding ε, is well-approximated by the first two terms of its asymptotic expansion [30]:

$$R^*(n, \varepsilon) \approx C - \sqrt{\frac{V}{n}} Q^{-1}(\varepsilon) =: \overline{R}(n, \varepsilon). \qquad (14.17)$$

Here, the channel capacity and channel dispersion are given by

$$C = \frac{1}{2} \log_2(1 + P) \qquad (14.18)$$

$$V = \frac{P}{2} \frac{P + 2}{(P + 1)^2} \log_2^2 e. \qquad (14.19)$$

The asymptotic expansion in (14.17) implies that longer packets can be encoded with a higher rate than shorter ones. Next, for $f \geq 0$ and $\varepsilon \in (0, 1)$, define the function $N(f, \varepsilon)$ as the solution $\overline{n} \geq 0$ to the equation

$$\overline{n}\overline{R}(\overline{n}, \varepsilon) = f. \qquad (14.20)$$

The function $N(f, \varepsilon)$ is the minimum number of channel uses necessary to deliver a message of f bits with an error probability not exceeding ε.

An important implication of (14.17) is that the conventional approach, where each message is encoded separately, is inefficient from a transmitter perspective. Instead, one can simply encode all messages jointly in one large packet and thereby meet the reliability constraints with a smaller average frame duration, an approach depicted in Figure 14.5(c). On the other hand, by encoding all messages jointly in one packet, each user needs to observe the whole frame and decode the messages destined to the other $K - 1$ users, which consumes a large amount of power at the users. Thus, the two approaches can be considered two extreme points on a tradeoff curve between average frame duration and average power consumption at the users. Figure 14.5(b) depicts a third approach: a compromise between the two extremes.

This section investigates the described tradeoff for a specific class of protocols. The sizes (in bits) of the messages, denoted by $|M_k|$, are random variables drawn independently from the set $\{0, \alpha_1, \ldots, \alpha_S\}$, for positive integers $\alpha_1 \leq \ldots \leq \alpha_S$, with probabilities $\{q, (1 - q)p_1, \ldots, (1 - q)p_S\}$. The messages themselves are chosen uniformly at random from the set $\{0, 1\}^{|M_k|}$. Both the messages and the message sizes are unknown at the users. We shall consider the class of protocols in which the transmitter uses L consecutive packets, where L can be a random integer which depends on the message sizes $\{|M_k|\}$. Each message is assigned to exactly one of these L packets, and we shall denote the set of users assigned to the ℓth packet by $\mathcal{U}_\ell \subseteq \{1, \ldots, K\}$, i.e., $\mathcal{U}_\ell \cap \mathcal{U}_{\ell'} = \emptyset$ for $\ell \neq \ell'$ and $\bigcup_\ell \mathcal{U}_\ell = \{1, \ldots, K\}$. The ℓth packet thereby encodes $\sum_{k \in \mathcal{U}_\ell} |M_k|$ message bits. Besides the messages, the transmitter also needs to encode metadata in some of the L packets. Let B_ℓ denote the number of bits of metadata that the transmitter encodes in the ℓth packet. Suppose that the transmitter spends N_ℓ channel uses on the ℓth packet. Assuming that the system uses optimal channel codes, the error probability when decoding the ℓth packet is well-approximated by

$$\mathcal{E}_\ell = Q^{-1}\left(\frac{N_\ell C - B_\ell - \sum_{k \in \mathcal{U}_\ell} |M_k|}{\sqrt{N_\ell V}}\right). \tag{14.21}$$

We are interested in understanding the tradeoff between the average frame duration

$$\bar{T} = \mathrm{E}\left[\sum_{\ell=1}^{L} N_\ell\right] \tag{14.22}$$

and the average power consumption at the users

$$\bar{P} = \mathrm{E}\left[\frac{1}{K}\sum_{\ell=1}^{L} |\mathcal{U}_\ell| N_\ell\right] \tag{14.23}$$

under the following constraint on the error probabilities of the users:

$$\max_{k \in \{1, \ldots, K\}} P\left[\text{user } k \text{ err}\,\middle|\,|M_k| \neq 0\right] \leq \epsilon. \tag{14.24}$$

A Lower Bound

We start by presenting a lower bound on the tradeoff curve between $\overline{T}$ and $\overline{P}$. In order to analyze this tradeoff, we aim at minimizing a weighted sum of the average frame duration and the average power consumption at the users:

$$\mathrm{E}\left[\sum_{\ell=1}^{L} N_\ell\right] + \beta \mathrm{E}\left[\frac{1}{K}\sum_{\ell=1}^{L}|\mathcal{U}_\ell|N_\ell\right] = \mathrm{E}\left[\sum_{\ell=1}^{L}\left(1 + \frac{\beta|\mathcal{U}_\ell|}{K}\right)N_\ell\right]. \tag{14.25}$$

In order to state the lower bound, we shall need the following definition

$$\widehat{N}(f) = \min_{\omega \in (0,\varepsilon_0)} \frac{N(f,\omega)}{\varepsilon_0 - \omega}. \tag{14.26}$$

Here, $\varepsilon_0 \in (0,1)$ is a parameter which can be optimized. It is easy to check that $|\varepsilon_0 - \varepsilon|_+ \widehat{N}(f) \leq N(f,\varepsilon)$ for all $\varepsilon \in (0,1)$ and $f \geq 0$. The expression $|\varepsilon_0 - \varepsilon|_+ \widehat{N}(f)$ can also be shown to be concave in f and convex in ε; properties that are used for the derivation of the lower bound on $\mathrm{E}[\overline{T}] + \beta \mathrm{E}[\overline{P}]$ for all $\beta > 0$ are presented next and proved in [3].

THEOREM 14.1   *We have*

$$\mathrm{E}[\overline{T}] + \beta\mathrm{E}[\overline{P}] \geq (1-q)K \min_{\boldsymbol{\varepsilon}} \phi_\beta(\mathbf{p} - \boldsymbol{\varepsilon}) \tag{14.27}$$

where the minimization is with respect to $\boldsymbol{\varepsilon} \in \mathbb{R}_+^S$ satisfying $\varepsilon_s \leq p_s$, for $s \in \{1,\ldots,S\}$, and $\mathbf{1}_S^\mathsf{T}\boldsymbol{\varepsilon} \leq \epsilon$, and where we have defined

$$\phi_\beta(\mathbf{x}) = \inf_{\zeta > 0} \frac{1}{\zeta}\widehat{N}(\zeta\boldsymbol{\alpha}^\mathsf{T}\mathbf{x})\left(1 + \frac{\beta\zeta\mathbf{1}_S^\mathsf{T}\mathbf{x}}{K}\right), \quad \mathbf{x} \in \mathbb{R}_+^S. \tag{14.28}$$

Here, we have also defined $\boldsymbol{\alpha} = (\alpha_1,\ldots,\alpha_S)$ and $\mathbf{1}_S$ as an S-dimensional all–one vector.

A Simple Genie-Aided Protocol

The lower bound provides us with a tradeoff curve that no protocol can beat. We shall next introduce a genie-aided protocol presented in [3]. This protocol only works provided that all message sizes $\{M_k\}_{k=1}^K$ are known to the users and thus provide us with an optimistic achievable tradeoff curve. Even though the tradeoffs achievable with genie-aided protocols are optimistic, it does allow us to compare different schemes for assigning messages to packets. The protocol can then be augmented with metadata in order to make it applicable. The key intuition behind our genie-aided protocol is that messages of similar sizes are grouped together.

First, for a set of users $\overline{\mathcal{U}} \subseteq \{1,\ldots,K\}$ and a parameter $\kappa \in \mathbb{N}$, we define a $(\overline{\mathcal{U}},\kappa,\epsilon)$ protocol as follows: The users $\overline{\mathcal{U}}$ are first divided into $G = \lceil|\overline{\mathcal{U}}|/\kappa\rceil$ disjoint sets $\{\overline{\mathcal{U}}_\ell\}_{\ell=1}^G$ satisfying $\bigcup_{\ell=1}^G \overline{\mathcal{U}}_\ell = \overline{\mathcal{U}}$ and

$$|\overline{\mathcal{U}}_\ell| = \begin{cases} |\overline{\mathcal{U}}|/G + 1, & \ell \in \{1,\ldots,\mathrm{mod}(|\overline{\mathcal{U}}|,G)\} \\ |\overline{\mathcal{U}}|/G, & \text{otherwise.} \end{cases} \tag{14.29}$$

Sequentially, for $\ell \in \{1,\ldots,G\}$, the transmitter jointly encodes the messages $\{M_k\}_{k\in\overline{\mathcal{U}}_\ell}$ and sends them to the decoders using $N(\sum_{k\in\overline{\mathcal{U}}_\ell}|M_k|,\varepsilon)$ channel uses.

For the genie-aided protocol, we define the sets $\mathcal{U}^{(s)} = \{k \in \mathcal{K} : |M_k| = \alpha_s\}$. Then, sequentially for each $s \in \{1, \ldots, S\}$, the transmitter delivers the messages $\{M_k\}_{k \in \mathcal{U}^{(s)}}$ using a $(\mathcal{U}^{(s)}, \kappa_s, \epsilon)$ protocol. Here, $\{\kappa_s\}_{s=1}^{S}$ are integer parameters, which can be optimized and determine the operating point on the tradeoff curve.

Augmenting the Genie-Aided Protocols with Metadata

In a working protocol, it is necessary for the transmitter to provide the users with a proper amount of metadata that allows each user to identify in what packet and how the desired message is encoded. The optimal approach depends on the number of users, the probabilities $q, p_1, \ldots, p_S$, and the realizable set of message sizes $\{0, \alpha_1, \ldots, \alpha_S\}$. In this section, we shall augment the genie-aided protocol with metadata by preceding in time the packets encoding messages with a header. In particular, all possible combinations of message sizes $\{|M_k|\}_{k=1}^{K}$ can be represented by at most $\lceil K \log_2(S+1) \rceil$ bits, which can be encoded in a header of approximately $N(\lceil K \log_2(S+1) \rceil, \epsilon_H)$ channel uses, meaning that the header can be decoded with an error probability not exceeding ϵ_H. Since each user needs to decode both the header and the packet containing the desired message, we let the genie-aided protocol use an error probability ϵ_D satisfying $(1 - \epsilon_D)(1 - \epsilon_H) = 1 - \epsilon$.

It is clear that this simplistic encoding scheme of metadata is far from optimal in all cases. The problem can be seen as a joint source–channel coding problem, where the goal is to encode the source $\{|M_k|\}_k$ using as few channel uses as possible. For example, in the case where q is near 1, i.e., only a small fraction of the messages are non-empty, it might easily be more efficient to first encode the subset of users with non-empty messages, and then encode the sizes of non-empty messages. As another example, consider the case where the number of users K is comparable or larger than the message sizes. In this case, the metadata can easily dominate the average power consumption. Here, it might be more efficient to first divide the K users into a number of disjoint user groups, which do not depend on the realized message sizes and are known by both the transmitter and all users on beforehand. The transmitter can then serve the users of each user group consecutively in time, and hence all users do not need to observe and decode metadata destined for users of other user groups.

Numerical Illustration

We depict the lower bound in Theorem 14.1, the genie-aided, and the genie-aided lower bound augmented with metadata in Figure 14.6. The plots show the average power consumption at the users as a function of the average frame duration. The parameters in these plots are given by $K = 16$, $q = p_1 = p_2 = \frac{1}{2}$, and the SNR is 10 dB for all users. Moreover, the message sizes (α_1, α_2) are given by $(64, 128)$ and $(256, 512)$ in the two subfigures. The error probability ϵ is fixed to 10^{-3}, and ϵ_D and ϵ_H are optimized. The plots illustrate tradeoffs, where one can choose to decrease the average frame duration at the expense of an increased average power consumption at the users. The optimal operating point on the tradeoff curve depends on the application at hand. We observe in both plots that there is a significant gap between the lower bound and the protocol augmented with control information, which illustrates the importance of the encoding scheme for the metadata.

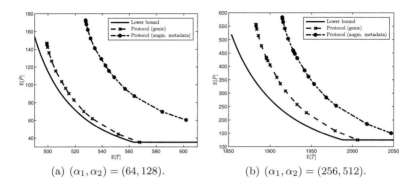

(a) $(\alpha_1, \alpha_2) = (64, 128)$.   (b) $(\alpha_1, \alpha_2) = (256, 512)$.

Figure 14.6 Average power consumption at the users as a function of the average frame duration for message sizes (α_1, α_2) equal to $(64, 128)$ in (a) and to $(256, 512)$ in (b). The figure depicts the lower bound in Theorem 14.1 and the tradeoff curves achievable by the genie-aided protocol and the protocol augmented with metadata.

14.5   Conclusion

In this chapter, we have made an attempt to further reconcile the long-standing gap between the information theory and the networking/protocol design community. The key to this reconciliation is the inclusion of protocol information/metadata within the information theoretic models. This is becoming particularly important for 5G wireless systems that will feature various services related to mMTC and Internet of Things (IoT). Those services are characterized by the transmission of small data portions, such that the size of the metadata becomes relevant and the transmission of metadata requires a significant amount of resources. We have discussed two multiuser setups that are relevant for 5G systems: the uplink (MAC) and the downlink (broadcast channel). For the uplink case, we have provided a discussion on various developments in the area of random access, a classical research area that is experiencing a revival due to new ideas and its great significance for the mMTC services in 5G systems, as well as low power wide area (LPWA) networks for IoT. The transmission of short packets to multiple users in the downlink has received much less attention, which does not mean it is not relevant or challenging. We have presented a study that illustrates the fundamental tradeoffs that emerge for the downlink case, providing a tractable model for data and protocol information.

References

[1] R. Gallager, "A perspective on multiaccess channels," *IEEE Trans. Inf. Theory*, vol. 31, no. 2, pp. 124–142, 1985.

[2] A. Ephremides and B. Hajek, "Information theory and communication networks: An unconsummated union," *IEEE Trans. Inf. Theory*, vol. 44, no. 6, pp. 2416–2434, Oct. 1998.

[3] K. F. Trillingsgaard and P. Popovski, "Downlink transmission of short packets: Framing and control information revisited," *IEEE Trans. Commun.*, vol. 65, no. 5, pp. 2048–2061, Feb. 2017.

[4] L. G. Roberts, "ALOHA packet system with and without slots and capture," *SIGCOMM Comput. Commun. Rev.*, vol. 5, no. 2, pp. 28–42, Apr. 1975.

[5] A. Laya, L. Alonso, and J. Alonso-Zarate, "Is the random access channel of LTE and LTE-A suitable for M2M communications? A survey of alternatives," *Commun. Surveys Tuts.*, vol. 16, no. 1, pp. 4–16, 2014.

[6] *Medium Access Control (MAC) protocol specification (Rel. 14)*, 3GPP Std. TS36.321 v14.9.0, 2018.

[7] *Medium Access Control (MAC) protocol specification (Rel. 15)*, 3GPP Std. TS38.321 v15.4.0, 2018.

[8] A. Ephremides, "The multi-access channel in a network: Stability and network coding issues," in *Multiple Access Channels*, E. Biglieri and L. Gyorfi, Eds. Amsterdam: IOS Press, 2007, pp. 287–298.

[9] J. Capetanakis, "Tree algorithms for packet broadcast channels," *IEEE Trans. Inf. Theory*, vol. 25, no. 5, pp. 505–515, Sep. 1979.

[10] S. Verdu, "Computation of the efficiency of the Mosely–Humblet contention resolution algorithm: A simple method," *Proc. IEEE*, vol. 74, no. 4, pp. 613–614, Apr. 1986.

[11] X. Chen, T. Y. Chen, and D. Guo, "Capacity of Gaussian many-access channels," *IEEE Trans. Inf. Theory*, vol. 63, no. 6, pp. 3516–3539, Jun. 2017.

[12] L. Liu, E. G. Larsson, W. Yu, et al., "Sparse signal processing for grant-free massive connectivity," *IEEE Signal Process. Mag.*, vol. 35, no. 5, pp. 88–89, Sep. 2018.

[13] L. Dai, B. Wang, Y. Yuan, et al., "Non-orthogonal multiple access for 5G: Solutions, challenges, opportunities, and future research trends," *IEEE Commun. Mag.*, vol. 53, no. 9, pp. 74–81, Sep. 2015.

[14] Z. Wu, K. Lu, C. Jiang, and X. Shao, "Comprehensive study and comparison on 5G NOMA schemes," *IEEE Access*, vol. 6, pp. 18 511–18 519, 2018.

[15] D. Danyev, B. Laczay, and M. Ruszinko, "Multiple access adder channel," in *Multiple Access Channels*, E. Biglieri and L. Gyorfi, Eds. Amsterdam: IOS Press, 2007, pp. 26–53.

[16] B. Lindström, "Determining subsets by unramified experiments," in *A Survey of Statistical Design and Linear Models*, J. N. Srivastava, Ed. Amsterdam: North-Holland, 1975.

[17] L. Gyorfi and B. Laczay, "Signature coding and information transfer for the multiple access adder channel," in *IEEE Inf. Theory Workshop*, San Antonic, TX, 2004, pp. 242–246.

[18] J. Goseling, C. Stefanovic, and P. Popovski, "Sign-compute-resolve for tree splitting random access," *IEEE Trans. Inf. Theory*, vol. 64, no. 7, pp. 5261–5276, Jul. 2018.

[19] Y. Polyanskiy, "A perspective on massive random-access," in *IEEE Int. Symp Inf. Theory (ISIT)*, Aachen, 2017, pp. 2523–2527.

[20] X. Chen and D. Guo, "Many-access channels: The Gaussian case with random user activities," in *IEEE Int. Symp. Inf. Theory (ISIT)*, Honolulu, 2014, pp. 3127–3131.

[21] O. Ordentlich and Y. Polyanskiy, "Low complexity schemes for the random access Gaussian channel," in *IEEE Int. Symp Inf. Theory (ISIT)*, Aachen, 2017, pp. 2528–2532.

[22] E. Paolini, C. Stefanovic, G. Liva, and P. Popovski, "Coded random access: Applying codes on graphs to design random access protocols," *IEEE Commun. Mag.*, vol. 53, no. 6, pp. 144–150, Jun. 2015.

[23] G. Liva, "Graph-based analysis and optimization of contention resolution diversity slotted ALOHA," *IEEE Trans. Commun.*, vol. 59, no. 2, pp. 477–487, Feb. 2011.

[24] E. Paolini, G. Liva, and M. Chiani, "Coded slotted ALOHA: A graph-based method for uncoordinated multiple access," *IEEE Trans. Inf. Theory*, vol. 61, no. 12, pp. 6815–6832, Dec. 2015.

[25] C. Stefanovic, E. Paolini, and G. Liva, "Asymptotic performance of coded slotted ALOHA with multi packet reception," *IEEE Commun. Lett.*, vol. 22, no. 1, pp. 105–108, Jan. 2018.

[26] S. Ghez, S. Verdú, and S. Schwartz, "Stability properties of slotted ALOHA with multi-packet reception capability," *IEEE Trans. Autom. Control*, vol. 33, no. 7, pp. 640–649, Jul. 1988.

[27] A. Zanella and M. Zorzi, "Theoretical analysis of the capture probability in wireless systems with multiple packet reception capabilities," *IEEE Trans. Commun.*, vol. 60, no. 4, pp. 1058–1071, Apr. 2012.

[28] F. Clazzer, E. Paolini, I. Mambelli, and C. Stefanovic, "Irregular repetition slotted ALOHA over the Rayleigh block fading channel with Capture," in *IEEE Int. Conf. Commun (ICC)*, 2017, pp. 1–6.

[29] H. Okada, Y. Igarashi, and Y. Nakanishi, "Analysis and application of framed ALOHA channel in satellite packet switching networks: FADRA method," *Electron. Commun. Japan*, vol. 60, pp. 60–72, Aug. 1977.

[30] Y. Polyanskiy, H. V. Poor, and S. Verdú, "Channel coding rate in the finite blocklength regime," *IEEE Trans. Inf. Theory*, vol. 56, no. 5, pp. 2307–2359, Apr. 2010.

[31] G. Durisi, T. Koch, and P. Popovski, "Toward massive, ultrareliable, and low-latency wireless communication with short packets," *Proc. IEEE*, vol. 104, no. 9, pp. 1711–1726, Sep. 2016.

[32] A. E. Gamal and Y.-H. Kim, *Network Information Theory*. Cambridge: Cambridge University Press, 2012.

[33] G. Caire, D. Tunninetti, and S. Verdu, "Suboptimality of TDMA in the low-power regime," *IEEE Trans. Inf. Theory*, vol. 50, pp. 608–620, Apr. 2004.

[34] M. Kim and H. Viswanathan, "Energy-efficient delay-tolerant communication: Revisiting optimality of superposition coding in broadcast channels," in *IEEE Global Comm. Conf. (GLOBECOM)*, Austin, IX, 2014, pp. 3526–3531.

[35] E. Dahlman, S. Parval, and J. Skold, *4G LTE/LTE-Advanced for Mobile Broadband*. New York: Academic Press, 2014.

15 Interference Management in Wireless Networks: An Information Theoretic Perspective

Ravi Tandon and Aydin Sezgin

15.1 Introduction

Interference is one of the key barriers in achieving higher spectral efficiency for wireless communications. In the last decade, there has been intense research activity in the area of interference management for wireless systems. The *optimal solution* to this challenge is ultimately connected to the characterization of fundamental performance bounds of wireless networks. Exact characterization of the information theoretic capacity region of an arbitrary wireless network, however, is in general an open problem. Most of the exact capacity region characterizations are typically limited to specific network configurations such as multiple-access networks, or special cases such as classes of broadcast and interference networks. Instead of exact capacity regions, an alternative path is to pursue approximate characterizations of capacity (or sum capacity) which is given by a high-power analysis.

In more detail, the information theoretic sum capacity of an interference network can be written as

$$C\,(\mathrm{SNR}) = \mathrm{DoF}\,\log_2\,(\mathrm{SNR}) + o\left(\log_2\,(\mathrm{SNR})\right), \tag{15.1}$$

where $C\,(\mathrm{SNR})$ denotes the capacity and SNR represents the ratio of the transmit power of a transmitter to the local thermal noise power at its respective receiver. The second term $o\left(\log_2\,(\mathrm{SNR})\right)$, which is basically an approximation error term, becomes negligible by definition relative to $\log_2\,(\mathrm{SNR})$ as the SNR becomes large. This approximate characterization becomes increasingly accurate as the signal and interference power increases while the noise power is low. The DoF can be obtained from (15.1) by

$$\mathrm{DoF} = \lim_{\mathrm{SNR}\to\infty} \frac{C\,(\mathrm{SNR})}{\log_2\,(\mathrm{SNR})}.$$

In addition to the DoF (degrees of freedom), the so-called generalized DoF (GDoF) is a finer metric to refine the results obtained in the DoF analysis. The GDoF are defined as [1]

$$\mathrm{GDoF} = \lim_{\substack{\mathrm{SNR,INR}\to\infty, \\ \mathrm{INR}=\mathrm{SNR}^\alpha}} \frac{C\,(\mathrm{SNR},\mathrm{INR})}{\log\,(\mathrm{SNR})}, \tag{15.2}$$

and provide the slope of the capacity C as a function of both the SNR and the interference-to-noise ratio

$$\text{INR} = \text{SNR}^{\alpha} \tag{15.3}$$

on a logarithmic scale. The parameter α defines the slope of increase of the INR as a function of the SNR. The special case of $\alpha = 1$ provides the DoF. The justification to consider the GDoF as a generalization of the DoF is that the DoF makes channels with different strengths (weak, strong, very strong, etc.) to equivalent channels due to the scaling with the same power. This, however, changes the nature of the problem, while the GDoF keeps the nature of the channel intact and is thus more suitable to analyze heterogeneous networks. The GDoF thus provides a first-order approximation of the capacity of the underlying interference channel (IC) which becomes exact asymptotically, and highlights other aspects of interference networks that are not made visible by the DoF. The DoF and GDoF are part of a general methodology of progressive refinement, i.e., a coarse capacity approximation followed by a finer capacity approximation, as illustrated in Figure 15.1. It can essentially be subdivided into **three progressive steps**, starting with the DoF to describe the capacity C (or capacity region) of a network. This is followed by the GDoF in order to achieve a refinement in the accuracy of the results.

As a representative example, consider the practically relevant K-user Gaussian interference network for which the input–output signal relationship can be described as:

$$y_k(t) = \sum_{j=1}^{K} \sqrt{Ph_{kj}(t)}x_j(t) + n_k(t) \quad 1 \le k \le K, \tag{15.4}$$

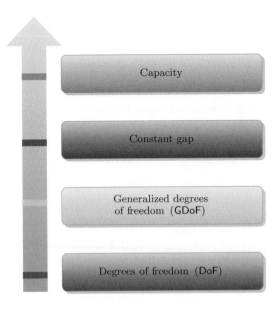

Figure 15.1 Approach of progressive refinement: Starting from coarse characterization (by DoF), to finer characterizations (through GDoF and constant gap results) leading ultimately to exact capacity.

where $y_k(t)$ is the received signal at user k and time t, and is composed of the desired signal $x_k(t)$ and disturbances, weighted by the respective fading channel parameters $h_{kj}(t)$, and the receiver noise $n_k(t)$. P denotes the transmit power. This model can be slightly modified (cf. [2]), such that it can be utilized for the extraction of new methodological insights with respect to the GDoF. Thus, we obtain for (skipping the time index)

$$y_k = \sum_{j=1}^{K} \sqrt{P} h_{kj} x_j + n_k = \sum_{j=1}^{K} \sqrt{P^{\alpha_{kj}}} g_{kj} x_j + n_k \quad 1 \le k \le K. \tag{15.5}$$

Here, g_{kj} denotes the channel coefficients of the fading channel from node j to node k. The channel strengths are modeled by α_{kj}. Thus, the channels h_{kj} have been decomposed into two components, attributed to fast fading (g_{kj}) and the path loss (α_{kj}).

The Gaussian channel model described above serves as the basis of the method referred to as the deterministic approach [3]. The deteriorating effects in the communication, such as interference and noise, are approximated as follows:

$$\mathbf{y}_k(t) = \sum_{j=1}^{K} \mathbf{D}^{q-n_{kj}} \mathbf{x}_j(t), \tag{15.6}$$

where n_{kj} denotes the channel strength, $\mathbf{x}_j(t)$ and $\mathbf{y}_k(t)$ are the binary representations of the transmit and receive signal, respectively, up to a precision of q bits. The matrix $\mathbf{D}$, given by

$$\mathbf{D} = \begin{bmatrix} 0 & 0 & \cdots & 0 \\ 1 & 0 & \cdots & 0 \\ 0 & 1 & \cdots & 0 \\ \vdots & \vdots & \ddots & \vdots \\ 0 & 0 & \cdots & 1 \end{bmatrix}, \tag{15.7}$$

is a so-called "downshift" matrix and models the impact of noise and the wireless fading channel. Now, the insights gained through the deterministic approach can be successfully transferred to the Gaussian channel itself. This has been proven extensively in numerous publications, as discussed next. For instance, the capacity of linear-high SNR deterministic networks has been studied in [3, 4]. To reiterate, the advantage of this linear-high SNR deterministic modeling is that it allows making statements and observations about achievable rates in a network, which are not possible, per se, in the noisy counterpart of the network. However, these observations and statements provide insights into the behavior of the system such that even concepts for the noisy network as well as capacity characterizations can be obtained. These insights can be made useful due to the interesting relation between the linear deterministic model of the network and its noisy counterpart, which can be proved for certain networks [3]. Using this model, one can develop strategies for the noisy channel that achieve rates within a constant gap of capacity upper bounds. This method has been applied successfully for the IC, for the many-to-one IC, and the so-called Y-channel in [5], [6], and [7, 8], respectively.

Thus, the final step of the approximative approach replaces the GDoF with finite SNR analysis and aims at improving the approximate characterization of the capacity region quantitatively and perhaps ideally (which is not predictable) obtain an exact characterization. To this end, the developed lower bounds (i.e., the conceptional approaches) and upper bounds (discussed later in the chapter) on the capacity/capacity regions need to be iteratively, through auxiliary parametrization, improved and systematically brought closer to each other.

Besides the linear-high SNR deterministic approximation, there are more general (semi-)deterministic models proposed by Costa and El Gamal in 1982 [9] and Telatar and Tse in 2007 [10]. With those models, one can obtain further insights for the noisy case. The general deterministic models are especially important for multiantenna systems [10], since the linear-high SNR deterministic model does not take into account the special properties of multiple-input multiple-output (MIMO) systems [11]. As a result, the capacity of the MIMO IC was characterized in [10] within a constant (independent of the SNR or channels) number of bits depending only on the number of receive antennas using the more general deterministic approaches.

15.1.1   Organization of the Chapter

This chapter is organized as follows. In Section 15.2, we discuss various classical interference management schemes, namely the approaches of treating interference as noise, interference avoidance, suppression, and cancellation. In Section 15.3, we present discussion on modern techniques, such as interference alignment, interference neutralization (and applications to multihop settings), as well as the use of structured codes. Section 15.4 focuses on the robustness aspects of the above scheme with respect to the availability of channel knowledge availability. In particular, we discuss the role of channel state information (CSI), and its impact on the feasibility of efficient interference management. We focus on various forms of CSI starting from no CSI, delayed CSI, alternating CSI, and other models that have been recently investigated, and different interference management schemes for such scenarios. In Section 15.5, we focus on the role of network and channel heterogeneity in interference management. In particular, we first discuss the conditions under which treating interference as noise (TIN) is information-theoretically optimal. We also discuss the recent framework of topological interference management (or TIM), to efficiently manage interference in partially connected wireless networks with minimal knowledge about the wireless channel. We next discuss a recently proposed approach of *interference-free orthogonal frequency division multiplexing (OFDM)*, which leverages heterogeneity in wireless channel characteristics (namely intersymbol interference [ISI] parameters) to achieve significant gains in spectral efficiency. Section 15.6 is focused on outer bounding techniques, with a discussion starting from cut-set bounds, as well as improved techniques, such as enhancement and genie-aided approaches, and virtual receivers.

15.2   Classical Interference Management Techniques

Considering interference management methods applied in practice nowadays leads to the interesting insight that robustness and simplicity are the main driving forces in the choice of which and how a specific scheme is applied. This will have an immediate effect on the devices in terms of price and functionality. In more detail, in practice three classical concepts are by now well established. With those concepts, one either aims at ignoring the interference by treating it simply as an additional random noise whenever it is received with sufficiently lower power compared to the power of the desired signal. Alternatively, one aims at avoiding interference by splitting the resources among the nodes in noninterfering or orthogonal segments. Thus, within each resource segment, the corresponding node receives the desired signal free of interference. Yet another approach aims at suppressing the interference either through spatial receive filtering or spatial beamforming in the case that multiple antennas are available at the nodes. Alternatively, spread spectrum methods can be utilized in which spreading sequences with good cross-correlation properties are assigned to the nodes.

While they might not be the desired choice in terms of elegance and beauty on how to manage interference, they are the first choice of a systems engineer as they ensure the status quo in allowing the feasibility of conveying data from one node A to another node B. Note that those concepts can be applied as a standalone solution, but they can also be combined and utilized in more sophisticated interference management methods such as interference cancellation. However, they also appear as part of modern interference management techniques (as will be clear in the following section), in which they are systematically placed to offer benefits without much coordination.

In the remainder of this section, we elaborate on those traditional and/or most frequently used methods in practice, and discuss their performance in terms of achievable rate, DoF, and GDoF.

15.2.1   Treating Interference as Noise

Suppose that we have concurrent transmissions within the same spectrum. Now, the presence of interference due to concurrent transmissions by multiple information sources might lead to deterioration in the overall quality of signal reception. As the interference has structure, one is tempted to infer useful information in order to remove it from the received signal. However, this might be challenging and require substantial effort and, depending on the achieved outcome, have insufficient benefits. In particular, the codebook, modulation order, and other details of the communication strategy have to be provided to the receiver by the interfering node either in an online fashion or prior to the communication. However, as long as the deteriorations in the overall quality of signal reception are acceptable, one can simply ignore that the interference is an information-bearing signal and instead treat it as unstructured and uncorrelated disturbance that contributes to the overall noise variance at the receiver [12]. The achievable rate obtained by this TIN approach can be stated as follows. Suppose that node k wants to communicate with its receiving counterpart in the presence of interference from other users, predominantly from user j. Then, the achievable rate of that user k is approximately given by

$$R_{\text{TIN}} = \left(1 - \alpha_{kj}\right) \log_2 \left(\text{SNR}\right) + o\left(\log_2 \left(\text{SNR}\right)\right), \tag{15.8}$$

where α_{kj} denotes the interference level as in (15.3) caused by transmitter j at receiver k, and $\alpha_{kk} = 1$. We observe that the rate monotonically decreases by the level of interference caused at the receiver. In certain cases, it might turn out that this scheme, despite its simplicity, is optimal from a GDoF or even from a capacity perspective. As such, there is no need to resort to schemes with higher complexity as those would only provide a bounded gain or no gain at all. In more detail, the optimality of TIN from a GDoF perspective for the noisy interference regime ($\alpha_{kj} \leq {}^1\!/_2$) was shown in [1] for the two-user IC. Later on, the authors of [13–15] were able to obtain conclusions on the optimality of TIN for nonasymptotic regimes, and hence obtain the capacity of the IC with weak (noisy) interference, the first capacity result after several decades of standstill.

If the interference is of comparable strength to the desired signal, one approach considered is to avoid the interference completely, as discussed next.

15.2.2   Interference Avoidance

It we cannot simply ignore the interference due to its comparably strong impact on the system performance, one approach is to split the available resources (time, frequency, code, space) into (orthogonal) segments and allocate the nodes such that they do not interfere with each other. This method of interference orthogonalization is referred to as interference avoidance. Instances of this are quite common in practice, such as time-division multiplex access (TDMA), frequency-division multiple access (FDMA), code-division multiple-access (CDMA), and space-division multiple access (SDMA). Consider Figure 15.2, in which the basic idea of those concepts is illustrated.

While TDMA and FDMA separate the nodes by orthogonal time slots and frequency bands, spreading sequences are utilized in CDMA to separate the users from each other. This surely depends on the properties of the sequences used and thus effects the achievable rates. If the nodes are active in different cell sectors within a cell of a cellular communication systems, spatial separation through antenna processing at the base stations and thus SDMA could be applied to avoid (or at least reduce) the

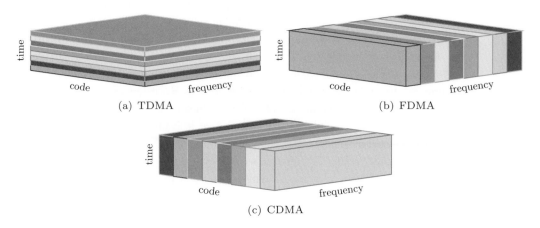

(a) TDMA   (b) FDMA

(c) CDMA

Figure 15.2 Classical interference avoidance/multiple-access schemes.

interference within the system. As such, the rate by avoiding interference for a user k is approximatively given by

$$R_{\text{IA}} = \frac{1}{K} \log_2 (\text{SNR}) + o \left(\log_2 (\text{SNR}) \right), \tag{15.9}$$

where K denotes the number of potential interferers, which is upper bounded by the overall number of users in the system. It was shown in [1] for the two-user IC that the orthogonalization of the users in time is optimal from a GDoF perspective in the case that the interference and the desired signal are at a comparable power level, i.e., $\alpha_{kj} = \alpha_{kk}$ for all $k, j, k \neq j$. If the transmitters lack full channel state information (CSIT) but know the connectivity within the network, it was shown in [16, 17] that for certain topologies orthogonal access schemes are optimal as well.

Another way of dealing with interference is to reduce its severity in a less dogmatic way of banning it to orthogonal subspace, be it time, frequency, code, or space, as will be discussed next.

15.2.3   Interference Suppression

In order to suppress the interference, the nodes have to be more reactive, for which some knowledge on the interference is needed. For instance, with multiple antennas available at the transmitter and/or receiver end of a wireless link, interference suppression by means of spatial processing can be employed. In more detail, if the receiver is equipped with multiple antennas it can use some of the spatial DoF available such that the effect of interference is completely eliminated or significantly reduced. Let $\mathbf{y}$ with

$$\mathbf{y} = \mathbf{H}_1 \mathbf{x}_1 + \mathbf{H}_2 \mathbf{x}_2 + \mathbf{z} \tag{15.10}$$

be the vector value received signal at a node interested in decoding the signal $\mathbf{x}_1$ carrying d_1 data streams, observed through channel $\mathbf{H}_1$ with additive noise $\mathbf{z}$. The number of antennas at the receive node, the interfering node, and the transmit node are given by N, M_I, and M_T, respectively. Suppose that the entries of the channels $\mathbf{H}_1$, $\mathbf{H}_2$ are realizations of an i.i.d. random variable, such that channels have full rank almost surely. Suppose further that $\mathbf{H}_2 \mathbf{x}_2$, represents the interference, which needs to be suppressed. This can be done at the interfering node by linear beamforming with $\mathbf{x}_2 = \mathbf{V}_2 \mathbf{d}_2$, where $\mathbf{V}_2$ represents the beamforming matrix and $\mathbf{d}_2$ is the data vector of length d_2 according to the number of its data streams. By choosing $\mathbf{V}_2$, such that it is orthogonal to the channel $\mathbf{H}_2$, i.e., $\mathbf{H}_2 \mathbf{V}_2 = \mathbf{0}$, with $\mathbf{V}_2 \neq \mathbf{0}$, the interfering node can completely suppress the interference otherwise visible at the receiving node. This can be achieved as long as the receiver has fewer antennas than the number of transmit antennas at the interfering node, i.e., $N < M_I$. Alternatively, the receiver can suppress the interference by linear post-processing with $\mathbf{U}\mathbf{y}$, such that

$$\mathbf{U}\mathbf{y} = \mathbf{U} \left(\mathbf{H}_1 \mathbf{x}_1 + \mathbf{H}_2 \mathbf{x}_2 + \mathbf{z} \right) = \mathbf{U}\mathbf{H}_2 \mathbf{x}_2 + \mathbf{U}\mathbf{z} \tag{15.11}$$

with $\mathbf{U}\mathbf{H}_2 \mathbf{x}_2 = \mathbf{0}$ and $\mathbf{U}\mathbf{H}_1 \neq \mathbf{0}$ as long as $d_2 < N$. In between those extremes we have the case that the interferer and the receiver jointly suppress the interference, i.e., $\mathbf{U}$ and $\mathbf{V}_2$ are designed such that

$$\mathbf{Uy} = \mathbf{U}\left(\mathbf{H}_1\mathbf{x}_1 + \mathbf{H}_2\mathbf{V}_2\mathbf{d}_2 + \mathbf{z}\right) = \mathbf{UH}_1\mathbf{x}_1 + \mathbf{Uz}, \tag{15.12}$$

with $\mathbf{UH}_2\mathbf{V} = \mathbf{0}$. Thus, by choosing $d_1 \leq N$ and $\mathbf{d}_2 \leq M_I$ appropriately, the suppression task can be split between transmitter and receiver [18, 19]. It is important to note that the interference suppression needs full cooperation among the receive and transmit antennas. In the case of distributed transmit antennas, for instance, the data must be provided to the nodes in advance, i.e., proactively.

The aforementioned methods of interference management are linear and thus of comparably reduced complexity. Nonlinear interference management schemes such as interference cancellation, discussed next, have higher complexity but can provide in general a significant performance gain.

15.2.4   Interference Cancellation

Suppose that a receiver equipped with N antennas is receiving a superposition of signals from K interfering users and one desired user.

Now, let

$$\mathbf{y} = \mathbf{h}x + \sum_{i=1}^{K} \mathbf{g}_i s_i + \mathbf{n} \tag{15.13}$$

be the received signal of the node equipped with N antennas with additive noise $\mathbf{n}$. The received signal contains the desired signal x obtained over the channel $\mathbf{h}$, which is subject to interference from interfering signals s_i via the channels $\mathbf{g}_i$, $1 \leq K$. If the receiver is able to obtain an exact copy of the interfering signals, then the interference could be completely removed without any loss. In practice, this is hardly given. Thus, the receiver has to utilize its receive antennas to detect and decode the interference signals. Once the receiver has an estimate $\hat{s}_i$ of the interfering signals, it can subtract it from the received signal to obtain

$$\tilde{\mathbf{y}} = \mathbf{h}x + \tilde{\mathbf{n}}, \tag{15.14}$$

where $\tilde{\mathbf{n}}$ contains the additive noise and some residual interference due to the imperfect cancellation step. The quality of the estimate and thus the amount of residual interference depends on the algebraic relation (linear dependence) between the channel vectors $\mathbf{g}_i$ and $\mathbf{h}$, $1 \leq i \leq K$. If the channels are rather co-linear, then the performance of the detection and decoding process might be poor, resulting in high residual interference. Generally speaking, there are two options to detect the interference signals, namely interference cancellation in parallel or successively. In the former, the receiver uses parallel branches of post-processing, in which at each branch i a postcoder $\mathbf{u}_i$ aims at reducing the effects of noise and the other interference signals s_j, $1 \leq j \leq K, j \neq i$, while improving the power level of s_i. There are various ways to design $\mathbf{u}_i$, among which linear methods like matched filtering, zero-forcing filtering, and minimum mean-square error filtering are the most common. The decoding of the interference signal i can be successful as long as the following condition on the communication rate,

$$R_i \le \log_2 \left(1 + \frac{P_i \mathbf{u}_i^H \mathbf{g}_i \mathbf{g}_i^H \mathbf{u}_i}{\mathbf{u}_i^H \left(\mathbf{hh}^H P_x + \sum_{j=1, j \ne i}^{K} \mathbf{g}_j \mathbf{g}_j^H P_j \right) \mathbf{u}_i} \right) \quad \forall i \qquad (15.15)$$

is fulfilled. Here, P_j denotes the power of the jth, $1 \le 1 \le K$, interference signal, while P_x is the power of the desired signal. The output of all parallel branches is then used in order to remove the interference signals from the received signal to obtain (15.14).

Alternatively, interference cancellation at the receiver can be done in a successive way. While the order of this procedure might have a strong impact on the performance, we assume for simplicity that the decoding order is done in the index order i of the interfering signals. To this end, the interference signal $i = 1$ is decoded first. The estimate of s_i is then used for cancellation from the received signal. Next, the signal s_2 is decoded and its contribution is removed from the remaining received signal. This procedure continues until the interference from user K is removed. Successive interference cancellation can be done successfully as long as

$$R_i \le \log_2 \left(1 + \frac{P_i \mathbf{u}_i^H \mathbf{g}_i \mathbf{g}_i^H \mathbf{u}_i}{\mathbf{u}_i^H \left(\mathbf{hh}^H P_x + \sum_{j=i+1}^{K} \mathbf{g}_j \mathbf{g}_j^H P_j \right) \mathbf{u}_i} \right) \quad \forall i \qquad (15.16)$$

is fulfilled.

As an alternative to canceling the interference at the receiver, one could aim to pre-cancel the interference already at the transmitter. To this end, the transmitter has to have knowledge of the interference deteriorating at the receiver. Furthermore, pre-cancellation might be the only option if the receiver does not have the capability to remove the interference due to lack of knowledge or available resources.

Now, suppose that the received signal is deteriorated by an additive interference term and noise. Let

$$y = x + s + z \qquad (15.17)$$

be the received signal at a node, where x represents the desired signal and z is the receiver noise. The interference is represented by s, which is known to the transmitter and unknown to the receiver. Such channels are referred to as channels with state, as illustrated in Figure 15.3.

Max Costa has shown in [20] that the capacity of such a channel is identical to the capacity of a channel without any interference term s. This is achieved by designing the transmit signal x as a superposition of the weighted interference term $1/\alpha s$ and the desired information u, with the optimal α given by

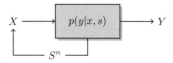

Figure 15.3 Channel with state: The state vector $S^n = (s_1, s_2, \ldots, s_n)$ is given to the transmitter but is not available at the receiver.

$$\alpha = 1 + \text{SNR}. \tag{15.18}$$

Thus, α balances nicely between addressing the impact of noise power versus the interference. The concept is referred to as dirty-paper coding or Costa-precoding for Gaussian channels and as Gelfand–Pinsker coding for general discrete memoryless channels [21]. Channels with state play an important role in communications in applications like storage with errors, DSL, wireless networks, and stenography [22–25]. For instance, consider the broadcast channel with one base station serving two users, where the overall transmit signal of the base station is given by

$$x = x_1 + x_2, \tag{15.19}$$

where x_1 and x_2 is the signal intended for users 1 and 2, respectively. The received signal at user 2 is given by

$$y_2 = x_2 + x_1 + z_2, \tag{15.20}$$

where x_1 is the interference known at the transmitter (here the base station). By dirty-paper encoding of x_2 with respect to x_1, we are able to cancel the impact of x_1 at user 2 and obtain

$$y_2 \approx x_2 + z_2. \tag{15.21}$$

Meanwhile, user 1 treats the signal x_2 as noise. Alternatively, the procedure can be used to pre-cancel the interference at user 1, while user 2 treats x_1 as noise. It can be shown that this scheme is indeed optimal. A practical (and suboptimal) realization of dirty-paper coding is given by the so-called Tomlinson–Harashima precoding [26, 27].

15.3   Modern Interference Management Techniques

15.3.1   Interference Alignment

In this section, we review the progress on the idea of interference alignment (IA) for interference management in wireless networks. Interference alignment is a linear precoding technique that attempts to align interfering signals in time, frequency, or space. The key idea is that users can coordinate their transmissions using linear precoding, such that the interference signal lies in a reduced dimensional subspace at each receiver. The concept of IA was originally introduced by Maddah-Ali et al. in [28] for the MIMO X channel. The ideas of IA for K-user ICs were fully developed by Cadambe and Jafar in [29], where it was shown that the sum capacity of the K-user IC scales as $\frac{K}{2}\log(P) + o(\log(P))$, showing the celebrated result that each user can get half the resources via IA.

We now briefly explain IA through an example of three-user IC presented in [29], where the concept of channel extensions is introduced. In particular, the idea is to transmit information from multiple transmitters over a certain number of channel extensions ($2n + 1$ for the example shown in Figure 15.4). The information symbols from transmitter i are denoted by the vectors s_i, for $i = 1, 2, 3$. For this example, transmitters 1 and

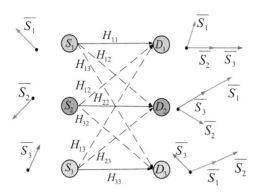

Figure 15.4 A three-user IC with achievable DoF = $\frac{3n+1}{2n+1}$, which converges to $3/2$ as $n \to \infty$. [29].

2 intend to send n symbols, whereas transmitter 3 sends $n + 1$ symbols to their respective receivers. The key idea is that the information symbols S_1, S_2, and S_3 are precoded (with precoding vectors, v_1, v_2, and v_3, respectively), such that the following alignment conditions are satisfied:

- Symbol vectors S_2, S_3 are aligned in an $(n + 1)$ dimensional subspace at receiver 1.
- Symbol vectors S_1, S_3 are aligned in an $(n + 1)$ dimensional subspace at receiver 2.
- Symbol vectors S_1, S_2 are aligned in an n dimensional subspace at receiver 3.

Linear IA schemes, i.e., those based on spatial beamforming, operate within the spatial dimensions provided by multiple antennas at the transmitting and receiving nodes, and seek to divide those spatial dimensions into separable subspaces to be occupied by interference and desired signals at each receiver. When the number of antennas at each node is insufficient, e.g., when each node is equipped with only one antenna, spatial interference alignment schemes are not feasible, instead, assuming knowledge of the upcoming time slots, coding over this entire time together can be beneficial.

This method has been generalized for the X network in [30, 31], where each source node has different messages for all destinations. The authors of [30] developed an asymptotic IA scheme for the $M \times N$-user SISO X network, which partially aligns the interference to asymptotically achieve the upper bound, which is $(M-1)Nn^\Gamma + N(n+1)^\Gamma$ over infinite symbol extensions. Specifically, they constructed a scheme to achieve $(M - 1)Nn^\Gamma + N(n + 1)^\Gamma$ DoF over an $(M - 1)Nn^\Gamma + N(n + 1)^\Gamma$ symbol extension of the channel, where $\Gamma = (M - 1)(N - 1)$ so that the achievable DoF are arbitrary close to $\frac{MN}{M+N-1}$ when $n \to \infty$. In [29], the work of the IC was under the assumption of time-varying channels; however, in [32] the authors studied this channel model with time-invariant channel coefficients and showed that the $K/2$ DoF can be achieved as in the previous case based on results from the Diophantine approximation in number theory that interference can be aligned using the properties of rational and irrational numbers and their relations. The authors in [32] built structural encoding for the IA; they also showed that the perfect alignment is achieved asymptotically based on partial IA.

However impressive the insights of the IA scheme, this scheme is challenging in practice. Therefore, some works [33, 34] have studied the practical issues for IA. In [34] the authors discussed the practical concerns to validate the scheme such as propagation models and the role of CSIT. They also propose algorithms for the design of the precoding and decoding matrices at finite SNR.

However, despite the theoretical optimality of the IA scheme in terms of achieving the DoF upper bound asymptotically via symbol extensions for the channel, the results of this scheme are fundamentally limited in practice. Therefore, some works [33, 34] have studied the practical issues for implementing the scheme. In [34] the authors discussed the practical concerns for implementing the scheme, such as the role of CSI. In fact, IA depends heavily on CSI and hence two competing CSI acquisition techniques have been discussed in [34]: channel reciprocity and channel feedback. Showing the limitations and benefits of each technique, they also highlighted the overhead of CSI acquisition for each. Another issue is the low-SNR performance; the authors of [35] have studied this case and showed that by relaxing the requirement of perfect alignment, and proposing a max-SINR algorithm that maximizes the per-stream SINR can outperform perfect IA in the low-SNR regime. It is worth mentioning that the low-SNR performance is not the only limitation for the IA scheme. In early works on IA, it was assumed there would be cooperation between users; however, this is fundamentally limited by constraints such as the number of antennas and the overhead resulting from exchanging information between users. As a result, for large networks this will inevitably lead to uncoordinated interference or colored noise at the receivers. Therefore, the authors of [33] addressed these problems by proposing modified algorithms for IA, taking into account the uncoordinated interference at the receivers.

The work on IA has been extended to various channel models such as the X channel in [31], cellular interference alignment [36, 37], where the practical considerations are taken into account such as clustering, relaying, feedback, and backhaul link cooperation, multihop networks [38–40], and the Y-channel [41] based on a signaling technique called signal space alignment. More specifically, consider three users and one relay node, where each user wants to send a message to all other users. This can be done by aligning a signal dedicated per two users at the relay in the multiple access channel (MAC) phase of communications. In the broadcast phase, the relay broadcasts these aligned signals to each receiver separately and then each user can subtract the self-interference to extract the desired message from the other two users.

15.3.2   Interference Neutralization

Interference neutralization refers to the distributed zero-forcing of interference when the interfering signal passes through multiple nodes before arriving at the unintended receivers. The terminology of interference neutralization was applied first in [4] for relay networks. The authors in [4] showed the potential benefits of this scheme over the well-known schemes such as IA and interference separation. The seminal work of [3] opened the door to deployment of such a transmission scheme in wireless networks, where the authors showed that by eliminating the noise the wireless interaction model can be simplified into a simple network model, which inspired several new insights for interference management. Guided by the works of [3, 4], the authors in [42] characterized the approximate capacity of certain interference relay networks (such as ZZ and ZS networks). This transmission technique has been investigated in other wireless channel models, such as MACs in [43], where the authors studied the three-user MAC with cognition capability of other users' messages and showed the usefulness of this scheme in terms of the DoF. Also, the optimal transmission strategies via interference neutralization for the instantaneous relay networks were studied in [44], where the authors provided an analytical framework for optimizing the transmission power from the relays to make the scheme feasible and showed how the performance of the neutralization scheme can be degraded if the transmitted power from the relay is limited and does not satisfy the quality of service (QoS) constraints in the network.

Interference neutralization was investigated by Rankov et al. in [45] for the multihop network specifically for $K \times R \times K$, and showed that interference neutralization is feasible if the number of relays R is lower bounded by $R \geq K(K-1)+1$. However, utilizing the IA technique with the interference neutralization scheme can achieve the cut-set bound on the DoF with fewer relays, more specifically $R = K$. Gou et al. in [38] proposed aligned interference neutralization, i.e., a novel combination of the two schemes, IA and interference neutralization for multihop networks. Specifically, they studied the $2 \times 2 \times 2$ interference network, with two sources, two relays, and two destinations, and showed that the cut-set bound of 2 DoF is achievable over infinite symbol extensions. We briefly discuss the aligned interference neutralization scheme for the $2 \times 2 \times 2$ network for which a sum DoF of

$$\text{DoF} = \frac{2L-1}{L} \xrightarrow[L \to \infty]{} 2 \tag{15.22}$$

is achievable [38]. Consider a layered $2 \times 2 \times 2$ multihop network as shown in Figure 15.5; each source node S_i has a message W_i to its corresponding destination node D_i, $\forall i \in \{1, 2\}$. There are no direct links between the sources and destinations, hence the messages from the sources are relayed over the relays $\{R_k\}_{k=1}^2$. In the first hop, the received signals at relays $\{R_k\}_{k=1}^2$ are as follows:

$$Y_{R_k}(t) = F_{k1}(t)X_1(t) + F_{k2}(t)X_2(t) + N_{R_k}(t), \tag{15.23}$$

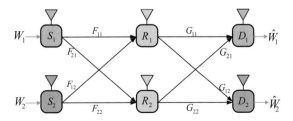

Figure 15.5 System model for the $2 \times 2 \times 2$ multihop network with single antennas at sources (S_1, S_2), relays (R_1, R_2), and destinations (D_1, D_2).

where $F_{ki}(t)$, $\forall k, i \in \{1, 2\}$ is the complex Gaussian channel coefficient from S_i to R_k, $X_i(t)$ is the transmitted signal from S_i, and $N_{R_k}(t)$ is the receiver circularly symmetric Gaussian noise with zero mean-unit variance at time slot t. In the second hop, $\{R_k\}_{k=1}^2$ transmit symbols $\{X_{R_k}\}_{k=1}^2$ to $\{D_k\}_{k=1}^2$. The received signal at D_k is given by:

$$Y_{D_k}(t) = G_{k1}(t)X_{R_1}(t) + G_{k2}(t)X_{R_2}(t) + N_{D_k}(t), \qquad (15.24)$$

where $G_{ki}(t)$, $\forall k, i \in \{1, 2\}$ is the complex Gaussian channel coefficient from R_i to D_k and $N_{D_k}(t)$ is the receiver circularly symmetric Gaussian noise with zero mean-unit variance at time slot t. All nodes are equipped with a single antenna; in addition, the transmitted signals from the nodes have an average power constraint P. The relays are assumed to be full-duplex nodes with perfect self-interference cancellation [46]. Perfect CSI is assumed at the transmitters, i.e., channel coefficients for receiver i are known instantaneously and without error. Specifically, the source nodes know the channels for the first hop only, relays know the channels for both hops, and destination nodes know the channels for the second hop only.

When considering L symbol extension of the network, the effective channel coefficients for the two hops will be viewed as MIMO channel matrices with block diagonal structure, i.e., $\mathbf{F}_{kj}$ and $\mathbf{G}_{kj} \in C^{L \times L}$, $k, j \in \{1, 2\}$. Let the transmitted symbols of sources $S_i, i \in \{1, 2\}$ be as follows:

$$\mathbf{s}_1 = [a_1 \ a_2 \ \dots \ a_L]^T$$
$$\mathbf{s}_2 = [b_1 \ b_2 \ \dots \ b_{L-1}]^T,$$

where $\{b_i\}_{i=1}^{L-1}$ are the information symbols sent from source node S_2, and $\{a_i\}_{i=1}^L$ are the information symbols sent from S_1. Source node S_1 sends $s_1(i)$ along with beamforming vector $\mathbf{v}_{1,i} \in C^{L \times 1}, i \in \{1, \dots, L\}$. Similarly, source node S_2 sends $s_2(i)$ along with beamforming vector $\mathbf{v}_{2,i} \in C^{L \times 1}, i \in \{1, \dots, L-1\}$. The precoding vectors are designed at the source nodes $\{S\}_{i=1}^2$ to satisfy the following two sets of conditions.

Interference Alignment Conditions

$$\mathbf{F}_{11}\mathbf{v}_{1,i+1} = \mathbf{F}_{12}\mathbf{v}_{2,i} \qquad (15.25)$$

$$\mathbf{F}_{21}\mathbf{v}_{1,i} = \mathbf{F}_{22}\mathbf{v}_{2,i}. \qquad (15.26)$$

The $(i+1)$th element of $\mathbf{x}_1$ is aligned with the ith element of $\mathbf{x}_2$ such that the information symbols from sources S_1 will be aligned with the information symbols sent from S_2 at relay R_1, except the first element of $\mathbf{x}_1$. Similarly, for relay R_2, the ith element of $\mathbf{x}_1$ is aligned with the ith element of $\mathbf{x}_2$, except the Lth of $\mathbf{x}_1$. $\mathbf{v}_{1,1} \in R^n$ is chosen to be the all one vectors, i.e., $[1, 1, \dots, 1]^t$.

Then each relay R_i will multiply the received signal with the inverse of the effective channel F_{R_i} to transmit in the second hop as follows:

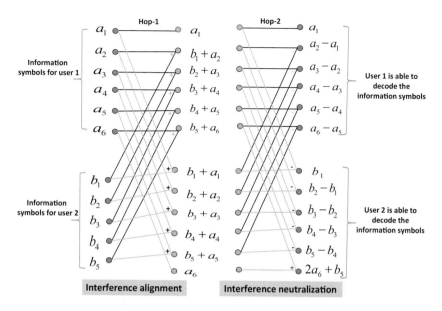

Figure 15.6 A proposed scheme in [38] achieves DoF $= \frac{2L-1}{L} = \frac{11}{6}$ with $L = 6$ symbol extensions.

$$\mathbf{x}_{R_1} = \mathbf{F}_{R_1}^{-1} \mathbf{y}_{R_1} = \begin{bmatrix} a_1 \\ b_1 + a_2 \\ \vdots \\ b_{L-1} + a_L \end{bmatrix} \tag{15.27}$$

$$\mathbf{x}_{R_2} = \mathbf{F}_{R_2}^{-1} \mathbf{y}_{R_2} = \begin{bmatrix} b_1 + a_1 \\ b_2 + a_2 \\ \vdots \\ b_{L-1} + a_{L-1} \\ a_L \end{bmatrix}. \tag{15.28}$$

Figure 15.6 shows an example for $L = 6$ symbol extensions. Source node S_1 sends information symbols $\{a_i\}_{i=1}^{6}$ while source node S_2 sends information symbols $\{b_i\}_{i=1}^{5}$. The resulting alignment of information symbols at both the relays are illustrated in the figure. The precoding vectors $\{\mathbf{v}_{R_i,j}\}_{j=1}^{L} \in C^{L \times 1}$, $i = 1, 2$ at the relay nodes $\{R\}_{i=1}^{2}$ are designed in the following subsection.

Interference Neutralization Conditions

$$\mathbf{G}_{11}\mathbf{v}_{R_1,i+1} = -\mathbf{G}_{12}\mathbf{v}_{R_2,i} \tag{15.29}$$

$$\mathbf{G}_{21}\mathbf{v}_{R_1,i} = -\mathbf{G}_{22}\mathbf{v}_{R_2,i}. \tag{15.30}$$

From the two previous conditions, D_1 can decode the first element of $\mathbf{x}_{R_1}$ and subtract it from the second element of $\mathbf{x}_{R_1}$ to decode the second information symbol and so on to

decode all the messages successively. Similarly, D_2 is applying the same decoding technique for $\mathbf{x}_{R_2}$. Figure 15.6 shows the required neutralization conditions for the second
hop in order to make sure that destination D_2 decodes its information symbols $\{b_i\}_{i=1}^5$
and destination D_1 also, decodes its information symbols $\{a_i\}_{i=1}^6$. The achievable DoF
for this example is 11/6.

The authors of [47] extended the work of the $2 \times 2 \times 2$ interference network to the
MIMO setting, and showed the achievability of the cut-set bound using a combination
of beamforming and aligned interference neutralization techniques. Also, the $2 \times 2 \times 2$
multi-hop network has been generalized into $K \times K \times K$ in [40], where the authors
showed that the cut-set bound on the DoF, which is K, can be achieved by a modified scheme called aligned network diagonalization, which is a nonlinear transmission
scheme. In [48], the impact of employing linear schemes in the multihop networks has
been studied by showing the limitations of such linear schemes.

15.3.3   Interference Management via Structured Codes

In the absence of spatial degrees of freedom, one approach to achieve alignment is
through the use of structured codes. In particular, lattice codes have been shown to be
an effective transmission strategy in order to overcome bottlenecks due to the constraint
(often referred to as the MAC-constraint) of decoding codewords from a superposition
of codewords, which arises naturally in two-way channels, multi-way channels, and
many other channels [8, 49–51]. In lattice coding, the codes are located on a grid so
that the sum of two codewords is a codeword, as shown in Figure 15.7. Consider Fig-
ure 15.7(a), where a one-dimensional lattice $\mathbb{Z}$ is drawn. The points $u_1, u_2 \in \mathbb{Z}$ represent
lattice codewords and in consequence $u_1 + u_2 \in \mathbb{Z}$ is also a codeword. This property is
not limited to a one-dimensional lattice and can be utilized in higher dimensions as well.
As another example, consider the two-dimensional lattice $\mathbb{Z}^2$ shown in Figure 15.7(b).
Here again, $u_1, u_2 \in \mathbb{Z}^2$ represent lattice codewords and, as such, the sum $u_1 + u_2 \in \mathbb{Z}^2$
is again a valid codeword.

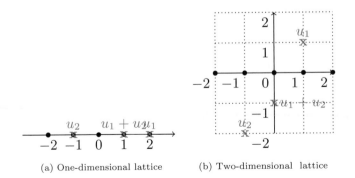

(a) One-dimensional lattice   (b) Two-dimensional lattice

Figure 15.7 One- and two-dimensional lattices.

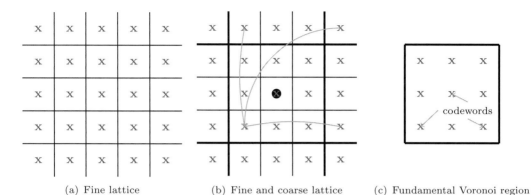

(a) Fine lattice   (b) Fine and coarse lattice   (c) Fundamental Voronoi region

Figure 15.8 Lattice codes.

The structure of a lattice code for communication is given as follows [52]. It contains a fine lattice Λ_f with elements $\mathbf{x} \in \Lambda_f$, as illustrated in Figure 15.8(a) with the respective boundaries around each lattice point. In addition to the fine lattice, a coarse lattice $\Lambda_c \subset \Lambda_f$ is employed as illustrated in Figure 15.8(b). Applying a modulo Λ_c operation on each point of the fine lattice maps it back to the so-called fundamental Voronoi region of the course lattice, as illustrated in Figure 15.8(c). The set of all points of the fine lattice inside the fundamental Voronoi region of the course lattice forms the nested lattice code. Note that the power constraint for transmission is satisfied by the choice of the volume of the fundamental Voronoi region and thus the choice of Λ_c. As an interesting fact, nested lattice codes have been shown to achieve the capacity of the P2P channel [52].

Now, suppose that we have a receiver that obtains a superposition of the desired signal plus some unwanted interference according to the system equation in (15.38). Suppose further that the interference signals are a realization of codewords obtained by a nested lattice code. For simplicity, assume we have two interferers in total which use a nested lattice code with rate $R_1 = R_2 = R$ and power P. Thus, the overall interference signal is given by $\mathbf{x}_1$ and $\mathbf{x}_2$. The receiver obtains $\mathbf{y}_r = \mathbf{x}_r + \mathbf{x}_1 + \mathbf{x}_2 + \mathbf{z}_r$, and wishes to decode the interference signals first by treating the desired signal $\mathbf{x}_r$ as additional noise to perform interference cancellation afterwards.

Consider Figure 15.10(a), which shows the individual interference vectors $\mathbf{x}_1$, $\mathbf{x}_2$, their superposition $\mathbf{x}_1 + \mathbf{x}_2$, as well as the overall signal $\mathbf{y}_r$ after adding the noise given by $\mathbf{x}_r + \mathbf{z}_r$. We observe that the received signal is moved out of the fundamental Voronoi region of the coarse lattice. The receiver now calculates $\mathbf{y}'_r = \mathbf{y}_r \bmod \Lambda_c$, which maps the received signal back to the Voronoi region as illustrated in Figure 15.10(b). Next, it decodes the signal $\mathbf{y}'_r$ to the nearest codeword and thus obtains $\mathbf{x}_\Sigma = (\mathbf{x}_1 + \mathbf{x}_2) \bmod \Lambda_c$ as long as a rate constraint on $\mathbf{x}_\Sigma$ is fulfilled. With the knowledge of $\mathbf{x}_\Sigma$, the receiver is able to cancel its impact in the following way

$$\mathbf{y}''_r = (\mathbf{y}_r - \mathbf{x}_\Sigma) \bmod \Lambda_c = (\mathbf{y}_r - (\mathbf{x}_1 + \mathbf{x}_2) \bmod \Lambda_c) \bmod \Lambda_c \tag{15.31}$$

$$= (\mathbf{y}_r - \mathbf{x}_1 + \mathbf{x}_2) \bmod \Lambda_c = (\mathbf{x}_r + \mathbf{z}_r) \bmod \Lambda_c. \tag{15.32}$$

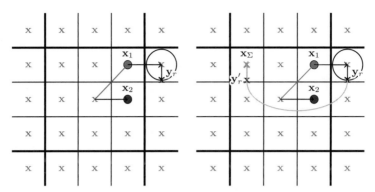

(a) Superposition of signals in the signal space

(b) Mapping the received signal by the modulo operation.

Figure 15.9 A two-dimensional example for computation at the relay (without dithers). The small and big squares represent the Voronoi regions of the fine and coarse lattices, respectively, and the × are find lattice points.

Now, the receiver can proceed to decode the desired signal $\mathbf{x}_r$ reliably as long as a certain rate constraint is fulfilled. The advantage that arises due to the use of lattice codes is that there is only a rate constraint on $\mathbf{x}_\Sigma$, but not on the individual signals $\mathbf{x}_1$, $\mathbf{x}_2$ separately. In consequence, fewer constraints have to be fulfilled which allows a reliable communication with higher rates.

15.4   Role of Channel Knowledge Availability in Interference Management

15.4.1   Role of CSIT and Various Models

In the previous section, we discussed several recently discovered interference manage-ment schemes, such as IA, interference neutralization, and variations. These schemes have been applied to a variety of multiuser networks that inherently face interference. One notable assumption behind the construction of such schemes is the availability of CSIT. This assumption can be difficult to justify in practical scenarios for sev-eral reasons: (1) the mode of CSI acquisition can involve significant overhead (which includes channel estimation and feedback for all transmitter/receiver pairs in the net-work); (2) the quality of CSI (which in turn depends on the quality and availability of feedback links) is almost never perfect; and (3) it may not be possible to share CSI arbitrarily across all users and may only be available locally. Motivated by the prac-tical significance of CSI availability, in this section, we will present an overview of interference management schemes by relaxing the assumptions on availability of CSI on various scales, such as time, space, delay, and/or quality. As we discuss below, variations on the availability of CSI significantly impact the nature of feasible inter-ference management schemes, and the associated information theoretic performance limits.

15.4.2   No CSIT

We first overview the extremal scenario in which there is no CSI available at any of the transmitters. Let us consider the K-user interference network with no CSIT. We assume local CSI at each of the receivers. For this network, when the channel gains H_{ij} are distributed i.i.d. for all (i, j), the total DoF of the network collapses to 1, i.e., the optimal DoF region is simply given by $\sum_i d_i \leq 1$, and can be achieved by TDMA. In fact, in this case, even full cooperation among the transmitters (i.e., converting the system to a broadcast network) does not increase the DoF. The intuitive reasoning for the collapse of DoF is as follows: Due to no CSI availability, the channel outputs at each of the receivers are statistically equivalent. Since the probability of decoding error at each receiver depends on the conditional probability of the channel outputs given the inputs of all transmitters, and each one of these conditional probabilities is the same across receivers, then if receiver j can decode its message so can any other receiver. Using this argument along with simple application of Fano's inequality leads to the upper bound of $\sum_i d_i \leq 1$. For more detail, we refer the reader to Section 15.6. This argument can be further generalized to the case of interference networks with multiple antennas. An in-depth study of a variety of multiuser MIMO networks (including interference, broadcast, X, and multihop interference networks) with no CSIT is presented in [53], where the optimal DoF is characterized in several scenarios. Further generalizations for the case of no CSIT were obtained in [54], where DoF results were obtained for the case of isotropic and independent fading for two-user MIMO interference network. The interesting conclusion of [54] is that the optimal DoF is achievable using random Gaussian codebooks (independent of channel sets), and it is impossible to increase the DoF using beamforming and alignment based approaches in the absence of CSIT. One path out of this valley is to deploy relay nodes, i.e., perform infrastructural improvements. Assuming that those relays have some knowledge in terms of CSI, they can help to boost the performance of the system considerably [55, 56]. Other alternatives are discussed next.

15.4.3   Delayed CSIT

The study of interference management techniques with completely outdated (or delayed) CSIT was initiated in the pioneering work [57]. The model of delayed CSI is the following: Consider time-varying fading channels, i.e., the channel coefficients $H_{ij}(t)$ take independent realizations over time indexed by t. In a delayed CSIT model, at time t, the transmitters only have access to the CSI of time up to time $(t - 1)$. Maddah-Ali and Tse showed that even completely outdated CSI at the transmitters can be significantly useful. It was shown that for the K-user broadcast channel with K single antenna users and K antennas at the transmitter, the optimal sum DoF of the system is given by $K/(1+1/2+\ldots+1/K)$. For large enough K, the optimal sum DoF scales as $\approx K/\log(K)$, which is a significant improvement compared to the case of no CSIT, for which the sum DoF collapses to 1. We explain the key idea behind the use of delayed CSI for multiuser broadcast channels through the following example.

Consider the two-receiver broadcast channel (BC) with single-antenna receivers, and transmitter with two antennas. With no CSI, the optimal DoF of this system is 1, whereas with delayed CSI, 4/3 DoF can be achieved, which is also information-theoretically optimal. To achieve 4/3 DoF, the transmitter sends two information symbols (a_1, a_2) intended for user 1 in the first time slot. Users 1 and 2 receive linear combinations $L_1(a_1, a_2), L_2(a_1, a_2)$ of the information symbols. In the second slot, the transmitter sends two information symbols (b_1, b_2) intended for user 2. Users 1 and 2 receive linear combinations $G_1(b_1, b_2), G_2(b_1, b_2)$ of the information symbols. The novel idea is to utilize unwanted symbols in a retrospective manner as follows: Over the first two time slots, user 1 received $G_1(b_1, b_2)$ and user 2 received $L_2(a_1, a_2)$, which are as such not useful. However, due to delayed CSI, the transmitter learns the channel coefficients of the first two time slots, and it can *recreate* L_2 and G_1 with a delay of two time slots. Subsequently, the transmitter can send $L_1 + G_2$ on one of its antennas, which will be seen at both the users. Hence, upon receiving this, the unwanted side information from the previous two time slots can be leveraged to remove the *interference* at the respective users and decode two symbols at each one. Thus, in three slots, we can transmit four total symbols, to achieve 4/3 DoF. The role of delayed CSI for the case of MIMO broadcast channels was investigated in [58], and tight characterizations were obtained for two users. The general characterization of the optimal DoF for multireceiver MIMO broadcast channels remains an open problem.

The role of delayed CSIT for interference networks was first investigated in [59], where it was shown that more than one DoF is achievable. Subsequently, there were several works [60–62] that provided improved DoF results for the K-user IC with delayed CSIT; however, all of these results did not achieve scaling of DoF with K, the number of users. A recent interesting work [63] is the first result to achieve DoF scaling for this problem. In particular, it is shown that $\sqrt{K}/2$ DoF are achievable for the SISO IC with delayed CSIT. The distributed nature of transmitters is one of the key challenge to improving the DoF with delayed CSIT. In particular, the interference at one receiver is a function of multiple transmitters, which only have access to their own information symbols. Thus, unlike the case of broadcast channels, each transmitter is unable to reconstruct the whole past interference. The interesting idea presented in [63] to tackle this issue is to carefully repeat the transmissions by the transmitters to enable more effective retrospective utilization of interference for decoding desired symbols at the receivers. To the best of the authors' knowledge, there are no existing nontrivial upper bounds on the DoF for the K-user interference channel with delayed CSIT, and proving (or disproving) the optimality of this scheme remains an open problem.

15.4.4   Alternating CSIT

Another interesting model known as alternating CSIT was introduced in [64], which captures the following aspect of CSI: that it can vary over time. Consider the MISO BC for the case in which perfect CSIT is available for one user and no CSIT is available

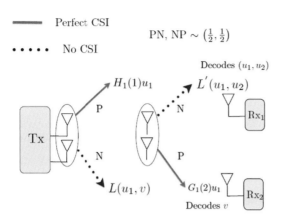

Figure 15.10 Scheme achieving 3/2 DoF for the MISO broadcast channel with alternating CSIT using two states (PN, NP), i.e., perfect CSI from one user and no CSI from the other user.

for the other user. The DoF for this setting was recently shown [65] to be 1. Now, staying within the assumption of full CSIT for one user and none for the other, suppose we allow the CSIT to vary, in the sense that half the time we have full CSIT for user 1 and none for user 2, and for the remaining half of the time we have full CSIT for user 2 and none for user 1. This is one example of what we call the alternating CSIT setting. In general terms, the defining feature of the alternating CSIT problem is a joint consideration of multiple CSIT states. We note that alternating CSIT may be practically unavoidable due to the time-varying nature of wireless networks. However, more interestingly, the form of CSIT may also be deliberately varied as a design choice, often with little or no additional overhead. For example, acquiring perfect CSIT for one user and none for the other for half the time and then switching the role of users for the remaining half of the time carries little or no additional overhead relative to the non-alternating case in which perfect CSIT is acquired for the same user for the entire time, while no CSIT is obtained for the other user. Moreover, another interesting aspect is that by coding across multiple CSIT states, higher performance (in terms of DoF) is often possible in comparison to treating them separately. We explain this idea through a simple example: Consider the case of two CSIT states (perfect CSI from one user and no CSI from the other user) occurring for equal fractions of time. In this case, by jointly coding across the PN and NP CSI states, one can achieve a higher sum DoF of 3/2 (compared to 1 which is the optimal for each individual state) as shown in Figure 15.10. The DoF region for the two-user MISO BC with alternating CSIT (where CSI can be either perfect, delayed, or not available) was characterized in [64] and extended to the case with security constraints in [66]. Other variations such as hybrid CSIT (in which CSI quality can vary across users, but remains invariant over time) have been considered in [67]. The DoF region of the two-user MIMO IC with hybrid CSIT was characterized in [68].

15.5   Recent Approaches Exploiting Network and Channel Heterogeneity

15.5.1   Optimality of TIN

Since the introduction of the IC by Ahlswede in [69] in 1974, some progress was made in the [70, 71] in the late 1970s and early 1980s by Carleial and Sato to characterize the capacity region for the very strong and strong interference case, respectively. With the introduction of the GDoF metric in [1], in which the strength of the interference compared to the desired signal is measured as

$$\alpha = \frac{\log(\text{INR})}{\log(\text{SNR})},$$

important insights have been gained on the characteristics of the capacity region of the IC in the asymptotic regime of high SNR. Roughly speaking, the parameter α links the strength of the INR to the SNR in a logarithmic way. It turned out that for $\alpha \le \frac{1}{2}$ the simple scheme of TIN is optimal in terms of GDoF and optimal within a constant gap of within 1 bit. The GDoF of the two-user Gaussian IC and the regime where TIN is GDoF-optimal ($\alpha \le 1/2$) is illustrated in Figure 15.11. As can be seen in the figure, the GDoF as a function of the α parameter shows an interesting W-behavior. While orthogonal interference regimes are optimal only at single points of interference levels α, the optimality of TIN is given for a whole regime from $0 \le \alpha \le 1/2$. The characterization of the GDoF required novel upper bounds which were based on providing additional side information to the receivers (genie-aided bounds) to enhance their capabilities.

The novel and tight upper bound on the GDoF of the two-user IC in [1] became a stepping stone to refined results presented in [13–15, 72] on the capacity characterization of the two-user IC at finite SNR in the so-called noisy-interference regime for single- and multiple-antenna systems, respectively. In more detail, the upper bound in [1] was

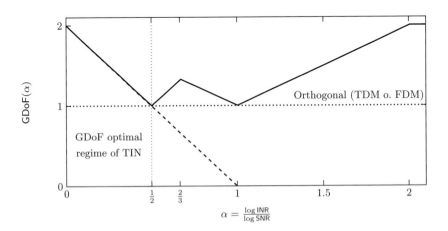

Figure 15.11 The GDoF characterization for the symmetric two-user Gaussian IC. The solid line represents the GDoF of the two-user IC and the dashed line is the achievable GDoF of TIN in this channel. The horizontal dotted line shows the performance of orthogonal interference avoidance schemes like TDM or FDM.

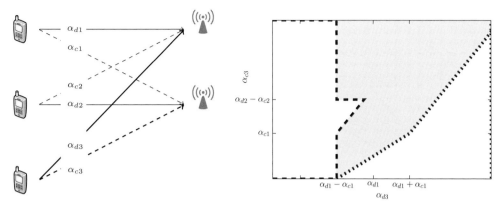

(a) Multiple access channel and point-to-point in- (b) TIN optimality/suboptimality regime for the PIMAC.
terference setup (PIMAC)

Figure 15.12 PIMAC system model and TIN optimality.

refined by allowing some correlation between the noise terms at the receivers. This upper
bound matches the achievable rate of TIN within the noisy-interference regime.

The result on the approximate optimality of TIN has motivated researchers to consider
more general setups. For instance, the MAC interfering with a point-to-point (P2P) link
referred to as PIMAC, as shown in Figure 15.12(a) was investigated in [73–75]. Consider
the IC formed by the nodes which are linked by the path denoted with α_{d1}, α_{d2} (desired
links) and α_{c1}, α_{c2} (interference links). Now, assuming that this IC is operating in the
noisy-interference regime, i.e., $\alpha_{c1} + \alpha_{c2} \leq \min(\alpha_{d1}, \alpha_{d2})$, it was shown in [73–75] that
a combination of TDMA and TIN achieves the capacity within a constant gap and thus
is GDoF optimal once a third user with a desired link of strength α_{d3} and interference
link α_{c3} is added to the system, thus forming the PIMAC. To be more precise, consider
the area plot in Figure 15.12(b), which illustrates the regimes in which TIN is optimal
(regimes indicated as dotted and dashed boxes) and where TIN is outperformed by other
more sophisticated schemes in terms of GDoF (gray shaded regime). As can be seen
from the figure, the regimes are nontrivially linked to the strength of the channel gains
α_{d3} and α_{c3}. In the dashed-box regime, it is optimal to keep user 3 silent and operate the
PIMAC in an IC mode with the remaining users. In the dotted-box regime, it is optimal
to keep user 1 silent and operate the PIMAC again in an IC mode with the remaining
users.

Consider Figure 15.13(a). In cases where (a) the desired link α_{d3} is stronger than the
desired link α_{d1} and (b) the interference caused by the third user (α_{c3}) is higher than
the value $\alpha_{d2} - \alpha_{c2}$, then the PIMAC can be operated as a P2P channel from user 3 to
receiver 1. This regime is illustrated by the solid dark-gray regime in the right top corner
of the figure. In consequence, the complexity of the GDoF optimal scheme reduces to
those used for interference–free channels. Now, consider Figure 15.13(b). The regime
marked by stripes is one in which the interference at receiving node 2 caused by user
3 in terms of α_{c3} is weaker than the interference due to user 1 (determined by α_{c1}).

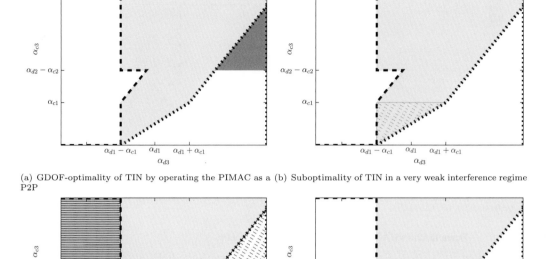

(a) GDOF-optimality of TIN by operating the PIMAC as a (b) Suboptimality of TIN in a very weak interference regime
P2P

(c) Optimality of TIN in regimes in which interference is not   (d) Optimality of TIN with all users active
very weak

Figure 15.13 PIMAC: GDoF-optimality and suboptimality of TIN. It is assumed that
$\alpha_{c1} + \alpha_{c2} \leq \min(\alpha_{d1}, \alpha_{d2})$.

Based on this condition, one could conclude that operating the PIMAC in an IC mode
with the receivers employing TIN in which only users 2 and 3 are active, while user
1 is silent should suffice to achieve the optimal GDoF. However, it turns out that in
this regime treating interference is suboptimal in terms of GDoF as IA methods can
be applied which strictly outperform TIN. An interesting insight is that this regime
does not vanish even though the overall interference caused to receiver 2 (determined
by α_{c1} and α_{c3}) can be arbitrarily small. Contrary to that regime is the behavior in
the regimes highlighted in Figure 15.13(c). In those regimes, although the interference
caused by user 3 cannot be considered as very weak, TIN is still optimal. In consequence, there is no need to deploy schemes with more computational complexity than
TIN, as the GDoF is already achieved by TIN. Finally, in Figure 15.13(d), regimes are
highlighted in which all users can communicate simultaneously with their receivers,
whereby TIN is applied. This scheme is a valid alternative to combining TDMA with
TIN in terms of achieving the optimal GDoF. An important finding in [76] is that a simple binary power control (on/off signaling) for TIN is sufficient to achieve the GDoF
optimality.

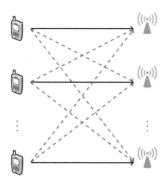

Figure 15.14 General K-user interference channel. Block solid lines represent desired links, while red dashed lines represent interference links.

Beside the PIMAC, the approximate capacity optimality of TIN in the general asymmetric K-user IC (as shown in Figure 15.14) has been studied in [77]. Interestingly, in this work it has been shown that TIN with power control achieves the capacity region of the K-user IC within a constant gap if for all transmitter–receiver pairs i, $1 \leq i \leq K$, the optimality condition

$$\text{SNR}|_{\text{dB}} \geq \max\{\text{observed INR}|_{\text{dB}}\} + \max\{\text{caused INR}|_{\text{dB}}\} \qquad (15.33)$$

holds. In words, the desired signal strength for each pair has to be at least equal to the strongest interference caused by this pair plus the strongest interference observed by this pair, with all values in dB scale. This can be expressed efficiently in terms of the α-parameters introduced earlier as follows:

$$\alpha_{ii} \geq \max_{j:j\neq i}\{\alpha_{ji}\} + \max_{k:k\neq i}\{\alpha_{ik}\}, \quad 1 \leq i,j,k \leq K, \qquad (15.34)$$

where α_{ji} denotes the channel strength level from transmitter i to receiver j. The result of [77] has been extended to several setups such as the compound IC in [78] and K-user parallel IC in [79]. Furthermore, when combined with zero-forcing, TIN was shown to remain optimal in the case of multiple-antenna systems [76]. Here, the strong interference can be blended out by applying zero-forcing, while the remaining interference is treated as noise, resulting in a nontrivial parameter regime in which TIN is GDoF optimal. The validity of the condition (15.33) was investigated for X channels in [80]. It was shown that when the condition above holds for a K-user IC obtained by setting some (to be more precise: $M \times N - K$) of the messages of the $M \times N$-X channel to null with $K = \min(M,N)$, then TIN is also optimal for the X channel. Thus, for all possible permutations of transmitter and receiver combinations, the condition in (15.33) has to be checked as follows. Let Π^M and Π^N be permutations of the transmitter and receiver indices, respectively. Then if for at least one choice of permutation pairs the following condition holds

$$\alpha_{\Pi_i^M \Pi_i^N} \geq \max_{j:j\neq i}\left\{\alpha_{\Pi_j^N \Pi_i^M}\right\} + \max_{k:k\neq i}\left\{\alpha_{\Pi_i^N \Pi_k^M}\right\},$$

$$1 \leq i \leq K, \quad 1 \leq j \leq N, \quad 1 \leq k \leq M, \qquad (15.35)$$

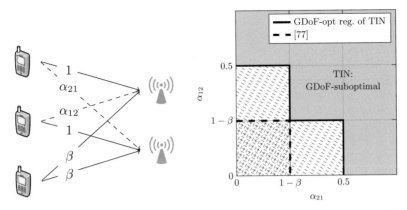

(a) X channel setup. The parameter (b) GDOF-optimality regimes: original
β fullfills $0.5 \leq \beta \leq 1$ from [77] and expanded region from [81]

Figure 15.15 A 3×2 X channel. Comparison of the regions obtained in [77] and [81].

then operating the X channel as a K-user IC with the communication pairs according to the corresponding Π^M and Π^N is GDoF-optimal.

It is important to note that the condition in (15.33) when applied to X channels is only sufficient but not necessary. This was highlighted in [81], in which the TIN optimality in an $M \times 2$ user X channel was considered. In more detail, the X channel shown in Figure 15.15(a) was considered. Assuming that β varies within the range $0.5 \leq \beta \leq 1$, the TIN optimality regime derived in [80] is illustrated in Figure 15.15(b) and given as the rectangular double-striped region at the left lower corner of the plot. In addition to this, the expanded GDoF-optimality regime derived in [81] is shown with the additional single-striped blocks on top and to the right of the double-striped region. Note that although the optimality conditions given in (15.35) is not fulfilled in the single-striped region, TIN is still GDoF optimal in the X channel. To achieve the optimal GDoF one simply turns off user 3 (with links of strength β) and operates the X channel as a two-user IC, in which the links with strength 1 and α_{12}, α_{21} represent the desired and interference links, respectively. Due to its simplicity, it was of interest to investigate the optimality of TIN in other networks as well. For instance, the optimality of TIN was investigated for the interference relay channel (IRC) in [82], shown in Figure 15.16(a). Here, the communication in an IC is supported by an additional relay with incoming links γ and outgoing links β. As one can intuitively agree, the GDoF of the basis network should be improved by utilizing the relay [83, 84]. However, the gain depends on the strength of the channel parameters β and γ. Consider, for instance, Figure 15.16(b), in which the GDoF of the IC and the IRC are shown. As one can see, the benefit of the relay in that particular case with $\beta = 0.3$ and $\gamma = 0.7$ is restricted to the weak interference regime. Furthermore, in the gray-shaded regime, we can simply turn off the relay as it does not provide any benefit when compared to the IC. Thus, in that regime, TIN is still optimal. The region in which TIN is optimal is, however, smaller for the IRC when compared to the IC. In fact, one can show that if the following condition holds

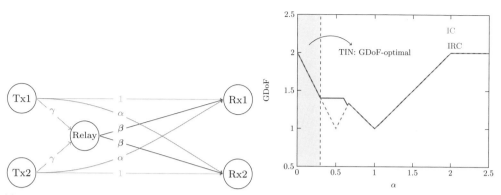

(a) Interference relay channel with symmetric channel gains

(b) GDoF of the IRC with $\beta = 0.3$, $\gamma = 0.7$. The GDoF of the IC is shown for comparison purposes.

Figure 15.16 The interference relay channel.

$$\max\{\alpha, \beta\} + \max\{\alpha, \gamma\} \leq 1, \tag{15.36}$$

then TIN is optimal in the IRC. Intuitively, this agrees fully with the condition for K-user ICs expressed verbally in (15.33). It turns out even if the interference is strong, applying TIN in the IRC might be still optimal. In more detail, if the following conditions

$$1 \leq \beta \leq \alpha \leq \gamma \quad \text{and} \quad \beta \leq \min\left\{\frac{\gamma}{2}, \gamma - \alpha\right\} \tag{15.37}$$

hold, then employing TIN at the receivers of the IRC is GDoF-optimal. Thus, the relay now enlarges the TIN-optimality regime. Thereby, the relay and the transmitters employ a cooperative interference cancellation scheme to reduce the interference level at the receivers below the noise level. Essentially, for sufficiently large γ according to the conditions above, the relay acts as a cognitive node with access to the messages of both transmitters. This aspect is illustrated in Figure 15.17, where the GDoF of the IRC is given for $\beta = 1.5$, $\gamma = 10$. It turns out that in that case if $\alpha \geq 1.5$, then TIN is GDoF-optimal. For comparison purposes, the GDoF of the IC is shown as well. As can be seen, substantial gains can be achieved by deploying a relay. Furthermore, bear in mind that in the regime $\alpha \geq 1.5$ successive decoding with its corresponding computational complexity is employed at the receivers for the IC to achieve the GDoF, while only TIN is needed for the IRC. Thus, the relay can help to reduce the computational load at the receivers, which is offloaded to the relay and thus the infrastructure provided by, e.g., the network providers. An important aspect to consider for the GDoF-optimality of TIN is the low level of coordination and requirements needed, thus making it a robust candidate for interference management both in theory and practice. In the next subsection, another robust method, referred to as topological interference management, is addressed.

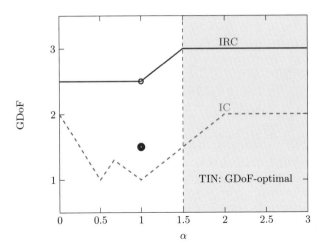

Figure 15.17 GDoF of IC and IRC with $\beta = 1.5$, $\gamma = 10$. The circle at $\alpha = 1$ illustrates the collapse of the GDoF at this singularity.

15.5.2   Topological Interference Management and Its Applications

Another variation of the interference management techniques is interference avoidance based on a coarse knowledge of the CSIT. Although our understanding of the information theoretic limits of the capacity of interference networks has become more advanced in recent decades, it is optimistically based on the availability of abundant CSIT. In this section, we highlight recent techniques that consider the case of minimal knowledge of channel condition at the transmitters. Starting from the degenerate setting of a fully connected interference channel with K transmitter–receiver pairs and no available CSIT even from the desired channels, it is easy to notice that the symmetric capacity of the network can be achieved by means of TDMA, where each user sends over a fraction $1/K$ of the total time. The achieved rate can be given for a certain level of SNR as $\frac{1}{K}\log(1 + K\text{SNR})$. Notice that this channel model is too pessimistic due to the assumption of fully connected topology meaning that each receiver can see interference from all the undesired links.

Consider now the case where the transmitters are allowed only one bit of channel knowledge for each undesired link in the network. A reasonable choice of the information carried on this bit is whether the channel is significant or weak. This can be obtained by a nominal measure of the received interference power at the receiver side for each undesired link and comparing them with the noise floor, where the bit is assigned a value 1 when the interference power is above the noise floor, and 0 otherwise. It is worthy to mention here that this one bit of channel knowledge is only attained once in the communication session, i.e., not per channel use, since it is a coarse measure, and not subject to change over time. The DoF measure can be obtained by setting the weak channels to zero, which is widely known as the *topological interference management* (TIM) problem for wireless networks [85, 86]. The assumption of the existence of weak channels in interference networks is observed due to the physical phenomena: path loss, shadowing, and fading. This assumption is the basis of frequency reuse already being used in cellular networks, and is expected to be even more utilized in the future due to the increasing complexity of the connectivity patterns owing to the use of directional antennas, indoor customer deployed networks, and the use of millimeter-wave bands, which experience high path loss and large variations in the transmitted power.

The TIM problem is known to have a strong connection with the index coding problem [87–91]. The antidote graph for index coding is formed by links providing side information (antidotes) to cancel interference from the destination, which give them a role opposite to the interfering links in wireless networks. Therefore, any TIM problem with a given topology is analogous to an index coding problem with an antidote graph complement to the topology graph of the corresponding TIM problem. Moreover, it was shown in [85] that the achievable linear DoF for the TIM problem is the same as the achievable linear DoF for the index coding problem. Let us take an example of a partially connected four-user interference channel in Figure 15.18(a), with the solid black lines representing the desired links, while the dotted lines represent the interfering links. A corresponding index coding problem is given in Figure 15.18(b). The symmetric DoF for two problems is 0.5 per user, achieved by means of IA. The *alignment graph* for the messages is given by the solid black lines in Figure 15.18(c), where a link exists between two messages if these two messages cause interference at the same unintended receiver in the TIM problem (no antidotes for the two messages at the same unintended receiver in the index coding problem). Also, the *conflict graph* is given in Figure 15.18(c) by the dotted lines existing between X_i and X_j if transmitter i interferes with receiver j in the IC problem (no antidotes for X_i at receiver j in the IC problem). Since there is no *internal conflict*, i.e., no link belongs to both the alignment and the conflict graphs, we align the messages X_1 and X_3 by precoding with a 2×1 vector V_1, and align the messages X_2 and X_4 by precoding with a 2×1 vector V_2, where V_1, and V_2 are two independent vectors. The receivers then project over the orthogonal direction of the aligned received interference. Therefore, we achieve a symmetric DoF of 0.5 per user.

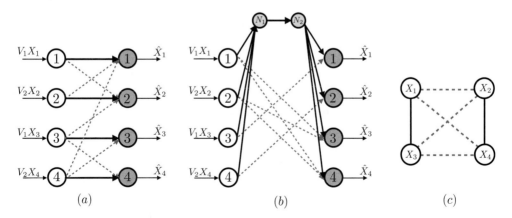

Figure 15.18 A partially connected interference channel (TIM problem) in (a), with the associated index coding problem in (b), and the alignment/conflict graph in (c).

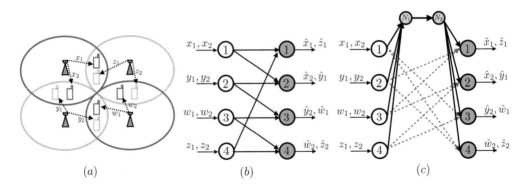

Figure 15.19 An example of four cells with two users falling in the boundaries each in (a), with the corresponding TIM and index coding problems in (b) and (c).

The TIM problem is also expected to receive tremendous interest due to its potential applications in the next generation of cellular systems [92–94] with the topology playing a big role especially when applying ideas of directional antennas and short-range millimeter waves. Consider the four cells with two boundary users each in Figure 15.19(a). With conventional frequency reuse (or TDMA) schemes, two noninterfering cells (cells with the same shading in the figure) can transmit over the same frequency band (in the same time for TDMA) achieving 0.5 DoF per cell, and hence 1/4 DoF per user. However, we can show that by exploiting the topology knowledge a DoF of 1/3 is achievable per user. We first notice that two users falling in the same boundary are statistically equivalent, i.e., they are seeing the same interference, and hence can be treated as one virtual user as shown in the TIM problem in Figure 15.19(b), and equivalently the index coding problem in Figure 15.19(c). By aligning the interference seen by each virtual receiver over a 3×1 precoding vector (e.g., x_2, and z_2 over V_1), we compress the interference into one dimension plus two dimensions for the two desired signals achieving a symmetric DoF of 1/3 per user.

The TIM problem has also been considered for the case of fast fading channels in [86, 95], where transmitters are only allowed to retransmit in order to neutralize the interference at nonintended parties. Also the conditions in order to achieve the best symmetric DoF of 0.5 per user (interference only occupying half the allowed dimensions) were studied. Another variation when only the topology is known but varying across time, which is also known as the alternating TIM problem, was considered in [96–98] to exploit synergetic benefits of coding across alternating topologies. The multilevel TIM problem was considered in [99] in order to characterize the capacity of partially connected intereference channels, where the network is decomposed into TIM and TIN components such that interference avoidance is used for significant interference links and TIN is used for weak interference links. In [100], transmitter cooperation was allowed, which is also known as the distributed broadcast setting. The TIM problem has also been considered in many different contexts, e.g., multiple-antenna scenarios in [101] for the MIMO interference channel, and in [102] for the MIMO broadcast channel, reconfigurable antennas in [103], and multihop networks in [104], where relaying and interference management are coupled.

15.5.3   Interference Management by Exploiting ISI Heterogeneity

In the previous section we discussed how higher DoF can be achieved by taking advantage of CSIT – be it, delayed, alternating, or mixed CSIT. In practice, however, CSIT is not always available. It therefore remains of importance to find an answer to the following problem: Is there a way to achieve positive DoF that scales linearly with the number of users K in the absence of CSIT?

Recent work [105] has shown that the answer to the above question is affirmative. It showed that, in the absence of CSIT, ISI heterogeneity (in terms of the variability in the number of channel impulse response (CIR) taps between transmitters and different receivers) can be exploited to achieve significant gains in spectral efficiency in the multiuser wideband IC. In particular, it is possible to achieve sum-DoF that scales linearly with K, for a K-user single-input single-output (SISO) interference channel with ISI. This section provides a brief summary of ideas presented in [105] and illustrates how positive sum-DoF (that scales linearly with the number of users) can be achieved by carefully aligning interfering symbols into a separate subspace from that occupied by the desired message symbols at the intended receiver.

Consider a K-user SISO interference channel with ISI as shown in Figure 15.20. Let $x_k[n]$, satisfying the transmission power constraint $\mathbf{E}[x_k^2[n]] = P$, be the transmitted signal from transmitter k at time n and let $\{h_{k,i}[l]\}_{l=1}^{L_{k,i}}$ be the channel impulse response from transmitter k to receiver i. $L_{k,i}$ denotes the number of effective channel taps between the kth transmitter and the ith receiver. The received signal, $y[n]$ in time slot n is

$$y_k[n] = \sum_{i=1}^{K} \sum_{l=1}^{L_{k,i}} h_{k,i}[l]x_i[n-l+1] + z_k[n], \tag{15.38}$$

where $z_k[n]$ is the i.i.d. complex Gaussian random channel noise variable seen by receiver k in the nth time slot. Each channel coefficient, $h_{k,i}[l]$, is assumed to be an

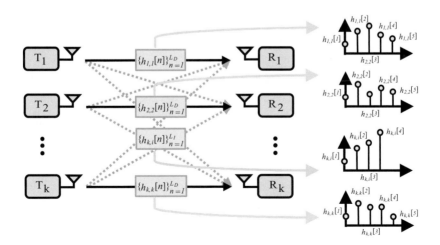

Figure 15.20 SISO interference channel with ISI.

independent complex Gaussian random variable with zero-mean and a variance that decreases exponentially as the number of channel taps increases.

The following theorem states the achievable sum-spectral efficiency for the above described IC with ISI. As will be clarified by subsequent examples, the principal idea, also called interference-free OFDM (IF-OFDM), to achieving positive sum-DoF is to create a noncirculant channel structure to allow decodability of the desired message symbols from transmitter k to receiver k while at the same time creating a circulant channel structure for all interfering channel symbols from transmitter i to receiver k in order to keep these interfering symbols in a separate subspace from that of the desired message symbols at the intended receiver's node. The interfering symbols are aligned in a separate subspace from that of the desired symbols by means of an output signal combiner matrix that creates a circulant matrix for the interfering symbols and a noncirculant matrix for the desired symbols as to preserve their linear algebraic decodability. To further allow the cancellation of the aligned interference at the receiver, the transmitted signal vector is precoded using complementary discrete time Fourier transform (DFT) vectors. At the receiver, after applying the combiner matrix to the received vector, complementary DFT vectors (i.e., vectors that are orthogonal to those applied in the precoding process) are applied to the received signal vectors. This step cancels out the circulant portion (i.e., the portion bearing the interference symbols) and leaves out the noncirculant portion (i.e., the portion bearing the desired message symbols).

THEOREM 15.1   *Consider a K-user wideband SISO interference channel with $L_{k,i}$. Let $L_D = \max_k L_{k,k}$ and $L_I = \max_k \max_i L_{k,i}$ for $k \neq i$, respectively. Then the achievable sum-DoF with completely no CSIT is*

$$d_{\Sigma} = \max \left\{ \sum_{k=1}^{K} \frac{(L_{k,k} - L_{k,i})^+}{(N + L_I - 1)}, 1 \right\}, \tag{15.39}$$

where $N = \max\{L_I, 2(L_D - L_I)\}$ and $(x)^+ \triangleq \max(x, 0)$.

EXAMPLE 15.2   Consider a K-user SISO IC with ISI with $L_{k,i} = L_I = 2$ and $L_{k,k} = L_D = 3$ for $i, k \in \{1, 2, \ldots, K\}$. The aim of this example is to show that a sum-DoF of $d_{\Sigma} = \frac{K}{4}$ is achievable. The key idea of IF-OFDM transmission is to create noncirculant channel structure for the desired wireless link in order to preserve decodability for the desired symbols at the intended receiver while at the same time creating circulant channel structure for the interfering wireless link. This is done using a combiner matrix. The interfering signals can be eliminated at the receiver by precoding the transmitted signal using a DFT matrix vector and applying its complementary DFT matrix vector to the output signal after the combiner matrix application step. Let $F_1 = \frac{1}{\sqrt{2}} \begin{bmatrix} 1 & 1 \end{bmatrix}^T$ be the beamforming vector carrying information symbol s_k, for $k \in \{1, 2, \ldots, K\}$. The beamforming vector, F_1, will be used for all transmitters. Let transmitter k send the signal vector X_k over two time slots as follows:

$$X_k = \begin{bmatrix} x_k[1] \\ x_k[2] \end{bmatrix} = F_1 s_k. \tag{15.40}$$

Therefore, by (15.38), the signal vector, Y_k seen at the intended receiver k during four time slots is given by

$$
Y_k = \begin{bmatrix} y_k[1] \\ y_k[2] \\ y_k[3] \\ y_k[4] \end{bmatrix} = \underbrace{\begin{bmatrix} h_{k,k}[1] & 0 \\ h_{k,k}[2] & h_{k,k}[1] \\ h_{k,k}[3] & h_{k,k}[2] \\ 0 & h_{k,k}[3] \end{bmatrix}}_{\mathbf{H_{k,k}}} \begin{bmatrix} x_k[1] \\ x_k[2] \end{bmatrix} + \sum_{i \neq k}^{K} \underbrace{\begin{bmatrix} h_{k,i}[1] & 0 \\ h_{k,i}[2] & h_{k,i}[1] \\ 0 & h_{k,i}[2] \\ 0 & 0 \end{bmatrix}}_{\mathbf{H_{k,i}}} \begin{bmatrix} x_i[1] \\ x_i[2] \end{bmatrix} + \begin{bmatrix} z_k[1] \\ z_k[2] \\ z_k[3] \\ z_k[4] \end{bmatrix}
$$

$$(15.41)$$

As noted earlier, the principal idea behind IF-OFDM transmission follows the following conditions:

- Ensure that the channel matrices $\mathbf{H_{k,i}}$ for the interfering links are kept circulant.
- Ensure that the channel matrices $\mathbf{H_{k,k}}$ for the desired links are kept noncirculant.

The idea enabling these two conditions to be satisfied is presented next: For receiver k, consider a binary combiner matrix $\tilde{\mathbf{D}}_\mathbf{k}$ of size 2×4, for $k \in \{1, 2, \ldots, K\}$, such that

$$
\tilde{\mathbf{D}}_\mathbf{k} = \begin{bmatrix} 1 & 0 & 1 & 0 \\ 0 & 1 & 0 & 0 \end{bmatrix}.
$$

$$(15.42)$$

Applying the linear combining matrix $\tilde{\mathbf{D}}_\mathbf{k}$ to Y_k in (15.41) leads to

$$
\bar{Y}_k = \tilde{\mathbf{D}}_\mathbf{k} Y_k = \begin{bmatrix} y_k[1] + y_k[3] \\ y_k[2] \end{bmatrix} = \underbrace{\begin{bmatrix} h_{k,k}[1] + h_{k,k}[3] & h_{k,k}[2] \\ h_{k,k}[2] & h_{k,k}[1] \end{bmatrix}}_{\bar{\mathbf{H}}_{\mathbf{k,k}}} \underbrace{\begin{bmatrix} x_k[1] \\ x_k[2] \end{bmatrix}}_{\bar{X}_k}
$$

$$
+ \sum_{i \neq k}^{K} \underbrace{\begin{bmatrix} h_{k,i}[1] & h_{k,i}[2] \\ h_{k,i}[2] & h_{k,i}[1] \end{bmatrix}}_{\bar{\mathbf{H}}_{\mathbf{k,i}}} \underbrace{\begin{bmatrix} x_i[1] \\ x_i[2] \end{bmatrix}}_{\bar{X}_i} + \underbrace{\begin{bmatrix} z_k[1] + z_k[3] \\ z_k[2] \end{bmatrix}}_{\bar{Z}_k}
$$

$$(15.43)$$

Clearly, all of the IC matrices $\bar{\mathbf{H}}_{\mathbf{k,i}}$ in (15.43) are complex circulant matrices of size 2×2. On the other hand, the desired 2×2 complex channel matrix $\bar{\mathbf{H}}_{\mathbf{k,k}}$ is not circulant. Using signal processing properties, it is known that a circulant matrix $\bar{\mathbf{H}}_{\mathbf{k,i}}$ can be diagonalized by the DFT as follows:

$$
\bar{\mathbf{H}}_{\mathbf{k,i}} = \begin{bmatrix} F_1 & F_2 \end{bmatrix} \begin{bmatrix} \lambda_{k,i}^1 & 0 \\ 0 & \lambda_{k,i}^2 \end{bmatrix} \begin{bmatrix} F_1 & F_2 \end{bmatrix}^H,
$$

$$(15.44)$$

where $F_1 = \frac{1}{\sqrt{2}} \begin{bmatrix} 1 & 1 \end{bmatrix}$ and $F_2 = \frac{1}{\sqrt{2}} \begin{bmatrix} 1 & -1 \end{bmatrix}$. F_1 and F_2 are the eigenvectors, and $\lambda_{k,i}^1$ and $\lambda_{k,i}^2$ are their associated eigenvalues, respectively. It should be noted that all the interference signals are aligned in the direction of F_1. This is because all the transmitters precode each symbol by applying the same beamforming vector, F_1, to create the transmitted vector, $X_k = F_1 s_k$. In linear algebraic terms, this can be thought of as

$$
\text{span}(\bar{\mathbf{H}}_{\mathbf{k,i}} F_1) = \text{span}(\bar{\mathbf{H}}_{\mathbf{k,j}} F_1),
$$

$$(15.45)$$

for all $j, i \in \{1, 2, \ldots, K\}/\{k\}$ and $j \neq i$. Equation (15.43) can, therefore, be rewritten as

$$\bar{Y}_k = \bar{\mathbf{H}}_{\mathbf{k},\mathbf{k}}F_1 s_k + \sum_{i \neq k} \bar{\mathbf{H}}_{\mathbf{k},\mathbf{i}}F_1 s_i + \bar{Z}_k \tag{15.46}$$

$$= \bar{\mathbf{H}}_{\mathbf{k},\mathbf{k}}F_1 s_k + \sum_{i \neq k} \lambda_{k,i}^1 F_1 s_i + \bar{Z}_k. \tag{15.47}$$

Using the fact that F_1 and F_2 are orthogonal to each other, in order to cancel out all the aligned interference at receiver k, we apply the receiver's beamforming vector F_2^H to $\bar{Y}_k$. This leads to the following interference-free output signal at the receiver:

$$\tilde{Y}_k = F_2^H \bar{Y}_k = F_2^H \bar{\mathbf{H}}_{\mathbf{k},\mathbf{k}}F_1 s_k + F_2^H \bar{Z}_k. \tag{15.48}$$

For the kth communication link, this example therefore achieves a rate of

$$R_k = \frac{1}{4} \log_2 \left(1 + \frac{2P|F_2^H \bar{\mathbf{H}}_{\mathbf{k},\mathbf{k}}F_1|^2}{3\sigma^2} \right), \tag{15.49}$$

where $E[|F_2^H \bar{Z}_k|^2] = \frac{3}{2\sigma^2}$, $E[|s_k|^2] = P$, and $F_2^H \bar{\mathbf{H}}_{\mathbf{k},\mathbf{k}}F_1 = \frac{-h_{k,k}[3]}{2}$. Since K data symbols are delivered using four time slots, at the above rate R_k, we thus achieve the following sum-DoF $d_{\sum} = \lim_{P \to \infty} \frac{\sum_{k=1}^{K} R_k(P)}{\log(P)} = \frac{K}{4}$

EXAMPLE 15.3   Consider a K-user SISO IC with ISI with $L_{k,k} = L_D = 3$ and $L_{k,i} = L_I = 4$. It is assumed that $h_{k,k}[l] \neq 0$ for $l \in \{1, 2, 3\}$. For $l \in \{3, 4\}$, it is assumed that $h_{k,i}[1] = h_{k,i}[2] = 0$ and $h_{k,i}[l] \neq 0$. This implies that $\text{supp}|(h_{k,k}[1], h_{k,k}[2], h_{k,k}[3])| = 3$ and $\text{supp}|(h_{k,i}[1], h_{k,i}[2], \ldots, h_{k,i}[4])| = 4$, where $\text{supp}(x)$ denotes the support set of x. This example explores sum-DoF for the case when the desired signal and the interference signals differ in arrival times at the receiver. This difference in arrival times stems from the propagation delay between the desired and the interfering links. The goal of this example is to show that each transmitter will be able to send one information symbol using five time slots, i.e., the scheme achieves sum-DoF of $d_{\sum} = \lim_{P \to \infty} \frac{\sum_{k=1}^{K} R_k(P)}{\log(P)} = \frac{K}{5}$. Consider a beamforming vector $F_1 = \frac{1}{\sqrt{2}} \begin{bmatrix} 1 & 1 \end{bmatrix}^T$ that carries the information symbol s_k, for $k \in \{1, 2, \ldots, K\}$. By equation (15.38), the received signal vector Y_k at receiver k over five time slots is

$$Y_k = \begin{bmatrix} y_k[1] \\ y_k[2] \\ y_k[3] \\ y_k[4] \\ y_k[5] \end{bmatrix} = \underbrace{\begin{bmatrix} h_{k,k}[1] & 0 \\ h_{k,k}[2] & h_{k,k}[1] \\ h_{k,k}[3] & h_{k,k}[2] \\ 0 & h_{k,k}[3] \\ 0 & 0 \end{bmatrix}}_{\mathbf{H}_{\mathbf{k},\mathbf{k}}} \begin{bmatrix} x_k[1] \\ x_k[2] \end{bmatrix} + \sum_{i \neq k}^{K} \underbrace{\begin{bmatrix} 0 & 0 \\ 0 & 0 \\ h_{k,i}[3] & 0 \\ h_{k,i}[4] & h_{k,i}[3] \\ 0 & h_{k,i}[4] \end{bmatrix}}_{\mathbf{H}_{\mathbf{k},\mathbf{i}}} \begin{bmatrix} x_i[1] \\ x_i[2] \end{bmatrix} + \begin{bmatrix} z_k[1] \\ z_k[2] \\ z_k[3] \\ z_k[4] \\ z_k[5] \end{bmatrix}.$$

$$\tag{15.50}$$

In order to obtain circulant channel matrices for the interfering links while keeping the desired link channel matrices noncirculant, we can apply a combiner matrix to the received signal, Y_k, to obtain $y_k[3] + y_k[5]$ and $y_k[4]$ as follows

$$\bar{Y}_k = \begin{bmatrix} y_k[3] + y_k[5] \\ y_k[4] \end{bmatrix} = \underbrace{\begin{bmatrix} h_{k,k}[3] & h_{k,k}[2] \\ 0 & h_{k,k}[3] \end{bmatrix}}_{\bar{\mathbf{H}}_{\mathbf{k,k}}} \underbrace{\begin{bmatrix} x_k[1] \\ x_k[2] \end{bmatrix}}_{\bar{X}_k}$$

$$+ \sum_{\substack{l \neq k}}^{K} \underbrace{\begin{bmatrix} h_{k,i}[1] & h_{k,i}[2] \\ h_{k,i}[2] & h_{k,i}[1] \end{bmatrix}}_{\bar{\mathbf{H}}_{\mathbf{k,i}}} \underbrace{\begin{bmatrix} x_i[1] \\ x_i[2] \end{bmatrix}}_{\bar{X}_i} + \underbrace{\begin{bmatrix} z_k[3] + z_k[5] \\ z_k[4] \end{bmatrix}}_{\bar{Z}_k} \quad (15.51)$$

Applying a decoding strategy similar to that used in Example 15.2, we achieve the sum-DoF of $\frac{K}{5}$.

It should be noted from Example 15.3 that, unlike in Example 15.2 where $L_D > L_I$, the IF-OFDM output signal combining idea, to create favorable channel matrices for the decodability of the desired symbols and the isolation of the interfering symbols into a separate subspace at the receiver, can also be applied to the scenario where $L_D < L_I$. In particular, this can be done when the support set for the desired link CIRs, $\{h_{k,k}[1], h_{k,k}[2], \ldots, h_{k,k}[L_D]\}$, is different from the support set for the interfering links CIRs, $\{h_{k,i}[1], h_{k,i}[2], \ldots, h_{k,i}[L_I]\}$. The reader is referred to [105] for further details on how, more generally, the idea of the combiner matrix allowing the isolation of the desired and interfering symbols into separate subspaces at the intended receiver can be coupled with the application of the complementary DFT matrix vectors in the precoding process both at the transmitter and at the receiver in order to cancel out interference symbols and still be able to decode the desired symbols and achieve sum-DoF result of (15.39) in a general OFDM setting.

The recently proposed novel idea in [105] shows that, for a K-user SISO interference channel with ISI, careful signal precoding for transmission, coupled with matrix combining, and received signal precoding can be devised by exploiting ISI channel link heterogeneity between desired and interfering channels. This enables the isolation of the information and the interfering symbols into separate subspaces at the intended receiver and, after interference cancellation, leads to sum-DoF that linearly scales with the number of users, K, in the absence of CSIT. Recent work in [106] has shown that a similar technique can be applied to K-cell interfering MACs to achieve linear scaling of sum-DoF with the number of cells. Interesting generalizations of this idea of ISI heterogeneity exploitation can be applied to other types of multiuser channel settings in the absence of CSIT.

15.6   Outer Bounding Techniques

Upper bounds are an essential part of the analysis and the characterization of optimality of the developed communication strategies. This section aims to highlight the essence of some frequently used bounding techniques by applying them to different topologies

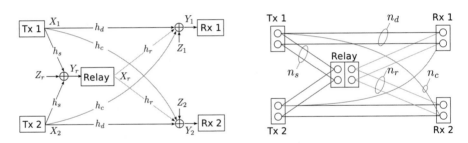

Figure 15.21 Symmetric interference relay channel: Gaussian channel and high-SNR linear deterministic modeling.

described by a simple channel model, the so-called linear deterministic channel model introduced in [3].

15.6.1   Cut-set Bounds

We start with the cut-set-bounds [107], which can be regarded as a classical bound. The essence of the cut-set bound is to split the system into two parts or sets $\mathcal{S}, \mathcal{S}^c$ and thereby measuring the flow of information between those two parts. All nodes within each set are allowed to fully cooperate, respectively. In general, there are many ways to apply such splits or cuts. The cut providing the least flow of information between the two parts of the system it is separating is of interest as it provides the least upper bound on the achievable rates. For instance, consider the so-called symmetric interference relay channel as illustrated in Figure 15.21 with two transmitters Tx1, Tx2, a relay, and two receivers Rx1, Rx2. The setup on the right serves as a high-SNR linear deterministic approximation of the Gaussian counterpart shown on the left. The relay supports the communication which takes place between Txi and Rxi, $i = 1, 2$. The received signals at receivers Rx1 and Rx2 of the high-SNR linear deterministic channel are given by

$$\mathbf{y}_1^n = \mathbf{D}^{q-n_d}\mathbf{x}_1^n \oplus \mathbf{D}^{q-n_c}\mathbf{x}_2^n \oplus \mathbf{D}^{q-n_s}\mathbf{x}_r^n, \qquad (15.52)$$

$$\mathbf{y}_2^n = \mathbf{D}^{q-n_{12}}\mathbf{x}_1^n \oplus \mathbf{D}^{q-n_{22}}\mathbf{x}_2^n \oplus \mathbf{D}^{q-n_s}\mathbf{x}_r^n, \qquad (15.53)$$

respectively, where $\mathbf{y}_i^n = \{\mathbf{y}_i(1), \ldots, \mathbf{y}_i(n)\}$ and $\mathbf{x}_i^n = \{\mathbf{x}_i(1), \ldots, \mathbf{x}_i(n)\}$ are the collections of the received and transmitted signals for all channel uses $1 \ldots n$. The parameters are defined similarly to (15.6) and (15.7). Consider, for instance, a cut which separates the network into the following sets $\mathcal{S} = \{\text{Tx1}, \text{Tx2}, \text{Relay}\}$ and $\mathcal{S}^c = \{\text{Rx1}, \text{Rx2}\}$. Then, a cut-set bound on the sum-rate $R_1 + R_2$ can be obtained with

$$R_{\mathcal{S} \to \mathcal{S}^c} \leq \max_{\Pr(\mathbf{x}_1, \mathbf{x}_2, \mathbf{x}_r)} I(\mathbf{x}_{\mathcal{S}}; \mathbf{y}_{\mathcal{S}^c} | \mathbf{x}_{\mathcal{S}^c}), \qquad (15.54)$$

where the maximization is with respect to the joint density function of the input alphabet of both users and relay for that cut, such that

$$R_1 + R_2 \leq \max_{\Pr(\mathbf{x}_1, \mathbf{x}_2, \mathbf{x}_r)} I(\mathbf{x}_1, \mathbf{x}_2, \mathbf{x}_r; \mathbf{y}_1, \mathbf{y}_2) \qquad (15.55)$$

$$= \max_{\Pr(\mathbf{x}_1, \mathbf{x}_2, \mathbf{x}_r)} H(\mathbf{y}_1, \mathbf{y}_2) - H(\mathbf{y}_1, \mathbf{y}_2 | \mathbf{x}_1, \mathbf{x}_2, \mathbf{x}_r) \qquad (15.56)$$

$$\overset{(a)}{\le} \max_{\Pr(\mathbf{x}_1,\mathbf{x}_2,\mathbf{x}_r)} H(\mathbf{y}_1,\mathbf{y}_2) \tag{15.57}$$

$$= \max_{\Pr(\mathbf{x}_1,\mathbf{x}_2,\mathbf{x}_r)} H(\mathbf{y}_1) + H(\mathbf{y}_2|\mathbf{y}_1), \tag{15.58}$$

where (a) follows since $\mathbf{y}_1$ and $\mathbf{y}_2$ are deterministic functions of $\mathbf{x}_1$, $\mathbf{x}_2$, and $\mathbf{x}_r$. The first term $H(\mathbf{y}_1)$ is upper bounded by i.i.d. Bern(1/2) inputs. Thus,

$$R_1 + R_2 \le \max_{\Pr(\mathbf{x}_1,\mathbf{x}_2,\mathbf{x}_r)} \max\{n_d, n_c, n_r\} + H(\mathbf{y}_2|\mathbf{y}_1). \tag{15.59}$$

The second term can be bounded as follows

$$H(\mathbf{y}_2|\mathbf{y}_1) = H(\mathbf{y}_2 \oplus \mathbf{y}_1|\mathbf{y}_1) \overset{(a)}{\le} H(\mathbf{y}_2 \oplus \mathbf{y}_1), \tag{15.60}$$

where (a) follows since conditioning does not increase entropy. Now, for $\mathbf{y}_2 \oplus \mathbf{y}_1$ we have

$$\mathbf{y}_2 \oplus \mathbf{y}_1 = \mathbf{D}^{q-n_d}\mathbf{x}_1^n \oplus \mathbf{D}^{q-n_c}\mathbf{x}_2^n \oplus \mathbf{D}^{q-n_s}\mathbf{x}_r^n \mathbf{D}^{q-n_{12}}\mathbf{x}_1^n \oplus \mathbf{D}^{q-n_{22}}\mathbf{x}_2^n \oplus \mathbf{D}^{q-n_s}\mathbf{x}_r^n \tag{15.61}$$

$$= \mathbf{D}^{q-n_d}\mathbf{x}_1^n \oplus \mathbf{D}^{q-n_c}\mathbf{x}_2^n \mathbf{D}^{q-n_{12}}\mathbf{x}_1^n \oplus \mathbf{D}^{q-n_{22}}\mathbf{x}_2^n, \tag{15.62}$$

which is a binary vector with at most $\max\{n_d, n_c\}$ nonzero elements, which maximizes $H(\mathbf{y}_2 \oplus \mathbf{y}_1)$ in case of i.i.d. Bern(1/2) components. Thus, we obtain

$$R_1 + R_2 \le \max\{n_d, n_c, n_r\} + \max\{n_d, n_c\}. \tag{15.63}$$

Other cut-set bounds can be obtained by choosing the set $\mathcal{S}$ and thus $\mathcal{S}^c$ differently.

While the cut-set bounds are usually simpler to obtain, it can be rather coarse meshed and thus result in overestimating the performance, i.e., it might not be a tight bound in terms of GDoF. The next types of methods allow a more punctuated way to obtain upper bounds and can be described as being small meshed when compared to cut-set bounds.

15.6.2   Enhancement and Genie-Aided Approaches

The idea of the genie-aided approach is, as the name suggests, that a genie is providing useful and well-chosen information to some nodes in an individual fashion, which is then utilized for deriving tight upper bounds.

One-Sided Genie

The concept in this method is to provide one node with additional information, referred to as genie information. Thus, this particular node is more capable than before. As such, the information could be utilized to improve the performance of the overall system. To illustrate this concept, consider the so-called two-user IC depicted in Figure 15.22(a). Here, transmitter Tx1 (Tx2) wishes to send information to the receiver Rx1 (Rx2) reliably. As Tx1 and Tx2 access the channel at the same time, the receivers obtain a superposition of the two signals. Thus, Tx1 causes interference at Rx2, while Tx2 causes interference at Rx1.

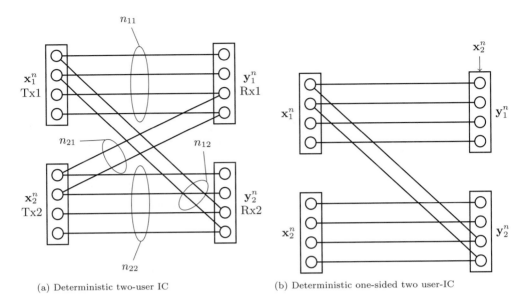

(a) Deterministic two-user IC   (b) Deterministic one-sided two user-IC

Figure 15.22 One-side genie approach in the deterministic IC.

The corresponding system equations are given by

$$y_1^n = D^{q-n_{11}} x_1^n \oplus D^{q-n_{21}} x_2^n, \tag{15.64}$$
$$y_2^n = D^{q-n_{12}} x_1^n \oplus D^{q-n_{22}} x_2^n, \tag{15.65}$$

respectively, where $y_i^n = \{y_i(1), \ldots, y_i(n)\}$ and $x_i^n = \{x_i(1), \ldots, x_i(n)\}$ are the collections of the received and transmitted signals for all channel uses $1 \ldots n$. We now provide x_2 to receiver Rx1 for free, such that Rx1 can remove its contribution from its received signal y_1^n. Thus, the system model reduces to the one shown in Figure 15.22(b).

As a result, we get

$$y_1^n = D^{q-n_{11}} x_1^n.$$

Now, we can derive the upper bound on the sum of the rates

$$
\begin{aligned}
n(R_1 + R_2) &= H(w_1) + H(w_2) \\
&= H(w_1) - H(w_1|y_1^n) + H(w_1|y_1^n) + H(w_2) - H(w_2|y_2^n) + H(w_2|y_2^n) \\
&\overset{(a)}{\leq} I(w_1; y_1^n) + I(w_2; y_2^n) + n\epsilon_1 + n\epsilon_2,
\end{aligned}
$$

where in (a) the definition of the mutual information and Fano's inequality was used. Applying the data processing inequality results in

$$n(R_1 + R_2 - \epsilon_1 - \epsilon_2) \leq I(x_1; y_1^n) + I(x_2; y_2^n).$$

We now provide x_2^n with the help of a genie to receiver Rx1 as mentioned in the discussion, which led to the system model in Figure 15.22(b), and obtain

$$n(R_1 + r_2 - \epsilon_1 - \epsilon_2) \le I(\mathbf{x}_1^n; \mathbf{y}_1^n, \mathbf{x}_2^n) + I(\mathbf{x}_2^n; \mathbf{y}_2^n)$$

$$\overset{(a)}{=} I(\mathbf{x}_1^n; \mathbf{x}_2^n) + I(\mathbf{x}_1^n; \mathbf{y}_1^n | \mathbf{x}_2^n) + I(\mathbf{x}_2^n; \mathbf{y}_2^n)$$

$$\overset{(b)}{=} I(\mathbf{x}_1^n; \mathbf{y}_1^n | \mathbf{x}_2^n) + I(\mathbf{x}_2^n; \mathbf{y}_2^n)$$

$$\overset{(c)}{=} H(\mathbf{y}_1^n | \mathbf{x}_2^n) + H(\mathbf{y}_2^n) - H(\mathbf{y}_2^n | \mathbf{x}_2^n)$$

$$= H(\mathbf{D}^{q-n_{11}} \mathbf{x}_1^n) + H(\mathbf{D}^{q-n_{11}} \mathbf{x}_1^n \oplus \mathbf{D}^{q-n_{22}} \mathbf{x}_2^n) - H(\mathbf{D}^{q-n_{12}} \mathbf{x}_1^n)$$

$$= (n_{11} - \max(n_{12}, n_{22}) - n_{12})n,$$

where the chain rule of mutual information and the fact that the transmit signals are stochastically independent were used in (a) and (b), respectively. Equality (c) holds due to the fact that in the deterministic model $\mathbf{y}_i$ is completely determined given $\mathbf{x}_1, \mathbf{x}_2$. Letting $n \to \infty (\epsilon_1, \epsilon_2 \to 0)$ leads to

$$R_1 + R_2 \le n_{11} + |n_{22} - n_{12}|^+.$$

A similar bound can be obtained by giving the corresponding interference signal for free to other nodes in the network. As we have seen, by providing some partial information to a particular node, an upper bound can be obtained, which could possibly be tight for certain parameter regimes.

Two-Sided Genie

While the one-sided genie approach helped to improve the interference burden of one receiver only, the two-sided genie provides side information to more than and node receiver. However, to get a useful bound, the side information has be carefully chosen in order not to reveal too much information. Starting with

$$n(R_1 + R_2 - \epsilon_1 - \epsilon_2) \le I(\mathbf{x}_1^n; \mathbf{y}_1^n) + I(\mathbf{x}_2; \mathbf{y}_2^n),$$

we provide the following side information:

$$\mathbf{s}_2^n = \mathbf{D}^{q-n_{21}} \mathbf{x}_2^n \tag{15.66}$$

$$\mathbf{s}_1^n = \mathbf{D}^{q-n_{12}} \mathbf{x}_1^n \tag{15.67}$$

to receivers Rx2 and Rx1, respectively, as shown in Figure 15.23. Thus, we do not provide the signal(s) causing the interference at the corresponding receiver. Rather, we provide to each receiver a view on how its desired signal is impairing the desired signals at the other receivers. Thus, we obtain

$$n(R_1 + R_2 - \epsilon_1 - \epsilon_2) \le I(\mathbf{x}_1^n; \mathbf{y}_1^n, \mathbf{s}_1^n) + I(\mathbf{x}_2^n; \mathbf{y}_2^n, \mathbf{s}_2^n)$$

$$\overset{(a)}{=} I(\mathbf{x}_1^n; \mathbf{s}_1^n) + I(\mathbf{x}_1^n; \mathbf{y}_1^n | \mathbf{s}_1^n) + I(\mathbf{x}_2^n; \mathbf{s}_2^n) + I(\mathbf{x}_2^n; \mathbf{y}_2^n | \mathbf{s}_2^n)$$

$$\overset{(b)}{=} H(\mathbf{s}_1^n) + H(\mathbf{y}_1^n | \mathbf{s}_1^n) - H(\mathbf{y}_1^n | \mathbf{s}_1^n, \mathbf{x}_1^n) + H(\mathbf{s}_2)$$
$$+ H(\mathbf{y}_2^n | \mathbf{s}_2^n) - H(\mathbf{y}_2^n | \mathbf{s}_2^n, \mathbf{x}_2^n)$$

$$\overset{(c)}{=} H(\mathbf{s}_1^n) + H(\mathbf{y}_1^n | \mathbf{s}_1^n) - H(\mathbf{s}_2^n) + H(\mathbf{s}_2^n) + H(\mathbf{y}_2^n | \mathbf{s}_2^n)$$
$$- H(\mathbf{s}_1^n)$$

$$= H(\mathbf{y}_1^n | \mathbf{s}_1^n) + H(\mathbf{y}_2^n | \mathbf{s}_2^n),$$

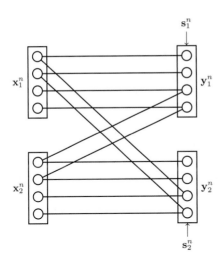

Figure 15.23 Two-sided genie deterministic IC.

where in (a) the channel rule of mutual information was used, in (b) the definition of the mutual information was used, and in (c) the side information was used to remove its contribution from (15.64) and (15.65), respectively. Letting $n \to \infty$

$$\Rightarrow R_1 + R_2 = \max(n_{21}, |n_{11} - n_{12}|^+) + \max(n_{12}, |n_{22} - n_{21}|^+).$$

This again provides an upper bound which depending on the parameter choices might be more tight than the previous bounds.

15.6.3   Virtual Receiver

In the concept of the virtual receiver, a new node is added to the network, however, without expanding the message set. This additional node allows then a larger flexibility in the manipulations of information measures to achieve a tight bound. For instance, in Figure 15.24(a) we deploy an additional receiver, Rx1v, which is interested in the same message as Rx1 and identical to Rx1 in terms of channel gains. Thus, for the sum-rate of this new system we obtain

$$n(2R_1 + R_2 - 2\epsilon_1 - \epsilon_2) \leq I(\mathbf{x}_1^n; \mathbf{y}_1^n) + I(\mathbf{x}_1^n; \mathbf{y}_1^n) + I(\mathbf{x}_2^n; \mathbf{y}_2^n).$$

In the next step, as shown in Figure 15.24(b), we now provide side information in the form of $\mathbf{s}_2$ to Rx2 and $\mathbf{x}_2$ to Rx1v, where $\mathbf{s}_2$ is defined as in (15.66). We proceed as follows:

$$n(2R_1 + r_2 - 2\epsilon_1 - \epsilon_2) \overset{(a)}{\leq} I(\mathbf{x}_1^n; \mathbf{y}_1^n) + I(\mathbf{x}_1^n; \mathbf{y}_1^n | \mathbf{x}_2^n) + I(\mathbf{x}_2^n; \mathbf{y}_2^n, \mathbf{s}_2^n)$$

$$= H(\mathbf{y}_1^n) - H(\mathbf{y}_1^n | \mathbf{x}_1^n) + H(\mathbf{y}_1^n | \mathbf{x}_2^n) - H(\mathbf{y}_1^n | \mathbf{x}_2^n, \mathbf{x}_1^n)$$

$$+ I(\mathbf{x}_2^n; \mathbf{s}_2^n) + I(\mathbf{x}_2^n; \mathbf{y}_2^n | \mathbf{s}_2^n)$$

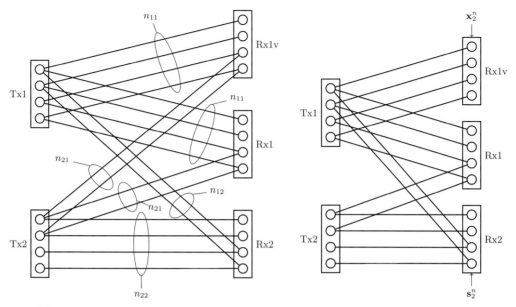

(a) Virtual receiver Rxlv is added to the system.

(b) Genie information is provided to the virtual receiver Rxlv as well as to receiver Rx2.

Figure 15.24 Concept of the virtual receiver.

$$\overset{(b)}{=} H(\mathbf{y}_1^n) - H(\mathbf{s}_2^n) + H(\mathbf{y}_1^n|\mathbf{x}_2^n) + H(\mathbf{s}_2^n) - H(\mathbf{s}_2^n|\mathbf{x}_2^n)$$
$$+ H(\mathbf{y}_2^n|\mathbf{s}_2^n) - H(\mathbf{y}_2^n|\mathbf{s}_2^n, \mathbf{x}_2^n)$$
$$\overset{(c)}{=} H(\mathbf{y}_1^n) + H(\mathbf{y}_1^n|\mathbf{x}_2^n) + H(\mathbf{y}_2^n|\mathbf{s}_2^n) - H(\mathbf{y}_2^n|\mathbf{x}_2^n)$$
$$\overset{(d)}{=} H(\mathbf{y}_1^n) + H(\mathbf{y}_1^n|\mathbf{x}_2^n) + H(\mathbf{y}_2^n|\mathbf{s}_2^n) - H(\mathbf{s}_1^n)$$
$$= n(\max(n_{11}, n_{21}) + n_{11} + \max(n_{12}, |n_{22}$$
$$- n_{21}|^+) - n_{12}),$$

where (a) follows from the independence of the messages and the chain rule of mutual information, (b) follows from the definition of $\mathbf{s}_2$ and that $\mathbf{y}_1^n$ is completely determined given $\mathbf{x}_1^n, \mathbf{x}_2^n$, (c) follows since $\mathbf{s}_2^n$ is a deterministic function of $\mathbf{x}_2^n$, and (d) follows from the definition of $\mathbf{s}_1^n$ in (15.67). Letting $n \to \infty$ results in

$$2R_1 + R_2 \le \max(n_{11}, n_{21}) + n_{11} + ||n_{22} - n_{21}|^+ - n_{12}|^+,$$

which is yet another useful bound that is tighter than other upper bounds for certain parameter regimes.

15.7   Outlook and Concluding Remarks

In this chapter, we presented a review of interference management techniques through an information theoretic perspective. The treatment of this topic was done with the goal of highlighting the various interference management techniques, their similarities, differences, and suitability based on the underlying network topology. In addition, we also discussed outer bounding methodologies (converse techniques) that can provably quantify the gap from optimality of various interference management schemes. The chapter discussed recent breakthroughs in the area as well as emerging topics whose aim is to make information theoretic approaches robust to channel knowledge availability, as well as the underlying network topology.

We conclude the chapter with a few open problems as well as a brief overview of other recent developments that are not completely covered in the previous sections. While the GDoF and the approximate capacity are known for the two-user IC, such characterizations remain elusive for general K-user interference networks. In the discussion of interference alignment and interference neutralization techniques, the number of channel extensions can be excessively large and require a large amount of channel diversity to achieve optimal DoF in an asymptotic sense (also see [108] and references therein for recent results in this direction).

Another important and practical aspect is the robustness of interference management schemes to finite-precision CSI. To this end, recent work on a novel converse technique (*aligned image sets*) [65] obtains new results and insights for these scenarios. For instance, one of the conclusions is that the DoF of broadcast and interference networks collapse-under finite-precision CSI (finites precision CSI refers to the setting in which the CSI precision *does not* improve with SNR). Generalizations of these ideas for obtaining improved bounds on GDoF and ultimately capacity is an interesting research direction. For the case in which CSI availability is delayed, there is a lack of outer bounding techniques for interference networks. Clearly, bounds for broadcast networks (which are optimal) are also valid outer bounds for interference networks however, these are not expected to be tight. Hence, an interesting open area is to obtain tight lower bounds for interference networks with delayed CSI.

Finally, another interesting emerging research area is that of cache-aided interference management. As storage costs go down, a viable alternative can be to exploit storage capabilities at transmitters/receivers to preemptively *cache* contents locally. This leads to novel problems related to joint consideration of storage, caching, and interference management. It is worth noting that some of the challenges (such as robustness aspects) carry over to the setting of cache-aided interference management, leading to an abundant set of interesting open problems in this area. We point the reader to some recent papers on this topic [109–111].

References

[1] R. H. Etkin, D. Tse, and H. Wang, "Gaussian interference channel capacity to within one bit," *IEEE Trans. Inform. Theory*, vol. 54, no. 12, pp. 5534–5562, Dec. 2008.

[2] T.Guo and S. A. Jafar, "Sum capacity of a class of symmetric SIMO Gaussian interference channels within o(1)," *IEEE Trans. Inform. Theory*, vol. 57, no. 4, pp. 1932–1958, 2011.

[3] A. S. Avestimehr, S. N. Diggavi, and D. N. C. Tse, "Wireless network information flow: A deterministic approach," *IEEE Trans. Inform. Theory*, vol. 57, no. 4, pp. 1872–1905, Mar. 2011.

[4] S. Mohajer, S. N. Diggavi, C. Fragouli, and D. Tse, "Transmission techniques for relay-interference networks," in *46th Ann. Allerton Conf. Commun. Control Comput.* Monticello, IL, Sept. 2008, pp. 467–474.

[5] G. Bresler and D. N. C. Tse, "The two-user Gaussian interference channel: A deterministic view," *Eur. Trans. Telecommun.*, vol. 8, no. 4, pp. 333–354, 2008.

[6] G. Bresler, A. Parekh, and D. N. C. Tse, "The approximate capacity of the many-to-one and one-to-many Gaussian interference channels," *IEEE Trans. Inform. Theory*, vol. 56, no. 9, pp. 4566–4592, Aug. 2010.

[7] A. Chaaban and A. Sezgin, "The capacity region of the linear shift deterministic Y-channel," in *2011 IEEE Int. Symp. Information Theory Proceedings (ISIT)*, St Petersburg Jul. 2011, pp. 26–29.

[8] A. Chaaban and A. Sezgin, "Multi-way communications: An information theoretic perspective," *Found. Trends Commun. Inform. Theory*, vol. 12, no. 3–4, pp. 185–371, 2015.

[9] A. El-Gamal and M. Costa, "The capacity region of a class of deterministic interference channels," *IEEE Trans. Inform. Theory*, vol. 28, no. 3, pp. 343–346, Mar. 1982.

[10] E. Telatar and D. Tse, "Bounds on the capacity region of a class of interference channels," in *2012 IEEE Int. Symp. Information Theory Proceedings (ISIT)*, Nice, 2007.

[11] M. Anand and P. R. Kumar, "On approximating Gaussian relay networks by deterministic networks," in *IEEE Information Theory Workshop*, Taormina, 2009.

[12] M. Charafeddine, A. Sezgin, Z. Han, and A. Paulraj, "Achievable and crystallized rate regions of the interference channel with interference as noise," *IEEE Trans. Wireless Commun.*, vol. 11, no. 3, pp. 1100–1111, 2012.

[13] A. S. Motahari and K. A. Khandani, "Capacity bounds for the Gaussian interference channel," *IEEE Trans. Inform. Theory*, vol. 55, no. 2, pp. 620–643, Feb. 2009.

[14] V. S. Annapureddy and V. V. Veeravalli, "Gaussian interference networks: Sum capacity in the low interference regime and new outer bounds on the capacity region," *IEEE Trans. Inform. Theory*, vol. 55, no. 6, pp. 3032–3050, Jun. 2009.

[15] X. Shang, G. Kramer, and B. Chen, "A new outer bound and the noisy-interference sum-rate capacity for Gaussian interference channels," *IEEE Trans. Inform. Theory*, vol. 55, no. 92, pp. 689–699, Feb. 2009.

[16] H. Yi, Xinping Sun, S. Jafar, and D. Gesbert, "TDMA is optimal for all-unicast DoF region of TIM if and only if topology is chordal bipartite."

[17] H. Maleki and S. Jafar, "Optimality of orthogonal access for one-dimensional convex cellular networks," *IEEE Commun. Lett.*, vol. 17, no. 9, pp. 1770–1773, 2013.

[18] A. Paulraj, R. Nabar, and D. Gore, *Introduction to Space-Time Wireless Communications*. Cambridge: Cambridge University Press, 2003.

[19] A. Sezgin, E. Jorswieck, and E. Costa, "LDC in MIMO Ricean channels: Optimal transmit strategy with MMSE detection," *IEEE Trans. Signal Process.*, vol. 56, no. 1, pp. 313–328, 2008.

[20] M. Costa, "Writing on dirty paper," *IEEE Trans. Inform. Theory*, vol. 29, no. 3, pp. 439–441, May 1983.

[21] M. Costa, "Coding for channel with random parameters," *Probl. Control Inform. Theory*, vol. 9, no. 1, pp. 19–31, 1980.

[22] G. Caire and S. Shamai, "On the achievable throughput of a multiantenna Gaussian broadcast channel," *IEEE Trans. Inform. Theory*, vol. 49, no. 3, Jul. 2003.

[23] C. Heegard and A. El Gamal, "Coding in a memory with defective cells," *Probl. Inf. Transm.*, vol. 10, no. 2, pp. 52–60, 1974.

[24] A. S. Cohen and A. Lapidoth, "The Gaussian watermarking game," *IEEE Trans. Inform. Theory*, vol. 48, no. 6, pp. 1639–1667, 2002.

[25] C. Heegard and A. El Gamal, "On the capacity of computer memories with defects," *IEEE Trans. Inform. Theory*, vol. 29, no. 5, pp. 731–739, 1983.

[26] M. Tomlinson, "New automatic equalizer employing modulo arithmetic," *Electron. Lett.*, vol. 7, no. 3, pp. 138–139, Mar. 1971.

[27] D. Wang, E. A. Jorswieck, A. Sezgin, and E. Costa, "Joint Tomlinson–Harashima precoding with diversity techniques for multiuser MIMO systems," in *IEEE Vehicular Technology Conference (VTC Spring)*, 2005.

[28] M. A. Maddah-Ali, A. S. Motahari, and A. K. Khandani, "Communication over MIMO X channels: Interference alignment, decomposition, and performance analysis," *IEEE Trans. Inform. Theory*, vol. 54, no. 8, pp. 3457–3470, Jul. 2008.

[29] V. R. Cadambe and S. A. Jafar, "Interference alignment and degrees of freedom of the K-user interference channel," *IEEE Trans. Inform. Theory*, vol. 54, no. 8, pp. 3425–3441, Jul. 2008.

[30] V. R. Cadambe and S. A. Jafar, "Degrees of freedom of wireless x networks," in *IEEE Int. Symp. Information Theory, 2008*, Toronto, 2008, pp. 1268–1272.

[31] S. A. Jafar and S. Shamai, "Degrees of freedom region of the MIMO X channel," *IEEE Trans. Inform. Theory*, vol. 54, no. 1, pp. 151–170, Jan. 2008.

[32] A. S. Motahari, S. Oveis-Gharan, M.-A. Maddah-Ali, and A. K. Khandani, "Real interference alignment: Exploiting the potential of single antenna systems," *IEEE Trans. Inform. Theory*, vol. 60, no. 8, pp. 4799–4810, Jun. 2014.

[33] S. W. Peters and R. W. Heath, "Interference alignment via alternating minimization," in *IEEE Inter. Conf. Acoust. Speech Signal Process.*, Taipei May 2009, pp. 2445–2448.

[34] O. El Ayach, S. W. Peters, and R. W. Heath, "The practical challenges of interference alignment," *IEEE Wireless Commun.*, vol. 20, no. 1, pp. 35–42, Mar. 2013.

[35] K. Gomadam, V. R. Cadambe, and S. A. Jafar, "A distributed numerical approach to interference alignment and applications to wireless interference networks," *IEEE Trans. Inform. Theory*, vol. 57, no. 6, pp. 3309–3322, May 2011.

[36] V. Ntranos, M. A. Maddah-Ali, and G. Caire, "Cellular interference alignment," *IEEE Trans. Inform. Theory*, vol. 61, no. 3, pp. 1194–1217, Jan. 2015.

[37] C. Suh, M. Ho, and D. N. Tse, "Downlink interference alignment," *IEEE Trans. Commun.*, vol. 59, no. 9, pp. 2616–2626, 2011.

[38] T. Gou, S. A. Jafar, C. Wang, S.-W. Jeon, and S.-Y. Chung, "Aligned interference neutralization and the degrees of freedom of $2 \times 2 \times 2$ interference channel," *IEEE Trans. Inform. Theory*, vol. 58, no. 7, pp. 4381–4395, Mar. 2012.

[39] I. Shomorony and A. S. Avestimehr, "Two-unicast wireless networks: Characterizing the degrees of freedom," *IEEE Trans. Inform. Theory*, vol. 59, no. 1, pp. 353–383, Aug. 2012.

[40] I. Shomorony and A. S. Avestimehr, "Degrees of freedom of two-hop wireless networks: Everyone gets the entire cake," *IEEE Trans. Inform. Theory*, vol. 60, no. 5, pp. 2417–2431, Mar. 2014.

[41] N. Lee, J.-B. Lim, and J. Chun, "Degrees of freedom of the MIMO Y channel: Signal space alignment for network coding," *IEEE Trans. Inform. Theory*, vol. 56, no. 7, pp. 3332–3342, Jun. 2010.

[42] S. Mohajer, S. N. Diggavi, and N. David, "Approximate capacity of a class of Gaussian relay-interference networks," in *IEEE Int. Symp. Inform. Theory (ISIT)*, Aug. 2009, pp. 31–35.

[43] A. Chaaban and A. Sezgin, "Interference alignment and neutralization in a cognitive 3-user MAC-interference channel: Degrees of freedom," in *12th Canadian Workshop on Information Theory (CWIT)*, Jun. 2011, pp. 2457–2461.

[44] Z. K. Ho and E. A. Jorswieck, "Instantaneous relaying: Optimal strategies and interference neutralization," *IEEE Trans. Signal Process.*, vol. 60, no. 12, pp. 6655–6668, Jul. 2012.

[45] B. Rankov and A. Wittneben, "Spectral efficient protocols for half-duplex fading relay channels," *IEEE J. Sel. Areas Commun.*, vol. 25, no. 2, Feb. 2007.

[46] D. Bharadia, E. McMilin, and S. Katti, "Full duplex radios," *ACM SIGCOMM Comput. Commun. Rev.*, vol. 43, no. 4, pp. 375–386, Aug. 2013.

[47] C. S. Vaze and M. K. Varanasi, "Beamforming and aligned interference neutralization achieve the degrees of freedom region of the $2 \times 2 \times 2$ MIMO interference network," in *Information Theory and Applications Workshop (ITA)*, Apr. 2012, pp. 199–203.

[48] I. Issa, S. L. Fong, and A. S. Avestimehr, "Two-hop interference channels: Impact of linear schemes," *IEEE Trans. Inform. Theory*, vol. 61, no. 10, pp. 5463–5489, Jul. 2015.

[49] A. Philosoph, T. Khisti, U. Erez, and R. Zamir, "Lattice strategies for the dirty multiple access channel," in *2012 IEEE Int. Symp. Information Theory Proceedings (ISIT)*, Nice, 2007, pp. 386–390.

[50] S. Sridharan, A. Jafarian, S. Vishwanath, S. A. Jafar, and S. Shamai, "A layered lattice coding scheme for a class of three user Gaussian interference channels," in *2008 46th Ann. Allerton Conf. Commun. Control Comput.*, Monticell, IL, Sept 2008, pp. 531–538.

[51] I. J. Baik and S. Y. Chung, "Network coding for two-way relay channels using lattices," in *2008 IEEE Int. Conf. Commun.*, Beijing May 2008, pp. 3898–3902.

[52] R. Zamir, "Lattices are everywhere," in *2009 Information Theory and Applications Workshop*, La Jolla, CA, Feb. 2009, pp. 392–421.

[53] C. S. Vaze and M. K. Varanasi, "The degree-of-freedom regions of MIMO broadcast, interference, and cognitive radio channels with no CSIT," *IEEE Trans. Inform. Theory*, vol. 58, no. 8, pp. 5354–5374, May 2012.

[54] Y. Zhu and D. Guo, "The degrees of freedom of isotropic MIMO interference channels without state information at the transmitters," *IEEE Trans. Inform. Theory*, vol. 58, no. 1, pp. 341–352, Jan. 2012.

[55] Y. Tian and A. Yener, "Guiding blind transmitters: Degrees of freedom optimal interference alignment using relays," *IEEE Trans. Inform. Theory*, vol. 59, no. 8, pp. 4819–4832, 2013.

[56] D. Frank, K. Ochs, and A. Sezgin, "A systematic approach for interference alignment in CSIT-less relay-aided X-networks," in *Wireless Commun. Netw. Conf. (WCNC), 2014*, Istanbul, 2014, pp. 1126–1131.

[57] M. A. Maddah-Ali and D. Tse, "Completely stale transmitter channel state information is still very useful," *IEEE Trans. Inform. Theory*, vol. 58, no. 7, pp. 4418–4431, July 2012.

[58] C. S. Vaze and M. K. Varanasi, "The degrees of freedom region of the two-user MIMO broadcast channel with delayed CSIT," in *2011 IEEE Int. Symp. Inform. Theory Proceedings*, St Petersburg July 2011, pp. 199–203.

[59] H. Maleki, S. A. Jafar, and S. Shamai, "Retrospective interference alignment over interference networks," *IEEE J. Sel. Topics Signal Process.*, vol. 6, no. 3, pp. 228–240, June 2012.

[60] M. J. Abdoli, A. Ghasemi, and A. K. Khandani, "On the degrees of freedom of K-user SISO interference and X channels with delayed CSIT," *IEEE Trans. Inform. Theory*, vol. 59, no. 10, pp. 6542–6561, Oct. 2013.

[61] M. Torrellas, A. Agustin, and J. Vidal, "On the degrees of freedom of the K-user MISO interference channel with imperfect delayed CSIT," in *IEEE Int. Conf. Acoust. Speech Signal Process. (ICASSP)*, Florence, May 2014, pp. 1155–1159.

[62] C. Hao and B. Clerckx, "Achievable sum DoF of the K-user MIMO interference channel with delayed CSIT," *IEEE Trans. Commun.*, vol. 64, no. 10, pp. 4165–4180, Oct. 2016.

[63] D. Castanheira, A. Silva, and A. Gameiro, "Retrospective interference alignment: Degrees of freedom scaling with distributed transmitters," *IEEE Trans. Inform. Theory*, vol. 63, no. 3, pp. 1721–1730, Mar. 2017.

[64] R. Tandon, S. A. Jafar, S. S. (Shitz), and H. V. Poor, "On the synergistic benefits of alternating CSIT for the MISO broadcast channel," *IEEE Trans. Inform. Theory*, vol. 59, no. 7, pp. 4106–4128, Jul. 2013.

[65] A. G. Davoodi and S. A. Jafar, "Aligned image sets under channel uncertainty: Settling conjectures on the collapse of degrees of freedom under finite precision CSIT," *IEEE Trans. Inform. Theory*, vol. 62, no. 10, pp. 5603–5618, 2016.

[66] E. Jorswieck, S. Tomasin, and A. Sezgin, "Broadcasting into the uncertainty: Authentication and confidentiality by physical-layer processing," *Proc. IEEE*, vol. 103, no. 10, pp. 1702–1724, 2015.

[67] S. Lashgari, R. Tandon, and S. Avestimehr, "MISO broadcast channel with hybrid CSIT: Beyond two users," *IEEE Trans. Inform. Theory*, vol. 62, no. 12, pp. 7056–7077, Dec. 2016.

[68] K. Mohanty, C. S. Vaze, and M. K. Varanasi, "The degrees of freedom region of the MIMO interference channel with hybrid CSIT," *IEEE Trans. Wireless Commun.*, vol. 14, no. 4, pp. 1837–1848, Apr. 2015.

[69] R. Ahlswede, "The capacity region of a channel with two senders and two receivers," *Ann. Probab.*, vol. 2, no. 5, pp. 805–814, Oct. 1974.

[70] A. B. Carleial, "A case where interference does not reduce capacity," *IEEE Trans. Inform. Theory*, vol. IT-21, no. 5, pp. 569–570, Sep. 1975.

[71] H. Sato, "The capacity of the Gaussian channel under strong interference," *IEEE Trans. Inform. Theory*, vol. IT-27, no. 6, pp. 786–788, Nov. 1981.

[72] B. Bandemer, A. Sezgin, and A. Paulraj, "On the noisy interference regime of the MISO Gaussian interference channel," in *Signals Syst. Comput., 2008 42nd Asilomar Conf.*, Pacific Grove, CA 2008, pp. 1098–1102.

[73] A. Chaaban and A. Sezgin, "Sub-optimality of treating interference as noise in the cellular uplink," in *2012 Int. ITG Workshop Smart Antennas (WSA)*, Dresden, 2012, pp. 238–242.

[74] A. Chaaban and A. Sezgin, "On the capacity of the 2-user Gaussian MAC interfering with a P2P link," in *Wireless Conf. 2011: Sustainable Wireless Technol. (European Wireless), 11th European*, 2011, pp. 1–6.

[75] S. Gherekhloo, A. Chaaban, C. Di, and A. Sezgin, "(Sub-) optimality of treating interference as noise in the cellular uplink with weak interference," *IEEE Trans. Inform. Theory*, vol. 62, no. 1, pp. 322–356, 2016.

[76] C. Geng and S. A. Jafar, "On the optimality of zero-forcing and treating interference as noise for K-user MIMO interference channels," in *2016 Int. Symp. Inform. Theory (ISIT)*, Barcelona 2016, pp. 2629–2633.

[77] C. Geng, N. Naderializadeh, A. S. Avestimehr, and S. A. Jafar, "On the optimality of treating interference as noise," *IEEE Trans. Inform. Theory*, vol. 61, no. 4, pp. 1753–1767, 2015.

[78] C. Geng and S. A. Jafar, "On the optimality of treating interference as noise: Compound interference networks," *IEEE Trans. Inform. Theory*, vol. 62, no. 8, pp. 4630–4653, 2016.

[79] H. Sun and S. A. Jafar, "On the optimality of treating interference as noise for k-user parallel Gaussian interference networks," *IEEE Trans. Inform. Theory*, vol. 62, no. 4, pp. 1911–1930, 2016.

[80] C. Geng, H. Sun, and S. A. Jafar, "On the optimality of treating interference as noise: General message sets," *IEEE Trans. Inform. Theory*, vol. 61, no. 7, pp. 3722–3736, Jul. 2015.

[81] S. Gherekhloo, A. Chaaban, and A. Sezgin, "Expanded GDoF-optimality regime of treating interference as noise in the $M \times 2$ X-channel," *IEEE Trans. Inform. Theory*, vol. 63, no. 1, pp. 355–376, 2017.

[82] S. Gherekhloo and A. Sezgin, "Optimality of treating interference as noise in the IRC: A GDoF perspective," in *Signals Syst. Comput., 2015 49th Asilomar Conf.*, Pacific Grove, CA, 2015, pp. 35–39.

[83] A. Chaaban and A. Sezgin, "On the generalized degrees of freedom of the Gaussian interference relay channel," *IEEE Trans. Inform. Theory*, vol. 58, no. 7, pp. 4432–4461, 2012.

[84] S. Gherekhloo, A. Chaaban, and A. Sezgin, "Cooperation for interference management: A GDoF perspective," *IEEE Trans. Inform. Theory*, vol. 62, no. 12, pp. 6986–7029, 2016.

[85] S. A. Jafar, "Topological interference management through index coding," *IEEE Trans. Inform. Theory*, vol. 60, no. 1, pp. 529–568, 2014.

[86] N. Naderializadeh and A. S. Avestimehr, "Interference networks with no CSIT: Impact of topology," *IEEE Trans. Inform. Theory*, vol. 61, no. 2, pp. 917–938, 2015.

[87] Y. Birk and T. Kol, "Informed-source coding-on-demand (ISCOD) over broadcast channels," in *INFOCOM'98. Seventeenth Ann. Joint Conf. IEEE Comput. Commun. Soc. Proc.*, vol. 3. Piscataway, NY: IEEE, 1998, pp. 1257–1264.

[88] H. Maleki, V. R. Cadambe, and S. A. Jafar, "Index coding: An interference alignment perspective," *IEEE Transactions on Information Theory*, vol. 60, no. 9, pp. 5402–5432, 2014.

[89] S. El Rouayheb, A. Sprintson, and C. Georghiades, "On the index coding problem and its relation to network coding and matroid theory," *IEEE Trans. Inform. Theory*, vol. 56, no. 7, pp. 3187–3195, 2010.

[90] M. Effros, S. El Rouayheb, and M. Langberg, "An equivalence between network coding and index coding," *IEEE Trans. Inform. Theory*, vol. 61, no. 5, pp. 2478–2487, 2015.

[91] A. S. Tehrani, A. G. Dimakis, and M. J. Neely, "Bipartite index coding," in *2012 Int. Symp. Inform. Theory Proceedings (ISIT)*, St Petersburg 2012, pp. 2246–2250.

[92] V. Ntranos, M. A. Maddah-Ali, and G. Caire, "Cellular interference alignment: Omni-directional antennas and asymmetric configurations," *IEEE Trans. Inform. Theory*, vol. 61, no. 12, pp. 6663–6679, 2015.

[93] Y. Gao, G. Wang, and S. A. Jafar, "Topological interference management for hexagonal cellular networks," *IEEE Trans. Wireless Commun.*, vol. 14, no. 5, pp. 2368–2376, 2015.

[94] V. Ntranos, M. A. Maddah-Ali, and G. Caire, "Cellular interference alignment," *IEEE Trans. Inform. Theory*, vol. 61, no. 3, pp. 1194–1217, 2015.

[95] N. Naderializadeh, A. El Gamal, and A. S. Avestimehr, "Topological interference management with just retransmission: What are the 'best' topologies?" in *2015 IEEE Int. Conf., Communications (ICC)*, London, 2015, pp. 4113–4119.

[96] H. Sun, C. Geng, and S. A. Jafar, "Topological interference management with alternating connectivity," in *2013 Int. Symp. Inform Theory*, Hong, Kong, 2013, pp. 399–403.

[97] S. Gherekhloo, A. Chaaban, and A. Sezgin, "Topological interference management with alternating connectivity: The Wyner-type three user interference channel," *arXiv:1310.2385*, 2013.

[98] S. Gherekhloo, A. Chaaban, and A. Sezgin, "Resolving entanglements in topological interference management with alternating connectivity," in *2014 IEEE Int. Symp. Inform. Theory (ISIT)*, Honolulu, HI, 2014, pp. 1772–1776.

[99] C. Geng, H. Sun, and S. A. Jafar, "Multilevel topological interference management," in *2013 IEEE Information Theory Workshop (ITW)*, Sevilla, 2013, pp. 1–5.

[100] X. Yi and D. Gesbert, "Topological interference management with transmitter cooperation," *IEEE Trans. Inform. Theory*, vol. 61, no. 11, pp. 6107–6130, 2015.

[101] H. Sun and S. A. Jafar, "Topological interference management with multiple antennas," in *2014 IEEE Int, Symp. Inform. Theory (ISIT)*, Honolulu, HI, 2014, pp. 1767–1771.

[102] P. Selvaprabhu, S. Chinnadurai, S. S. Song, and M. H. Lee, "Topological interference alignment for MIMO interference broadcast channel," in *2016 Int. Conf. Inform. Commun. Technol. Convergence (ICTC)*, 2016, pp. 1068–1072.

[103] H. Yang, N. Naderializadeh, A. S. Avestimehr, and J. Lee, "Topological interference management with reconfigurable antennas," in *2016 IEEE Int. Symp. Inform. Theory (ISIT)*, Bascelona 2016, pp. 555–559.

[104] I. Shomorony and S. Avestimehr, "Multihop wireless networks: A unified approach to relaying and interference management," *Found. Trends. Netw.*, vol. 8, no. 3, pp. 149–280, 2014.

[105] N. Lee, "Interference-free OFDM: Rethinking OFDM for interference networks with inter-symbol interference," *CoRR*, vol. abs/1609.02517, 2016. [Online]. Available: http://arxiv.org/abs/1609.02517.

[106] Y.-S. Jeon, N. Lee, and R. Tandon, "Degrees of freedom and achievable rate of wide-band multi-cell multiple access channels with no CSIT."

[107] T. M. Cover and J. A. Thomas, *Elements of Information Theory*. Chichester: Wiley, 2006.

[108] C. T. Li and A. Ozgur, "Channel diversity needed for vector space interference alignment," *IEEE Trans. Inform. Theory*, vol. 62, no. 4, pp. 1942–1956, 2016.

[109] R. Tandon and O. Simeone, "Harnessing cloud and edge synergies: Towards an information theory of fog radio access networks," *IEEE Commun. Maga.*, vol. 54, no. 8, pp. 44–50, 2016.

[110] N. Naderializadeh, M. A. Maddah-Ali, and A. S. Avestimehr, "Fundamental limits of cache-aided interference management," *IEEE Trans. Inform. Theory*, vol. 63, no. 5, pp. 3092–3107, 2017.

[111] J. Hachem, U. Niesen, and S. N. Diggavi, "A layered caching architecture for the interference channel." in *2016 IEEE Int. Symp. Inform. Theory*, Barcelona, 2016.

16 Cooperative Cellular Communications

Benjamin M. Zaidel, Michèle Wigger, and Shlomo Shamai (Shitz)

16.1 Introduction

One of the main technical challenges of future cellular systems is to improve the handling of coexisting communications beyond simple orthogonal access (in time, frequency, or other dimensions). In other words, improved interference mitigation techniques both for uplink and downlink are necessary in order to meet the requirements on data rates. To this end, 5G will include advanced cooperative communication techniques as one of its main evolutions. In particular, 5G will on one hand improve the existing cooperation mechanisms in 4G, e.g., by better exploiting the possibilities of device-to-device (D2D) communications between mobile users or improving the relaying strategies at the helping terminals such as pico or femto base stations (BSs). On the other hand, 5G will also include new infrastructural elements that create new opportunities for cooperative communications. Examples are clustered multipoint processing or cloud radio-access network (C-RAN) architectures, as discussed in Chapter 6, which shift encoding and decoding previously done at the BSs to central processors. Another example is cache storage memories distributed across the network that can be used to prefetch popular content in advance so as to reduce network load and delay at peak-traffic periods. Recent works, most prominently [1], have shown that when employing a smart prefetching strategy, the prefetched contents can enable opportunities for *multicast*, and thus for cooperative coding. See also Chapters 5 and 17 for more information on networks with cache memories.

The idea of using cooperative communication strategies for interference mitigation dates back a few decades in the information theoretic literature. We cite here the first results on the standard *relay channel* by van der Meulen [2] and by El Gamal and Cover [3], on the multiple-access channel (MAC) with cooperative encoders by Willems [4], and on the broadcast channel (BC) with cooperative decoders by Dabora and Servetto [5]. These works showed that cooperative communication can be used to:

- exchange transmitted signals at the transmitting ends, so as to mitigate interference when creating new transmit signals, and to exchange decoded data at the receiving ends, so as to mitigate interference before decoding additional data;
- communicate over alternative, more favorable paths; and

● exchange data at the transmitting ends to allow for *coordinated multipoint communication (CoMP)*, i.e., distributed multiantenna transmission, and to exchange observed signals at the receiving ends to allow for distributed multiantenna reception.

Even though the benefits of cooperation have been known for long in theory, in practice their implementation is not widely spread. Reasons for this include the increased signaling overhead and the increased encoding and decoding complexity of cooperative communication strategies compared to non-cooperative strategies, as well as additional synchronization requirements and additional delays caused by multihop communications. The points we raised are, however, less of a barrier for 5G than for previous standards, thanks to improved chip technologies and faster signal processing. Moreover, the need for higher data rates seems to be more urgent than when preparing previous standards due to desired streaming applications such as virtually augmented movies and interactive video games.

This chapter presents a detailed overview of the cooperative communication techniques in cellular systems and the benefits they attain over standard non-cooperative techniques from an information theoretic perspective. For ease of exposition, the focus is on *Wyner*-type cellular models [6, 7] where interference is only short-ranged. This means that signals sent by BSs (or mobile users) are only received at mobile users (or BSs) in the same or in adjacent cells. Wyner's model allows in principle that the cells be arranged in two dimensions, e.g., in a *hexagonal cell-array model*. However, this chapter mostly focuses on Wyner's *linear cell-array* model where cells are arranged on a line, e.g., along a railway line or a highway or in corridors inside a building. The linear cell-array model often allows for closed-form expressions of information theoretic quantities like *capacity*, *minimum energy per bit*, or *low- and high-signal-to-noise ratio (SNR) slopes*, thus providing a clean basis for our discussion on the benefits of cooperation.

It is worth noting that the analytical tractability of the aforementioned Wyner-type models comes at the expense of modeling limitations. More specifically, these models cannot account for user mobility or different types of BSs, and the precise locations of the various mobiles within the cells are only accounted for to a limited extent. Still, Wyner-type models can be accurate when users are densely distributed in cells, which is expected for future 5G networks, and when the focus of the analysis is on sum-rate [8]. We refer the reader to the summary and outlook section for ideas on how to obtain results on more refined models.

This chapter is organized as follows. At first, the Wyner-type model for both uplink and downlink is reviewed. Subsequently, system architectures for multicell processing are discussed, where BSs can either communicate to a central processor or they can fully cooperate in overlapping or non-overlapping clusters. Then, variations on the basic setting are discussed, where mobile users can cooperate directly or through relays, or when they have access to cache memories where information can be prefetched. The chapter is concluded with a short summary and outlook section. A more elaborate account of some of the topics discussed here can be found in [9], which has been used as a primary source of reference in the compilation of this chapter.

16.2   The Wyner-Type Model

16.2.1   Uplink Channel Model

In this section we describe the underlying principles of the Wyner-type linear cell-array model, starting with the uplink channel. The natural extension of the model to the more common two-dimensional hexagonal cell array is addressed in Section 16.5. The linear model assumes that the system comprises M cells arranged in a linear array; a setting suitable for describing, e.g., communications along highways, railways, or indoor corridors.

Wyner-type system models typically rely on three fundamental properties: finite-horizon intercell interference, full symmetry among users and cells, and full synchronism of the signals received at all BSs. The finite-horizon assumption accounts for the fact that signals arriving from distant cells typically decay significantly, and their received power lies below the background noise level. A common approximation is therefore to treat such signals as part of the additive white Gaussian noise (AWGN). Specifically, we will assume henceforth that intercell interference at any BS is limited to arrive from L_r cells to the right and L_ℓ cells to the left (BSs close to the edges of the linear array will experience less interference).

The symmetry property implies an equal number of active users K in each of the cells, and that all users experience the same channel characteristics toward their local BS, and toward neighboring BSs. Precisely, the channel characteristics may depend only on the *relative* distance between the local cell and the receiving BS. For simplicity, it is assumed here that both users and BSs employ single transmit and receive antennas. We note, however, that extensions of the model to account for multiple antennas or, equivalently, code-division multiple-access (CDMA), have also been considered in the literature, see, e.g., [10–12].

Finally, the assumed full synchronism dictates that the symbol epochs of all users in all cells coincide at *all* BSs at which the respective signals are received. Thus, any processing or propagation distance-related delays are totally ignored. Furthermore, in addition to being symbol-synchronous, frame synchronism is also assumed among all users; i.e., all users are assumed to start the transmission of their codewords simultaneously (see, e.g., [13, 14] for a discussion on the impact of asynchronism on the capacity of Gaussian MACs). While clearly unrealistic, this abstraction considerably simplifies the analysis, and allows us to focus on the more fundamental information theoretic aspects particular to cellular communications.

Summarizing the aforementioned properties, the received signal at BS m at discrete time $t \in \{1, \ldots, n\}$ can be expressed as (see the illustration in Figure 16.1)

$$y_m(t) = \sum_{l=-L_\ell}^{L_r} \mathbf{h}_{m,l}^\mathsf{T}(t)\mathbf{x}_{m+l}(t) + z_m(t), \quad m \in \{1, \ldots, M\}, \tag{16.1}$$

where $\mathbf{x}_m(t) \in \mathbb{C}^K$ is the $K \times 1$ vector of symbols transmitted by the users in cell m, $\mathbf{h}_{m,l}(t) \in \mathbb{C}^K$ is the vector of channel gains to cell m from users located l cells apart, and $z_m(t) \sim \mathcal{CN}(0, 1)$ denotes the proper complex AWGN at the receiver of BS m. All

$$L_\ell = L_r = 1, \ K = 3$$

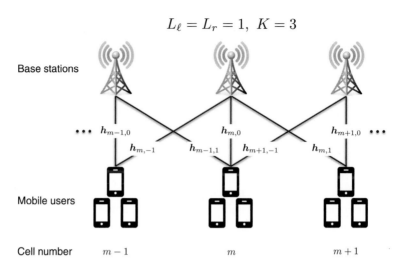

Figure 16.1 The Wyner-type linear cell-array with $L_\ell = L_r = 1$, and $K = 3$.

noise samples are assumed to be independent and identically distributed (i.i.d.) across time and space. We assume equal per-user *per-block* power constraints

$$\frac{1}{n} \sum_{t=1}^{n} \mathbb{E}\left\{ |[\mathbf{x}_m(t)]_k|^2 \right\} \leq \frac{P}{K}, \quad m \in \{1, \dots, M\}, \ k \in \{1, \dots, K\}, \tag{16.2}$$

where P denotes the respective *per-cell* total power constraint, and $[\mathbf{a}]_k$ denotes the kth component of the vector $\mathbf{a}$. *Full* channel state information (CSI) is assumed at the receiving ends, while the users are assumed to be only informed about the transmission strategy; i.e., the choice of codebooks, transmission rates, and transmit powers.

Omitting the time index for simplicity, the channel model (16.1) can also be more compactly rewritten in a vector-matrix form as

$$\mathbf{y} = \mathsf{H}_M \mathbf{x} + \mathbf{z}, \tag{16.3}$$

where

$$\mathbf{y} = [y_1, \dots, y_M]^\mathsf{T},$$
$$\mathbf{x} = [\mathbf{x}_1^\mathsf{T}, \dots, \mathbf{x}_M^\mathsf{T}]^\mathsf{T},$$
$$\mathbf{z} = [z_1, \dots, z_M]^\mathsf{T},$$

and H_M is the $M \times MK$ matrix whose mth row is given by

$$\left[\mathbf{h}_{m,-(m-1)}^\mathsf{T}, \mathbf{h}_{m,-(m-2)}^\mathsf{T}, \dots, \mathbf{h}_{m,0}^\mathsf{T}, \mathbf{h}_{m,1}^\mathsf{T}, \dots, \mathbf{h}_{m,(M-m)}^\mathsf{T} \right],$$

with $\mathbf{h}_{m,k}^\mathsf{T} = 0$ whenever $k \notin [-L_\ell, L_r]$. This immediately implies that H_M is a finite-band matrix, and takes on a *block-Toeplitz* form in the absence of fading, where the channel coefficients are set as deterministic constants. This model is reminiscent of the model for a time-varying single-user channel with *intersymbol interference* (ISI). In fact, techniques used to analyze channels with ISI [15] can also be used to analyze Wyner-type models, especially in the limiting regime where the number of cells $M \to \infty$.

Considering the limiting regime $M \to \infty$ allows us to obtain a more insightful and analytically tractable characterization of information theoretic performance measures of interest – in particular, because it allows us to avoid the inevitable edge effect of the linear cell-array model. In this limiting regime, all cells essentially experience (statistically) the same intercell interference pattern, and edge cells have a vanishing impact on system-level performance (see, e.g., [6, 16]). An alternative approach to avoid edge effects is to consider a slightly modified system model where the cells are arranged on a *circle* (see, e.g., [7, 11, 17]). The circular model has the advantage that full symmetry among all cells is achieved already for $M \geq 3$, and it lends itself more easily to full analytical treatment for $M < \infty$ in certain regimes.

Special Cases

Several special cases of the general Wyner-type model (16.1) are often considered in the literature. The first and foremost is the *Gaussian Wyner model*, which assumes a symmetric intercell interference horizon, where $L_r = L_\ell = L$, and fixed (nonfading) channel gains. Furthermore, the channel gains are assumed to be *cell-homogeneous* [9], where

$$\mathbf{h}_{m,k}(t) = \alpha_k \mathbf{1}_K \, , \ m \in \{1, \ldots, M\} \, , \ -L \leq k \leq L \, , \ t \in \{1, \ldots, n\} \, , \tag{16.4}$$

and *symmetric* with $\alpha_k = \alpha_{-k} \in [0, 1]$, $\alpha_0 = 1$ (the quantities $\{\alpha_k\}_{k=-L,\ldots,L}$ are taken here as deterministic constants). $\mathbf{1}_K$ denotes here the K-dimensional column vector of 1s. Such assumptions may be interpreted as representing a setup where the users are concentrated close to the cell centers. The setting originally suggested by Wyner [6] corresponds to $L = 1$, which by denoting $\alpha_1 \triangleq \alpha$ yields

$$y_m(t) = \mathbf{1}_K^\mathsf{T} \mathbf{x}_m(t) + \alpha \mathbf{1}_K^\mathsf{T} \mathbf{x}_{m+1}(t) + \alpha \mathbf{1}_K^\mathsf{T} \mathbf{x}_{m-1}(t) + z_m(t) \, , \quad m \in \{1, \ldots, M\} \, . \tag{16.5}$$

This system model is illustrated in Figure 16.2.

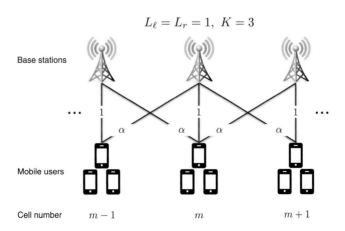

$$L_\ell = L_r = 1, \ K = 3$$

Base stations

Mobile users

Cell number $m - 1$ m $m + 1$

Figure 16.2 The Gaussian Wyner linear cell-array model with $L_\ell = L_r = 1$ and $K = 3$.

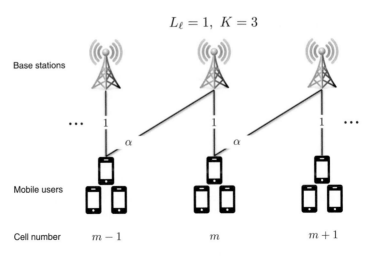

$$L_\ell = 1, \quad K = 3$$

Figure 16.3 The Gaussian soft-handoff model with $L_\ell = 1$, $L_r = 0$ and $K = 3$.

The *Gaussian soft-handoff model* (Figure 16.3) is a similar model, whose main difference lies in the fact that intercell interference is only assumed to arrive from *one side* of the linear array, say, from the left. We thus set $L_\ell = L$, $L_r = 0$, and $\mathbf{h}_{m,k}(t) = \alpha_k \mathbf{1}_K$ for $-L \leq k \leq 0$, $m \in \{1, \dots, M\}$, with $\alpha_0 = 1$. For the particular case of $L = 1$, as considered, e.g., in [17, 18], one gets

$$y_m(t) = \mathbf{1}_K^\mathsf{T} \mathbf{x}_m(t) + \alpha \mathbf{1}_K^\mathsf{T} \mathbf{x}_{m-1}(t) + z_m(t), \quad m \in \{1, \dots, M\}, \tag{16.6}$$

where again we denote $\alpha_1 \triangleq \alpha$. This model is illustrated in Figure 16.3, and for $L = 1$ may be interpreted as representing a "soft-handoff" scenario, where the users are located close to the cell edges, and their signals are received at the two nearest BSs (hence its name).

Both aforementioned models can be modified in a natural manner to account for flat fading. The *fading Wyner model* corresponds to $\mathbf{h}_{m,k}(t) = \alpha_k \tilde{\mathbf{h}}_{m,k}(t)$, $m \in \{1, \dots, M\}$, $-L \leq k \leq L$, $t \in \{1, \dots, n\}$, where $\tilde{\mathbf{h}}_{m,k}(t) \in \mathbb{C}^K$ are random vectors with i.i.d. entries, independent over m and k, and distributed according to some joint distribution π_k. The marginals of π_k are assumed to have zero mean and satisfy $\mathbb{E}\left\{ |[\tilde{\mathbf{h}}_{m,k}(t)]_q|^2 \right\} = 1$, $q \in \{1, \dots, K\}$. Similar to the Gaussian model, statistical symmetry is assumed where $\pi_k = \pi_{-k}$ and $\alpha_k = \alpha_{-k}$. We restrict the discussion to stationary and ergodic fading processes [16, 19],[1] and further particularize to Rayleigh fading. The entries of $\tilde{\mathbf{h}}_{m,k}(t)$ are thus assumed to be zero-mean proper complex Gaussian random variables with unit variance. The *fading soft-handoff model* (see, e.g., [17, 20–22]) is defined in an analogous manner to the corresponding Gaussian model, but with $\mathbf{h}_{m,k}(t) = \alpha_k \tilde{\mathbf{h}}_{m,k}(t)$ for $-L \leq k \leq 0$.

---

[1] A quasi-static fading model, where the channel gains are randomly chosen but are otherwise constant during the transmission of a codeword, can also be easily accommodated into this setting.

16.2.2   Downlink Channel Model

The downlink Wyner-type model is the dual of the uplink model. Focusing on the compact vector-matrix form (cf. (16.3)), the received signals in the downlink can be expressed as[2]

$$\mathbf{y} = \mathsf{H}_M^\dagger \mathbf{x} + \mathbf{z} \,, \tag{16.7}$$

where

$$\mathbf{y} = [\mathbf{y}_1^\mathsf{T}, \dots, \mathbf{y}_M^\mathsf{T}]^\mathsf{T},$$

$\mathbf{y}_m$, $m \in \{1, \dots, M\}$, is the $K \times 1$ vector of signals received by the users in cell m, $\mathbf{x}$ is the $M \times 1$ vector of signals transmitted by the M BSs, the matrix H_M is of the same dimensions as in (16.3), and $\mathbf{z}$ is the $MK \times 1$ vector of zero-mean complex AWGNs at the receivers of all users in all cells. The additive noises are assumed to be i.i.d. with unit variances; i.e., $\mathbf{z} \sim \mathcal{CN}(\mathbf{0}, \mathsf{I}_{KM})$. A *per-BS* (equivalently, per-cell) *per-block* power constraint will be assumed henceforth:

$$\frac{1}{n} \sum_{t=1}^{n} \mathbb{E}\left\{ |[\mathbf{x}(t)]_m|^2 \right\} \leq P \,, \quad m \in \{1, \dots, M\} \,. \tag{16.8}$$

Full CSI is assumed at the transmitting ends, encompassing here the BSs as well as any joint cooperative preprocessing system entity. The users are assumed, in contrast, to be only cognizant of their own channels, and of the transmission strategy employed in the downlink (see, e.g., [17, 23]).

Special Cases

Completely analogous special cases to the ones specified for the uplink channel are commonly considered for the downlink. Specifically, particularizing to $L = 1$, the signal received by user k in cell m according to the *downlink* Gaussian Wyner model reads [23] (cf. (16.5))

$$[\mathbf{y}_m(t)]_k = x_m(t) + \alpha x_{m+1}(t) + \alpha x_{m-1}(t) + [\mathbf{z}_m(t)]_k \,, \ m \in \{1, \dots, M\}, \ k \in \{1, \dots, K\}. \tag{16.9}$$

Similarly, we get for the corresponding Gaussian soft-handoff model

$$[\mathbf{y}_m(t)]_k = x_m(t) + \alpha x_{m+1}(t) + [\mathbf{z}_m(t)]_k \,, \ m \in \{1, \dots, M\}, \ k \in \{1, \dots, K\}. \tag{16.10}$$

The respective flat-fading models are defined again analogously to the corresponding uplink models [17].

16.2.3   System Architectures for Multicell Processing

To facilitate cooperative multicell processing, the system architecture must be appropriately designed. In particular, backhaul links between the BSs and system entities that implement cooperative processing must be established, with large enough bandwidth

---

[2] We reuse here some of the symbols defined for the uplink model whenever their role is clear from the context, as in '[9].

to utilize the benefits of cooperation. Inter-BS links are also a viable tool for cooperative communications, allowing for information sharing between neighboring BSs. In this chapter we mainly focus on two typical architectural models.

Central Processor with Finite-Capacity Backhaul

In this C-RAN setting *all* BSs are connected to a central processor (CP) via links of (generally) *finite*-capacity C (bit/s/Hz). In the uplink, the CP is responsible for performing joint decoding of the messages transmitted by *all users in all cells* based on the signals received at all BSs. In the downlink, the CP performs joint encoding and preprocessing of the signals transmitted by all BSs to all users [24, 25]. The limiting case where $C \to \infty$ corresponds to full (ideal) joint multicell processing (MCP) [6, 16, 17, 23]. The uplink can thus be viewed as a distributed (either Gaussian or fading) multiple-input multiple-output (MIMO) MAC, and the well-known information theoretic results for the capacity region of the MIMO MAC readily apply. Analogously, the downlink can be viewed as a distributed MIMO BC, for which dirty-paper coding (DPC) is an optimal transmission strategy [26]. This chapter presents the most basic results on C-RAN architectures with Wyner-type models (see Section 16.4). Further details can be found in Chapter 6.

We note here that another potential form of cooperative communications relies on *local* connectivity between neighboring BSs via links of finite capacity C (bit/s/Hz). Such connectivity can be modeled as either bi- or unidirectional [27, 28]. The setting may be regarded as corresponding to certain practical scenarios, as already facilitated, e.g., by the inter-BS X2-links in LTE-advanced 4G cellular systems [29]. The type of processing employed at the BSs depends on the nature of the information shared among them. The case where $C \to \infty$ naturally coincides with ideal full joint MCP. This system architecture can be combined with the C-RAN architecture described in the previous paragraph – see, e.g., [30, 31]. Local connectivity for BSs will not be further detailed in this chapter. (Chapter 4 analyzes this scenario in depth.) However, some results can be deduced from the dual setup with local connectivity between mobile users discussed in Section 16.6.

Clustered MCP

Another practically inclined line of models assumes that cooperative processing is restricted to a *cluster* (or subset) of cells. More specifically, a cluster processor is assigned to handle the encoding and decoding functions of messages associated with users operating within the cluster. Clusters may overlap in general as, e.g., in [32, 33], although the simpler case of non-overlapping clusters is also of clear practical interest (see, e.g., [34, 35]). Note that with clustered MCP out-of-cluster interference becomes a *crucial factor*. This puts the setting within the difficult framework of the *MIMO interference channel*, the capacity region of which is still generally unknown (see the discussion in [36, 37]). For simplicity, we focus here on unlimited backhaul connectivity between the cluster's BSs and the cluster processor, while noting that restricted settings with limited connectivity may apply as well [38].

16.2.4   Variations on the Basic Setting

The basic setting has been extended to incorporate more advanced architectures or to better account for practical constraints. Advanced architectures include, for example, cooperation links between mobile users [28, 39–42], mobile users acting as helper relays for each other [40, 43], dedicated relay terminals [44], or cache memories at user terminals [45]. Further practical constraints that have been addressed in the literature are unreliability of backhaul links [46, 47] or dynamic user activity [48, 49].

User Cooperation

Orthogonal noninterfering communication links of given capacities, such as Bluetooth or microwave links, enable mobile users to cooperate. For example, for the downlink, mobile users can exploit these links to exchange quantized versions of their received signals or parts of their decoded messages so as to enable joint decoding or successive interference cancellation, and in the uplink, they can exploit them to exchange parts of messages or quantized versions of transmit signals so as to enable joint beamforming or DPC. Notice that when the capacity of the cooperation links is sufficiently large, the scenario coincides with a scenario where all mobile users can fully cooperate in their transmission or reception.

Orthogonal cooperation links between users were introduced in a multi-access scenario by Willems [4], and subsequently studied in [50–54]. This type of cooperation is often also referred to as *conferencing*.

Relaying

Relaying refers to a scenario where helper terminals *overhear* signals transmitted by mobile users, i.e., they observe a noisy version of these transmit signals, and use these observations to assist the communication from these users to the BS. The most popular relaying strategies are decode-and-forward (DF), compress-and-forward (CF) [3], and amplify-and-forward (AF) [55]. Helper terminals can either be other mobile users in the network or dedicated relay terminals.

The advantage of relaying compared to conferencing is that it does not require additional communication resources (e.g., bandwidth). Its disadvantage is that its performance can be reduced, e.g., because CF and AF protocols enhance the noise level at the receiving BSs and DF imposes additional decoding constraints.

Caching Protocols

In future 5G networks, BSs or user terminals are equipped with dedicated storage memories where they can prefetch popular content so as to reduce delay and network traffic in peak-traffic periods. The main challenge is that the content has to be cached before knowing which files the users will request in the peak-traffic period. A conventional approach is to store the same popular contents in the cache memories of the users. This allows the receivers to locally retrieve the contents without burdening the network. However, further caching gains, so-called *global caching gains*, are possible if different contents are stored at different users [1]. Specifically, a careful design of the cache contents creates coding opportunities to simultaneously serve multiple users during the peak-traffic periods. Subsequent related works [56–63] extended this idea to *noisy* broadcast channels, for which they also identified further possible caching gains. Similar ideas for fully connected interference networks were studied in [64–66].

16.3   Optimum Full Joint Multicell Processing

To better assess the impact of cooperative centralized processing in cellular systems, it is important first to consider, as a reference, the ultimate performance limits. Such metrics correspond to an idealized setting where all BSs are connected to the CP via backhaul links of unlimited capacity ($C \to \infty$). The uplink can thus be viewed as a distributed MIMO MAC, while the downlink corresponds to a distributed MIMO BC. For the sake of simplicity, and for ease of comparison, we assess performance in terms of the average total achievable per-cell sum-rate (throughput) in bit/s/Hz. We will also consider the large number of cells limit ($M \to \infty$), where the underlying symmetry property of Wyner-type models dictates that all cells are (essentially) equivalent,[3] and that the maximum per-cell sum-rate is achievable with equal rate allocation for all users [6]. Performance in Gaussian channels will be discussed first, and a discussion on the impact of flat-fading channels will then follow.

16.3.1   Gaussian Channels

Uplink

The per-cell uplink sum-rate capacity with full joint MCP was derived by Wyner in [6]. Referring to (16.3), and considering the MIMO MAC interpretation while recalling that the channel transfer matrix H_M is fixed and deterministic in the Gaussian regime, the per-cell uplink sum-capacity is given by the normalized input–output mutual information:

$$
\begin{aligned}
R^M_{\mathrm{ul,MCP}}(P) &= \frac{1}{M} I\{\mathbf{x}; \mathbf{y}\} \\
&\overset{(a)}{=} \frac{1}{M} \log_2 \det\left(I_M + \frac{P}{K} H_M H_M^\dagger \right) \\
&= \frac{1}{M} \sum_{m=1}^{M} \log_2\left(1 + \frac{P}{K}\lambda_m \left(H_M H_M^\dagger\right) \right) \\
&= \frac{1}{M} \int_0^\infty \log_2\left(1 + \frac{P}{K} x \right) \, dF^M_{HH^\dagger}(x),
\end{aligned}
\tag{16.11}
$$

where (a) follows from the optimality of Gaussian codebooks in this setting (see, e.g., [6]) and the underlying input constraints, $\left\{\lambda_m \left(H_M H_M^\dagger\right)\right\}_{m=1,...,M}$ are the eigenvalues of $H_M H_M^\dagger$, and

---

[3] Recall that any potential edge effects have a vanishing impact as $M \to \infty$.

$$F_{\mathsf{HH}^\dagger}^M(x) \triangleq \frac{1}{M} \sum_{m=1}^M \mathbb{1}\left(\lambda_m\left(\mathsf{H}_M\mathsf{H}_M^\dagger\right) \le x\right) \tag{16.12}$$

denotes the empirical eigenvalue distribution of $\mathsf{H}_M\mathsf{H}_M^\dagger$.

Note that (16.11) is obtained while all users in all cells transmit simultaneously in the same frequency band, which corresponds to a frequency reuse factor of 1 (see the discussion in the following). As for intracell scheduling, the sum-capacity (16.11) can be attained by employing a "CDMA-like" scheme, where all intracell users transmit simultaneously with an equal power P/K for the whole transmission block, while occupying the whole available bandwidth. This scheme shall be referred to henceforth as the "wideband" (WB) transmission scheme following [6]. Alternatively, the sum-capacity can also be achieved with intracell time-division multiple-access (TDMA) scheduling, where the users transmit for a fraction $1/K$ of the time with power P (the *total* available power for each cell). We note here that the optimality of intracell TDMA stems from the relaxed per-*block* power constraint, and the fact that all intracell users experience *identical* channel gains; otherwise, this conclusion might no longer hold [9] (see also references therein).

Considering the Gaussian Wyner model (16.4), and assuming intracell TDMA scheduling (without loss of optimality), we let $K = 1$ and observe that the channel transfer matrix takes a Toeplitz form with the first column given by $[1, \alpha_1, \alpha_2, \ldots, \alpha_L, \mathbf{0}_{M-L-1}]^\mathsf{T}$. Thus, in the large number of cells limit ($M \to \infty$) the per-cell sum-capacity (16.11) boils down to a simple integral expression following Szegö's theorem [67]:

$$R_{\mathrm{ul,MCP}}(P) = \int_0^1 \log_2\left(1 + P\left(1 + 2\sum_{k=1}^L \alpha_k \cos(2\pi k\theta)\right)^2\right) d\theta . \tag{16.13}$$

The same result can in fact be obtained by interpreting the channel dictated by (16.4) as equivalent to a *time-invariant* ISI channel with finite memory, as considered, e.g., in [15]. Here, the signals received at the different BSs are taken as the time samples of the channel output, while the signals transmitted by the different users are interpreted as the channel inputs at discrete times. A similar analysis for the MIMO setting can be found in [68]. The Gaussian soft-handoff model can be treated in a completely analogous manner. In particular, for $L = 1$ the corresponding integral expression for the per-cell sum-capacity takes a closed explicit form [9, 17, 18]:

$$R_{\mathrm{ul,MCP}}(P) = \log_2\left(\frac{1 + (1 + \alpha^2)P + \sqrt{1 + 2(1 + \alpha^2)P + (1 - \alpha^2)^2 P^2}}{2}\right) . \tag{16.14}$$

It is useful to compare the per-cell sum-capacity of optimum joint MCP to the corresponding achievable throughput with more traditional approaches for cellular systems. A common scheme for spectral resource allocation among the cells is *spatial frequency reuse*. Accordingly, adhering to either the linear or circular cell-array model, two BSs utilize the same set of frequency bands at the same time only if they are separated $F \ge 1$ cells apart, where F denotes the *frequency reuse factor*. Transmissions in each cell thus occupy $1/F$ of the available spectrum, and the respective spectral density can be proportionally scaled to retain the overall average power constraint. Note that spatial frequency reuse is equivalent in terms of the achievable throughput to an *intercell time-sharing* (ICTS) protocol [19] subjected to the per-block power constraint (16.2).

Focusing on the Gaussian Wyner model, then standard *single-cell processing* (SCP) corresponds to local decoding of the transmitted *intracell* messages, while treating all out-of-cell interference as noise. This leads to the following total achievable per-cell sum-rate [9]:

$$R_{\text{ul,SCP}}(P,F) = \begin{cases} \frac{1}{F}\log_2\left(1 + \frac{FP}{1+2FP\sum_{k=1}^{\lfloor L/F\rfloor}\alpha_{kF}^2}\right) & \text{if } F \le L, \\ \frac{1}{F}\log_2(1 + FP) & \text{if } F > L. \end{cases} \tag{16.15}$$

Again, the sum-rate (16.15) can be achieved either with the WB scheme, or with intracell TDMA scheduling, as was the case for optimum joint MCP.

Note that when the frequency reuse factor is not large enough ($F \le L$), each BS still experiences out-of-cell interference with average power $2FP\sum_{k=1}^{\lfloor L/F\rfloor}\alpha_{kF}^2$, and the system becomes *interference limited*. In such a case, SCP performance can be enhanced if the receivers at the BSs are also informed of the codebooks employed by the interfering users in adjacent cells. This setup allows for joint decoding of both useful and interfering signals, which becomes beneficial when the interference is strong enough (see, e.g., [19, 69, 70]). As a straightforward example, we consider here the particular case of $F = L = 1$, and the simple *adjacent cell decoder* (ACD) suggested in [19]. The latter either fully decodes the interference, or simply treats interference as noise, whichever is beneficial in terms of the achievable throughput. This yields the following per-cell sum-rate [19]:

$$R_{\text{ul,ACD}}(P) = \max\left\{\log_2\left(1 + \frac{P}{1+2\alpha^2 P}\right),\right.$$
$$\left.\min\left\{\tfrac{1}{3}\log_2\left(1 + (1+2\alpha^2)P\right), \tfrac{1}{2}\log_2\left(1 + 2\alpha^2 P\right)\right\}\right\}. \tag{16.16}$$

Numerical results for the Gaussian Wyner model with $F = L = 1$ and $\alpha = \frac{1}{2}$ are shown in Figure 16.4. The figure shows the per-cell spectral efficiency in bit/s/Hz of SCP with interference treated as noise, the ACD, ICTS (equivalent to taking $F = 2$ in (16.15)), and optimum joint MCP. The results are plotted as a function of the system average *transmit* $\frac{E_b}{N_0}$, obtained via the relation $R_{\{\cdot\}}(P) = R_{\{\cdot\}}\left(\frac{E_b}{N_0}\right)$, where $P = R_{\{\cdot\}}\left(\frac{E_b}{N_0}\right)\cdot\frac{E_b}{N_0}$. Here $R_{\{\cdot\}}(P)$ stands for the per-cell sum-rate of the corresponding scheme. The spectral efficiency of the single-user AWGN channel (with power constraint P) is also shown for reference. The results clearly demonstrate that optimum joint MCP leads to a dramatic performance enhancement. In particular, note that it achieves the *optimum unit high-SNR slope*. The high-SNR slope S_∞ is also known as *multiplexing gain* or *degrees of freedom* (DoF), and it describes the logarithmic growth of the sum-rate R_{sum} of a system in the high-SNR regime. For large input powers ($P \gg 1$) the cell's sum-rate grows as $R_{\text{sum}} \approx S_\infty \cdot \log_2(P)$. The unit high-SNR slope achieved

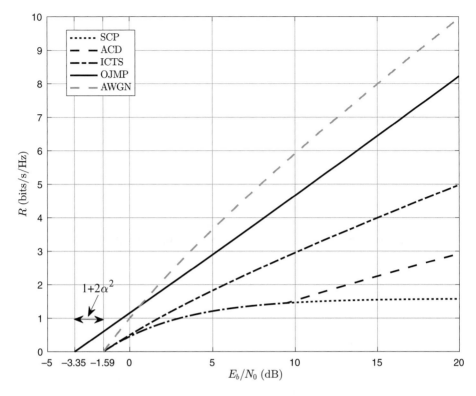

Figure 16.4 Uplink spectral efficiency: SCP vs. MCP ($M \to \infty$, $L = 1$, $\alpha = \frac{1}{2}$).

by MCP corresponds thus to the same rate as over a single user AWGN channel with the total available cell-power P, as if no out-of-cell interference was present. In contrast, naive SCP exhibits interference-limited behavior, the ACD achieves the high-SNR slope $\mathcal{S}_{\infty,\text{ACD}} = \frac{1}{3}$, and ICTS achieves the high-SNR slope $\mathcal{S}_{\infty,\text{ICTS}} = \frac{1}{F} = \frac{1}{2}$. The latter observation emphasizes the severe penalty incurred by forced orthogonalization among the cells, as dictated by the frequency reuse scheme (equivalently, ICTS).

Note also that optimum joint MCP decreases the minimum transmit $\frac{E_b}{N_0}$ that enables reliable communication, which for the Gaussian Wyner model boils down to [9]

$$\left(\frac{E_b}{N_0}\right)_{\min} = \frac{\log_e 2}{1 + 2\sum_{k=1}^{L}\alpha_k^2},$$

(16.17)

where the $1 + 2\sum_{k=1}^{L}\alpha_k^2$ factor designates the full *array gain* of the system. The presence of out-of-cell interference does, however, affect the low-SNR slope, $\mathcal{S}_0 \triangleq -\log_e 2 \frac{2(\dot{R}(0))^2}{\ddot{R}(0)}$, which for $L = 1$ reads (recall that $\mathcal{S}_0 = 2$ for the single user AWGN channel)

$$\mathcal{S}_0 = \frac{2(1 + 2\alpha^2)^2}{1 + 12\alpha^2 + 6\alpha^4}.$$

(16.18)

Analogous expressions can be straightforwardly obtained for the Gaussian soft-handoff model [9].

Downlink

The per-cell downlink sum-capacity of optimum joint MCP for the Gaussian Wyner model is derived by observing that the system can be interpreted as a distributed MIMO BC with *individual per antenna* power constraints. It is well known that DPC is the capacity-achieving strategy for this channel [26]; however, direct computation of the sum-capacity is nontrivial due to the nonconvex nature of the underlying optimization problem. This difficulty has been circumvented in [17], where it was shown that the per-cell sum-capacity can be expressed as the solution to the following min–max optimization problem:

$$R_{\text{dl,MCP}}^{M}(P) = \frac{1}{M} \min_{\Lambda_M} \max_{\mathsf{D}_M} \log_2 \left(\frac{\det \left(\Lambda_M + P \, \mathsf{H}_M \mathsf{D}_{MK} \mathsf{H}_M^\dagger \right)}{\det \Lambda_M} \right), \qquad (16.19)$$

where Λ_M and D_{MK} are $M \times M$ and $MK \times MK$ diagonal matrices satisfying $\text{Tr}(\Lambda_M) \leq 1$ and $\text{Tr}(\mathsf{D}_{MK}) \leq 1$, respectively. The result relies on the *Lagrangian uplink–downlink duality principle* derived in [71, 72], which addresses, in particular, the individual per-BS power constraints typical to cellular systems.[4] Furthermore, it was shown in [17] that for the Gaussian Wyner and soft-handoff models the per-cell sum-capacity coincides with the corresponding uplink sum-capacity ((16.13) and (16.14)) as $M \to \infty$. Both intracell TDMA scheduling and the WB scheme are optimal in the latter setting. The above conclusions continue to hold when $M < \infty$ for the circular array model. We finally note that the achievable per-cell downlink sum-rate with SCP coincides with (16.15), and thus the main observations made with respect to the uplink continue to hold for the downlink as well.

16.3.2   Ergodic Flat-Fading Channels

Uplink

For flat-fading ergodic channels, the underlying assumption on the available CSI at both transmitting and receiving ends implies that the uplink per-cell sum-capacity is given by

$$R_{\text{ul,MCP}}^{\text{erg},M}(P) = \frac{1}{M} I\{\mathbf{x}; \mathbf{y}|\mathsf{H}_M\} = \mathbb{E}\{R_{\text{ul,MCP}}^{M}(P)\} = \mathbb{E}\left\{ \frac{1}{M} \log_2 \det \left(\mathsf{I}_M + \frac{P}{K} \mathsf{H}_M \mathsf{H}_M^\dagger \right) \right\},$$

$$(16.20)$$

where $R_{\text{ul,MCP}}^{M}(P)$ is given in (16.11), and the expectation is taken with respect to the distribution of the channel matrix H_M. Standard information theoretic arguments lead to the conclusion that $R_{\text{ul,MCP}}^{\text{erg},M}(P)$ is achievable by the WB scheme (combined with optimum joint MCP). However, in contrast to the Gaussian setting, intracell TDMA is generally suboptimal in the presence of fading. Focusing on the large number of cells limit, $M \to$

---

[4] We note here that the idea of employing DPC in the cellular downlink was first suggested in [23]. However, the scheme proposed therein relied on the generally suboptimal zero-forcing dirty paper (ZF-DP) strategy [73], combined with a *total sum-power* constraint over all BSs.

∞, [16, lemma 4] states that the maximum achievable per-cell sum-rate with intracell TDMA is upper bounded by $R_{\text{ul,MCP}}^{\text{erg},M}(P)$ with equality achieved when $K = 1$ (i.e., when only a single user is active in each cell). Intuitively, this result can be explained by the notion of *multiuser diversity*. When many users are active in each cell, with high probability at least some of the users are likely to experience good channel conditions leading to high throughput.

Unfortunately, calculating (16.20) analytically turns out to be a formidable task even in the large number of cells limit. The main difficulty stems from the fact that classical random matrix theory results often used in information theoretic analyses do not apply to Wyner-type channel models (see, e.g., [74, 75]). Note in particular that the matrix $H_M H_M^\dagger$ is a *finite* (narrow) band matrix, where the number of random entries only grows linearly with the matrix dimensions (in contrast, say, to Wishart matrices where the number of random variables grows quadratically with the dimensions, leading to the famous Marčenko-Pastur limiting empirical eigenvalue distribution [74]).

A limiting result for (16.20) as $M \to \infty$ has been derived in [21] for the soft-handoff model, and expressed in terms of the *top Lyapunov exponent* of the product of a sequence of certain random matrices via a version of the Thouless formula.[5] The result relies on the study of random Schrödinger operators, and proves the convergence of the average per-cell sum-capacity as $M \to \infty$ (see [21] for the details). That said, this limit result does not lend itself to practical computation, and it is therefore convenient to resort to lower and upper bounds on the sum-capacity.

A notable exception is the soft-handoff model with $L = 1$, $\alpha = 1$, and $K = 1$ (corresponding to intracell TDMA). For this setting a closed form *analytical* expression has been derived in [17] for the average per-cell sum-rate based on a result by Narula [76] (referring to the two-tap time-varying ISI channel). Accordingly, as $M \to \infty$, the average per-cell sum-rate with intracell TDMA converges to

$$R_{\text{ul,MCP}}^{\text{erg}}(P)\Big|_{K=1} = \int_1^\infty \log_2 x \, \frac{\log_e(x) \, e^{-x/P}}{\text{Ei}\left(\frac{1}{P}\right) P} \, dx \,, \tag{16.21}$$

where $\text{Ei}(x) = \int_1^\infty e^{-xt}/t \, dt$ is the exponential integral function.

Analytical upper and lower bounds on the average achievable per-cell sum-rate with intracell TDMA for the fading Wyner model with $L = 1$ were derived in [16], based on the limiting moments of the eigenvalue distribution of $H_M H_M^\dagger$. More generally, it can be shown via information theoretic arguments (analogous to the proof of [21, proposition IV.3]) that for the general fading Wyner-type model, the limiting sum-capacity is lower and upper bounded as [9]

$$\frac{M}{M + L_\ell + L_r} \mathbb{E}\left\{ \tilde{R}_{\text{ul,MCP}}^M \left(\frac{M + L_\ell + L_r}{M} P \right) \right\} \leq R_{\text{ul,MCP}}^{\text{erg}} \leq \mathbb{E}\left\{ \tilde{R}_{\text{ul,MCP}}^M (P) \right\} \,. \tag{16.22}$$

---

[5] See also [20] for a study of the non-ergodic setting for which results of the type of the central limit theorem and large deviations were obtained. In particular, $R_{\text{ul,MCP}}^M(P)$ was shown therein to converge almost surely to a deterministic limit as $M \to \infty$ (the same type of convergence is, in fact, also established in [21]).

Here, $\tilde{R}^M_{\text{ul,MCP}}(P)$ is the instantaneous sum-rate expression for a slightly modified model with L_ℓ and L_r empty cells added on the right and left boundaries of the system, respectively.

An insightful upper bound on the sum-capacity in the presence of fading can be obtained by considering the limit of large number of users per cell $K \to \infty$. The limit is taken here while fixing the *total per-cell* transmit power P. The key observation is that, by the underlying assumptions on the fading processes, the entries of the matrix $\frac{1}{K}H_M H_M^\dagger$ converge (a.s.) as $K \to \infty$ to their mean values (by the strong law of large numbers). The matrix $\frac{1}{K}H_M H_M^\dagger$ thus approaches a *deterministic* Toeplitz matrix. Again, Szegö's theorem [67] can be employed to conclude that for the linear Wyner model the per-cell sum-capacity (16.20) converges as $M \to \infty$ in the presence of Rayleigh fading to [16]

$$R^{\text{erg}}_{\text{ul,MCP}}(P) = \log_2\left(1 + P\left(1 + 2\sum_{k=1}^{L}\alpha_k^2\right)\right). \qquad (16.23)$$

This result clearly indicates that when the number of users in each cell is large enough, MCP not only *fully mitigates* intercell interference, but also achieves the *full array gain* of the system: $1 + 2\sum_{k=1}^{L}\alpha_k^2$. Comparing (16.23) and (16.13), it is concluded that fading turns out to be *beneficial* in this regime. The phenomenon can be attributed again to the notion of multiuser diversity.

Note also that the large K sum-capacity limit constitutes an upper bound on the sum-capacity for any finite K, which follows by applying Jensen's inequality in (16.20). The reader must be cautioned, however, *not* to conclude that fading is beneficial in general for any finite K. To demonstrate this, consider the case of $K = 1$ and $\alpha_k = 0$ $\forall k$. Clearly, the per-cell sum-capacity corresponds here to the standard point-to-point channel for which it is well known that fading cannot increase capacity (see [16, 17] for a more elaborate discussion). That said, numerical results in [16, 17] based on Monte Carlo simulations indicate that the upper bound (16.23) becomes quite tight already for moderate values of K.

The limiting sum-capacity (16.23) corresponds to a unit high-SNR slope, $\mathcal{S}^{\text{erg}}_{\infty,\text{ul,MCP}} = 1$, as was the case in the absence of fading. This observation holds, in fact, also for $K < \infty$ as shown in [21]. Turning to the low-SNR regime, it can be shown that the minimum transmit $\frac{E_b}{N_0}$ that enables reliable communication is identical to (16.17), implying that fading does not incur any loss in this respect. The low-SNR slope reads [9, 12]:

$$\mathcal{S}^{\text{erg}}_{0,\text{ul,MCP}} = \frac{2K}{1 + K} \quad \forall K \geq 1. \qquad (16.24)$$

Particularizing to $L = 1$, and comparing this expression to (16.18), we conclude that fading becomes beneficial if the number of users in each cell exceeds a certain threshold (which decreases with α [9]). As $K \to \infty$ one gets $\mathcal{S}^{\text{erg}}_{0,\text{ul,MCP}} \to 2$, which is the corresponding slope for the single-user AWGN channel. A more in-depth characterization of the achievable sum-rate in extreme SNR regimes with joint MCP can be found in [22].

We conclude this section by noting that with SCP and reuse factor F, the per-cell sum-rate is obtained by averaging the analogous expressions to (16.15) with respect to the channel gains. In contrast to joint MCP, intracell TDMA may be advantageous over the WB scheme with SCP for $F \leq L$ when the intercell interference is large enough [19]. This stems from the fact that intracell TDMA reduces the overall intercell interference power. The reader is also referred to [77] for additional observations regarding the performance of intracell TDMA in the framework of the soft-handoff model.

Downlink

The downlink average per-cell ergodic sum-capacity for Wyner-type models is given (analogously to the uplink) by

$$R_{\text{dl,MCP}}^{\text{erg},M}(P) = \mathbb{E}\left\{ R_{\text{dl,MCP}}^{M}(P) \right\}, \tag{16.25}$$

where $R_{\text{dl,MCP}}^{M}(P)$ is obtained as in (16.19). As in the Gaussian setting, the sum-capacity is attained via DPC. Similar to the corresponding uplink expression (16.20), computation of (16.25) is difficult; particularly in view of the min–max optimization problem involved. We therefore resort to the limit of large number of intracell users ($K \gg 1$) for the sake of analytical tractability. We also focus on the large number of cells limit ($M \to \infty$).

Starting with the soft-handoff model, we consider for simplicity the case of $L = 1$ and $\alpha = 1$. For this setting, lower and upper bounds on the ergodic per-cell sum-capacity were derived in [17]. Referring to the min–max optimization problem (16.19), a lower bound on $R_{\text{dl,MCP}}^{\text{erg},M}(P)$ can be obtained by *fixing* a diagonal matrix D_{MK}, and minimizing over the diagonal matrices Λ_M that satisfy the unit trace constraint. Fixing D_{MK} is equivalent to a particular choice of power allocation among the users in the *dual uplink* (see [17, 71]). Consider a "threshold crossing" strategy, according to which only users received in the dual uplink with fade power levels exceeding a certain threshold L_{th} (at *all* receiving BSs) are allowed to transmit. Thus, as $K \to \infty$, the number of active users in each cell crystallizes to $K_0 \triangleq K e^{-2L_{\text{th}}}$, and allocating an equal power $1/(K_0 M)$ to all active users meets the constraint $\text{Tr}(\mathsf{D})_{MK} \leq 1$ in (16.19). In such a case $\Lambda_M = \mathsf{I}_M$ is a minimizing noise covariance matrix for (16.19). If one further chooses L_{th} so that $K_0 \to \infty$ as $K \to \infty$, then the strong law of large numbers can be invoked for the set of active users (in a similar manner to the arguments leading to (16.23)). Specifically, setting $L_{\text{th}} = \frac{1-\epsilon}{2} \log_e K$, where $0 < \epsilon < 1$, yields $K_0 = K^\epsilon$. This eventually leads to the following lower bound on the ergodic downlink per-cell sum-capacity [17]:

$$R_{\text{dl,MCP}}^{\text{erg}}(P) \geq \log_2 \left(1 + P \left((1 - \epsilon) \log_e K + 2 \right) \right). \tag{16.26}$$

It is important to note here that the power allocation and scheduling scheme employed in the dual uplink to produce (16.26) also translates to a *downlink* DPC-based transmission scheme that achieves the same per-cell sum-rate. This can be accomplished using uplink–downlink transformations derived in [78, 79].

A corresponding upper bound on the ergodic average per-cell sum-capacity can be derived by fixing $\Lambda_M = \mathsf{I}_M$, and optimizing over D_{MK} in (16.19). Further bounding the channel fade levels by the strongest gains received at each BS (where the maximum is taken over all intracell users), and observing that the maximum of K i.i.d. exponentially distributed random variables scales as $\log_e K$ for $K \gg 1$ (neglecting lower-order terms [80]), one eventually gets [17]

$$R_{\mathrm{dl,MCP}}^{\mathrm{erg}}(P) \leq \log_2\left(1 + 2P\log_e K\right). \qquad (16.27)$$

The upper and lower bounds in (16.26) and (16.27) turn out to be rather tight, and for $\epsilon \ll 1$ the gap between the bounds is less than 1 bit/s/Hz in the high-SNR region. By similar arguments, it can be shown that the ergodic per-cell sum-capacity for the fading Wyner model (with $L = 1$) can be lower and upper bounded as [17]

$$\log_2\left(1 + P\left((1-\epsilon)\log_e K + 1 + 2\alpha^2\right)\right)$$
$$\leq R_{\mathrm{dl,MCP}}^{\mathrm{erg}}(P) \leq \log_2\left(1 + (1+2\alpha^2)P\log_e K\right). \qquad (16.28)$$

From (16.26)–(16.28) we conclude that the downlink ergodic per-cell sum-capacity scales as $\log_2 \log_e K$ for $K \gg 1$. This scaling law is a manifestation of multiuser diversity, and results from a scheduling scheme that only serves the users with the most favorable channel conditions. The same scaling law is also observed in the single-cell setting with multiple receive antennas at the BS (e.g., [80]). Note that this comes in contrast to the corresponding result for the uplink (16.23), where no such multiuser diversity gains are observed. The striking difference stems from the underlying assumption that in the uplink no instantaneous CSI is available to the users, whereas the threshold-crossing scheduling scheme considered above for the dual uplink *does* require threshold-level crossing indication feedback from the receiving BSs. By the uplink–downlink duality principle, the same scaling law can in principle be achieved in the uplink if the channel coherence time permits appropriate CSI feedback to the users. We finally note that DPC is not a requirement for achieving the $\log_2 \log_e K$ scaling law for $K \gg 1$. In fact, it was shown in [81] that the same scaling law can be achieved via a simple linear zero-forcing (ZF) strategy, combined with intracell TDMA scheduling, whereby at any given time only a single user with the maximum channel gain is served in each cell.

Turning to extreme-SNR characterization, all aforementioned bounds lead to the conclusion that the high-SNR slope of joint MCP in the downlink equals unity. The low-SNR parameters were shown in [17] to satisfy

$$\frac{\log_e 2}{2\log_e K} \leq \left(\frac{E_b}{N_0}\right)_{\min} \leq \frac{\log_e 2}{(1-\epsilon)\log_e K + 2} \qquad (16.29)$$

and

$$S_{0,\mathrm{dl,MCP}}^{\mathrm{erg}} = 2 \qquad (16.30)$$

for the soft-handoff model, and

$$\frac{\log_e 2}{(1+2\alpha^2)\log_e K} \leq \left(\frac{E_b}{N_0}\right)_{\min} \leq \frac{\log_e 2}{(1-\epsilon)\log_e K + 1 + 2\alpha^2} \qquad (16.31)$$

and

$$S_{0,\mathrm{dl,MCP}}^{\mathrm{erg}} = 2 \qquad (16.32)$$

for the Wyner model.

16.4 Central Processor with Finite-Capacity Backhaul

We now consider a more realistic scenario for joint MCP. The idealized setting considered in Section 16.3 assumed that the signals received at the BSs are fully available at the CP. However, in practical systems communicating high-accuracy samples of the received signals to the CP necessitates large bandwidths, and an extensive deployment of optical fibers to support them. It is therefore of great interest to understand the impact of realistic constraints on the capacity of the available backhaul links on the achievable throughput. Comparison of the results with the ultimate performance limits of Section 16.3 will provide valuable insights into the potential performance enhancement of joint MCP that can be *actually achieved* in practice. It will also elucidate the manner in which constraints on the backhaul capacity affect performance.

We thus assume here a C-RAN architecture where BSs are connected to the CP via *finite capacity* backhaul links of C bit/s/Hz, as depicted in Figure 16.5, and review the main results of [24]. The backhaul links are assumed here to be error-free, and of fixed capacity known both at the BSs and at the CP. The impact of *unreliable* backhaul links in a non-ergodic setting, where each link either has a state-dependent capacity, or fails completely with a certain probability, was studied in [46, 47].

16.4.1 Uplink

It is useful to start the discussion with a simple upper bound on the achievable per-cell average sum-rate, which may constitute a benchmark result. This is the so-called *cut-set bound* [82], which reads

$$R_{\text{ul,UB}}^{M}(P, C) = \min\left\{C, R_{\text{ul,MCP}}^{M}(P)\right\}, \tag{16.33}$$

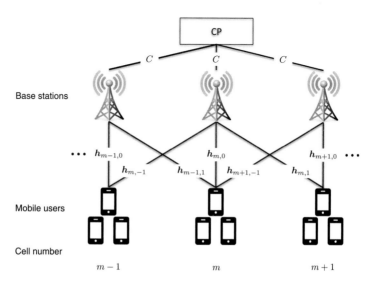

Figure 16.5 Joint multicell processing with limited capacity backhaul links.

where $R^M_{\text{ul,MCP}}(P)$ is the average per-cell sum-capacity with unlimited backhaul links (16.11). For the corresponding ergodic fading setting, one gets a similar bound by replacing $R^M_{\text{ul,MCP}}(P)$ with $R^{\text{erg},M}_{\text{ul,MCP}}(P)$ (cf. (16.20)).

In this section, we restrict the discussion to two types of practically oriented MCP strategies, and investigate how close they can get to the upper bound (16.33). The first strategy is that of *oblivious BSs*, where the BSs are assumed to be ignorant of the users' codebooks, and have to compress the received signal before communicating with the CP. Such BSs are relatively simple, and require only a radio transceiver and minimal baseband processing capabilities [9]. The oblivious strategy was first considered for the *nomadic* regime [83], where nomadic terminals send information to a remote destination via agents (BSs in the current setting) using codebooks that are unknown to the agents. The second strategy that we consider here involves *informed BSs*. Here, it is assumed that the BSs have full *decoding* capabilities, and are aware of the users' codebooks. To simplify the analysis, we assume henceforth for both settings that Gaussian codebooks are employed. It is important to note in this respect that this transmission strategy is by no means claimed to be optimal, and in fact a discrete input may outperform Gaussian signaling in such bottleneck settings, as shown, e.g., in [83].

Gaussian Channels

In the Gaussian setting, we focus for simplicity and without loss of sum-rate optimality on intracell TDMA; namely, $K = 1$ (see the discussion in the previous section). Starting with oblivious BSs, we consider a CF strategy, as proposed in [24, 83]. The underlying idea is to perform *distributed Wyner–Ziv compression*[6] at the BSs, while leveraging the fact that the signals received at different BSs are *correlated*. This allows for a more efficient use of the limited backhaul links. The setting falls within the framework of distributed source coding and the "CEO problem" [86].[7] Accordingly, the BSs compress the received signals to C bit/s/Hz, and send an appropriate function of the compressed index to the CP. A straightforward strategy at this point is decompression of the received signals at the CP followed by decoding of the messages based on the decompressed signals. However, the MCP approach suggested in [24] relies on *joint* decompression and decoding at the CP, which may potentially lead to enhanced performance.

---

[6] Wyner–Ziv compression [84] is a coding scheme for achieving the rate-distortion function of lossy source coding with side information at the decoder. The original setting in [84] involves a discrete memoryless source which is to be communicated to the decoder via a noiseless link at a prescribed distortion. The underlying assumption is that a side information sequence (correlated to the source) is available to the decoder *but not to the encoder*, and the reconstruction of the source relies on the *whole* side information sequence. See also Chapter 2 and [85, chapter 11.4].

[7] The chief executive officer (CEO) problem refers to a setting consisting of a single random source, which is to be communicated to a decoder (the CEO) via many sensors (or "agents"). The sensors observe noisy versions of the source, and need to forward their observations to the CEO. The CEO then uses the observations to estimate the source. Communication between the sensors and the CEO is done via unidirectional links of finite capacities. The objective in this setting is to minimize the average distortion of the estimation at the CEO, given the finite capacities of those links. See Chapter 2, [85, chapter 12.4], and [9, appendix C] for a more elaborate review of this setting.

Considering the large number of cells limit ($M \to \infty$), it can be shown that the following per-cell sum-rate is achievable by the CF strategy described above [24]:

$$R_{\mathrm{ul,obl}}(P, C) = \lim_{M \to \infty} \frac{1}{M} \log_2 \det \left(\mathsf{I}_M + (1 - 2^{-r^*}) P \, \mathsf{H}_M \mathsf{H}_M^\dagger \right), \qquad (16.34)$$

where r^* is the solution of the equation

$$F(P, r^*) = C - r^*, \qquad (16.35)$$

and

$$F(P, r) \triangleq \lim_{M \to \infty} \frac{1}{M} \log_2 \det \left(\mathsf{I}_M + (1 - 2^{-r}) P \, \mathsf{H}_M \mathsf{H}_M^\dagger \right). \qquad (16.36)$$

Comparing this result to (16.11), it is observed that the finite-capacity backhaul induces an effective *SNR loss* of $1/(1 - 2^{-r^*})$. Letting $C \to \infty$, one gets $r^* \to \infty$, which immediately implies, as expected, that $R_{\mathrm{ul,obl}}(P, \infty) = R_{\mathrm{ul,MCP}}(P)$. However, for $C < \infty$ the parameter r^* can be interpreted as the rate "wasted" for *compressing the noise* embedded in the received signals [83]. Note that this loss is unavoidable since in the absence of local decoding capabilities, the BSs cannot separate the useful signals from the additive noise. The per-cell sum-rate is thus equal to the capacity of the backhaul link minus the portion r^* wasted on noise compression.

We further note that in the absence of intercell interference ($\alpha = 0$), the per-cell achievable rate reduces to

$$R_{\mathrm{ul,obl}}(P, C) = \log_2 \left(1 + P \frac{1 - 2^{-C}}{1 + P 2^{-C}} \right). \qquad (16.37)$$

Here, an SNR penalty of $\frac{1 + P 2^{-C}}{1 - 2^{-C}}$ is observed compared to the achievable rate with ideal backhaul links, which is simply $\log_2(1 + P)$.

Particularizing to the Gaussian Wyner model with $L = 1$, we get [24]

$$R_{\mathrm{ul,obl}}(P, C) = F(P, r^*)$$
$$= \int_0^1 \log_2 \left(1 + P(1 - 2^{-r^*})(1 + 2\alpha \cos(2\pi\theta))^2 \right) d\theta, \qquad (16.38)$$

where r^* is the unique solution to the fixed point equation $F(P, r^*) = C - r^*$. Similarly, for the corresponding soft-handoff model ($L = 1$) the per-cell achievable rate reads [24]

$$R_{\mathrm{ul,obl}}(P, C) = F(P, r^*)$$
$$= \log_2 \left(\frac{1 + (1 + \alpha^2)P + 2\alpha^2 2^{-C} P^2 + \sqrt{1 + 2(1 + \alpha^2)P + \left((1 - \alpha^2)^2 + 4\alpha^2 2^{-C} \right) P^2}}{2 \left(1 + 2^{-C} P \right) \left(1 + \alpha^2 2^{-C} P \right)} \right). \qquad (16.39)$$

Again, both expressions reduce to (16.13) and (16.14), respectively, for $C \to \infty$.

Turning to the extreme-SNR characterization of (16.34), then clearly, for any fixed backhaul capacity C, the per cell sum-rate approaches C as the SNR increases without bound. Since this rate is finite, the immediate implication is that the high-SNR slope is

$\mathcal{S}_{\infty,\text{obl}} = 0$. As shown in [24], to retain the high-SNR characteristics of the unlimited backhaul setting the backhaul capacity must scale as

$$C(P) \underset{P \to \infty}{\simeq} \log_2 P + \Theta(P), \tag{16.40}$$

where $\Theta(P) \to \infty$ as $P \to \infty$ at arbitrary rate [24], and we used the fact that $\mathcal{S}_\infty = 1$ for the unlimited backhaul setting. Considering the low-SNR regime, it was shown in [24] that the minimum $\frac{E_b}{N_0}$ that enables reliable communication and the low-SNR slope are given by

$$\left(\frac{E_b}{N_0} \right)_{\text{min,obl}} = \left(\frac{\tilde{E}_b}{N_0} \right)_{\text{min}} \frac{1}{1 - 2^{-C}} \tag{16.41}$$

and

$$\mathcal{S}_{0,\text{obl}} = \tilde{\mathcal{S}}_0 \frac{1}{1 + \tilde{\mathcal{S}}_0 \frac{2^{-C}}{1 - 2^{-C}}}, \tag{16.42}$$

where $\left(\frac{\tilde{E}_b}{N_0} \right)_{\text{min}}$ and $\tilde{\mathcal{S}}_0$ denote here the corresponding parameters for the unlimited backhaul setting (cf. (16.11)). Note that (16.41) indicates that the limited backhaul capacity induces an SNR loss factor of $\frac{1}{1 - 2^{-C}}$, which quantifies, e.g., to 0.5 dB for $C \approx 3.2$ bit/s/Hz. We further note that the above extreme-SNR characterization was derived in [24] for the general result (16.34), and thus it is not restricted to either the Wyner model, or the soft-handoff model.

We have seen so far that when the BSs are constrained to be oblivious, the impact of limited backhaul capacity is an effective capacity loss, which can be attributed to the rate wasted on noise compression at the BSs. It is now interesting to investigate what might be the impact of endowing the BSs with decoding capabilities, allowing for partial separation of desired signals from noise at the BS, and for more efficient use of the backhaul links.

To simplify the discussion, we consider here a rate-splitting scheme according to which each user splits its message into two parts [24]: One part is intended to be decoded only at the CP (by means of joint MCP), and the second part is to be decoded locally at the BS. The transmitted signal is a superposition of the two codewords associated with the above two sub-messages, and the local decoder at the BS treats the signal associated with the sub-message decoded at the CP as additional noise. We denote henceforth the power allocated to the sub-message intended for decoding at the CP by βP, and the power allocated to the locally decoded sub-message is thus $(1 - \beta)P$, where $0 \leq \beta \leq 1$. The decoded messages are losslessly communicated to the CP via the backhaul links. Furthermore, the signals associated with the locally decoded sub-messages (the respective codewords) can be canceled out from the received signals, and the CF scheme described above for the oblivious setting can then be performed with respect to the remaining signals. Clearly, forwarding the locally decoded messages reduces the bandwidth available for compression.

Let $\bar{R}_{\text{d}}(\beta)$ denote the achievable rate of local decoding (while *ignoring* the limited backhaul capacity). Following [19], while relying on the MAC rate region, local decoding within the Wyner model (taking $L = 1$) can be performed in three ways [24]:

decoding all (three) messages received at the BS; decoding only the local message while treating the rest as noise; decoding only the messages received from adjacent cells. This yields the following expression for the achievable rate (cf. (16.16)):

$$\bar{R}_d(\beta) = \max \left\{ \log_2 \left(1 + \frac{(1-\beta)P}{1+(\beta+2\alpha^2)P} \right), \right.$$
$$\left. \min \left\{ \frac{1}{2} \log_2 \left(1 + \frac{(1-\beta)2\alpha^2 P}{1+\beta(1+2\alpha^2)P} \right), \frac{1}{3} \log_2 \left(1 + \frac{(1+2\alpha^2)(1-\beta)P}{1+\beta(1+2\alpha^2)P} \right) \right\} \right\}.$$
$$(16.43)$$

The corresponding expression for the soft-handoff model ($L = 1$) is given by

$$\bar{R}_d(\beta) = \max \left\{ \log_2 \left(1 + \frac{(1-\beta)P}{1+(\beta+\alpha^2)P} \right), \frac{1}{2} \log_2 \left(1 + \frac{(1-\beta)(1+\alpha^2)P}{1+\beta(1+\alpha^2)P} \right) \right\}.$$
$$(16.44)$$

But, clearly, the rate of the locally decoded messages cannot exceed the backhaul capacity C. Thus, the *actual* local decoding rate reads

$$R_d(\beta) = \min \left\{ \bar{R}_d(\beta), C \right\}.$$
$$(16.45)$$

With that in mind, the overall achievable rate with informed BSs is given by [24]

$$R_{ul,inf}(P, C) = \max_{0 \leq \beta \leq 1} \left\{ F(\beta P, r^*) + R_d(\beta) \right\},$$
$$(16.46)$$

where r^* is the solution to

$$F(\beta P, r^*) = C - R_d(\beta) - r^*.$$
$$(16.47)$$

Here, the function $F(\beta P, r^*)$ is given by (16.36), which particularizes to (16.38) and (16.39) for the Wyner and the soft-handoff models, respectively. The achievable rate (16.46) can be interpreted as the sum of the rate of the locally decoded messages and the rate $F(\beta P, r^*)$ achieved using the CF strategy of the oblivious setting. The maximization in (16.46) facilitates the fact that the achievable rate can be optimized with respect to the power splitting between the locally decoded codewords, and the codewords decoded at the CP. Note in particular that in the absence of intercell interference ($\alpha = 0$), the rate (16.46) is maximized for $\beta = 0$. This outcome should come as no surprise since in the absence of intercell interference joint MCP provides no gain, and local decoding at the BSs is optimal.

It was further observed in [24] that a useful strategy is, in fact, time sharing between the two extreme approaches; namely, either fully decode the messages locally at the BSs, or fully decode them jointly at the CP (as in the oblivious setting). This finally yields the following achievable per-cell sum-rate with informed BSs:

$$R_{ul,inf}^{ts}(P, C) = \max_{r \geq r^*} \left\{ R_d(0) + (C - R_d(0)) \frac{F(P, r) - R_d(0)}{F(P, r) + r - R_d(0)} \right\},$$
$$(16.48)$$

where r^* is calculated according to (16.35).

Turning to the extreme-SNR characterization of (16.48), we observe the following. In the high-SNR regime the benefits of local decoding vanish, since (16.43) and (16.44) clearly imply that the high-SNR slope for $\beta = 0$ is either $\frac{1}{3}$ or $\frac{1}{2}$, respectively, even for *unlimited* backhaul capacity. This comes in contrast to the oblivious setting where the high-SNR slope is unaffected if the backhaul capacity is appropriately increased with the SNR.

For the low-SNR regime, it was shown in [24] that the low-SNR parameters corresponding to (16.48) satisfy

$$\left(\frac{E_b}{N_0}\right)_{\min}^{\text{ts}} = \frac{1}{\lambda_o \left(\left(\frac{E_b}{N_0}\right)_{\min,\text{d}}\right)^{-1} + (1-\lambda_o)\left(\left(\frac{\tilde{E}_b}{N_0}\right)_{\min,\text{obl}}\right)^{-1}}, \tag{16.49}$$

and

$$S_0^{\text{ts}} = \frac{\left(\left(\frac{E_b}{N_0}\right)_{\min}^{\text{ts}}\right)^{-2}}{\lambda_o \left(S_{0,\text{d}}\right)^{-1}\left(\left(\frac{E_b}{N_0}\right)_{\min,\text{d}}\right)^{-2} + (1-\lambda_o)\left(\tilde{S}_{0,\text{obl}}\right)^{-1}\left(\left(\frac{\tilde{E}_b}{N_0}\right)_{\min,\text{obl}}\right)^{-2}}, \tag{16.50}$$

with λ_o being the optimized time-sharing parameter

$$\lambda_o = 1 - \frac{C}{r^*}, \tag{16.51}$$

and $r^* = \max\{C, \tilde{r}\}$, where $\tilde{r}$ is the unique solution of

$$2^{-\tilde{r}}\left(1 + \tilde{r}\log_e 2\right) = \frac{2\alpha^2}{1+2\alpha^2} \tag{16.52}$$

for the Wyner model, and

$$2^{-\tilde{r}}\left(1 + \tilde{r}\log_e 2\right) = \frac{\alpha^2}{1+\alpha^2} \tag{16.53}$$

for the soft-handoff model (both with $L = 1$). In (16.49) and (16.50) the $(\cdot)_{\text{obl}}$ notation refers to the low-SNR parameters of the oblivious setting, but with r^* as described above replacing the backhaul capacity C. The notation $(\cdot)_{\text{d}}$ designates the low-SNR parameters for local decoding (16.45). The latter are given by

$$\left(\frac{E_b}{N_0}\right)_{\min,\text{d}} = \log_e 2, \quad S_{0,\text{d}} = \frac{2}{1+4\alpha^2} \tag{16.54}$$

for the Wyner model, and

$$\left(\frac{E_b}{N_0}\right)_{\min,\text{d}} = \log_e 2, \quad S_{0,\text{d}} = \frac{2}{1+2\alpha^2} \tag{16.55}$$

for the soft-handoff model.

Some numerical results for the achievable average per-cell sum-rate with limited backhaul capacity are shown in Figure 16.6. The figure shows achievable sum-rates for the Wyner model ($L = 1$) as a function of the intercell interference factor α, for $P = 10\,\text{dB}$ and $C = 6$ bit/s/Hz. In addition to the sum-rates according to (16.38) and (16.48), corresponding, respectively, to oblivious and informed BSs, the figure also shows for reference the sum-capacity with unlimited backhaul links. The figure clearly

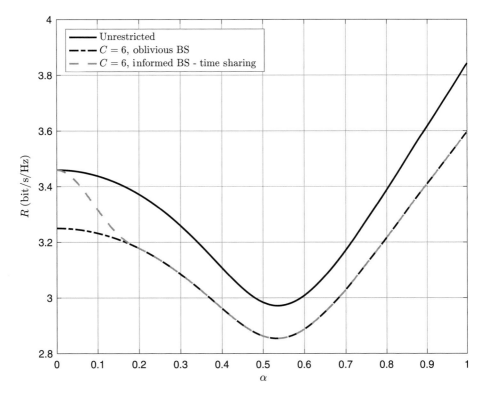

Figure 16.6 Average achievable per-cell sum-rates for the Gaussian Wyner model with $L = 1$ and $K = 1$, plotted as a function of the intercell interference factor α. The results correspond to $P = 10$ dB and $C = 6$ bit/s/Hz [9, figure 5.1].

demonstrates the performance degradation induced by restricting the backhaul capacity. It is also evident that local decoding is beneficial only when the interference level drops below a certain threshold. The figure shows that since for $C = 6$ bit/s/Hz (and $P = 10$ dB) the cut-set upper bound (16.33) is governed by the joint MCP sum-capacity, the achievable rate with local decoding coincides with the upper bound only at a single point, corresponding to $\alpha = 0$ (no intercell interference), where local decoding is optimal.

As shown in [24], if the backhaul capacity is further reduced so that the cut-set bound is governed by the backhaul capacity, then the upper bound can be achieved via local decoding for a certain positive range of intercell interference levels. In contrast, the oblivious CF approach cannot achieve the upper bound for finite values of C.

We finally note that the achievable sum-rates with restricted backhaul capacity may be potentially improved if instead of using Gaussian codebooks and the simple local decoding scheme described above, a transmission scheme that relies on *structured* codes is employed. The underlying idea is that instead of locally decoding the individual *messages* of the users whose signals are received at the BSs, each BS decodes a *function* of the corresponding transmitted codewords. Such a scheme is suggested for the Wyner model setting in [87] (see also [88] and references therein for the underlying theory). More specifically, the proposed scheme relies on nested lattice codes, which have the property that a linear combination of codewords with *integer* coefficients followed by a modulo-lattice operation is still equal to a codeword from the same codebook. This property is practically appealing, since it implies that the complexity of decoding a single codeword or a weighted sum of codewords is the same. One must bear in mind, however, that in the Wyner model setting, the intercell interference factors (comprising the "weights" of the aforementioned linear combination) are, in general, *not* integers. Still, each BS may decode an appropriately chosen linear combination of codewords with integer coefficients, and treat the remaining signal as Gaussian noise.

Some improvement for the achievable throughput with such lattice codes may be obtained if additional "private" messages intended to be fully locally decoded at the BSs are superimposed on the lattice codewords. The idea here is to reduce the deleterious impact of non-integer channel coefficients on the achievable rates. As shown in [87], the lattice-based scheme may outperform the oblivious scheme for sufficiently low or high intercell interference factors. The reader is referred to [87] and [9] for a more detailed account.

Ergodic Flat-Fading Channels

We start the discussion by recalling the observation made for the ideal setting that with joint MCP intracell TDMA is no longer optimal in the presence of fading. This observation continues to hold for constrained backhaul links as well. We therefore consider instead the WB scheme with $K \geq 1$, and focus for conciseness on oblivious BSs. As shown in [24], the ergodic per-cell sum-rate for this setting in the large number of cells limit ($M \to \infty$) is given by

$$R_{\text{ul,obl}}^{\text{erg}}(C, P) = F(P, r^*)$$

$$= \lim_{M \to \infty} \max_{\substack{r_m(\mathsf{H}_M):\mathbb{C}^{M \times MK} \to \mathbb{R}^+ \\ \text{s.t. } \mathbb{E}\{r_m(\mathsf{H}_M)\}=r^*}} \mathbb{E}\left\{ \frac{1}{M} \log_2 \det\left(\mathsf{I}_M \right. \right.$$

$$\left. \left. + \frac{P}{K} \operatorname{diag}\left(1 - 2^{-r_m(\mathsf{H}_M)}\right)_{m=1}^{M} \mathsf{H}_M \mathsf{H}_M^{\dagger} \right) \right\},$$

(16.56)

where the mappings $\{r_m(\mathsf{H}_M)\}_{m=1}^{M}$ are chosen to satisfy (16.35). $r_m(\mathsf{H}_M)$, $m = 1, \ldots, M$, should be interpreted here as the rate invested by BS m for compressing its received noise when the channel matrix realization is H_M.

To simplify the discussion, a lower bound on (16.56) can be obtained by taking $r_m(\mathsf{H}_M) = r^*$, $m = 1, \ldots, M$, for all realizations of H_M. This bound becomes tight for large K, and its gap from (16.56) is quite small already for $K = 2$ [24]. Still, calculating (16.56) remains a formidable task (see the discussion in Section 16.3.2), and we focus here on the limit of a large number of users per cell ($K \gg 1$) in order to present closed-form analytical results (as in Section 16.3.2).

Assuming Rayleigh fading, and starting with the Wyner model with $L = 1$, the ergodic per-cell sum-rate for $K \gg 1$ boils down to

$$R_{\text{ul,obl}}^{\text{erg}}(C, P) = \log_2\left(1 + (1 + 2\alpha^2)P\frac{(1 - 2^{-C})}{1 + (1 + 2\alpha^2)P2^{-C}}\right). \qquad (16.57)$$

Comparing this result to (16.37), we see that for large K the achievable sum-rate is the same as in a single-user single-BS setting, but with *enhanced power* $(1 + 2\alpha^2)P$. Note that a completely analogous result was obtained for the ideal setting (cf. (16.23)), and the two achievable sum-rates coincide for $C \to \infty$. The effective SNR loss factor due to limited backhaul capacity can be quantified as

$$\frac{1 + (1 + 2\alpha^2)P2^{-C}}{1 - 2^{-C}}, \quad K \gg 1. \qquad (16.58)$$

The corresponding sum-rate in the soft-handoff model was shown in [24] to be

$$R_{\text{ul,obl}}^{\text{erg}}(C, P) = \log_2\left(1 + (1 + \alpha^2)P\frac{(1 - 2^{-C})}{1 + (1 + \alpha^2)P2^{-C}}\right). \qquad (16.59)$$

Both (16.57) and (16.59) achieve the cut-set upper bound when either C or P increases, while the other is held fixed.

Turning to the low-SNR regime, while allowing for an *arbitrary number* of users per-cell K, it can be shown that for the Wyner model the minimum $\frac{E_b}{N_0}$ that enables reliable communication and the low-SNR slope read [24]

$$\left(\frac{E_b}{N_0}\right)_{\text{min,obl}} = \frac{\log_e 2}{(1 + 2\alpha^2)(1 - 2^{-C})} \qquad (16.60)$$

and

$$S_{0,\text{obl}} = \frac{2(1 - 2^{-C})}{1 + \frac{1}{K} + \left(1 - \frac{1}{K}\right)2^{-C}}. \qquad (16.61)$$

The corresponding parameters for the soft-handoff model are given by

$$\left(\frac{E_b}{N_0}\right)_{\text{min,obl}} = \frac{\log_e 2}{(1 + \alpha^2)(1 - 2^{-C})} \qquad (16.62)$$

and

$$S_{0,\text{obl}} = \frac{2(1 - 2^{-C})}{1 + \frac{1}{K} + \left(1 - \frac{1}{K}\right)2^{-C}}. \qquad (16.63)$$

The impact of limited backhaul capacity is clearly observed to be an increase in the minimum $\frac{E_b}{N_0}$ required for reliable communication, and a corresponding decrease in the low-SNR slope. Both effects vanish asymptotically with the increase of C.

16.4.2   Downlink

We now turn to discuss the impact of constrained backhaul links on the downlink. For the sake of conciseness, we restrict the discussion to Gaussian channels, while considering the soft-handoff model with $L = 1$ and $M \to \infty$, and review two simple downlink transmission schemes, as suggested in [25]. The first scheme assumes *oblivious BSs* with no *encoding* capabilities; namely, all encoding functions must be performed at the CP. The second scheme allows *local* encoding capabilities at the BS (but no codebook information sharing between different BSs), and the encoding tasks are thus shared by the BS and the CP.[8] In such a case we refer to the BS as an *informed* BSs. These two schemes can be viewed as the downlink analogs of the two schemes discussed for the uplink in the previous section. For both schemes we focus on intracell TDMA ($K = 1$), which achieves the optimal per-cell sum-rate with unlimited backhaul.

While not encompassing the full scope of the setting, we believe that the results presented in the sequel do capture some of the fundamental characteristics of the downlink backhaul constrained setting, and provide useful insights into practical implications of joint downlink MCP. To set a reference, we note that the uplink–downlink duality principle implies that the cut-set bound in (16.33) continues to hold also for the downlink (recall that the uplink and downlink per-cell sum-capacities are the same in the absence of fading – see Section 16.3).

With oblivious BSs, the simple transmission scheme suggested in [25] assumes that the CP first performs joint DPC for all users, which would have been optimal for $C \to \infty$ (as in Section 16.3.1). This encoding process produces, per BS, a sequence of n Gaussian symbols whose transmission would have ideally facilitated DPC in the absence of any constraints on the backhaul. Next, to address the backhaul constraints, the CP quantizes each such sequence using a Gaussian quantization codebook with 2^{nC} codewords, which in turn produces a corresponding compression codeword (of length n symbols). The *index* of this codeword is then sent to the respective BS via its backhaul link. Finally, each BS reconstructs the locally received quantized codeword, and simply transmits its symbols toward the users over n channel uses.

Note that the above procedure must take into account that BSs inevitably also forward quantization noise. Thus, to address the underlying per-BS power constraint (16.8), the DPC scheme employed by the CP must produce Gaussian symbols of *reduced variances* so that the Gaussian symbols actually transmitted by the BSs indeed have variance P. Following standard rate-distortion theory arguments (see [25, appendix A]), the resulting channel can be viewed as an equivalent soft-handoff model (16.10), but with reduced effective SNR:

$$\tilde{P} = \frac{P}{1 + \frac{1+(1+\alpha^2)P}{2^C - 1}} \cdot \tag{16.64}$$

The finite backhaul capacity thus induces an SNR degradation factor of $1 + \frac{1+(1+\alpha^2)P}{2^C - 1}$, which vanishes for $C \to \infty$. The resulting per-cell rate then reads

---

[8] See [25] for additional, more sophisticated schemes with local encoding capabilities at the BSs.

$$R_{\text{dl,obl}}(P, C) = \log_2 \left(\frac{1 + (1 + \alpha^2)\tilde{P} + \sqrt{1 + 2(1 + \alpha^2)\tilde{P} + (1 - \alpha^2)\tilde{P}^2}}{2} \right), \quad (16.65)$$

which is identical to (16.14), but with P replaced by the effective SNR (16.64).

Considering the high-SNR characteristics of this result, it is observed that letting C scale as $r \log_2 P$ in (16.65), one obtains the high-SNR slope $S_{\infty,\text{dl,obl}} = \min\{r, 1\}$. Thus, with $C \sim \log_2 P$ the maximum unit high-SNR slope can be attained. Clearly, the high-SNR slope is zero whenever C is fixed, and as shown in [25], the limiting rate falls short of the cut-set bound C in this regime unless $\alpha = 0$. The low-SNR characterization of (16.65) coincides with (16.41) and (16.42), which further facilitates the correspondence between the oblivious uplink and downlink schemes in extreme-SNR regimes.

Turning to *informed* BSs with local encoding capabilities, the scheme suggested in [25] relies on the following observation. For the soft-handoff model with $L = 1$, each user receives interference only from a single BS located in the cell with preceding index (cf. (16.10)). Additionally, if the BS in a given cell knew the signal transmitted by the adjacent BS that interferes with the local user, it could use DPC to effectively eliminate this interference. The idea is therefore to send each BS a *quantized* version of the codeword transmitted by the (adjacent) interfering BS, along with the message intended to the local user. The BS can then locally employ DPC with respect to the quantized interfering signal. Note that the latter procedure requires the CP to reproduce the transmitted signals of each BS. However, this is indeed possible if one further assumes that the CP is fully informed regarding the encoding functions and quantization codebooks employed by all BSs (as well as the messages intended to all users). We finally note that in view of the rate constraint on the backhaul, the sum of the quantization rate and the rate of the messages sent to the users may not exceed the total backhaul capacity C.

Taking the above considerations into account, then as shown in [25], the suggested local encoding scheme achieves the following per-cell rate:

$$R_{\text{dl,inf}}(P, C) = \begin{cases} C & \text{if } C \leq \log_2\left(1 + \frac{P}{1+\alpha^2 P}\right) \\ R'_{\text{dl,inf}}(P, C) & \text{otherwise,} \end{cases} \quad (16.66)$$

where

$$R'_{\text{dl,inf}}(P, C) = \log_2 \left(1 - \frac{2^C}{\alpha^2 P} + \sqrt{1 + \frac{2^{C+1}}{\alpha^2}\left(2 + \frac{1}{P}\right) + \frac{2^{2C}}{\alpha^4 P^2}} \right) - 1 \quad (16.67)$$

for $\alpha > 0$, while for $\alpha = 0$

$$R'_{\text{dl,inf}}(P, C) = \log_2(1 + P). \quad (16.68)$$

Note that as $C \to \infty$, the rate $R_{\text{dl,inf}}(P, C)$ approaches $\log_2(1+P)$, which corresponds to a vanishing quantization noise and perfect elimination of intercell interference via DPC. However, this result implies that the suggested scheme fails to utilize the underlying array gain (a factor of $1 + \alpha^2$ in this setting), which is indeed achieved by joint MCP with unlimited backhaul capacity. In contrast, if $C \leq \log_2\left(1 + \frac{P}{1+\alpha^2 P}\right)$, or when $\alpha = 0$, then the suggested scheme is optimal (recall that the first constraint in (16.66)

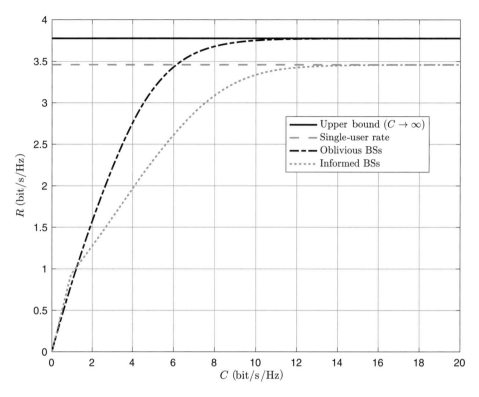

Figure 16.7 Per-cell achievable downlink rates for the soft-handoff model with $L = 1$ and $K = 1$, plotted as a function of the backhaul capacity C. The results are shown for $P = 10\,\mathrm{dB}$ and $\alpha = 1$.

corresponds to the rate achieved with straightforward local encoding and treating all interference as noise).

An additional observation [25] is that letting $C = r \log_2 P$ yields a high-SNR slope of $\mathcal{S}_{\infty,\mathrm{dl,inf}} = \min\left\{\frac{r}{2}, 1\right\}$, and hence the optimal unit slope is achieved by having $C \sim 2 \log_2 P$. This is *double* the rate required for the oblivious setting. Turning to the low-SNR regime, it is shown in [25] that the local encoding scheme achieves the following low-SNR parameters:

$$\left(\frac{E_b}{N_0}\right)_{\mathrm{min,dl,inf}} = \log_e 2 \tag{16.69}$$

and

$$\mathcal{S}_{0,\mathrm{inf}} = \frac{2}{1 + 2\alpha^2 2^{-C}}. \tag{16.70}$$

The minimum $\frac{E_b}{N_0}$ that enables reliable communication is the same here as in a *single-user setting* independently of C (in contrast to the oblivious scheme), while the same holds for the low-SNR slope only when $C \to \infty$.

Figures 16.7 and 16.8 show some numerical results for the achievable rates presented above. Figure 16.7 shows the achievable rates as a function of the backhaul capacity

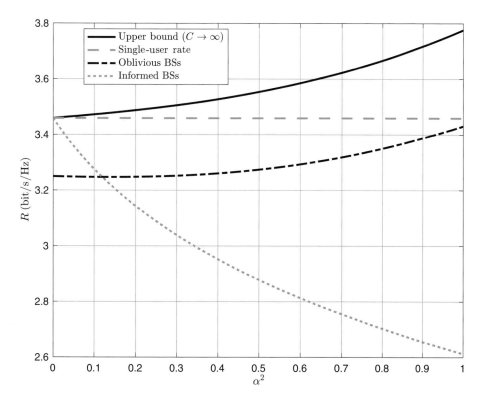

Figure 16.8 Per-cell achievable downlink rates for the soft-handoff model with $L = 1$ and $K = 1$, plotted as a function of the squared intercell interference factor α^2. The results are shown for $P = 10\,\mathrm{dB}$ and $C = 6$ bit/s/Hz.

for $P = 10\,\mathrm{dB}$ and $\alpha = 1$. The results indicate that indeed a backhaul capacity of about 2–3 times the per-cell transmission rate suffices to get quite close to the unlimited backhaul upper bound. The results also indicate that for high intercell interference factors the oblivious scheme outperforms the simple local encoding scheme, except for very low values of C (for which local encoding is beneficial). Figure 16.8 compares the two transmission schemes for a fixed backhaul capacity of $C = 6$ bit/s/Hz, for different values of the intercell interference factor α. The results show that local decoding with informed BSs turns out to be superior when the intercell interference level is small.

16.5   Clustered Multicell Processing

16.5.1   Non-Overlapping Clusters

In the previous section we considered a constrained setting where the CP is connected to the *entire* set of BSs, but via backhaul links of limited capacity. In the current section we turn to discuss another form of practical restriction. Here, the restriction involves the

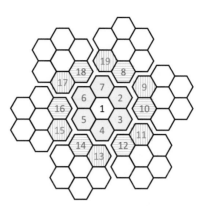

Figure 16.9 A simple two-dimensional cellular network model [34, 35].

number of BSs whose signals can be jointly processed to facilitate cooperative process-ing. More specifically, we now assume that the network is divided into clusters, each comprising a relatively small number of BSs, where the received, or transmitted, sig-nals are jointly processed. For ease of exposition, we focus on the uplink, and consider the Wyner *two-dimensional* (2D) hexagonal cell-array model [6]. The latter relies on the same principles as the linear model considered in previous sections, but exhibits a richer structure of intercell interference, which provides more insights into the impact of clustering in cellular networks. The interference induced by signals transmitted from neighboring clusters is a key factor in any clustering architecture, and as indicated in Section 16.2.3 it puts the setting in the framework of the MIMO interference channel model (see, e.g., [36, 37]).

Straightforwardly treating out-of-cluster interference as noise might lead to an *interference-limited* behavior. To address that, we focus here on a rate-splitting ("lay-ering") strategy inspired by [89] which allows for partial decoding of out-of-cluster interference, as suggested in [34, 35]. For simplicity, the system cells are divided into identical *non-overlapping* clusters comprising seven cells, as shown in Figure 16.9. Each cluster employs a *local* CP that jointly processes the signals received at the cluster's BSs. No cooperation between clusters is assumed. We note here that the achievable throughputs with the above clustering scheme provide lower bounds on the achievable throughputs with more general clustering approaches, including in particular overlap-ping clusters and various approaches for sharing information between different clusters. Such schemes are deferred to the following section.

We assume here that only adjacent cell interference is present, and characterized by a single parameter $\alpha \in (0, 1]$.[9] The discussion is restricted to AWGN channels, while assuming a single active user per cell; namely, an intracell TDMA strategy, which is known to be sum-rate optimal in the absence of fading. Referring to Figure 16.9, and considering cell 1 without loss of generality, the received signal at the local BS (at some arbitrary time instance t) is given by

---

[9] This corresponds to setting $L = 1$ in the linear Wyner model.

$$y_1(t) = x_1(t) + \alpha \sum_{j=2}^{7} x_j(t) + z_1(t), \tag{16.71}$$

where $x_j(t)$ denotes the signal transmitted by the user in cell j, and $z_j(t) \sim \mathcal{CN}(0, 1)$ denotes the zero-mean AWGN at the receiver of the jth BS. The transmitted signals are assumed to be subject to a per-user power constraint of P.

In view of the full symmetry of the system model, we focus henceforth (again, without loss of generality) on the central cluster in Figure 16.9, comprising cells $1, 2, \ldots, 7$. Note that the system cells from which signals are received at the central cluster's CP can be divided into four sets, while referring to the cell numbers in Figure 16.9. The set $\mathcal{I} \triangleq \{2, 3, 4, 5, 6, 7\}$ comprises all *intracluster* cells, except for the central cell (cell 1), which we consider as a separate set. The set $\mathcal{J} \triangleq \{8, 10, 12, 14, 16, 18\}$ comprises out-of-cluster cells from which each of the signals is received at *two* of the cluster's BSs. Finally, the set $\mathcal{K} \triangleq \{9, 11, 13, 15, 17, 19\}$ comprises the out-of-cluster cells from which each of the signals is received only at a single BS within the cluster.

According to the aforementioned rate-splitting strategy, each user in the *noncentral* cells of each cluster splits its messages into three *independently encoded* layers. The transmitted signal is formed as a superposition of the three codewords corresponding to each of the three layers. Layer 1, transmitted at rate R_1, is intended to be decoded in all clusters in which the corresponding user's signal is received (i.e., both in the local cluster and in neighboring clusters). Layer 2, transmitted at rate R_2, is intended to be decoded only in clusters in which the corresponding user's signal is received by *at least two BSs* (i.e., Layer 2 is only decoded out of the signals originating in cell groups $\mathcal{J}$ and $\mathcal{I}$). Finally, Layer 3, transmitted at rate R_3, is intended to be decoded only by the local cluster's CP (i.e., only from cell group $\mathcal{I}$). Users in central cells (i.e., cell 1 when considering the central cluster in Figure 16.9) employ a *single-layered transmission*, since by the underlying channel model (16.71) their signals do not induce interference in neighboring clusters. The target performance measure is taken as the average throughput per cell, defined as:

$$R_{\text{avg}} \triangleq \frac{1}{7}(R_c + 6R_{nc}) = \frac{1}{7}(R_c + 6(R_1 + R_2 + R_3)), \tag{16.72}$$

where R_c and R_{nc} denote the achievable sum-rates of users in central cells and in noncentral cells within each cluster, respectively.

Note that the above transmission strategy lets us treat the signal of each noncentral transmitter as a superposition of the signals of (up to) three independent *virtual* transmitters. The signals corresponding to layers that are not decoded by the cluster CP are treated as AWGN. We assume throughout that complex Gaussian random codebooks are employed for all layers, and that the symbols transmitted by each of the virtual transmitters can be treated as independent zero-mean proper complex Gaussian random variables, identically distributed within a layer, and with variances corresponding to the chosen power allocation scheme. This approach, combined with treating undecoded interference as AWGN, is justified by its robustness, and the fact that Gaussian signaling achieves the sum-capacity of the Gaussian MIMO interference channel in certain regimes, as shown, e.g., in [36, 37].

A fundamental feature of the layered transmission strategy applied here is that *the same rate-splitting and power allocation scheme* is imposed on all users in all noncentral cells throughout the system. This is since each such cell can belong to each of the three generic types of cell groups (i.e., $\mathcal{K}$, $\mathcal{J}$, and $\mathcal{I}$), depending on the point of view of the cluster it is observed from. Denoting by P_i, $i = 1, \ldots, 3$, the power allocated to Layer i transmissions in each noncentral cell, the underlying power constraint for the channel in (16.71) implies that any power allocation scheme must satisfy

$$P_1 + P_2 + P_3 \leq P. \tag{16.73}$$

Omitting the time index, we denote by $x_k^{(i)}$ the signal transmitted by the virtual transmitter corresponding to Layer i in cell k. Similarly, $\mathbf{x}_{\mathcal{G}}^{(i)}$, respectively $\mathbf{x}_{\mathcal{G}}^{(S)}$, denote the column vector of signals transmitted by the virtual transmitters corresponding to Layer i, respectively the layers in the set S, from the cells included in the set $\mathcal{G}$ (with entries numerically ordered first by layer and then by cell number). The signal observed by the central cluster's CP can thus be written as (cf. (16.71)):

$$\mathbf{y} = \begin{bmatrix} 1 & \alpha \mathbf{1}_6^T \\ \alpha \mathbf{1}_6 & \mathrm{Circ}(\mathbf{a}) \end{bmatrix} \mathbf{x} + \begin{bmatrix} 0 & \mathbf{0}_6^T \\ \mathbf{0}_6 & \alpha \mathsf{I}_6 \end{bmatrix} \begin{bmatrix} 0 \\ \mathbf{x}_{\mathcal{K}}^{(1)} \end{bmatrix} + \begin{bmatrix} 0 & \mathbf{0}_6^T \\ \mathbf{0}_6 & \mathrm{Circ}(\mathbf{b}) \end{bmatrix} \left(\begin{bmatrix} 0 \\ \mathbf{x}_{\mathcal{J}}^{(1)} \end{bmatrix} + \begin{bmatrix} 0 \\ \mathbf{x}_{\mathcal{J}}^{(2)} \end{bmatrix} \right) + \mathbf{n}, \tag{16.74}$$

where $\mathbf{y} = [y_1, \ldots, y_7]^T$, $\mathbf{x} = [x_1, \ldots, x_7]^T$ (with x_i denoting the *overall* signal transmitted by the user in cell i), $\mathrm{Circ}(\mathbf{v})$ denotes the circulant matrix whose first column is $\mathbf{v}$, $\mathbf{a} = [1, \alpha, 0, 0, 0, \alpha]^T$, $\mathbf{b} = [\alpha, 0, 0, 0, 0, \alpha]^T$, $\mathbf{0}_\ell$ is the $\ell \times 1$ vector of zeros, and $\mathbf{n} = [n_1, \ldots, n_7]^T$ denotes the *effective* noise vector. The latter is given by

$$\mathbf{n} = \begin{bmatrix} 0 \\ \alpha \left(x_8^{(3)} + x_9^{(2)} + x_9^{(3)} + x_{10}^{(3)} \right) \\ \alpha \left(x_{10}^{(3)} + x_{11}^{(2)} + x_{11}^{(3)} + x_{12}^{(3)} \right) \\ \alpha \left(x_{12}^{(3)} + x_{13}^{(2)} + x_{13}^{(3)} + x_{14}^{(3)} \right) \\ \alpha \left(x_{14}^{(3)} + x_{15}^{(2)} + x_{15}^{(3)} + x_{16}^{(3)} \right) \\ \alpha \left(x_{16}^{(3)} + x_{17}^{(2)} + x_{17}^{(3)} + x_{18}^{(3)} \right) \\ \alpha \left(x_{18}^{(3)} + x_{19}^{(2)} + x_{19}^{(3)} + x_8^{(3)} \right) \end{bmatrix} + \mathbf{z}, \tag{16.75}$$

where $\mathbf{z} \sim \mathcal{CN}(\mathbf{0}, \mathsf{I}_7)$ is the vector of AWGNs at the BS receivers. With the underlying assumption on the codebook construction, the covariance matrix of the effective noise is given by

$$\Lambda_N = \mathbb{E}\left\{ \mathbf{n}\mathbf{n}^\dagger \right\} = \mathsf{I}_7 + \alpha^2 (P_2 + P_3) \begin{bmatrix} 0 & \mathbf{0}_6^T \\ \mathbf{0}_6 & \mathsf{I}_6 \end{bmatrix} + \alpha^2 P_3 \begin{bmatrix} 0 & \mathbf{0}_6^T \\ \mathbf{0}_6 & \mathrm{Circ}(\mathbf{c}) \end{bmatrix}, \tag{16.76}$$

where $\mathbf{c} = [2, 1, 0, 0, 0, 1]^T$. The first term in (16.76) is due to the AWGN at the receivers, the second term is due to undecoded signals received from cells forming the set $\mathcal{K}$, and the third term corresponds to undecoded signals received from cells forming the set $\mathcal{J}$.

The underlying channel in (16.74) is a MIMO MAC, where the effective noise incorporates all undecoded out-of-cluster signals (in addition to AWGN). Interpreting each of the layers transmitted by the users as an independent virtual transmitter, the total number of virtual transmitters sums up to 37. However, the inherent symmetry properties of the system model imply that all virtual transmitters corresponding to the same layer within any of the sets of cells $\mathcal{I}$, $\mathcal{J}$, or $\mathcal{K}$ are in fact completely equivalent, and employ the same rates. Hence, the channel can be viewed as an equivalent MIMO MAC with *seven* multiple antenna users. Six users employ six transmit antennas, and correspond to the virtual transmitters of the decoded layers in cell groups $\mathcal{K}$, $\mathcal{J}$, and $\mathcal{I}$. The seventh user is the user in cell 1, who employs a single transmit antenna. By this interpretation, the transmit covariance matrices must be restricted to be multiples of the identity, to facilitate the symmetry properties, and the fact that virtual transmitters in different cells do not cooperate.

As shown in [34], the maximum achievable average per-cell throughput is given by the solution of the following optimization problem:[10]

$$\underset{\{R_c,R_i,P_i\}}{\text{maximize}} \quad \tfrac{1}{7}\left(R_c + 6(R_1 + R_2 + R_3)\right)$$

subject to:

$$R_c + \sum_{\mathcal{S}_\mathcal{K},\mathcal{S}_\mathcal{J},\mathcal{S}_\mathcal{I}} R_i$$
$$\leq I\left\{x_1, \mathbf{x}_\mathcal{K}^{(\mathcal{S}_\mathcal{K})}, \mathbf{x}_\mathcal{J}^{(\mathcal{S}_\mathcal{J})}, \mathbf{x}_\mathcal{I}^{(\mathcal{S}_\mathcal{I})}; \mathbf{y} \mid \mathbf{x}_\mathcal{K}^{(\mathcal{S}_\mathcal{K}^c)}, \mathbf{x}_\mathcal{J}^{(\mathcal{S}_\mathcal{J}^c)}, \mathbf{x}_\mathcal{I}^{(\mathcal{S}_\mathcal{I}^c)}\right\} \qquad (16.77)$$
$$\sum_{\mathcal{S}_\mathcal{K},\mathcal{S}_\mathcal{J},\mathcal{S}_\mathcal{I}} R_i$$
$$\leq I\left\{\mathbf{x}_\mathcal{K}^{(\mathcal{S}_\mathcal{K})}, \mathbf{x}_\mathcal{J}^{(\mathcal{S}_\mathcal{J})}, \mathbf{x}_\mathcal{I}^{(\mathcal{S}_\mathcal{I})}; \mathbf{y} \mid x_1, \mathbf{x}_\mathcal{K}^{(\mathcal{S}_\mathcal{K}^c)}, \mathbf{x}_\mathcal{J}^{(\mathcal{S}_\mathcal{J}^c)}, \mathbf{x}_\mathcal{I}^{(\mathcal{S}_\mathcal{I}^c)}\right\}$$
$$\text{for all } \mathcal{S}_\mathcal{I} \subseteq \{1,2,3\}, \mathcal{S}_\mathcal{J} \subseteq \{1,2\}, \mathcal{S}_\mathcal{K} \subseteq \{1\},$$

where $\sum_{\mathcal{S}_\mathcal{K},\mathcal{S}_\mathcal{J},\mathcal{S}_\mathcal{I}} R_i$ represents the sum-rate of all virtual transmitters corresponding to the constraining mutual information expression, $\{P_1,P_2,P_3\}$ satisfy (16.73), and entries corresponding to empty sets should be ignored. For random Gaussian codebooks, the constraints in (16.77) can be shown to take the following general form

$$aR_1 + bR_2 + cR_3 + dR_c \leq \log_2\left(\frac{A(P_1,P_2,P_3) - B(P_1,P_2,P_3)}{C(P_1,P_2,P_3) - D(P_1,P_2,P_3)}\right), \qquad (16.78)$$

where a,b,c,d are integers, and $A(P_1,P_2,P_3), \ldots, D(P_1,P_2,P_3)$ are *posynomials*.[11] The optimization problem (16.77) is hence *nonconvex*. However, it is still possible to efficiently obtain an approximate solution to the problem, by applying the *complementary geometric programming* (CGP) algorithm introduced in [90] (see [34] for a more elaborate discussion on the application of this algorithm in the current setting).

---

[10] We note here that the proof of this result relies on information theoretic considerations, and does not depend on the assumed Gaussianity of either the transmitted signals or noise [34].

[11] A posynomial is a function $f : \mathbb{R}^n \to \mathbb{R}$ of the form $f(\mathbf{t}) = \sum_{m=1}^{M} \gamma_m t_1^{\beta_{1m}} t_2^{\beta_{2m}} \cdots t_n^{\beta_{nm}}$, where $\gamma_m > 0$, and $\beta_{jm} \in \mathbb{R}$, $m = 1, \ldots, M, j = 1, \ldots, n$. Each term in the sum is referred to as a posynomial term or *monomial*.

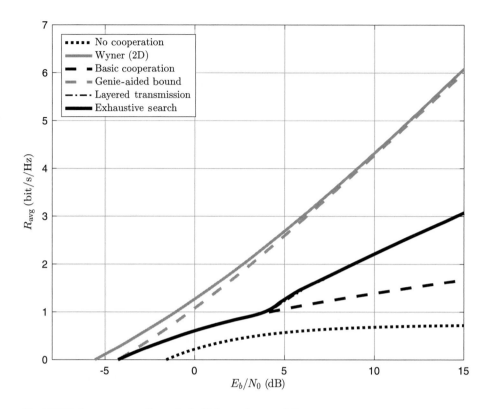

Figure 16.10 Average per-cell spectral efficiency for $\alpha = 0.5$.

Figure 16.10 shows the achievable per-cell spectral efficiency of the layered rate-splitting transmission scheme as a function of the average transmit $\frac{E_b}{N_0}$, for a fixed interference factor of $\alpha = 0.5$. The figure shows both the approximate solution to (16.77), as obtained using the CGP algorithm, as well as corresponding results obtained by means of exhaustive search over the feasible set of optimization variables. Both results exhibit an excellent match.

To further assess the impact of clustering, two lower and upper bounds are also included in Figure 16.10. The lower bounds correspond to the case of *no cooperation* (which boils down to SCP – cf. (16.15)), and to basic "naive" cooperation where joint processing of intracluster transmissions is employed, while *all* out-of-cluster interference is treated as noise (this corresponds to setting $P_1 = P_2 = 0, P_3 = P$). The first of the two upper bounds correspond to full optimum joint MCP [6] and reads:[12]

$$R_{\text{ul,MCP,2D}}(P) = \int_0^1 \int_0^1 \log_2\left(1 + (1 + 2\alpha F_2(\theta_1, \theta_2))^2 KP\right) d\theta_1 \, d\theta_2, \qquad (16.79)$$

where

$$F_2(\theta_1, \theta_2) \triangleq \cos(2\pi\theta_1) + \cos(2\pi\theta_2) + \cos(2\pi(\theta_1 + \theta_2)). \qquad (16.80)$$

---

[12] This is the analog of (16.13) for the 2D Wyner model.

Table 16.1 Clustered multicell processing with non-overlapping clusters: extreme-SNR characterization for nonfading channels.

Scheme	$\left(\frac{E_b}{N_0}\right)_{min}$	$\left(\frac{E_b}{N_0}\right)_{min}\bigg|_{\alpha=0.5}$	S_0	$S_0\big|_{\alpha=0.5}$	S_∞
Layered transmission	$\frac{\log_e 2}{1+\frac{24}{7}\alpha^2}$	$-4.28\,(\text{dB})$	$\frac{2}{7}\frac{(7+24\alpha^2)^2}{336\alpha^4+192\alpha^3+180\alpha^2+7}$	0.498	$\frac{3}{7}$
Genie-aided bound	$\frac{\log_e 2}{1+\frac{24}{7}\alpha^2}$	$-4.28\,(\text{dB})$	$\frac{2}{7}\frac{(7+24\alpha^2)^2}{204\alpha^4+144\alpha^3+144\alpha^2+7}$	0.655	1
Optimum joint MCP	$\frac{\log_e 2}{1+6\alpha^2}$	$-5.57\,(\text{dB})$	$2\frac{(1+6\alpha^2)^2}{90\alpha^4+48\alpha^3+36\alpha^2+1}$	0.578	1

The second upper bound reflects a "genie-aided" achievable throughput, where the genie provides as side information all out-of-cluster interfering signals (the setting thus corresponds to a *single isolated cluster*). The results clearly indicate that the rate-splitting scheme significantly enhances performance as one moves away from the low-interference regime. In the latter case, treating interference as noise is known to be optimal [36, 37], and the spectral efficiency coincides with the basic cooperation lower bound. The performance enhancement over SCP is clearly apparent for the whole $\frac{E_b}{N_0}$ range, and becomes more pronounced for high $\frac{E_b}{N_0}$, as the latter scheme is interference limited.

The extreme-SNR characterization of the clustered layered rate-splitting transmission strategy is summarized in Table 16.1. Note that the clustering scheme induces a loss in the minimum $\frac{E_b}{N_0}$ that enables reliable communication when compared to optimum joint MCP (16.79), as the latter achieves the full array diversity gain factor of $1 + 6\alpha^2$ in the low-SNR regime (the loss amounts to 1.29 dB for $\alpha = 0.5$, see Figure 16.10). However, the minimum $\frac{E_b}{N_0}$ coincides with the corresponding value of the "genie-aided" upper bound, implying that the genie cannot improve the array gain in this regime. Both low- and high-SNR slopes are, however, lower than the corresponding slopes of the upper bounds, indicating the inherent performance loss due to *clustered* multicell processing in the absence of fading. We finally note in this respect that, as it turns out, when Rayleigh fading is introduced the clustered layered transmission scheme can indeed achieve the *maximum possible high-SNR slope*, in the large number of users per-cell limit. The reader is referred to [35] for a more detailed account on the impact of fading.

16.5.2 Overlapping Clusters

The discussion in this section focuses on the Gaussian Wyner linear cell-array model with $L_\ell = L_r = 1$ in Figure 16.2, and on the Gaussian soft-handoff model with $L_\ell = 1$ and $L_r = 0$ in Figure 16.3. Like in the previous section, in the uplink messages are decoded based on clusters of receive antennas, and in the downlink each transmit signal encodes a cluster of messages. The difference is that here different clusters overlap.

More specifically, a regular cluster organization is considered where in the uplink the BS of a given cell m decodes its desired messages based on its own received signal as well as based on the received signals at the BSs in cells $m-1, m-2, \ldots, m-\tau_\ell$ and in cells $m+1, m+2, \ldots, m+\tau_r$. In the downlink, the BS of a given cell m encodes the messages intended for the mobile users in this cell m as well as the messages for the mobile users in cells $m-1, m-2, \ldots, m-\tau_\ell$ and in cells $m+1, m+2, \ldots m+\tau_r$.

For simplicity, this section assumes intracell TDMA scheduling and focuses on the high-SNR slope. The goal here is to understand the influence of the cluster size, i.e., of the parameters τ_ℓ and τ_r, on the high-SNR slope of the system. The results in the following subsections are derived in [33].

Gaussian Soft-Handoff Model

Uplink

Consider first the uplink where each BS can decode its intended message based on its own receive antenna as well as based on the receive antennas of the BSs in cells $m-1, \ldots, m-\tau_\ell$ and in cells $m+1, \ldots, m+\tau_r$. As the following result shows, the parameters τ_ℓ and τ_r do not have the same impact on the high-SNR slope. Due to the asymmetry in the network it can be more beneficial to access more antennas in left-neighboring cells than in right-neighboring cells. In fact, the high-SNR slope is:

$$\mathcal{S}_{\infty,\text{ul,OvCI}} = 1 - \left\lceil \frac{M - (\tau_\ell + 1)}{\tau_\ell + \tau_r + 2} \right\rceil \cdot \frac{1}{M}. \tag{16.81}$$

The high-SNR slope in (16.81) is achieved by deactivating every $\tau_\ell + \tau_r + 2$-th transmitter and by appropriately combining successive interference cancellation from the right and from the left at the receiver side.

Notice that the asymmetry in the roles of τ_ℓ and τ_r with respect to the high-SNR slope in (16.81) is only a boundary effect and vanishes in the limit as the number of cells M tends to infinity. In fact, the limiting high-SNR slope reads

$$\lim_{M \to \infty} \mathcal{S}_{\infty,\text{ul,OvCI}} = \frac{\tau_\ell + \tau_r + 1}{\tau_\ell + \tau_r + 2}. \tag{16.82}$$

Downlink

Consider now the downlink where each BS can construct its transmit signal based on the message intended for the mobile user in its own cell as well as the messages intended for the mobiles in cells $m-1, \ldots, m-\tau_\ell$ and in cells $m+1, \ldots, m+\tau_r$. The high-SNR slope of this downlink setup is

$$\mathcal{S}_{\infty,\text{dl,OvCI}} = 1 - \left\lceil \frac{M - (\tau_\ell + 1)}{\tau_\ell + \tau_r + 2} \right\rceil \cdot \frac{1}{M}. \tag{16.83}$$

The high-SNR slope in (16.83) is achieved by deactivating every $\tau_\ell + \tau_r + 2$-th transmitter and by appropriately combining DPC from the right and from the left.

The result in (16.83) indicates that it is more valuable (in terms of high-SNR slope) for a BS to have access to the messages intended for mobile users in cells to its left than to the messages for mobile users in cells to its right. But similarly as for the uplink, this is only a boundary effect that vanishes as $M \to \infty$. We also notice that the high-SNR slopes for the uplink and downlink in (16.81) and (16.83) coincide.

El Gamal et al. [91] considered a related scenario where the message clusters can be designed freely. The only constraint is that each BS can access no more than $\tau_\ell + \tau_r + 1$ messages, but one can freely choose which ones. Under this relaxed constraint, the asymptotic high-SNR slope as $M \to \infty$ was shown to be $\frac{2(\tau_\ell + \tau_r + 1)}{2(\tau_\ell + \tau_r + 1) + 1}$, and thus larger than the asymptotic high-SNR slope in (16.82). The optimal high-SNR slope in this modified setup is attained by designing clusters that contain messages of $\tau_\ell + \tau_r + 1$ consecutive cells, but the cluster at the BS of cell m does not necessarily include the message intended for the mobile in cell m.

Gaussian Wyner Linear Cell-Array Model

For simplicity, we will focus on the symmetric scenario $\tau_\ell = \tau_r = \tau$ and assume that all principal submatrices of the channel matrix H_M in (16.3) and (16.7) are of full rank. Recall that the matrix H_M is Toeplitz and for $L_\ell = L_r = 1$ it has entry 1 on the diagonal and entry α on the first upper and lower diagonals. For most values of the cross-channel gain α all principal submatrices of H_M have full rank. (In fact, the set of α values for which this is not the case has Lebesgue measure 0.)

Uplink

For the uplink of the Gaussian Wyner linear cell-array model, when each receiving BS can decode based on its own receive signal as well as based on the receive signals of the BSs in cells $m - 1, \ldots, m - \tau$ and in cells $m + 1, \ldots, m + \tau$, the high-SNR slope is:

$$\mathcal{S}_{\infty,\mathrm{ul},\mathrm{OvCl}} = 1 - \left\lfloor \frac{M}{\tau + 2} \right\rfloor \cdot \frac{1}{M}. \tag{16.84}$$

The high-SNR slope in (16.84) is achieved by deactivating the last transmitter/receiver pair in every block of $\tau + 2$ transmitter/receiver pairs, and by appropriately combining successive interference cancellation from the right and from the left at the receiver side. This result continues to hold (with probability 1) even when the channel coefficients are not all equal to the same value α (see Figure 16.2), but rather each of the nonzero channel coefficients is drawn according to a continuous distribution.

The finite-SNR performance of this system was considered in [32], where Levy and Shamai provided upper and lower bounds on the per-cell capacity. The lower bound is based on rate splitting and the upper bound on genie-aided arguments.

Downlink

For the downlink of the Gaussian Wyner linear cell-array model, when the BS of cell m can produce its transmit signal based on the message for the mobile in its own cell and the messages for the mobiles in cells $m - 1, \ldots, m - \tau$ and $m + 1, \ldots, m + \tau$, the high-SNR slope is

$$\mathcal{S}_{\infty,\mathrm{dl},\mathrm{OvCl}} = 1 - \left\lfloor \frac{M}{\tau + 2} \right\rfloor \cdot \frac{1}{M}. \tag{16.85}$$

The high-SNR slope in (16.85) is achieved by deactivating the last transmitter/receiver pair in every block of $\tau + 2$ transmitter/receiver pairs, and by appropriately combining DPC from the right and from the left at the transmitter side. Note that also for the Gaussian Wyner linear cell-array model, the high-SNR slopes for the uplink and downlink coincide. Bande et al. [92] showed that a larger high-SNR slope per user can be attained over the downlink of the Gaussian Wyner linear cell-array model when each BS can choose which $2\tau + 1$ messages it wishes to learn.

Concluding Remarks

The high-SNR slope results in this section show duality of uplink and downlink even in a scenario with overlapping clusters of BSs. The results presented here further extend to scenarios with clusters both at the transmitter and at the receiver side (used to model at the same time cooperation between BSs and between mobile users) (see [33]).

Uplink and downlink performances of networks with clustered encoding or decoding have also been studied for other network models [92–97]. For example, in [93] the authors completely characterized the set of networks and transmitter side information that for a finite number of users K have optimum high-SNR slope equal to 1 or $1 - \frac{1}{K}$. In [97], a network with clustered encoding is presented where the high-SNR slope is even increased by adding an interference link to the network.

16.6   User Cooperation

User cooperation can be employed between users in the same cell (*intracell cooperation*) or between users in adjacent cells (*intercell cooperation*). For the Wyner-type model considered in this chapter, intercell cooperation allows a mobile user in a given cell m to directly communicate with the mobile users in the two adjacent cells $m - 1$ and $m + 1$. See Figure 16.11 for a depiction of this setup for the Gaussian Wyner model (adaptation to the corresponding soft-handoff model is straightforward). Cooperation can still be established among mobile users in cells that are further away by means of a

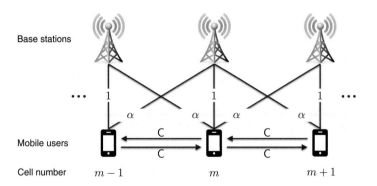

Figure 16.11 Gaussian Wyner model with cooperation links between mobiles of capacity C (bit/s/Hz).

multihop communication from mobile user to mobile user. For practical considerations (delay and complexity), the number of such hops is limited to a maximum number of $\kappa_{\max}$ hops.

The setup is then described in more detail as follows. For each round $j = 1, \ldots, \kappa_{\max}$, a mobile user in a given cell m can send a cooperation signal to the mobile users in the two adjacent cells $m - 1$ and $m + 1$, where this cooperation signal is calculated as a function of the mobile user's own message (uplink) or received signal (downlink) and the cooperation signals obtained in the $j-1$ previous rounds. The total of the cooperation signals sent in the $\kappa_{\max}$ rounds over each of the cooperation links is subject to a total capacity constraint C (bit/s/Hz).

16.6.1   Gaussian Soft-Handoff Model

Consider the Gaussian soft-handoff model ($L_\ell = 1$ and $L_r = 0$) with SCP at the BSs, intracell TDMA scheduling, and intercell cooperation between mobile users (see Figure 16.11). Both for the uplink as well as for the downlink, the high-SNR slope per-user with intercell cooperation over $\kappa_{\max}$ rounds is [28]

$$\mathcal{S}_{\infty,ul,coop} = \mathcal{S}_{\infty,dl,coop} = \begin{cases} \frac{1}{2} + \mu_\infty, & \text{if } \mu_\infty \leq \frac{\kappa_{\max}}{2(1+\kappa_{\max})} \\ \frac{2\kappa_{\max}+1}{2\kappa_{\max}+2}, & \text{if } \mu_\infty > \frac{\kappa_{\max}}{2(1+\kappa_{\max})}, \end{cases} \tag{16.86}$$

where μ_∞ is the high-SNR slope of the cooperation links, i.e., $C = \mu_\infty \log_2 P$.

In the uplink, the above high-SNR slope is achieved by silencing every $2\kappa_{\max} + 2$-th transmitter – which splits the network into non-interfering subnetworks with $2\kappa_{\max} + 1$ transmitter/receiver pairs – and then combining DPC from left and right at the transmitter side to attain a high-SNR slope of $2\kappa_{\max} + 1$ in each of the resulting subnetworks. This requires that mobile users use the cooperation links to share quantized versions of their transmit signals with their immediate right or left neighbors. In the downlink, the desired high-SNR slope is achieved by silencing the same set of transmitters and by combining successive interference cancellation from the left and from the right at the receiving mobile users. This requires that mobile users use the cooperation links to share their decoded signals with their right or left neighbors.

The high-SNR slope result in (16.86) exhibits that for small cooperation capacity C with $\mu_\infty \leq 1/4$, one cooperation round attains the largest high-SNR slope that is possible when the number of cooperation rounds is not limited. In contrast, for larger values of C, constraining the number of cooperation rounds limits the achievable high-SNR slope. In particular, for a fixed number of cooperation rounds $\kappa_{\max} < \infty$, the high-SNR slope with user cooperation saturates at $\frac{2\kappa_{\max}+1}{2\kappa_{\max}+2}$ even when the capacity of the cooperation links is unbounded, i.e., $C \to \infty$.

16.6.2   Gaussian Wyner Linear Cell-Array Model

Consider the uplink of the Gaussian Wyner linear cell-array model (assuming $L_\ell = L_r = 1$) with full joint MCP at the BSs and with intercell or intracell cooperation at the mobiles. For the case of intercell cooperation only and when TDMA is employed for mobile users in the same cell, [40] proposes the following strategy. Each mobile user of cell m sends a part of its message that is of rate $\frac{C}{\kappa_{max}}$ to the mobiles in cells $m - \kappa_{max}, \ldots, m-1$ and in cells $m+1, \ldots, m+\kappa_{max}$ by employing multihop communication from mobile user to mobile user. The shared message part of the active mobile user in cell m is then jointly transmitted by the $2\kappa_{max} + 1$ active mobile users in cells $m - \kappa_{max}, \ldots, m+\kappa_{max}$ over the uplink to the BSs that employ full joint MCP. As the number of cells $M \to \infty$, the rate achieved per-cell can be expressed in an integral form similar to (16.13), given by [40]:

$$R_{ul,inter} = \max_{P_c,P_p,H_c} \min \left\{ \int_0^1 \log_2 \left(1 + P_p H(f)^2 + P_c H(f)^2 |H_c(f)|^2\right) df, \right.$$
$$\left. \int_0^1 \log_2(1 + P_p H(f)^2) \, df + \frac{C}{\kappa_{max}} \right\}. \qquad (16.87)$$

Here, P_c and P_p denote, respectively, the powers each mobile user allocates for the transmission of the common message parts and the private message part (with $P_c + P_p = P$); $H_c(f)$ is the frequency response of the $2 + \kappa_{max} + 1$-taps filter employed by the set of $2\kappa_{max} + 1$ mobile users to jointly transmit a common message part; and $H(f) = 1 + 2\alpha \cos(2\pi f)$ is the frequency response of the Gaussian Wyner linear cell-array network when this network is interpreted as an ISI channel. As expected, the achievable per-cell rate in (16.87) increases with the number of cooperation rounds κ_{max}, because a larger value of κ_{max} enlarges the optimization domain of $H_c(f)$. The optimal choice for $H_c(f)$ approximates the waterfilling solution. The minimum $\frac{E_b}{N_0}$ that enables reliable communications is approximately given by [40]

$$\left(\frac{E_b}{N_0}\right)_{min,ul,inter} \approx \frac{\log_e 2}{(1 + 2\alpha)^2(1 - 8\alpha\pi^2/(3(1 + 2\alpha)\kappa_{max}^2))}. \qquad (16.88)$$

Consider now a scenario with intracell cooperation with J mobiles in each cell and with a maximum of $\kappa_{max} = J$ cooperation rounds, where the proximity of the mobile users means that all users in a cell can receive the cooperation signal sent by a mobile user in the same cell. Assume that each mobile user shares a part of its message that is of rate not exceeding C with all other mobile users in the cell, and that these shared message parts are jointly transmitted by all the mobiles in a cell. The described strategy achieves the per-cell rate [40]:

$$R_{ul,intra} = \max_{P_c,P_p} \min \left\{ \int_0^1 \log_2 \left(1 + (P_p + JP_c)H(f)^2\right) df, \right.$$
$$\left. \int_0^1 \log_2(1 + P_p H(f)^2) \, df + \frac{C}{\kappa_{max}} \right\}. \qquad (16.89)$$

16.6.3 Concluding Remarks

Achievable rates for other interference networks with conferencing were given in [98–101]. The work in [42] considers a state-dependent MAC with distributed state information and where transmitters can cooperate over a conferencing link. It is shown in [42] that in such a scenario the transmitters should use the cooperation links not only to exchange parts of their messages but also part of their knowledge about the channel state.

16.7 Relaying

Consider the uplink of the Gaussian Wyner linear cell-array model where relays aid the transmission from the mobile users to the BSs. These relays can either be other mobile users or dedicated relay terminals. Two scenarios are distinguished: a scenario without a direct communication link from the mobile users to the corresponding BSs where the communication needs to be multihop, and a scenario where such a direct link does exist. The former scenario is motivated by setups where relay terminals are on the direct path from mobile users to BSs. For simplicity, the focus here is on TDMA, where during each time frame a single mobile user per cell actively transmits its message to the BS. The two scenarios are illustrated in Figure 16.12. The dashed connections indicate the direct path communications from mobile users to BSs, which are only present in the second scenario.

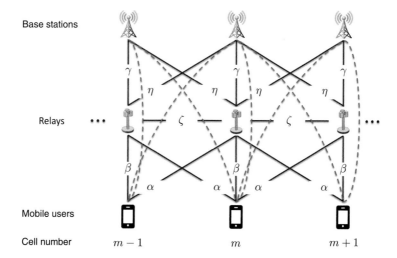

Figure 16.12 The Gaussian Wyner linear cell-array model with relays. The direct mobile-to-relay and relay-to-BS channel gains are denoted, respectively, by β and γ. The corresponding intercell interference channel gains are denoted, respectively, by α and η. Also, ζ denotes the inter-relay channel gain. The dashed connections indicate the direct paths from mobile users to BSs, which are present only in the second scenario that we consider but not in the first (the corresponding channel gains are omitted for simplicity).

16.7.1   Multihop Networks

For the Gaussian Wynerr linear cell-array model, a reasonable assumption is that relays that are located toward the center of a cell observe the signals sent by the mobile users in the same cell (with some channel gain β), as well as the signals sent by the mobile users and the relays in the two adjacent cells (with channel gains α and ζ, respectively). For this scenario and where there are no direct paths from mobile users to BSs (see Figure 16.12 without the dashed connections), [44] analyzes the uplink rates achieved by AF and CF strategies at the relays when the BSs perform either SCP or full joint MCP. The resulting rate-expressions are presented in the form of solutions to fixed-point equations.

Numerical investigation shows that for both AF and CF strategies, full joint MCP significantly improves performance over SCP. Moreover, for full joint MCP, CF improves over AF for large relay transmission powers and for large inter-relay interference. A specific numerical example for a unit noise variance at all terminals, transmit power of 20 dB at the mobile users, direct channel gains $\beta = \gamma = 1$, and interference channel gains $\alpha = \eta = \sqrt{0.2}$ and $\zeta = \sqrt{0.1}$, is presented in Figure 16.13. The per-cell rate is plotted as function of the ratio of relay-to-mobile power for the AF and CF strategies under both SCP and full joint MCP at the BSs.

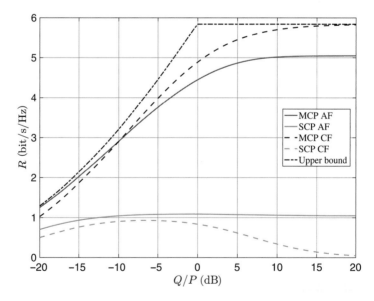

Figure 16.13 Per-cell achievable uplink rates, plotted as a function of the ratio of relay power Q versus user power P. The results are shown for a setup with unit Gaussian noise variances, mobile user power $P = 20$ dB, direct mobile-to-relay and relay-to-BS channel gains $\beta = \gamma = 1$, and crossover channel gains $\alpha = \eta = \sqrt{0.2}$ and $\zeta = \sqrt{0.1}$. Relays are assumed to use their full power (see [44] for the beneficial impact of optimized relay power allocation for SCP with CF). A "cut-set like" upper bound on the achievable rate [44] is also shown for comparison.

16.7.2 Networks with Direct Paths

As opposed to the previous section, here it is assumed that the signal transmitted by a mobile user in a given cell is also observed at the BS in the same cell and at the BSs in the two adjacent cells, as well as at the relay in the same cell (see the dashed connections in Figure 16.12). The relays apply an AF strategy with either *orthogonal* or *nonorthogonal* access, where in the former protocol the transmitting mobile user and the relay terminal transmit in different bandwidths or time frames, which is not the case for the second protocol.

Closed-form expressions for the rates achieved by AF under both protocols were given in [39]. When the number of cells $M \to \infty$, these rate expressions take integral forms:

$$\lim_{M \to \infty} R_{\text{ul,AF}} = \int_0^1 \log_2 \left(P \frac{r_0 + 2 \sum_{k=1}^3 r_k \cos(k\pi\theta)}{s_0 + \sum_{k=1}^2 s_k \cos(k\pi\theta)} \right) d\theta , \qquad (16.90)$$

where the expressions for the parameters r_0, r_1, r_2, r_3 and s_0, s_1, s_2 are different under orthogonal and nonorthogonal access and can be found in [39].

Numerical examinations in [39] show that under the full joint MCP considered here, even in the advantageous scenario where the helping relay is located in the direct line-of-sight path from the transmitting mobile to the BS, AF protocols deteriorate the performance compared to direct transmission when the SNR of the links from the mobile users to the relays is large. This is because the AF protocol inherently introduces additional noise to the system. AF protocols are beneficial at low SNR. In particular, the orthogonal access protocol outperforms the nonorthogonal access protocol at very low SNR, while at medium SNR nonorthogonal access AF performs best. In contrast to these results for MCP, under SCP at the BSs both AF protocols outperform direct transmission for almost all values of SNR (see again [39]).

For the fading Wyner linear cell-array model, [43] analyzed the performance of DF with orthogonal access, i.e., with mobile users transmitting in different time frames or bandwidths than the relaying terminal. At low SNR and for full joint MCP at the BSs, the minimum $\frac{E_b}{N_0}$ that enables reliable communications is

$$\left(\frac{E_b}{N_0} \right)_{\text{min,ul,inter}} = 2 \log_e 2 . \qquad (16.91)$$

16.8 Caching Protocols

Consider the downlink of the Gaussian soft-handoff model and the Gaussian Wyner linear cell-array model with mobile users having cache memories where they can pre-store contents. Most previous works on interference networks with receiver caching considered fully connected interference networks where for successful delivery it suffices that each demanded message is stored at least at one of the BSs of the network. In

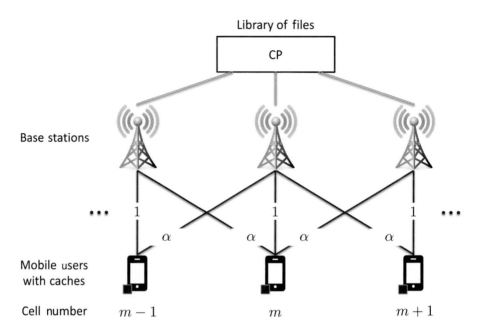

Figure 16.14 C-RAN caching architecture for the Gaussian Wyner linear cell-array model.

sparsely connected Wyner-type models, this condition is not sufficient and the C-RAN architecture in Figure 16.14 is proposed to overcome this limitation.

Let a CP have access to a large library with D files, each consisting of nR bits, and let it be connected with M BSs, which in their turn communicate with M mobile users over one of the two aforementioned Wyner-type networks. Each mobile user eventually wishes to download one of the files in the library, but it is a priori not clear which one. In fact, communication takes place in three phases, where the first phase takes place well in advance, before mobile users have decided which files to download. During this first phase, the server prefetches information about files in the library in the users' cache memories. This communication typically happens in periods of low network congestion (e.g., at night) and is thus modeled as error- and noise-free. During the second phase the users decide which file to download and tell their demands to the BSs in their own cell and in the adjacent cells, which then download the corresponding files from the CP. In the third phase the BSs *cooperatively* deliver the requested files to the mobile users, where they exploit the fact that the mobile users have prestored information in their cache memories.

Applying naive cache placement and delivery strategies, with $n\mu$ bits of cache memory at each user, n being the block length of communication, it is possible to achieve a rate

$$R_{\text{naive}} = R_{\text{no\_cache}} + \frac{\mu}{D}, \tag{16.92}$$

where $R_{\text{no\_cache}}$ is the maximum rate achieved when users have no cache. The following results from [45] show that by carefully choosing the cache placement and delivery strategies, it is possible to achieve rates that go far beyond the naive additive gain of $\frac{\mu}{D}$. In fact, for sufficiently large interference channels it is possible to attain the same high-SNR slope per user as over an interference-free network. In other words, in this regime caching allows us to eliminate all interference without sacrificing the local caching gain, i.e., the benefit that some information is locally stored at the mobile users.

More specifically, for the Gaussian soft-handoff model, [45] shows that the cache-aided high-SNR slope per user $S_{\infty,\text{dl,cache}}$ is bounded as:

$$\min\left\{\frac{2}{3} + \frac{3\mu_\infty}{2D}, 1 + \frac{\mu_\infty}{D}\right\} \leq S_{\infty,\text{dl,cache}} \leq \min\left\{\frac{2}{3} + \frac{3\mu_\infty}{D}, 1 + \frac{\mu_\infty}{D}\right\}, \quad (16.93)$$

where μ_∞ denotes the high-SNR slope of the cache size. That means, for high SNR, each mobile user has a cache memory of size of $n \cdot \mu_\infty \log_2 P$ bits. The lower bound on the high-SNR slope is achieved by combining coded caching with broadcast communication at the BSs with cache-aided interference cancellation at the mobiles. Notice that this strategy achieves the performance of an interference-free network with caches when $\frac{\mu_\infty}{D} \geq \frac{2}{3}$:

$$S_{\infty,\text{dl,cache}} = 1 + \frac{\mu_\infty}{D}, \qquad \text{for} \quad \mu_\infty > \frac{2}{3}D. \quad (16.94)$$

For the Gaussian Wyner linear cell-array model, [45] bounds the cache-aided high-SNR slope per user as:

$$\min\left\{\frac{2}{3} + \frac{4\mu_\infty}{3D}, 1 + \frac{\mu_\infty}{D}\right\} \leq S_{\infty,\text{dl,cache}} \leq \min\left\{\frac{2}{3} + \frac{6\mu_\infty}{D}, 1 + \frac{\mu_\infty}{D}\right\}. \quad (16.95)$$

The lower bound on the high-SNR slope is achieved by a scheme that periodically silences some of the BSs and lets the active BSs broadcast coded caching packets. To distribute the load equally among the BSs, different instances of this scheme are time-shared, where a different set of BSs is silenced in each of these instances. This strategy achieves the performance of an interference-free network with caches when $\frac{\mu_\infty}{D} \geq 1$:

$$S_{\infty,\text{dl,cache}} = 1 + \frac{\mu_\infty}{D}. \quad (16.96)$$

Figure 16.15 illustrates the upper and lower bounds in (16.93) and (16.95). Upper bounds are illustrated with dashed lines and lower bounds with solid lines.

A similar multistage system has been considered in [102], however for noise-free broadcast systems. Other models for cellular networks with cache memories were studied in [103–106]. An appealing option in cellular systems is to equip BSs with cache memories. In C-RAN architectures this leads to the framework of *Fog-RANs* (F-RANs), as discussed in Chapter 6.

16.9   Summary and Outlook

This chapter discusses the influence that various kinds of cooperation between mobile users or between BSs have on the rates that can be achieved in future 5G networks.

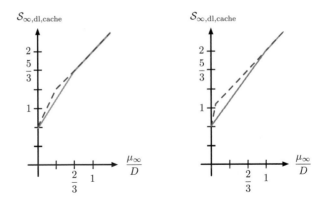

Figure 16.15 Upper and lower bounds on the cache-aided high-SNR slope per user. The figure on the left-hand side depicts bounds for the Gaussian soft-handoff model, and the figure on the right-hand side depicts bounds for the Gaussian Wyner linear cell-array model.

As illustrated in this chapter, when focusing on Wyner-type models, analytical closed-form solutions can be obtained for the most important cooperation scenarios. For some scenarios, however, this model might be too simple. In fact, it can only model short-range interference between neighboring transmitters and receivers. Moreover, it assumes some regularity in the interference pattern, which hinders its accounting for different locations of the mobiles within the cells. See, e.g., [107–109] for some discussion on this aspect. When there is a large number of mobile users in each cell, an averaging effect takes place so that user locations can be considered almost constant, thus rendering the Wyner model more accurate [8]. However, in future 5G networks, irregularities of interference patterns are not only caused by different user locations within the cells, but also by the tiered network architecture involving various kinds of BSs (macro, pico, femto BSs) that have different signaling and interference strengths.

First attempts to overcome the aforementioned modeling limitations in general are based on the ideas of introducing user activity patterns [48, 49], or modeling user locations using Poisson point processes, and analyzing the performance of such systems using tools from stochastic geometry. A stochastic geometry analysis of cellular networks with cooperative communications has, for example, been performed in [110, 111].

Future research directions of information theoretic studies on Wyner-type networks include also the analysis of irregular interference patterns, of frequency-selective channels, for which [68] provides a mature framework, or of coding delays, for example by means of the finite-blocklength coding framework introduced in [112]. Concerning coding delays, a recent study [113] has examined the rate-tradeoff between two types of coexisting data traffics, one with more stringent delay constraints than the other. The results show that on a Wyner soft-handoff network, depending on the regime of operation, imposing a strict delay on one of the traffics does not harm the overall sum-rate. An interesting future research direction is to further refine this analysis and extend it to more involved network models.

Irrespective of the precise network model, for the future it will be interesting to analyze all the mentioned extensions also from a more system-wide perspective that additionally accounts for signaling overheads and synchronization issues.

Acknowledgments

The work of B. M. Zaidel and S. Shamai (Shitz) was supported by the Heron Consortium of the Israel Innovation Authority, The work of S. Shamai (Shitz) has also been supported by the European Union's Horizon 2020 Research and Innovation Programme, under grant agreement no. 694630. The work of M. Wigger was supported by the same program under grant agreement no. 715111.

References

[1] M. A. Maddah-Ali and U. Niesen, "Fundamental limits of caching," *IEEE Trans. Inf. Theory*, vol. 60, no. 5, pp. 2856–2867, May 2014.

[2] E. C. van der Meulen, "Three-terminal communication channels," *Adv. Appl. Prob.*, vol. 3, pp. 120–154, 1971.

[3] T. Cover and A. El Gamal, "Capacity theorems for the relay channel," *IEEE Trans. Inf. Theory*, vol. 25, no. 5, pp. 572–584, Sep. 1979.

[4] F. Willems, "The discrete memoryless multiple access channel with partially cooperating encoders (corresp.)," *IEEE Trans. Inf. Theory*, vol. 29, no. 3, pp. 441–445, May 1983.

[5] R. Dabora and S. Servetto, "Broadcast channels with cooperating decoders," *IEEE Trans. Inf. Theory*, vol. 52, no. 12, pp. 5438–5454, Dec. 2006.

[6] A. D. Wyner, "Shannon-theoretic approach to a Gaussian cellular multiple-access channel," *IEEE Trans. Inf. Theory*, vol. 40, no. 6, pp. 1713–1727, Nov. 1994.

[7] S. V. Hanly and P. A. Whiting, "Information-theoretic capacity of multi-receiver networks," *Telecommun. Syst.*, vol. 1, pp. 1–42, 1993.

[8] J. Xu, J. Zhang, and J. G. Andrews, "On the accuracy of the Wyner model in cellular networks," *IEEE Trans. Wireless Commun.*, vol. 10, no. 9, pp. 3098–3109, Sep. 2011.

[9] O. Simeone, N. Levy, A. Sanderovich, et al., "Cooperative wireless cellular systems: An information-theoretic view," *Found. Trends Commun. Inf. Theory*, vol. 8, no. 1–2, pp. 1–177, 2012.

[10] B. M. Zaidel, S. Shamai (Shitz), and S. Verdú, "Multi-cell uplink spectral efficiency of coded DS-CDMA with random signatures," *IEEE J. Sel. Areas Commun.*, vol. 19, no. 8, pp. 1556–1569, Aug. 2001.

[11] D. Aktas, M. N. Bacha, J. Evans, and S. V. Hanly, "Scaling results on the sum capacity of cellular networks with MIMO links," *IEEE Trans. Inf. Theory*, vol. 52, no. 7, pp. 3264–3274, Jul. 2006.

[12] O. Somekh, B. M. Zaidel, and S. Shamai (Shitz), "Spectral efficiency of joint multiple cell-site processors for randomly spread DS-CDMA systems," *IEEE Trans. Inf. Theory*, vol. 53, no. 7, pp. 2625–2637, Jul. 2007.

[13] R. Cheng and S. Verdú, "The effect of asynchronism on the capacity of Gaussian multiple-access channels," *IEEE Trans. Inf. Theory*, vol. 38, no. 1, pp. 2–13, Jan. 1992.

[14] S. Verdú, "The capacity region of the symbol-asynchronous Gaussian multiple-access channel," *IEEE Trans. Inf. Theory*, vol. 35, no. 4, pp. 733–751, Jul. 1989.

[15] W. Hirt and J. L. Massey, "Capacity of the discrete-time Gaussian channel with intersymbol interference," *IEEE Trans. Inf. Theory*, vol. 34, pp. 380–388, May 1988.

[16] O. Somekh and S. Shamai (Shitz), "Shannon-theoretic approach to a Gaussian cellular multi-access channel with fading," *IEEE Trans. Inf. Theory*, vol. 46, no. 4, pp. 1401–1425, Jul. 2000.

[17] O. Somekh, B. M. Zaidel, and S. Shamai (Shitz), "Sum rate characterization of joint multiple cell-site processing," *IEEE Trans. Inf. Theory*, vol. 53, no. 12, pp. 4473–4497, Dec. 2007.

[18] S. Jing, D. N. C. Tse, J. Hou, et al., "Multi-cell downlink capacity with coordinated processing," in *Proc. 2007 Inf. Theory Appl. Workshop (ITA)*, San Diego, CA, Jan. 29–Feb. 2, 2007.

[19] S. Shamai (Shitz) and A. D. Wyner, "Information-theoretic considerations for symmetric, cellular, multiple-access fading channels. Parts I & II," *IEEE Trans. Inf. Theory*, vol. 43, no. 6, pp. 1877–1911, Nov. 1997.

[20] N. Levy, O. Zeitouni, and S. Shamai (Shitz), "Central limit theorem and large deviations of the fading Wyner cellular model via product of random matrices theory," *Probl. Inf. Transm.*, vol. 45, no. 1, pp. 5–22, 2009.

[21] N. Levy, O. Zeitouni, and S. Shamai (Shitz), "On information rates of the fading Wyner cellular model via the Thouless formula for the strip," *IEEE Trans. Inf. Theory*, vol. 56, no. 11, pp. 5495–5514, Nov. 2010.

[22] N. Levy, O. Somekh, S. Shamai (Shitz), and O. Zeitouni, "On certain large random Hermitian Jacobi matrices with applications to wireless communications," *IEEE Trans. Inf. Theory*, vol. 55, no. 4, pp. 1534–1554, Apr. 2009.

[23] S. Shamai (Shitz) and B. M. Zaidel, "Enhancing the cellular downlink capacity via co-processing at the transmitting end," in *Proc. 2001 IEEE 53rd Veh. Techn. Conf. (VTC)*, vol. 3, Rhodes, May 6–9, 2001, pp. 1745–1749.

[24] A. Sanderovich, O. Somekh, H. V. Poor, and S. Shamai (Shitz), "Uplink macro diversity of limited backhaul cellular network," *IEEE Trans. Inf. Theory*, vol. 55, no. 8, pp. 3457–3478, Aug. 2009.

[25] O. Simeone, O. Somekh, H. V. Poor, and S. Shamai (Shitz), "Downlink multicell processing with limited-backhaul capacity," *EURASIP J. Adv. Signal Process.*, vol. 2009, pp. 3:1–3:10, Feb. 2009.

[26] H. Weingarten, Y. Steinberg, and S. Shamai (Shitz), "The capacity region of the Gaussian multiple-input multiple-output broadcast channel," *IEEE Trans. Inf. Theory*, vol. 52, no. 9, pp. 3936–3964, Sep. 2006.

[27] O. Simeone, O. Somekh, V. H. Poor, and S. Shamai (Shitz), "Local base station cooperation via finite-capacity links for the uplink of linear cellular networks," *IEEE Trans. Inf. Theory*, vol. 55, no. 1, pp. 190–204, Jan. 2009.

[28] M. Wigger, R. Timo, and S. Shamai (Shitz), "Conferencing in Wyner's asymmetric interference network: Effect of number of rounds," *IEEE Trans. Inf. Theory*, vol. 63, no. 2, pp. 1199–1226, Feb. 2017.

[29] J. G. Andrews, A. Ghosh, J. Zhang, and R. Muhamed, *Fundamentals of LTE*.   Boston, MA: Prentice-Hall, 2010.

[30] C.-Y. Wang, M. Wigger, and A. Zaidi, "On achievability for downlink cloud radio access networks with base station cooperation," in *2017 IEEE Wireless Commun. and Netw. Conf. (WCNC)*, San Francisco, CA, March 2017.

[31] S.-H. Park, O. Simeone, and S. S. (Shitz), "Uplink sum-rate analysis of C-RAN with interconnected radio units," in *Proc. 2017 IEEE Inf. Theory Workshop (ITW)*, Kaohsiung, Taiwan, Nov. 6–10, 2017, pp. 171–175.

[32] N. Levy and S. Shamai (Shitz), "Clustered local decoding for Wyner-type cellular models," *IEEE Trans. Inf. Theory*, vol. 55, no. 11, pp. 4967–4985, Nov. 2009.

[33] A. Lapidoth, N. Levy, S. Shamai (Shitz), and M. Wigger, "Cognitive Wyner networks with clustered decoding," *IEEE Trans. Inf. Theory*, vol. 60, no. 10, pp. 6342–6367, Oct. 2014.

[34] G. Katz, B. M. Zaidel, and S. Shamai (Shitz), "On layered transmission in clustered cooperative cellular architectures," in *Proc. 2013 IEEE Int. Symp. Inf. Theory (ISIT)*, Istanbul, Jul. 7–12, 2013, pp. 1162–1166.

[35] G. Katz, B. M. Zaidel, and S. Shamai (Shitz), "On layered transmission in flat-fading clustered cooperative cellular architectures," in *Proc. 2014 IEEE 28th Conv. Electrical and Electronics Engineers Israel (IEEEI)*, Israel, Dec. 3–5, 2014.

[36] V. S. Annapureddy and V. V. Veeravalli, "Sum capacity of MIMO interference channels in the low interference regime," *IEEE Trans. Inf. Theory*, vol. 57, no. 5, pp. 2565–2581, May 2011.

[37] X. Shang, B. Chen, G. Kramer, and H. V. Poor, "Capacity regions and sum-rate capacities of vector Gaussian interference channels," *IEEE Trans. Inf. Theory*, vol. 56, no. 10, pp. 5030–5044, Oct. 2010.

[38] S. Shamai (Shitz) and M. Wigger, "Rate-limited transmitter-cooperation in Wyner's asymmetric interference network," in *Proc. 2011 IEEE Int. Symp. Inf. Theory (ISIT)*, St Petersburg, Jul. 31–Aug. 5, 2011, pp. 429–433.

[39] O. Simeone, O. Somekh, Y. Bar-Ness, and U. Spagnolini, "Uplink throughput of TDMA cellular systems with multicell processing and amplify-and-forward cooperation between mobiles," *IEEE Trans. Wireless Commun.*, vol. 6, no. 8, pp. 2942–2951, Aug. 2007.

[40] O. Simeone, O. Somekh, G. Kramer, H. V. Poor, and S. Shamai (Shitz), "Throughput of cellular systems with conferencing mobiles and cooperative base stations," *EURASIP J. Wireless Commun. and Netw.*, vol. 2008, pp. 1–14, Feb. 2008.

[41] O. Simeone, O. Somekh, G. Kramer, H. V. Poor, and S. Shamai (Shitz), "Uplink sum-rate analysis of a multicell system with feedback," in *Proc. 46th Ann. Allerton Conf. Commun. Control Comp.*, Monticello, IL, Sep. 2008, pp. 1030–1036.

[42] Z. Goldfeld, H. H. Permuter, and B. M. Zaidel, "The finite state MAC with cooperative encoders and delayed CSI," *IEEE Trans. Inf. Theory*, vol. 60, no. 10, pp. 6181–6203, Oct. 2014.

[43] O. Simeone, O. Somekh, Y. Bar-Ness, and U. Spagnolini, "Throughput of low-power cellular systems with collaborative base stations and relaying," *IEEE Trans. Inf. Theory*, vol. 54, no. 1, pp. 459–467, Jan. 2008.

[44] O. Somekh, O. Simeone, H. V. Poor, and S. Shamai (Shitz), "Cellular systems with non-regenerative relaying and cooperative base stations," *IEEE Trans. Wireless Commun.*, vol. 9, no. 8, pp. 2654–2663, Aug. 2010.

[45] M. Wigger, R. Timo, and S. Shamai (Shitz), "Complete interference mitigation through receiver-caching in Wyner's networks," in *Proc. 2016 IEEE Inf. Theory Workshop (ITW)*, Cambridge, Sep. 2016, pp. 335–339.

[46] O. Simeone, E. Erkip, and S. Shamai (Shitz), "Robust transmission and interference management for femtocells with unreliable network access," *IEEE J. Sel. Areas Commun.*, vol. 28, no. 9, pp. 1469–1478, Dec. 2010.

[47] O. Simeone, O. Somekh, E. Erkip, H. V. Poor, and S. Shamai (Shitz), "Robust communication via decentralized processing with unreliable backhaul links," *IEEE Trans. Inf. Theory*, vol. 57, no. 7, pp. 4187–4201, Jul. 2011.

[48] O. Somekh, O. Simeone, H. V. Poor, and S. Shamai (Shitz), "Throughput of cellular uplink with dynamic user activity and cooperative base-stations," in *Proc. 2009 IEEE Inf. Theory Workshop (ITW)*, Taormina, Oct. 2009, pp. 610–614.

[49] N. Levy and S. Shamai (Shitz), "Information theoretic aspects of users' activity in a Wyner-like cellular model," *IEEE Trans. Inf. Theory*, vol. 56, no. 5, pp. 2241–2248, May 2010.

[50] A. Sendonaris, E. Erkip, and B. Aazhang, "User cooperation diversity – Part I: System description," *IEEE Trans. Commun.*, vol. 51, no. 11, pp. 1927–1938, Nov. 2003.

[51] A. Sendonaris, E. Erkip, and B. Aazhang, "User cooperation diversity – Part II: Implementation aspects and performance analysis," *IEEE Trans. Commun.*, vol. 51, no. 11, pp. 1939–1948, Nov. 2003.

[52] M. A. Wigger and G. Kramer, "Three-user MIMO MACs with cooperation," in *Proc. 2009 IEEE Inf. Theory Workshop (ITW)*, Volos, Jun. 10–12, 2009, pp. 221–225.

[53] S. I. Bross, A. Lapidoth, and M. Wigger, "Dirty-paper coding for the Gaussian multiaccess channel with conferencing," *IEEE Trans. Inf. Theory*, vol. 58, no. 9, pp. 5640–5668, Sep. 2012.

[54] O. Simeone, O. Somekh, G. Kramer, H. V. Poor, and S. Shamai, "Three-user Gaussian multiple access channel with partially cooperating encoders," in *42nd Asilomar Conf. Signals, Syst. and Comp.*, Pacific Grove, CA Oct. 2008, pp. 85–89.

[55] B. Schein and R. Gallager, "The Gaussian parallel relay network," in *Proc. 2000 IEEE Int. Symp. Inf. Theory (ISIT)*, Sorrento, Jun. 25–30, 2000, p. 22.

[56] S. Saeedi Bidokhti, M. Wigger, and R. Timo, "Noisy broadcast networks with receiver caching," *IEEE Trans. Inf. Theory*, vol. 64, no. 11, pp. 6996–7016, May 2018.

[57] S. Saeedi Bidokhti, M. Wigger, and A. Yener, "Benefits of cache assignment on degraded broadcast channels," *IEEE Trans. Inf. Theory*, vol. 65, no. 11, pp. 6999–7019, Nov. 2019.

[58] P. Hassanzadeh, E. Erkip, and A. T. J. Llorca, "Distortion-memory tradeoffs in cache-aided wireless video delivery," in *Proc. 53rd Ann. Allerton Conf. Commun. Control Comp.*, Monticello, IL, Oct. 2015.

[59] A. Ghorbel, M. Kobayashi, and S. Yang, "Content delivery in erasure broadcast channels with cache and feedback," *IEEE Trans. Inf. Theory*, vol. 62, no. 11, pp. 6407–6422, Nov. 2016.

[60] W. Huang, S. Wang, N. Ding, F. Yang, and W. Zhang, "The performance analysis of coded cache in wireless fading channel," *arXiv:1504.01452v1*, 2015.

[61] S. Wang, X. Tian, and H. Liu, "Exploiting the unexploited of coded caching for wireless content distribution," in *Proc. IEEE 2015 Int. Conf. Comp., Netw. Commun. (ICNC)*, 2015, pp. 700–706.

[62] S. Yang, K.-H. Ngo, and M. Kobayashi, "Content delivery with coded caching and massive MIMO in 5G," in *Proc. 9th Intern. Symp. Turbo Codes Iterative Inf. Proc. (ISTC)*, Brest, Sep. 2016, pp. 370–374.

[63] J. Zhang and P. Elia, "Fundamental limits of cache-aided wireless BC: Interplay of coded-caching and CSIT feedback," *IEEE Trans. Inf. Theory*, vol. 63, no. 15, pp. 3142–3160, May 2017.

[64] J. S. Pujol Roig, F. Tosato, and D. Gündüz, "Interference networks with caches at both ends," in *Proc. IEEE Int. Conf. Commun. (ICC)*, Paris, May 21–25, 2017.

[65] M. A. Maddah-Ali and U. Niesen, "Cache-aided interference channels," in *Proc. 2015 IEEE Int. Symp. Inf. Theory (ISIT)*, Hong Kong, Jun. 14–19, 2015.

[66] N. Naderializadeh, M. A. Maddah-Ali, and A. S. Avestimehr, "Fundamental limits of cache-aided interference management," *IEEE Trans. Inf. Theory*, vol. 63, no. 5, pp. 3092–3107, May 2017.

[67] R. M. Gray, "Toeplitz and circulant matrices: A review," *Found. Trends Commun. Inf. Theory*, vol. 2, no. 3, Dec. 2006.

[68] L. H. Brandenburg and A. D. Wyner, "Capacity of the Gaussian channel with memory: The multivariate case," *Bell Syst. Tech. J.*, vol. 53, no. 5, pp. 745–778, May–Jun. 1974.

[69] Y. Liu and E. Erkip, "On the sum capacity of K-user cascade Gaussian Z-interference channel," in *Proc. 2011 IEEE Int. Symp. Inf. Theory (ISIT)*, St. Petersburg, Jul. 31–Aug. 5, 2011, pp. 1382–1386.

[70] L. Zhou and W. Yu, "On the capacity of the K-user cyclic Gaussian interference channel," *IEEE Trans. Inf. Theory*, vol. 59, no. 1, pp. 154–165, Jan. 2013.

[71] W. Yu, "Uplink–downlink duality via minimax duality," *IEEE Trans. Inf. Theory*, vol. 52, no. 2, pp. 361–374, Feb. 2006.

[72] W. Yu and T. Lan, "Minimax duality of Gaussian vector broadcast channels," in *Proc. 2004 IEEE Int. Symp. Inf. Theory (ISIT'04)*, Chicago, IL, Jun. 27–Jul. 2, 2004, p. 177.

[73] G. Caire and S. Shamai (Shitz), "On the achievable throughput of multiantenna Gaussian broadcast channels," *IEEE Trans. Inf. Theory*, vol. 49, no. 7, pp. 1691–1706, Jul. 2003.

[74] A. M. Tulino and S. Verdú, "Random matrix theory and wireless communications," *Found. Trends Commun. Inf. Theory*, vol. 1, no. 1, pp. 1–182, 2004.

[75] R. Couillet and M. Debbah, *Random Matrix Methods for Wireless Communications*. Cambridge: Cambridge University Press, 2011.

[76] A. Narula, "Information theoretic analysis of multiple-antenna transmission diversity," PhD Thesis, Massachusetts Institute of Technology (MIT), Jun. 1997.

[77] Y. Liang and A. Goldsmith, "Symmetric rate capacity of cellular systems with cooperative base stations," in *Proc. IEEE GLOBECOM 2006*, San Francisco, CA, Nov. 27–Dec. 1, 2006.

[78] S. Vishwanath, N. Jindal, and A. Goldsmith, "Duality, achievable rates, and sum-rate capacity of Gaussian MIMO broadcast channels," *IEEE Trans. Inf. Theory*, vol. 49, no. 10, pp. 2658–2668, Oct. 2003.

[79] P. Viswanath and D. Tse, "Sum capacity of the vector Gaussian channel and uplink–downlink duality," *IEEE Trans. Inf. Theory*, vol. 49, no. 9, pp. 1912–1921, Aug. 2003.

[80] M. Sharif and B. Hassibi, "On the capacity of MIMO broadcast channel with partial side information," *IEEE Trans. Inf. Theory*, vol. 51, no. 2, pp. 506–522, Feb. 2005.

[81] O. Somekh, O. Simeone, Y. Bar-Ness, A. M. Haimovich, and S. Shamai (Shitz), "Cooperative multicell zero-forcing beamforming in cellular downlink channels," *IEEE Trans. Inf. Theory*, vol. 55, no. 7, pp. 3206–3219, Jul. 2009.

[82] T. M. Cover and J. Thomas, *Elements of Information Theory*, 2nd ed. Chichester: Wiley, 2006.

[83] A. Sanderovich, S. Shamai (Shitz), Y. Steinberg, and G. Kramer, "Communication via decentralized processing," *IEEE Trans. Inf. Theory*, vol. 54, no. 7, pp. 3008–3023, Jul. 2008.

[84] A. D. Wyner and J. Ziv, "The rate-distortion function for source coding with side information at the decoder," *IEEE Trans. Inf. Theory*, vol. IT-22, no. 1, pp. 1–10, Jan. 1976.

[85] A. El Gamal and Y.-H. Kim, *Network Information Theory*. Cambridge: Cambridge University Press, 2011.

[86] T. Berger, Z. Zhang, and H. Viswanathan, "The CEO problem," *IEEE Trans. Inf. Theory*, vol. 42, no. 3, pp. 887–902, May 1996.

[87] B. Nazer, A. Sanderovich, M. Gastpar, and S. Shamai (Shitz), "Structured superposition for backhaul constrained cellular uplink," in *Proc. 2009 IEEE Int. Symp. Inf. Theory (ISIT)*, Seoul, Jun. 28–Jul. 3, 2009.

[88] B. Nazer and M. Gastpar, "Compute-and-forward: Harnessing interference through structured codes," *IEEE Trans. Inf. Theory*, vol. 57, no. 10, pp. 6463–6486, Oct. 2011.

[89] T. S. Han and K. Kobayashi, "A new achievable rate region for the interference channel," *IEEE Trans. Inf. Theory*, vol. 27, no. 1, pp. 49–60, Jan. 1981.

[90] M. Avriel and A. C. Williams, "Complementary geometric programming," *SIAM J. Appl. Math.*, vol. 19, no. 1, pp. 125–141, Jul. 1970.

[91] A. El Gamal, V. S. Annapureddy, and V. V. Veeravalli, "Interference channels with coordinated multipoint transmission: Degrees of freedom, message assignment, and fractional reuse," *IEEE Trans. Inf. Theory*, vol. 60, no. 6, pp. 3483–3498, Jun. 2014.

[92] M. Bande, A. El Gamal, and V. V. Veeravalli, "Degrees of freedom in wireless interference networks with cooperative transmission and backhaul load constraints," *IEEE Trans. Inf. Theory*, vol. 65, no. 9, pp. 5816–5832, Sep. 2019.

[93] A. Lapidoth, S. Shamai (Shitz), and M. Wigger, "On cognitive interference networks," in *Proc. 2007 IEEE Inf. Theory Workshop (ITW)*, Lake Tahoe, NV, Sep. 2–7, 2007.

[94] D. Gesbert, S. Hanly, H. Huang, et al., "Multi-cell MIMO cooperative networks: A new look at interference," *IEEE J. Sel. Areas Commun.*, vol. 28, no. 9, pp. 1380–1408, Dec. 2010.

[95] A. El Gamal, V. S. Annapureddy, and V. V. Veeravalli, "Degrees of freedom of interference channels with comp transmission and reception," *IEEE Trans. Inf. Theory*, vol. 58, no. 9, pp. 5740–5760, Sep. 2012.

[96] S.-J. Kim, S. Jain, and G. B. Giannakis, "Backhaul-constrained multi-cell cooperation using compressive sensing and spectral clustering," in *Proc. IEEE 13th Intern. Workshop Proc. Adv. Wireless Commun. (SPAWC)*, Cesme, Jun. 17–20, 2012, pp. 65–69.

[97] C. Wang, S. A. Jafar, S. Shamai, and M. Wigger, "Interference, cooperation and connectivity: A degrees of freedom perspective," in *Proc. IEEE Int. Symp. Inf. Theory (ISIT)*, St. Petersburg, Jul. 31–Aug. 5, 2011.

[98] Y. Cao and B. Chen, "An achievable rate region for interference channels with conferencing," in *Proc. 2007 IEEE Int. Symp. Inf. Theory (ISIT)*, Nice, Jun. 24–29, 2007, pp. 1251–1255.

[99] I. H. Wang and D. N. C. Tse, "Interference mitigation through limited receiver cooperation," *IEEE Trans. Inf. Theory*, vol. 57, no. 5, pp. 2913–2940, May 2011.

[100] S. Shamai and M. Wigger, "Rate-limited transmitter-cooperation in Wyner's asymmetric interference network," in *Proc. 2011 IEEE Int. Symp. Inf. Theory (ISIT)*, July 2011, pp. 425–429.

[101] I. Marić, R. D. Yates, and G. Kramer, "Capacity of interference channels with partial transmitter cooperation," *IEEE Trans. Inf. Theory*, vol. 53, no. 10, pp. 3536–3548, Oct. 2007.

[102] J. Hachem, N. Karamchandani, and S. N. Diggavi, "Coded caching for multi-level popularity and access," *IEEE Trans. Inf. Theory*, vol. 63, no. 5, pp. 3108–3141, May 2017.

[103] S. P. Shariatpanah, S. A. Motahari, and B. H. Khalaj, "Multi-server coded caching," *IEEE Trans. Inf. Theory*, vol. 62, no. 12, pp. 7253–7271, Dec. 2016.

[104] Y. Ugur, Z. H. Awan, and A. Sezgin, "Cloud radio access networks with coded caching," in *Proc. 20th Int. ITG Workshop Smart Antennas (WSA 2016)*, Munich, Mar. 9–11, 2016.

[105] S.-H. Park, O. Simeone, and S. Shamai (Shitz), "Joint optimization of cloud and edge processing for fog radio access networks," *IEEE Trans. Wireless Commun.*, vol. 15, no. 11, pp. 7621–7632, Nov. 2016.

[106] M. Peng, K. Z. S. Yan, and C. Wang, "Fog computing based radio access networks: Issues and challenges," *IEEE Network*, vol. 30, no. 4, pp. 46–53, Jul./Aug. 2016.

[107] J. G. Andrews, F. Baccelli, and R. K. Ganti, "A tractable approach to coverage and rate in cellular networks," *IEEE Trans. Commun.*, vol. 59, no. 11, pp. 3122–3134, Nov. 2011.

[108] H. S. Dhillon, R. K. Ganti, F. Baccelli, and J. G. Andrews, "Modeling and analysis of K-tier downlink heterogeneous cellular networks," *IEEE J. Sel. Areas Commun.*, vol. 30, no. 3, pp. 550–560, Apr. 2012.

[109] T. D. Novlan, H. S. Dhillon, and J. G. Andrews, "Analytical modeling of uplink cellular," *IEEE Trans. Wireless Commun.*, vol. 12, no. 6, pp. 2669–2679, Jun. 2013.

[110] K. Huang and J. G. Andrews, "An analytical framework for multicell cooperation via stochastic geometry and large deviations," *IEEE Trans. Inf. Theory*, vol. 59, no. 4, pp. 2501–2516, Apr. 2013.

[111] F. Baccelli and A. Giovanidis, "A stochastic geometry framework for analyzing pairwise-cooperative cellular networks," *IEEE Trans. Wireless Commun.*, vol. 14, no. 2, pp. 794–808, Feb. 2015.

[112] Y. Polyanskiy, V. Poor, and S. Verdú, "Channel coding rate in the finite blocklength regime," *IEEE Trans. Inf. Theory*, vol. 56, no. 5, pp. 2307–2359, May 2010.

[113] H. Nikbakht, M. Wigger, and R. Timo, "Mixed delay constraints in Wyner's soft-handoff network," in *Proc. 2018 IEEE Int. Symp. Inf. Theory (ISIT)*, Vail, CO, Jun. 17–22, 2018.

17 Service Delivery in 5G

Jaime Llorca, Antonia Tulino, and Giuseppe Caire

17.1 Introduction

We are at the dawn of an era in networking that will be shaped by the digitization and interconnection of virtually every asset in the physical world, with the goal of automating and optimizing physical systems and processes, as well as augmenting human knowledge, cognition, and life experiences. 5G networks are expected to become the digital fabric underpinning this new transformational era [1].

With the increasing penetration of cloud resources (e.g., via edge cloud or fog computing technologies) [2, 3], 5G networks will become highly distributed computing infrastructures able to host content and applications close to information sources and end users, providing rapid response, analysis, and delivery of augmented information in real time. Unlike traditional information services, in which users consume information that is produced or stored at a given source and is delivered via a communications network, this new class of real-time computation services, also referred to as *augmented information (AgI)* services [4, 5], provides end users with information that results from the *real-time processing* of source streams via possibly multiple functions distributed throughout the network. AgI services can be classified into two main categories, depending on whether the end consumers are humans or machines:

1. **Augmented experience services** enable the consumption of multimedia streams that result from the combination of multiple live sources and contextual information of real-time relevance. Examples include telepresence, real-time computer vision, virtual classrooms/labs/offices, and augmented/virtual reality [6, 7].
2. **Automation services** enable the real-time processing of information sourced at sensing devices in physical infrastructures such as homes, offices, factories, in order to deliver instructions that optimize and control the automated operation of associated physical systems. Examples include smart homes/buildings/cities, automated transportation, and industrial internet services [8].

In 5G, via what is referred to as *network slicing* [9, 10], services or group of services with similar requirements take the form of independent virtual networks within the shared physical infrastructure, as illustrated in Figure 17.1. Network slicing builds on the advent of network function virtualization (NFV) and software defined networking (SDN), which allow the deployment of network services in the form of elastic software functions instantiated over general purpose servers at multiple cloud locations, and

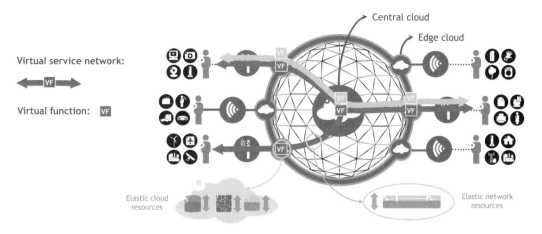

Virtual service network:

Virtual function:

Figure 17.1 With 5G network slicing, services, or group of services with similar performance requirements, form independent elastic virtual networks, each consuming a slice of the cloud network's physical infrastructure.

interconnected via a programmable network fabric [11, 12]. Hence, slices can flexibly and elastically consume storage, computation, and networking resources according to changing network conditions and service requirements. This way, operators can host a large variety of services over a common infrastructure, reducing both capital and operational expenses, while delivering high-quality experiences.

A key aspect driving both efficiency and performance is the actual placement of the service functions, the routing of network flows through the appropriate function instances, and the associated allocation of cloud and network resources. Traditional information services have dealt with the efficient flow of information from data sources to destinations, but considered the location of sources, destinations, and processing elements as static functions, mostly based on rigid hardware deployments. In contrast, the efficient delivery of AgI services over cloud-integrated 5G networks requires jointly and dynamically deciding where to execute each service function, how to schedule network flows for both routing and computation, and how to elastically allocate computation, storage, and networking resources. Altogether, the configuration and control of AgI services in cloud networks poses new fundamental challenges that call for novel solutions in the crossings of network information theory, optimization, and control.

This chapter takes a look at the crucial problem of *optimal configuration and control of AgI services over cloud-integrated networks*. After reviewing state-of-the-art approaches for capacity characterization and service delivery in traditional communication networks in Section 17.2, we introduce general models for AgI services and cloud-integrated networks in Section 17.3. We then describe how to characterize the cloud network capacity region via properly constructed multi-commodity-chain cloud network flow equations in Section 17.4. Section 17.5 presents throughput-optimal algorithms for dynamic flow scheduling and joint computation/communication resource allocation. Finally, Section 17.6 sheds light into interesting future research directions.

17.2   State of the Art: Traditional Communication Networks

The network layer capacity region (or stability region) of traditional communication networks is defined as the set of traffic rates that can be stabilized by any control strategy that affects routing, scheduling, and resource allocation, where stability refers to the bounded occupancy of the network's underlying queuing system [13]. Such a capacity region can be characterized by the solution to the associated multi-commodity flow (MCF) problem, in which commodities representing source–destination traffic demands are to be routed through a network with communication capacity constraints. A routing and scheduling policy is said to be *throughput-optimal* if it can stabilize any input rate matrix in the interior to the capacity region.

Stochastic network optimization methods such as *Lyapunov optimization* are well-known techniques for the systematic design of throughput-optimal routing and scheduling policies that do not require any knowledge of the traffic arrival rates. A first renowned application of Lyapunov drift control in multihop networks is the backpressure (BP) algorithm [14], which, by minimizing an upper bound on the Lyapunov drift of the underlying queuing system, allows network nodes to make local routing and scheduling decisions based on differential queue backlog information. The BP algorithm achieves routing without ever designing an explicit route, it is oblivious of traffic arrival rates, and can therefore adapt to time-varying network conditions. By further adding a penalty term (e.g., related to the network resource allocation cost) to the Lyapunov drift expression, the Lyapunov drift-plus-penalty (LDP) control methodology [15, 16] allows the design of throughput-optimal algorithms that also minimize overall average resource cost. A number of applications of LDP control have yielded efficient routing and resource allocation algorithms for traditional multihop communication networks [15–19].

Throughout this chapter, in order to characterize the capacity of modern cloud-integrated networks, we will describe how to generalize the MCF model to include unique aspects of AgI services in distributed computing networks, namely flow chaining, flow scaling, and joint computation/communication resource allocation, and how to extend the LDP control methodology for the design of throughput- and cost-optimal policies that jointly schedule communication and computation resources for the delivery of AgI services.

17.3   Network and Service Models

In this section, we describe how to model generic AgI services and cloud-integrated networks, and introduce the multi-commodity-chain cloud network flow model that will be used in the following sections for the characterization of the cloud network capacity region and the design throughput-optimal cloud network control policies.

17.3.1   Cloud Network Model

In the following, we use the term *cloud network* to refer to a network with both communication and computation resources (see e.g., [11, 20–24]). A cloud network can be modeled by a directed graph $\mathcal{G} = (\mathcal{V}, \mathcal{E})$, where vertices represent distributed computing locations (e.g., core cloud nodes, edge cloud nodes, compute-enabled base stations, or end devices with embedded computing resources), and edges represent network links between computing locations.

Cloud network node $u \in \mathcal{V}$ has compute capacity c_u in processing resource units (e.g., CPUs, virtual machines, servers), and the cost of allocating one processing resource unit at node u is given by w_u. Analogously, link (u, v) has transmission capacity c_{uv} in bandwidth resource units (e.g., RAN resource blocks, wavelengths), and the cost per bandwidth resource unit is given by w_{uv}. Finally, γ_{uv} denotes the bandwidth resource units required to route one flow unit through link (u, v).[1]

17.3.2   AgI Service Model

Recall that the term *AgI service* refers to a generalization of traditional information services, in which source flows need to get processed by a set of service functions before being delivered to their corresponding destinations [4, 5].

In line with, e.g., [4, 5, 20–24], a generic service $\phi \in \Phi$ can be described by a directed acyclic graph $\mathcal{T}^\phi(\mathcal{O}, \mathcal{I})$, where vertices represent service functions, and directed edges describe the flow of information among the service functions, as shown in Figure 17.2.

The leaves (vertices with no incoming edges) of the service graph represent the source functions. A source function may simply be an *ingress* function indicating the arrival of packet flows at a particular network location, or an actual producer function such as a camera capturing a video stream or other type of sensing devices. Source flows must then go through the set of service functions specified by the service graph until creating

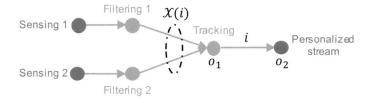

Figure 17.2 Example of a service graph for a video-tracking service, in which flows from two video sensors go through filtering and tracking functions before being displayed at a consumer device.

---

[1] For ease of exposition, we assume static links of fixed capacity and discuss extensions to networks with time-varying link states (e.g., wireless links) in Section 17.6.1.

the end flow that gets consumed by the consumption functions represented by the roots (vertices with no outgoing edges) of the service graph. A consumption function may simply be an *egress* function indicating the delivery of a final flow at a given destination, or an actual consumer function such as the display of a video stream.

For each service flow, represented by edge $i = (o_1, o_2) \in \mathcal{I}$, we use $\mathcal{X}(i)$ to denote the set of incoming edges to $o_1 \in \mathcal{O}$, i.e., the set of input flows required to generate flow $i \in \mathcal{I}$ via function $o_1 \in \mathcal{O}$. In addition, $\eta^{(i)}$ denotes the resource requirement to process flow $i \in \mathcal{I}$, in resource units per flow unit (e.g., CPUs per bps).

A key feature of AgI services is that flows can change size as they go through service function processing (e.g., after video transcoding). This is modeled by a flow scaling factor $\xi^{(i,j)}$ denoting the flow size ratio between output flow i and input flow $j \in \mathcal{X}(i)$.

Finally, we denote by $\mathcal{S}^\phi \subset \mathcal{I}$ the set of source flows, represented by the set of outgoing edges from source functions in $\mathcal{T}^\phi$, and by $\mathcal{Q}^\phi \subset \mathcal{I}$ the set of end flows, represented by the set of incoming edges to consumption functions in $\mathcal{T}^\phi$.

17.3.3   Multicommodity-Chain Flow Model

In the multicommodity-chain (MCC) flow model, a commodity refers to the flow of packets between two consecutive service functions used to meet the service demand of a given destination, and it is denoted by the tuple (d, ϕ, i), indicating the packets of flow $i \in \mathcal{I}$ used to meed the demand of destination $d \in \mathcal{V}$ for service $\phi \in \Phi$ (see, e.g., [20, 22–24]).

Consider a time slotted system with slots normalized to integral units $t \in \{0, 1, 2, \cdots\}$. We denote by $a_u^{(d,\phi,i)}(t)$ the exogenous arrival rate of commodity (d, ϕ, i) at node u during time slot t, and by $\lambda_u^{(d,\phi,i)}$ its expected value. We assume that $a_u^{(d,\phi,i)}(t)$ is independently and identically distributed (i.i.d.) across time slots, and that $a_u^{(d,\phi,i)}(t) = 0$ for $i \notin \mathcal{S}^\phi$, i.e., there are no exogenous arrivals of intermediate commodities.

Cloud network nodes buffer packets according to their commodities, with $Q_u^{(d,\phi,i)}(t)$ denoting the amount of commodity (d, ϕ, i) packets in the queue of node u at the beginning of time slot t.

In order to model the computation, production, and consumption of service flows at a cloud network node, [20] introduced the notion of a *cloud-augmented graph*, which results from augmenting each node in $\mathcal{G}$ with the gadget in Figure 17.3, where $p(u)$, $s(u)$, and $q(u)$ indicate processing, source, and demand functionalities at node u, respectively. Flows arriving at node u can get forwarded to a neighboring node $p(u)$ to get processed by a service function. When a service flow goes through processing edge $(u, p(u))$, it consumes computation resources at node u. After service processing, the output flow gets back to node u for either consumption or forwarding to a neighbor node via a free-cost edge of high enough capacity $(p(u), u)$.

A cloud network control policy can then be described by the following computation and communication flow scheduling and resource allocation decision variables:

- $\mu_{uv}^{(d,\phi,i)}(t)$: rate of commodity (d, ϕ, i) on link (u, v) at time t;
- $\mu_{up(u)}^{(d,\phi,i)}(t)$: rate of commodity (d, ϕ, i) being processed at node u at time t;

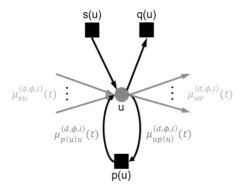

Figure 17.3 Cloud-augmented graph for node $u \in \mathcal{V}$, with network edges represented in gray, and additional source, demand, and processing edges in black. Flow scheduling variables indicate how packets of a given commodity can arrive to node u via transmission from neighbor nodes, as well as via local service processing.

- $\mu_{p(u)u}^{(d,\phi,i)}(t)$: rate of commodity (d,ϕ,i) being generated at node u at time t;
- $y_u(t)$: number of processing resource units allocated at node u at time t;
- $y_{uv}(t)$: number of transmission resource units allocated on link (u,v) at time t.

In the MCC flow model, cloud network queues build up from the transmission of packets from incoming neighbors *and* from the local processing of packets via service functions, resulting in the following MCC *queuing dynamics* and *service chaining* equations:

$$Q_u^{(d,\phi,i)}(t+1) \leq \left[Q_u^{(d,\phi,i)}(t) - \sum_{(u,v)} \mu_{uv}^{(d,\phi,i)}(t) - \mu_{up(u)}^{(d,\phi,i)}(t) \right]^+$$

$$+ \sum_{(v,u)} \mu_{vu}^{(d,\phi,i)}(t) + \mu_{p(u)u}^{(d,\phi,i)}(t) + a_i^{(d,\phi,i)}(t), \quad \forall u,d,\phi,i,t, \qquad (17.1)$$

$$\mu_{p(u)u}^{(d,\phi,i)}(t) = \xi^{(i,j)} \mu_{up(u)}^{(d,\phi,j)}(t), \quad \forall u,d,\phi,i,j \in \mathcal{X}(i),t, \qquad (17.2)$$

where $[x]^+$ denotes $\max\{x,0\}$, and $Q_d^{(d,\phi,i)}(t) = 0, \forall d,\phi,i \in \mathcal{Q}^\phi,t$.

Observe that the queuing dynamics (17.1) together with the service chaining constraints (17.2) assure that the packets of commodity (d,ϕ,i) exogenously arriving at node u at time t, the packets of commodity (d,ϕ,i) transmitted over the incoming links of node u at time t, and the packets of commodity $(d,\phi,j), j \in \mathcal{X}(i)$, locally processed at node u at time t scaled by $\xi^{(i,j)}$, all build up at the queue of commodity (d,ϕ,i) at node u at time $t+1$. An example of a four-node cloud network queuing system is illustrated in Figure 17.4.

Finally, processing and transmission maximum flow rates are dictated by the associated computation and communication resource allocation decisions:

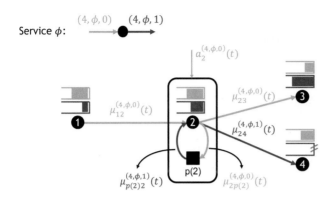

Figure 17.4 MCC cloud network queuing model for the delivery of a single-function service for destination node 4. Packets of the source commodity $(4, \phi, 0)$ enter at node 2 and get buffered at the light gray queues as they get forwarded through the network. When commodity $(4, \phi, 0)$-packets get processed at node 2, they build up as packets of commodity $(4, \phi, 1)$, which then get buffered at the dark gray queues as they get forwarded through the network, until exiting at node 4.

$$\sum_{(d,\phi,i)} \mu_{up(u)}^{(d,\phi,i)}(t)\eta^{(i)} \leq y_u(t), \quad \forall u, t, \tag{17.3}$$

$$\sum_{(d,\phi,i)} \mu_{uv}^{(d,\phi,i)}(t)\gamma_{uv} \leq y_{uv}(t), \quad \forall (u, v), t. \tag{17.4}$$

Note that the MCC flow model generalizes traditional multicommodity flow by including the characterization of key new features for the delivery of AgI services over distributed computing networks, namely, *flow chaining, flow scaling, and joint computation/communication resource allocation.*

17.4   Cloud Network Capacity Region

For a given cloud network $\mathcal{G}$ and set of offered services Φ, the cloud network capacity region $\Lambda(\mathcal{G}, \Phi)$ is defined as the closure of all average input rate vectors $\lambda = \{\lambda^{(d,\phi,i)}\}$ that can be stabilized by a cloud network control algorithm, whose decisions conform to the cloud network and service structure $\{\mathcal{G}, \Phi\}$ specified in the previous section.

Note that this is a network layer notion of capacity that considers all choices of the decision variables $\{\mu(t), y(t)\}$. This is distinct from the information theoretic notion of network capacity, which includes optimization over all possible coding strategies and involves many of the unsolved problems of network information theory [25].

Also, note that the cloud network capacity region depends not only on the network topology $\mathcal{G}$, but also on the service structure Φ, i.e., the set of services with associated flow chaining/scaling parameters.

The following theorem from [23, 24] characterizes the cloud network capacity region:

THEOREM 17.1   *The cloud network capacity region* $\Lambda(\mathcal{G}, \Phi)$ *consists of all average input rates* λ *for which there exist MCC flow variables* $f_{uv}^{(d,\phi,i)}$, $f_{p(u)u}^{(d,\phi,i)}$, $f_{up(u)}^{(d,\phi,i)}$, *together with nonnegative numbers* $\alpha_{uv,k}$, $\alpha_{u,k}$, $\beta_{uv,k}^{(d,\phi,i)}$, $\beta_{u,k}^{(d,\phi,i)}$, *such that*

$$\sum_{(v,u)} f_{vu}^{(d,\phi,i)} + f_{p(u)u}^{(d,\phi,i)} + \lambda_u^{(d,\phi,i)} \le \sum_{(u,v)} f_{uv}^{(d,\phi,i)} + f_{up(u)}^{(d,\phi,i)},$$

$$\forall u, d, \phi, i \notin \mathcal{Q}^{\phi} \text{ and } \forall u = d, d, \phi, i \in \mathcal{Q}^{\phi} \quad (17.5\text{a})$$

$$f_{p(u)u}^{(d,\phi,i)} = \xi^{(i,j)} f_{up(u)}^{(d,\phi,j)}, \quad \forall u, d, \phi, i, j \in \mathcal{X}(i) \quad (17.5\text{b})$$

$$f_{up(u)}^{(d,\phi,i)} \le \frac{1}{\eta^{(i)}} \sum_{k=1}^{c_u} k\, \alpha_{u,k}\, \beta_{u,k}^{(d,\phi,i)}, \quad \forall u, d, \phi, i \quad (17.5\text{c})$$

$$f_{uv}^{(d,\phi,i)} \le \frac{1}{\gamma_{uv}} \sum_{k=1}^{c_{uv}} k\, \alpha_{uv,k}\, \beta_{uv,k}^{(d,\phi,i)}, \quad \forall (u, v), d, \phi, i \quad (17.5\text{d})$$

$$f_{up(u)}^{(d,\phi,i)} = 0, \quad \forall u, d, \phi, i \in \mathcal{Q}^{\phi} \quad (17.5\text{e})$$

$$f_{p(u)u}^{(d,\phi,i)} = 0, \quad \forall u, d, \phi, i \in \mathcal{S}^{\phi} \quad (17.5\text{f})$$

$$f_{up(u)}^{(d,\phi,i)} \ge 0, f_{uv}^{(d,\phi,i)} \ge 0, \quad \forall u, (u, v), d, \phi, i \quad (17.5\text{g})$$

$$\sum_{k=1}^{c_u} \alpha_{u,k} \le 1, \sum_{k=1}^{c_{uv}} \alpha_{uv,k} \le 1, \quad \forall u, (u, v) \quad (17.5\text{h})$$

$$\sum_{(d,\phi,i)} \beta_{u,k}^{(d,\phi,i)} \le 1, \sum_{(d,\phi,i)} \beta_{uv,k}^{(d,\phi,i)} \le 1, \quad \forall u, (u, v), k. \quad (17.5\text{i})$$

Furthermore, the minimum average resource cost required for cloud network stability is given by

$$\bar{h}^* = \min_{\{\alpha_{u,k}, \alpha_{uv,k}, \beta_{u,k}^{(d,\phi,i)}, \beta_{uv,k}^{(d,\phi,i)}\}} \bar{h}, \quad (17.6)$$

where

$$\bar{h} = \sum_u \sum_k \alpha_{u,k} w_{u,k} + \sum_{(u,v)} \sum_k \alpha_{uv,k} w_{uv,k}. \quad (17.7)$$

Proof.   A detailed proof of Theorem 17.1 can be found in [24]. ∎

In Theorem 17.1, (17.5a) and (17.5b) describe *generalized flow conservation* constraints and *service chaining constraints*, respectively, which are key new elements for cloud network stability. Specifically, the generalized flow conservation constraints in (17.5a) assure the conservation of the combined computation/communication flow of a given commodity at every cloud network node. That is, the total outgoing flow, composed of the outgoing transport flows, the processing flow leaving node u toward its processing element $p(u)$, and the outgoing egress/consumption flow, must be larger or equal to the total incoming flow, composed of the incoming transport flows, the source/ingress flow, and the processing flow coming in from $p(u)$. Note that the final

Service graph   Cloud network flow solution

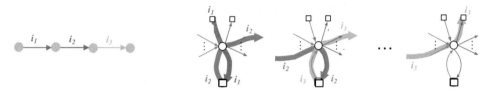

Figure 17.5 Illustration of a cloud network flow solution with MCC flows satisfying generalized flow conservation and flow chaining constraints.

commodities are assumed to be consumed at their respective destinations, where (17.5a) does not apply. On the other hand, the service chaining constraints in (17.5b) assure that in order to have positive flow of a given commodity coming out of the processing element of a cloud network node, a flow of each of the input commodities in the service graph, scaled by its corresponding flow scaling factor, must be present at the input of the processing element. An example of a cloud network flow solution satisfying generalized flow conservation and service chaining constraints is illustrated in Figure 17.5.

Constraints (17.5c) and (17.5d) describe computation and communication capacity constraints, where the probability values $\alpha_{u,k}$, $\alpha_{uv,k}$, $\beta_{u,k}^{(d,\phi,i)}$, $\beta_{uv,k}^{(d,\phi,i)}$ define a *stationary randomized policy* as follows:

- $\alpha_{u,k}$: the probability that k processing resource units are allocated at node u;
- $\alpha_{uv,k}$: the probability that k transmission resource units are allocated at link (i,j);
- $\beta_{u,k}^{(d,\phi,i)}$: the probability that node u processes commodity (d,ϕ,i), conditioned on the allocation of k processing resource units at node u;
- $\beta_{uv,k}^{(d,\phi,i)}$: the probability that link (u,v) transmits commodity (d,ϕ,i), conditioned on the allocation of k transmission resource units at link (u,v).

Hence, Theorem 17.1 demonstrates that, for any input rate $\lambda \in \Lambda(\mathcal{G}, \Phi)$, there exists a stationary randomized policy that uses fixed probabilities to make transmission and processing decisions at each time slot, which can support the given λ, while minimizing overall average cloud network cost. However, the difficulty in directly solving for the parameters that characterize such a stationary randomized policy and the requirement on the knowledge of λ, motivates the design of online dynamic cloud network control solutions with matching performance guarantees.

17.5   Dynamic Cloud Network Control

In this section, we describe distributed dynamic cloud network control (DCNC) strategies that drive processing and transmission flow scheduling and resource allocation decisions with global performance guarantees. Such DCNC algorithms were derived in [22–24] by extending LDP control to stabilize MCC queuing systems, resulting in the first throughput-optimal algorithms for the class of MCC cloud network flow problems that generalize traditional multicommodity network flow by including flow chaining, flow scaling, and joint communication/computation resource allocation.

17.5.1 Cloud Network Lyapunov Drift-Plus-Penalty

Let $\mathbf{Q}(t)$ represent the vector of queue backlog values of all the commodities at all the cloud network nodes. The cloud network *Lyapunov drift* is defined as

$$\Delta\left(\mathbf{Q}\left(t\right)\right) \triangleq \frac{1}{2}\mathbb{E}\left\{ \|\mathbf{Q}\left(t+1\right)\|^2 - \|\mathbf{Q}\left(t\right)\|^2 \middle| \mathbf{Q}\left(t\right) \right\}, \tag{17.8}$$

where $\| \cdot \|$ indicates the Euclidean norm, and the expectation is taken over the ensemble of all exogenous arrival realizations.

The one-step Lyapunov drift-plus-penalty (LDP) is then defined as

$$\Delta\left(\mathbf{Q}\left(t\right)\right) + V\mathbb{E}\left\{ h(t) \middle| \mathbf{Q}\left(t\right) \right\}, \tag{17.9}$$

where V is a nonnegative control parameter that determines the degree to which resource cost minimization is emphasized.

After squaring both sides of (17.1) and following standard LDP manipulations (see e.g., [16]), the LDP can be upper bounded as

$$\Delta\left(\mathbf{Q}\left(t\right)\right) + V\mathbb{E}\left\{ h(t) \middle| \mathbf{Q}\left(t\right) \right\} \leq V\mathbb{E}\left\{ h(t) \middle| \mathbf{Q}\left(t\right) \right\}$$
$$+ \mathbb{E}\left\{ \Gamma(t) + Z(t) \middle| \mathbf{Q}(t) \right\} + \boldsymbol{\lambda}^\dagger \mathbf{Q}(t), \tag{17.10}$$

where

$$\Gamma(t) \triangleq \frac{1}{2}\sum_u \sum_{(d,\phi,i)} \left\{ \left[\sum_{(u,v)} \mu_{uv}^{(d,\phi,i)}(t) + \mu_{up(u)}^{(d,\phi,i)}(t) \right]^2 \right.$$
$$\left. + \left[\sum_{(v,u)} \mu_{vu}^{(d,\phi,i)}(t) + \mu_{p(u)u}^{(d,\phi,i)}(t) + a_u^{(d,\phi,i)}(t) \right]^2 \right\},$$

$$Z(t) \triangleq \sum_u \sum_{(d,\phi,i)} Q_u^{(d,\phi,i)}(t) \left[\sum_{(v,u)} \mu_{vu}^{(d,\phi,i)}(t) + \mu_{p(u)u}^{(d,\phi,i)}(t) \right.$$
$$\left. - \sum_{(u,v)} \mu_{uv}^{(d,\phi,i)}(t) - \mu_{up(u)}^{(d,\phi,i)}(t) \right].$$

DCNC algorithms are derived via the minimization of different metrics extracted from the cloud network's LDP upper bound in (17.10), leading to throughput-optimal flow scheduling and resource allocation policies that differ in the actual cost–delay trade-off performance [24]. In the following, we describe DCNC-L, which, while being the most basic DCNC algorithm presented in [24], illustrates the key elements driving joint processing and transmission flow scheduling and resource allocation.

17.5.2   Linear Dynamic Cloud Network Control (DCNC-L)

For ease of exposition, we describe the DCNC-L algorithm assuming *single-input single-output* service chains, with $i = \mathcal{X}(j) \in \mathcal{I}$ denoting the input flow needed to generate flow $j \in \mathcal{I}$. Extensions to more generic service models are described in [24, section VIII].

DCNC-L is designed to minimize, at each time slot, the linear metric $Z(t) + Vh(t)$ obtained from the right-hand side of (17.10) subject to instantaneous flow chaining and capacity constraints ((17.2) and (17.3)). This minimization can be decomposed into independent subproblems for each processing and transmission interface. Specifically, the subproblem that affects processing flow scheduling and resource allocation decisions at node u is given by

$$\min \qquad V \sum_u w_u\, y_u(t) - \sum_{(d,\phi,i)} Z_u^{(d,\phi,i)}(t) \qquad (17.11a)$$

$$\text{s.t.} \qquad (17.2),(17.3) \qquad (17.11b)$$

where

$$Z_u^{(d,\phi,i)}(t) \triangleq \mu_{up(u)}^{(d,\phi,i)}(t)\left[Q_u^{(d,\phi,i)}(t) - \xi^{(i,j)}Q_u^{(d,\phi,j)}(t)\right], \quad i = \mathcal{X}(j).$$

Analogously, the subproblem that affects transmission flow scheduling and resource allocation decisions at link (u,v) is given by

$$\min \qquad V \sum_{(u,v)} w_{uv}\, y_{uv}(t) - \sum_{(d,\phi,i)} Z_{uv}^{(d,\phi,i)}(t) \qquad (17.12a)$$

$$\text{s.t.} \qquad (17.4) \qquad (17.12b)$$

where

$$Z_{uv}^{(d,\phi,i)}(t) \triangleq \mu_{uv}^{(d,\phi,i)}(t)\left[Q_u^{(d,\phi,i)}(t) - Q_v^{(d,\phi,i)}(t)\right].$$

The goal of minimizing (17.11a) and (17.12a) at each time slot is to greedily push the cloud network queues toward a lightly congested state, while minimizing cloud network resource usage regulated by the control parameter V. Observe that (17.11a) and (17.12a) are linear with respect to $\mu_{up(u)}^{(d,\phi,i)}(t)$ and $\mu_{uv}^{(d,\phi,i)}(t)$. Hence, the minimizations can be achieved via *max-weight-matching* [26], leading to the following distributed flow scheduling and resource allocation policy:

Local processing decisions: At the beginning of each time slot t, node u observes its local queue backlogs and performs the following operations:

1. Compute the *processing utility weight* of each commodity:

$$W_u^{(d,\phi,i)}(t) = \begin{cases} \frac{1}{\eta^{(i)}}\left[Q_u^{(d,\phi,i)}(t) - \xi^{(i,j)}Q_u^{(d,\phi,j)}(t)\right]^+, & \forall d,\phi, i \notin \mathcal{Q}^\phi, j = \mathcal{X}^{-1}(i) \\ 0 & \text{otherwise} \end{cases}$$

Note that the processing utility weight $W_u^{(d,\phi,i)}(t)$ is indicative of the potential benefit of processing commodity (d,ϕ,i) at node u at time t, in terms of local congestion reduction. Observe that a large value of $\xi^{(i,j)}$ reduces the priority of processing (d,ϕ,i).

2. Compute the max-weight commodity:

$$(d, \phi, i)^* = \arg \max_{(d,\phi,i)} \left\{ W_u^{(d,\phi,i)}(t) \right\}.$$

3. Allocate processing resources:

$$y_u(t) = \begin{cases} c_u & \text{if } W_{uv}^{(d,\phi,i)^*}(t) - Vw_{uv} > 0 \\ 0 & \text{otherwise} \end{cases}.$$

4. Assign processing flow rates:

$$\mu_{up(u)}^{(d,\phi,i)^*}(t) = \frac{y_u}{\eta^{(i)}},$$

$$\mu_{up(u)}^{(d,\phi,i)}(t) = 0, \quad \forall (d, \phi, i) \neq (d, \phi, i)^*.$$

Local transmission decisions: At the beginning of each time slot t, node u observes its local queue backlogs and those of its neighbors, and performs the following operations for each of its outgoing links:

1. Compute the *transmission utility weight* of each commodity:

$$W_{uv}^{(d,\phi,i)}(t) = \left[Q_u^{(d,\phi,i)}(t) - Q_v^{(d,\phi,i)}(t) \right]^+.$$

2. Compute the max-weight commodity:

$$(d, \phi, i)^* = \arg \max_{(d,\phi,i)} \left\{ W_{uv}^{(d,\phi,i)}(t) \right\}.$$

3. Allocate transmission resources:

$$y_{uv}(t) = \begin{cases} c_{uv} & \text{if } W_{uv}^{(d,\phi,i)}(t) - Vw_{uv} > 0 \\ 0 & \text{otherwise} \end{cases}.$$

4. Assign transmission flow rates:

$$\mu_{uv}^{(d,\phi,i)^*}(t) = \frac{y_{uv}}{\gamma_{uv}}$$

$$\mu_{uv}^{(d,\phi,i)}(t) = 0, \quad \forall (d, \phi, i) \neq (d, \phi, i)^*.$$

Observe that DCNC-L incurs low computational complexity at each cloud network node, where the number of computations scales only linearly with the total number of commodities. Enhanced versions of the DCNC-L algorithm are provided in [24]. In particular, an extension to DCNC-L, termed DCNC-Q, is derived via the minimization of a quadratic metric from the LDP bound, leading to improved cost–delay tradeoff performance at the expense of increased computational complexity. In addition, the use of a shortest transmission-plus-processing distance bias is shown to improve the cost–delay tradeoff of both DCNC-L and DCNC-Q, but especially of DCNC-L, in low-congestion scenarios [23, 24].

Figure 17.6 from [23] shows the performance of different DCNC algorithms for the control of two service chains over the Abilene US Continental network. Figures 17.6(a)

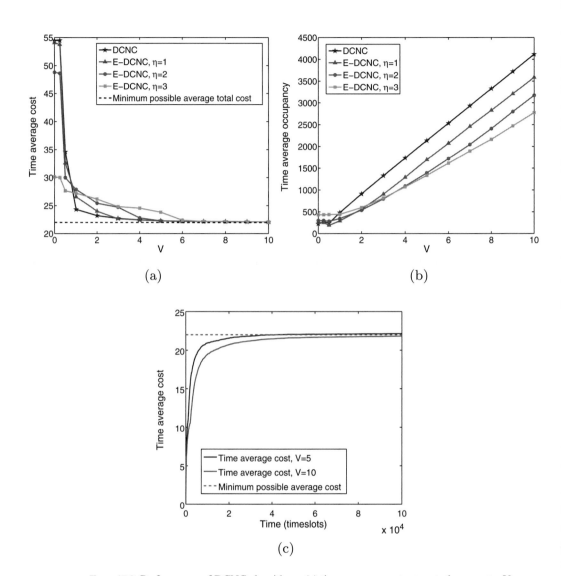

Figure 17.6 Performance of DCNC algorithms: (a) time average cost vs control parameter V; (b) time average total occupancy vs control parameter V; (c) time average cost evolution over time.

and 17.6(b) demonstrate the tradeoff between average cost and average total queue backlog (occupancy) as a function of the control parameter V that drives the cost–delay tradeoff. In Figure 17.6(a) and 17.6(b), E-DCNC is used to denote the enhanced version of DCNC-L (just DCNC in the figures) that includes the shortest transmission-plus-processing distance bias (with η indicating the weight of the bias term). These results clearly demonstrate the $[O(1/V), O(V)]$ cost–delay tradeoff as suggested by the theoretical analysis in [23, 24]. Figure 17.6(c) shows the evolution of the average cloud network cost over time, illustrating how a larger control parameter V allows achieving a closer to optimal solution, at the expense of slower time convergence. A more extensive set of results comparing the performance of different DCNC algorithms in illustrative network settings can be found in [24].

17.6   Extensions

17.6.1   Time-Varying Topologies

The studies covered so far in this chapter have assumed static network topology and channel states. However, driven by the increasingly mobile and wireless nature of end user devices and access technologies, 5G networks hosting and delivering AgI services will be in part characterized by time-varying topology states.

Along these lines, [4, 5, 27] extended the LDP analysis of [23, 24] to wireless computing networks. In particular, under the assumption that only statistical channel state information (CSI) is available at the transmitters, [4] characterized the wireless computing network capacity region, and developed a dynamic wireless computing network control algorithm that guarantees throughput-optimality via the use of the *broadcast approach* [28, 29] in the transmission flow scheduling decisions. The work in [27] then studied the impact of the availability of precise CSI on the capacity region and associated control policies.

Another important aspect to consider for the delivery of video-based services under time-varying wireless channels is the design of adaptive scheduling policies that determine the video rates to be delivered to each user with the objective of fairly maximizing the quality of users' video streaming experience [30–32].

17.6.2   Reconfiguration-Aware Control

Most existing works for dynamic routing, scheduling, and resource allocation assume negligible reconfiguration time and cost. While recent advances in virtualization technologies provide increased elasticity in the allocation of compute and network resources, reconfiguration times and costs are still nonnegligible, and can vary significantly depending, e.g., on the actual computing platform (virtual machines, containers, etc.).

In the presence of finite nonnegligible reconfiguration delays and costs, it is key to strike the right balance between allocating just enough resources to support the current traffic load and minimize the number of reconfigurations. This becomes an especially challenging task in an online setting under unknown time-varying demands.

The recent work of [33] started the study of dynamic cloud network control strategies under nonnegligible reconfiguration delay and cost. The authors showed that while the capacity region remains unchanged irrespective of the actual reconfiguration delay/cost values, a reconfiguration-agnostic policy may fail to guarantee throughput-optimality and minimum cost under nonzero reconfiguration delay/cost. They developed an adaptive dynamic cloud network control policy that allows network nodes to make local flow scheduling and resource allocation decisions while controlling the frequency of reconfiguration in order to support any input rate in the capacity region and achieve arbitrarily close to minimum cost for any finite reconfiguration delay/cost values.

Interesting future directions include analyzing the impact of demand predictions in the performance of reconfiguration-aware cloud network control algorithms.

17.6.3   Mixed-Cast Traffic

Many emerging AgI services, e.g., augmented reality, are characterized by a multicast upstream phase, where data streams useful for multiple users are uploaded to edge cloud nodes for processing, and a unicast downstream phase, in which personalized unicast streams are delivered to end users.

In this context, one of the main challenges is to develop a queueing model that captures the ability to create copies of data items inside the cloud network, so that mechanisms such as multicast and caching can be accurately modeled.

The work of [34] developed the first throughput-optimal algorithm for the control of service function chains under any mix of unicast and multicast traffic. The proposed policy, named universal computing network control (UCNC), builds on the recently proposed universal max-weight policy for throughput-optimal control of mixed-cast traffic in communication networks to determine both routes and processing locations for packets upon their arrival at a distributed computing network. UCNC guarantees that packets (1) are processed by a specified chain of service functions; (2) follow cycle-free routes between consecutive functions, and; (3) are delivered to their corresponding set of destinations via proper packet duplications.

Interesting future directions include addressing the design of mixed-cast traffic control policies for more generic service models, i.e., beyond service chains.

17.6.4   End-to-End Service Optimization

A number of recent works have also studied the *static* version of the service control problem, in which the goal is to find the placement of service functions and the routing of network flows for a given set of (a-priori known) average service demands.

Earlier versions of this problem include the service function placement problem, formulated in [35] as a generalization of facility location and generalized assignment, and for which efficient algorithms with bi-criteria approximation guarantees were also given in [35]. However, the model in [35] does not capture service chaining nor flow routing optimization. Another line of work studied the joint service placement and routing problem using tools from virtual network embedding, e.g., [36, 37], but without providing algorithms with approximation guarantees.

The works of [20, 21] introduced a flow-based model for optimizing the distribution (placement and routing) of services with arbitrary chaining relationships over capacitated cloud networks. Services are described via a directed acyclic graph, and the function placement and flow routing are determined by solving a minimum-cost network flow problem with generalized flow conservation and service chaining constraints. The work of [38] then leveraged the LDP-based analyses of [22–24] to design accelerated dual subgradient algorithms with approximation guarantees for such static cloud network flow problems under unicast service chains.

Interesting open problems for which efficient approximations are not known include the optimization of services described by generic service graphs (beyond service chains) and/or that contain mixed-cast traffic flows.

References

[1]   M. Weldon, *The Future of x Network*. Boca Raton, FL: CRL Press.

[2]   P. Mach and Z. Becvar, "Mobile edge computing: A survey on architecture and computation offloading," *IEEE Commun. Surv. Tutorials*, vol. 19, no. 3, pp. 1628–1656, 2017.

[3]   Y C. Hu, M. Patel, D. Sabella, N. Sprecher, and V. Young., "Mobile edge computing: A key technology towards 5G," ETSI White Paper, 11, 2015.

[4]   H. Feng, J. Llorca, A. M. Tulino, and A. F. Molisch, "On the delivery of augmented information services over wireless computing networks," in *IEEE Int. Conf. Commun. (ICC)*, Paris, May 2017.

[5]   H. Feng, J. Llorca, A. M. Tulino, and A. Molisch, "Optimal control of wireless computing networks," *IEEE/ACM Trans. Wireless Commun.*, vol. 26, no. 1, pp. 506–519, 2018.

[6]   A. B. Craig, *Understanding Augmented Reality: Concepts and Applications*. Oxford: Newnes, 2013.

[7]   M. S. Elbamby, C. Perfecto, M. Bennis, and K. Doppler, "Towards low-latency and ultra-reliable virtual reality," *IEEE Netw.*, vol. 32, no. 2, pp. 78–84, 2018.

[8]   "Industrial internet consortium." [Online]. Available at. www.iiconsortium.org.

[9]   X. Foukas, G. Patounas, E. A., and M. K. Marina, "Network slicing in 5G: Survey and challenges," *IEEE Commun. Mag.*, vol. 55, no. 5, pp. 94–100, 2017.

[10] M. Richart, J. Baliosian, J. Serrat, and J. Gorricho, "Resource slicing in virtual wireless networks: A survey," *IEEE Trans. Netw. Service Manage.*, vol. 13, no. 3, pp. 462–476, 2016.

[11] "The programmable cloud network: A primer on SDN and NFV," Bell Labs Strategic White Paper, June 2013.

[12] J. G. Herrera and J. F. Botero, "Resource allocation in NFV: A comprehensive survey," *IEEE Trans. Netw. Service Manage.*, vol. 13, 2016.

[13] L. Georgiadis, L. Tassiulas, and M. J. Neely, "Resource allocation and cross-layer control in wireless networks," *Found. Trends Networking*, vol. 1, no. 1, pp. 1–144, 2006.

[14] L. Tassiulas and A. Ephremides, "Stability properties of constrained queueing systems and scheduling policies for maximum throughput in multihop radio networks," *IEEE Trans. Auto. Control*, vol. 37, no. 12, pp. 1936–1948, Dec. 1992.

[15] L. Georgiadis, M. J. Neely, and L. Tassiulas, *Resource Allocation and Cross-Layer Control in Wireless Networks*. Delft: Now Publishers Inc, 2006.

[16] M. J. Neely, Stochastic Network Optimization with Application to Communication and Queueing Systems. Williston, VT: Morgan and Claypool Publishers, 2010.

[17] M. J. Neely, "Energy optimal control for time-varying wireless networks," *IEEE Trans. Inform. Theory*, vol. 52, no. 7, pp. 2915–2934, Jul. 2006.

[18] M. J. Neely, "Optimal backpressure routing for wireless networks with multi-receiver diversity," in *2006 40th Ann. Conf. Inform. Sci. Syst.*, vol. 3, March 2006, pp. 18–25.

[19] H. Feng and A. F. Molisch, "Diversity backpressure scheduling and routing with mutual information accumulation in wireless ad-hoc networks," *arXiv:1305.5588*, 2013.

[20] M. Barcelo, J. Llorca, A. M. Tulino, and N. Raman, "The cloud service distribution problem in distributed cloud networks," in *IEEE Int. Conf. Commun. (ICC)*, London, 2015, pp. 1–5.

[21] M. Barcelo, A. Correa, J. Llorca, et al. "IoT-cloud service optimization in next generation smart environments," *IEEE J. Sel. Areas Commun.*, vol. 34, no. 12, pp. 4077–4090, 2016.

[22] H. Feng, J. Llorca, A. M. Tulino, and A. Molisch, "Dynamic network service optimization in distributed cloud networks," in *IEEE INFOCOM SWFAN Workshop*, April 2016.

[23] H. Feng, J. Llorca, A. M. Tulino, and A. Molisch, "Optimal dynamic cloud network control," in *IEEE Int. Conf. Commun. (ICC)*, Kuala Lumpar, May 2016.

[24] H. Feng, J. Llorca, A. M. Tulino, and A. Molisch, "Optimal dynamic cloud network control," *IEEE/ACM Trans. Networking*, vol. 26, no. 5, pp. 2118–2131, 2018.

[25] T. M. Cover and J. A. Thomas, *Elements of Information Theory*. Chichester: Wiley, 1991.

[26] J. Kleinberg and E. Tardos, *Algorithm Design*. Upper Saddle Riwer, NJ: Pearson, Addison-Wesley, 2006.

[27] H. Feng, J. Llorca, A. M. Tulino, and A. Molisch, "Impact of channel state information on wireless computing network control," *IEEE ASILOMAR*, Pacific Grove, CA, 2017.

[28] S. Shamai and A. Steiner, "A broadcast approach for a single-user slowly fading MIMO channel," *IEEE Trans. Inform. Theory*, vol. 49, no. 10, pp. 2617–2635, 2003.

[29] A. M. Tulino, G. Caire, and S. Shamai, "Broadcast approach for the sparse-input random-sampled MIMO Gaussian channel," in *Proc. IEEE Int. Symp. Inform. Theory (ISIT)*, Honolulu, HI, 2014.

[30] K. Miller, D. Bethanabhotla, G. Caire, and A. Wolisz, "A control-theoretic approach to adaptive video streaming in dense wireless networks," *IEEE Trans. Multimed.*, vol. 17, no. 8, pp. 1309–1327, 2015.

[31] D. Bethanabhotla, G. Caire, and M. J. Neely, "Adaptive video streaming for wireless networks with multiple users and helpers," *IEEE Trans. Commun.* vol. 63, no. 1, pp. 268–285, 2014.

[32] D. Bethanabhotla, G. Caire, and M. J. Neely, "Wiflix: Adaptive video streaming in massive MU-MIMO wireless networks," *IEEE Trans. Wireless Commun.* vol. 15, no. 6, pp. 4088–4103, 2016.

[33] C. Wang, J. Llorca, A. M. Tulino, and J. T., "Dynamic cloud network control under reconfiguration delay and cost," *arXiv:1802.06581*, 2017.

[34] J. Zhang, A. Sinha, J. Llorca, A. M. Tulino, and E. Modiano, "Optimal control of distributed computing networks under mixed cast traffic flows," in *Proc. INFOCOM, Paris*; May 2018.

[35] R. Cohen, L. Lewin-Eytan, J. Naor, and D. Raz, "Near optimal placement of virtual network functions," in *Proc. IEEE INFOCOM*, Kowloon, April 2015.

[36] M. C. Luizelli, L. R. Bays, L. S. Buriol, M. P. Barcellos, and L. P. Gaspary, "Piecing together the NFV provisioning puzzle: Efficient placement and chaining of virtual network functions," in *IFIP/IEEE Int. Symp. Integrat. Netw. Manage.*, Ottawa, 2015, pp. 98–106.

[37] F. Bari, S. R. Chowdhury, R. Ahmed, R. Boutaba, and O. C. M. B. Duarte, "On orchestrating virtual network functions in NFV," *IEEE Trans. Netw. Service Manag.* DOI: 10.11091 CNSM, 2015, 7367338.

[38] H. Feng, J. Llorca, A. M. Tulino, D. Raz, and A. F. Molisch, "Approximation algorithms for the NFV service distribution problem," in *IEEE INFOCOM*, Atlanta GA, 2017.

18 A Broadcast Approach to Fading Channels under Secrecy Constraints

Shaofeng Zou, Yingbin Liang, Lifeng Lai, H. Vincent Poor, and Shlomo Shamai (Shitz)

18.1 Introduction

In wireless networks, communication signals are transmitted via the open medium of free space, and hence can be easily eavesdropped upon by any receiver within range. The major challenge of secure wireless communication is due to this broadcast nature of wireless radio. One way to address this challenge is via physical layer security, which is based on the information theoretic ideas of Wyner [1] that characterize the ability of the physical channel to achieve secure communication. This approach has the potential to significantly reduce requirements on the infrastructure and improve communication flexibility and dynamics without the inherent use of secret keys. Before we proceed, we first take a careful look at Wyner's approach to secure information transmission, which is based on the wiretap channel.

18.1.1 Basic Wiretap Channel

The possibility that a physical communication channel could, in itself, provide secure information transmission was first demonstrated by Wyner in his seminal work [1] via the wiretap channel depicted in Figure 18.1. In this channel, a transmitter wishes to send a message to a legitimate receiver while keeping this message as secret as possible from an eavesdropper. The eavesdropper is assumed to have unlimited computing power and complete knowledge of the communication protocols used by the transmitter and legitimate receiver, including coding schemes and transmission strategies. The transmitter and legitimate receiver do not share pre-deployed secret keys. Wyner's security approach makes use of inherent randomness of the physical medium to provide advantage to the legitimate receiver over the eavesdropper. In particular, stochastic coding is applied to exploit the difference in the physical channel characteristics between the legitimate receiver and the eavesdropper.

Wyner's approach to secrecy is analyzed via an information theoretic quantity called the *equivocation rate*, which was introduced by Shannon in [2] as a precise measure of the secrecy level of a message at an eavesdropper. Mathematically, the equivocation rate is defined to be the entropy rate of the message W given the n-length channel output Z^n at the eavesdropper, i.e.,

$$\frac{1}{n}H(W|Z^n). \tag{18.1}$$

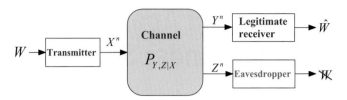

Figure 18.1 The wiretap channel.

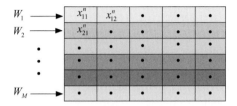

Figure 18.2 A wiretap codebook.

The equivocation rate represents the amount of randomness in the message that remains after the eavesdropper has observed its received version of the transmitted message. Thus, higher equivocation indicates lower likelihood that the message may be inferred by the eavesdropper. This measure of secrecy level enables secrecy to be considered under Shannon's general framework of information theory [3], and hence provides an analytical basis for characterizing the fundamental limits on communication rates given a certain secrecy constraint. An important aspect of this characterization is the secrecy capacity, i.e., the largest communication rate such that the legitimate receiver can successfully decode the message while the eavesdropper does not obtain any information about the message, in the limit as $n \to \infty$, i.e., asymptotically perfect secrecy is achieved.

Consider Wyner's wiretap channel, a single use of which is characterized by a transition probability distribution $P_{YZ|X}$ with X being the channel input, and Y and Z being the respective channel outputs at the receiver and the eavesdropper. The channel is used n independent times, and ultimately we consider asymptotics as $n \to \infty$. The secrecy capacity is characterized by Wyner in [1] for the degraded case, in which the variables X, Y, and Z form a Markov chain in that order, a condition that we write as $X \to Y \to Z$. In this case, the secrecy capacity is given by

$$C_s = \max_{P_X}[I(X;Y) - I(X;Z)]. \tag{18.2}$$

We note that the secrecy capacity is determined by the difference between the communication rate to the legitimate receiver, i.e., $I(X;Y)$, and the rate to the eavesdropper, i.e., $I(X;Z)$. To achieve the secrecy capacity, the transmitter first establishes a codebook in which the codewords are divided into a number of bins. Figure 18.2 depicts such a codebook, in which each bin consists of the cells in the same row with each cell representing a codeword. The transmitter then employs a stochastic coding approach that maps each message W to a bin, and randomly selects one of the codewords in the bin for transmission. It was shown by Wyner that if the number of bins and the size of each bin are chosen carefully based on channel statistics, stochastic coding achieves perfect secrecy, i.e., the eavesdropper does not learn any information about the message.

It can be seen from (18.2) that the secrecy capacity is in general smaller than the capacity of the channel. A misinterpretation of this fact is that in order to achieve secure communication, we sacrifice some of the reliable communication rate. However, this is not true. In Wyner's binning scheme, the transmitter randomly selects one of the code-words in the bin to introduce randomness to confuse the eavesdropper. In fact, the index of the codeword within the bin can also be used to carry a message. However, this message cannot be kept secure from the eavesdropper. In this way, the total communication rate is still the same as the capacity of the channel, and furthermore, part of the transmitted message is kept secure from the eavesdropper. From this point of view, secrecy is provided as an additional benefit without sacrificing the total communication rate. Of course, the benefit does not come for free, because the codebook should be designed with the binning structure. We refer to a scheme that uses one part of a message to protect another part of the message as *embedded coding*.

Wyner's analysis was further generalized by Csiszár and Körner in [4] for the general wiretap channel that is not necessarily degraded, for which the secrecy capacity is given by

$$C_s = \max_{P_{UX}P_{YZ|X}} [I(U;Y) - I(U;Z)], \tag{18.3}$$

where U is an auxiliary random variable satisfying the Markov chain condition $U \rightarrow X \rightarrow (Y,Z)$. By introducing U, the transmitter creates a *prefix channel* from the auxiliary input U to the actual channel input X, and hence more randomness is introduced for secrecy enhancement. Although such a prefix channel does not help for the degraded channel, it enhances the secrecy rate in general, as demonstrated in [5].

Wyner's secrecy scheme and performance analysis have been generalized to various wireless network models in many recent studies. Reviews of these studies can be found in [6–10].

18.1.2   Secure Communication with Channel State Uncertainty

In order to guarantee security, the design parameters for secure encoding, i.e., the size of the codebook and the number of codewords in each bin, must be chosen based on the physical states of the channels to the legitimate receiver and the eavesdropper. Hence, the transmitter's knowledge about these channel states critically determines how accurate the design parameters for secure coding may be chosen and how well the secure coding schemes perform.

However, wireless channels are typically time-varying in nature, which brings significant design challenges. In general, the transmitter's channel state information (CSI) is obtained by training and feedback, i.e., the transmitter sends out a training sequence, and the receiver first estimates the channel state, and then feeds this information back to the transmitter (see [9] for recent studies with channel state uncertainty). In many wireless scenarios, the channel state is difficult to track due to the following reasons. First, the channel may change very fast, and it is not easy to track the channel state accurately. Second, the feedback needs a control channel, which may not be available or has only limited bandwidth for low-rate transmission. Third, and very importantly, in secure transmissions, eavesdroppers have no incentive to feed back their channel states unless they also are participants in the network for receiving other (non-secure) messages.

To design secure communication schemes with channel state uncertainty, three major types of wireless systems categorized based on different system requirements, i.e., secrecy-oriented, rate-oriented, and state-oriented systems, are usually considered. The design for each system will depend both on how the channel uncertainty affects the three major interacting issues of rate, secrecy, and state, and on how these three issues are prioritized in each system.

Secrecy-oriented systems require (asymptotic) perfect secrecy in Shannon's sense to be guaranteed for all information bits transmitted, and do not tolerate any information leakage no matter what channel state occurs. Hence, security schemes for such systems should be robust to various communication environments. Secrecy-oriented systems can be modeled by the compound wiretap channel [5, 11–14], in which the channel to the legitimate receiver and the channel to the eavesdropper each has a number of possible states. The system requirement is that the source message must be transmitted to the legitimate receiver reliably and must be kept perfectly secure from the eavesdropper regardless of which pair of channel states occur. The transmitter does not know the channel states to the legitimate receiver and the eavesdropper, and hence construction of transmission schemes needs to address the design challenge that any state may occur.

Rate-oriented systems require a certain target rate to be achieved at the legitimate receiver. In contrast to the secrecy-oriented systems, rate-oriented systems tolerate a certain amount of information leakage to the eavesdropper, i.e., secrecy outage. Hence, perfect secrecy may not be achieved. The probability of information leakage is an important performance measure here. The goal is to design communication schemes to minimize the outage probability [15–21]. Rate-oriented systems can be modeled in terms of a fading channel whose state is random and time-varying, and the system transmission is subject to stringent delay constraints.

State-oriented systems require achievement of as high a rate as the legitimate receiver's channel supports, and as high a secrecy level as the eavesdropper's channel allows [15, 22, 23]. In such systems, there is neither a perfect secrecy constraint as in secrecy-oriented systems nor a target-rate constraint as in rate-oriented systems. Transmitted information bits may not be protected at the same secrecy level, and may not be equally likely to be decoded by the legitimate receiver. A major challenge in the design lies in that security schemes must be universal due to channel uncertainty, while the performance must automatically adapt to achieve the best secrecy performance supported by the channel state that occurs.

18.1.3   Objective of this Chapter

After the deployment of 4G, the 5G network has become a major topic of interest in both academia and industry. The security of the 5G network is a matter of key importance that requires careful consideration. Among the novel approaches proposed to enhance the security of the 5G network, physical layer security is considered as a strong candidate for many applications. Recent surveys and monographs [6–10, 24–27] provide a comprehensive review from various angles of physical layer security. The challenges that physical layer security faces with the 5G network are also summarized in [28–31].

As we have discussed in the last subsection, the CSI is very important in the design of a secure communication scheme. The wiretap channel with channel state uncertainty is also considered as a relevant ingredient for 5G systems. In [32] and [9], the authors present a survey with a focus on physical layer security with imperfect CSI. However, in some cases, a feedback link is not available, i.e., there is no CSI. For example, the fast-fading wiretap channel with no CSI at either the transmitter or the receivers was studied in [33, 34]. The block-fading wiretap channel with no CSI at the transmitter was studied in [35]. The wiretap channels with no CSI were also generalized to the multiple-antenna case in [36–38].

In this chapter, we aim to present a detailed state-of-the-art review of physical layer security with no CSI at the transmitter with an emphasis on using the broadcast approach in [39, 40] to tackle the block-fading wiretap channels, aiming at providing insights into using the broadcast approach in [39, 40] to reformulate the secure communication problems with channel state uncertainty into secure broadcasting problems, designing novel schemes and further inspiring new applications.

Toward this end, we first provide an overview of the broadcast approach in [39, 40]. The broadcast approach was first proposed for wireless communication with channel state uncertainty but without a secrecy constraint. The basic idea is to view the wireless channel as a broadcast channel with a number of receivers each corresponding to a different channel state. We then introduce the information theoretic studies of the application of the broadcast approach to the fading wiretap channel [35] and the Gaussian fading channel with secrecy outside a bounded range [41]. In the fading wiretap channel, both the legitimate receiver's and eavesdropper's channels are fading. In the Gaussian fading channel with secrecy outside a bounded range, there is only one receiver, which is considered as both a legitimate receiver and an eavesdropper, i.e., part of the message needs to be decoded and another part of the message needs to be kept secure. The receiver's channel is fading, and as the channel state gets better, more information is required to be decoded and less information is required to be secure. The information required to be decoded by the receiver with state h_0 should be kept secure from the receiver with state worse than h_0 by Δ, which is referred to as secrecy outside a bounded range. The performance measure of interest for both problems is the expected secrecy rate averaged over a long time period.

18.2   Broadcast Approach for Gaussian Slowly Fading Channels

The broadcast approach was introduced in [39, 40] for wireless communication with channel state uncertainty but without a secrecy constraint. This approach facilitates adaptation of the transmission rate to the actual unknown channel state without having any feedback link from the receiver to provide the CSI.

Consider the Gaussian fading channel, whose input–output relationship for one channel use is given by

$$Y = HX + W, \tag{18.4}$$

where Y is the output at the receiver and X is the input at the transmitter. The channel is corrupted by not only additive complex Gaussian random noise W with zero-mean and unit variance, but also a multiplicative random fading gain coefficient H, which is assumed to be continuous. The noise variables are assumed to be independent and identically distributed (i.i.d.) over channel uses. The fading coefficient H is assumed to experience block fading, i.e., it is constant within a coding block and changes ergodically across blocks, and is assumed to be known at the receiver (by channel estimation) but not known to the transmitter. The block length is assumed to be sufficiently long such that one codeword can be successfully transmitted if properly constructed. If the CSI were perfectly known at the transmitter, the channel capacity would be $C = \mathbb{E}_H \log(1 + H^2 P)$, where P is the average power constraint over each block on the channel input.

However, the actual CSI might not be available due to lack of feedback resources or delay constraints. The broadcast approach [39, 40] was introduced to address this problem. The basic idea is to view the fading channel as a degraded Gaussian broadcast channel with a continuum of receivers each experiencing a different fading coefficient h. The variable h is then interpreted as a continuous index. The transmitter can be viewed as transmitting in parallel infinitely many information streams parametrized by h, where the power assigned to the stream index by h is $P(h)dh \geq 0$, with $\int P(h)dh = P$. The receiver experiencing a realization h is able to decode its own data stream and the all the data streams indexed by $u < h$. Those data streams indexed by $u > h$ cannot be decoded and are considered as additional additive noise. Thus, the incremental differential rate is given by

$$
\begin{aligned}
dR(h) &= \log\left(1 + \frac{h^2 P(h)dh}{1 + h^2 \int_{u>h} P(u)du}\right) \\
&= \frac{h^2 P(h)dh}{1 + h^2 \int_{u>h} P(u)du}.
\end{aligned}
\tag{18.5}
$$

Then the achievable rate at fading level h is $R(h) = \int_{v \leq h} dR(v)$. The expected achievable rate over all fading realizations is $\mathbb{E}_H R(H)$. Optimization of the expected achievable rate over power allocation function $P(h)$ is of interest:

$$\max_{P(h)} \mathbb{E}_H R(H). \tag{18.6}$$

Studying the expected achievable rate without CSI can be put into the framework of variable-to-fixed coding as in [42], for which the number of observed channel symbols (blocklength) is prespecified, but the number of reliably recovered information bits depends on channel conditions. The variable-to-fixed channel capacity (see definition 3

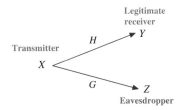

Figure 18.3 An illustration of the fading wiretap channel.

in [42]) is achieved by this broadcast approach for the scalar Gaussian fading channel [42, example 9].

18.3 Fading Wiretap Channel

In this section, we introduce the results of using the broadcast approach to address the fading wiretap channel in [35]. Consider the fading wiretap channel, in which both the legitimate receiver's and eavesdropper's channels are corrupted by additive Gaussian noise and random multiplicative fading coefficients. The transmission schemes must satisfy a delay constraint such that messages should be transmitted within one block, i.e., coding across blocks [43] is not allowed. The overall goal of designed achievable schemes for this problem is that the legitimate receiver decodes more information as its channel gets better, and out of the information decoded at the legitimate receiver, more information is kept secure from the eavesdropper as the eavesdropper's channel state gets worse. Moreover, the performance measure of interest is the delay-limited expected secrecy rate averaged over a long time period, which is different from the outage performance studied in [15, 20, 21] that focused on the delay-limited rate only over a short time period (e.g., one coherence block). Two types of broadcast approaches are developed for two simpler fading wiretap channels in which only one of the legitimate channel and eavesdropper's channel is fading, respectively. These two approaches are then combined to study the general scenario in which both channels are fading. The idea is to view the fading channel as a degraded broadcast channel with infinitely many legitimate receivers and/or eavesdroppers with a continuum of channel states.

18.3.1 Information Theoretic Model

Consider a fading wiretap channel (see Figure 18.3). The transmitter has one confidential message intended for the legitimate receiver and would like to keep the message secure from the eavesdropper. The legitimate receiver's channel and the eavesdropper's channel are corrupted both by additive complex Gaussian noise and random multiplicative fading gain coefficients. The channel input–output relationship for one channel use is given by

$$Y = HX + U,$$
$$Z = GX + V, \tag{18.7}$$

where X is the channel input, Y and Z are channel outputs at the legitimate receiver and eavesdropper, respectively, U and V are additive complex Gaussian random variables with zero-mean and unit variance, and H and G are random fading gain coefficients.

As in the previous section, the fading gains H and G are assumed to experience block fading, i.e., they are constant within a coding block and change ergodically across blocks. The blocklength is assumed to be sufficiently large such that one codeword can be successfully transmitted if properly constructed. The channel input is subject to an average power constraint P over each block. The noise variables are assumed to be independent per channel use within each block. The instantaneous CSI is assumed to be unknown to the transmitter, and each receiver knows its own channel state. The transmitted message is required to be decoded within one block, i.e., satisfies the delay constraint.

Each message is required to be decoded at the legitimate receiver with a small probability of error, and needs to be kept as secure as possible from the eavesdropper. Again, the measure of security is based on the equivocation rate:

$$\frac{1}{n}H(W|Z^n),$$

where Z^n is the channel output at the eavesdropper within one block. Using the fact that the mutual information $I(W; Z^n) = H(W) - H(W|Z^n)$, we can say that the message W is secure from the eavesdropper if

$$I(W; Z^n) \le n\epsilon_n, \tag{18.8}$$

where $\epsilon_n \to 0$ as $n \to \infty$. It is not required that all messages transmitted over the channel be secure from the eavesdropper. The performance measure is the secrecy rate, i.e., the rate of the messages that are kept secure from the eavesdropper. The secrecy rate is averaged over a large number of blocks. The goal is to achieve a secrecy rate as high as the legitimate receiver's channel supports, and as the eavesdropper's channel permits without instantaneous CSI at the transmitter.

18.3.2   Fading Legitimate Receiver's Channel

We first present the results in [35] for the case in which only the legitimate receiver experiences a block-fading channel, i.e., H is random and G is fixed. To convey the central idea, we start with the case with discrete legitimate channel states.

Assume that the legitimate receiver's channel can take L fading states, i.e., $|H_1| \le |H_2| \le \cdots \le |H_L|$. In [35], the authors generalize the broadcast approach introduced in [39, 40] for the fading channel without a secrecy constraint. More specifically, the entire message is split into L parts, $W_1, \ldots, W_L$, so that the legitimate receiver can decode the first ℓ messages if its channel realization is H_ℓ, for $\ell = 1, \ldots, L$. The eavesdropper is kept ignorant of all messages. In this sense, the fading wiretap channel here can be viewed as a *degraded* broadcast channel with multiple legitimate receivers and one eavesdropper [44].

A secrecy rate tuple $(R_1, \ldots, R_L)$ is *achievable* if there exists a coding scheme that encodes the messages $W_1, \ldots, W_L$ at the rate tuple $(R_1, \ldots, R_L)$ such that for $l = 1, \ldots, L$, the legitimate receiver decodes W_l with a small probability of error if its channel realization is H_l, and all messages $W_1, \ldots, W_L$ are kept secure from the eavesdropper. The following theorem specifies the achievable secrecy rate tuples that can be achieved by a broadcast approach.

THEOREM 18.1 *[35, theorem 1] For the fading wiretap channel with the legitimate receiver having L fading states $H_1, \ldots, H_L$, and the eavesdropper has one fixed fading state G, where $|G| < |H_1| \leq |H_2| \leq \cdots \leq |H_L|$, the following secrecy rate tuples $(R_1, \ldots, R_L)$ are achievable:*

$$R_\ell = \log\left(1 + \frac{|H_\ell|^2 P_\ell}{1 + |H_\ell|^2 \sum_{k=\ell+1}^{L} P_k}\right)$$
$$- \log\left(1 + \frac{|G|^2 P_\ell}{1 + |G|^2 \sum_{k=\ell+1}^{L} P_k}\right), \quad \ell = 1, \ldots, L, \tag{18.9}$$

where P_ℓ denotes the transmission power assigned for transmitting W_ℓ and satisfies the power constraint $\sum_{\ell=1}^{L} P_\ell \leq P$.

This problem reduces to the Gaussian wiretap channel if the legitimate receiver has only one possible state, i.e., $L = 1$. Then, Theorem 18.1 is equivalent to the optimal secrecy capacity of the Gaussian wiretap channel.

By the degradedness condition, the messages decoded by a receiver with worse channel state should also be decodable by a receiver with a better channel state. Hence, the legitimate receiver at channel state H_ℓ should be able to decode $W_1, \ldots, W_\ell$. Then, the total rate that is achievable when the legitimate receiver is at channel state H_ℓ is

$$\sum_{j=1}^{\ell} R_j = \sum_{j=1}^{\ell} \left[\log\left(1 + \frac{|H_j|^2 P_j}{1 + |H_j|^2 \sum_{k=j+1}^{L} P_k}\right) - \log\left(1 + \frac{|G|^2 P_j}{1 + |G|^2 \sum_{k=j+1}^{L} P_k}\right) \right]$$
$$= \left[\sum_{j=1}^{\ell} \log\left(1 + \frac{|H_j|^2 P_j}{1 + |H_j|^2 \sum_{k=j+1}^{L} P_k}\right) \right] - \log\left(1 + \frac{|G|^2 \sum_{k=1}^{\ell} P_k}{1 + |G|^2 \sum_{k=\ell+1}^{L} P_k}\right).$$
$$\tag{18.10}$$

The average achievable secrecy rate is

$$\sum_{\ell=1}^{L} P(H_\ell) \sum_{j=1}^{\ell} R_j.$$

This average secrecy rate can be further optimized over the power allocation function P_ℓ subject to $\sum_{\ell=1}^{L} P_\ell \leq P$.

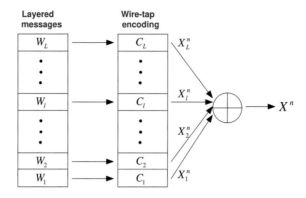

Figure 18.4 Codebook design for the broadcast approach.

The basic idea of the achievable scheme is based on designing a subcodebook for each message W_ℓ, employing random binning within each subcodebook, and then assembling them (see Figure 18.4). More specifically, for each message W_ℓ, a subcodebook $\mathcal{C}_\ell$ is designed with $2^{n\widetilde{R}_\ell}$ codewords, where $\widetilde{R}_\ell = \log\left(1 + \frac{|H_\ell|^2 P_\ell}{1+|H_\ell|^2 \sum_{k=\ell+1}^L P_k}\right)$. These codewords are further divided into 2^{nR_ℓ} bins, where

$$R_\ell = \log\left(1 + \frac{|H_\ell|^2 P_\ell}{1 + |H_\ell|^2 \sum_{k=\ell+1}^L P_k}\right) - \log\left(1 + \frac{|G|^2 P_\ell}{1 + |G|^2 \sum_{k=\ell+1}^L P_k}\right). \quad (18.11)$$

In order to transmit message $(w_1, \ldots, w_L)$, for each message w_ℓ, the transmitter uniformly and randomly chooses a codeword x_ℓ^n from the w_ℓth bin in the subcodebook $\mathcal{C}_\ell$, then sends $x^n = \sum_{\ell=1}^L x_\ell^n$ into the channel. For detailed derivations of the error probability of decoding and equivocation rate, we refer the reader to [35].

The result for discrete fading channels can be generalized to the continuous fading wiretap channel, i.e., the fading coefficient H can take continuous values. For each channel state $H = h$, let $s = |h|^2$. Using the broadcast approach for a Gaussian fading channel without secrecy constraint as in [39, 40], the message is divided into infinitely many layers of messages indexed by s. For each layer s of the message, the transmitter allocates power $\rho(s)ds$, where $\int_0^\infty \rho(s)ds \leq P$. Denote $\Sigma(s) = \int_s^\infty \rho(x)dx$, i.e., the power allocated to the layers of messages intended for legitimate channel states better than s. It is clear that $\rho(s) = -\Sigma'(s)$.

The average secrecy rate for the continuous legitimate channel state follows by applying Theorem 18.1:

$$R = \log e \max_{\Sigma(x)} \int_{|G|^2}^\infty (1 - F(x)) \left[\frac{-x\Sigma'(x)}{1 + x\Sigma(x)} + \frac{|G|^2\Sigma'(x)}{1 + |G|^2\Sigma(x)}\right] dx, \quad (18.12)$$

where $f(\cdot)$ is the probability density function of the fading state s, and $F(\cdot)$ is the cumulative distribution function of s. The achievable scheme is designed similarly to the case with a discrete channel state. The detailed proof can be found in [35].

To maximize the average achievable secrecy rate, consider the following optimization problem:

$$\max_{\Sigma(x)} \int_{|G|^2}^{\infty} (1 - F(x)) \left[\frac{-x\Sigma'(x)}{1 + x\Sigma(x)} + \frac{|G|^2 \Sigma'(x)}{1 + |G|^2 \Sigma(x)} \right] dx$$

$$\text{subject to} \quad 0 \le \Sigma(x) \le P, \ \Sigma'(x) \le 0, \quad \text{for } x \ge 0. \tag{18.13}$$

This is a problem of the constrained calculus of variation. Applying the techniques in [45], a necessary condition on the structure of the optimal power allocation function can be characterized in the following theorem (see [35] for a detailed proof).

THEOREM 18.2 *[35, theorem 2] Let*

$$\eta(x) = \frac{1 - F(x) - (x - |G|^2)f(x)}{xf(x)(x - |G|^2) - (1 - F(x))|G|^2}. \tag{18.14}$$

An optimal solution to (18.13), *if one exists, has the following structure. There exist* $0 \le x_1 < y_1 < x_2 < y_2 < \cdots < x_m < y_m$, *such that* $\eta(x)$ *is strictly decreasing over* $[x_i, y_i]$ *for* $i = 1, \ldots, n$, $\eta(x_1) = P$, $\eta(y_n) = \eta(x_0) = 0$, $\eta(y_i) = \eta(x_{i+1})$ *for* $i = 1, \ldots, m - 1$, *and*

$$\Sigma^*(x) = \begin{cases} P & 0 \le x \le x_1; \\ \eta(x) & x_i \le x \le y_i, \text{ for } i = 1, \ldots, m; \\ \eta(y_i) = \eta(x_{i+1}), & y_i < x < x_{i+1}, \text{ for } i = 1, \ldots, m - 1; \\ 0 & y_m \le x. \end{cases} \tag{18.15}$$

Consider an example when the legitimate channel experiences Rayleigh fading, i.e., $|H|^2$ is exponentially distributed with parameter σ_1. In this case, $\eta(x)$ in (18.14) is

$$\eta(x) = \frac{\sigma_1 - x + |G|^2}{x(x - |G|^2) - \sigma_1 |G|^2}. \tag{18.16}$$

Solving $\eta(x_1) = P$ and $\eta(y_1) = 0$,

$$y_1 = \sigma_1 + |G|^2,$$

$$x_1 = \frac{(P|G|^2 - 1) + \sqrt{(P|G|^2 - 1)^2 + 4P(P\sigma_1|G|^2 + |G|^2 + \sigma_1)}}{2P}. \tag{18.17}$$

It can be verified that $|G|^2 < x_1 < y_1$, $\eta(x)$ is strictly decreasing over $[x_1, y_1]$, because the numerator of $\eta(x)$ is decreasing, and the denominator of $\eta(x)$ is increasing over $[x_1, y_1]$. Since x_1 and y_1 are both unique solutions to $\eta(x_1) = P$ and $\eta(y_1) = 0$, respectively, and $\eta(x)$ is strictly decreasing over $[x_1, y_1]$, the optimal $\Sigma^*(x)$ is thus given by

$$\Sigma^*(x) = \begin{cases} P & 0 \le x \le x_1; \\ \eta(x) & x_1 \le x \le y_1; \\ 0 & y_1 \le x. \end{cases} \tag{18.18}$$

Since the above $\Sigma^*(x)$ is the unique function that satisfies the conditions given in Theorem 18.2, it is the only possible optimal solution for the power allocation function.

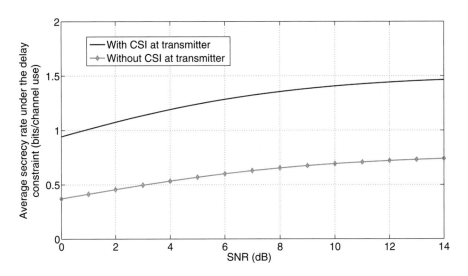

Figure 18.5 Comparison of the average secrecy rates: only the channel to the legitimate receiver is fading.

Thus, we can obtain the optimized average secrecy rate via a broadcast approach for the Rayleigh fading channel.

A comparison of the average secrecy rate with the case when the CSI is available at the transmitter is provided in Figure 18.5 [35]. $|H|^2$ is exponentially distributed with the parameter $\sigma_1 = 2$. The eavesdropper's channel state is at $|G|^2 = \frac{1}{2}$. The average secrecy rates achieved via the broadcast approach and the rates achievable when the legitimate receiver's CSI is known at the transmitter are plotted. With the legitimate receiver's CSI at the transmitter, the average secrecy rate (which is also the capacity) can be obtained by averaging the secrecy rate for each channel state over the state distribution and optimizing over all possible power allocation over the channel states as follows:

$$\bar{R} = \max_{P(s):E_s[P(s)]\leq P} \int_{|G|^2}^{\infty} \left[\log\left(1 + sP(s)\right) - \log\left(1 + |G|^2 P(s)\right)\right] \rho(s)\mathrm{d}s, \qquad (18.19)$$

where the optimizing power allocation can be obtained by using the Lagrange multiplier method as in [35].

Due to the lack of CSI at the transmitter, the transmitter's power is spread over many layers of messages in order to accommodate possibly occurring channel states, which results in a lower average secrecy rate than the case with CSI known at the transmitter.

18.3.3   Fading Eavesdropper's Channel

Consider the case in which only the eavesdropper experiences a block-fading channel, i.e., H is fixed and G is random. The transmitter is assumed to not know the CSI. We start with the case with the discrete eavesdropper's channel states.

Assume that the eavesdropper has a finite number of channel states, i.e., G may take L states, $G_1, \ldots, G_L$, with $|G_1|^2 < |G_2|^2 < \cdots < |G_L|^2 < |H|^2$. If the eavesdropper has a better state than $|H|^2$, then no message can be transmitted securely. The following results are applicable by considering only the eavesdropper's states that are worse than $|H|^2$.

A second type of broadcast approach is developed for this case in [35]. The message is split into L layers of submessages $W_1, W_2, \ldots, W_L$. In this case, the fading wiretap channel can be viewed as a degraded broadcast channel with one legitimate receiver and multiple eavesdroppers.

A secrecy rate tuple $(R_1, \ldots, R_L)$ is achievable if there exists a coding scheme that encodes $W_1, \ldots, W_L$ at the rate tuple $(R_1, \ldots, R_L)$ such that the legitimate receiver can decode all messages with a small probability of error, and message W_ℓ is kept secure from the eavesdropper if the eavesdropper's channel state is G_ℓ for $\ell = 1, \ldots, L$. Via the broadcast approach, the achievable rate tuples are characterized in the following theorem.

THEOREM 18.3 *[35, theorem 3] Consider the fading wiretap channel with the legitimate receiver having a fixed channel state H and the eavesdropper possibly having one of L fading states $G_1, \ldots, G_L$ with $|G_1|^2 < |G_2|^2 < \cdots < |G_L|^2 < |H|^2$. The following secrecy rate tuples $(R_1, \ldots, R_L)$ are achievable:*

$$R_\ell = \log\left(1 + |G_{\ell+1}|^2 P\right) - \log\left(1 + |G_\ell|^2 P\right), \quad for \quad \ell = 1, \ldots, L-1; \quad (18.20)$$

$$R_L = \log\left(1 + |H|^2 P\right) - \log\left(1 + |G_L|^2 P\right). \quad (18.21)$$

By the degradedness condition of the reformulated broadcast channel, the messages that are secure from the eavesdropper with the state G_j are also secure from the eavesdropper with the state G_ℓ if $|G_j| > |G_\ell|$, and all $W_\ell, \ldots, W_L$ are secure from the eavesdropper at the state G_ℓ if $(R_1, \ldots, R_L)$ is achievable. Hence, the total rate of the messages that are secure from the eavesdropper at the channel state G_l is given by

$$R_\ell + R_{\ell+1} + \cdots + R_L = \log\left(1 + |H|^2 P\right) - \log\left(1 + |G_\ell|^2 P\right). \quad (18.22)$$

The secrecy rate in (18.22) is equal to the secrecy capacity of a wiretap channel with state pair (H, G_ℓ). Thus, this type of broadcast approach achieves the best secrecy rate that the instantaneous channel allows, although the transmitter does not know the CSI. The only loss due to this broadcast approach is that some lower-layer messages may not be kept secure from the eavesdropper. This is in contrast to the case with only the legitimate receiver having a fading channel, for which all messages are kept secure from the eavesdropper, but the achieved secrecy rate might not be optimal.

Although the legitimate receiver does not know the eavesdropper's CSI, the broadcast approach is still able to keep some layers of submessages secure from the eavesdropper. However, without the eavesdropper's CSI, the transmitter only knows the probability that a certain layer of submessage is secure from the eavesdropper.

The achievable scheme is based on embedded coding [46]. A codebook is constructed which contains $2^{n \log(1+|H|^2 P)}$ codewords x^n. Each codeword is indexed by L layers of submessages $(w_1, \ldots, w_L)$ and a random index q, where

$$q = 1, 2, \ldots, 2^{n \log(1+|G_1|^2 P)},$$

$$w_1 = 1, 2, \ldots, 2^{n[\log(1+|G_2|^2 P) - \log(1+|G_1|^2 P)]},$$

$$w_2 = 1, 2, \ldots, 2^{n[\log(1+|G_3|^2 P) - \log(1+|G_2|^2 P)]},$$

$$\vdots$$

$$w_{L-1} = 1, 2, \ldots, 2^{n[\log(1+|G_L|^2 P) - \log(1+|G_{L-1}|^2 P)]},$$

$$w_L = 1, 2, \ldots, 2^{n[\log(1+|H|^2 P) - \log(1+|G_L|^2 P)]}. \tag{18.23}$$

To transmit a message tuple $(w_1, \ldots, w_L)$, the transmitter randomly and uniformly generates q, and transmits $x^n(q, w_1, w_2, \ldots, w_L)$. Following steps similar to those in [7, section 2.3], the decoding error and equivocation rate can be bounded properly which shows that the rate tuples in Theorem 18.3 are achievable.

The results for the case with discrete eavesdropper's states can be generalized to the continuous case. In this case, the broadcast approach views the model as a degraded broadcast channel with one legitimate receiver and infinitely many eavesdroppers with a continuum of channel states. The messages are encoded correspondingly into a continuum of layers. For each state $G = g$, let $u = |g|^2$, and use u as an index for the layer of the message that needs to be kept secure from the eavesdropper with state g. Following Theorem 18.3, the following average secrecy rate is achievable:

$$R = Q(|H|^2) \log \left(1 + |H|^2 P \right) - \int_0^{|H|^2} q(u) \log(1 + uP) du, \tag{18.24}$$

where $q(\cdot)$ and $Q(\cdot)$ are the probability density function and cumulative distribution function of $|G|^2$, respectively.

Figure 18.6 provides a clear illustration of the layers of messages that are secure from the eavesdropper. Based on this result, the probabilistic secrecy, i.e., the probability that a given secrecy rate R is achievable, denoted by $\Pr(R)$, can be characterized [35].

A comparison of the average secrecy rate with the case when the CSI is available at the transmitter is provided in Figure 18.7 [35]. The distribution of $u = |G|^2$ is exponential with the parameter $\sigma_2 = 0.5$, i.e., $p(u) = \frac{1}{\sigma_2} e^{-u/\sigma_2}$. The legitimate receiver's channel state is at $|H|^2 = 2$. The average secrecy rates achieved via the broadcast approach and the rates achievable when the eavesdropper's CSI is known at the transmitter are plotted. It is clear that the rates corresponding to the two cases are very close, suggesting that the knowledge of the eavesdropper's CSI does not provide much advantage to achieve better secrecy rates. This is not surprising, as the broadcast approach already achieves the maximum possible secrecy rate for each block. The small gap between the two rates is because with the CSI, the transmitter can adapt its power allocation over the channel states to achieve a better rate.

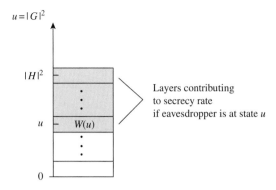

Figure 18.6 An illustration of the layers of messages that are secure from the eavesdropper.

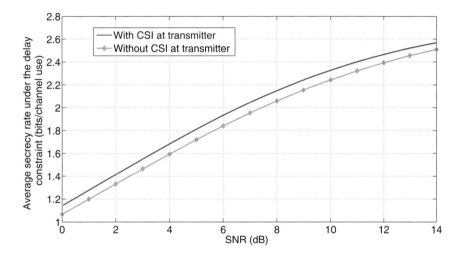

Figure 18.7 Comparison of the average secrecy rates: only the channel to the eavesdropper is fading.

18.3.4 Fading Legitimate and Eavesdropper's Channels

We now consider the most general case in which both the legitimate channel and eavesdropper's channel are experiencing block fading, i.e., H and G are random.

First, assume that both the legitimate receiver and the eavesdropper have finite numbers of channel states, i.e., H and G take one of $H_1, \ldots, H_L$ values and one of $G_1, \ldots, G_K$ values, respectively, where $|H_1| < \cdots < |H_L|$ and $|G_1| < \cdots < |G_K|$. For each $1 \le \ell \le L$, use K_ℓ to denote the largest index of the state level of G that is below H_ℓ, i.e., $K_\ell = \max_{|G_k| \le |H_\ell|} k$. For this setting, a broadcast approach combines the two broadcast approaches developed for the two previous cases with fading legitimate channel only and fading eavesdropper's channel only. The entire message is first split into a number of components $W_{\ell[1,K_\ell]}$ for $1 \le \ell \le L$, where $W_{\ell[1,K_\ell]}$ denotes $W_{\ell 1}, \ldots, W_{\ell K_\ell}$.

A secrecy rate tuple $\{R_{\ell[1,K_\ell]}\}_{\ell=1,\ldots,L}$ is achievable if there exists a coding scheme that encodes the messages $W_{\ell[1,K_\ell]}$ at the rates $R_{\ell[1,K_\ell]}$ for $1 \le \ell \le L$ such that if the legitimate receiver's channel is at H_ℓ and the eavesdropper's channel is at G_k for $1 \le \ell \le L$ and $1 \le k \le K_\ell$, then the legitimate receiver decodes the message $W_{\ell k}$ with a vanishing probability of error and the eavesdropper is kept ignorant of the message $W_{\ell k}$.

Through a combination of the two broadcast approaches for the two previous cases, the following rate tuples are achievable.

THEOREM 18.4   *[35, theorem 4] For the fading wiretap channel with the legitimate receiver having one of L fading states $H_1,\ldots,H_L$ with $|H_1| < \cdots < |H_L|$ and the eavesdropper having one of K fading states $G_1,\ldots,G_K$ with $|G_1| < \cdots < |G_K|$, the following secrecy rate tuples $(R_{1,[1,K_1]},\ldots,R_{L[1,K_L]})$ are achievable:*

$$R_{\ell k} = \begin{cases} \log\left(1 + \dfrac{|G_{k+1}|^2 P_\ell}{1+|G_{k+1}|^2 \sum_{j=\ell+1}^{L} P_j}\right) - \log\left(1 + \dfrac{|G_k|^2 P_\ell}{1+|G_k|^2 \sum_{j=\ell+1}^{L} P_j}\right), \\ \qquad for\ 1 \le \ell \le L, \quad 1 \le k \le K_\ell - 1 \\[2mm] \log\left(1 + \dfrac{|H_\ell|^2 P_\ell}{1+|H_\ell|^2 \sum_{j=\ell+1}^{L} P_j}\right) - \log\left(1 + \dfrac{|G_{K_\ell}|^2 P_\ell}{1+|G_{K_\ell}|^2 \sum_{j=\ell+1}^{L} P_j}\right), \\ \qquad for\ 1 \le \ell \le L, \quad k = K_\ell \end{cases} \tag{18.25}$$

where P_ℓ denotes the transmission power assigned to state ℓ and satisfies the power constraint $\sum_{\ell=1}^{L} P_\ell \le P$.

An argument similar to that used in the case with only a fading legitimate channel is that the messages that can be decoded by the legitimate receiver with state H_j can also be decoded by those with state better than H_j. Here, the legitimate receiver with state H_ℓ can decode all the messages $W_{1[1,K_1]},\ldots,W_{\ell,[1,K_\ell]}$ for $\ell = 1,\ldots,L$. Also an argument similar to that used in the case with only the fading eavesdropper's channel is that the messages that are secure from the eavesdropper with state G_j are also secure from those eavesdroppers with worse state than G_k, and all $W_{1[k,K_1]},\ldots,W_{L[k,K_L]}$ are secure from the eavesdropper at state G_k.

For the case with both legitimate receiver and eavesdropper having fading channels, the design of the achievable scheme via the broadcast approach does not require that the legitimate receiver know the CSI. However, not knowing the CSI only enables the transmitter to know the probability that certain layers of messages are secure, which is referred to as probabilistic secrecy.

The achievable scheme via a combination of the two broadcast approaches is to design a subcodebook for each H_ℓ, for $1 \le \ell \le L$, and then assemble them as in the case with only a fading legitimate channel. More specifically, for each state of the legitimate chan-

nel H_ℓ, $1 \le \ell \le L$, design a subcodebook C_ℓ, which contains $2^{n\log\left(1+\frac{|H_\ell|^2 P_\ell}{1+|H_\ell|^2 \sum_{j=\ell+1}^{L} P_j}\right)}$ codewords x_ℓ^n indexed by $(q_\ell, w_{\ell 1}, w_{\ell 2}, \ldots, w_{\ell K_\ell})$, where

$$q_\ell = 1, 2, \ldots, 2^{n \log\left(1 + \frac{|G_1|^2 P_\ell}{1 + |G_1|^2 \sum_{j=\ell+1}^{L} P_j}\right)},$$

$$w_{\ell 1} = 1, 2, \ldots, 2^{n\left[\log\left(1 + \frac{|G_2|^2 P_\ell}{1 + |G_2|^2 \sum_{j=\ell+1}^{L} P_j}\right) - \log\left(1 + \frac{|G_1|^2 P_\ell}{1 + |G_1|^2 \sum_{j=\ell+1}^{L} P_j}\right)\right]},$$

$$w_{\ell 2} = 1, 2, \ldots, 2^{n\left[\log\left(1 + \frac{|G_3|^2 P_\ell}{1 + |G_3|^2 \sum_{j=\ell+1}^{L} P_j}\right) - \log\left(1 + \frac{|G_2|^2 P_\ell}{1 + |G_2|^2 \sum_{j=\ell+1}^{L} P_j}\right)\right]},$$

$$\vdots$$

$$w_{\ell(K_\ell - 1)} = 1, 2, \ldots, 2^{n\left[\log\left(1 + \frac{|G_{K_\ell}|^2 P_\ell}{1 + |G_{K_\ell}|^2 \sum_{j=\ell+1}^{L} P_j}\right) - \log\left(1 + \frac{|G_{K_\ell - 1}|^2 P_\ell}{1 + |G_{K_\ell - 1}|^2 \sum_{j=\ell+1}^{L} P_j}\right)\right]},$$

$$w_{\ell K_\ell} = 1, 2, \ldots, 2^{n\left[\log\left(1 + \frac{|H_\ell|^2 P_\ell}{1 + |H_\ell|^2 \sum_{j=\ell+1}^{L} P_j}\right) - \log\left(1 + \frac{|G_{K_\ell}|^2 P_\ell}{1 + |G_{K_\ell}|^2 \sum_{j=\ell+1}^{L} P_j}\right)\right]}. \qquad (18.26)$$

To send a message tuple $w_{1[1,K_1]}, \ldots, w_{L,[L,K_L]}$, the transmitter first randomly and uniformly generates q_ℓ, and chooses the subcodeword $x_\ell^n(q_\ell, w_{\ell 1}, \ldots, w_{\ell K_\ell})$ for $\ell = 1, \ldots, L$. Then the transmitter sends $x^n = \sum_{\ell=1}^{L} x_\ell^n(q_\ell, w_{\ell 1}, \ldots, w_{\ell K_\ell})$. Following steps similar to those in [7, section 2.3], it can be shown that there exists a codebook with asymptotically small probability of error and equivocation rates.

The results can be generalized to the case with continuous channel states. For each channel state pair $(H, G) = (h, g)$, let $(s, u) = (|h|^2, |g|^2)$, and use (s, u) to index layers of messages. For each layer s, the transmitter allocates power $\rho(s)ds$, and we use $\Sigma(s)$ to denote the total power allocated to the layers with better channel states than s. Hence, $\Sigma(s) = \int_s^\infty \rho(x)dx$, and $\rho(s) = -\Sigma'(s)$.

By viewing this model as a degraded broadcast channel with an infinite number of legitimate receivers with a continuum of channel states and infinite number of eavesdroppers with a continuum of channel states, the following average secrecy rate is achievable [35]:

$$R = \log e \max_{\Sigma(x)} \int_0^\infty dx(1 - F(x))\Sigma'(x)\left[\frac{-xQ(x)}{1 + x\Sigma(x)} + \int_0^x du \frac{uq(u)}{1 + u\Sigma(x)}\right], \qquad (18.27)$$

where $F(\cdot)$ and $Q(\cdot)$ are cumulative distribution functions of s and u, respectively. An illustration of the layers that contribute to the secrecy rate $R(s, u)$ is shown in Figure 18.8.

By solving an optimization problem with respect to the power allocation function $\Sigma(x)$, we can obtain the best achievable average secrecy rate using the broadcast approach, which is to solve the following problem:

$$\max_{\Sigma(x)} \int_0^\infty S(x, \Sigma(x), \Sigma'(x))dx$$

$$\text{subject to} \quad 0 \le \Sigma(x) \le P, \quad \Sigma'(x) \le 0, \quad \text{for } x \ge 0, \qquad (18.28)$$

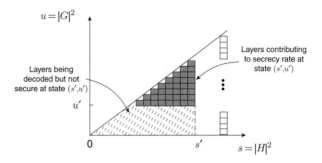

Figure 18.8 An illustration of the layers of messages that are decodable at the legitimate receiver and secure from the eavesdropper.

where

$$S(x, \Sigma(x), \Sigma'(x)) = (1 - F(x))Q(x)\frac{-x\Sigma'(x)}{1 + x\Sigma(x)} + (1 - F(x))\Sigma'(x)\int_0^x \frac{uq(u)}{1 + u\Sigma(x)}du. \quad (18.29)$$

It is shown in [35] that the optimal solution to the above optimization problem needs to satisfy the following structure. There exist $0 \le x_1 < y_1 < x_2 < y_2 < \cdots < x_m < y_m$, and a function $\eta(x)$, such that $\eta(x)$ satisfies

$$\frac{(1 - F(x))Q(x)}{(1 + x\eta(x))^2} = \frac{xf(x)Q(x)}{1 + x\eta(x)} - f(x)\int_0^x \frac{uq(u)}{1 + u\eta(x)}du \quad (18.30)$$

and is strictly decreasing over $[x_i, y_i]$ for $i = 1, \ldots, m$, $\eta(x_1) = P$, $\eta(y_m) = 0$, $\eta(y_i) = \eta(x_{i+1})$ for $i = 1, \ldots, m - 1$, and an optimal $\Sigma^*(x)$ is given by

$$\Sigma^*(x) = \begin{cases} P & 0 \le x \le x_1; \\ \eta(x) & x_i \le x \le y_i, \\ & \text{for} \quad i = 1, \ldots, m; \\ \eta(y_i) = \eta(x_{i+1}), & y_i < x < x_{i+1}, \\ & \text{for} \quad i = 1, \ldots, m - 1; \\ 0 & y_m \le x. \end{cases} \quad (18.31)$$

We next consider an example in [35] in which both the legitimate receiver and eavesdropper are Rayleigh fading channels, i.e., s and u are exponentially distributed as characterized by

$$f(x) = \frac{1}{\sigma_1}e^{-\frac{x}{\sigma_1}} \quad \text{and} \quad F(x) = 1 - e^{-\frac{x}{\sigma_1}}, \quad x \ge 0, \quad (18.32)$$

$$q(x) = \frac{1}{\sigma_2}e^{-\frac{x}{\sigma_2}} \quad \text{and} \quad Q(x) = 1 - e^{-\frac{x}{\sigma_2}}, \quad x \ge 0, \quad (18.33)$$

where σ_1 and σ_2 are parameters for the exponential distributions of s and u, respectively.

The Euler condition (18.30) now becomes

$$\frac{1 - e^{-\frac{x}{\sigma_2}}}{(1 + x\Sigma(x))^2} - \frac{x(1 - e^{-\frac{x}{\sigma_2}})}{\sigma_1(1 + x\Sigma(x))} + \frac{1}{\sigma_1\sigma_2}\int_0^x \frac{ue^{-\frac{u}{\sigma_2}}}{1 + u\Sigma(x)}du = 0. \quad (18.34)$$

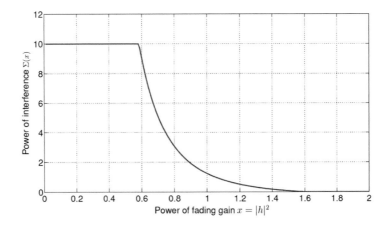

Figure 18.9 An optimal function $\Sigma(x)$ for Rayleigh fading channels with $P = 10\,\text{dB}$ and $\sigma_1 = \sigma_2 = 1$.

Consider the case with $\sigma_1 = \sigma_2 = 1$. Following from the above condition, if $\Sigma(y_1) = 0$, then y_1 satisfies

$$2 - 2e^{-y_1} - y_1 = 0,$$

whose root can be computed numerically and is approximately equal to

$$y_1 = 1.5936.$$

Using the condition (18.34), it is easy to find a $\Sigma^*(x)$ function that satisfies the necessary condition given above. The function $\Sigma^*(x)$ for the case with $P = 10\,\text{dB}$, and $\sigma_1 = \sigma_2 = 1$ is plotted as an example in Figure 18.9. In this example, $\Sigma^*(x)$ is strictly decreasing over the interval $[x_1, y_1]$, which implies that the optimal solution is unique if it exists.

A comparison of the average secrecy rate for the case when the CSI is available at the transmitter is provided in Figure 18.10 [35]. The random variables $s = |H|^2$ and $u = |G|^2$ are independent and are both exponentially distributed with the parameters $\sigma_1 = 2$ and $\sigma_2 = \frac{1}{2}$. In Figure 18.7, the average secrecy rates achieved via the broadcast approach and the rates achievable when both channels' CSI is known at the transmitter are plotted. We see that the CSI provides a great advantage in achieving better average secrecy rates in this case.

18.4   Fading Channel with Secrecy Outside a Bounded Range

In this section, we present the results of using the broadcast approach to solve the problem of secure communication over a fading channel with secrecy outside a bounded range.

Consider a fading channel in which a transmitter sends information to a receiver, and the channel from the transmitter to the receiver is corrupted by additive Gaussian

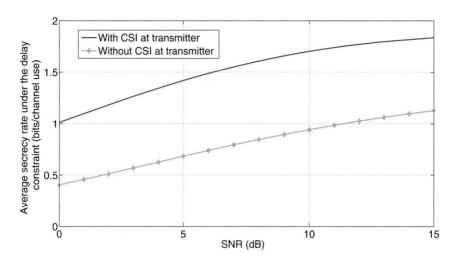

Figure 18.10 Comparison of the average secrecy rates: the channels to both the legitimate receiver and the eavesdropper are fading.

noise and random multiplicative fading coefficient H. The only receiver is considered as both legitimate and eavesdropping simultaneously, i.e., part of the message needs to be decoded and part of the message needs to be secure. If the channel state is better, then more information is required to be decoded by the receiver, and if the channel state is worse, more information is required to be kept secure from the receiver. Furthermore, the information intended to be decoded by the receiver if it has state h_0 should be kept secure from the receiver if it has state worse than h_0 by Δ, i.e., $|H| < |h_0| - \Delta$, for all $|h_0| \geq \Delta$, which is referred to as secrecy outside a bounded range [41]. The amount of information it can decode and that is being kept secure depends on its channel state. The overall goal of the designed achievable scheme is that more information is decoded and less information is secure from the receiver as the receiver's channel state gets better [41, 47, 48]. Similarly, the performance measure of interest is the delay-limited expected secrecy rate averaged over a long time period. This problem can be viewed as a degraded broadcast channel with infinitely many receivers with a continuum of channel states via the broadcast approach. The decoding and secrecy requirements are layered decoding and layered secrecy, and with secrecy outside a bounded range [41, 47, 48].

18.4.1 Information Theoretic Model

We consider a fading channel in which a transmitter sends information to a receiver. The channel input–output relationship for one channel use is given by

$$Y = HX + U, \tag{18.35}$$

where Y is the channel output, X is the channel input, H is random fading gain coefficient, and U is complex Gaussian random noise with zero-mean and unit variance.

The fading gain coefficient H is assumed to experience block fading as before, and the blocklength is assumed to be sufficiently large such that one codeword can be successfully transmitted if properly constructed. The channel input is subject to an average power constraint P over each block:

$$\frac{1}{n} \sum_{i=1}^{n} \mathbb{E}[|X_i|^2] \leq P, \tag{18.36}$$

where i denotes the symbol time (i.e., channel use) index, and n is the blocklength. The noise variable U is assumed to be i.i.d. over channel uses within each block. The instantaneous CSI is assumed to be unknown to the transmitter, and known to the receiver. The transmitted message is required to be decoded within one block, i.e., satisfies a delay constraint, and coding across blocks is not allowed.

It is required that more information is decoded by the receiver if it has a better channel state, and more information is kept secure from the receiver if it has a worse channel state. It is further required that the secrecy is outside a bounded range [41]. More specifically, the information intended to be decoded by the receiver if it has channel state h_0 should be kept secure from the receiver if it has channel state worse than h_0 by Δ, i.e., $|H| < |h_0| - \Delta$, for all $|h_0| \geq \Delta$. Here Δ is the secrecy range.

We consider two scenarios. In the first scenario, the receiver has finitely many channel states, i.e., H can take one of $H_1, \ldots, H_L$ values, where $|H_1| < \cdots < |H_L|$. It is assumed that the message intended to be decoded by the receiver if it has channel state H_k should be kept secure from the receiver if it has channel state worse than H_k by two levels of channel quality, i.e., $|H| \leq |H_{k-2}|$, and Δ is two levels of channel quality. This model is then generalized to the case in which Δ is arbitrary m levels of channel quality, where $m \geq 2$, i.e., the message intended to be decoded by the receiver if it has channel state H_k should be kept secure from the receiver if it has channel state worse than H_{k-m}.

In the second scenario, the fading coefficient H can take continuous values. The message at the transmitter is divided into infinitely many layers. It is required that the layers of the messages intended to be decoded by the receiver if it has a better state than h_0 be secure from the receiver if its state is worse than h_0 by Δ, i.e., $|H| < |h_0| - \Delta$.

18.4.2   Discrete Channel States

Before we introduce the results for the fading channel, we first introduce the results for the problem of degraded broadcast channel with secrecy outside a bounded range in [41]. In this problem, an L-receiver degraded broadcast channel is considered. A transmitter sends information to L receivers through a discrete memoryless broadcast channel. The channel is assumed to be degraded, i.e., the following Markov chain condition holds:

$$X \to Y_L \to Y_{L-1} \to \cdots \to Y_1, \tag{18.37}$$

where Y_ℓ is the channel output at receiver ℓ. Hence, the channel quality gradually degrades from receiver L to receiver 1. There are in total L messages $W_1, W_2, \ldots, W_L$ intended for L receivers with the following decoding and secrecy requirements. Receiver

k is required to decode messages $W_1, W_2, \ldots, W_k$, for $k = 1, 2, \ldots, L$, and to be kept secure of $W_{k+2}, \ldots, W_L$, for $k = 1, \ldots, L - 2$. For this problem, the secrecy capacity region is characterized as follows:

PROPOSITION 18.5   (41, theorem 1): *Consider the L-receiver degraded broadcast channel with secrecy outside a bounded range. The secrecy capacity region consists of rate tuples $(R_1, R_2, \ldots, R_L)$ satisfying*

$$R_1 \leq I(U_1; Y_1), \tag{18.38a}$$

$$\sum_{j=2}^{k} R_j \leq \sum_{j=2}^{k} I(U_j; Y_j | U_{j-1}), \quad for \ 2 \leq k \leq L, \tag{18.38b}$$

$$\sum_{j=l}^{k} R_j \leq \left(\sum_{j=l-1}^{k} I(U_j; Y_j | U_{j-1}) \right) - I(U_k; Y_{l-2} | U_{l-2}), \ for \ 3 \leq l \leq k \leq L, \tag{18.38c}$$

for some $P_{U_1 \ldots U_K}$ satisfying the following Markov chain condition:

$$U_1 \to U_2 \to \cdots \to U_K \to Y_K \to \cdots \to Y_2 \to Y_1. \tag{18.39}$$

Using the broadcast approach, the problem of a fading channel can be viewed as a degraded broadcast channel with L receivers, with each receiver k experiencing the fading coefficient H_k, for $1 \leq k \leq L$. The transmitter splits the entire message into L submessages, $W_1, \ldots, W_L$. By the decoding and secrecy requirements, message k should be decoded by receiver k, and should be kept secure from receiver $k - 2$. Due to the degradedness condition, receiver k can decode messages $W_1, \ldots, W_k$, and cannot decode messages $W_{k+2}, \ldots, W_L$. For each message i, the transmitter assigns power P_i, such that $\sum_{i=1}^{L} P_i \leq P$.

The rate tuple $(R_1, \ldots, R_L)$ is achievable if there exists a coding scheme that encodes $W_1, \ldots, W_L$ at the rate $(R_1, \ldots, R_L)$ such that for $k = 1, \ldots, L$, the receiver can decode W_k with small probability of error if its channel state is H_k, and $W_{k+2}, \ldots, W_L$ are kept secure from it.

Via the broadcast approach, this fading channel can be reformulated into a degraded broadcast channel with layered decoding and layered secrecy and with secrecy outside a bounded range as in [41]. The following theorem characterizes the secrecy rate tuples that are achievable.

THEOREM 18.6   *For the fading channel where the receiver has L fading states $H_1, \ldots, H_L$, where $H_1 < \cdots < H_L$, the following secrecy rate tuples $(R_1, \ldots, R_L)$ are achievable:*

$$R_1 \leq \log \left(1 + \frac{|H_1|^2 P_1}{|H_1|^2 \sum_{i=2}^{L} P_i + 1} \right), \tag{18.40a}$$

$$\sum_{j=2}^{k} R_j \leq \sum_{j=2}^{k} \log \left(1 + \frac{|H_j|^2 P_j}{1 + |H_j|^2 \sum_{i=j+1}^{L} P_i} \right), \ for \ 2 \leq k \leq L, \tag{18.40b}$$

$$\sum_{j=\ell}^{k} R_j \leq \left[\sum_{j=\ell-1}^{k} \log \left(1 + \frac{|H_j|^2 P_j}{1 + |H_j|^2 \sum_{i=j+1}^{L} P_i} \right) \right] - \log \left(1 + \frac{|H_{\ell-2}|^2 \sum_{i=\ell-1}^{k} P_i}{1 + |H_{\ell-2}|^2 \sum_{i=k+1}^{L} P_i} \right),$$

$$\text{for } 3 \leq \ell \leq k \leq L. \qquad (18.40c)$$

In the above region, the bounds (18.40a) and (18.40b) are due to the decoding requirements, i.e., the receiver with state H_k should decode messages $W_1, \ldots, W_k$, for $1 \leq k \leq L$. The bounds (18.40c) are due to the secrecy requirements, i.e., messages $W_\ell, \ldots, W_k$ need to be kept secure from the receiver with state $H_{\ell-2}$ for $3 \leq l \leq k \leq L$. Furthermore, the bounds (18.40c) can be written as

$$\sum_{j=\ell}^{k} R_j \leq \sum_{j=\ell-1}^{k} \left[\log \left(1 + \frac{|H_j|^2 P_j}{1 + |H_j|^2 \sum_{i=j+1}^{L} P_i} \right) - \log \left(1 + \frac{|H_{\ell-2}|^2 P_j}{1 + |H_{\ell-2}|^2 \sum_{i=j+1}^{L} P_i} \right) \right]$$

$$\stackrel{\Delta}{=} \sum_{j=\ell-1}^{k} A_j, \qquad (18.41)$$

which has a clear intuitive interpretation. The term A_j corresponds to the rate of message W_j that can be kept secure from the receiver with state $H_{\ell-2}$ given the knowledge of $W_1, \ldots, W_{j-1}$. Those rates A_j for $\ell - 1 \leq j \leq k$ can all be counted toward $\sum_{j=\ell}^{k} R_j$ in accordance to the secrecy requirement of keeping $W_\ell, \ldots, W_k$ secure from the receiver with state $H_{\ell-2}$.

Due to the degradedness condition, the total rate satisfying the secrecy constraint that is achievable when the receiver is at channel state H_ℓ is $\sum_{j=1}^{\ell} R_j$. Then the average achievable secrecy rate is

$$\sum_{\ell=1}^{L} \left[P(H_\ell) \sum_{j=1}^{\ell} R_j \right].$$

This average secrecy rate can be further optimized with respect to the power allocation P_ℓ subject to a power constraint $\sum_{\ell=1}^{L} P_\ell \leq P$.

A very important property of the designed scheme is that it is adaptive to the actual unknown channel state. More specifically, if the channel state is better, then more messages can be decoded, and fewer messages are kept secure from the receiver. This adaptive property does not require the transmitter to know the instantaneous CSI.

The achievable scheme follows from the one in [41], which is based on superposition coding, random binning, and rate splitting and sharing. More specifically, for each message, one layer of codebook is designed, i.e., layer k corresponds to W_k, for $1 \leq k \leq L$. Within each layer, random binning is employed, i.e., the codewords in each layer are divided into a number of bins, where the bin number contains the information of the corresponding message.

Furthermore, rate splitting and sharing is employed to enlarge the achievable region. More specifically, within the kth layer, the message is split into two parts $W_{k,1}, W_{k,2}$. The message $W_{k,1}$ serves as embedded coding, which is a random source in addition to the random binning to protect $W_{k,2}$ and the higher-layer messages from the receiver with state H_{k-1}, i.e., the messages $W_{k,2}, W_{k+1,1}, W_{k+1,2}, \ldots, W_{L,1}, W_{L,2}$ are secure from the receiver with state H_{k-1}. Furthermore, the receiver with state better than H_{k-1} can also decode $W_{k,2}$ because of the degradedness condition. Thus, the message $W_{k,2}$ satisfies both the decoding and secrecy requirements for message W_{k+1}, and hence the rate of $W_{k,2}$ can be counted toward the rate of either W_k or W_{k+1}. By such a rate-sharing strategy, the achievable region is enlarged.

The motivation for adding the ingredient of the rate splitting and sharing is due to an important observation that although those layers within the bounded range Δ are not required to be kept secure, part of their rate is kept secure anyway, and hence can be shared with higher-layer messages, which helps to enlarge the achievable rate region. We note that such a rate splitting and sharing strategy is critical to achieve the secrecy capacity region for the degraded broadcast channel with secrecy outside a bounded range in [41].

Based on such an achievable scheme, we obtain an achievable region in terms of $R_{k,1}$ and $R_{k,2}$, i.e., the rate of $W_{k,1}$ and $W_{k,2}$, for $1 \le k \le L$. Define $R_k = R_{k-1,2} + R_{k,1}$ for $3 \le k \le K - 1$, $R_2 = R_{2,1}$ and $R_K = R_{K-1,2} + R_{K,1} + R_{K,2}$. A novel inductive Fourier–Motzkin elimination algorithm that eliminates the rate pairs $R_{k-1,2}$ and $R_{k,1}$ for $3 \le k \le K$ one at each step is designed in [41]. The region $\mathcal{R}_k$ after eliminating $R_{k-1,2}$ and $R_{k,1}$ possesses a common structure. By doing this recursively, we obtain the region as shown in (18.40). Further details can be found in [41].

We next present the result for the case in which Δ equals an arbitrary number of m levels of channel quality, where $m \ge 2$. The following secrecy rate region can be achieved using a scheme similar to that used in Theorem 18.6.

THEOREM 18.7 *For the fading channel with receiver having L fading states $H_1, \ldots, H_L$, and secrecy outside m levels of channel quality, where $H_1 < \cdots < H_L$, the following secrecy rate tuples are achievable:*

$$R_1 \le \log\left(1 + \frac{|H_1|^2 P_1}{|H_1|^2 \sum_{i=2}^{L} P_i + 1}\right),$$

$$\sum_{j=2}^{k} R_j \le \sum_{j=2}^{k} \log\left(1 + \frac{|H_j|^2 P_j}{1 + |H_j|^2 \sum_{i=j+1}^{L} P_i}\right), \quad for\ 2 \le k \le L, \tag{18.42}$$

$$\sum_{j=\ell}^{k} R_j \le \left[\sum_{j=\ell-m+1}^{k} \log\left(1 + \frac{|H_j|^2 P_j}{1 + |H_j|^2 \sum_{i=j+1}^{L} P_i}\right)\right] \tag{18.43}$$

$$- \log\left(1 + \frac{|H_{\ell-m}|^2 \sum_{i=\ell-m+1}^{k} P_i}{1 + |H_{\ell-m}|^2 \sum_{i=k+1}^{L} P_i}\right),$$

$$for\ m + 1 \le \ell \le k \le L. \tag{18.44}$$

To understand how well the broadcast approach performs, we compare its performance to an outer bound, which is derived by considering the case with no secrecy constraint, i.e., Δ equals infinitely many levels of channel quality. This is equivalent to the fading channel without a secrecy constraint as in [39, 40]. Then the average capacity in (18.6), which is also the variable-to-fixed channel capacity [42], can serve as an outer bound for the problem with a secrecy constraint.

18.4.3 Continuous Channel States

In this subsection, we consider the case with continuous channel states, i.e., H can take continuous values.

It is required that the messages intended to be decoded by the receiver if it has a better state than h_0 be secure from the receiver if it has state worse than h_0 by Δ, i.e., $|H| < |h_0| - \Delta$, for all $|h_0| \geq \Delta$. For each channel state $H = h$, let $s = |h|^2$.

Motivated by the broadcast approach for the Gaussian fading channel without a secrecy constraint in [39], the message is divided into infinitely many layers of messages indexed by s. For each layer s of message, the transmitter allocates power $\rho(s)ds$, where $\rho(s)$, satisfying

$$\int_0^\infty \rho(s)ds \leq P$$

is the power allocation function. Denote

$$\Sigma(s) = \int_s^\infty \rho(x)dx,$$

which is the power allocated to the layers of messages intended for the receiver with channel state better than s. It is clear that $\rho(s) = -\Sigma'(s)$.

It is assumed that the message indexed by s is required to be decoded by the receiver with state $|h| = \sqrt{s}$ and to be kept secure from the receiver with state worse than $\sqrt{s} - \Delta$. We use $R(s)ds$ to denote the incremental differential rate for the layer indexed by s. Using the broadcast approach, the fading channel with continuous channel states can be viewed as a degraded broadcast channel with infinitely many receivers with a continuum of channel states.

THEOREM 18.8 *Consider the fading channel with continuous channel states, i.e., H can take continuous values. Any incremental differential rate $R(s)ds$ satisfying the following constraints is achievable:*

$$\int_0^t R(s)ds \leq \int_0^t \frac{x\rho(x)dx}{1 + x\Sigma(x)}, \qquad \text{for } t \geq 0, \tag{18.45a}$$

$$\int_{t_1}^{t_2} R(s)ds \leq \int_{(\sqrt{t_1}-\Delta)^2}^{t_2} \frac{x\rho(x)dx}{1 + x\Sigma(x)} - \log\left(1 + \frac{(\sqrt{t_1} - \Delta)^2 \left(\Sigma((\sqrt{t_1} - \Delta)^2) - \Sigma(t_2)\right)}{1 + (\sqrt{t_1} - \Delta)^2 \Sigma(t_2)}\right),$$

$$\text{for } \Delta^2 \leq t_1 \leq t_2 \leq \infty. \tag{18.45b}$$

To maximize the average secrecy rate, it suffices to solve the following optimization problem:

$$\max_{\rho(s)} \int_0^\infty \rho(s)ds \left(\int_0^s R(t)dt\right), \tag{18.46}$$

subject to the constraints in (18.45) and $\int_0^\infty \rho(x) \leq P$, where $p(s)$ is the probability distribution function of s.

In order to understand how well the broadcast approach performs, we let $\Delta = \infty$. Then the fading channel with secrecy outside a bounded range is equivalent to the fading channel without any secrecy constraint, as in [40, 42]. Then the average capacity in (18.6), which is also the variable-to-fixed channel capacity [42], can serve as an outer bound for the problem with secrecy constraint.

18.5   Discussion

In this chapter, we have provided an overview of recent information theoretic studies of using physical layer security to solve secure communication problems with an emphasis on the broadcast approach.

As we commented in Section 18.1.1, secrecy is provided as an additional benefit without sacrificing the reliable communication rate for the basic wiretap channel. In the scenarios considered in this chapter and even more complicated scenarios, this might not hold anymore. Then an interesting question worth exploring is: If the receiver does not want to compromise on the achievable communication rate (i.e., without sacrificing the reliable communication rate), what secret rate can we get as a benefit of physical layer security?

Under the class of communication models with channel state uncertainty, there are many open problems that require further exploration. For example, it is of interest to consider a compound scenario. For this model, instead of a single legitimate receiver and/or a single eavesdropper, there are multiple legitimate receivers and eavesdroppers, each of them having a random state unknown to the transmitter. The decoding and secrecy requirements for the messages are more complicated with more receivers. This model is much more flexible for modeling practical networks with clusters of receivers. It is also interesting to study the scenario in which the eavesdroppers collude to obtain a better estimate of the message.

The performance measure of interest in this chapter is the expected secrecy rate. Although the variable-to-fixed channel capacity [42] can serve as an outer bound, this bound in general is not tight since it does not take the secrecy constraint into consideration. It is thus of interest to provide tighter outer bounds for a comprehensive understanding of how well the broadcast approach performs. Moreover, for the problem of the fading wiretap channel, not all the information is secure from the eavesdropper. Thus it is also of interest to study the tradeoff between the average leakage rate and the average secrecy rate. Furthermore, the secrecy rate discussed in this chapter can only guarantee with "high" probability that a data block is secure, as the information theoretic security is an average measure. It is also desired to obtain an accurate characterization of the probability that a data block is secure.

It would also be of interest to study the broadcast approach with a relaxed delay constraint, in which coding over a few blocks is allowed. Some of the ideas in [49] might be worth exploring for the case with secrecy constraints. Furthermore, it is also of interest to explore the case with a stringent delay constraint. In this chapter, we have focused on the secrecy rate averaged over all random channel realizations. This rate is achievable in an ergodic manner, i.e., in average over a large number of blocks. However, in a delay-constrained scenario [22], a certain rate must be guaranteed at each single block, and hence averaging over multiple blocks is not possible.

Another interesting topic is the heterogeneous scenario with multiple information flows subject to different delay constraints. This scenario demonstrates the combination of two extreme cases: a single block delay constraint and no delay constraint (i.e., the ergodic case). In this scenario, careful design for layering these information flows is needed. It is reasonable to assign the messages with more stringent constraints to lower-level layers so that these data can be decoded at the legitimate receiver in most channel states.

This chapter has focused on characterizing the information theoretic performance limits based on random coding arguments. The design of secrecy coding schemes for realistic environments is still challenging (see [50] for an overview). It is also of further interest to design practical coding schemes such as low-density parity-check (LDPC) codes [51, 52] and polar codes [53–58] to achieve secure communication.

18.6   Acknowledgments

The work of Y. Liang was supported in part by NSF grant CCF-1801846. The work of L. Lai was supported in part by NSF grants ECCS-1660140, CCF-1665073g, and CNS-1824553. The work of H. V. Poor was supported in part by the National Science Foundation under grants CCF-0939370, CCF-1513915, and CCF-1908308. The work of S. Shamai was supported in part by the European Union's Horizon 2020 Research and Innovation Programme under grant 694630 and in part by the United States – Israel Binational Science Foundation under grant BSF-2018710.

References

[1] A. D. Wyner, "The wire-tap channel," *Bell Syst. Tech. J.*, vol. 54, no. 8, pp. 1355–1387, Oct. 1975.

[2] C. E. Shannon, "Communication theory of secrecy systems," *Bell Syst. Tech. J.*, vol. 28, pp. 656–715, 1949.

[3] C. E. Shannon, "A mathematical theory of communication," *Bell Syst. Tech. J.*, vol. 27, pp. 379–423, 623–656, 1948.

[4] I. Csiszár and J. Körner, "Broadcast channels with confidential messages," *IEEE Trans. Inf. Theory*, vol. 24, no. 3, pp. 339–348, May 1978.

[5] Y. Liang, G. Kramer, H. V. Poor, and S. Shamai (Shitz), "Compound wire-tap channels," *EURASIP J. Wireless Commun. Netw., Special Issue on Wireless Physical Layer Security*, 2009.

[6] M. Bloch and J. Barros, *Physical-Layer Security: From Information Theory to Security Engineering*. Cambridge: Cambridge University Press, 2011.

[7]   Y. Liang, H. V. Poor, and S. Shamai (Shitz), "Information theoretic security," *Found. Trends Commun. Inform. Theory*, vol. 5, no. 4-5, pp. 355–580, 2008.

[8]   M. Baldi and S. Tomasin, *Physical and Data-Link Security Techniques for Future Communication Systems*. New York: Springer, 2016.

[9]   A. Hyadi, Z. Rezki, and M.-S. Alouini, "An overview of physical layer security in wireless communication systems with CSIT uncertainty," *IEEE Access*, vol. 4, pp. 6121–6132, 2016.

[10] H. V. Poor and R. F. Schaefer, "Wireless physical layer security," *Proc. Natl. Acad. Sci. U.S.A.*, vol. 114, no. 1, pp. 19–26, Jan. 2017.

[11] E. Ekrem and S. Ulukus, "Degraded compound multi-receiver wiretap channels," *IEEE Trans. Inf. Theory*, vol. 58, no. 9, pp. 5681–5698, 2012.

[12] A. Khisti, "On the MISO compound wiretap channel," in *Proc. Inform. Theory Applications Workshop (ITA)*, La Jolla, CA, Jan 2010, pp. 1–7.

[13] T. Liu, V. Prabhakaran, and S. Vishwanath, "The secrecy capacity of a class of parallel Gaussian compound wiretap channels," in *Proc. IEEE Int. Symp. Inform. Theory (ISIT)*, Toronto, Jul 2008, pp. 116–120.

[14] M. Kobayashi, Y. Liang, S. Shamai, and M. Debbah, "On the compound MIMO broadcast channels with confidential messages," in *Proc. IEEE Int. Symp. Inform. Theory (ISIT)*, Seoul, Jun. 2009, pp. 1283–1287.

[15] Y. Liang, H. V. Poor, and S. Shamai (Shitz), "Secure communication over fading channels," *IEEE Trans. Inform. Theory, Special Issue on Information Theoretic Security*, vol. 54, no. 6, pp. 2470–2492, Jun. 2008.

[16] Y. Liang, H. V. Poor, and S. Shamai (Shitz), "Secure communication under channel uncertainty," in *Securing Wireless Communications at the Physical Layer*. New York: Springer, 2009.

[17] O. Gungor, J. Tan, C. E. Koksal, H. El-Gamal, and N. B. Shroff, "Secrecy outage capacity of fading channels," *IEEE Trans. Inf. Theory*, vol. 59, no. 9, pp. 5379–5397, 2013.

[18] X. Zhou, M. R. McKay, B. Maham, and A. Hjorungnes, "Rethinking the secrecy outage formulation: A secure transmission design perspective," *IEEE Commun. Lett.*, vol. 15, no. 3, pp. 302–304, 2011.

[19] X. Chen, L. Lei, H. Zhang, and C. Yuen, "On the secrecy outage capacity of physical layer security in large-scale MIMO relaying systems with imperfect CSI," in *Proc. IEEE Int. Conf. Commun. (ICC)*, Sydney, 2014, pp. 2052–2057.

[20] P. Parada and R. Blahut, "Secrecy capacity of SIMO and slow fading channels," in *Proc. IEEE Int. Symp. Inform. Theory (ISIT)*, Adelaide, 2005, pp. 2152–2155.

[21] M. Bloch, J. Barros, M. R. D. Rodrigues, and S. W. McLaughlin, "Wireless information-theoretic security," *IEEE Trans. Inf. Theory*, vol. 54, no. 6, pp. 2515–2534, 2008.

[22] K. Khalil, M. Youssef, O. O. Koyluoglu, and H. El Gamal, "On the delay limited secrecy capacity of fading channels," in *Proc. IEEE Int. Symp. Inform. Theory (ISIT)*, Seoul, Jun.–Jul. 2009.

[23] Y. Liang, H. V. Poor, and L. Ying, "Wireless broadcast networks: Reliability, security and stability," in *Proc. Inform. Theory Applications Workshop (ITA)*, La Jolla, CA, Jan. 2008.

[24] R. Liu and W. Trappe, *Securing Wireless Communications at the Physical Layer*. New York. Springer, 2010.

[25] X. Zhou, L. Song, and Y. Zhang, *Physical Layer Security in Wireless Communications*. Boca Raton, FL: CRC Press, 2013.

[26] E. A. Jorswieck, A. Wolf, and S. Gerbracht, "Secrecy on the physical layer in wireless networks," in *Trends in Telecommunications Technologies*. London: InTech, 2010.

[27] A. Mukherjee, S. A. A. Fakoorian, J. Huang, and A. L. Swindlehurst, "Principles of physical layer security in multiuser wireless networks: A survey," *IEEE Commun. Surv. Tutorials*, vol. 16, no. 3, pp. 1550–1573, 2014.

[28] S. Hidano, M. Pečovský, and S. Kiyomoto, "New security challenges in the 5G network," in *Proc. Int. Symp. Intell. Computation Appl.* 2015, pp. 619–630.

[29] N. Yang, L. Wang, G. Geraci, et al., "Safeguarding 5G wireless communication networks using physical layer security," *IEEE Commun. Mag.*, vol. 53, no. 4, pp. 20–27, 2015.

[30] W. Trappe, "The challenges facing physical layer security," *IEEE Commun. Mag.*, vol. 53, no. 6, pp. 16–20, 2015.

[31] Y. Zou, J. Zhu, X. Wang, and L. Hanzo, "A survey on wireless security: Technical challenges, recent advances, and future trends," *Proc. IEEE*, vol. 104, no. 9, pp. 1727–1765, 2016.

[32] B. He, X. Zhou, and T. D. Abhayapala, "Wireless physical layer security with imperfect channel state information: A survey," *ZTE Commun.*, vol. 11, no. 3, pp. 11–19, Sep. 2013.

[33] P.-H. Lin and E. Jorswieck, "On the fast fading Gaussian wiretap channel with statistical channel state information at the transmitter," *IEEE Trans. Inf. Forensics Security*, vol. 11, no. 1, pp. 46–58, 2016.

[34] P. Mukherjee and S. Ulukus, "Fading wiretap channel with no CSI anywhere," in *Proc. IEEE Int. Symp. Inform. Theory (ISIT)*, Istanbul, 2013, pp. 1347–1351.

[35] Y. Liang, L. Lai, H. V. Poor, and S. Shamai (Shitz), "A broadcast approach for fading wiretap channels," *IEEE Trans. Inf. Theory*, vol. 60, no. 2, pp. 842–858, Feb. 2014.

[36] G. Brante, H. Alves, R. D. Souza, and M. Latva-aho, "Secrecy analysis of transmit antenna selection cooperative schemes with no channel state information at the transmitter," *IEEE Trans. Commun.*, vol. 63, no. 4, pp. 1330–1342, 2015.

[37] S.-C. Lin and P.-H. Lin, "On secrecy capacity of fast fading multiple-input wiretap channels with statistical CSIT," *IEEE Trans. Inf. Forensics Security*, vol. 8, no. 2, pp. 414–419, 2013.

[38] T.-Y. Liu, P. Mukherjee, S. Ulukus, S.-C. Lin, and Y.-W. P. Hong, "Secure degrees of freedom of MIMO Rayleigh block fading wiretap channels with no CSI anywhere," *IEEE Trans. Wireless Commun.*, vol. 14, no. 5, pp. 2655–2669, 2015.

[39] S. Shamai (Shitz), "A broadcast strategy for the Gaussian slowly fading channel," in *Proc. IEEE Int. Symp. Inform. Theory (ISIT)*, Ulm, June 1997, p. 150.

[40] S. Shamai (Shitz) and A. Steiner, "A broadcast approach for a single-user slowly fading MIMO channel," *IEEE Trans. Inf. Theory*, vol. 49, no. 10, pp. 2617–2635, Oct. 2003.

[41] S. Zou, Y. Liang, L. Lai, H. V. Poor, and S. Shamai, "Degraded broadcast channel with secrecy outside a bounded range," *arXiv:1609.06353*, 2016.

[42] S. Verdú and S. Shamai (Shitz), "Variable-rate channel capacity," *IEEE Trans. Inf. Theory*, vol. 56, no. 6, pp. 2651–2667, Jun. 2010.

[43] P. K. Gopala, L. Lai, and H. El Gamal, "On the secrecy capacity of fading channels," *IEEE Trans. Inf. Theory*, vol. 54, no. 10, pp. 4687–4698, 2008.

[44] E. Ekrem and S. Ulukus, "Secrecy capacity of a class of broadcast channels with an eavesdropper," *EURASIP J. Wireless Commun. Netw.*, vol. 2009, pp. 1:1–1:29, Mar. 2009.

[45] J. Gregory and K. Pericak-Spector, "New methods of solving general constrained calculus of variations problems involving PDEs," *Utilitas Mathematica*, vol. 58, pp. 215–224, Nov. 2000.

[46] H. D. Ly, T. Liu, and Y. Blankenship, "Security embedding codes," *IEEE Trans. Inf. Forensics Security*, vol. 7, no. 1, pp. 148–159, Feb. 2012.

[47] S. Zou, Y. Liang, L. Lai, and S. Shamai (Shitz), "An information theoretical approach to secrecy sharing," *IEEE Trans. Inf. Theory*, vol. 61, no. 6, pp. 3121–3136, 2015.

[48] S. Zou, Y. Liang, L. Lai, H. V. Poor, and S. Shamai, "Broadcast networks with layered decoding and layered secrecy: Theory and applications," *Proc. IEEE*, vol. 103, no. 10, pp. 1841–1856, 2015.

[49] P. A. Whiting and E. M. Yeh, "Broadcasting over uncertain channels with decoding delay constraints," *IEEE Trans. Inf. Theory*, vol. 52, no. 3, pp. 904–921, 2006.

[50] W. K. Harrison, J. Almeida, M. R. Bloch, S. W. McLaughlin, and J. Barros, "Coding for secrecy: An overview of error-control coding techniques for physical-layer security," *IEEE Signal Proc. Mag.*, vol. 30, no. 5, pp. 41–50, 2013.

[51] V. Rathi, M. Andersson, R. Thobaben, J. Kliewer, and M. Skoglund, "Performance analysis and design of two edge-type LDPC codes for the BEC wiretap channel," *IEEE Trans. Inf. Theory*, vol. 59, no. 2, pp. 1048–1064, 2013.

[52] A. Subramanian, A. Thangaraj, and M. Bloch, "Strong secrecy on the binary erasure wiretap channel using large girth LDPC codes," *IEEE Trans. Inf. Forensics Security*, vol. 6, no. 3, pp. 585–594, Sep. 2011.

[53] O. O. Koyluoglu and H. El Gamal, "Polar coding for secure transmission and key agreement," *IEEE Trans. Inf. Forensics Security*, vol. 7, no. 5, pp. 1472–1483, Oct. 2012.

[54] E. Hof and S. Shamai, "Secrecy-achieving polar-coding," in *Proc. IEEE Inform. Theory Workshop (ITW)*, La Jolla, CA Aug. 2010.

[55] B. Duo, P. Wang, Y. Li, and B. Vucetic, "Secure transmission for relay-eavesdropper channels using polar coding," in *Proc. IEEE Int. Conf. Commun. (ICC)*, Sydney, Jun. 2014, pp. 2197–2202.

[56] S. A. A. Fakoorian and A. L. Swindlehurst, "On the optimality of polar codes for the deterministic wiretap channel," in *Proc. Asilomar Conf. Signals, Syst. Comput.*, Pacific Grove, CA Nov. 2013, pp. 2089–2093.

[57] E. Sasoglu and A. Vardy, "A new polar coding scheme for strong security on wiretap channels," in *Proc. IEEE Int. Symp. Inform. Theory (ISIT)*, Istanbul, 2013.

[58] J. del Olmo and J. R. Fonollosa, "Strong secrecy on a class of degraded broadcast channels using polar codes," *arXiv:1607.07815*, 2016.

19 Cognitive Cooperation and State Management: An Information Theoretic Perspective

Anelia Somekh-Baruch, Yingbin Liang, Haim Permuter, and Shlomo Shamai (Shitz)

19.1 Introduction

This chapter focuses on central aspects of improving spectral efficiency in wireless networks, such as 5G and beyond, via cognition, giving rise to cooperation in terms of messages, channel states, and interference mitigation. Cognition in this setting reflects the capability to acquire other user's information about the network transmissions (messages) and/or interference that impacts the network performance. In particular, such interference can be associated with other types of communications sharing the same resources, such as device-to-device wireless links as well as other networks. Furthermore, the models we address account also for state-dependent channels, where states again may reflect the aforementioned cognitive interference, or in fact, an actual channel state (such as fading, and the like). The focus of this chapter is the information theoretic perspective of those cognitive elements, which actually, based on advanced technology, do not only provide bounds in terms of achievable and outer capacity regions, but reflect also what can actually be achieved in practice. The very simple models considered and analyzed in this chapter provide an example of the impact of cognition on communication rates. Another issue is the impact of cognition not only rate-wise but delay-wise, which is an important factor in cellular technology (from the information theoretic perspective); see, for example, [1]. Other practical issues such as asynchronism can also affect the communication rates significantly [2, 3]. Although receiving much attention recently ([4–6]), many of the fundamental problems of cognitive multiterminal networks remain unsolved.

Cooperation between users in communication networks such as in multiple accesses channels (MAC), broadcast channels (BC), interference channels (IC) can significantly increase the achievable throughput. This fact has been observed in many seemingly different scenarios: conferencing encoders and decoders, channels with cribbing encoders, feedback systems, state-dependent channels with states available to several users, and more. Many coding schemes have been suggested in the literature for the various scenarios, and all culminate in one common theme: Cooperation resources are exploited to generate statistical dependence between the inputs of the various users. Theoretically, this means that the set of distributions over which the achievable region is evaluated is larger, resulting in higher achievable rates. In the first section of this chapter, a brief overview of the main cooperative setting, their fundamental limits, and some ideas on how to achieve these are given, putting the focus on models that are relevant to current and future wireless systems.

State-dependent channels with different state cognition at different nodes in network communications is a classical setting, and different aspects of this have been addressed almost since the introduction of information theory. In the second section of this chapter, we focus on cognition of states and messages in multiuser channels. The simple models we address which are directly relevant to wireless cellular systems, account for a two-user state-dependent cognitive MAC. In this model, there are two transmitters and a single receiver. The primary transmitter wishes to transmit one message and a cognitive secondary user that observes both the primary user's message (which can be viewed as a common message) and the state, transmits an additional message. The receiver needs to decode both messages reliably. The second model being studied consists of a two-user state-dependent cognitive interference channel (IC), It generalizes the above cognitive MAC to the case in which the messages are intended for two different receivers, keeping the roles of the two transmitters as in the first model. The third model being analyzed also involves a cognitive IC, but in this case the state is degenerate and the cognition of the secondary user is limited only to the message. In this setup also the common message should be decoded by the two receivers of the IC. The fourth model introduces a three-user (state-independent) cognitive communication channel that generalizes instances of the previous models.

One major technology to improve the throughput and spectral efficiency of wireless networks is interference management, which drives evolution of wireless networks from one generation to another. By now it has been recognized that interference management is a central factor in wireless cellular technology, and in fact this aspect dictates future views and technologies, as such 5G cellular systems. Techniques to deal with interference, in all up-to-date cellular networks, follow the basic principle of orthogonalizing transmissions in time, frequency, code, and space, which yields time division multiple access (TDMA), frequency division multiple access (FDMA), code division multiple access (CDMA), and more advanced orthogonal frequency division multiplex (OFDM) and zero-forcing, and/or orthogonalized multi-input multiple-output (MIMO) technologies (see, e.g., [7–9] and reference therein). However, orthogonal schemes typically do not reach the best spectral efficiency and are not information-theoretically optimal in general. On the other hand, various advanced nonorthogonal interference cancellation schemes are proposed, motivated by information theoretic designs. Some of the famous applications refer to interference cancellation CDMA, third-generation approaches, which mimic the Han–Kobayashi [10] classical strategies used for the IC, that is the receiver decodes interference partially, and then subtracts it from the output to reduce its deleterious impact. Such a scheme was recently exploited in a downlink nonorthogonal multiple access (NOMA) scheme for interference management [11–13]. However, such successive interference cancellation requires users to share codebooks and hence can be very complex to implement in practice, especially when transmissions are not within the same network domain. It is then desirable to develop new low-cost interference control approaches for wireless networks, and the following properties of

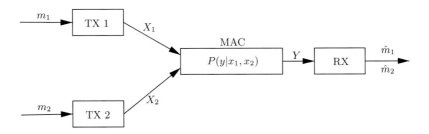

Figure 19.1 Multiple access channel as an information theoretic model for uplink communication.

interference signals offer such a promise. In fact, cellular interference signals may reflect coded information sent to certain intended nodes, and hence such signals as codeword sequences are typically noncausally known by various nodes in the network. In cellular systems this accounts for base station inflicted interference, where that base station can inform other base stations about its interference via the backhaul network or inform access points in the cell via wired or specially dedicated links. In general, nodes in a wireless network that possess noncausal interference information should be able to exploit it to assist cancellation of such interference [14–16], and may also be referred to as helpers. Such models are of relevance not only for cellular systems, but reflect also the situation observed in future wireless networks such as those focused on the Internet of Things (IoT) [17–19]. In the third part of this chapter, we focus on such approaches with helpers being cognizant of interference only, and thus the helper's signal can assist the interference cancellation at the receiver rather than cooperatively encoding transmitted information in the typical user/relay cooperation [20–22] addressed in the former two parts of the chapter. The models account also for multiple helpers [15, 23], where the cooperation among helpers is based on their common knowledge about the state/interference, and the helpers are kept ignorant of the actual transmitted messages by the communicating nodes. The third part of this chapter provides an overview of the type of models that capture the cooperation via interference cognition, and examines recent information theoretic characterization of the performance limits for such types of models.

19.2   Cooperative Uplink

Uplink communication is modeled using the (MAC) as shown in Figure 19.1. The two transmitters (aka encoders), TX 1 and TX 2, have two independent messages that need to be sent via the memoryless channel.

There are two main cooperation modes between the encoders (transmitters) in the MAC, and both were introduced by Willems [24–26]. In the first, the two encoders use a rate-limited cooperation link to cooperate and share as much of their private messages as possible. In the second approach, named cribbing, each encoder "listens" to and obtains the output of the other. Capacity regions for the two approaches were established separately by Willems. Furthermore, the cribbing setting was generalized in [27] to partial

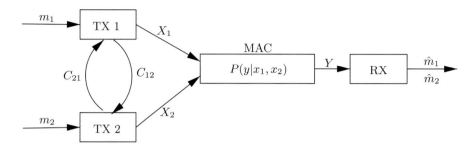

Figure 19.2 Multiple access channel with conferencing link cooperation.

cribbing, which means that each of the two encoders obtains a deterministic function of the other encoder's output. The partial cribbing is especially important in the continuous alphabet, such as in the Gaussian MAC, because perfect cribbing in a continuous alphabet means full cooperation between the encoders regardless of the cribbing delay. We now introduce in detail the two cooperation modes: MAC with conferencing encoders and MAC with cribbing. Then we discuss how to combine them and finally extend the results in the presence of state information.

19.2.1   Cooperation via Conferencing

In this setting the transmitters/encoders can cooperate with each other via a limited-rate link. The setting is depicted in Figure 19.2. Transmitter 1 can transmit to Transmitter 2 a message at rate c_{12} and Transmitter 2 can transmit to Transmitter 1 at a rate c_{21}. Originally, Willems considered the case where there can be multiple transmissions between the encoders (namely, the cooperative bits are sent in k rounds); in terms of capacity this does not make any difference and for simplicity of the code we assume that the cooperation is done at once. The formal definition of the coding scheme is the following:

DEFINITION 19.1 (*Cooperating encoder code*)   *A* $(2^{nR_1}, 2^{nR_2}, 2^{nC_{12}}, 2^{nC_{21}}, n)$ *code with cooperating encoders consists of four encoding functions*

$$f_{12} : \{1, \ldots, 2^{nR_1}\} \mapsto \{1, \ldots, 2^{nC_{12}}\},$$
$$f_{21} : \{1, \ldots, 2^{nR_2}\} \times \{1, \ldots, 2^{nC_{12}}\} \mapsto \{1, \ldots, 2^{nC_{21}}\},$$
$$f_1 : \{1, \ldots, 2^{nR_1}\} \times \{1, \ldots, 2^{nC_{21}}\} \mapsto \mathcal{X}_1^n,$$
$$f_2 : \{1, \ldots, 2^{nR_2}\} \times \{1, \ldots, 2^{nC_{12}}\} \mapsto \mathcal{X}_2^n, \tag{19.1}$$

and a decoding function,

$$g : \mathcal{Y}^n \mapsto \{1, \ldots, 2^{nR_1}\} \times \{1, \ldots, 2^{nR_2}\}. \tag{19.2}$$

The average probability of error for $(2^{nR_1}, 2^{nR_2}, 2^{nC_{12}}, n)$ *code is defined as*

$$P_e^{(n)} = \frac{1}{2^{n(R_1+R_2)}} \sum_{m_1, m_2} \Pr\{g(Y^n, S^n) \neq (m_1, m_2) | (m_1, m_2) \text{ sent}\}. \tag{19.3}$$

A rate (R_1, R_2) is said to be achievable for cooperating MAC with cooperation link C_{12}, C_{21} if there exists a sequence of $(2^{nR_1}, 2^{nR_2}, 2^{nC_{12}}, 2^{nC_{21}}, n)$ codes with $P_e^{(n)} \to 0$. The capacity region of the MAC is the closure of all achievable rates. The following theorem describes the capacity region of cooperating MAC.

THEOREM 19.2 (Capacity region of MAC with limited-rate link cooperation [24, 25]) *The capacity region of the MAC with two-way cooperating encoders as shown in Figure 19.2 is the closure of the set of rates that satisfy*

$$R_1 \leq I(X_1; Y|X_2, U) + C_{12} \tag{19.4}$$

$$R_2 \leq I(X_2; Y|X_1, U) + C_{21} \tag{19.5}$$

$$R_1 + R_2 \leq \min \left\{ \begin{array}{c} I(X_1, X_2; Y|U) + C_{12} + C_{21} \\ I(X_1, X_2; Y) \end{array} \right\}, \tag{19.6}$$

for some joint distribution of the form

$$P(u)P(x_1|u)P(x_2|u)P(y|x_1, x_2), \tag{19.7}$$

where U is an auxiliary random variable with bounded cardinality.

Willems showed that to achieve the capacity region the encoders should use the cooperation link in order to share parts of their private messages and then use a coding scheme for the ordinary MAC, which was found earlier by Slepian and Wolf [28]. Willems' model allows interactive communication between the encoders; however, he showed that the optimality is achieved even in a single round of communication between the encoders [24]. The results has been extended recently to the case when the cooperation may be absent [29].

19.2.2   Cooperation via Cribbing

The second kind of cooperation was also introducd by Willems [24]. *Cribbing encoder* refers to the case where the encoder knows perfectly the other output encoder, possibly with delay or look-ahead. The work by Willems on MACs with cribbing encoders has been extended to the IC [30], and to state-dependent MACs [31]. However, for the Gaussian case, where the encoder output is of a continuous alphabet, the cribbing idea is not an interesting case [32] since it implies full cooperation between the encoders regardless of the delay of the cribbing. This is due to the fact that in a single epoch a noiseless continuous signal may transmit an infinite amount of information. Motivated by this fact, [27] introduced the idea of "partial cribbing," where one encoder only knows a quantized version, or, more generally, a deterministic function of the coded output of other encoder. Partial cribbing is an extension of perfect cribbing and a more realistic case. The most realistic case would be noisy cribbing, which is an open problem.

Here we present two kinds of partial cribbing: causal and strictly causal. Causal partial cribbing means that at time i the encoder observes (and uses) the partial cribbing signal without delay, i.e., Z_i. Strictly causal partial cribbing means that at time i the encoder observes the partial cribbing with a delay, i.e., Z_{i-1}. We consider the capacity region for two different cases according to the causality or the strict causality of the cribbing

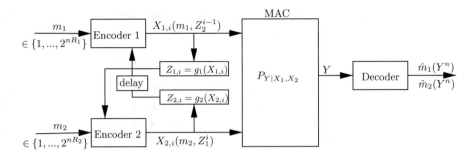

Figure 19.3 Partial (deterministic-function) cribbing. Each encoder observes a deterministic function of the other encoder with or without delay. Encoder 1 observes the cribbing in a strictly causal way, i.e., with delay, and Encoder 2 observes the cribbing causally, i.e., without delay. The setting corresponds to Case B in this chapter.

Case A: The cribbing for both encoders is *strictly causal*.

Case B: The cribbing for one encoder is *causal* and for the other encoder is *strictly causal*.

The setting that is depicted in Figure 19.3 is the case where one encoder has causal partial cribbing and the other strictly causal partial cribbing, namely Case B. Let's define more formally the settings.

DEFINITION 19.3 (*Partial cribbing code*)   *A* $(2^{nR_1}, 2^{nR_2}, n)$ *code with partial cribbing, as shown in Figure 19.3, consists at time i of an encoding function at Encoder 1*

$$\text{Case A, B,} f_{1,i} : \{1, \ldots, 2^{nR_1}\} \times \mathcal{Z}_2^{i-1} \mapsto X_{1,i}, \tag{19.8}$$

and an encoding function at Encoder 2 that changes according to the following case settings

$$\text{Case A } f_{2,i} : \{1, \ldots, 2^{nR_2}\} \times \mathcal{Z}_1^{i-1} \mapsto X_{1,i},$$
$$\text{Case B } f_{2,i} : \{1, \ldots, 2^{nR_2}\} \times \mathcal{Z}_1^{i} \mapsto X_{1,i}, \tag{19.9}$$

and a decoding function,

$$g : \mathcal{Y}^n \mapsto \{1, \ldots, 2^{nR_1}\} \times \{1, \ldots, 2^{nR_2}\}. \tag{19.10}$$

Let us define the following regions $\mathcal{R}_A, \mathcal{R}_B$, which are contained in $\mathbb{R}_+^2$, namely, contained in the set of nonnegative two-dimensional real numbers:

$$\mathcal{R}_A = \left\{ \begin{array}{l} R_1 \leq H(Z_1|U) + I(X_1; Y|X_2, Z_1, U), \\ R_2 \leq H(Z_2|U) + I(X_2; Y|X_1, Z_2, U), \\ R_1 + R_2 \leq I(X_1, X_2; Y|U, Z_1, Z_2) + H(Z_1, Z_2|U), \\ R_1 + R_2 \leq I(X_1, X_2; Y), \text{ for} \\ P(u)P(x_1, z_1|u)P(x_2, z_2|u)P(y|x_1, x_2). \end{array} \right. \tag{19.11}$$

The region $\mathcal{R}_B$ is defined with the same set of inequalities as in (19.11), but the joint distribution is of the form

$$P(u)P(x_1, z_1|u)P(x_2, z_2|z_1, u)P(y|x_1, x_2). \tag{19.12}$$

THEOREM 19.4 (Capacity region of MAC with partial cribbing [27]) *The capacity regions of the MAC with strictly causal (Case A) and mixed causal and strictly causal (Case B) partial cribbing as described in Definition 19.3 are $\mathcal{R}_A$, $\mathcal{R}_B$, respectively. It is enough to restrict the alphabet of U, to satisfy $|\mathcal{U}| \leq \min(|\mathcal{Y}| + 3, |\mathcal{X}_1||\mathcal{X}_2| + 2))$.*

The main idea that is used in solving MACs with partial cribbing encoders is that the message should be split into two parts: One part should be used for generating cooperation between the encoders. When the cribbing is strictly causal an explicit cooperation is generated. Namely, the cribbing encoder explicitly decodes the message of the previous block from the other encoder and uses it in the generation of the codeword. In the case of causal cribbing, the cribbing encoder decodes the previous block from the other encoder, and on top of it uses the knowledge of the current cribbed signal to generate a correlation between the outputs of the encoders.

Since the explicit cooperation between the encoders involves messages from the current block and from the previous block, a special technique called backward decoding was introduced. The main idea of backward decoding is that it starts the decoding from the last block, and decodes simultaneously the messages of the current block and the message of the previous block, which is used for generating the cooperation between the encoders.

EXAMPLE 19.5 (Gaussian MAC with quantized cribbing) In this example we consider the additive Gaussian noise MAC, i.e., $Y = X_1 + X_2 + W$, where W is a memoryless Gaussian noise with variance N, i.e., $W \sim \text{Norm}(0, N)$. We assume power constraints P_1 and P_2 on the inputs from Encoder 1 and Encoder 2, respectively. If the encoders do not cooperate then the capacity is given by

$$R_1 \leq \frac{1}{2} \log \left(1 + \frac{P_1}{N} \right)$$

$$R_2 \leq \frac{1}{2} \log \left(1 + \frac{P_2}{N} \right)$$

$$R_1 + R_2 \leq \frac{1}{2} \log \left(1 + \frac{P_1 + P_2}{N} \right). \tag{19.13}$$

If there is perfect cribbing from Encoder 1 to Encoder 2, either with delay or without, the capacity is the same as if Encoder 2 knows the message of Encoder 1, since Encoder 1 can send the message in one epoch time. Hence, the capacity is the union over $0 \leq \rho \leq 1$ of the regions

$$R_2 \leq \frac{1}{2} \log \left(1 + \frac{P_2}{N}(1 - \rho^2) \right)$$

$$R_1 + R_2 \leq \frac{1}{2} \log \left(1 + \frac{P_1 + 2\rho\sqrt{P_1 P_2} + P_2}{N} \right). \tag{19.14}$$

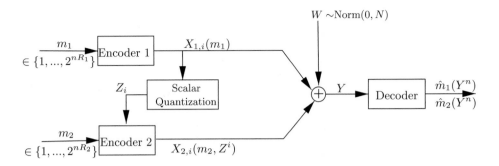

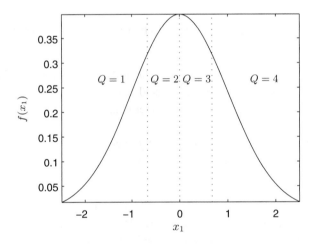

Figure 19.4 Gaussian MAC with quantized cribbing. The cribbing that Encoder 2 observes is the quantized signal from Encoder 1. There exist power constraints $\sum_{i=1}^{n} E[X_{1,i}^2] \leq P_1$ and $\sum_{i=1}^{n} E[X_{1,i}^2] \leq P_2$.

Figure 19.5 The two-bit quantizer's boundaries are designed such that if the input signal has a normal distribution with variance $P_1 = 1$ the output values from the quantizer have equal probability. The input to the two-bit quantizer is X_1 and the output is $Q \in \{1, 2, 3, 4\}$.

Now, let us consider the case where Encoder 2 observes a quantized version of the signal from Encoder 1 without delay. The setting is depicted in Figure 19.4. We assume that the quantizer is a scalar quantizer designed such that under a Gaussian input with variance $P_1 = 1$ the discrete values have the same probability (see Figure 19.5 for an example of a two-bit quantizer).

19.2.3   Combining Conferencing and Cribbing

Here we discuss combining cooperation and partial cribbing and use them simultaneously to obtain better performance and a larger capacity region. In a MAC with combined cooperation and partial cribbing, depicted in Figure 19.6, Encoder 1 and

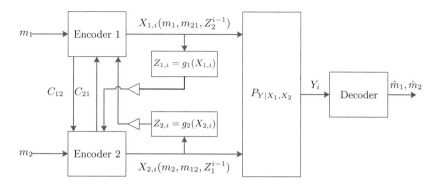

Figure 19.6 MAC with combined cooperation and partial cribbing. Encoder 1 and Encoder 2 obtain messages M_{21} and M_{12} prior to transmission. The cribbing is done strictly causally by both encoders. This setting corresponds to Case A.

Encoder 2 obtain messages M_{21} and M_{12} prior to transmission. As for the cribbing part, we address two cases:

1. **Case A**, the cribbing is done strictly causally by both encoders, i.e., Encoder 1 forms $X_{1,i}$ as a function of (M_1, M_{21}, Z_2^{i-1}) and Encoder 2 forms $X_{2,i}$ as a function of (M_2, M_{12}, Z_1^{i-1}) where $Z_{1,i}$ and $Z_{2,i}$ are deterministic functions of $X_{1,i}$ and $X_{2,i}$, respectively.
2. **Case B**, the cribbing is done strictly causally by Encoder 1 and causally by Encoder 2, i.e., Encoder 1 forms $X_{1,i}$ as a function of (M_{21}, Z_2^{i-1}) and Encoder 2 forms $X_{2,i}$ as a function of (M_{12}, Z_1^i).

The settings of combined cooperation and cribbing considered in this paper give the fundamental limits and insights on how to design optimal coding for communication systems where the users have cognition capabilities, and therefore "listen" to each other's signals and, in addition, cooperate with each other via dedicated links. We show that combining cribbing and cooperation is straightforward since it does not require any additional auxiliary random variable (RV) compared with either cribbing or cooperation exclusively. Therefore, the combination of cooperation and cribbing should be considered in future cooperative wireless cellular systems.

DEFINITION 19.6 (*Code for combined cooperation and partial cribbing*) *A* $(2^{nR_1}, 2^{nR_2}, 2^{nC_{12}}, 2^{nC_{21}}, n)$ *code for the MAC with combined cooperation and partial cribbing, as shown in Figure 19.6, consists of encoding functions at Encoder 1 and Encoder 2*

$$f_{12} : \{1, \ldots, 2^{nR_1}\} \mapsto \{1, \ldots, 2^{nC_{12}}\}, \tag{19.15}$$

$$f_{21} : \{1, \ldots, 2^{nR_2}\} \mapsto \{1, \ldots, 2^{nC_{21}}\}, \tag{19.16}$$

$$f_{1,i} : \{1, \ldots, 2^{nR_1}\} \times \{1, \ldots, 2^{nC_{21}}\} \times \mathcal{Z}_2^{i-1} \mapsto \mathcal{X}_{1,i}, \tag{19.17}$$

$$f_{2,i}^A : \{1, \ldots, 2^{nR_2}\} \times \{1, \ldots, 2^{nC_{12}}\} \times \mathcal{Z}_1^{i-1} \mapsto \mathcal{X}_{2,i}, \tag{19.18}$$

$$f_{2,i}^B : \{1, \ldots, 2^{nR_2}\} \times \{1, \ldots, 2^{nC_{12}}\} \times \mathcal{Z}_1^i \mapsto \mathcal{X}_{2,i}, \tag{19.19}$$

where $i \in \{1, \ldots, n\}$, and a decoding function

$$g : \mathcal{Y}^n \mapsto \{1, \ldots, 2^{nR_1}\} \times \{1, \ldots, 2^{nR_2}\}. \tag{19.20}$$

Let us define the regions $\mathcal{R}^A$ and $\mathcal{R}^B$ that are contained in the set of nonnegative two-dimensional real numbers, which we henceforth denote by $\mathbb{R}_+^2$.

$$\mathcal{R}^A = \begin{cases} R_1 \le I(X_1; Y | X_2, Z_1, U) + H(Z_1 | U) + C_{12}, \\ R_2 \le I(X_2; Y | X_1, Z_2, U) + H(Z_2 | U) + C_{21}, \\ R_1 + R_2 \le I(X_1, X_2; Y | U, Z_1, Z_2) + H(Z_1, Z_2 | U) + C_{12} + C_{21}, \\ R_1 + R_2 \le I(X_1, X_2; Y), \text{ for} \\ P(u) P(x_1 | u) \mathbb{1}_{z_1 = g_1(x_1)} P(x_2 | u) \mathbb{1}_{z_2 = g_2(x_2)} P(y | x_1, x_2). \end{cases} \tag{19.21}$$

The region $\mathcal{R}^B$ is defined with the same set of inequalities as in (19.21), but the joint distribution is of the form

$$P(u) P(x_1 | u) \mathbb{1}_{z_1 = g_1(x_1)} P(x_2 | u, z_1) \mathbb{1}_{h_2 = f(x_2)} P(y | x_1, x_2). \tag{19.22}$$

THEOREM 19.7 (Capacity region of the MAC with combined cooperation and partial cribbing [33]) *The capacity regions of the MAC with combined cooperation and strictly causal (Case A) and mixed strictly causal and causal (Case B) partial cribbing, as described in Definition 19.6, are $\mathcal{R}^A$ and $\mathcal{R}^B$, respectively.*

We note that $H(Z_1 | U) = I(Z_1; X_1 | U)$, which practically represents the capacity between Encoder 1 and Encoder 2. Thus the cribbing term $I(Z_1; X_1 | U)$ plays the same role (in a quantitative sense) as the cooperation link term, C_{12}; both are capacities between Encoder 1 and Encoder 2. Similarly, the role of $I(Z_2; X_2 | U)$ in C_{21} and of $I(Z_1, Z_2; X_1, X_2 | U)$ in $C_{12} + C_{21}$. Hence, the important feature here is the mutual information on the cooperation, whether the cooperation is done by cribbing or by dedicated links, since they both behave similarly.

Surprisingly, the capacity region of the MAC with combined cooperation and partial cribbing can be expressed using only one auxiliary RV, similar to the capacity regions of the MAC with cooperation and with partially cribbing encoders. The reason is that in an optimal coding scheme, the encoders use both cooperation and partial cribbing to generate a common message between the encoders. This implies that if for the MAC with partial cribbing we have a "good code," namely a code that achieves the capacity region, then by performing minor modifications, namely increasing the common message rate, we can construct a "good code" for the MAC in which cooperation and partial cribbing are combined. The coding techniques we use in this chapter include block Markov coding (introduced by Willems), joint typicality decoding, backward decoding, and double rate splitting, the last of which is necessary because we need to split the original message twice: One part will be obtained through the cooperation link and the other part using partial cribbing.

EXAMPLE 19.8 (Gaussian MAC with combined cooperation and quantized cribbing [33]) We now consider a Gaussian MAC where $Y = X_1 + X_2 + W$ and $W \sim N(0, N)$, depicted in Figure 19.7.

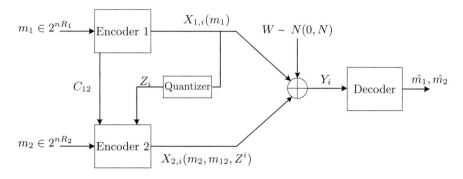

Figure 19.7 Gaussian MAC with one-sided combined cooperation and quantized cribbing. Message M_{12} is sent prior to transmission and Z_i is known causally at Encoder 2.

We assume that the power constraint over the outputs of Encoder 1 and Encoder 2 is

$$\frac{1}{n} E\left[\sum_{i=1}^{n} X_{l,i}^2\right] \leq P_l, \text{ for } l = 1, 2. \tag{19.23}$$

Prior to transmission, Encoder 1 sends a message M_{12} to Encoder 2. In addition, Encoder 2 cribs causally from Encoder 1 and obtains Z_i, which is a scalar quantization of the signal $X_{1,i}$. First, we examine an inner bound to the capacity region. Since Encoder 2 can ignore the cribbed symbols and the messages that it obtains from Encoder 1, the capacity region of the Gaussian MAC without cooperation and quantized cribbing is contained in the capacity region with combined cooperation and quantized cribbing. Hence, we have the following inner bound:

$$R_1 \leq \frac{1}{2} \log(1 + \frac{P_1}{N}),$$
$$R_2 \leq \frac{1}{2} \log(1 + \frac{P_2}{N}),$$
$$R_1 + R_2 \leq \frac{1}{2} \log(1 + \frac{P_1 + P_2}{N}). \tag{19.24}$$

On the other hand, an outer bound is obtained when there is full cooperation or perfect cribbing, i.e., Encoder 2 obtains the message m_1 before sending X_2. The capacity region in this case is

$$R_2 \leq \frac{1}{2} \log(1 + \frac{P_2}{N}(1 - \rho^2)),$$
$$R_1 + R_2 \leq \frac{1}{2} \log(1 + \frac{P_1 + 2\rho\sqrt{P_1 P_2} + P_2}{N}). \tag{19.25}$$

We set the following distributions:

$$X_1 = \lambda U + X_1', \tag{19.26}$$
$$X_2 = \bar{\lambda} U + X_2', \tag{19.27}$$

where

$$U \sim N(0, P_0) \ , \ P_0 = \left(\sqrt{\bar{\beta}_1 P_1} + \sqrt{\bar{\beta}_2 P_2} \right)^2,$$

$$P_{X_2'|Z,U}(x_2'|z,u) = \bar{\rho} P_{X_2''}(x_2') + \rho P_{X_1'|Z,U}(x_2'|z,u),$$

$$X_1' \sim N(0, \beta_1 P_1),$$

$$X_2'' \sim N(0, \beta_2 P_2),$$

$$\lambda = \sqrt{\frac{\bar{\beta}_1 P_1}{P_0}} \ , \ \bar{\lambda} = 1 - \lambda,$$

$$\beta_1, \beta_2, \rho \in [0, 1]. \tag{19.28}$$

The intuition behind the choice of these distributions is as follows. The common message, signified as U, is obtained via the rate-limited link and the two encoders cooperate to send that common message. Since the cooperation and cribbing are one-sided, only Encoder 2 can help Encoder 1 send its private message. The idea behind the choice of $P_{X_2'|Z,U}(x_2'|z,u)$ is that Encoder 2 will send $\bar{\rho}$ of the time its private message and ρ of the time the estimation of Encoder 1's private message, X_1', conditioned on the cribbing Z and the cooperation U.

19.2.4   Conferencing with Channel State Information

Now, we discuss the case of cooperation in the presence of state information. We consider the problem of MAC with cooperating encoder, where different partial state information is known at each encoder and perfect state information is known at the decoder. The setting of the problem is depicted in Figure 19.8. The state of the channel is given by the pair (S_1, S_2), where Encoder 1 knows S_1, Encoder 2 knows S_2, and the decoder knows the pair (S_1, S_2). The cooperation links C_{12} and C_{21} may increase the capacity region by transmission of the state information that is missing to the encoders and by sharing parts of the private messages (m_1, m_2).

Now we define the code of MAC with two-way cooperation where different state information is available at each encoder and full state information is available at the receiver, as depicted in Figure 19.8.

DEFINITION 19.9 (*Code of cooperating encoders with state information*)   *A* $(2^{nR_1}, 2^{nR_2}, 2^{nC_{12}}, 2^{nC_{21}}, n)$ *code with two-way cooperating encoders, where each encoder has partial state information, consists of four encoding functions*

$$f_{12} : \{1, \ldots, 2^{nR_1}\} \times \mathcal{S}_1^n \mapsto \{1, \ldots, 2^{nC_{12}}\},$$

$$f_{21} : \{1, \ldots, 2^{nR_2}\} \times \{1, \ldots, 2^{nC_{12}}\} \times \mathcal{S}_2^n \mapsto \{1, \ldots, 2^{nC_{21}}\},$$

$$f_1 : \{1, \ldots, 2^{nR_1}\} \times \{1, \ldots, 2^{nC_{21}}\} \times \mathcal{S}_1^n \mapsto \mathcal{X}_1^n,$$

$$f_2 : \{1, \ldots, 2^{nR_2}\} \times \{1, \ldots, 2^{nC_{12}}\} \times \mathcal{S}_2^n \mapsto \mathcal{X}_2^n, \tag{19.29}$$

and a decoding function,

$$g : \mathcal{Y}^n \times \mathcal{S}_1^n \times \mathcal{S}_2^n \mapsto \{1, \ldots, 2^{nR_1}\} \times \{1, \ldots, 2^{nR_2}\}. \tag{19.30}$$

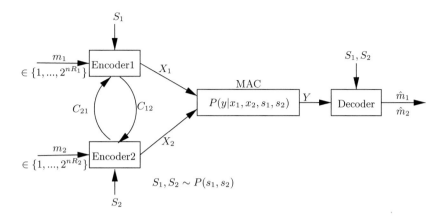

Figure 19.8 MAC with cooperation where different partial state information is known to each encoder, and full state information is known to the decoder.

THEOREM 19.10 (Capacity region of MAC with cooperating encoders and state information) *The capacity region of the MAC with two-way cooperating encoders and with partial state information as shown in Figure 19.8 is the closure of the set of rates that satisfy*

$$C_{12} \geq I(U; S_1|S_2)$$

$$C_{21} \geq I(V; S_2|S_1, U)$$

$$R_1 \leq I(X_1; Y|X_2, S_1, S_2, U, V) + C_{12} - I(U; S_1|S_2)$$

$$R_2 \leq I(X_2; Y|X_1, S_1, S_2, U, V) + C_{21} - I(V; S_2|S_1, U)$$

$$R_1 + R_2 \leq \min \begin{cases} I(X_1, X_2; Y|S_1, S_2, U, V) + C_{12} + C_{21} - I(U; S_1|S_2) - I(V; S_2|S_1, U) \\ I(X_1, X_2; Y|S_1, S_2) \end{cases}$$

for some joint distribution of the form

$$P(s_1, s_2)P(u|s_1)P(v|s_2, u)P(x_1|s_1, u, v)P(x_2|s_2, u, v)P(y|x_1, x_2, s_1, s_2), \qquad (19.31)$$

where U and V are auxiliary RVs with bounded cardinality.

In the achievability proof of the theorem we use double-binning, which was introduced by Liu et al. [34, 35] to achieve secrecy capacity in the broadcast channel. Here, the double-binning is needed since one layer of binning will be used for transmitting a common message between the encoders and an additional layer of binning is needed for choosing a specific typical sequence using side information, as done in the Wyner–Ziv problem [36] and two-way source coding [37]. In a double-binning coding scheme we have special bins that contain other bins rather than codewords, and we call such a special bin a *superbin*, as depicted in Figure 19.9.

We would like to emphasize that in the two way cooperation problem we discuss here, the ordering of operations on the conferencing links (first Transmitter 1, then Transmitter 2) is imposed in the problem definition in (19.29). This is different from the definition

a bin
(contains codewords)

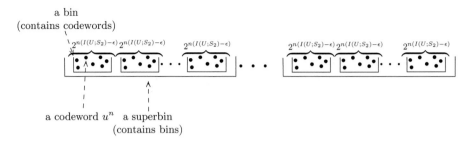

a codeword u^n a superbin
(contains bins)

Figure 19.9 Double-binning for the achievability of Theorem 19.10. Double-binning consists of two-layer bins, where in the first layer we have bins that contain codewords and in the second layer we have superbins that contain bins.

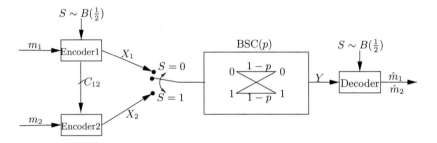

Figure 19.10 An example of a MAC with one-way cooperation and state information at one encoder and the decoder. The state S controls the switch. When $S = 0$, $Y = X_1 + Z$ and when $S = 1$, $Y = X_2 + Z$, and $Z \sim B(p)$.

of conferencing that is in Willems' work [25], which allows for any ordering and for multiple rounds of conferencing. In order to deal with multiple rounds of conferencing one should use techniques for two-way source coding developed by Kaspi [37] for the state cooperation and combine it with message cooperation.

EXAMPLE 19.11 (Comparison to message-only and state-only cooperation) Consider the example given in Figure 19.10, where the state of the channel controls the switch that determines which input goes through a binary symmetric channel (BSC) with parameter p. When $S = 0$, the binary input X_1 goes through and when $S = 1$ the binary input X_2 goes through, hence the output of the channel Y is given by

$$Y = \bar{S}X_1 \oplus SX_2 \oplus Z, \tag{19.32}$$

where $Z \sim \text{Bernouli}(\frac{1}{2})$ and is independent of S, the symbol $\oplus$ denotes XOR, and $\bar{S}$ denotes $1 - S$. We also have the constraint on the portion of 1s at the encoders, namely for any pair of codeword (x_1^n, x_2^n), $\frac{1}{n}\sum_{i=1}^{n} x_{1,i} \leq p_1$, and $\frac{1}{n}\sum_{i=1}^{n} x_{2,i} \leq p_2$. The capacity region is given by Theorem 19.10 where additional constraints $\mathbb{E}[X_1] \leq p_1$ and $\mathbb{E}[X_2] \leq p_2$ are needed ($\mathbb{E}[\cdot]$ denotes expectation). We obtain from Theorem 19.10 that the capacity region is the set of all rate-pairs (R_1, R_2) that satisfy

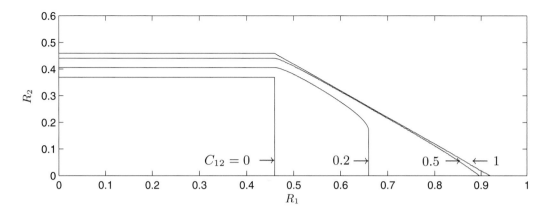

Figure 19.11 Capacity region of the example depicted in Figure 19.10 for several values of C_{12}, i.e, $C_{12} = [0, 0.2, 0.5, 1]$ and $p_z = 0.01, p_1 = p_2 = 0.25$.

$$R_1 \leq \frac{1}{2}H_b(p_1 * p_z) - \frac{1}{2}H_b(p_z) + C_{12} - I(U; S)$$

$$R_2 \leq \frac{1}{2}H(X_2 \oplus Z|S = 1, U) - \frac{1}{2}H_b(p_z)$$

$$R_1 + R_2 \leq \frac{1}{2}H_b(p_1 * p_z) + \frac{1}{2}H(X_2 \oplus Z|S = 1) - H_b(p_z) \qquad (19.33)$$

for some conditional distributions $P(u|s)$ and $P(x_2|u)$, where $I(U; S) \leq C_{12}$. The term $H_b(p)$ denotes the binary entropy function, which is defined for $0 \leq p \leq 1$ as $H_b(p) = -p \log p - (1 - p) \log(1 - p)$. The term $p * q$ denotes the parameter of a Bernoulli distribution that results from convolving mod-2 two Bernoulli distributions with parameters p and q, i.e., $p * q = (1 - p)q + (1 - q)p$.

Figure 19.11 illustrates the influence of the cooperation rate on the capacity region. It shows the capacity regions for several rates of cooperation $C_{12} = [0, 0.2, 0.5, 1]$ where $p_z = 0.01, p_1 = p_2 = 0.25$. One can see that when the cooperation rate is small an increase in the cooperation rate significantly influences the capacity region; however, for a large cooperation rate, such as $C_{12} > 0.5$, an increase in the cooperation rate hardly influences the capacity region.

Comparison to two different kinds of cooperation: In the setting analyzed in this chapter, we assumed a cooperation link that may use both the message and the state information. Recent works assumed similar settings where the cooperation depends only on the state [38], as depicted in Figure 19.12, or on the message only [39, 40] as depicted in Figure 19.13.

For the first case, where the cooperation may use only the state information (Figure 19.12), the capacity region was derived in [38] and may be written as

$$C_{12} = I(U; S) \qquad (19.34)$$

$$R_1 \leq I(X_1; Y|X_2, S, U) \qquad (19.35)$$

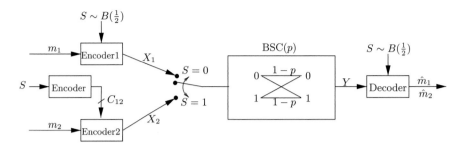

Figure 19.12 State cooperation. An example where the cooperation is limited-rate state information and is independent of the message.

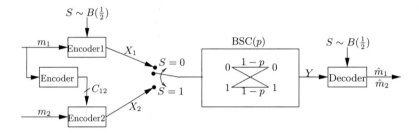

Figure 19.13 Message cooperation. An example where the cooperation is a function of the message only, and then after the cooperation stage the channel state is available to Encoder 1.

$$R_2 \leq I(X_2; Y|X_1, S, U) \tag{19.36}$$

$$R_1 + R_2 \leq I(X_1, X_2; Y|S, U), \tag{19.37}$$

for some joint distribution of the form

$$P(s, u)P(x_1|s, u)P(x_2|u)P(y|x_1, x_2, s). \tag{19.38}$$

For the second case, where the cooperation may use only the message (Fig. 19.13), the capacity region was considered in [39, 40] and may be written as

$$R_1 \leq I(X_1; Y|X_2, S, U) + C_{12} \tag{19.39}$$

$$R_2 \leq I(X_2; Y|X_1, S, U) \tag{19.40}$$

$$R_1 + R_2 \leq \min \left\{ \begin{array}{c} I(X_1, X_2; Y|S, U) + C_{12}, \\ I(X_1, X_2; Y|S) \end{array} \right\}, \tag{19.41}$$

for some joint distribution of the form

$$P(s)P(u)P(x_1|s, u)P(x_2|u)P(y|x_1, x_2, s), \tag{19.42}$$

where U and V are auxiliary RVs with bounded cardinality.

Figure 19.14 depicts the capacity regions obtained for a cooperation link $C_{12} = 0.5$ for the three settings:

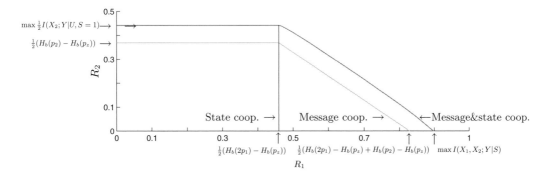

Figure 19.14 The regions of three settings with a cooperation link $C_{12} = 0.5$. The line cs corresponds to the case where the cooperation is based only on the state information as depicted in Figure 19.12. The line cm corresponds to the case where the cooperation is based only on the message and not on the state as depicted in Figure 19.13. Finally, the line csm is the one that corresponds to the setting of this chapter where the cooperation may use both the state and the message as depicted in Figure 19.10.

1. state-cooperation, where the cooperation is based only on the state information (Figure 19.12);
2. message-cooperation, where the cooperation is based only on the message (Figure 19.13);
3. message–state cooperation, where the cooperation may use both the state and the message (Figure 19.10).

In this example, one can note from Figure 19.14 that state cooperation and message cooperation have different gains that are combined in the message–state cooperation. State cooperation increases the capacity region only in the direction of R_2 since only Encoder 2 receives the state information via the cooperation link. Message cooperation increases the capacity region only in the direction of R_1 since only the message from Encoder 1 is transmitted via the cooperation link. However, message–state cooperation increases the capacity region in the direction of both R_1 and R_2 since it combines the advantages of both message and state cooperation.

19.2.5   Cribbing with Channel State Information

Consider a MAC with partial cribbing and causal or noncausal channel state information (CSI), as depicted in Figure 19.15. This channel depends on the CSI (S_1, S_2) sequence that is known to the decoder, and each encoder $w \in \{1, 2\}$ has noncausal or causal access to one state component $S_w \in \mathcal{S}_w$. Each encoder w sends a message M_w over the channel. Encoder 2 is cribbing Encoder 1; the cribbing is strictly causal, partial, and controlled by S_1. Namely, the cribbed signal at time i, denoted by Z_i, is a deterministic function of $X_{1,i}$ and $S_{1,i}$. The cribbed information is used by Encoder 2 to assist Encoder 1.

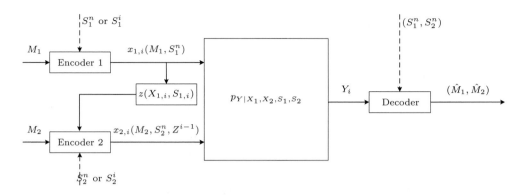

Figure 19.15 State-dependent MAC with two state components and one side cribbing. The CSI is known causally, i.e., S_1^i, or noncausally S_1^n. The cribbing in this figure is *strictly* causal – $X_2 = x_{2,i}(M_2, S_2^n, Z^{i-1})$.

There are four cases considered in this setting that combines causal CSI and noncausal CSI with strictly causal partial cribbing and causal partial cribbing.

DEFINITION 19.12 (*Code for MAC*)   *A (R_1, R_2, n) code $\mathcal{C}_n$ for the state-dependent MAC with partial cribbing and two state components is defined by*

$$x_1^n : [1 : 2^{nR_1}] \times S_1^n \to \mathcal{X}_1^n$$
$$\hat{m}_1 : \mathcal{Y}^n \times S_1^n \times S_2^n \to [1 : 2^{nR_1}]$$
$$\hat{m}_2 : \mathcal{Y}^n \times S_1^n \times S_2^n \to [1 : 2^{nR_2}],$$

and the encoding of $x_{2,i}$, for $1 \le i \le n$ has four different possibilities according to the setting of the problem:

$$x_{2,i} : [1 : 2^{nR_2}] \times S_2^i \times Z^{i-1} \to \mathcal{X}_2 \qquad \text{causal CSI, strictly causal cribbing}$$
$$x_{2,i} : [1 : 2^{nR_2}] \times S_2^i \times Z^i \to \mathcal{X}_2 \qquad \text{causal CSI, causal cribbing}$$
$$x_{2,i} : [1 : 2^{nR_2}] \times S_2^n \times Z^{i-1} \to \mathcal{X}_2 \qquad \text{noncausal CSI, strictly causal cribbing}$$
$$x_{2,i} : [1 : 2^{nR_2}] \times S_2^n \times Z^i \to \mathcal{X}_2 \qquad \text{noncausal CSI, causal cribbing}$$

THEOREM 19.13 (MAC with cribbing and CSI [41, 42])   *The capacity region for discrete memoryless MAC CSI and partial cribbing as in Figure 19.15 is given by the set of rate pairs (R_1, R_2) that satisfy*

$$R_1 \le I(X_1; Y|X_2, Z, S_1, S_2, U) + H(Z|S_1, U) - I(U; S_1|S_2) \tag{19.43}$$
$$R_2 \le I(X_2; Y|X_1, S_1, S_2, U) \tag{19.44}$$
$$R_1 + R_2 \le I(X_1, X_2; Y|Z, S_1, S_2, U) + H(Z|S_1, U) - I(U; S_1|S_2) \tag{19.45}$$
$$R_1 + R_2 \le I(X_1, X_2; Y|S_1, S_2) \tag{19.46}$$

- *for probability mass functions (pmfs) of the followings forms:*
- *causal CSI and strictly causal cribbing: $p(u)p(x_1|u, s_1)p(x_2|u, s_2)$,*
- *causal CSI, causal cribbing: $p(u)p(x_1|u, s_1)p(x_2|z, u, s_2)$,*

- *noncausal CSI and strictly causal cribbing: $p(x_1, u|s_1)p(x_2|u, s_2)$, that satisfies $I(U; S_1|S_2) \leq H(Z|S_1, U)$,*
- *noncausal CSI, causal cribbing: $p(x_1, u|s_1)p(x_2|u, s_2, z)$, that satisfies $I(U; S_1|S_2) \leq H(Z|S_1, U)$.*

The capacity is achieved by a scheme based on *cooperative-bin-forward*. This scheme allows cooperation between the encoders without the need to decode a part of the message by the encoder that receives the cribbed signal. The transmission is divided into blocks and each possible cribbed signal is mapped to a bin. The bin index is used by the encoders to choose the cooperation codeword in the next transmission block. In *causal CSI* settings [41] the cooperation is independent of the state. In *noncausal CSI* settings [42] dependency between the encoder transmission and the state can increase the transmission rates. The first encoder implicitly conveys partial state information to the second encoder. In particular, it uses the states of the next block and selects a cooperation codeword accordingly and the encoder's transmission depends on the cooperation codeword and therefore also on the states.

19.3 Cognition of State and Messages

In this section we focus on cognition of state and message in multiuser channels. We begin by describing the model of a two-user state-dependent cognitive MAC. In this model there are two transmitters, a primary one and a secondary (cognitive) one, and a single receive. The primary transmitter wishes to transmit a message, and the secondary transmitter observes both the state and the primary transmitters message and transmits an additional message. Since the primary transmitters message is known to both transmitters, it an be viewed as a common message. The receiver needs to decode both messages reliably. The second model consists of a two-user state-dependent cognitive interference channel; it generalizes the above cognitive MAC to the case in which the messages are intended for two different receivers, keeping the roles of the two transmitters as in the first model. The third model also involves a cognitive IC, but in this case the state is degenerate and the cognition of the secondary user is limited only to the message. In this setup also the common message should be decoded by the two receivers of the IC. The fourth model introduces a three-user cognitive communication channel model without state which generalizes instances of the previous models.

19.3.1 Cooperative Multiple Access Encoding with States Available at One Transmitter

This section describes the channel models that were studied in [43] and generalizes the Gel'fand–Pinsker channel [44] with noncausal state information as well as Shannon's channel [45] with causal state information by encompassing the setup of a cognitive memoryless MAC.

A stationary memoryless state-dependent MAC is defined by the channel conditional probability distribution $W_{Y|S,X_1,X_2}$ from $\mathcal{S} \times \mathcal{X}_1 \times \mathcal{X}_2$ to $\mathcal{Y}$ and a distribution Q_S on the set

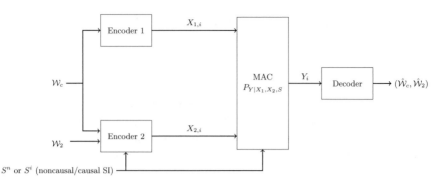

Figure 19.16 Asymmetric state-dependent MAC with a common message.

$\mathcal{S}$. Denote by $X_1^n = (X_{1,1}, \ldots, X_{1,n})$ and $X_2^n = (X_{2,1}, \ldots, X_{2,n})$ the inputs of Transmitters 1 and 2 to the channel, respectively. Let Y^n designate the output of the channel.

From the stationarity and memorylessness assumptions, one has

$$P_{Y^n|S^n,X_1^n,X_2^n}(y^n|s^n,x^n,\tilde{x}^n) = \prod_{i=1}^{n} W_{Y|S,X_1,X_2}(y_i|s_i,x_i,\tilde{x}_i). \tag{19.47}$$

The symbols $S_i, X_{1,i}, X_{2,i}$, and Y_i stand for the channel state, the channel inputs produced by two distinct encoders, and the channel output, at time index i, respectively. The channel states S^n are assumed to be i.i.d., where S_i is distributed according to Q_S for all i.

The term generalized Gel'fand–Pinsker (GGP) channel will designate the setup in which Encoder 2 observes the CSI noncausally, and the term asymmetric causal state-dependent channel will designate the setup in which Encoder 2 observes the states causally.

The private message $\mathcal{W}_2$ and the private message $\mathcal{W}_c$ are independent RVs uniformly distributed over the sets $\{1, \ldots, M_2\}$ and $\{1, \ldots, M_c\}$, respectively, where $M_2 = \lfloor e^{nR_2} \rfloor$ and $M_c = \lfloor e^{nR_c} \rfloor$.

An (e^{R_c}, e^{R_2}, n) code for the channel with noncausal state information as depicted in Figure 19.16 (where S^n is the input to Encoder 2), consists of two encoders $\varphi_n^{(1)}, \varphi_n^{(2)}$ and a decoder ψ_n: The first encoder, which is unaware of the CSI is defined by the mapping

$$\varphi_n^{(1)} : \{1, \ldots, M_c\} \to \mathcal{X}_1^n. \tag{19.48}$$

The second encoder observes the CSI noncausally, and is defined by a mapping

$$\varphi_n^{(2)} : \{1, \ldots, M_c\} \times \{1, \ldots, M_2\} \times \mathcal{S}^n \to \mathcal{X}_2^n. \tag{19.49}$$

The decoder is a mapping

$$\psi_n : \mathcal{Y}^n \to \{1, \ldots, M_c\} \times \{1, \ldots, M_2\}. \tag{19.50}$$

An (e^{R_c}, e^{R_2}, n) code for the asymmetric causal state-dependent channel is defined similarly with the exception that the second encoder is defined by a sequence of mappings

$$\varphi_{n,i}^{(2)} : \{1, \ldots, M_c\} \times \{1, \ldots, M_2\} \times \mathcal{S}^i \to \mathcal{X}_2 \quad i = 1, \ldots, n, \qquad (19.51)$$

and at time index i the channel input is given by $X_{2,i} = \varphi_{n,i}^{(2)}(\mathcal{W}_c, \mathcal{W}_2, S^i)$. This channel is depicted in Figure 19.16, where S^i is the input to Encoder 2 at time instant i.

The capacity region of the channel is defined as the set of rate-pairs (R_c, R_2) that enable reliable decoding of both messages by the decoder.

Note that a MAC with a common message [28] whose capacity region is given as follows is a channel model that generalizes the GGP with no state: This is given by ([28]) as the closure of the union of rate-pairs (R_c, R_2) satisfying

$$R_2 \le I(X_2; Y | X_1, Z),$$
$$R_c + R_2 \le I(X_1, X_2; Y), \qquad (19.52)$$

for some $P_{Z, X_1, X_2, Y} = P_Z P_{X_1 | Z} P_{X_2 | Z} W_{Y | X_1, X_2}$.

Finite Input Alphabet Cognitive MAC with Noncausal State Information

The following theorem ([43]) establishes a single-letter expression characterizing the capacity region of the finite input alphabet GGP channel.

THEOREM 19.14 *The capacity region of the finite input alphabet GGP channel, $\mathcal{C}$, is the closure of the union of all rate-pairs (R_c, R_2) satisfying*

$$R_2 \le I(U; Y | X_1) - I(U; S | X_1)$$
$$R_c + R_2 \le I(U, X_1; Y) - I(U, X_1; S), \qquad (19.53)$$

for some joint distribution $P_{S, X_1, U, X_2, Y}$ on $\mathcal{S} \times \mathcal{X}_1 \times \mathcal{U} \times \mathcal{X}_2 \times \mathcal{Y}$ having the form

$$P_{S, X_1, U, X_2, Y} = Q_S P_{X_1} P_{U, X_2 | S, X_1} W_{Y | S, X_1, X_2}, \qquad (19.54)$$

where $|\mathcal{U}| \le |\mathcal{S}| \cdot |\mathcal{X}_1| \cdot |\mathcal{X}_2|$.

The coding scheme that achieves the capacity region of Theorem 19.14 was introduced in [43] and is described next. Fix a distribution P_{S, X_1, U, X_2} satisfying (19.54). The uninformed user employs ordinary i.i.d. $\sim P_{X_1}$ random coding to transmit at rate $R_1 = I(X_1; Y)$. The informed user transmits additional information using a binning scheme at rate $R_2' = I(U; Y, X_1) - I(U; S, X_1) = I(U; Y | X_1) - I(U; S | X_1)$ by treating X_1 as part of the state given that X_1 is already available at the decoder. That can be done since the information sent by the uninformed encoder has been decoded first. Further, the information sent by the informed encoder at rate R_2' can be shared between private message $\mathcal{W}_2$ and common message $\mathcal{W}_c$.

Note that feedback to the informed encoder does not increase the capacity region of Theorem 19.14. In other words, the capacity region remains unchanged if the uninformed encoder is a mapping of the form (19.48), whereas before producing the ith channel input symbol, the informed encoder observes the previous channel outputs, Y^{i-1}, and is defined by a sequence of mappings $\varphi_n^{(2)} = \{\varphi_n^{(2,i)}\}_{i=1}^n$ with

$$\varphi_n^{(2,i)} : \{1, \ldots, M_c\} \times \{1, \ldots, M_2\} \times \mathcal{S}^n \times \mathcal{Y}^{i-1} \to \mathcal{X}_2. \qquad (19.55)$$

Note also that when the channel does not depend on the states, i.e., $W_{Y|S,X_1,X_2} = W_{Y|X_1,X_2}$, the expression for the capacity region reduces to (19.52).

The important special case of common message transmission only (no private message) is also described in [43].

COROLLARY 19.15   *The common message capacity of the finite input alphabet GGP channel is given by*

$$C = \max \left[I(U, X_1; Y) - I(U, X_1; S) \right], \tag{19.56}$$

where the maximum is over all the joint measures $P_{S,X_1,U,X_2,Y}$ on $\mathcal{S} \times \mathcal{X}_1 \times \mathcal{U} \times \mathcal{X}_2 \times \mathcal{Y}$ having the form

$$P_{S,X_1,U,X_2,Y} = Q_S P_{X_1} P_{U,X_2|S,X_1} W_{Y|S,X_1,X_2}, \tag{19.57}$$

where $|\mathcal{U}| \le |\mathcal{S}| \cdot |\mathcal{X}_1| \cdot |\mathcal{X}_2|$.

It is noted that there exists a maximizing distribution with X_2 that is a deterministic function of (S, X_1, U).

In the case of an ordinary single-user GGP channel, the trivial upper bound on the capacity $\max_{P_{X|S}} I(X; Y|S)$ accounts for the case of state information available non-causally to the encoder as well as to the decoder. This bound was generalized in [43] in the following theorem to an outer bound on the capacity region of the GGP channel. Providing such an outer bound (for an already characterized capacity region) is useful in the proof of the converse part of the coding theorem for the explicit expression to the capacity region of the Gaussian GGP channel, because this upper bound is achievable in the Gaussian case.

THEOREM 19.16   *The closure of the convex hull of the set of rate-pairs satisfying*

$$R_2 \le I(X_2; Y|S, X_1)$$
$$R_c + R_2 \le I(X_1, X_2; Y|S) - I(S; X_1|Y) \tag{19.58}$$

for some distribution $P_{S,X_1,X_2,Y} = Q_S P_{X_1} P_{X_2|S,X_1} W_{Y|S,X_1,X_2}$ is an outer bound on the capacity region of the GGP channel.

An interesting subclass of GGP channels is the *memoryless parallel channel with noncausal asymmetric side information*. It is a GGP channel with $Y = (Y_1, Y_2)$ and

$$W_{Y_1,Y_2|S,X_1,X_2} = W_{Y_1|X_1,S} W_{Y_2|X_2,S}. \tag{19.59}$$

In words, this is a GGP channel with two outputs $Y_{1,1}, \ldots, Y_{1,n}$ and $Y_{2,1}, \ldots, Y_{2,n}$ that are both observed by the receiver. If, in addition, one has

$$W_{Y_2|X_2,S} = W_{Y_2|X_2}, \tag{19.60}$$

the parallel channel is termed degenerate.

The following theorem establishes the fact that in the case of a degenerate parallel GGP channel the CSI does not help.

THEOREM 19.17 *The capacity region of the degenerate parallel GGP channel is equal to the capacity region obtained without transmitter CSI; i.e.,*

$$R_2 \le C_2$$
$$R_c + R_2 \le C_1 + C_2, \tag{19.61}$$

where C_1 is the capacity of the channel $W_{Y_1|X_1,S}$ obtained without transmitter CSI, and C_2 is the capacity of the channel $W_{Y_2|X_2}$.

Note that from Theorem 19.17 it follows that for the second encoder, sending information about the common message is always advantageous over sending information about the channel states.

Finite Input Alphabet Cognitive MAC with Causal State Information

There are many practical applications for which the state sequence is not known in advance, but rather in a causal manner, as in (19.51).

THEOREM 19.18 *The capacity region of the finite input alphabet causal asymmetric state-dependent channel is given by the closure of the set of rate-pairs (R_2, R_c) satisfying*

$$R_2 \le I(U; Y|X_1)$$
$$R_c + R_2 \le I(U, X_1; Y), \tag{19.62}$$

for some joint distribution $P_{S,X_1,U,X_2,Y}$ on $\mathcal{S} \times \mathcal{X}_1 \times \mathcal{U} \times \mathcal{X}_2 \times \mathcal{Y}$ having the form

$$P_{S,X_1,U,X_2,Y} = Q_S P_{X_1,U} P_{X_2|S,X_1,U} W_{Y|S,X_1,X_2}, \tag{19.63}$$

where $|\mathcal{U}|$ satisfies $|\mathcal{U}| \le |\mathcal{S}| \cdot |\mathcal{X}_1| \cdot |\mathcal{X}_2| + 1$.

Specializing the result of Theorem 19.18 to the case where there is only a transmission of a common message, the following result is obtained in [43].

COROLLARY 19.19 *The common message capacity of the finite input alphabet causal asymmetric state-dependent channel is given by*

$$\max I(U; Y), \tag{19.64}$$

where the maximum is over all the joint measures $P_{S,X_1,U,X_2,Y}$ on $\mathcal{S} \times \mathcal{X}_1 \times \mathcal{U} \times \mathcal{X}_2 \times \mathcal{Y}$ having the form

$$P_{S,X_1,U,X_2,Y} = Q_S P_U P_{X_1|U} P_{X_2|S,U,X_1} W_{Y|S,X_1,X_2}, \tag{19.65}$$

where X_1 is a deterministic function of U, and $|\mathcal{U}| \le |\mathcal{S}| \cdot |\mathcal{X}_1| \cdot |\mathcal{X}_2| + 2$.

Note that the single-letter expression for the common message capacity can be deduced by a direct application of the formula derived by Shannon [45] for a channel with input alphabet $\mathcal{X}_1 \times \mathcal{X}_2$ if the CSI is available to both of the encoders.

The Binary GGP Channel

Consider the following example of a binary channel which is also studied in [43]. The output of the channel Y_i is given by the equation

$$Y_i = X_{1,i} \oplus X_{2,i} \oplus S_i \oplus N_i, \tag{19.66}$$

where all the RVs are binary $\{0, 1\}$, and $\oplus$ denotes modulo-2 addition. The interference process, $S_i, i \geq 1$, and the noise process, $N_i, i \geq 1$, are assumed to be independent i.i.d. Bernoulli(P_s) and Bernoulli(P_n) processes, respectively. The power constraints are given by:

$$\frac{1}{n} \sum_{i=1}^{n} X_{1,i}^2 \leq P_1 \quad \text{and} \quad \frac{1}{n} \sum_{i=1}^{n} X_{2,i}^2 \leq P_2. \tag{19.67}$$

The capacity region in this case is given by the closure of the union of all rate-pairs, (R_c, R_2), satisfying (19.53) for some joint measure $P_{S,X_1,U,X_2,Y}$ taking the form (19.54) with $E(X_1) \leq P_1$ and $E(X_2) \leq P_2$.

Next, consider the case in which $P_n = 0$; that is, the noise-free case, whose capacity region can be expressed in a simpler form which does not involve an auxiliary RV.

THEOREM 19.20   *The capacity region of the noise-free binary GGP channel is given by the convex hull of the union of the set of rate-pairs (R_c, R_2) which satisfy*

$$R_2 \leq H(X_2|X_1, S)$$
$$R_c + R_2 \leq I(X_1; Y) + H(X_2|X_1, S) \tag{19.68}$$

for some joint measure $P_{S,X_1,X_2,Y} = Q_S P_{X_1} P_{X_2|S,X_1}$ with $E(X_1) \leq P_1$ and $E(X_2) \leq P_2$, and where $Y = X_1 + X_2 + S$.

In Figure 19.17, the capacity region is plotted for several values of P_s and for $P_1 = P_2 = 0.2, P_n = 0$.

The Gaussian GGP Channel

The Gaussian GGP channel is defined by the following input–output relationship:

$$Y_i = X_{1,i} + X_{2,i} + S_i + N_i. \tag{19.69}$$

The noise processes, $S_i, i \geq 1$ and $N_i, i \geq 1$, are assumed to be zero-mean Gaussian i.i.d. with $E(S_i^2) = Q$ and $E(N_i^2) = N$. The process $N_1, \ldots, N_n$ is independent of (X_1^n, X_2^n, S^n). The power constraints are as in (19.67).

An application of the single-letter expression derived for the finite alphabet case to the Gaussian GGP channel $W_{Y|S,X_1,X_2}(y|s,x,x') = \frac{1}{\sqrt{2\pi N}} e^{-(y-s-x-x')^2/2N}$ with the individual power constraint (which is obtained using standard techniques [46]) yields the capacity region $\mathcal{C}(P_1, P_2, Q, N)$ as the closure of the union of rate-pairs (R_c, R_2) satisfying (19.53) where the allowed joint distribution of S, X_1, U, X_2, Y satisfies (19.54) and

$$E(X_1^2) \leq P_1, \quad E(X_2^2) \leq P_2. \tag{19.70}$$

This results in the following explicit characterization for the capacity region of this channel for the individual power constraints, which was presented in [43].

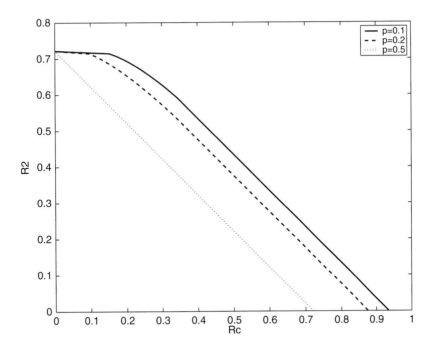

Figure 19.17 The capacity region of the noise-free binary GGP channel for $P_1 = P_2 = 0.2$.

THEOREM 19.21   *The capacity region $\mathcal{C}(P_1, P_2, Q, N)$ of the Gaussian GGP channel under individual power constraints is given by the union of the rate-pairs satisfying*

$$R_2 \le \frac{1}{2} \log \left(1 + \frac{P_2(1 - \rho_{12}^2 - \rho_{2s}^2)}{N} \right)$$

$$R_c + R_2 \le \frac{1}{2} \log \left(1 + \frac{P_2(1 - \rho_{12}^2 - \rho_{2s}^2)}{N} \right)$$

$$+ \frac{1}{2} \log \left(1 + \frac{\left(\sqrt{P_1} + \rho_{12}\sqrt{P_2} \right)^2}{P_2(1 - \rho_{12}^2 - \rho_{2s}^2) + \left(\sqrt{Q} + \rho_{2s}\sqrt{P_2} \right)^2 + N} \right) \qquad (19.71)$$

for some $\rho_{12} \in [0, 1], \rho_{2s} \in [-1, 0]$ such that

$$\rho_{12}^2 + \rho_{2s}^2 \le 1. \qquad (19.72)$$

The GGP channel capacity formula for the Gaussian channel was calculated explicitly by Costa in [47]. The proof relied on a binning scheme, termed "dirty paper coding," which was shown to achieve the same reliably transmitted rate as though the interference S^n were not there. Hence, Costa's problem, as well as its multiuser counterpart [44], were solvable because the trivial operative upper bound is achievable. The upper bound of Theorem 19.16 plays the role of the operative bound and constitutes the core of the converse part in [43].

Thus, one can generalize the interference cancellation of Costa to the GGP channel asymmetric setup by showing that the upper bound of Theorem 19.16 is actually achievable in the Gaussian GGP channel. In view of (19.58), this implies that the subtracted term, $I(S; X_1|Y)$, can be interpreted as the inevitable rate loss incurred due to the fact that S is known only to the second transmitter (and not to both). Indeed, any information that X_1 conveys to Y about S is an inevitable waste of rate.

The proof is based on showing that for the Gaussian channel in (19.53), attention can be restricted to the jointly Gaussian (S, X_1, X_2) without loss of generality, where an optimal choice for U is

$$U = X_2 + \alpha_{opt}S \tag{19.73}$$

with

$$\alpha_{opt} = \frac{P_2P_1Q - P_1\sigma_{2s}^2 - P_1N\sigma_{2s} - \sigma_{12}^2Q}{P_2P_1Q + P_1NQ - P_1\sigma_{2s}^2 - \sigma_{12}^2Q}, \tag{19.74}$$

where different values of $\sigma_{12} = E(X_1X_2)$ and $\sigma_{2s} = E(X_2S)$ are chosen to achieve different points that lie in (or, on the border of) the capacity region.

Table 19.1 presents a summary of the capacity region $\mathcal{C}(P_1, P_2, Q, N)$ expressions that can be deduced in several extreme cases. For infinite Q, the common message capacity degenerates to that of Costa's channel without the uninformed user, as expected. The capacity region takes a triangular shape because the amount of information that can be reliably transmitted by the uninformed user becomes negligible. A similar phenomenon occurs when $P_1 = 0$.

For $Q = 0$, there is no side information because the only noise present is $N_i, i \geq 1$. The informed encoder can therefore decide which portion of its power, ΔP_1, to devote to transmission of its own message, whereas the remaining power, $(1 - \Delta)P_1$, is allocated to coherent transmission (with the uninformed encoder) of the common message.

If the power of the informed encoder, P_2, is zero, it cannot transmit information, or help the uninformed user by partially canceling the interference. Hence, the common message capacity corresponds to a situation where the effective noise is $S_i + N_i, i \geq 1$ and the power used for transmission is P_1.

Specializing the results pertaining to the Gaussian channel to the case where only a common message is transmitted; that is, $\max_{\rho_{12}, \rho_{2s}} [I(X_1, X_2; Y|S) - I(X_1; S|Y)]$, where (X_1, X_2, S) are jointly Gaussian with $E(X_i^2) = P_i, i = 1, 2$, the following expression is obtained.

THEOREM 19.22   *The common message capacity of the Gaussian GGP channel under individual power constraints is given by the following formula.*

$$C(P_1, P_2, Q, N) =$$

$$\begin{cases} \frac{1}{2}\log\left(1 + \frac{P_1}{Q}\right) + \frac{1}{2}\log\left(1 + \frac{P_2}{N}\right) & \text{if } \frac{P_1(P_2+N)^2}{(P_1+Q)} \leq P_2Q \\ \max_{\rho\in[-1,0]} \frac{1}{2}\log\left(1 + \frac{\left(\sqrt{P_1}+\sqrt{P_2}\sqrt{1-\rho^2}\right)^2}{\left(\sqrt{Q}+\sqrt{P_2}\cdot\rho\right)^2+N}\right) & \text{o.w.} \end{cases} \tag{19.75}$$

Table 19.1 Extreme case analysis.

Regime	Behavior of $\mathcal{C}(P_1, P_2, Q, N)$
$Q \to \infty$	$R_2 \leq \frac{1}{2} \log \left(1 + \frac{P_2}{N} \right),$ $R_c + R_2 \leq \frac{1}{2} \log \left(1 + \frac{P_2}{N} \right)$
$P_1 = 0$	$R_2 \leq \frac{1}{2} \log \left(1 + \frac{P_2}{N} \right),$ $R_c + R_2 \leq \frac{1}{2} \log \left(1 + \frac{P_2}{N} \right)$
$Q = 0$	$R_2 \leq \frac{1}{2} \log \left(1 + \frac{P_2 \Delta}{N} \right),$ $R_c + R_2 \leq \frac{1}{2} \log \left(1 + \frac{(\sqrt{P_1} + \sqrt{(1-\Delta)P_2})^2}{N} \right)$ $+ \frac{1}{2} \log \left(1 + \frac{P_2 \Delta}{N} \right)$
$P_2 = 0$	$R_2 = 0,\ R_c \leq \frac{1}{2} \log \left(1 + \frac{P_1}{N+Q} \right)$

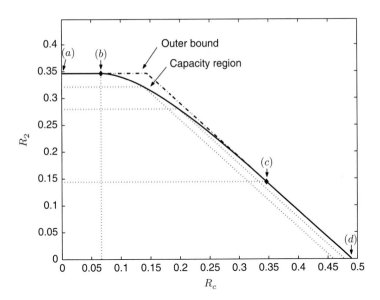

Figure 19.18 Capacity region for $P_1 = \frac{1}{2}, P_2 = N = 1, Q = \frac{3}{2}, \mathcal{C}(\frac{1}{2}, 1, \frac{3}{2}, 1)$, and outer bound.

where, in fact, the maximization over ρ can be limited to either $\rho = -1$, $\rho = 0$, or a real root ρ of a fourth-order polynomial that satisfies $\rho \in [-1, 0]$.

The set of parameters (P_1, P_2, Q, N) such that

$$\frac{P_1(P_2 + N)^2}{P_1 + Q} \geq P_2 Q \tag{19.76}$$

is referred to as the *silent regime* and its complement will be referred to as the *active regime*.

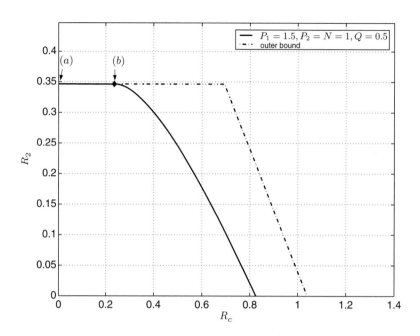

Figure 19.19 Capacity region for $P_1 = \frac{3}{2}, P_2 = N = 1, Q = \frac{1}{2}, C(\frac{1}{2}, 1, \frac{3}{2}, 1)$, and the outer bound.

The Silent Regime

The optimal values of σ_{12} and σ_{2s} as far as the common message capacity is concerned in the silent regime satisfy

$$P_1 \sigma_{2s}^2 + Q \sigma_{12}^2 = P_1 P_2 Q, \qquad (19.77)$$

or equivalently,

$$\rho_{12}^2 + \rho_{2s}^2 = 1. \qquad (19.78)$$

This is also equivalent to

$$E\left(X_2 - \hat{X}_2^{lin}(X_1, S)\right)^2 = 0, \qquad (19.79)$$

where $\hat{X}_2^{lin}(X_1, S)$ is the optimal linear estimator (in the MMSE sense) of X_2 given X_1 and S

$$\hat{X}_2^{lin}(X_1, S) = \frac{\sigma_{12}}{P_1} X_1 + \frac{\sigma_{2s}}{Q} S. \qquad (19.80)$$

Equation (19.79) implies that in the silent regime

$$X_2 = \hat{X}_2^{lin}(X_1, S) = \frac{\sigma_{12}}{P_1} X_1 + \frac{\sigma_{2s}}{Q} S, \qquad (19.81)$$

and thus,

$$Y = X_1 \left(1 + \frac{\sigma_{12}}{P_1}\right) + S \left(1 + \frac{\sigma_{2s}}{Q}\right) + V, \qquad (19.82)$$

where $V \sim \mathcal{N}(0, N)$; by calculating the optimal value of α (19.74) while accounting for (19.77), one obtains that

$$\alpha_{opt}^{silent} = -\frac{\sigma_{2s}}{Q}$$

$$U_{opt}^{silent} = X_2 - \frac{\sigma_{2s}}{Q}S = \frac{\sigma_{12}}{P_1}X_1. \tag{19.83}$$

Hence, in the silent regime of parameters, the common message capacity (19.75) formula is equal to

$$\max_{\sigma_{12}, \sigma_{2s}} I(U, X_1; Y) - I(U, X_1; S)|_{U = \frac{\sigma_{12}}{P_1}X_1}$$

$$= \max_{\sigma_{12}, \sigma_{2s}} I\left(X_1; X_1\left(1 + \frac{\sigma_{12}}{P_1}\right) + S\left(1 + \frac{\sigma_{2s}}{Q}\right) + V\right), \tag{19.84}$$

with σ_{12}, σ_{2s} satisfying (19.78). Inspecting (19.84), it is easy to verify that a simpler selection of U,

$$U_{opt}^{silent} = 0 \tag{19.85}$$

yields the same achievable rate and hence is also optimal.

The fact that in the silent regime the common message capacity is equal to (19.84) suggests that in this regime, in order to achieve capacity, the informed encoder can allocate all its power to decreasing the interference and enhancing the signal of the uninformed encoder. No power is allocated to the transmission of additional information, which is why this region is referred to as silent.

Given (19.81) and (19.85), the informed encoder does not need to employ a binning scheme and simply transmits the n-vector $\tilde{x}$ whose ith symbol is given by

$$\tilde{x}_i = x_i \frac{\sigma_{12}}{P_1} + s_i \frac{\sigma_{2s}}{Q}, \tag{19.86}$$

where x_i is the primary user's ith transmitted symbol $\sigma_{2s} = \sqrt{P_2 Q} \cdot \rho$, where ρ is the maximizer in (19.75) and $\rho_{12} = \sqrt{1 - \rho_{2s}^2}$.

The Active Regime

This parameter regime is referred to as active because here the informed encoder needs to tradeoff among three goals: decreasing the interference, enhancing the signal of the uninformed encoder, and transmitting additional information (as opposed to the silent regime where no additional information is transmitted). When keeping the other parameters fixed, the higher the interference Q, the greater the share the power that the informed user allocates to the additional information at the expense of interference reduction and enhancement of the uninformed user's signal. In this regime as well, the maximizing distribution of (X_1, X_2, S) is jointly Gaussian, but with different parameters

$$\sigma_{12}^{active} = -\sigma_{2s}^{active} = \frac{P_1(P_2 + N)}{P_1 + Q}; \tag{19.87}$$

that is,

$$\rho_{12}^{active} = \frac{P_1(P_2 + N)}{\sqrt{P_1 P_2}(P_1 + Q)}, \rho_{2s}^{active} = -\frac{P_1(P_2 + N)}{\sqrt{Q P_2}(P_1 + Q)}. \tag{19.88}$$

The resulting α_{opt} (see (19.74)) when using the correlations (19.87) is given by

$$\alpha_{opt}^{active} = \frac{P_2}{P_2 + N},\tag{19.89}$$

which is equal to the optimal α in Costa's setup [47]. As mentioned above, the choice of correlations (19.87) results in a surprising phenomenon which only occurs in the active regime. The highest achievable common message rate is $\frac{1}{2}\log\left(1 + \frac{P_1}{Q}\right) + \frac{1}{2}\log\left(1 + \frac{P_2}{N}\right)$, the same as that of a decoder that observes both $Y_1 = X_1 + S$ and $Y_2 = X_2 + V$ rather than $Y = X_1 + X_2 + S + V$. Consequently, even if the decoder is constrained to see only the sum of the channel outputs, the upper bound of the Gaussian degenerate parallel channel with an asymmetric noncausal CSI (see Theorem 19.17) can actually be achieved.

The random scheme is as described after Theorem 19.14 with Gaussian P_{S,X_1,X_2} and the covariance matrix

$$\begin{pmatrix} Q & 0 & \sigma_{2s}^{active} \\ 0 & P_1 & \sigma_{12}^{active} \\ \sigma_{2s}^{active} & \sigma_{12}^{active} & P_2 \end{pmatrix},$$

where $\sigma_{12}^{active}, \sigma_{2s}^{active}$ are defined in (19.87), and

$$U = X_2 + \alpha_{opt}^{active} S \tag{19.90}$$

(see (19.89)).

Numerical Results

In Figure 19.20 the common message capacity is plotted as a function of Q for fixed values of P_1, P_2, N which were chosen to form two groups (the first group consists of $(P_1 = 2, P_2 = N = 1)$, $(P_1 = 4, P_2 = N = 2)$, and $(P_1 = 6, P_2 = N = 3)$, and the second has $(P_1 = 5, P_2 = 2, N = 1)$, $(P_1 = 10, P_2 = 4, N = 2)$, and $(P_1 = 20, P_2 = 8, N = 4)$). The common message capacity values for both $Q = 0$ and for $Q \to \infty$ are equal for all the members of each of these groups. The diamonds are used to indicate the transition points between the silent regime and the active regime $Q = -\frac{P_1}{2} + \frac{\sqrt{P_1(P_1 P_2 + 4(N + P_2)^2)}}{2\sqrt{P_2}}$.

19.3.2   Cognitive State-Dependent Interference Channel

The GGP channel model can be viewed as a special case of a cognitive state-dependent IC, as depicted in Figure 19.21, which was discussed in [48].

The definition of the channel model is similar to (19.47), with the modification that in the IC case, the channel has two outputs Y_1 and Y_2, and each of the channel output sequences is observed by a distinct receiver.

The IC is said to be *asymmetric cognitive* if only Encoder 2 knows the channel state sequence, and if additionally the encoders for the channel are defined by the mappings

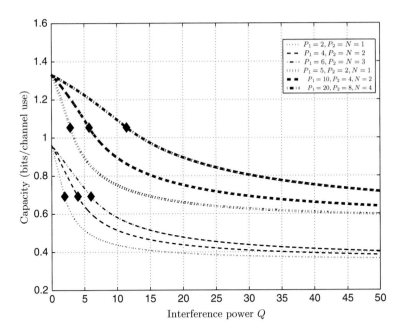

Figure 19.20 Common message capacity as a function of the interference power Q.

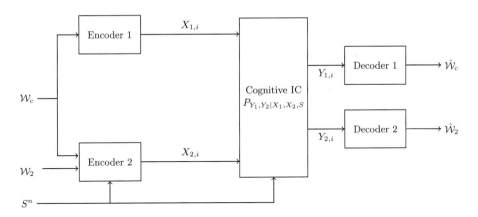

Figure 19.21 Asymmetric state-dependent interference channel with a common message.

$$\varphi_{1,n} : \{1, \ldots, M_1\} \to \mathcal{X}_1^n$$
$$\varphi_{2,n} : \{1, \ldots, M_1\} \times \{1, \ldots, M_2\} \times \mathcal{S}^n \to \mathcal{X}_2^n. \tag{19.91}$$

A decoder for the asymmetric cognitive IC is defined by the mappings

$$\psi_{1,n} : \mathcal{Y}_1^n \to \{1, \ldots, M_1\}$$
$$\psi_{1,n} : \mathcal{Y}_2^n \to \{1, \ldots, M_2\}. \tag{19.92}$$

In other words, Decoder i should decode messages i, for $i \in \{1, 2\}$.

An achievable region for the finite alphabet cognitive asymmetric noncausal CSI interference channel is given in the following theorem [48].

THEOREM 19.23 *(Inner bound) The closure of the convex hull of the set of rate-pairs that satisfy*

$$R_1 \leq I(U, X_1; Y_1) - I(U, X_1; S)$$
$$R_2 \leq I(V; Y_2) - I(V; U, X_1, S),$$
(19.93)

for some joint law

$$P_{S, X_1, U, V, X_2, Y_1, Y_2} = P_S P_{X_1} P_{U, V, X_2 | S, X_1} P_{Y_1, Y_2 | S, X_1, X_2},$$
(19.94)

is achievable for the finite alphabet cognitive asymmetric noncausal CSI IC.

The random-coding scheme relies on the generalized binning principle in [49].

Generation of Codebooks

Let $M_1 = M_{1,a} \times M_{1,b}$. Encoder 1's codebook is denoted by $\{x_1, \ldots, x_{M_{1,a}}\}$, where the codewords are selected randomly and independently with i.i.d. entries drawn from distribution P_{X_1}. For each codeword, x_ℓ, $\ell = 1, \ldots, M_{1,a}$, another codebook is drawn (which is referred to as a U-codebook) consisting of $M_{1,b} \times J_1$ auxiliary vectors, denoted $\{u_{\ell,k,j}\}$ $j \in \{1, \ldots, J_1\}, k \in \{1, \ldots, M_{1,b}\}$. This drawing is done independently and with i.i.d. entries given x_ℓ, with the ith symbol distributed according to $P_{U|X_1}(\cdot|x_i)$ given x_i where x_i is the ith symbol of x_ℓ. Each codeword in Encoder 1's codebook is therefore associated with a U-codebook of auxiliary codewords.

Further, a codebook (referred to as the V-codebook) which consists of $M_2 \times J_2$ i.i.d. independent codewords $\{v_{m,\tilde{j}}\}, m \in \{1, \ldots, M_2\}, \tilde{j} \in \{1, \ldots, J_2\}$ is drawn where each symbol is distributed according to P_V.

Encoding

Let $m_1 = (\ell, k)$ and m_2 denote the message indices to be transmitted. Encoder 1 transmits x_ℓ. Encoder 2 searches the kth bin of the ℓth U-codebook for a vector $u = u_{\ell,k,j_0}$ that is jointly typical with (x_ℓ, s). Subsequently, it searches the V-codebook for a vector $v = v_{m,\tilde{j}_0}$ that is jointly typical with (x_ℓ, u, s). Encoder 2 transmits a vector that is i.i.d. conditioned on (x_ℓ, u, s, v) with conditional marginal distribution $P_{X_2|X_1, S, U, V}$.

In the event that such j_0 or $\tilde{j}_0$ are not found, or if the observed state sequence s is not typical, an error is declared.

Decoding

Upon observing y_1, decoder 1 searches for indices (ℓ', k', j') such that $x_{\ell'}, u_{\ell', k', j'}$ are jointly typical with $y_{(i)}$ and outputs $\hat{m}_1 = (\ell', k')$. Decoder 2, which observes y_2, searches for a pair of indices $(m', \tilde{j}')$ such that $v_{m', \tilde{j}'}$ is typical with y_2 and outputs $\hat{m}_2 = m'$.

If either of the decoders fail to find appropriate codewords, or if the typical codewords are not unique, an error is declared.

The achievable region described in Theorem 19.23 with no states reduces to the achievable region presented in [50], which is given by the closure of the convex hull of the union of rate-pairs that satisfy

$$R_1 \leq I(U, X_1; Y_1)$$
$$R_2 \leq I(V; Y_2) - I(V; U, X_1) \tag{19.95}$$

for some $P_{X_1,U,V,X_2,Y_1,Y_2} = P_{X_1,U,V,X_2} P_{Y_1,Y_2|X_1,X_2}$.

The following result presents a rather straightforward outer bound on the capacity region.

THEOREM 19.24   *(Outer bound) The set of achievable rate-pairs of the cognitive IC with asymmetric CSI is contained in the closure of the set of rate-pairs (R_1, R_2) that satisfy*

$$R_1 \leq I(U, X_1; Y_1) - I(U, X_1; S)$$
$$R_2 \leq I(X_2; Y_2|X_1, S) \tag{19.96}$$

for some joint law $P_{S,X_1,U,V,X_2,Y_1,Y_2}$ satisfying (19.94).

The bound on R_1 is derived similarly to [49] and the bound on R_2 follows from a genie-aided Decoder 2 which observe the other user's signal as well as the states sequence.

Another outer bound is the capacity region of the same channel, where the receivers are cognizant of the state sequence ([48]).

THEOREM 19.25   *(Outer bound) The set of achievable rate-pairs of the cognitive IC with asymmetric CSI is contained in the closure of the set of rate-pairs (R_1, R_2) that satisfy*

$$R_1 \leq I(U, X_1; Y_1|S)$$
$$R_2 \leq I(X_2; Y_2|X_1, S)$$
$$R_1 + R_2 \leq I(X_2; Y_2|U, X_1, S) + I(U, X_1; Y_1|S) \tag{19.97}$$

for some joint law $P_{S,X_1,U,V,X_2,Y_1,Y_2}$ satisfying (19.94).

Example: The Gaussian Cognitive Interference Channel with Asymmetric CSI

This section deals with the Gaussian cognitive IC. The channel outputs at Receivers 1 and 2 at time instant i are given, respectively, by

$$Y_{1,i} = X_{1,i} + aX_{2,i} + S_{1,i} + N_{1,i}$$
$$Y_{2,i} = bX_{1,i} + X_{2,i} + S_{2,i} + N_{2,i}, \tag{19.98}$$

where a and b are real numbers, $\{N_{1,i}\}_{i=1}^{\infty}$ and $\{N_{2,i}\}_{i=1}^{\infty}$ are independent i.i.d. unit variance Gaussian processes, $\{(S_{1,i}, S_{2,i})\}_{i=1}^{\infty}$ is an i.i.d. Gaussian process that is independent of $\{(N_{1,i}, N_{2,i})\}_{i=1}^{\infty}$. The RVs $S_{1,i}$ and $S_{2,i}$ may be correlated.

The state sequence $S_i \triangleq (S_{1,i}, S_{2,i}), i = 1, \ldots, n$ is assumed to be known to the second transmitter noncausally, and the transmitters are subject to the power constraints in (19.67).

First, it is assumed that the channel states are known to User 2 only.

It is easy to realize that at least when $|a| \leq 1$, if both transmitters know $S^n = (S_1^n, S_2^n)$ noncausally, they can cancel the interference S^n. Consequently the capacity region reduces to that of the channel without the interference S^n as in [50]. To understand why this holds, assume for convenience $S = S_1 = S_2$, and consider the encoding scheme used in [50] which also employs dirty paper coding with respect to S^n: Encoder 2 allocates a fraction β of its power to transmit the signal $X_{2,1}$ to help Encoder 1, and a fraction $(1-\beta)$ to transmit its own message by the signal $X_{2,2}$. The signal transmitted by Encoder 2 is given by $X_2 = X_{2,1} + X_{2,2}$. The signal $X_1 + aX_{2,1}$ is dirty paper coded against S. Then, $X_{2,2}$ is dirty paper coded against the effective interference $bX_1 + X_{2,1} + S$.

The importance of cooperation for successful interference cancellation can be seen by the fact that while interference cancellation is achievable in this cognitive regime with users observing S, it is impossible in the noncognitive Gaussian IC [51]. In what follows it is shown that when only the cognitive transmitter knows S^n, complete cancellation of the interference S^n is impossible (unless S_1^n is 0). A single-letter outer bound to the capacity of an IC of the form in (19.109) with full asymmetric CSI is given.

THEOREM 19.26 *(Outer bound) The set of achievable rate-pairs of the cognitive IC (19.109) with $|a| \leq 1$ and where $S^n = (S_1^n, S_2^n)$ is known noncausally to Transmitter 2 is contained in the closure of the set of rate-pairs (R_1, R_2) that satisfy*

$$R_1 \leq I(U, X_1; Y_1) - I(U, X_1; S)$$
$$R_2 \leq I(X_2; Y_2 | U, X_1, S) \tag{19.99}$$

for some joint law $P_{S, X_1, U, V, X_2, Y_1, Y_2}$ satisfying (19.94).

Next, the capacity region for the Gaussian case can be found by using Theorem 19.26. Denote without loss of generality

$$S_{2,i} = dS_{1,i} + \tilde{S} \tag{19.100}$$

where d is some real number and $S_{1,i}$ and $\tilde{S}_i$ are uncorrelated, and let

$$Q = E\left[S_1^2\right], \quad \tilde{Q} = E\left[\tilde{S}^2\right], \tag{19.101}$$

$$\tilde{\sigma}_{2s} = E[X_2 \tilde{S}], \quad \tilde{\rho}_{2s} = \frac{\tilde{\sigma}_{2s}}{\sqrt{P_2 \tilde{Q}}}. \tag{19.102}$$

Having defined the alternative noises, $S_1, \tilde{S}$, the alternative channel representation of (19.109) can be applied

$$Y_{1,i} = X_{1,i} + aX_{2,i} + S_{1,i} + N_{1,i} \tag{19.103}$$
$$Y_{2,i} = bX_{1,i} + X_{2,i} + dS_{1,i} + \tilde{S}_i + N_{2,i}. \tag{19.104}$$

Let

$$C(\Delta, \rho, P_1, P_2, Q, a)$$

$$= \frac{1}{2} \log \left(1 + \frac{\left(\sqrt{P_1} + \sqrt{1 - \Delta - \rho^2 a} \sqrt{P_2} \right)^2}{a^2 P_2 \Delta + (\sqrt{P_2} a \rho + \sqrt{Q})^2 + 1} \right), \qquad (19.105)$$

THEOREM 19.27 *The capacity region of the cognitive asymmetric Gaussian IC with noncausal CSI and $|a| \leq 1$ is given by the set of rate-pairs satisfying*

$$R_1 \leq \max_{\Delta \geq [\gamma, 1], \rho \in [-\sqrt{1-\Delta}, 0]} \left[C(\Delta, \rho, P_1, P_2, Q, a) \right.$$

$$\left. + \frac{1}{2} \log \left(1 + \Delta a^2 P_2 \right) \right] - \frac{1}{2} \log \left(1 + \gamma a^2 P_2 \right)$$

$$R_2 \leq \frac{1}{2} \log \left(1 + \gamma P_2 \right), \qquad (19.106)$$

for some $\gamma \in [0, 1]$.

The rate region of Theorem 19.27 is also achievable when $|a| > 1$, in which case it is not necessarily tight.

The proof of the converse part of Theorem 19.27 relies on the entropy power inequality, also used in [50]. To prove the direct part of Theorem 19.27 one should choose $P_{X_1, X_2, U, V|S_1, \tilde{S}} = P_{X_1} P_{X_2, U, V|S_1, \tilde{S}, X_1}$ which is Gaussian, with

$$E[X_1 X_2] = \sigma_{12} = \frac{\rho_{12}}{\sqrt{P_1 P_2}}, \quad E[X_2 S_1] = \sigma_{2s} = \frac{\rho_{2s}}{\sqrt{P_2 Q}}$$

$$E[X_2 \tilde{S}] = \tilde{\sigma}_{2,s} = 0, \quad U = aX_2 + \alpha S_1 + Z, \qquad (19.107)$$

where Z is a zero-mean Gaussian RV with variance $P_z > 0$, independent of $(X_1, X_2, S_1, \tilde{S}, U)$, and α is given by (19.74), i.e., it is identical to the optimal value of α for the MAC [49] because α only affects the term $I(U, X_1; Y_1) - I(U, X_1; S)$ (for $U = X_2 + \alpha S$) in (19.93), but it does not affect the term $I(V; Y_2) - I(V; U, X_1, S)$.

It is also worth noting that the capacity depends on the interference S_1 but it does not depend on the fraction of the interference S_2 which is uncorrelated with S_1, i.e., $\tilde{S}$ (see (19.100)). Hence, interference cancellation of the noise $\tilde{S}$ is achieved.

Figure 19.22 plots the capacity region for $P_1 = \frac{1}{2}, P_2 = 1$ and several values of Q. The achievable rates decrease as a function of Q. In the high-interference regime, $Q \to \infty$, only the cognitive encoder functions, making it an interference-inflicted broadcast setting, for which the interference can be absolutely removed [52]. This result also persists for strong interferences $1 < |a| < \infty$.

19.3.3 Cognitive Interference Channels with a Common Message

The channel in Figure 19.23 was investigated in [53]. It differs from the model in [48] in that the common message needs to be decoded by both decoders, and it is state-free.

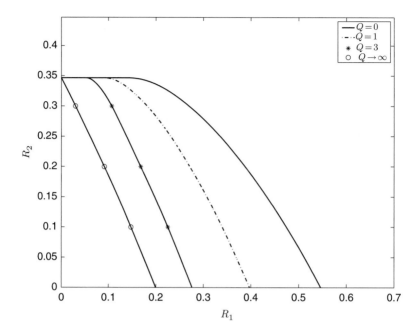

Figure 19.22 Capacity region of the asymmetric cooperative state-dependent Gaussian interference channel for $P_1 = \frac{1}{2}, P_2 = 1$.

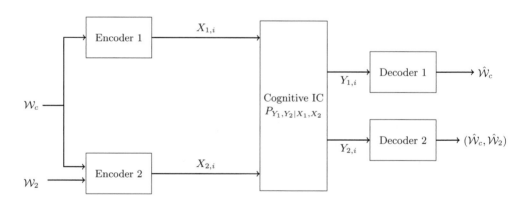

Figure 19.23 Cognitive interference channel.

THEOREM 19.28   *The capacity region of the cognitive IC is given by the union of the convex hull of nonnegative rate-pairs (R_1, R_3) which satisfy*

$$R_1 \leq I(U, X_1; Y)$$
$$R_2 \leq I(X_2; Z|X_1) \tag{19.108}$$
$$R_1 + R_2 \leq \min\{I(U, X_1; Y), I(U, X_1; Z)\} + I(X_2; Z|U, X_1),$$

where the auxiliary RV U is bounded in cardinality by $|\mathcal{U}| \leq |\mathcal{X}_1| \cdot |\mathcal{X}_2| + 1$.

In this section, the Gaussian cognitive IC is considered. The channel outputs at Receivers 1 and 2 at time i are given, respectively, by

$$Y_i = X_{1,i} + aX_{2,i} + N_{1,i}$$
$$Z_i = bX_{1,i} + X_{2,i} + N_{2,i}, \tag{19.109}$$

where $\{N_{1,i}\}_{i=1}^{\infty}$ and $\{N_{2,i}\}_{i=1}^{\infty}$ are independent memoryless unit variant Gaussian processes, and a and b are real constants. Assume that the transmitters are subject to the power constraints in (19.67). Consider the cases with $|a| \geq 1$ and $|a| < 1$, separately.

THEOREM 19.29 *For the Gaussian cognitive IC if $|a| \geq 1$, then the capacity region is given by the set of nonnegative rate-pairs (R_1, R_2) which satisfy*

$$
\begin{aligned}
&R_1 \geq 0, R_2 \geq 0 \\
&R_2 \leq \tfrac{1}{2} \log\left(1 + (1 - \rho^2)P_2\right) \\
&R_1 + R_2 \leq \tfrac{1}{2} \log\left(1 + b^2 P_1 + P_2 + 2b\rho\sqrt{P_1 P_2}\right) \\
&R_1 + R_2 \leq \tfrac{1}{2} \log\left(1 + P_1 + a^2 P_2 + 2a\rho\sqrt{P_1 P_2}\right)
\end{aligned}
\tag{19.110}
$$

where the logarithmic function is to base 2.

The capacity region for the case $|a| < 1$ is given in the following.

THEOREM 19.30 *For the Gaussian cognitive IC with a confidential message, if $|a| < 1$, the capacity region is given by the union of rate-pairs which satisfy*

$$
\begin{aligned}
&R_1 \leq \min\left\{\tfrac{1}{2} \log\left(1 + \frac{P_1 + \rho^2 a^2 P_2 + 2\rho a \sqrt{\beta P_1 P_2}}{1 + (1 - \rho^2)a^2 P_2}\right), \tfrac{1}{2} \log\left(1 + \frac{b^2 P_1 + \rho^2 P_2 + 2\rho b \sqrt{\beta P_1 P_2}}{1 + (1 - \rho^2)P_2}\right)\right\}, \\
&R_2 \leq \tfrac{1}{2} \log\left(1 + (1 - \rho^2)P_2\right)
\end{aligned}
\tag{19.111}
$$

for some $\rho \in [-1, 1]$.

Note that in (19.111), if $a > 0$ and $b > 0$, it is sufficient to consider $\beta = 1$ whereas if $a < 0$ and $b < 0$, it is sufficient to consider $\beta = 0$.

Note that the regions (19.110) and (19.111) are valid for both positive and negative values of a and b.

19.3.4 Cognition and Cooperation in Interfered MACs

A discrete memoryless multiple-access Z-interference channel (MA-CZIC) is defined by the input alphabets $(\mathcal{X}_1, \mathcal{X}_2, \mathcal{X}_3)$ and output alphabets $(\mathcal{Y}, \mathcal{Z})$ and by the transition probabilities $P_{Y|X_1 X_2 X_3}$ and $P_{Z|X_3}$. The channel outputs are generated as follows:

$$\Pr\left(y^n, z^n | x_1^n, x_2^n, x_3^n\right) = \prod_{t=1}^{n} p(y_t | x_{1,t}, x_{2,t}, x_{3,t}) p(z_t | x_{3,t}). \tag{19.112}$$

An $(2^{nR_1}, 2^{nR_2}, 2^{nR_3}, n)$ code for the MA-CZIC consists of three encoder and two decoders.

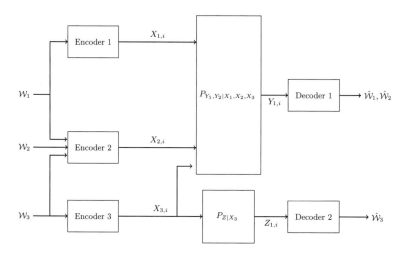

Figure 19.24 Multiple-access cognitive Z-interference channel with full unidirectional cooperation.

Encoders 1, 3 are defined by the deterministic mappings

$$f_1 : \{1, \ldots, M_1\} \to \mathcal{X}_1^n \tag{19.113}$$

$$f_3 : \{1, \ldots, M_3\} \to \mathcal{X}_3^n. \tag{19.114}$$

The structure of Encoder 2 depends on the application in that it can operate strictly causally, causally, or observe the entire message W_1 prior to the beginning of transmission (full unidirectional cooperation). The three options are given by the mappings $f_{2,k}^{(sc)}$, $f_{2,k}^{(c)}$, and f_2, respectively:

$$f_{2,k}^{(sc)} : \{1, \ldots, M_2\} \times \{1, \ldots, M_3\} \times \mathcal{X}_1^{k-1} \to \mathcal{X}_2 \qquad k = 1, \ldots, n, \tag{19.115}$$

$$f_{2,k}^{(c)} : \{1, \ldots, M_2\} \times \{1, \ldots, M_3\} \times \mathcal{X}_1^k \to \mathcal{X}_2 \qquad k = 1, \ldots, n, \tag{19.116}$$

$$f_2 : \{1, \ldots, M_1\} \times \{1, \ldots, M_2\} \times \{1, \ldots, M_3\} \to \mathcal{X}_2^n. \tag{19.117}$$

The primary and secondary decoders, denoted g_1 and g_2, respectively, are given by

$$g_1 : \mathcal{Y}^n \to \{1, \ldots, M_1\} \times \{1, \ldots, M_2\}, \tag{19.118}$$

$$g_3 : \mathcal{Z}^n \to \{1, \ldots, M_3\}. \tag{19.119}$$

The channels with a causal Encoder 2 and with full unidirectional cooperation are depicted in Figures 19.24 and 19.25, respectively.

Decoder 1 is expected to decode the messages W_1, W_2 reliably, and Decoder 2 is expected to decode W_3 reliably.

Characterizing the capacity region of the channel is a challenging task, since except for special cases, even the capacity region of the Z-interference channel is yet unknown. In [54, 55] the following inner bounds were given for the causal and strictly causal channels.

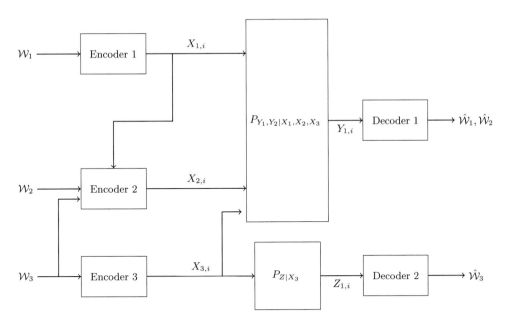

Figure 19.25 Multiple-access cognitive Z-interference channel with a causal Encoder 2.

THEOREM 19.31   *The closure of the convex hull of the set of all rate-triples* (R_1, R_2, R_3) *satisfying*

$$R_1 \le H(X_1|V) \tag{19.120a}$$

$$R_2 \le I(U; Y|VLX_1) - I(U; X_3|VL) \tag{19.120b}$$

$$R_1 + R_2 \le I(VUX_1; Y|L) - I(U; X_3|VL) \tag{19.120c}$$

$$R_3 \le I(X_3; Z|L) + \min\{I(L; Y), I(L; Z)\} \tag{19.120d}$$

for some probability distribution of the form

$$P_{VLUX_1X_2X_3} = P_V P_L P_{X_3|L} P_{X_1|V} P_{UX_2|VLX_3} \tag{19.121}$$

is achievable for the MA-CZIC with a strictly causal cribbing encoder.

   The same result holds for a causal cribbing encoder with the exception that the probability distribution takes on the form

$$P_{VLUX_1X_2X_3} = P_V P_L P_{X_3|L} P_{X_1|V} P_{U|VLX_3} P_{X_2|UVLX_3X_1}. \tag{19.122}$$

The following theorem provides a single-letter expression for an achievable region of the MA-CZIC with full unidirectional cooperation (common message).

THEOREM 19.32   *The closure of the convex hull of the set of all rate-triples* (R_1, R_2, R_3) *satisfying*

$$R_2 \leq I(U; Y|LX_1) - I(U; X_3|LX_1) \tag{19.123a}$$

$$R_1 + R_2 \leq I(X_1 U; Y|L) - I(U; X_3|LX_1) \tag{19.123b}$$

$$R_3 \leq I(X_3; Z|L) + \min\{I(L; Y), I(L; Z)\} \tag{19.123c}$$

for some probability distribution of the form

$$P_{LUX_1 X_2 X_3} = P_{X_1} P_L P_{X_3|L} P_{UX_2|X_1 LX_3} \tag{19.124}$$

is achievable for the MA-CZIC with full unidirectional cooperation.

Following are a few comments regarding the achievability region (19.120a)–(19.120d). In the coding scheme, both Encoder 1 and Encoder 2 use Block–Markov superposition encoding, whereas the primary decoder uses backward decoding [56]. In this scheme, the RV V represents the "resolution information" [26]; i.e., the current block information used for encoding the preceding block. Encoder 3 employs rate-splitting, where the RV L represents the part of W_3 that can be decoded by both the primary and secondary decoders. This can be observed by the term $\min\{I(L; Y), I(L; Z)\}$, which appears in (19.120d). The complementary part of W_3, while fully decoded by the secondary decoder, serves as a channel state for the primary channel in the form of X_3. The cognitive encoder (Encoder 2) also uses Gel'fand–Pinsker binning[44] of U against X_3 to reduce interference, assuming an already successful decoding of V and L at the primary decoder, as can be seen in (19.120b).

Note also that the achievable region (19.120a)–(19.120d) is consistent with previously studied special cases:

- The region achievable for the cognitive Z-interference channel, studied in [57], with User 2 and User 3 as the cognitive and noncognitive users, respectively, can be obtained by setting $R_1 = 0$ and $X_1 = 0$; one can also set $V = \emptyset$. The equations then reduce to

$$R_2 \leq I(U; Y|L) - I(U; X_3|L) \tag{19.125}$$

$$R_3 \leq I(X_3; Z|L) + \min\{I(L; Y), I(L; Z)\}. \tag{19.126}$$

- Removing Encoder 2 by setting $R_2 = 0$ and $X_2 = 0$, one can also set $U = \emptyset$ and one gets the classical Z-interference channel with two users, Encoder 1 and Encoder 3. In this case Y is dependent on V only through X_1, since $V \rightarrow X_1 \rightarrow Y$, and $I(VUX_1; Y|L) = I(X_1; Y|L)$. This yields

$$R_1 \leq I(X_1; Y|L) \tag{19.127}$$

$$R_3 \leq I(X_3; Z|L) + \min\{I(L; Y), I(L; Z)\}. \tag{19.128}$$

- An interesting setup occurs when setting $W_2 = 0$ without removing Encoder 2. This models a relay channel, where Encoder 2 is a relay which has no message of its own and learns the information of the transmitter by cribbing (modeling excellent signal to noise ratio (SNR) conditions on this path). This model relates to [58] if the structure of the primary user (X_3) is replaced by i.i.d. state symbols known noncausally at the relay, as in [58].

- Removing Encoder 3 by setting $R_3 = 0$ and $X_3 = 0$, one can also set $L = \emptyset$, and the expression reduces to

$$R_1 \leq H(X_1|V) \tag{19.129}$$

$$R_2 \leq I(U; Y|VX_1) \tag{19.130}$$

$$R_1 + R_2 \leq I(VUX_1; Y). \tag{19.131}$$

By setting $U = X_2$ one gets the achievable region of the MAC with Encoder 2 as the cribbing encoder [26].

- By setting $L = 0$, removing inequality (19.120d), and replacing X_3 by S, whose given probability distribution is not to be optimized, the region reduces to the one in [56].
- It is worth noting that in the case of a Gaussian channel, Encoder 2 can become completely cognizant of the message W_1 from a single sample of X_1. This special case can be made nontrivial by adding a noisy channel or some deterministic function (a quantizer, for instance) between X_1 and Encoder 2.
- Finally, observe the case where Encoder 1's output may be viewed as two parts $X_1 = (X_{1a}, X_{1b})$ where only the first part of the input affects the channel; i.e., $P_{Y|X_1X_2X_3} = P_{Y|X_{1a}X_2X_3}$. In this case, if the second part X_{1b} is rich enough (e.g., continuous alphabet) Encoder 1 will be able to transfer an infinite amount of data to Encoder 2, to the extent of conveying the entire message W_1. This is equivalent to the case of full cooperation from Encoder 1 to Encoder 2; i.e., the case where Encoder 2 has full knowledge of Encoder 1's data W_1. Hence, the cooperative state-dependent MAC where the state is known noncausally at the cognitive encoder [49] may also be considered a special case of the MA-CZIC, when X_3 is replaced with an i.i.d. state S.

The outer bounds are as follows.

THEOREM 19.33 *Achievable rate-triples (R_1, R_2, R_3) for the MA-CZIC with a strictly causal cribbing encoder belonging to the closure of the convex hull of all rate-triples that satisfy*

$$R_1 \leq H(X_1|V) \tag{19.132a}$$

$$R_2 \leq I(U; Y|VLX_1) - I(U; Z|VL) \tag{19.132b}$$

$$R_1 + R_2 \leq I(VUX_1; Y|L) - I(VU; Z|L) \tag{19.132c}$$

$$R_3 \leq I(X_3; Z|L) + \min\{I(L; Y), I(L; Z)\} \tag{19.132d}$$

for some probability distribution of the form

$$P_{VLUX_1X_2X_3} = P_{X_3} P_V P_{L|X_3V} P_{X_1|V} P_{UX_2|VLX_3}. \tag{19.133}$$

The same result holds for a causal cribbing encoder with the exception that the probability distribution takes the form

$$P_{VLUX_1X_2X_3} = P_{X_3} P_V P_{L|X_3V} P_{X_1|V} P_{UX_2|VLX_3X_1}. \tag{19.134}$$

The alphabets' cardinalities of the auxiliary RVs L, V, U can be bounded in a standard manner in terms of the alphabet sizes of X_1, X_2, X_3.

The following theorem provides a single-letter expression for an outer bound on the capacity region of the MA-CZIC with full unidirectional cooperation.

THEOREM 19.34   *Achievable rate-triples (R_1, R_2, R_3) for the MA-CZIC with full unidirectional cooperation belong to the closure of the convex hull of rate-regions given by*

$$R_1 + R_2 \leq I(VUX_1; Y|L) - I(VU; Z|L) \tag{19.135a}$$
$$R_2 \leq I(U; Y|LVX_1) - I(U; Z|LV) \tag{19.135b}$$
$$R_3 \leq I(X_3; Z|L) + \min\{I(L; Y), I(L; Z)\} \tag{19.135c}$$

for some probability distribution of the form

$$P_{LVUX_1X_2X_3} = P_V P_{X_3} P_{L|VX_3} 1_{\{X_1 = f(V)\}} P_{UX_2|VLX_3}. \tag{19.136}$$

The inequalities (19.135a)–(19.135c) are identical to (19.132b)–(19.132d), where the probability distribution form in (19.136) is a special case of (19.133). Thus, the outer bound established for R_2, R_3, and the sum-rate $R_1 + R_2$ in the *strictly causal* MA-CZIC, also holds for the case of full unidirectional cooperation. However, one would expect the outer bounds on R_2 and the sum-rate $R_1 + R_2$ to be smaller for the channel with the cribbing encoder, thus implying that the outer bound for the MA-CZIC is generally not tight.

Special Cases

Consider the special case of a *more-capable* MA-CZIC channel; i.e., a channel that satisfies the inequality $I(X_3; Y) \geq I(X_3; Z)$ for all probability distributions of the form $P_V P_{X_1|V} P_{X_3} P_{X_2|VX_3} P_{Y|X_1X_2X_3} P_{Z|X_3}$. For this class of channels the capacity region has the following closed-form expression:

THEOREM 19.35   *The capacity region of the more-capable strictly causal MA-CZIC channel is the closure of the convex hull of the set of all rate-triples (R_1, R_2, R_3) satisfying*

$$R_1 \leq H(X_1|V) \tag{19.137a}$$
$$R_2 \leq I(X_2; Y|VX_1X_3) \tag{19.137b}$$
$$R_1 + R_2 \leq I(X_1X_2; Y|X_3) \tag{19.137c}$$
$$R_3 \leq I(X_3; Z) \tag{19.137d}$$

for some probability distribution of the form

$$P_{VX_1X_2X_3} = P_V P_{X_1|V} P_{X_3} P_{X_2|VX_3}. \tag{19.138}$$

In the causal case the region (19.137a)–(19.137d) holds with a probability distribution of the form

$$P_{VX_1X_2X_3} = P_V P_{X_1|V} P_{X_3} P_{X_2|VX_1X_3}. \tag{19.139}$$

Unfortunately, the requirement that the MA-CZIC is *more-capable* implies that the receiver Y has better reception of the signal X_3 than its designated receiver Z, which is somewhat optimistic.

Analysis of maximal rates achieved for either one of the transmitters can be found in [55]. In [55] the case of partial cribbing is considered, where Encoder 2 views X_1 through a deterministic function,

$$h : \mathcal{X}_1 \to \mathcal{Y}_2, \tag{19.140}$$

instead of obtaining X_1 directly. This cribbing scheme is motivated by continuous input alphabet MA-CZIC, since perfect cribbing results in the degenerated case of full cooperation between the encoders and requires an infinite capacity link. Inner and outer bounds on the capacity region are provided in [55].

19.4   Cooperation via Distributed Interference Cognition

In this section, we present a new framework for interference management in wireless networks, arising recently based on the following key perspective of the interference. In fact, interference signals in nature contain coded information sent to certain intended nodes, and hence such signals as codeword sequences are typically *noncausally* known by various nodes in the network. For example, an interferer clearly knows the interference signal that it causes to other users noncausally, because such interference is the codeword that this interfererer transmits to its intended receivers. As another example, if the interferer is a base station, it can easily inform other base stations about its interference via the backhaul network or inform access points in the cell via wired links. The major observation here is that *interference cognition* naturally exists or can be established at very low costs in networks. Thus, nodes that possess noncausal interference information should be able to exploit it to assist cancellation of such interference.

One major advantage of such an idea is that the design of interference cancellation is handled mainly at the interferer or the helper side, which are typically powerful nodes (such as base stations and access points) in networks and can hence easily take the extra load of interference cancellation. Since the design is on one side, it can be made more efficient and does not require sharing codebooks as in successive interference cancellation in the Han–Kobayashi scheme. Moreover, the design can be made transparent to nodes being interfered with. This is very useful in cognitive networks (e.g., [59–61]) and Internet of Things (IoT) networks [17–19]. With the interference to primary nodes being canceled by the interferer itself or helper nodes, the access of primary channels can be made simultaneous and transparent from primary networks.

Differently from cognitive or relay channels, e.g., [20–22, 62], where nodes have cognition of transmitted information, here a helper has cognition of only the interference. Thus, the helper's signal mainly assists in interference cancellation at the receiver rather than cooperatively encoding transmitted information in the typical user cooperation and relay schemes. In particular, when there are multiple helpers, the cooperation among helpers is based on their common knowledge about the state/interference, whereas in typical relay networks the cooperation among relays is based on their common knowledge of transmitted messages. We next provide an overview of the type of models that capture the cooperation via interference cognition, and recent information theoretic characterization of the performance limits for such types of models.

19.4.1   Helper-Assisted Point-to-Point State-Dependent Channel

The basic model that captures user cooperation via interference cognition is the following helper-assisted state-dependent channel, which was initially proposed in [14] and was further studied in [15]. To define the model more formally, the transmitter designs an encoder $f : \mathcal{W} \to \mathcal{X}^n$ to map a message $w \in \mathcal{W}$ to a codeword $x^n \in \mathcal{X}^n$, which is transmitted over the channel, and is corrupted by an independent and identically distributed (i.i.d.) state sequence S^n. The state sequence is assumed to be known at a helper noncausally, but at neither the transmitter nor the receiver. Then the helper designs an encoder $f_0 : \mathcal{S}^n \to \mathcal{X}_0^n$ to map a state sequence $s^n \in \mathcal{S}^n$ to a codeword $x_0^n \in \mathcal{X}_0^n$ and sends it over the channel. The channel is characterized by the transition probability distribution $P_{Y|X_0,X,S}$. The receiver designs a decoder $g : \mathcal{Y}^n \to \mathcal{W}$ to map a received sequence y^n into a message $\hat{w} \in \mathcal{W}$.

We assume that the message is uniformly distributed over the set $\mathcal{W}$, and then define the average probability of error for a length-n code as follows:

$$P_e = \frac{1}{|\mathcal{W}|} \sum_{w=1}^{|\mathcal{W}|} Pr\{\hat{w} \neq w\}. \tag{19.141}$$

We say a rate R is *achievable* if there exist a sequence of message sets $\mathcal{W}^{(n)}$ with $|\mathcal{W}^{(n)}| = 2^{nR}$ and encoder–decoder tuples $(f_0^{(n)}, f^{(n)}, g^{(n)})$, such that the average probability of error $P_e^{(n)} \to 0$ as $n \to \infty$. We wish to find the *capacity* of the above channel, which is defined as the largest achievable rate via a certain transmission scheme by the transmitter and the helper.

In this model, the helper assists the state cancellation by transmitting certain help signals via its cross-link channel to the receiver. The major feature here is that the state is known at a different node helper from the sender of the message, so that encoding of the message and precoding of the state cannot be designed jointly. This is in contrast to the classical state-dependent channel, e.g., [44, 47], where the transmitter itself also knows the state so that it can employ random binning/dirty paper coding to encode the message and state jointly to cancel the state interference. Thus, such a helper-assisted state-dependent channel captures a new interference management scenario, in which an interferer knows (noncausally) the interference that it causes to other users and can/should exploit such information for canceling the interference. Such an idea can be illustrated by a device-to-device (D2D) assisted cellular network (e.g., [63–66]). Here, the cellular base station knows the interference that it causes to D2D users noncausally, because such interference is the signal that the base station sends to a cellular user. Hence, the helper-assisted channel captures such interference as the noncausal state sequence, and the base station serves as a helper to exploit such information about the interference (i.e., state) and send a helper signal to assist D2D users to cancel the interference. Although simply reversing the state can cancel the state interference directly, such a scheme is not preferred in general because (1) reversing the state also reduces the cellular communication rate; and (2) state power can sometimes be very large so that reversing the state is not energy efficient. Therefore, it is desirable to design more intelligent schemes that may precode the state information to deal with the interference.

The helper-assisted state-dependent channel was first proposed and studied in [14], which established lower and upper bounds on the capacity. In particular, the lower bound was derived based on lattice coding. The capacity was established in the asymptotic regime as the state power goes to infinity. More recently, in [16], a lower bound on the capacity was derived based on a scheme which integrates the precoding of the state into a help signal using single-bin dirty paper coding and the direct subtraction of state. We present such a lower bound as follows.

PROPOSITION 19.36 *([16]): For the state-dependent point-to-point channel with a helper, the following rate is achievable:*

$$R \leqslant \min\{I(UX;Y) - I(U;S), I(X;Y|U)\} \tag{19.142}$$

for some distribution $P_{X_0US}P_XP_{Y|X_0XS}$.

Intuitively, we interpret the above achievable rate in (19.142) as follows. In the first term

$$I(U,X;Y) - I(U;S) = I(U,X;Y) - I(U,X;S),$$

(U,X) play the role of the auxiliary variable in the Gel'fand–Pinsker scheme. The second term

$$I(X;Y|U) = I(U,X;Y) - I(U;Y)$$

is the result of coding via (U,X), but paying the price needed to convey U to the receiver.

We next consider the Gaussian case of the channel, where the channel output equals the summation of the channel input from the transmitter X, the input X_0 from the helper, the Gaussian distributed state variable $S \sim \mathcal{N}(0, Q)$, and the independent Gaussian noise variable $N_z \sim \mathcal{N}(0, 1)$, and is given by

$$Z = X + X_0 + S + N_z. \tag{19.143}$$

The channel inputs X and X_0 are subject to the average power constraints P and P_0, respectively. Proposition 19.36 can be applied to obtain a lower bound on the capacity for the above Gaussian channel.

PROPOSITION 19.37 *([16]): For the state-dependent Gaussian channel with a helper, the following rate is achievable*

$$R \leqslant \max_{\substack{(\alpha,\beta) \ s.t. \\ -\sqrt{\frac{P_0}{Q}} \leqslant \beta \leqslant \sqrt{\frac{P_0}{Q}}}} \min\{R_1(\alpha, \beta), R_2(\alpha, \beta)\}, \tag{19.144}$$

where

$$R_1(\alpha, \beta) = \frac{1}{2} \log \frac{P_0'(P_0' + (1 + \beta)^2 Q + P + 1)}{P_0' Q(\alpha - 1 - \beta)^2 + P_0' + \alpha^2 Q}, \tag{19.145a}$$

$$R_2(\alpha, \beta) = \frac{1}{2} \log \left(1 + \frac{P(P_0' + \alpha^2 Q)}{P_0' Q(\alpha - 1 - \beta)^2 + P_0' + \alpha^2 Q} \right), \tag{19.145b}$$

and $P_0' = P_0 - \beta^2 Q$.

The above lower bound is obtained by substituting the following joint Gaussian distribution for the RVs in Proposition 19.36:

$$X_0 = X_{00} + \beta S$$
$$U = X_{00} + \alpha S, \tag{19.146}$$

where X_{00} is independent of S and $X_{00} \sim \mathcal{N}(0, P_0')$ with $P_0' = P_0 - \beta^2 Q$ and $-\sqrt{\frac{P_0}{Q}} \leq \beta \leq \sqrt{\frac{P_0}{Q}}$. It can be seen that in (19.146), the helper's input X_0 contains two parts with X_{00} designed using single-bin dirty paper coding, and βS serving for state subtraction. The parameter β captures the tradeoff between the two schemes.

In Proposition 19.37, we set $\alpha = \frac{(1+\beta)P_0'}{P_0'+1}$ that optimizes $R_1(\alpha, \beta)$ and $\beta = \rho_{0S}\sqrt{\frac{P_0}{Q}}$ (to better illustrate the result, where $-1 \leq \rho_{0S} \leq 1$), and obtain the following lower bound.

COROLLARY 19.38   *(Lower bound 1 [16]): For the state-dependent Gaussian channel with a helper, the following rate R is achievable*

$$R = \max_{-1 \leq \rho_{0S} \leq 1} \min\{R_1(\rho_{0S}), R_2(\rho_{0S})\}, \tag{19.147}$$

where

$$R_1(\rho_{0S}) = \frac{1}{2} \log \left(1 + \frac{P}{Q + 2\rho_{0S}\sqrt{P_0 Q} + P_0 + 1} \right) + \frac{1}{2} \log(1 + P_0 - \rho_{0S}^2 P_0) \tag{19.148a}$$

$$R_2(\rho_{0S}) = \frac{1}{2} \log \left(1 + \frac{P((1 + P_0(1 - \rho_{0S}^2))^2 + (1 - \rho_{0S}^2)P_0(\sqrt{Q} + \rho_{0S}\sqrt{P_0})^2)}{(Q + 2\rho_{0S}\sqrt{P_0 Q} + P_0 + 1)(1 + P_0 - \rho_{0S}^2 P_0)} \right). \tag{19.148b}$$

In Proposition 19.37, we set $\beta = \alpha - 1$ that maximizes $R_2(\alpha, \beta)$, and obtain the following lower bound.

COROLLARY 19.39   *(Lower bound 2 [16]): For the state-dependent Gaussian channel with a helper, the following rate R is achievable*

$$R = \min \left\{ \frac{1}{2} \log \frac{P_0'(P_0' + \alpha^2 Q + P + 1)}{P_0' + \alpha^2 Q}, \frac{1}{2} \log(1 + P) \right\}. \tag{19.149}$$

for some $\alpha \in \Omega_\alpha = \{\alpha : 1 - \sqrt{\frac{P_0}{Q}} \leq \alpha \leq 1 + \sqrt{\frac{P_0}{Q}}\}$.

We further present two upper bounds on the capacity.

PROPOSITION 19.40 *(Upper bound 1 [14]): The capacity of the state-dependent Gaussian channel with a helper is upper bounded as*

$$C \leqslant \max_{-1 \leqslant \rho_{0S} \leqslant 1} \frac{1}{2} \log\left(1 + \frac{P}{Q + 2\rho_{0S}\sqrt{P_0 Q} + P_0 + 1}\right) + \frac{1}{2} \log(1 + P_0 - \rho_{0S}^2 P_0).$$

$$(19.150)$$

The following upper bound follows from the capacity of the channel between the transmitter and receiver without state, which clearly serves as an upper bound on the capacity of the state-dependent channel.

PROPOSITION 19.41 *(Upper bound 2) The capacity of the state-dependent Gaussian channel with a helper is upper bounded as*

$$C \leqslant \frac{1}{2} \log(1 + P).$$

$$(19.151)$$

We observe that the lower bound 1 matches the upper bound 1 in Proposition 19.40. Hence, we characterize the following capacity result for certain channel parameters.

THEOREM 19.42 *(Capacity in regime 1 [16]) For the state-dependent Gaussian channel with a helper, suppose ρ_{0S}^* maximizes $R_1(\rho_{0S})$ in (19.148a). If the channel parameters satisfy the following condition:*

$$R_1(\rho_{0S}^*) \leqslant R_2(\rho_{0S}^*),$$

$$(19.152)$$

where $R_2(\rho_{0S})$ is given in (19.148b), then the channel capacity $C = R_1(\rho_{0S}^*)$.

We note that for channels that satisfy the condition (19.152), the capacity $R_1(\rho_{0S}^*)$ is less than the capacity of the channel without the state. Thus, in such cases, the state interference cannot be fully canceled by any scheme.

We further observe that lower bound 2 matches the upper bound 2 in Proposition 19.41. Thus, we characterize the capacity for certain channel parameters.

THEOREM 19.43 *(Capacity in regime 2 [16]) For the state-dependent Gaussian channel with a helper, if the channel parameters satisfy the following condition:*

$$P_0'^2 \geq \alpha^2 Q(P + 1 - P_0'),$$

$$(19.153)$$

where $P_0' = P_0 - (\alpha - 1)^2 Q$ holds for some $\alpha \in \Omega_\alpha = \{\alpha : 1 - \sqrt{\frac{P_0}{Q}} \leq \alpha \leq 1 + \sqrt{\frac{P_0}{Q}}\}$, then the channel capacity $C = \frac{1}{2} \log(1 + P)$.

Hence, under the condition (19.153), the state-dependent channel achieves the capacity of the channel without state, and thus the state can be fully canceled even if the state-cognitive node (i.e., the helper) does not know the message.

We next demonstrate the capacity results via numerical results.

In Figure 19.26, we choose $P = 5$, and $Q = 12$, and plot how the lower bounds 1 and 2 in Corollaries 19.38 and 19.39 and the upper bounds 1 and 2 in Propositions 19.40 and 19.41 change as the helper's power P_0 increases. It can be seen that if $P_0 \leq 2.5$, then

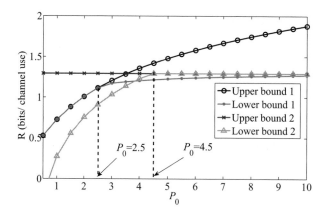

Figure 19.26 Lower and upper bounds on the capacity for the state-dependent channel with a helper.

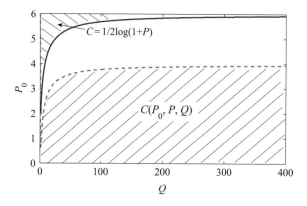

Figure 19.27 Ranges of parameters for which the capacity is characterized.

the lower bound 1 and the upper bound 1 match each other, which verifies the capacity result in Theorem 19.42. If $P_0 \geq 4.5$, the lower bound 2 and the upper bound 2 match each other, which verifies the capacity result in Theorem 19.43. We can also see that the channel capacity increases if the helper's power P_0 increases. However, for sufficiently large P_0, the state is perfectly canceled, and the capacity of the channel without state is achieved. We further note that the state can be fully canceled even when $P_0 < Q$ (e.g., $4.5 \leq P_0 \leq 10$). Clearly, the state is fully canceled not simply by state subtraction, but also by the dirty paper coding in these cases.

In Figure 19.27, we set $P = 5$, and plot the range of the channel parameters (Q, P_0) for which the capacity is characterized. Each parameter pair (Q, P_0) determines a point in the figure. The lower shaded area includes channel parameters satisfying (19.152), and hence the capacity is characterized by a function of P, P_0, and Q. The upper shaded area includes channel parameters that satisfy (19.153), i.e., P_0 is large enough compared to Q, and hence the state is fully canceled, i.e., the capacity of the channel without state is achieved.

19.4.2   Helper-Assisted State-Dependent MAC

We consider the state-dependent MAC with a helper, in which Transmitter 1 sends a message W_1, and Transmitter 2 sends a message W_2 to the receiver. Each Transmitter k for $k = 1, 2$ designs an encoder $f_k : \mathcal{W} \to \mathcal{X}_k^n$ to map a message $w_k \in \mathcal{W}_k$ to a codeword $x_k^n \in \mathcal{X}_k^n$. The two inputs x_1^n and x_2^n are transmitted over the MAC to a receiver, which is corrupted by an i.i.d. state sequence S^n. The state sequence is known to a helper noncausally, but known neither by the transmitters nor by the receiver. Hence, to assist the receiver to cancel the state interference, the helper designs an encoder $f_0 : \mathcal{S}^n \to \mathcal{X}_0^n$ to map the state sequence $s^n \in \mathcal{S}^n$ into a codeword $x_0^n \in \mathcal{X}_0^n$. The channel transition probability is given by $P_{Y|X_0 X_1 X_2 S}$. The decoder $g : \mathcal{Y}^n \to (\mathcal{W}_1, \mathcal{W}_2)$ at the receiver maps the received sequence y^n into two messages $\hat{w}_k \in \mathcal{W}_k$ for $k = 1, 2$.

We define the average probability of error for a length-n code as

$$P_e^{(n)} = \frac{1}{|\mathcal{W}_1||\mathcal{W}_2|} \sum_{w_1=1}^{|\mathcal{W}_1|} \sum_{w_2=1}^{|\mathcal{W}_2|} Pr\{(\hat{w}_1, \hat{w}_2) \neq (w_1, w_2)\}. \tag{19.154}$$

We say that a rate-pair (R_1, R_2) is *achievable* if there exists a sequence of message sets $\mathcal{W}_k^{(n)}$ with $|\mathcal{W}_k^{(n)}| = 2^{nR_k}$ for $k = 1, 2$, and encoder–decoder tuples $(f_0^{(n)}, f_1^{(n)}, f_2^{(n)}, g^{(n)})$ such that the average error probability $P_e^{(n)} \to 0$ as $n \to \infty$. We define the *capacity region* to be the closure of the set of all achievable rate pairs (R_1, R_2).

We also consider the state-dependent Gaussian channel, in which the output at the receiver equals the summation of the channel inputs X_0, X_1, and X_2, which are subject to the average power constraints P_0, P_1, and P_2, the Gaussian distributed state $S \sim \mathcal{N}(0, Q)$ and the Gaussian distributed noise variable $N \sim \mathcal{N}(0, 1)$.

$$Y = X_0 + X_1 + X_2 + S + N. \tag{19.155}$$

The analysis of the state-dependent point-to-point channel with a helper was further generalized to the state-dependent MAC model with a helper in [16]. First, the following inner bound on the capacity region was derived based on the scheme that integrates direct state cancellation and single-bin dirty paper coding.

PROPOSITION 19.44   *([16]): For the state-dependent Gaussian MAC with a helper, an inner bound on the capacity region consists of rate-pairs (R_1, R_2) satisfying:*

$$R_1 \leqslant \min\{f(\alpha, \beta, P_1), g(\alpha, \beta, P_1)\} \tag{19.156a}$$

$$R_2 \leqslant \min\{f(\alpha, \beta, P_2), g(\alpha, \beta, P_2)\} \tag{19.156b}$$

$$R_1 + R_2 \leqslant \min\{f(\alpha, \beta, P_1 + P_2), g(\alpha, \beta, P_1 + P_2)\} \tag{19.156c}$$

for some real constants α and β satisfying $-\sqrt{\frac{P_0}{Q}} \leqslant \beta \leqslant \sqrt{\frac{P_0}{Q}}$. In the above bounds,

$$f(\alpha, \beta, P) = \frac{1}{2} \log \frac{P_0'(P_0' + (1 + \beta)^2 Q + P + 1)}{P_0' Q(\alpha - 1 - \beta)^2 + P_0' + \alpha^2 Q}, \tag{19.157}$$

$$g(\alpha, \beta, P) = \frac{1}{2} \log \left(1 + \frac{P(P_0' + \alpha^2 Q)}{P_0' Q(\alpha - 1 - \beta)^2 + P_0' + \alpha^2 Q}\right), \tag{19.158}$$

where $P_0' = P_0 - \beta^2 Q$.

The upper bounds for the point-to-point channel can also be generalized to provide the following outer bound for the MAC model.

PROPOSITION 19.45 *([16]): An outer bound on the capacity region of the state-dependent Gaussian MAC with a helper consists of rate-pairs (R_1, R_2) satisfying:*

$$R_1 \leqslant \min \left\{ \frac{1}{2} \log(1 + \frac{P_1}{Q + 2\rho_{0S}\sqrt{P_0 Q} + P_0 + 1}) + \frac{1}{2} \log(1 + P_0 - \rho_{0S}^2 P_0), \right.$$
$$\left. \frac{1}{2} \log(1 + P_1) \right\} \tag{19.159a}$$

$$R_2 \leqslant \min \left\{ \frac{1}{2} \log(1 + \frac{P_2}{Q + 2\rho_{0S}\sqrt{P_0 Q} + P_0 + 1}) + \frac{1}{2} \log(1 + P_0 - \rho_{0S}^2 P_0), \right.$$
$$\left. \frac{1}{2} \log(1 + P_2) \right\} \tag{19.159b}$$

$$R_1 + R_2 \leqslant \min \left\{ \frac{1}{2} \log(1 + \frac{P_1 + P_2}{Q + 2\rho_{0S}\sqrt{P_0 Q} + P_0 + 1}) + \frac{1}{2} \log(1 + P_0 - \rho_{0S}^2 P_0), \right.$$
$$\left. \frac{1}{2} \log(1 + P_1 + P_2) \right\} \tag{19.159c}$$

for some ρ_{0S} that satisfies $-1 \leqslant \rho_{0S} \leqslant 1$.

We follow the same strategy in the point-to-point channel to compare the inner and outer bounds, i.e., we analyze the bounds (19.156a)–(19.156c) in the inner bound separately, and characterize conditions on the channel parameters (P_0, P_1, P_2, Q) under which these bounds meet the bounds (19.159a)–(19.159c) in the outer bound, respectively. In this way, we characterize separate segments on the capacity region boundary under the corresponding conditions on the channel parameters. Intersections of conditions for the individual and sum-rates then collectively characterize channel parameters under which multiple segments on the capacity boundary are obtained. Based on such an idea, we partition the channel parameters into appropriate cases, and characterize segments on the capacity region boundary for all these cases.

THEOREM 19.46 *(Segment of capacity region [16]) The channel parameters (P_0, P_1, P_2, Q) can be partitioned into the sets $\mathcal{A}_1, \mathcal{B}_1, \mathcal{C}_1$, where*

$$\mathcal{A}_1 = \{(P_0, P_1, P_2, Q) : f(\alpha_1, \beta_1, P_1) \leqslant g(\alpha_1, \beta_1, P_1)\}$$
$$\mathcal{C}_1 = \{(P_0, P_1, P_2, Q) : P_0'^2 \geqslant \alpha^2 Q(P_1 + 1 - P_0')$$
$$\text{where } P_0' = P_0 - (\alpha - 1)^2 Q, \text{ for some } \alpha \in \Omega_\alpha\}$$
$$\mathcal{B}_1 = (\mathcal{A}_1 \cup \mathcal{C}_1)^c.$$

If $(P_0, P_1, P_2, Q) \in \mathcal{A}_1$, then $R_1 = f(\alpha_1, \beta_1, P_1)$ captures one segment of the capacity region boundary, where the state cannot be fully canceled. If $(P_0, P_1, P_2, Q) \in \mathcal{C}_1$, then $R_1 = \frac{1}{2} \log(1 + P_1)$ captures one segment of the capacity region boundary, where the state is fully canceled. If $(P_0, P_1, P_2, Q) \in \mathcal{B}_1$, R_1 segment of the capacity region boundary is not characterized.

The channel parameters (P_0, P_1, P_2, Q) can alternatively be partitioned into the sets $\mathcal{A}_2, \mathcal{B}_2, \mathcal{C}_2$, *where*

$$\mathcal{A}_2 = \{(P_0, P_1, P_2, Q) : f(\alpha_2, \beta_2, P_2) \leqslant g(\alpha_2, \beta_2, P_2)\}$$

$$\mathcal{C}_2 = \{(P_0, P_1, P_2, Q) : P_0'^2 \geq \alpha^2 Q(P_2 + 1 - P_0')$$

$$\text{where } P_0' = P_0 - (\alpha - 1)^2 Q, \text{ for some } \alpha \in \Omega_\alpha\}$$

$$\mathcal{B}_2 = (\mathcal{A}_2 \cup \mathcal{C}_2)^c.$$

If $(P_0, P_1, P_2, Q) \in \mathcal{A}_2$, then $R_2 = f(\alpha_2, \beta_2, P_2)$ captures one segment of the capacity region boundary, where the state cannot be fully canceled. If $(P_0, P_1, P_2, Q) \in \mathcal{C}_2$, then $R_2 = \frac{1}{2}\log(1 + P_2)$ captures one segment of the capacity region boundary, where the state is fully canceled.

Furthermore, the channel parameters (P_0, P_1, P_2, Q) can also be partitioned into the sets $\mathcal{A}_3, \mathcal{B}_3, \mathcal{C}_3$, *where*

$$\mathcal{A}_3 = \{(P_0, P_1, P_2, Q) :$$
$$f(\alpha_3, \beta_3, P_1 + P_2) \leqslant g(\alpha_3, \beta_3, P_1 + P_2)\}$$

$$\mathcal{C}_3 = \{(P_0, P_1, P_2, Q) : P_0'^2 \geq \alpha^2 Q(P_1 + P_2 + 1 - P_0')$$

$$\text{where } P_0' = P_0 - (\alpha - 1)^2 Q, \text{ for some } \alpha \in \Omega_\alpha\}$$

$$\mathcal{B}_3 = (\mathcal{A}_3 \cup \mathcal{C}_3)^c.$$

If $(P_0, P_1, P_2, Q) \in \mathcal{A}_3$, then $R_1 + R_2 = f(\alpha_3, \beta_3, P_1 + P_2)$ captures one segment of the sum-capacity, where the state cannot be fully canceled. If $(P_0, P_1, P_2, Q) \in \mathcal{C}_3$, then $R_1 + R_2 = \frac{1}{2}\log(1 + P_1 + P_2)$ captures one segment of the sum-capacity, where the state is fully canceled.

The above theorem partitions the channel parameters in three ways to characterize the segments on the capacity region with respect to the optimal rates R_1, R_2, and $R_1 + R_2$, respectively. Clearly, intersection of three sets (with each from one partition) characterizes all segments on the capacity region boundary. For example, if a given channel parameter tuple satisfies $(P_0, P_1, P_2, Q) \in (\mathcal{C}_1 \cap \mathcal{C}_2 \cap \mathcal{A}_3)$, then following Theorem 19.46, line segments characterized by $R_1 = \frac{1}{2}\log(1 + P_1)$, $R_2 = \frac{1}{2}\log(1 + P_2)$, and $R_1 + R_2 = f(\alpha_3, \beta_3, P_1 + P_2)$ are on the capacity region boundary. Since parameters α and β that achieve these segments are not the same, the intersection of these segments is not on the capacity region boundary.

It is clear that all possible intersections of sets to which the channel parameters can belong have at most $3^3 = 27$ cases. However, since $(P_0, P_1, P_2, Q) \in \mathcal{C}_3$ must imply that they also belong to $\mathcal{C}_1$ and $\mathcal{C}_2$, the total number of cases turns out to be $3^2 \times 2 + 1 = 19$. For one of these cases, we characterize the full capacity region, in which the state is fully canceled, and hence, the capacity region of the Gaussian MAC without state is achieved.

THEOREM 19.47   *(Full capacity in one case [16]) If (P_0, P_1, P_2, Q) satisfies*

$$P_0'^2 \geq \alpha^2 Q(P_1 + P_2 + 1 - P_0'), \tag{19.160}$$

where $P_0' = P_0 - (\alpha - 1)^2 Q$ for some $\alpha \in \Omega_\alpha$, then the capacity region of the state-dependent Gaussian MAC contains (R_1, R_2) satisfying

$$R_1 \leqslant \frac{1}{2} \log(1 + P_1)$$

$$R_2 \leqslant \frac{1}{2} \log(1 + P_2)$$

$$R_1 + R_2 \leqslant \frac{1}{2} \log(1 + P_1 + P_2),$$

which achieves the capacity region of the Gaussian MAC without state.

Two special sets of channel parameters in the above case are interesting. First, if $P_0 \geq Q$, then $\alpha = 0 \in \Omega_\alpha$ and the condition clearly holds. Here, the helper has sufficient power to directly cancel the state. Second, if $P_1 + P_2 + 1 \leq P_0 < Q$, then the condition holds for $\alpha = 1 \in \Omega_\alpha$ for arbitrarily large Q. This implies that if the helper's power is above a certain threshold, then the state can always be canceled for arbitrary state power Q (even for infinite Q).

19.4.3   Helper-Assisted State-Dependent Parallel Channel

In this section, we consider the state-dependent parallel network with a common state-cognitive helper. Here, K transmitters wish to send K messages to their corresponding receivers over state-corrupted parallel channels, and a helper wishes to help these receivers to cancel state interference by knowing the state information noncausally.

More specifically, each transmitter (say transmitter k) designs an encoder $f_k : \mathcal{W}_k \to \mathcal{X}_k^n$ to map a message $w_k \in \mathcal{W}_k$ to a codeword $x_k^n \in \mathcal{X}_k^n$ for $k = 1, \ldots, K$. The K inputs $x_1^n, \ldots, x_K^n$ are transmitted over K parallel channels, respectively. Each receiver (say Receiver k) is interfered by an i.i.d. state sequence S_k^n for $k = 1, \ldots, K$, which is known at a common helper, but at none of the transmitters $1, \ldots, K$ and receivers $1, \ldots, K$. Thus, the helper designs an encoder $f_0 : \{S_1^n, \ldots, S_K^n\} \to \mathcal{X}_0^n$ to map the state sequences $(s_1^n, \ldots, s_K^n) \in \mathcal{S}_1^n \times \ldots \times \mathcal{S}_K^n$ to a codeword $x_0^n \in \mathcal{X}_0^n$. The entire channel transition probability is given by $P_{Y_0|X_0} \prod_{k=1}^K P_{Y_k|X_0,X_k,S_k}$. Each receiver k designs a decoder $g_k : \mathcal{Y}_k^n \to \mathcal{W}_k$ to map a received sequence y_k^n into a message $\hat{w}_k \in \mathcal{W}_k$ for $k = 1, \ldots, K$.

We define the average probability of error for a length-n code as

$$P_e^{(n)} = \frac{1}{|\mathcal{W}_1| \ldots |\mathcal{W}_K|} \sum_{w_1=1}^{|\mathcal{W}_1|} \cdots \sum_{w_k=1}^{|\mathcal{W}_k|} Pr\{(\hat{w}_1, \ldots, \hat{w}_K) \neq (w_1, \ldots, w_K)\}. \tag{19.161}$$

We say a rate tuple $(R_1, \ldots, R_K)$ is *achievable* if there exists a sequence of message sets $\mathcal{W}_k^{(n)}$ with $|\mathcal{W}_k^{(n)}| = 2^{nR_k}$ for $k = 0, 1, \ldots, K$, and encoder–decoder tuples $(f_0^{(n)}, f_1^{(n)}, \ldots, f_K^{(n)}, g_1^{(n)}, \ldots, g_K^{(n)})$ such that the average error probability $P_e^{(n)} \to 0$ as $n \to \infty$. We define the *capacity region* to be the closure of the set consisting of all achievable rate tuples $(R_1, \ldots, R_K)$.

We consider the following three types of Gaussian channel models.

Gaussian Model I

In this model, the channel outputs at the two receivers are corrupted by the same but differently scaled states. The outputs at the two receivers for one channel use are given by

$$Y_1 = X_0 + X_1 + S + Z_1 \tag{19.162a}$$

$$Y_2 = bX_0 + X_2 + aS + Z_2, \tag{19.162b}$$

where Z_1 and Z_2 are independent Gaussian noise variables $Z_1 \sim \mathcal{N}(0,1)$ and $Z_2 \sim \mathcal{N}(0,1)$, and S is the Gaussian state variable $S \sim \mathcal{N}(0,Q)$. We assume that the channel inputs X_1 and X_2 are subject to the average power constraints P_1 and P_2, respectively. In the channel model, the constant a represents the channel gain of the state sequence in the second subchannel with respect to the first subchannel, and the constant b is the gain of the helper signal in the second subchannel with respect to the first subchannel. Here, the helper's power and the state power can be arbitrary.

This model was recently studied in [67], and we next introduce the capacity characterization from [67]. First, an achievable region can be derived based on an achievable scheme similar to that for the point-to-point case, which integrates direct state cancellation and single-bin dirty paper coding.

PROPOSITION 19.48   *(Inner bound [67]): For Gaussian model I, an inner bound on the capacity region consists of rate-pairs (R_1, R_2) satisfying:*

$$R_1 \leq \min \left\{ f_{1,1}(\alpha, \beta, P_1), g_{1,1}(\alpha, \beta, P_1) \right\} \tag{19.163a}$$

$$R_2 \leq \min \left\{ f_{a,b}(\alpha, \beta, P_2), g_{a,b}(\alpha, \beta, P_2) \right\} \tag{19.163b}$$

where, α and β are real constants satisfying $-\sqrt{\frac{P_0}{Q}} \leq \beta \leq \sqrt{\frac{P_0}{Q}}$, and

$$f_{a,b}(\alpha, \beta, P) = \frac{1}{2} \log \frac{P_0' \left(b^2 P_0' + (a + b\beta)^2 Q + P + 1 \right)}{P_0' Q (b\alpha - a - b\beta)^2 + P_0' + \alpha^2 Q}, \tag{19.164a}$$

$$g_{a,b}(\alpha, \beta, P) = \frac{1}{2} \log \left(1 + \frac{P \left(P_0' + \alpha^2 Q \right)}{P_0' Q (b\alpha - a - b\beta)^2 + P_0' + \alpha^2 Q} \right), \tag{19.164b}$$

where $P_0' = P_0 - \beta^2 Q$.

The scheme involves successive cancellation of the channel state, i.e., each receiver first decodes the helper's signal, uses it to cancel the state, and then decodes the desired signal, which yields the first term in each rate of the above bound. More specifically,

We next present an outer bound which applies the point-to-point channel capacity and the upper bound derived for the point-to-point channel with a helper in [14].

PROPOSITION 19.49   *(Upper bound [14]): An outer bound on the capacity region of the state-dependent parallel Gaussian channel with a helper consists of rate-pairs (R_1, R_2) satisfying:*

$$R_1 \leq \min\left\{\frac{1}{2}\log\left(1 + \frac{P_1}{P_0 + 2\rho_{0S}\sqrt{P_0 Q} + Q + 1}\right)\right.$$
$$\left. + \frac{1}{2}\log\left((1 - \rho_{0S}^2)P_0 + 1\right), \frac{1}{2}\log(1 + P_1)\right\} \tag{19.165a}$$

$$R_2 \leq \min\left\{\frac{1}{2}\log\left(1 + \frac{P_2}{b^2 P_0 + 2ab\rho_{0S}\sqrt{P_0 Q} + a^2 Q + 1}\right)\right.$$
$$\left. + \frac{1}{2}\log\left((1 - \rho_{0S}^2)b^2 P_0 + 1\right), \frac{1}{2}\log(1 + P_2)\right\} \tag{19.165b}$$

for some ρ_{0S} that satisfies $-1 \leq \rho_{0S} \leq 1$.

We then optimize α and β in Proposition 19.48, and compare the rate bounds with the outer bounds in Proposition 19.49 to characterize the points or segments on the capacity region boundary, in a similar way to that for the state-dependent MAC with a helper. To this end, we first optimize the bounds for R_1 and R_2 respectively, and then provide conditions on channel parameters such that these bounds match the outer bound. Based on the conditions, we partition the channel parameters into the sets, in which different segments of the capacity region boundary can be obtained.

We next focus on an example case with $a = b$ so that R_1 and R_2 are optimized by the same set of coefficients α and β when $P_0'^2 \geq \alpha^2 Q(P_1 + 1 - P_0')$. In this way, both R_1 and R_2 can achieve the corresponding point-to-point channel capacity, i.e., the state is fully canceled. The following theorem presents this result.

THEOREM 19.50 *If $a = b$ and $P_0'^2 \geq \alpha^2 Q(P_1 + 1 - P_0')$, where $P_0' = P_0 - (\alpha - 1)^2 Q$, for some $\alpha \in \Omega_\alpha$, then the capacity region of the state-dependent parallel Gaussian channel with a helper and under the same but differently scaled states contains (R_1, R_2) satisfying*

$$R_1 \leq \frac{1}{2}\log(1 + P_1)$$
$$R_2 \leq \frac{1}{2}\log(1 + P_2).$$

Gaussian Model II

In this model, the helper assists both transmitter–receiver pairs, but only one receiver is interfered by a state sequence. The channel input–output relationship is given by

$$Y_1 = X_0 + X_1 + S + N_1, \tag{19.166a}$$
$$Y_2 = X_0 + X_2 + N_2. \tag{19.166b}$$

The assumptions for the noise variables Z_1 and Z_2, the state S, and the channel inputs X_0, X_1, X_2 are the same as Gaussian model I.

This model in the regime with the state power approaching infinity, i.e., $Q \to \infty$, has been studied in [15]. We next introduce the results in [15] as follows.

The authors of [15] developed the following achievable scheme to obtain an inner bound on the capacity for the discrete memoryless channel. Here, since the helper assists

Receiver 1 to deal with the state, it necessarily causes interference to Receiver 2. Hence, to maximize the transmission rate to Receiver 2, the helper should simply be silent. But without the helper's assistance, Receiver 1 gets zero rate due to infinite state power. Thus, it is desirable to retain Receiver 2's point-to-point rate and still achieve a certain positive rate for Receiver 1 simultaneously, which requires the helper to still assist Receiver 1 without causing interference to Receiver 2. The design is to split the helper's signal into two auxiliary RVs U and V, where U represents the help signal for Receiver 1 to cancel the state while treating V as noise, and V serves to help Receiver 2 to cancel the interference caused by U. Since there is no state interference at Receiver 2, U is decoded only at Receiver 1. The following achievable region can be obtained based on such an achievable scheme.

PROPOSITION 19.51 *For the discrete memoryless channel of model II, an achievable region consists of the rate-pair (R_1, R_2) satisfying*

$$R_1 \leqslant I(X_1; Y_1 U) \tag{19.167a}$$

$$R_1 \leqslant I(X_1 U; Y_1) - I(U; S_1) \tag{19.167b}$$

$$R_2 \leqslant I(X_2; Y_2 V) \tag{19.167c}$$

$$R_2 \leqslant I(X_2 V; Y_2) - I(V; U S_1) \tag{19.167d}$$

for some distribution $P_{S_1 U V X_0 X_1 X_2} = P_{S_1} P_{UVX_0|S_1} P_{X_1} P_{X_2}$, where U and V are auxiliary RVs.

Following Proposition 19.51, the following rate region can also be achieved, which consists of rate-pair (R_1, R_2) satisfying

$$R_1 \leqslant I(X_1; Y_1|U), \tag{19.168a}$$

$$R_2 \leqslant I(X_2; Y_2|V), \tag{19.168b}$$

for some distribution $P_{S_1 U V X_0 X_1 X_2} = P_{S_1} P_{UVX_0|S_1} P_{X_1} P_{X_2}$, where U and V are auxiliary that satisfy

$$I(U; Y_1) \geqslant I(U; S_1), \tag{19.169a}$$

$$I(V; Y_2) \geqslant I(V; U S_1). \tag{19.169b}$$

In the above achievable region, we choose the following joint Gaussian distribution for random variables and obtain the inner bound for Gaussian model II:

$$U = X_{01} + \alpha S_1, \quad V = X_{02} + \beta X_{01}$$

$$X_0 = X_{01} + X_{02}$$

$$X_{01} \sim \mathcal{N}(0, P_{01}), \quad X_{02} \sim \mathcal{N}(0, P_{02})$$

$$X_1 \sim \mathcal{N}(0, P_1), \quad X_2 \sim \mathcal{N}(0, P_2),$$

where X_{01}, X_{02}, X_1, X_2, and S_1 are independent.

PROPOSITION 19.52   (*Inner bound for Gaussian model II [15]*): *For Gaussian model II, in the regime when $Q_1 \to \infty$, an inner bound on the capacity region consists of rate-pairs (R_1, R_2) satisfying:*

$$R_1 \leq \frac{1}{2} \log \left(1 + \frac{P_1}{(1 - \frac{1}{\alpha})^2 P_{01} + P_{02} + 1} \right) \tag{19.170a}$$

$$R_2 \leq \frac{1}{2} \log \left(1 + \frac{P_2}{1 + \frac{(\beta - 1)^2 P_{02} P_{01}}{P_{02} + \beta^2 P_{01}}} \right), \tag{19.170b}$$

where $P_{01}, P_{02} \geq 0$, $P_{01} + P_{02} \leq P_0$, $0 < \alpha \leq \frac{2P_{01}}{1 + P_0 + P_1}$, and $P_{02}^2 + 2\beta P_{01} P_{02} \geq \beta^2 P_{01}(P_{02} + P_2 + 1)$.

The following proposition provides an outer bound on the capacity region.

PROPOSITION 19.53   (*Outer bound for Gaussian model II [15]*): *For the Gaussian channel of model II with $W_0 = \phi$, in the regime when $Q_1 \to \infty$, an outer bound on the capacity region consists of rate-pairs (R_1, R_2) satisfying:*

$$R_1 \leq \min \left\{ \frac{1}{2} \log(1 + P_0), \frac{1}{2} \log(1 + P_1) \right\} \tag{19.171a}$$

$$R_2 \leq \frac{1}{2} \log(1 + P_2) \tag{19.171b}$$

$$R_1 + R_2 \leq \frac{1}{2} \log(1 + P_0 + P_2). \tag{19.171c}$$

In the above rate region, (19.171a) provides the best single-user rate of Receiver 1 with the dedicated help of the helper, (19.171b) represents the point-to-point capacity for Receiver 2, and (19.171c) suggests that due to the shared common helper, there is still a rate constraint on the sum-rate, although the two transmitters communicate over parallel channels.

Comparing the inner and outer bounds given in Propositions 19.52 and 19.53, respectively, we characterize two segments of the boundary of the capacity region, over which the two bounds meet.

THEOREM 19.54   (*Segments of capacity boundary for Gaussian model II [15]*) Consider the Gaussian channel of model II with $W_0 = \phi$, in the regime when $Q_1 \to \infty$ the rate points on the line A–B (see Figure 19.28) are on the capacity region boundary. More specifically, if $\frac{1}{2}(1 + P_0 + P_1) \geq \frac{P_0^2}{P_0 + P_2 + 1}$, points A and B are characterized as

$$\text{Point } A : \left(0, \frac{1}{2} \log(1 + P_2) \right)$$

$$\text{Point } B : \left(\frac{1}{2} \log \left(1 + \frac{4P_1 P_0^2}{(1 + P_0 + P_1)^2(1 + P_0 + P_2) - 4P_1 P_0^2} \right), \frac{1}{2} \log(1 + P_2) \right).$$

$$\tag{19.172}$$

If $\frac{1}{2}(1 + P_0 + P_1) < \frac{P_0^2}{P_0+P_2+1}$, points A and B are characterized as

$$\text{Point A}: \left(0, \frac{1}{2}\log(1 + P_2)\right)$$

$$\text{Point B}: \left(\frac{1}{2}\log\left(1 + \frac{P_1(P_0 + P_2 + 1)}{P_0 + (P_0 + 1)(P_2 + 1)}\right), \frac{1}{2}\log(1 + P_2)\right).$$

(19.173)

Furthermore, the rate points on the line C–D (see Figure 19.28) are also on the capacity region boundary. If $P_1 \geqslant P_0 + 1$, the points C and D are characterized as

$$\text{Point C}: \left(\frac{1}{2}\log(1 + P_0), \frac{1}{2}\log\left(1 + \frac{P_2}{P_0 + 1}\right)\right)$$

$$\text{Point D}: \left(\frac{1}{2}\log(1 + P_0), 0\right),$$

(19.174)

as illustrated in Figure 19.28 (a).

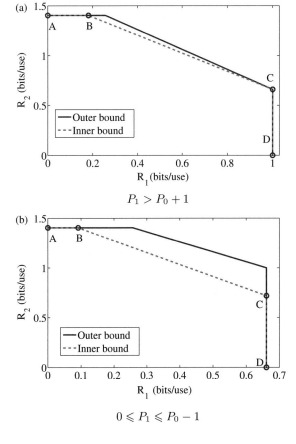

Figure 19.28 Segments of the capacity boundary for the Gaussian channel of model II.

If $P_1 \leqslant P_0 - 1$, the points C and D are characterized as

$$\text{Point } C : \left(\frac{1}{2} \log(1 + P_1), \frac{1}{2} \log \left(1 + \frac{P_2}{P_1 + 2} \right) \right)$$

$$\text{Point } D : \left(\frac{1}{2} \log(1 + P_1), 0 \right),$$

(19.175)

as illustrated in Figure 19.28(b).

Theorem 19.54 has a few important implications. (1) It suggests that via the proposed coding scheme the helper can effectively assist Receiver 1 (which achieves a positive rate) without causing interference to Receiver 2 (which achieves the point-to-point capacity) as indicated by the capacity result for the line A–B. (2) The helper can help Receiver 1 to deal with the state whereas Receiver 2 treats the helper's signal as noise to achieve the line C–D in Theorem 19.54. If Receiver 1's rate is maximized, such a scheme is optimal as guaranteed by the outer bound. (3) If $Q_1 \to \infty$, and $P_1 \geqslant P_0 + 1$, then the sum-capacity is characterized as $\frac{1}{2} \log(1 + P_0 + P_2)$, and is achieved by point C as illustrated in Figure 19.28(a).

Gaussian Model III

In this model, there are multiple transmitter–receiver pairs assisted by a common helper, and each receiver is corrupted by a state sequence. The state sequences are distributed independently. The channel outputs at receivers $1, \ldots, K$ for one symbol time are given by

$$Y_0 = X_0 + N_0, \tag{19.176a}$$

$$Y_k = X_0 + X_k + S_k + N_k, \quad \text{for} \quad k = 1, \ldots, K. \tag{19.176b}$$

In the above three models, $N_1 \ldots, N_K$ are independently Gaussian distributed noise variables, i.e., $N_0, \ldots, N_K \sim \mathcal{N}(0, 1)$, and $S_1, \ldots, S_K$ are independently Gaussian distributed state variables, i.e., $S_k \sim \mathcal{N}(0, Q_k)$ for $k = 1, \ldots, K$. The channel inputs $X_0, X_1, \ldots, X_K$ are subject to the average power constraints $P_0, P_1, \ldots, P_K$.

This model in the infinite state power regime, i.e., $Q \to \infty$, has been studied in [15]. We next introduce the results in [15] as follows. We first consider the scenario with two receivers, and then extend the result to the more general scenario with more than two receivers.

The following proposition provides an outer bound on the capacity region.

PROPOSITION 19.55 *(Outer bound for Gaussian model III [15]): For the Gaussian model III with $K = 2$, in the regime when $Q_1, Q_2 \to \infty$, an outer bound on the capacity region consists of rate pairs (R_1, R_2) satisfying:*

$$R_1 \leqslant \frac{1}{2} \log(1 + P_1) \tag{19.177}$$

$$R_2 \leqslant \frac{1}{2} \log(1 + P_2) \tag{19.178}$$

$$R_1 + R_2 \leqslant \frac{1}{2} \log(1 + P_0). \tag{19.179}$$

Due to the fact that one common helper assists multiple receivers, it may not be possible for it to fully cancel the two independent high-power states simultaneously. Thus, although the two transmitters communicate over parallel channels, the above outer bound still includes a sum rate constraint determined by the helper's power. This also suggests that the following time-sharing scheme may be desirable to achieve the sum rate upper bound. Namely, transmitter 1 transmits only over the fraction γ of the total time duration with power constraint $\frac{P_1}{\gamma}$, and transmitter 2 transmits only over the fraction $1 - \gamma$ of time with power constraint $\frac{P_2}{1-\gamma}$. During each slot, the helper fully assists only one transmitter. Hence, for each time slot, the following result for the point-to-point channel at the infinite state power regime can be applied. In particular, we express the achievable rate of the point-to-point channel as follows:

$$R(P, P_0) := \begin{cases} \frac{1}{2}\log(1 + P_0), & P \geqslant P_0 + 1 \\ \frac{1}{2}\log(1 + \frac{4P_0P}{4P_0 + (P_0 - P - 1)^2}), & P_0 - 1 \leqslant P \leqslant P_0 + 1 \\ \frac{1}{2}\log(1 + P), & P \leqslant P_0 - 1, \end{cases} \qquad (19.180)$$

where P and P_0 are power constraints respectively at the transmitter and the helper. The following achievable region then follows by applying the above achievable rate.

PROPOSITION 19.56   (*Inner bound for Gaussian model III [15]*): *For Gaussian model III with $K = 2$, in the regime with $Q_1, Q_2 \to \infty$, an inner bound on the capacity region consists of rate-pairs (R_1, R_2) satisfying:*

$$R_1 \leqslant \gamma R\left(\frac{P_1}{\gamma}, P_0\right) \qquad (19.181a)$$

$$R_2 \leqslant (1 - \gamma)R\left(\frac{P_2}{1 - \gamma}, P_0\right), \qquad (19.181b)$$

where $0 \leqslant \gamma \leqslant 1$ is the time-sharing coefficient, and the function $R(\cdot, \cdot)$ is defined in (19.180).

It can be seen from (19.180) that if $P \geqslant P_0 + 1$, the single-user rate is maximized to be $\frac{1}{2}\log(1 + P_0)$. We also observe that the time-sharing scheme achieves the sum-rate upper bound (19.179), in which case each transmitter simultaneously achieves the best single-user rate $\frac{1}{2}\log(1 + P_0)$ over their transmission fraction of time. This is because both of their powers get boosted over a certain fraction of time, although neither power is larger than $P_0 + 1$. The following theorem characterizes the sum capacity of the channel for the scenario described above.

THEOREM 19.57   (*Sum capacity for Gaussian model III [15]*) *For Gaussian model III with $K = 2$, in the regime with $Q_1, Q_2 \to \infty$, if $P_1 + P_2 \geqslant P_0 + 1$, then the sum-capacity equals $\frac{1}{2}\log(1 + P_0)$. The rate points that achieve the sum capacity (i.e., on the capacity region boundary) are characterized as $(R_1, R_2) = \left(\gamma R(\frac{P_1}{\gamma}, P_0), (1 - \gamma)R(\frac{P_2}{1-\gamma}, P_0)\right)$ for $\gamma \in \left(\max(1 - \frac{P_2}{P_0+1}, 0), \min(\frac{P_1}{P_0+1}, 1)\right)$.*

Thus, under certain parameters, the following full capacity region can be obtained.

COROLLARY 19.58   (*Capacity region for Gaussian model III under certain channel parameters [15]*) *For the Gaussian channel of model III with $K = 2$ and $W_0 = \phi$, in the regime with $Q_1, Q_2 \to \infty$, if P_1, $P_2 \geqslant P_0 + 1$, then the capacity region consists of the rate pair (R_1, R_2) satisfying $R_1 + R_2 \leqslant \frac{1}{2} \log(1 + P_0)$.*

The above results have also been further generalized to the case with more than two receivers, which we summarize as follows.

PROPOSITION 19.59   *For Gaussian model III with $K \geqslant 2$, in the regime when $Q_1, \ldots, Q_K \to \infty$, an outer bound on the capacity region consists of rate tuples $(R_0, \ldots, R_K)$ satisfying:*

$$R_k \leqslant \frac{1}{2} \log(1 + P_k) \quad \text{for } k = 1, \ldots, K$$

$$\sum_{k=1}^{K} R_k \leqslant \frac{1}{2} \log(1 + P_0).$$

To generalize the time-sharing scheme, we divide the entire transmission time into K slots, and then the helper assists one receiver during each slot. Then the following proposition generalizes Proposition 19.56 to the case with more than two receivers.

PROPOSITION 19.60   *For Gaussian model III with $K \geqslant 2$, in the regime when $Q_1, \ldots, Q_K \to \infty$, an inner bound consists of rate tuples $(R_1, \ldots, R_K)$ satisfying:*

$$R_k \leqslant \gamma_k R\left(\frac{P_k}{\gamma_k}, P_0\right), \quad \text{for } k = 1, \ldots, K$$

for some γ_k that satisfy

$$\sum_{k=1}^{K} \gamma_k = 1, \text{ and } \gamma_k \geqslant 0 \quad \text{for } k = 1, \ldots, K,$$

where $R(\cdot, \cdot)$ is the function defined in (19.180).

By comparing the inner and outer bounds, the following result on the capacity as an extension of Theorem 19.57 can be obtained.

THEOREM 19.61   *For Gaussian model III with $K \geqslant 2$, in the regime with $Q_1, \ldots, Q_k \to \infty$, the sum-capacity equals $\frac{1}{2} \log(1 + P_0)$. The rate points that achieve the sum capacity on the capacity region boundary are characterized as*

$$R_k \leqslant \frac{\gamma_k}{2} \log(1 + P_0) \quad k = 1, \ldots, K,$$

where

$$\sum_{k=1}^{K} \gamma_k = 1, \quad \gamma_k \geqslant 0$$

$$\frac{P_k}{\gamma_k} \geqslant P_0 + 1 \quad \text{for } k = 1, \ldots, K.$$

19.4.4   Helper-Assisted State-Dependent Broadcast Channel

In this section, we consider the state-dependent broadcast channel with a helper, in which a transmitter designs an encoder $f : (\mathcal{W}_1, \mathcal{W}_2) \to \mathcal{X}^n$ to map two independent messages $w_1 \in \mathcal{W}_1$ and $w_2 \in \mathcal{W}_2$ to a codeword $x^n \in \mathcal{X}^n$. The input x^n is transmitted over the broadcast channel. The state sequence S^n consists of i.i.d. components, and is assumed to be known at a helper, but not at either the transmitter or the receivers. Thus, in order to assist the receivers, the helper designs an encoder $f_0 : S^n \to \mathcal{X}_0^n$ to map the state sequences $s^n \in S^n$ to a codeword $x_0^n \in \mathcal{X}_0^n$. The channel transition probability is given by $P_{Y_1 Y_2 | X_0 X S}$. We here focus on the Gaussian model with the outputs at the two receivers for one channel use given by

$$Y_1 = a(X_0 + S) + X + Z_1, \tag{19.182a}$$
$$Y_2 = X_0 + S + X + Z_2, \tag{19.182b}$$

where Z_1 and Z_2 are Gaussian distributed noise variables $Z_1 \sim \mathcal{N}(0, N_1)$, $Z_2 \sim \mathcal{N}(0, N_2)$, and S is Gaussian distributed state variable $S \sim \mathcal{N}(0, Q)$. Then, each one of the receivers designs a decoder $g_k : \mathcal{Y}_k^n \to \mathcal{W}_k$ to map a received sequence y_k^n into a message $\hat{w}_k \in \mathcal{W}_k$ for $k = 1, 2$. The channel inputs X_0 and X are subject to the average power constraints P_0 and P, respectively.

We define the average probability of error for a length-n code as

$$P_e^{(n)} = \frac{1}{|\mathcal{W}_1||\mathcal{W}_2|} \sum_{w_1=1}^{|\mathcal{W}_1|} \sum_{w_2=1}^{|\mathcal{W}_2|} Pr\{(\hat{w}_1, \hat{w}_2) \neq (w_1, w_2)\}. \tag{19.183}$$

We say that a rate pair (R_1, R_2) is *achievable* if there exists a sequence of message sets $\mathcal{W}_k^{(n)}$ with $|\mathcal{W}_k^{(n)}| = 2^{nR_k}$ for $k = 1, 2$, and encoder–decoders $(f_0^{(n)}, f^{(n)}, g_1^{(n)}, g_2^{(n)})$ such that the average error probability $P_e^{(n)} \to 0$ as $n \to \infty$. We define the *capacity region* to be the closure of the set of all achievable rate pairs (R_1, R_2).

For ease of analysis, we consider the following model which is equivalent to that given by (19.182a) and (19.182b)

$$Y_1 = X_0 + S + \frac{1}{a}(X + Z_1), \tag{19.184a}$$
$$Y_2 = X_0 + S + X + Z_2. \tag{19.184b}$$

This model has been studied in [23], and we next introduce the results of [23]. Without loss of generality, we assume that $N_1 \geqslant N_2$, which implies that in the original broadcast channel without state, Receiver 1's channel quality is worse than that of Receiver 2. The following proposition provides an outer bound on the capacity region.

PROPOSITION 19.62   (*Outer bound on the capacity [23]): For the state-dependent Gaussian broadcast channel with a helper and with $N_1 \geqslant N_2$, an outer bound on the capacity region consists of rate-pairs (R_1, R_2) satisfying:*

$$R_1 \leqslant \min\left\{\frac{1}{2}\log(1 + \frac{P_1}{P - P_1 + N_1}), \frac{1}{2}\log(1 + \frac{a^2 P_0}{N_1})\right.$$

$$+ \frac{1}{2} \log \left(1 + \frac{P_0 + 2\sqrt{P_0 Q} + \frac{1}{a^2}(P + N_1)}{Q}\right)\Bigg\}, \tag{19.185a}$$

$$R_2 \leqslant \min \left\{\frac{1}{2}\log(1 + \frac{P - P_1}{N_2}), \frac{1}{2}\log(1 + \frac{P_0}{N_2})\right.$$

$$\left. + \frac{1}{2}\log\left(1 + \frac{P_0 + 2\sqrt{P_0 Q} + P + N_2}{Q}\right)\right\}, \tag{19.185b}$$

where $0 \leqslant P_1 \leqslant P$.

We observe that each bound above includes the capacity region of the Gaussian broadcast channel without state as well as the best single-user rate with the helper dedicated to help each receiver.

To derive the following achievable region, we adopt superposition coding for the transmitter to broadcast two messages and the dirty coding for the helper to assist the receivers.

PROPOSITION 19.63   *For the state-dependent broadcast channel with a helper, an inner bound on the capacity region consists of rate-pairs* (R_1, R_2) *satisfying:*

$$R_1 \leqslant I(V; Y_1 | U), \tag{19.186a}$$

$$R_2 \leqslant I(X; Y_2 | UV), \tag{19.186b}$$

$$R_1 + R_2 \leqslant I(X; Y_2 | U), \tag{19.186c}$$

for some distribution $P_S P_{U|S} P_{X_0|SU} P_V P_{X|V}$, *where* $I(U; Y_k) \geqslant I(U; S)$ *for* $k = 1, 2$.

Proposition 19.63 yields the following achievable rate region for the Gaussian channel by choosing the jointly Gaussian distribution as follows:

$$U = X_{00} + \alpha S, \ X_0 = X_{00} + \beta S$$
$$X = V + X', \ X_{00} \sim \mathcal{N}(0, P_{00})$$
$$V \sim \mathcal{N}(0, P_1), \ X' \sim \mathcal{N}(0, P - P_1),$$

where X_{00}, V, X', and S are independent.

PROPOSITION 19.64   *(Inner bound on capacity region [23]): For the state-dependent Gaussian broadcast channel in scenario II with* $N_1 \geqslant N_2$, *an inner bound on the capacity region consists of rate pairs* (R_1, R_2) *satisfying:*

$$R_1 \leqslant \frac{1}{2}\log\left(1 + \frac{P_1}{\frac{(1+\beta-\alpha)^2 a^2 P_{00} Q}{P_{00} + \alpha^2 Q} + P - P_1 + N_1}\right), \tag{19.187a}$$

$$R_2 \leqslant \frac{1}{2}\log\left(1 + \frac{P - P_1}{\frac{(1+\beta-\alpha)^2 P_{00} Q}{P_{00} + \alpha^2 Q} + N_2}\right), \tag{19.187b}$$

$$R_1 + R_2 \leqslant \frac{1}{2}\log\left(1 + \frac{P}{\frac{(1+\beta-\alpha)^2 P_{00} Q}{P_{00} + \alpha^2 Q} + N_2}\right), \tag{19.187c}$$

where $P_{00} + \beta^2 Q \leqslant P_0$, $P_{00} \geqslant 0$, $0 \leqslant P_1 \leqslant P$,

$$\alpha^2 Q \frac{P + N_1}{a^2} + \alpha^2 P_{00} Q - 2\alpha(1 + \beta) P_{00} Q \leqslant P_{00}^2, \quad and$$

$$\alpha^2 Q(P + N_2 + P_{00}) - 2\alpha(1 + \beta) P_{00} Q \leqslant P_{00}^2.$$

The capacity region for two ranges of channel parameters can then be obtained by comparing the above outer and inner bounds.

THEOREM 19.65   (*Capacity region in the high state power regime [23]): For the state-dependent Gaussian broadcast channel in scenario II with $N_1 \geqslant N_2$, if $P_0 \geqslant \max\{P + N_2, \frac{P+N_1}{a^2}\}$, the capacity region consists of rate-pairs (R_1, R_2) satisfying:*

$$R_1 \leqslant \frac{1}{2} \log\left(1 + \frac{P_1}{P - P_1 + N_1}\right), \tag{19.188a}$$

$$R_2 \leqslant \frac{1}{2} \log\left(1 + \frac{P - P_1}{N_2}\right). \tag{19.188b}$$

It can be seen from Theorem 19.65 that the capacity region of the corresponding channel without state is achieved (i.e., the state is fully canceled) if the helper's power is larger than a certain threshold, even if the state power is large.

If the helper's power is smaller than a threshold, the following theorem presents a capacity characterization.

THEOREM 19.66   (*Capacity region in the low state power regime [23]): For the state-dependent Gaussian broadcast channel with a helper and with $N_1 \geqslant N_2$, if $P_0 < \max\{P + N_2, \frac{P+N_1}{a^2}\}$, and*

$$\sqrt{Q} \leqslant \max_{0 \leqslant P_{00} \leqslant P_0} \sqrt{P_0 - P_{00}} + \frac{P_{00}}{\sqrt{\max\{\frac{P+N_1}{a^2}, P + N_2\} - P_{00}}}, \tag{19.189}$$

the capacity region consists of rate-pairs (R_1, R_2) satisfying:

$$R_1 \leqslant \frac{1}{2} \log(1 + \frac{P_1}{P - P_1 + N_1}), \tag{19.190a}$$

$$R_2 \leqslant \frac{1}{2} \log(1 + \frac{P - P_1}{N_2}). \tag{19.190b}$$

It can be seen from Theorem 19.66 that if the helper's power is not large enough, it can fully cancel the state that has limited power. Interestingly, such cancellation is not simply by subtracting the state, but by dirty paper coding, because the state power can still be larger than the helper's power. One example channel has its parameters given by $\max\{P + N_2, \frac{P+N_1}{a^2}\} = 7.5$, $P_0 = 5$, and $Q = 9$.

Acknowledgments

The work of H. Permuter, A. Baruch, and S. Shamai was supported by the Heron via the Israel Ministry of Economy and Science. The work of Y. Liang was supported by the U.S. National Science Foundation under Grant CCF-1801846.

References

[1] H. Nikbakht, M. Wigger, and Shlomo Shamai (Shitz), "Mixed delay constraints in Wyner's soft-handoff network," *arXiv,1801.05681*, 2018.

[2] M. Yemini, A. Somekh-Baruch, and A. Leshem, "Asynchronous transmission over single-user state-dependent channels," *IEEE Trans. Inf. Theory*, vol. 61, no. 11, pp. 5854–5867, Nov. 2015.

[3] M. Yemini, A. Somekh-Baruch, and A. Leshem, "On the multiple access channel with asynchronous cognition," *IEEE Trans. Inf. Theory*, vol. 62, no. 10, pp. 5643–5663, Oct. 2016.

[4] A. Goldsmith, S. A. Jafar, I. Maric, and S. Srinivasa, "Breaking spectrum gridlock with cognitive radios: An information theoretic perspective," *Proc. IEEE*, vol. 97, no. 5, pp. 894–914, May 2009.

[5] A. Jovicic and P. Viswanath, "Cognitive radio: An information-theoretic perspective," *IEEE Trans. Inf. Theory*, vol. 55, no. 9, pp. 3945–3958, Sep., 2009.

[6] K. Nagananda, P. Mohapatra, C. R. Murthy, and S. Kishore, "Multiuser cognitive radio networks: An information-theoretic perspective," *Int. J. Adv. Eng. Sci. Appl. Math.*, vol. 5, no. 1, pp. 43–65, 2013.

[7] D. Tse and P. Viswanath, *Fundamentals of Wireless Communication*.   Cambridge: Cambridge University Press, 2005.

[8] A. Goldsmith, *Wireless Communications*.   Cambridge: Cambridge University Press, 2005.

[9] J. Li, X. Wu, and R. Laroia, *OFDMA Mobile Broadband Communications: A Systems Approach*.   Cambridge: Cambridge University Press, 2013.

[10] T. S. Han and K. Kobayashi, "A new achievable rate region for the interference channel," *IEEE Trans. Inf. Theory*, vol. 27, no. 1, pp. 49–60, Jan. 1981.

[11] Y. Saito, A. Benjebbour, Y. Kishiyama, and T. Nakamura, "System-level performance evaluation of downlink non-orthogonal multiple access (NOMA)," in *Proc. IEEE Intl. Symp. Personal, Indoor and Mobile Radio Communications (PIMRC)*, London, Sep. 2013.

[12] Q. Li, H. Niu, A. Papathanassiou, and G. Wu, "5G network capacity: Key elements and technologies," *IEEE Veh. Technol. Mag.*, vol. 9, pp. 71–78, 2014.

[13] P. Xu, Z. Ding, X. Dai, and H. V. Poor, "NOMA: An information theoretic perspective," *arXiv: 1504.07751*, 2015.

[14] S. Mallik and R. Koetter, "Helpers for cleaning dirty papers," in *Proc. IEEE Int. ITG Conf. Source Channel Coding (SCC)*, Ulm, Jan. 2008.

[15] R. Duan, Y. Liang, A. Khisti, and S. Shamai, "Parallel Gaussian networks with a common state-cognitive helper," *IEEE Trans. Inf. Theory*, vol. 61, no. 12, pp. 6680–6699, Dec. 2015.

[16] Y. Sun, R. Duan, Y. Liang, A. Khisti, and S. Shamai, "Capacity characterization for state-dependent Gaussian channel with a helper," *IEEE Trans. Inf. Theory*, vol. 62, no. 12, pp. 7123–7134, Dec. 2016.

[17] K. Ashton, "That 'internet of Things' thing," *RFiD Journal*, 2009.

[18] L. Atzori, A. Iera, and G. Morabiton, "The internet of things: A survey," *Computer Netw.*, vol. 54, pp. 2787–2805, 2010.

[19] J. Gubbi, R. Buyya, S. Marusic, and M. Palaniswami, "Internet of Things (IoT): A vision, architectural elements, and future directions," *Fut. Gen. Comput. Syst.*, vol. 29, pp. 1645–1660, 2013.

[20] N. Devroye, P. Mitran, and V. Tarokh, "Achievable rates in cognitive radio channels," *IEEE Trans. Inf. Theory*, vol. 52, no. 5, pp. 1813–1827, May 2006.

[21] A. Sendonaris, E. Erkip, and B. Aazhang, "User cooperation diversity – part I: System description," *IEEE Trans. Commun.*, vol. 51, no. 11, pp. 1927–1938, Nov. 2003.

[22] A. Sendonaris, E. Erkip, and B. Aazhang, "User cooperation diversity – part II: Implementation aspects and performance analysis," *IEEE Trans. Commun.*, vol. 51, no. 11, pp. 1939–1948, Nov. 2003.

[23] R. Duan, Y. Liang, and S. Shamai (Shitz), "Dirty interference cancellation for Gaussian broadcast channels," in *Proc. IEEE Inf. Theory Workshop (ITW)*, Hobart, Nov. 2014.

[24] F. M. J. Willems, "Information-theoretical results for the discrete memoryless multiple access channel," PhD Dissertation, Katholieke Universiteit Leuven, 1982.

[25] F. M. J. Willems, "The discrete memoryless multiple channel with partially cooperating encoders," *IEEE Trans. Inf. Theory*, vol. 29, no. 6, pp. 441–445, 1983.

[26] F. Willems and E. van der Meulen, "The discrete memoryless multiple-access channel with cribbing encoders," *IEEE Trans. Inf. Theory*, vol. IT-31, no. 3, pp. 313–327, May 1985.

[27] H. H. Permuter and H. Asnani, "Multiple access channel with partial and controlled cribbing encoders," *IEEE Trans. Inf. Theory*, vol. 59, pp. 2252–2266, 2013.

[28] D. Slepian and J. K. Wolf, "A coding theorem for multiple access channels with correlated sources," *Bell Syst. Tech. J.*, vol. 52, pp. 1037–1076, Sep. 1973.

[29] W. Huleihel and Y. Steinberg, "Channels with cooperation links that may be absent," *IEEE Trans. Inf. Theory*, vol. 63, no. 29, pp. 5886–5906, 2017.

[30] S. I. Bross, Y. Steinberg, and S. Tinguely, "The causal cognitive interference channel," in *Int. Zurich Sem. Commun. (IZS)*, Zurich March 3–5, 2010.

[31] S. I. Bross and A. Lapidoth, "The state-dependent multiple-access channel with states available at a cribbing encoder," in *2010 IEEE 26th Conv. Electric. Electron. Eng. Israel (IEEEI 2010)*, 2010.

[32] F. M. Willems, "The multiple-access channel with cribbing encoders revisited," tutorial lecture at MSRI, Berkeley, Workshop Mathematics of Relaying and Cooperation in Communication Networks Apr. 10–12, 2006.

[33] T. Kopetz, H. H. Permuter, and S. S. Shamai, "Multiple access channels with combined cooperation and partial cribbing," *IEEE Trans. Inf. Theory*, vol. 62, no. 2, pp. 825–848, 2016.

[34] R. Liu, I. Maric, P. Spasojevic, and R. D. Yates, "Discrete memoryless interference and broadcast channels with confidential messages: Secrecy capacity regions," *IEEE Trans. Inf. Theory*, vol. 54, no. 6, pp. 2493–2507, 2008.

[35] R. Liu and V. Poor, "Secrecy capacity region of a multi-antenna Gaussian broadcast channel with confidential messages," *IEEE Trans. Inf. Theory*, vol. 55, no. 3, pp. 1235–1249, 2009.

[36] A. D. Wyner and J. Ziv, "The rate-distortion function for source coding with side information at the decoder," *IEEE Trans. Inf. Theory*, vol. 22, no. 1, pp. 1–10, 1976.

[37] A. H. Kaspi, "Two-way source coding with a fidelity criterion," *IEEE Trans. Inf. Theory*, vol. 31, no. 6, pp. 735–740, 1985.

[38] Y. Cemal and Y. Steinberg, "The multiple-access channel with partial state information at the encoders," *IEEE Trans. Inf. Theory*, vol. 51, no. 11, pp. 3992–4003, 2005.

[39] A. Haghi, R. Khosravi-Farsani, M. R. Aref, and F. Marvasti, "The capacity region of fading multiple access channels with cooperative encoders and partial CSIT," in *Proc. IEEE Int. Symp. Inf. Theory (ISIT)*, Austin, TX, 2010.

[40] A. Haghi, R. Khosravi-Farsani, M. R. Aref, and F. Marvasti, "The capacity region of *p*-transmitter/*q*-receiver multiple access channels with common information." [Online]. Available at: ee.sharif.ir/ali_haghi/IT SUBMIT.pdf.

[41] R. Kolte, A. Özgür, and H. H. Permuter, "Cooperative binning for semideterministic channels," *IEEE Trans. Inf. Theory*, vol. 62, no. 3, pp. 1231–1249, 2016.

[42] I. B. Gattegno, H. H. Permuter, S. Shamai, and A. Özgür, "Cooperative binning for semi-deterministic channels with non-causal state information," arXiv:1703.08099, 2017.

[43] A. Somekh-Baruch, S. Shamai (Shitz), and S. Verdú, "Cooperative multiple-access encoding with states available at one transmitter," *IEEE Trans. Inf. Theory*, vol. 54, no. 10, pp. 4448–4469, 2008.

[44] S. Gel'fand and M. Pinsker, "Coding for channels with random parameters," *Probl. Contr. Inf. Theory*, vol. 9, no. 1, pp. 19–31, Jan. 1980.

[45] C. E. Shannon, "Channels with side information at the transmitter," *IBM J. Res. Develop.*, pp. 289–293, 1958.

[46] R. G. Gallager, *Information Theory and Reliable Communication*. New York: Wiley, 1968.

[47] M. H. M. Costa, "Writing on dirty paper," *IEEE Trans. Inform. Theory*, vol. 29, no. 3, pp. 439–441, May 1983.

[48] A. Somekh-Baruch, S. Shamai (Shitz), and S. Verdú, "Cognitive interference channels with state information," in *Proc. IEEE Int. Symp. Inf. Theory (ISIT)*, Toronto Jul. 2008, pp. 1353–1357.

[49] A. Somekh-Baruch, S. S. (Shitz), and S. Verdú, "Cooperative multiple-access encoding with states available at one transmitter," *IEEE Trans. Inf. Theory*, vol. 54, no. 10, pp. 4448–4469, 2008.

[50] W. Wu, S. Vishwanath, and A. Arapostathis, "Capacity of a class of cognitive radio channels: Interference channels with degraded message sets," *IEEE Trans. Inf. Theory*, vol. 53, no. 11, pp. 4391–4399, Nov. 2007.

[51] Y. H. Kim, A. Sutivong, and S. Sigurjónsson, "Multiple user writing on dirty paper," in *Proc. Int. Symp. Inf. Theory*, Chicago, IL, 2004, p. 534.

[52] Y. Steinberg and S. Shamai (Shitz), "Achievable rates of the broadcast channel with states known at the transmitter," in *Proc. Int. Symp. Inf. Theory*, Adelaide, 2005.

[53] Y. Liang, A. Somekh-Baruch, H. V. Poor, S. Shamai (Shitz), and S. Verdú, "Capacity of cognitive interference channels with and without secrecy," *IEEE Trans. Inf. Theory*, vol. 55, no. 2, pp. 604–619, Feb. 2009.

[54] J. Shimonovich, A. Somekh-Baruch, and S. Shamai (Shitz), "Cognitive cooperative communications on the multiple access channel," in *2013 IEEE Inf. Theory Workshop (ITW)*, Sept. 2013, pp. 1–5.

[55] J. Shimonovich, A. Somekh-Baruch, and S. Shamai (Shitz), "Cognition and cooperation in interfered multiple access channels," *Entropy, special issue on Network Information Theory*, vol. 19, no. 7, pp. 1–33, Jul. 2017.

[56] S. Bross and A. Lapidoth, "The state-dependent multiple-access channel with states available at a cribbing encoder," in *Electri. Electron. Eng. Israel (IEEEI), 2010 IEEE 26th Conv.*, 2010, pp. 665–669.

[57] N. Liu, I. Maric, A. Goldsmith, and S. Shamai (Shitz), "Capacity bounds and exact results for the cognitive Z-interference channel," *IEEE Trans. Inf. Theory*, vol. 59, no. 2, pp. 886–893, 2013.

[58] A. Zaidi, S. Kotagiri, J. Laneman, and L. Vandendorpe, "Cooperative relaying with state available noncausally at the relay," *IEEE Trans. Inf. Theory*, vol. 56, no. 5, pp. 2272–2298, 2010.

[59] J. Mitola, "Cognitive radio: Making software radios more personal," *IEEE Personal Commun.*, vol. 6, pp. 13–18, Aug. 1999.

[60] S. Haykin, "Cognitive radio: Brain-empowered wireless communications," *IEEE J. Sel. Areas Commun.*, vol. 23, pp. 201–220, Feb. 2005.

[61] I. F. Akyildiz, W.-Y. Lee, M. C. Vuran, and S. Mohanty, "Next generation/dynamics spectrum access/cognitive radio wireless networks: A survey," *Comput. Netw.*, vol. 50, pp. 2127–2159, 2006.

[62] T. M. Cover and A. A. El Gamal, "Capacity theorems for the relay channel," *IEEE Trans. Inf. Theory*, vol. 25, no. 5, pp. 572–584, Sep. 1979.

[63] K. Doppler, M. Rinne, C. Wijting, C. Ribeiro, and K. Hugl, "Device-to-device communication as an underlay to LTE-advanced networks," *IEEE Commun. Mag.*, vol. 47, no. 12, pp. 42–49, 2009.

[64] G. Fodor and N. Reider, "A distributed power control scheme for cellular network assisted D2D communications," in *Proc. IEEE Global Telecommun. Conf. (GLOBECOM 2011)*, Houston, TX, Dec. 2011.

[65] L. Lei, Z. Zhong, C. Lin, and X. Shen, "Operator controlled device-to-device communications in LTE-advanced networks," *IEEE Trans. Wireless Commun.*, vol. 19, no. 3, pp. 96–104, 2012.

[66] H.-J. Su, P.-T. Tu, B. Su, and H.-B. Tseng, "Device-to-device communication with dirty paper coded simultaneous transmission," in *Proc. IEEE Conf. Veh. Technol. VTCfall*, Boston, MA, Sep. 2015.

[67] M. Dikshtein, R. Duan, Y. Liang, and S. Shamai (Shitz), "State-dependent parallel Gaussian channels with a state-cognitive helper," in *Proc. Int. Zurich Sem. Inf. Commun.*, Zurich, 2018.

Index